BREEDING OILSEED CROPS FOR SUSTAINABLE PRODUCTION

BREEDING OILSEED CROPS FOR SUSTAINABLE PRODUCTION

Opportunities and Constraints

Edited by

SURINDER KUMAR GUPTA

Sher-e-Kashmir University of Agricultural Sciences & Technology Chatha, Jammu (J&K), India

ELSEVIER

AMSTERDAM • BOSTON • HEIDELBERG • LONDON
NEW YORK • OXFORD • PARIS • SAN DIEGO
SAN FRANCISCO • SINGAPORE • SYDNEY • TOKYO

Academic Press is an Imprint of Elsevier

Academic Press is an imprint of Elsevier
125, London Wall, EC2Y 5AS, UK
525 B Street, Suite 1800, San Diego, CA 92101-4495, USA
225 Wyman Street, Waltham, MA 02451, USA
The Boulevard, Langford Lane, Kidlington, Oxford OX5 1GB, UK

British Library Cataloguing-in-Publication Data
A catalogue record for this book is available from the British Library

Library of Congress Cataloging-in-Publication Data
A catalog record for this book is available from the Library of Congress

ISBN: 978-0-12-801309-0

For information on all Academic Press publications
visit our website at http://store.elsevier.com/

Publisher: Nikki Levy
Acquisition Editor: Nancy Maragioglio
Editorial Project Manager: Billie-Jean Fernandez
Production Project Manager: Julia Haynes
Designer: Mark Rogers

Typeset by Thomson Digital

Contents

11. Olives

AURORA DÍAZ

12. Soybean

ADITYA PRATAP, SURINDER KUMAR GUPTA, JITENDRA KUMAR, SUHEL MEHANDI, VANKAT R. PANDEY

13. Omics – A New Approach to Sustainable Production

SAJAD MAJEED ZARGAR, NANCY GUPTA, MUSLIMA NAZIR, RAKEEB AHMAD MIR, SURINDER KUMAR GUPTA, GANESH KUMAR AGRAWAL, RANDEEP RAKWAL

14. Forecasting Diseases and Insect Pests for a Value-Added Agroadvisory System

AMRENDER KUMAR, CHIRANTAN CHATTOPADHYAY, BIMAL KUMAR BHATTACHARYA, VINOD KUMAR, AMRENDRA KUMAR MISHRA

15. Designer Oil Crops

MUKHLESUR RAHMAN, MONIKA MICHALAK DE JIMÉNEZ

16. Genetic Improvement of Rapeseed Mustard through Induced Mutations

VINOD CHOUDHARY, SANJAY J. JAMBHULKAR

17. Pollination Interventions

UMA SHANKAR, DHARAM P. ABROL

18. Breeding Oilseed Crops for Climate Change

ABDULLAH A. JARADAT

19. Possibilities of Sustainable Oil Processing

BERTRAND MATTHÄUS

20. Integrated Pest Management

DHARAM P. ABROL, UMA SHANKAR

List of Contributors

Dharam P. Abrol Faculty of Agriculture, Division of Entomology, Sher-e-Kashmir University of Agricultural Sciences & Technology Chatha, Jammu, Shalimar, J&K, India

Ganesh Kumar Agrawal Research Laboratory for Biotechnology and Biochemistry (RLABB), Kathmandu, Nepal

Mothilal Alagirisamy All India Co-ordinated Research Project on Groundnut, Division of Plant Breeding and Genetics, Regional Research Station, Tamil Nadu Agricultural University, Vridhachalam, India

Basharat Ali Institute of Crop Science and Zhejiang Key Laboratory of Crop Germplasm, Zhejiang University, Hangzhou, China

Vincent Arondel Membrane Biogenesis Laboratory, UMR5200 CNRS, University of Bordeaux, France

Bimal Kumar Bhattacharya Crop Inventory and Agro-ecosystem Division, Space Applications Centre, Indian Space Research Organization (ISRO), Ahmedabad, India

Rajani Bisen Project Coordinating Unit, All India Coordinated Research Project (Sesame and Niger), Indian Council of Agricultural Research (ICAR), JNKVV Campus, Jabalpur, India

Chirantan Chattopadhyay National Centre for Integrated Pest Management, Indian Council of Agricultural Research (ICAR), New Delhi, India

Vinod Choudhary Nuclear Agriculture & Biotechnology Division, Bhabha Atomic Research Centre, Mumbai, India

Peng Cui Institute of Crop Science and Zhejiang Key Laboratory of Crop Germplasm, Zhejiang University, Hangzhou, China

Monika Michalak de Jiménez Department of Plant Sciences, North Dakota State University, Fargo, ND, USA

Aurora Díaz Unidad de Hortofruticultura, Instituto Agroalimentario de Aragón (IA2) (CITA-Universidad de Zaragoza), Av. Montañana, Zaragoza, Spain

Maho-Yalen J. Edson Department of Biological Sciences, Higher Teachers' Training College, University of Yaounde 1, Yaounde, Cameroon

Youmbi Emmanuel Department of Plant Biology, University of Yaounde 1, Yaounde; Tissue Culture Laboratory, Centre Africain de Recherche sur Bananiers et Plantains (CARBAP), Njombé, Cameroun

Muhammad A. Farooq Institute of Crop Science and Zhejiang Key Laboratory of Crop Germplasm, Zhejiang University, Hangzhou, China

Ngando-Ebongue G. Frank Selection and Genetic Improvement Section, Specialized Centre for Oil Palm Research of La Dibamba, Institute of Agricultural Research for Development (IRAD), Douala, Cameroon

Rafaqat A. Gill Institute of Crop Science and Zhejiang Key Laboratory of Crop Germplasm, Zhejiang University, Hangzhou, China

Ntsomboh-Ntsefong Godswill Selection and Genetic Improvement Section, Specialized Centre for Oil Palm Research of La Dibamba, Institute of Agricultural Research for Development (IRAD), Douala; Department of Plant Biology, University of Yaounde 1, Yaounde, Cameroon

Nancy Gupta School of Biotechnology, Sher-e-Kashmir University of Agricultural Sciences & Technology of Jammu, Chatha, Jammu, J&K, India

Rameshwer Dass Gupta Sher-e-Kashmir University of Agricultural Sciences & Technology of Jammu, Chatha, Jammu (J&K), India

Surinder Kumar Gupta Division of Plant Breeding & Genetics, Faculty of Agriculture, Sher-e-Kashmir University of Agricultural Sciences & Technology Chatha, Jammu (J&K), India

Faisal Islam Institute of Crop Science and Zhejiang Key Laboratory of Crop Germplasm, Zhejiang University, Hangzhou, China

Surabhi Jain Project Coordinating Unit, All India Coordinated Research Project (Sesame and Niger), Indian Council of Agricultural Research (ICAR), JNKVV Campus, Jabalpur, India

Sanjay J. Jambhulkar Nuclear Agriculture & Biotechnology Division, Bhabha Atomic Research Centre, Mumbai, India

Abdullah A. Jaradat USDA-ARS and Department of Agronomy and Plant Genetics, University of Minnesota, MN, USA

Yalcin Kaya Engineering Faculty, Department of Genetic and Bioengineering, Trakya University, Edirne, Turkey

Tabi-Mbi Kingsley Selection and Genetic Improvement Section, Specialized Centre for Oil Palm Research of La Dibamba, Institute of Agricultural Research for Development (IRAD), Douala; Department of Plant Biology, University of Yaounde 1, Yaounde, Cameroon

Amrender Kumar Indian Agricultural Research Institute, Indian Council of Agricultural Research (ICAR), New Delhi, India

Jitendra Kumar Crop Improvement Division, Indian Institute of Pulses Research, Kanpur, India

Vinod Kumar Directorate of Rapeseed-Mustard Research, Indian Council of Agricultural Research (ICAR), Sewar, Bharatpur, Rajasthan, India

Bell J. Martin Department of Plant Biology, University of Yaounde 1, Yaounde, Cameroon

Bertrand Matthäus Max Rubner-Institut, Federal Research Institute for Nutrition and Food, Department for Quality and Safety of Cereals, Working Group for Lipid Research, Detmold, Germany

Suhel Mehandi Crop Improvement Division, Indian Institute of Pulses Research, Kanpur, India

Rakeeb Ahmad Mir School of Biosciences and Biotechnology, BGSB University, Rajouri, India

Amrendra Kumar Mishra Indian Agricultural Research Institute, Indian Council of Agricultural Research (ICAR), New Delhi, India

Ullah Najeeb Department of Plant and Food Sciences, Faculty of Agriculture and Environment, University of Sydney, Eveleigh, NSW, Australia

Muslima Nazir Department of Botany, Faculty of Science, Jamia Hamdard University, Jamia Nagar, New Delhi, India

Nandini Nimbkar Department of Genetics and Plant Breeding, Nimbkar Agricultural Research Institute, Phaltan, Maharashtra, India

Anand Kumar Panday Project Coordinating Unit, All India Coordinated Research Project (Sesame and Niger), Indian Council of Agricultural Research (ICAR), JNKVV Campus, Jabalpur, India

Vankat R. Pandey Crop Improvement Division, Indian Institute of Pulses Research, Kanpur, India

Aditya Pratap Crop Improvement Division, Indian Institute of Pulses Research, Kanpur, India

Mukhlesur Rahman Department of Plant Sciences, North Dakota State University, Fargo, ND, USA

Randeep Rakwal Research Laboratory for Biotechnology and Biochemistry (RLABB), Kathmandu, Nepal; Organization for Educational Initiatives, University of Tsukuba, Tennoudai, Tsukuba, Ibaraki, Japan

A.R.G. Ranganatha Project Coordinating Unit, All India Coordinated Research Project (Sesame and Niger), Indian Council of Agricultural Research (ICAR), JNKVV Campus, Jabalpur, India

Uma Shankar Faculty of Agriculture, Division of Entomology, Sher-e-Kashmir University of Agricultural Sciences & Technology Chatha, Jammu, Shalimar, J&K, India

Shikha Sharma Project Coordinating Unit, All India Coordinated Research Project (Sesame and Niger), Indian Council of Agricultural Research (ICAR), JNKVV Campus, Jabalpur, India

Vrijendra Singh Department of Genetics and Plant Breeding, Nimbkar Agricultural Research Institute, Phaltan, Maharashtra, India

Suriya A.C.N. Perera Division of Genetics and Plant Breeding, Coconut Research Institute, Lunuwila, Sri Lanka

Ajambang-Nchu Walter Selection and Genetic Improvement Section, Specialized Centre for Oil Palm Research of La Dibamba, Institute of Agricultural Research for Development (IRAD), Douala, Cameroon; Plant Molecular Biology Laboratory, Department of Plant Breeding and Biotechnology, Bogor Agricultural University, Bogor, Indonesia

Ling Xu Zhejiang Province Key Laboratory of Plant Secondary Metabolism and Regulation, College of Life Sciences, Zhejiang Sci-Tech University, Hangzhou, China

Sajad Majeed Zargar Centre for Plant Biotechnology, Division of Biotechnology, Sher-e-Kashmir University of Agricultural Sciences & Technology of Kashmir, Shalimar, Srinagar, J&K, India

Weijun Zhou Institute of Crop Science and Zhejiang Key Laboratory of Crop Germplasm, Zhejiang University, Hangzhou, China

Preface

Oilseed crops are grown under varied agroclimatic situations ranging from tropical to temperate regions of the world and are vital commodity in trade and commerce. World population continues to increase, thus creating an increasing demand of oil and its varied products. Despite the fact that technological advances made in all the major crops, the need and opportunities to increase the production and oil yield are as great as they have ever been. It has been possible only due to the increase in area under each crop as well as high-yielding varieties. Current breeding effort worldwide are focused on sustainable production and higher oil yield per unit area of land with a view to maximizing returns. Besides oil yield, breeding populations with many traits such as fatty acids, vitamins, high carotene etc. are identified in various oil crops for industrial/pharmaceutical purposes. Technological advances have also been made in each crop to create value addition to make the production sustainable.

The book includes 20 chapters, which have been well prepared by leading scientists of the world with vast experience and whose contributions are well known over the world. Chapters 1 and 2 deal with new strategies for oilseed production and breeding for sustainable production: heavy metal tolerance while Chapters 3–12 deal with breeding brassicas, sunflower, groundnut, sesame, niger, safflower, coconut, oilpalm, and olive for sustainable production. Chapter 13 describes a new approach – OMICS for sustainable production followed by a chapter on forecasting diseases and insect-pests for value added agroadvisory system (Chapter 14). Designer oilcrops (Chapter 15) is the most important chapter, which describes various technological advances till date to make production sustainable. Chapters 16 and 17 describe genetic improvement through mutation breeding and pollination interventions, respectively. Breeding for climate change followed by oil technology is presented in Chapters 18 and 19 and integrated pest management in Chapter 20. Above all, breeding oil crops for climate change and designer oil crops have added new dimensions in this book.

I am highly indebted to all my contributors especially Professor W.J. Zhou, Crop Science Institute, Hangzhou, China, Abdullah A. Jaradat, USDA-ARS, University of Minnesota, USA, Mukhlesur Rahman, North Dakota State University, USA, and Bertrand Matthäus, Max Rubner-Institute, Federal Research Institute for Nutrition and Food, Detmold, Germany for their ready response. I am indeed grateful to Nancy Maragioglio, Senior Acquisitions Editor, Julia Haynes, Senior Project Manager, S&T Books, Elsevier, Academic Press for making every effort to make this book valuable for readers. Lastly, I owe a lot to my wife Dr Neena Gupta and both my kids, Kavya and Kanav for their patience during the preparation of this manuscript.

Editor
Surinder Kumar Gupta

CHAPTER

1

Strategies for Increasing the Production of Oilseed on a Sustainable Basis

Rameshwer Dass Gupta, Surinder Kumar Gupta***

*Sher-e-Kashmir University of Agricultural Sciences & Technology of Jammu, Chatha, Jammu (J&K), India

**Division of Plant Breeding & Genetics, Faculty of Agriculture, Sher-e-Kashmir University of Agricultural Sciences & Technology Chatha, Jammu (J&K), India

INTRODUCTION

Among the oilseed crops, soybean is the major contributor to the world's oilseed economy followed by rapeseed mustard, cotton, peanut, and sunflower. The most important tropical oilseeds are the coconut, palm kernels, and groundnut. The major oilseed-producing areas are in temperate zones. America and Europe together account for more than 60% of the world production of oilseed whereas substantially small production (<5%) comes from tropical areas such as Africa, Malaysia, and Indonesia. Both oilseed and oil production have consistently increased over the years to meet the ever increasing demand for vegetable oils. Among the oilseeds, soybean is the chief oilseed crop. *Brassica* spp. is the second largest oilseed crop after soybean (*Glycine max* (L.) Merr.), surpassing peanut (*Arachis hypogaea* L.), sunflower (*Helianthus annuus* L.), and cottonseed (*Gossypium hirsutum* L.) over the last two decades (FAO, 2010; Agricultural Outlook 2010–19). Palms are grown predominantly in the tropical areas of the world as perennial trees and are an important source of vegetable oil. About two-thirds of the total fat oil production is supplied by oilseeds, with palm oil having a maximum share of 33%. Copra, cotton, palm, peanut, rapeseed, soybean, and sunflower are the oilseed crops, which dominate the international markets for trade purposes.

India has agroecological conditions that favor growing nine oilseed crops (Hegde and Sudhakara Babu, 2000). These oilseed crops consist of seven edible oilseeds – groundnut, rapeseed mustard, soybean, sunflower, sesame, safflower, and niger – and two nonedible

Breeding Oilseed Crops for Sustainable Production. http://dx.doi.org/10.1016/B978-0-12-801309-0.00001-X

sources – castor and linseed. Apart from this, a wide range of other minor oilseeds and oil-bearing tree species are grown in India.

India is, in fact, the largest producer of castor, sesame, safflower, and niger and the second largest producer of groundnut and rapeseed mustard. Oilseed constitutes the second largest agricultural commodity after cereals in the country, sharing 14% of gross cropped area and accounting for about 5% of gross national product and about 10% of the value of all agricultural products. Nearly 14 million people are involved in the production of oilseed and another 1 million in its processing.

In spite of India's premier position in terms of cropped area and production, its oilseed productivity is one of the lowest. Moreover, the contribution of India to the world's annual oilseed production is less than 10%. Besides this, the per capita availability of oil is calculated to be about 9.2 kg against the world average of 11 kg. A recommended per capita requirement of 14 kg has been specified by India's Council of Medical Research. Furthermore, the demand for edible oils both in quantity and quality is increasing due to an increase in population, which was 1.21 billion in 2011 (Luna, 2011) and is presently more than 1.25 billion (Gupta, 2014). Because the population is annually increasing at 1.8% and the oil consumption rate is increasing at 6%, the per capita consumption of oils and fats has doubled during the past 15 years. Annual demand has been growing at a rate of about 6% over the last 13 years touching nearly 12.5 million tons during 1998–1999.

In light of the above, India witnessed an alarming edible oil scenario during the early 1980s due to low production and productivity, and heavy import of edible oil. Between 1981 and 1986, India imported edible oil worth Rs. 38,870,000,000. Owing to a heavy outflow of foreign exchange for edible oils, next only to petroleum products, the government of India set up the Technology Mission on Oilseeds during May 1986. Due to the concerted efforts of this mission, a quantum jump in oilseed production, from 10.83 million tons to 25.21 million tons (1998–1999), was achieved, with productivity increasing from 570 kg ha^{-1} to 932 kg ha^{-1} (Sudhakar et al., 2000).

During 2002 the consumption of vegetable oil for the population of 1027 million was 9 million tons at an estimated per capita consumption of 8.74 kg, whereas its total consumption was 12.3 million tons at a per capita consumption of 10.7 kg during 2010 when the population was 1120 million. The demand for edible oil both in quantity and quality is increasing due to the increase in population and improvement in the standard of living in addition to a growing industrial demand. The annual demand which was 125 million tons (Gupta, 2009) has now risen to over 130 million tons, whereas production is barely 75 million tons. To meet this requirement for edible oil consumption, oilseed output is required to be 42.24 million tons. An integrated oilseed development program was initiated in different Indian states, with more than 3000 oilseed societies involving 11 million farmers and 25 million hectares of land, with the purpose of increasing oilseed productivity (Gupta, 2009). However, despite these efforts, oilseed productivity still continues to be as low as 944 kg ha^{-1} compared with the world level of 1632 kg ha^{-1}.

Presently, there is not much scope to expand the cultivable area used for oilseed. The continuous shortage of cooking oil would suggest that the Oilseeds Technology Mission, and the growing of oil palms, has had little impact. These energy-rich crops suffer from a number of constraints as they are grown in poor environments and are susceptible to pests and diseases. Moreover, farmers prefer to grow high-yield varieties of cereals and thus earn more profit.

In light of the above, the improved technology developed to boost the output of the major oilseed crops to meet the country's needs for edible oil needs to be used by farmers both in letter and spirit. Adopting these new technologies and strategies will not only meet the country's need for edible oils, but also assist in the production of oilseed on a sustainable basis.

EXTENDING IRRIGATION FACILITIES

Oilseed crops are cultivated largely under rain-fed conditions and due to erratic rainfall the crops suffer from moisture stress at different growth stages resulting in low yields. The production of oilseed can therefore be stabilized and significantly increased by extending irrigation facilities to oilseed. The water requirement of oilseed in general is low and the crop responds remarkably to limited irrigation facilities provided at critical growth periods, specifically at sowing, flowering, and seed/pod formation. In light of the above, wherever it becomes possible, irrigation facilities are required to be extended. Contrary to this, where it is not possible to create irrigation facilities, moisture conservation practices should be followed.

IMPORTANT MOISTURE CONSERVATION PRACTICES

The following are important moisture conservation practices, which can help to improve and sustain the productivity of oilseed crops:

1. *Mulch.* Mulching consists of either stirring the upper soil surface layer to a depth of about 2.0–2.5 cm or spreading litter, plant leaves, stubble, or straw on the soil with the objective of conserving soil moisture and reducing water loss through evaporation (Gupta and Banerjee, 1991). Apart from these organic sources, mulching can also be done using plastic sheets, gravels, etc., known as inorganic mulching. Mulch is mostly produced by running a shallow harrow over the soil, which breaks up the soil surface down to a depth of 2.5–3.0 cm. Mulching is an agronomic input with the potential to ameliorate stress by reducing soil erosion, lowering moisture evaporation, preventing crust formation, and improving water infiltration. It facilitates water storage in the soil for crops to use throughout the growing season, enhancing water use efficiency and, thereby, sustaining crop productivity. Mulching also provides several benefits to crop productivity by checking weeds and buffering the heat energy and nutrient status of the soil. As organic mulch sources break down, they increase the soil organic matter content, which in the long run improves soil structure, water holding capacity, and fertility. Apart from these benefits, use of mulch has also been found to increase the yield of oilseed crops. Use of wheat straw mulch at 5 t ha^{-1} has helped to advance the germination of summer groundnut in the Saurashtra region of Gujarat by 2–3 days and has conserved soil moisture and controlled weed growth, increasing yield by 40% (Reddy and Patti, 1998). Furthermore, it was found that a combination of a criss-cross method of sowing and wheat straw mulch could double the summer groundnut yield in the Saurashtra region. Mulching brings marked improvement in growth components such as plant height, dry matter accumulation. Soil mulch (or dust mulch), stubble mulch, straw mulch, plastic mulch, and vertical mulch are the various types of mulching. The greatest advantage of using these kinds of mulches is to overcome the deleterious effect of water stress during a plant's stages of growth (Kumar et al., 2006).

2. *Timely weed control.* In rain-fed situations, water is often the most critical factor in determining the potential yield of oilseed and other crops. As weeds can compete with oilseed crops for soil moisture and plant nutrients, effective weed control is important, especially before sowing. This is because weed growth at the early stages of crop growth has a considerable effect on the availability of soil moisture to crops. The reduction in oilseed crop yields varies between 20% and 60%, depending on the type of soil, season, and intensity of weed infestation. Integrated weed management (i.e., the use of herbicides in conjunction with other control methods mechanical, cultural, or agronomic) should be followed (Dubey et al., 2011) for soybean crops. It is important to keep weeds under check until at least 25–30 days after sowing so that they can establish well and withstand competition from weeds without being affected adversely. Thinning crop plants soon after germination, thereby retaining an optimum population, and removing weeds by one or two intercultivation methods can reduce the adverse influence of weeds on both crop growth and yield.
3. *Seed selection and seed rate.* The seeds used for sowing should have at least 80–85% germination capacity. Quality and certified seeds should always be used. Seed rate which depends upon the germination percentage, seed size, and time of sowing should be selected according to the recommendations given on the packet and practices formulated by the state agricultural universities, state departments of agriculture, and Indian Council of Agricultural Research stations. For example, the G.B. Pant University of Agriculture and Technology (Pantnagar) has recommended using 75–80 kg ha^{-1} of seed in the case of soybean, giving 80–85% germination. Under late sowing conditions the seed rate may be raised to 100 kg ha^{-1}.
4. *Proper timing and methods of sowing.* Broadcasting seeds at the convenience of farmers, with little or no thought for time and conditions, has been another reason for poor crop stands. Crops can be affected not only by droughts and rain but also by pests and diseases during their important stages of growth if sowing is not done at the right time. Hence, adherence to scheduled times of sowing and plowing in rows to deposit seeds in moist zones ensures better germination, crop growth, and yield. Furthermore, the optimum depth of seed placement avoids consumption of seeds by birds – something that is very common in broadcast crop seeds.

   In northern India, where irrigation facilities are available, soybean can be sown during the second fortnight of June. On the other hand, where irrigation facilities are not available, it should be sown during the first fortnight of July.
5. *Crop and varietal selection.* Most competing profitable oilseed crops for a region and season can be selected as per their ecological requirements. For instance, safflower and soybean crops for the postrainy season in the Malwa Plateau of Madhya Pradesh are competitive and profitable compared with rabi sorghum, gram, and wheat. Under dryland situations, crops and their varieties have to be selected to suit the effective growing season as determined by rainfall and soil type (Hegde and Sudhakara Babu, 2000).

   According to Kachroo and Sharma (2008), instead of enhancing oilseed crop acreage, especially sunflower, the motive must be to increase productivity by using good-quality seeds of high-yielding varieties and by adopting the latest technology. Likewise, to harvest a good soybean yield, farmers should grow only improved and recommended varieties according to their location (Saxena, 2008). Varietal selection should be based on the growing area and varietal characters. Most soybean varieties are resistant to disease. Soybean varieties suitable for different agroclimatic regions of India are shown in Table 1.1. The

TABLE 1.1 Suitable Soybean Varieties for Different Regions of India

Region	Recommended varieties
Hilly regions of northern India (lower and mid-hills)	VL Soya-2, 21, 47, PK-327, Pusa-16, 20, 24, PS-1042, 1092
North Indian plains (Punjab, Haryana, Delhi, eastern and western Uttar Pradesh, and western Bihar)	Bragg, PK-262, 327, 416, PS-564, 1024, 1347, Pusa-24, 37, SL-96, 525, 688, Birsa Soya-1
Central India (Bundelkhand, Rajasthan, Gujarat, western and northern Maharashtra)	Durga, Gaurav, Gujarat Soybean-1, 2, JS-80, 21, 75, 46, 76, 205, Pusa-16, 22, 24, 37, Punjab-1, PS-1029
Northeast India (Assam, West Bengal, parts of Bihar, Meghalaya)	Birsa Soybean-1, Bragg, Pusa-16, 22, 24, JS-335, Raus-S
South India (Karnataka, Tamil Nadu, Andhra Pradesh, South Maharashtra)	CO-1, Devis, Monetta, Hardee, Pusa-37, 40, PS-1029, MACS-450, JS-335

Source: Saxena (2008).

performance of various groundnut cultivars grown in the red lateritic and alluvial soils of West Bengal, Orissa, and Assam are shown in Table 1.2. Recently, Pant Rai-20, a bold-seeded variety of mustard, has been released by the State Variety Release Committee of Uttarakhand (Bhajan et al., 2013). This variety is the result of the concerted efforts of G.B. Pant University of Agriculture and Technology (Pantnagar). It is a medium-maturing variety under Uttarakhand conditions with a greater yield than Kranti.

6. *Crop rotation.* The growing order in which chosen cultivated crops follow one another, in a set cycle, on the same field, over a definite period of time is termed crop rotation. The period may vary from 1 year to 3 years or even more. Rotations are planned carefully after considering the nature of the soil, climate, irrigation, demand in the market, and price. The main advantages of appropriate rotations are that the soil is managed properly regarding tillage and soil fertility is replenished by including a restorative leguminous

TABLE 1.2 Performance of Groundnut Cultivars in the Eastern States of India

	Kharif			Orissa		Assam
Cultivars	**Red lateritic soil**	**Alluvial soil**	**Rabi**	**Kharif**	**Rabi**	**Rabi**
TG23	–	–	2210	–	–	–
ICGS44	2312	2555	2920	750–1200	2500	–
AK12–24	1474	1577	–	650–1000	2500	–
ICGS5	1780	1940	1910	–	–	–
TG24	1981	2095	–	–	–	–
ICGS11	1929	2053	1920	750–1200	2691	2650
JI24	1831	1908	2250	–	–	2421
Girnar	2331	1918	2400	1070–1200	2200	2350
DRG12	–	–	–	–	2915	–
RSHY1	–	–	–	–	2040	–

Source: Ghosh (1999).

crop in between the exhaustive cereal or fibrous crops. For example, growing a rabi summer groundnut crop in rice fallows has been found to be very effective in replenishing the nitrogen content of the soil. Similarly, when a soybean oilseed crop is rotated between a maize and wheat-cropping sequence during the kharif (monsoon) season, the nitrogen content of the soil is increased. This is explained due to both of these crops belonging to the leguminous family, which is known to fix atmospheric nitrogen. Apart from these advantages, the possibility of some soils developing soil toxicity is removed by crop rotation and a healthier soil condition is therefore maintained. Soil microorganisms are able to play their full part in enriching the soil because of their action on soil organic matter. Crop rotation provides a good chance to eradicate weeds and help destroy fungal diseases and insect pests. In the red soils of the Telangana region of Andhra Pradesh, growing of sunflower rotated with groundnut has been found to be a promising crop sequence since 2000. In this cropping sequence the farmers of the area grow one rain-fed crop followed by a second crop watered from wells, tanks, or canal systems. The summer sunflower has emerged as a potential crop in this area after the harvest of cereals, pulses, oilseed castor, and groundnut (Reddy, 2000). Of these two oilseed crops, growing of groundnut was found to be more profitable in the sunflower and groundnut sequence.

GROWING HEAT AND DROUGHT-RESISTANT MUSTARD VARIETIES

Traditionally grown in Rajasthan, Haryana, and Madhya Pradesh, heat and drought-tolerant mustard varieties developed by public sector institutions during the last few years have made a debut in the southern states of Tamil Nadu and Karnataka in the current rabi (dry) season (Anon., 2014). Two such heat and drought-resistant varieties, Pusa-21 and Pusa-29 developed by the Indian Agricultural Research Institute (IARI), were sown on a trial basis in Tamil Nadu and Karnataka. These varieties possess not only the ability to tolerate higher temperatures in October, but also possess low erucic acid, which lessens pungency in the oil and is considered healthy. If these new varieties are accepted by the farmers of southern India, the country's annual mustard production is expected to rise during the next few years, which may reduce the dependence on edible oil imports.

The country's annual mustard production has been around 6.6–8.0 million tons over the last 5 years. This sustained production is attributed to early-sown, heat and drought-tolerant varieties such as Pusa (25, 27, and 28), Vijay, Mahak, and Agrani, developed by the IARI. It is important to mention that most of the country's mustard oil consumption is in the eastern and northern parts of the country.

Keeping in mind these points, farmers are encouraged to grow heat and drought-resistant varieties of mustard wherever conditions are suitable to sustain mustard production.

INTEGRATED NUTRIENT MANAGEMENT

Integrated nutrient management consists of the judicious use of chemical fertilizers in combination with organic manures, industrial and crop wastes, and biological nitrogen-fixing organisms.

Use of Inorganic Fertilizers

Once soil nutrient supplies are depleted through crop removal, the only method of replenishment is through use of external sources. The introduction of high-yielding varieties of various crops, especially rice and wheat, which are highly responsive to inorganic fertilizers, resulted in spectacular increases in their yield. In fact, fertilizers were the major components of the green revolution. However, excessive use of fertilizers led to several other problems such as pest and disease occurrence, nutrient deficiencies in phosphorus, sulfur, and micronutrients as well as physical problems. Hence, a balanced use of nitrogen–phosphorus–potassium (NPK) fertilizers in the ratio 4:2:1 is required. It would be much better if recommended levels of fertilizers were based on soil test results.

Nutrient Requirements of Oilseed

Groundnuts

Groundnuts require large amounts of primary (N, P, K), secondary (Ca, Mg, S), and micronutrients (B, Mo, Fe, and Mn). Although groundnuts are capable of meeting 60–80% of their own N requirement through atmospheric nitrogen fixation by Rhizobia in its root nodules (Gupta et al., 2012), in general the response of groundnuts vary from 37.5 kg ha^{-1} to 50.0 kg ha^{-1} under Maharashtra soil conditions and from 12.5 kg ha^{-1} to 25.0 kg ha^{-1} in the soils of Gujarat. Contrary to this, the N, P_2O_5, and K_2O requirement of groundnuts lies in the ranges 10–30, 15–40, and 0–45 kg ha^{-1}, respectively, in the groundnut-growing areas of Andhra Pradesh, Tamil Nadu, Maharashtra; Madhya Pradesh; and Rajasthan (Ghosh et al., 2005).

Soybeans

Like groundnuts, the N requirement of soybeans is also higher due to the presence of higher protein content. However, soybeans, which belong to the leguminous family, is capable of fixing atmospheric nitrogen to the extent of 49–130 kg ha^{-1}. In general, an application of 30 kg–ha^{-1} of N was found to be the optimum for obtaining a high yield in rainy season soybean crops. Phosphorus is critical to the flowering and pod development stages of soybean. Application of 26.2 kg ha^{-1} of P was found to be optimum for attaining a high yield of kharif soybean in the loamy, sand soil at Ludhiana (Singh et al., 2002). Potassic fertilizers are used in very low amounts. Generally, there is no response in soybean growth to added potassic fertilizers in north Indian soils due to the dominance of illite clay minerals in these soils.

Mustard

Large amounts of N, P_2O_5, and K_2O (144, 35, and 40 kg ha^{-1}) are required for rapeseed mustard, which is mainly grown in Rajasthan, Uttar Pradesh, Haryana, Madhya Pradesh, and Gujarat (Mandal et al., 2002). In Punjab, particularly in the Hoshiarpur district, gobhi sarson under a higher fertilizer dose (N, P_2O_5, K_2O in the ratio 125:45:30) gave better performance than under recommended doses (100:30:30) giving a yield margin of 2.3 q ha^{-1} (quintals per hectare).

Application of N at 112.5 kg ha^{-1} increased significantly the number of branches, secondary branches per siliqua, number of seeds per siliqua, and finally the seed yield of gobhi sarson (*B. napus*) (Thakur et al., 2008). Improved varieties of Indian mustard have been reported to respond to nitrogen applications up to 120 kg ha^{-1} (Kumar and Chauhan, 2005). Like pulses, the productivity of mustard (Varnua) responded significantly to 20 kg K_2O ha^{-1} in the subtropical soils of Kangra (Himachal Pradesh) and 60 kg K_2O ha^{-1} in those of temperate soils (Sharma et al., 1998). It has been found experimentally that producing 1 ton of Bhawani toria (*B. rapa* var. *napus*) grown at the Regional Research Station (Dhaula Kuan), representing the subhumid, subtropical zone of Himachal Pradesh, requires application of 45 kg N, 3 kg P, and 31 kg K per hectare (Suri et al., 2002).

Balanced use of fertilizers has been reported to increase seed yields by 37–38% (Bhajan and Bhati, 2009) in the case of mustard and toria grown in Uttarakhand and Uttar Pradesh. In another study, Bhajan et al. (2007) found that fertilizers contributed to a 37–73% increase in the yield of mustard by adoption of a full package of practices.

Mustard grown in a sandy, clay loam soil at Meerut also responded to higher fertilizer doses. The interaction between N, P, and K also showed that the Pusa Bold variety of mustard responded well to 120 kg N + 60 kg P_2O_5 + 60 kg K_2O ha^{-1}, whereas Varnua responded well to 80 kg N + 40 kg P_2O_5 + 40 kg K_2O – producing the same yield level as mentioned in *Fertilizer News* (Mandal et al., 2002).

Sunflowers and Safflowers

The estimated N, P_2O_5, and K_2O requirement for sunflowers has been found to be in the order of 63.3, 19.1, and 126.0 kg ha^{-1}, respectively, whereas the amount of these nutrients is 38.8, 8.4, and 22.0 kg ha^{-1} for safflower oilseed crops (Hegde and Sudhakara Babu, 2009). To get economic yields of kharif sunflowers in the red loam soil of the Coimbatore region, application of NPK fertilizer (50, 60, and 40 kg ha^{-1}) was necessary along with the addition of FYM at 10 t ha^{-1} (Mandal et al., 2002). However, in rabi sunflower cultivation, application of the recommended dose of NPK, as stated earlier, was found to be ideal.

According to Vishwakarma et al. (2007), sunflowers require 120 kg N ha^{-1}, 60 kg P_2O_5 ha^{-1}, 40 kg K_2O ha^{-1}, and 45 kg S ha^{-1}. A split application of N results in high growth, yield attributes, seed yield, and other quality parameters, when compared with a full basal application of N. Safflowers should be fertilized with 40 kg N ha^{-1}, 40 kg P_2O_5 ha^{-1}, and 35 kg K_2O ha^{-1} as a basal application.

Padmavathi and Mahavishnan (2007) have found that the response of safflowers to fertilizers mainly depends on the availability of moisture in the soil. Under rain-fed conditions, all fertilizers should be added as a basal dose by drill. They should be placed in seed furrows, below the soil. In the traditional single-cropped rabi tracts of Maharashtra and Karnataka, the application of fertilizers 2–3 weeks prior to the optimum planting time is recommended for maximum efficiency under receding soil moisture.

Under irrigated conditions 50% of N and 100% P and K fertilizers should be added at the time of sowing with the remaining 50% N being top-dressed after 5 weeks – at the time of first irrigation (Padmavathi and Mahavishnan, 2007).

Sesame, Linseed, and Niger

The major sesame-growing states are Gujarat, Rajasthan, Madhya Pradesh, Tamil Nadu, Maharashtra, Andhra Pradesh, West Bengal, and Karnataka. Like other member of the oilseed family, sesame also removes a high quantity of plant nutrients (Mandal et al., 2002). Economic yields of sesame requires the application of 62, 24, and 64 kg ha^{-1} of N, P, and K. Mostly there is a positive interaction with N and P. Thus, combined and optimum applications of N and P always ameliorate sesame yield. The application of 20 kg K_2O ha^{-1} increased yield ranging from 55–60% under the subtropical soil conditions of Himachal Pradesh (Sharma et al., 1998). At Parbhani, on black cotton soil with a slightly alkaline pH (7.6), the highest grain yield of summer sesame was obtained with a combined application of 120 kg N and 175 kg P_2O_5 ha^{-1}. This was statistically on par with application of 120 kg N and 175 kg P_2O_5 ha^{-1}, which was recorded by application of 120 kg N and 50 kg P_2O_5 ha^{-1}.

Linseed is mainly cultivated in three states in India: Madhya Pradesh, Uttar Pradesh, and Maharashtra. To get a maximum yield of linseed (i.e., to the extent of getting 16 q ha^{-1}), 96 kg of N, 13 kg of P, and 72 kg of K ha^{-1} are required to be added. However, Dixit and Sharma (1993) found that the response of linseed crops to added K was only measurable up to 60 kg K_2O ha^{-1} under Himachal soil conditions. Potassium applications of up to 60 kg K_2O ha^{-1} also proved beneficial to linseed crops grown in the temperate zone soils of Kangra in Himachal Pradesh (Sharma et al., 1998). Crop response to applied K is attributed to the light, textured nature of these soils, to poor organic matter, to low cation exchange capacity, and to their being poorly buffered with respect to K saturation.

Castor

Under unirrigated conditions, castor crops should be fertilized with 40 kg N, P_2O_5, and K_2O ha^{-1} basally and 10 kg N ha^{-1} as a top dressing. Use of a single super phosphate (SSP) not only fulfills the requirement for P but also supplies Ca, Mg, and S to the crop. However, for irrigated castor raised in North Gujarat a split application of 120 kg N ha^{-1} (40 kg basal + 30 kg N at 30 and 70 days after sowing + 20 kg N at 100 days after sowing) is recommended as it was found to give the maximum return per rupee invested (Anon., 1997, 2000).

Use of Inorganics and Organics

Gobi sarson and toria are the major rabi crops grown on the mid-hills of the northwestern Himalayan state of Himachal Pradesh. These crops significantly responded to 120 kg N ha^{-1} in manured plots and 80 kg N ha^{-1} in unmanured plots on acidic silty clay loam soil (a type of alfisol) in the mid-hills of Palampur (Mankotia and Sharma, 1996). In another study, Mankotia and Sharma (1997) observed that the seed yield of gobi, sarson, and toria increased considerably, with the highest yield of 1733 kg ha^{-1} being obtained with 160 kg N ha^{-1} + 5 t FYM ha^{-1} (on a dry weight basis). This was about 54% higher than with a sole application of 160 kg N ha^{-1}. The yield advantage (percent increase in seed yield) due to the application of FYM was 25, 37, and 42 kg seed ha^{-1} at 40, 80, and 120 kg N ha^{-1}, respectively. The application of FYM results in better yields because it facilitates improved utilization of both applied and native mineralized soil N. The impact of FYM application on the response of crops to added P was similar that of N. The highest yield of 1546 kg ha^{-1} was obtained with an application of 35 kg P ha^{-1} + 5 t FYM ha^{-1} (Mandal et al., 2002).

Use of Inorganic Fertilizers and Biofertilizers

Indian mustard is responsive to chemical fertilizers. In view of the escalating prices of chemical fertilizers, there is dire need for alternative sources of nitrogenous and phosphatic fertilizers, especially biofertilizers. Nonsymbiotic bacteria such as *Azotobacter* and *Azospirillum* are potential biofertilizers which could replace nitrogenous fertilizers as a result of phosphate-dissolving bacteria replacing phosphatic fertilizers. Interaction between biofertilizers and N levels was found to be significant on the sandy loam soil of Gurgaon (Haryana), which revealed that an inoculated crop receiving 30 kg N ha^{-1} gave a grain yield equivalent to 60 kg ha^{-1} (Mandal et al., 2002). Similar results were obtained with *Azotobacter* inoculation at Hisar (Haryana).

Use of Phosphorus

A consistent increase in soybean yield due to N application has been found in the presence of applied phosphatic fertilizers. The highest seed yield was obtained with an application of 60 kg N + 120 kg P_2O_5. At different locations of the Pithorgarh district, in the hilly areas and foothills of Kumaon (Uttar Pradesh), an application of N at 40 kg ha^{-1} and P_2O_5 at 80 kg ha^{-1} enhanced the yield of soybean crops and, therefore, net return. In the clayey soil (deep vertisol) at Raipur (Madhya Pradesh), a similar response to the combined application of N and P was observed (Patel and Chandravanshi, 1996).

Application of Lime to Control the Deficiency of Calcium

Application of lime should be the main approach to the management of acid soils since it improves base saturation, inactivates Al, Fe, and Mn when present in excess, reduces P fixation, and improves both biological N fixation and mineralization of N.

Application of lime at 1–2.5 t ha^{-1}, depending upon the intensity of soil acidity, is required to be applied 3–4 weeks prior to sowing groundnut crops, since the P content of soils in the eastern and northeastern states of India is very low (<10 mg kg^{-1} soil). After reclamation of acid soils, gypsum should be added at 250 kg ha^{-1} at the time of sowing along with other fertilizers in order to supply Ca and S to groundnut crops (Ghosh, 1999).

Sustainable Productivity and Soil Quality

The results of a long-term, integrated nutrient management experiment (1985–2010) revealed that the accumulation of organic carbon was higher in every soil profile under different management practices (Rao, 2012). In another field experiment that looked at the organic farming of groundnut crops between 2002 and 2010, soil organic carbon content in the top 15 cm of soil was found to increase at a comparatively higher rate in organic plots than in inorganic plots, although this was not reflected in pod yield.

Soil organic carbon content in the top 15 cm of soil was found to increase from initial levels of 2.19 kg after 6 years of application of 10 t FYM ha^{-1} to groundnuts grown on red loamy soils at Tirupati (Andhra Pradesh), whereas it increased to 3.19 kg only in a plot to which inorganic fertilizers were applied in the same period of time (Annual Report, 2008).

Sulfur Deficiency in Soils and Its Role in Sustainable Oilseed Crop Production

In the early 1990s it was found that soils in about 130 districts of various states were suffering from some degree of S deficiency. In 2007, based on soil research results, it was estimated that soils in over 250 districts of India suffered from varying degrees of

S deficiency (Kavitha et al., 2014). Close to 40–45% of the soil samples analyzed were found to be of low S status, requiring an application of S for sustainable management. This percentage increases in the cases of soils in which oilseed crops are grown. This is attributed to the following reasons:

1. Sulfur is essential for protein synthesis, primarily because S is a constituent of three S-containing amino acids (cystine, homocysteine, and methionine) which are the building blocks of proteins. About 90% of plant S is present in these three amino acids.
2. Sulfur is essential for the synthesis of oils. This is why adequate sulfur is so crucial for oilseed crops and the activation of enzymes, which help in biochemical reactions within the plant.
3. Sulfur improves the protein content and oil percentage in seeds.
4. The effect of sulfur application on yield, and the various attributes of different oilseed crops, was studied in S-deficient soils under various Indian climatic conditions. It was found that:
   a. Application of S at 60 kg ha^{-1} with gypsum into the groundnut-rice cropping system recorded the highest cumulative grain yield under Orissa soil conditions (Gupta et al., 2012). The highest oil content (47.1%) was also recorded after application of S at the same rate with gypsum. Oil content in the seeds of groundnut crops was also observed.
   b. Application of 15, 30, and 45 kg S ha^{-1} with superphosphates in the raya–wheat cropping system, at different locations in Punjab, increased *raya* seed yield significantly over a control crop. However, the yield increase in wheat was not significant.
   c. Oilseed crops such as sesame, safflower, linseed, and soybean were also found to be very responsive to S. Approximately 12 kg S are required to produce 1 t of oilseed.
   d. Application of 10 kg S ha^{-1} under the rain-fed conditions of Andhra Pradesh, and 20 kg S ha^{-1} under irrigated conditions, via elemental S, showed an increase in castor yield when compared with a control crop.
   e. Linseed crops responded markedly to S addition on the S-deficient soils of Uttar Pradesh, Rajasthan, and Punjab (Gupta et al., 2012).
   f. Application of 40 kg S ha^{-1} along with 112.5 kg N resulted in a significantly higher number of primary and secondary branches per plant (Thakur et al., 2008), as well as increases in the number of siliquae per plant, seeds per siliqua, and seed yield of gobhi sarson (*B. napus* ssp. *oleifera* var. *annua*) under the mid-hill conditions of Himachal Pradesh. However, oil content was not affected by S application. Contrary to this, oil content increased significantly with increasing levels of S in the case of rain-fed mustard on inceptisol soils in the mid-hill intermediate zones of Jammu and Kashmir (Sharma and Jalali, 2001).
   g. A significant response in mustard crop was observed with 50 kg S ha^{-1} for seeds and 75 kg S ha^{-1} for straw and oil yields. The highest seed (2.47 t ha^{-1}) and oil yields (0.99 t ha^{-1}) were recorded for 75 kg S ha^{-1} in the cold arid region of Ladakh in Jammu and Kashmir (Thakur, 2008).

Application of 20–40 kg S ha^{-1} enhanced the S content of Indian mustard under the typic ustochrept soil of Gujarat. However, the highest S content in grain and straw (0.78 and 0.32%, respectively) from Indian mustard was recorded after 20 kg S ha^{-1} application (Mehta et al., 2013).

Although the optimum rates of S to be added depend upon such factors as the S status of soil, yield potential, nitrogen level in coarse textured soils with low sulfate retention

TABLE 1.3 Nitrogen, Phosphate, Potassium, and Sulfur Uptake

		Total uptake (kg ha^{-1})			
Crop	**Yield (kg ha^{-1})**	N	P	K	S
Groundnut	1900	121	19	–	15
Mustard	2596	131	25	133	45
Soybean	2500	125	43	101	22
Sunflower	2308	114	26	141	17
Safflower	1909	74	07	35	24
Sesame	1200	62	24	64	14
Linseed	1600	96	13	72	09

Source: Pal and Singh (2010).

capacity, and S uptake, a rule of thumb is that S uptake per ton of oilseed grain production is 5–20 kg ha^{-1}, with an average of 12 kg ha^{-1} (Pal and Singh, 2010).

Crop uptake of S and its comparison to NPK uptake are shown in Table 1.3.

Micronutrients

A field experiment was carried out to study the effect of molybdenum (Mo) on yield, quality, and nutrient content in Indian mustard (*B. juncea* L.) on the loamy sand (typic ustochrept) soil of Anand (Gujarat) at three abundances (0, 1, and 2 kg ha^{-1}) using ammonium molybdate (Mehta et al., 2013). The results indicated that Mo application increased Mo content in mustard. Although the oil content of Indian mustard grains was not affected much by the application of Mo it did improve the available Mo content of the soil. In light of the impressive benefit gained by applying boron (B) and zinc (Zn) (i.e., of increasing oilseed productivity) (Hegde and Sudhakara Babu, 2002), there is a dire need to harness their role extensively. Of the various oilseed crops, soybean and groundnut – being leguminous in nature – are relatively sensitive to deficiencies in Mo, Zn, Mn, and Fe. Application of 0.5 kg Mo ha^{-1} has been found to remove Mo deficiency and increase soybean yields.

The biomass produced by soybean is higher than wheat and the average uptake of Fe, Mn, Zn, and Cu by a single crop of soybean is higher than that of wheat (Ghosh et al., 2001). These crops also differed significantly in their ability to remove micronutrients from the soil. The uptake of Fe and Mn was much higher than that of Zn and Cu because of their initial concentrations in the soil.

SEED INOCULATION WITH *RHIZOBIUM* CULTURE

Of the nine-oilseed crops only two – soybean and groundnut, belong to the leguminous family. Like other legumes they also fix atmospheric nitrogen. To fix more atmospheric nitrogen, seed inoculation is done using an appropriate *Rhizobium* culture. The seeds of soybean are inoculated with *Bradyrhizobium japonicum* at 5 kg. Treated seeds should be kept under shade and then immediately sown. It has been found that application of 30 kg N and 60 kg P_2O_5 ha^{-1} along with seed treatment with *Rhizobium* gave significantly higher grain yield of soybean (2995 kg ha^{-1}) and fixed more N than when P was not applied along with seed inoculation at Raipur (Singh et al., 2006).

A good crop of soybean can yield 30–35 q ha^{-1} and generally does not require additional N application. Soybean is able to fix 200–250 kg N ha^{-1} from the atmosphere and after harvest of this crop it leaves 35–40 kg N ha^{-1} for the succeeding rabi season crop (Saxena, 2008).

In Manipur, the results of application of *Rhizobium* culture have been found to be encouraging when it is inoculated in groundnut seeds before sowing. Rhizobial inoculation not only increased the availability of phosphorus in the groundnut rhizosphere but its uptake was also increased by groundnuts grown in four different soils of Tamil Nadu (Vidhyasekaran et al., 2013). *Rhizobium* spp. from groundnuts were found to release phosphorus from water-insoluble tricalcium phosphate *in vitro*. The bacterium was found to produce phosphatase enzymes in appreciable quantities.

The effect of coinoculaton of *B. japonicum* and *Pseudomonas* spp. on nitrogen and phosphorus availability to soybean grown under a pot trial was studied (Gautam and Pant, 2002). The results indicated a significant increase in modulation, atmospheric nitrogen fixation, phosphate solubilization in soil and dry matter of soybean (*Glycine max* L. Merr.) with single superphosphate and rock phosphate of 26.6 kg P ha^{-1}. Rock phosphate proved better for nodulation and growth of soybean when used in combination with any phosphate-solubilizing bacterial strains.

In West Bengal, multilocational trials using improved strains of *Bradyrhizobium* – IGR6 and IGR40 – have shown an increase in pod yield of groundnuts of 150–170 kg ha^{-1}. These strains are tolerant to commonly used seed-treating fungicides such as Bavistin and Thiram (Ghosh, 1999).

Inoculation with Other Organisms

Inoculation of castor seed with nonsymbiotic N fertilizers such as *Azospirillum*, *Azotobacter*, and *Clostridium* plus a mixed culture of P-solubilizing organisms led to enhanced growth, dry matter, as well as N and P uptake by castor plants grown in sandy soil. It is noteworthy that of all these cultures *Azospirillum* was found more effective than the others, while P was solubilized by organisms even from rock phosphate (Raghavaiah, 2005).

For irrigated castor in North Gujarat, seed treatment with *Azospirillum* or phosphorus-solubilizing bacteria at 50 g seed kg^{-1} and application of recommended fertilizers (75–50–0 g NPK ha^{-1}) in conjunction with castor cake 1 t ha^{-1} or 5 t FYM ha^{-1} has been recommended to increase and sustain castor yield.

In Andhra Pradesh, 50% recommended doses of fertilizer + *Azospirillum* seed treatment + 25% N through FYM showed higher castor yields under rain-fed conditions (Raghavaiah, 2005).

Rock phosphate at 132 kg P ha^{-1} using *P. striata* could also be a more economic and alternative source of phosphorus. Use of *P. striata* with both sources of phosphorus (superphosphate and rock phosphate) proved more effective. Phosphate-dissolving organisms like *Aspergillus awamori* are found in the medium black soils (pH 7.2–7.7) of Sehore (Madhya Pradesh).

INTEGRATED PEST MANAGEMENT

Integrated pest management (IPM) is a broad ecological approach to managing insect pest problems rather than their complete eradication. The IPM approach, in fact, is the best mix of all available methods and techniques that are used for total crop management in order to keep the pest population below the economic threshold level. Greater emphasis is placed on the use of resistant varieties of crops to pests (diseases, cultural control, fertilizer and soil water management, and use of botanical pesticides).

Cultural Control

Cultural methods such as crop rotation, deep tillage, removal of weeds, and adjusting sowing/harvesting time should be followed to maximize prevention of pests and diseases.

Fertilizer and Soil Water Management

Soils rich in organic matter with balanced status of NPK and micronutrients can withstand pest attacks. Appropriate soil moisture is also important for limiting pest attacks and maintaining good plant growth.

Botanical Pesticides

Botanical pesticides that have been found less persistent in the environment and safe for mammals/other nontarget organisms should preferably be used to control various insect pests. Commercial products/formulations prepared from such trees as neem (*Azadirachta indica*), drek (*Melia azedarach*), and bael (*Aegle marmelos*) should preferably be used over synthetic or chemical pesticides to control insect pests.

Today, neem is the most promising source of botanical origin biopesticides. Neem owes its toxic attributes to a number of bitter compounds called meliacins such as azadirachtin, nimbin, and salanin, of which azadirachtin is the most potent. These products affect pests by functioning as feeding and oviposition deterrents, insect regulators, and toxicants. Any pest escaping one effect may get killed by the other effects. Neem has been found effective against more than 200 pests including locusts, hairy caterpillars, pod borers, stem borers, beetles, jassids, aphids, and mealy bugs (Chander and Munshi, 1995; Gupta and Singh, 2008).

The crushed roots of marigold (*Tagetes erecta*) provide good control of root knot nematodes when applied to soil. The seed extract of grapefruit (*Citrus paradisi*) can be used to control potato beetle. Garlic acts as a repellent for control of various insect pests. The root extract of drumstick (*Moringa oleifera*) inhibits the growth of several bacteria.

Botanical pesticides such as rotenone, pyrethrum, ryania, and nicotine were in vogue before the advent of synthetic pesticides. However, these botanicals were thereafter relegated to insignificant positions in pest control. Pyrethrum is extracted from the dried flowers of chrysanthemums (*Chrysanthemum cinerariefolium*). It has a rapid knockdown effect on flying insects and is effective against household pests, etc.

Main Insect Pests of Oilseed and Their Control

Oilseed is damaged by a number of pests, of which mustard aphid is the most serious, followed by mustard sawfly and painted bug (Reena et al., 2010). Similarly, groundnuts are attacked by such insect pests as groundnut aphid, leaf miner, stem borer, red hairy caterpillar, and white grub. Sesame is mostly attacked by a few major insect pests such as gall fly, til hawk moth, shoot webbers, capsule borers, aphids, thrips, leaf hoppers, and pod bugs.

Linseed is damaged by gall midge maggots that feed on buds and flowers. Jassids, white flies, thrips, and hairy caterpillars are the other insects that cause damage to the linseed crop. Safflower aphids cause serious damage by sucking cell sap. It too is attacked by bud fly, caterpillars, and leaf thrips.

To check insect pests, the following strategies require to be adopted:

1. Early sowing of mustard and groundnuts during the season escapes damage from mustard groundnut aphids.
2. Deep summer plowing helps expose the white grubs resting in the soil which cause lot of damage to groundnuts.
3. Growing of aphid-resistant varieties of mustard and thrip-resistant varieties of groundnuts such as Robut 33-1, Kadiri-3, and ICCS 8603 are essential.
4. Fields should be maintained weed free.
5. Setting up of light traps up at night to monitor and attract leaf miners, white grubs, and hairy caterpillars. Pheromone traps should be installed to prevent leaf miner infestation of groundnut.
6. Grow gall fly–resistant varieties of sesame. Field plowing during winter suppresses shoot webber activity in sesame.
7. Linseed crop timely sown escapes peak season damage by gall midges.
8. Always use recommended crop varieties that are resistant to various insect pests. Selection should be on the basis of main characteristics.

Use of Biofungicide for Control of Root Rot Diseases of Oilseed

As an alternative to chemical control throughout the world, plant pathologists have developed various microorganisms to control soil-borne pathogens. Of these *Trichoderma viride*, a fungus was found to be the most effective in controlling root rot in oilseed such as groundnuts, sesame and sunflowers. Apart from controlling root rot fungus (*Microphomina phaseolina*), *T. viride* is also known to produce certain growth-promoting substances, which resulted in vigorous growth in oilseeds and pulses (Ranganathan and Raguchander, 1995).

This biofungicide is used as a dry-seed dresser and is compatible with biofertilizers such as rhizobium and azospirillum. It is not harmful to animal and birds. The fungus is present in all soils without causing hazards or pollution.

In light of the above, farmers in areas where this biofungicide is available should use it instead of chemical fungicides.

INTERCROPPING

Growing two or more crops simultaneously without distinct row arrangement is called "mixed farming." Contrary to this, cultivation of two or more crops simultaneously in the same field with distinct row arrangement is known as "intercropping." Crop intensification has both time and space dimensions as a result of intercrop competition during all or part of crop growth and therefore farmers have to manage more than one crop in the same field at a time.

It is noteworth that India's farmers have been practicing mixed cropping since time immemorial (Gupta et al., 2012). It is attributed to the fact that this farming system assists in enhancing the crop productivity of land, provides good land cover to protect soil from water and wind erosion, increases employment opportunities, and gives more return. For example, intercropping of groundnuts with pigeon peas (2:1) is found more remunerative in Orissa (Ghosh, 1999). Similarly, intercropping of maize–groundnuts (1:1) in Arunachal and rice–groundnuts (3:1) in West Bengal has provided more net returns than when maize and rice are grown as sole crops.

A field experiment was carried out during the kharif season to study the effect of intercropping on the soil fertility and economics of sunflower and companion legumes such as soybeans, cowpeas, green grams, and rajmash (Khan et al., 2006). The results indicated that there was considerable improvement in residual soil N as a result of sunflower–legume combinations. All intercropped treatments had made better use of available soil P and K. Intercropping of sunflower–soybean (1:1) produced the highest net return (Rs. 14,850 ha^{-1}) with a benefit–cost ratio of 1:41. The next most promising treatment was sunflower–cowpea (1:1), followed by sunflower–green gram (1:1). According to Sharma and Sharma (2001), intercropping of sunflowers and groundnuts (1:2) had detrimental effects on the yield of sunflowers due to competition for growth-limiting factors (i.e., light, nutrients, and available space). However, intercropping of sunflowers and groundnuts (1:1) gave higher productivity (i.e., 3136 kg ha^{-1} than normal sunflowers where productivity was 2801 kg ha^{-1}.

Diversification

The profitability of oilseed solely from primary products such as seed and oil will not be sustainable. Besides primary product oil, oilseed crops provide immense scope for diversified uses with high-value speciality products and derivatives. From the vegetable oil consumption point of view, either for edible or for fuel purposes, the situation is envisaged toward valuing oil for its intrinsic value for calorie or for desired fatty acid, which is beyond the scope of individual crops as perceived now. De oils with requisite blends can meet the expectation, and to that extent the potential of individual oilseed crops would be for oil yield or for desired fatty acid, and not for oil from specific crops. Thus, the present wide diversity of oilseed crops may narrow down to a few high oil-yielding crops. As for the unique nonoil value aspects of specific aroma or nonoil uses (medicinal, ornamental, or other uses), individual oilseed crops could be grown for speciality purposes irrespective of productivity level.

Major opportunities for oilseed crop diversification and value addition include their introduction as catch crops in paddy fallows to utilize residual moisture and fertility; component crops in major wide-spaced field crops such as sugarcane, pigeon pea, cotton, and maize for sunflowers; main crops with groundnut, soybean, finger millet, pigeon pea, cluster bean, short-duration pulses, etc., for castor and sunflowers; with chickpea and coriander for safflowers; rabi castor under limited irrigation protection; and sunflowers for the Indo-Gangetic plains of

Punjab, Haryana, Western Uttar Pradesh in spring, and Bihar, Odisha, and West Bengal in rabi/summer. Soybeans also offer an opportunity for rice–wheat cropping system in the north.

References

Agricultural Outlook 2010. http://stats.oecd.org/index.aspx?queryid=23342-2019.

Annual Report, 2008. Regional Agricultural Research Station, Tirupati, ANGRAU.

Anon., 1997. Annual Progress Report (Castor). Directorate of Oilseeds Research, Hyderabad, pp. 83.

Anon., 2000. Annual Progress Report (Castor). Directorate of Oilseeds Research, Hyderabad, pp. 83–84.

Anon., 2014. Heat-resistant mustard debuts in Southern India. Agric. Today 17 (1), 19.

Bhajan, R., Bhati, U., 2009. Rapeseed-mustard: potential crop for diverse ecosystem. Indian Farmers Digest 42 (10), 14–16.

Bhajan, R., Singh, A.K., Singh, S.S., Sharma, C.L., Chaturvedi, 2007. Strategies and varietal options for increasing rapeseed-mustard production in Uttarakhand and Uttar Pradesh. Indian Farmers Digest 40 (8), 37–40.

Bhajan, R., Pant, U., Pant, D.P., Singh, A.K., 2013. Pant Rai-20: a bold seeded variety of mustard. Indian Farmers Digest 46 (12), 17–20.

Biswas, B.C., Das, S., 1999. Role of fertilizer in doubling oilseeds productivity in India. Fertilizer News 44 (10), 23–26.

Chander, S., Munshi, A.D., 1995. Ecofriendly botanical pesticides. Intensive Agric. 33 (1–2), 37–39.

Dixit, S.P., Sharma, P.K., 1993. Indian Journal Agric. Science. (Original not seen). In: Dev, G., Acharya, C.L. (Eds.), Cited from Use of potassium in Himachal Pradesh agriculture. Potash and Phosphate Institute of Canada–India Programme, Gurgaon, India, 1998, pp. 40–43.

Dubey, R., Dubey, D.P., Singh, D., Bhatt, C., 2011. Effective weed management strategies in soybean. Indian Farmers Digest 44 (9), 16–19.

FAO. 2010. faostat.fao.org/site/339/default.aspx.

Gautam, P., Pant, L.M., 2002. Effect of coinoculation of *Pseudomonas* species and *Bradyrhizobium japonicum* on nitrogen and phosphorus availability to soybean (*Glycine max*). Indian J. Agric. Sci. 72 (4), 248–251.

Ghosh, P.K., 1999. Groundnut in eastern and northeastern states: present status and future strategies. Indian Farming 49 (8), 11–17.

Ghosh, P.K., Manna, M.C., Hati, K.M., Mandal, K.G., Bandyopadhyay, K.K., Misra, A.K., Tripathi, A.K., Chaudhary, R.S., Acharya, C.L., 2001. International Arachis Newsletter, pp. 51–53.

Ghosh, P.K., Singh, R.K., Bandyopadhyay, K.K., Misra, A.K., Manna, M.C., 2005. Role of integrated plant nutrient supply for sustainable production in groundnut-based cropping system in India. Fertilizers News 50 (3), 45–48, 51–53.

Gupta, Y.P., 2009. In yellow revolution, seeds of a mission. The States Times 14 (246), 6.

Gupta, M.P., 2014. Challenges of food security. The Daily Excelsior 50 (77), 6.

Gupta, R.D., Banerjee, S.K., 1991. Problems and Management of Soil and Forest Resources of North West Himalayas. Mahajan Book Centre, Jammu, India.

Gupta, R.D., Singh, H., 2008. Indiscriminate use of pesticides in agriculture – public health issues and their control. Indian Farmers Digest 41 (1), 8–13.

Gupta, R.D., Arora, S., Gupta, S.K., 2012. Withering yellow revolution in the Indian context. In: Gupta, S.K. (Ed.), Technological Innovations in Major Oil Crop, vol. 2, Springer Science and Business Media, LLC.

Hegde, D.M., Sudhakara Babu, S.N., 2000. Prospect of sustainable oilseeds production in India. Agrinews (an information update from Department of Agriculture and Cooperation, Ministry of Agriculture, Government of India) 3(1), 3–4.

Hegde, D.M., Sudhakara Babu, S.N., 2002. Nutrient management in oilseed crops. In: Tandon, H.L.S. (Ed.), Nutrient Management Recommendations for Pulses and Oilseeds. Fertilizer Development and Consultation Organization, New Delhi.

Hegde, D.M., Sudhakara Babu, S.N., 2009. Declining factor productivity and improving nutrient use efficiency in oilseeds. Indian J. Agron. 54 (1), 1–8.

Kachroo, J., Sharma, M., 2008. Sunflower scenario in India. SKUAST J. Res. 7 (1), 122–131.

Kavitha, P., Siva Prasad, P.N., Reddy, S.K., 2014. Deficiency symptoms of sulphur in different crops and their management. Agrobios Newsl. 12 (10), 31–33.

Khan, M.I., Shah, M.H., Raja, W., Teeli, N.A., 2006. Effect of intercropping on soil fertility and economics of sunflower (*Helianthus annuus* L.) and companion legumes under temperate condition of Kashmir. Environ. Ecol. 245 (1), 171–173.

Kumar, A., Chauhan, J.S., 2005. Status and future thrust areas of rapeseed-mustard research in India. Indian J. Agric. Sci. 75 (11), 621–635.

Kumar, N., Chandra, S., Srivastava, A.K., 2006. Mulching: an effective tool for sustainable agriculture in rainfed farming. Indian Farmers Digest 39 (6), 30–33.

Luna, R.K., 2011. How India can reap demographic dividend. The Tribune 131 (103), 11.

Mandal, K.G., Ghouri, P.K., Wanjari, R.H., Hati, K.M., Bandyopadhyay, K.K., Mishra, A.K., 2002. Practical implication of nutrient × nutrient interaction to boost oilseeds productivity in India. Fertilizer News 47 (7), 13–18, 21–26.

Mankotia, B.S., Sharma, H.L., 1996. J. Oilseeds Res. 13(1), 27–28. (Original not seen). Cited from Practical implication of nutrient × nutrient interaction to boost oilseeds productivity in India. Fertilizer News 47(7), 13–18, 21–26.

Mankotia, B.S. and Sharma, H.L. 1997. Indian J. Agric. Sci. 67(3), 106–109. (Original not seen). Cited from Practical implication of nutrient × nutrient interaction to boost oilseeds productivity in India. Fertilizer News 47 (7), 13–18, 21–26.

Mehta, B.L., Meena, M.C., Patel, K.P., Ramani, V.P., 2013. Direct and residual effect of molybdenum and sulphur on yield and quality of Indian mustard and pearl millet in typic ustochrept soil of Gujarat. J. Indian Soc. Soil Sci. 61 (4), 359–364.

Padmavathi, P., Mahavishnan, K., 2007. Technologies for sustainable safflower production. Indian Farmers Digest 40 (8), 23–25.

Pal, Y., Singh, R.P., 2010. Sulphur: a key nutrient for oilseed crops. Indian Farmers Digest 43 (2), 17–18.

Patel, S.R., Chandravanshi, B.R., 1996. Indian J. Agron. 41(1), 601–03. (Original not seen). Cited from Practical implication of nutrient × nutrient interaction to boost oilseeds productivity in India. Fertilizer News 47(7), 13–18, 21–26.

Raghavaiah, C.V., 2005. Nutrient management in castor (*Ricinus communis* L.) and castor-based cropping systems. Indian J. Fert. 1 (3), 23–24, 27–33.

Ranganathan, K., Raguchander, T., 1995. Trichoderma – a mycofungi for the control of root rot diseases of pulses and oilseeds. Intensive Agric. 33 (1–2), 11.

Rao, P.C., 2012. Sustainability productivity and soil quality through soil organic matter maintenance. J. Indian Soc. Soil Sci. 60 (Suppl.), 46–48.

Reddy, B.N., 2000. Sunflower–groundnut: a promising crop sequence for red soils of Andhra Pradesh. Agrinews (an information update from Department of Agriculture and Cooperation, Ministry of Agriculture, Government of India) 3(1), 7.

Reddy, P.S., Patti, D., 1998. Indian oilseeds: present status and future needs. Indian J. Agric. Sci. 68 (8 Special Issue), 453–459.

Reena, Sinha, B.K., Prasad, S., 2010. Integrated insect pest management for dryland farming systems. In: Sharma et al. (Ed.), Technological Innovations in Dryland Farming System. IBDC Publishers, Lucknow, India.

Saxena, S.C., 2008. How to get good yield from soybean. Indian Farmers Digest 41 (6), 47–50.

Sharma, M.P., Jalali, V.K., 2001. Response of rainfed mustard (*B. juncea*) to sulphur and phosphorus on Inceptisol in midhill-intermediate zone of Jammu and Kashmir. Indian J. Agric. Sci. 71 (3), 195–196.

Sharma, P.K., Mahajan, K.K., Sud, R.D., Dixit, S.P., 1998. Response of pulse and oilseed crops to applied potassium in subtropical and wet temperate zone of Himachal Pradesh. In: Dev, G., Acharya, C.L. (Eds.), Use of Potassium in Himachal Pradesh Agriculture. Proceeding of Workshop held at HPKV, Palampur. Potash and Phosphate Institute of Canada-India Programme, Gurgaon, India.

Singh, D., Singh, H., Sita, R., 2002. Growth and yield of spring sunflower and groundnut as influenced by different cropping systems and rates of nitrogen. Ann. Plant Soil Res. 3 (2), 222–225.

Singh, R.K., Ghosh, P.K., Mishra, A.K., Mandal, K.G., Hati, K.M., 2006. Integrated plant nutrient supply for sustainable production in soybean-based cropping system. Indian J. Fert. 1 (11), 25–32.

Sudhakar, P.D., Singh, S., Rastogi, R., 2000. Oilseeds: current scenario and future strategy. Agrinews 3 (1), 1, 3.

Suri, V.K., Verma, T.S., Jaipaul, 2002. Nutrient efficiencies and their evaluation in Bhawani toria (*B. rapa* var. *napus*) on farmers fields for prescription based fertilizer recommendations. Indian J. Agric. Sci. 72 (8), 464–466.

Thakur, S.K., 2008. Effect of phosphorus and sulphur on yield and their uptake by Indian mustard (*Brassica juncea* I.) under cold arid region of Ladakh. In: Extended Summaries – National Seminar on INM, 2008.

Thakur, K.S., Kumar, A., Manuja, S., 2008. Effect of nitrogen and sulphur application on yield and yield attributes of gobhi sarson varieties under midhill conditions of Himachal Pradesh. In: Extended Summaries – National Seminar on INM, 2008, pp. 28–29.

Vidhyasekaran, P., Balaraman, N., Delveekasundaram, M., Viswanathan, G., Rangaswami, G., 2013. Phosphate dissolving activity of *Rhizobium* spp. from groundnut. Department of Microbiology, Tamil Nadu Agricultural University, India.

Vishwakarma, V.K., Sharma, M., Kumar, J., 2007. Balanced fertilization for higher yield of oilseed crops. Indian Farmers Digest 40 (8), 13–14.

CHAPTER

2

Breeding Oil Crops for Sustainable Production: Heavy Metal Tolerance

Muhammad A. Farooq, Basharat Ali, Rafaqat A. Gill, Faisal Islam, Peng Cui, Weijun Zhou

Institute of Crop Science and Zhejiang Key Laboratory of Crop Germplasm, Zhejiang University, Hangzhou, China

INTRODUCTION

Since ancient times, oilseed plants have been used to produce edible oils but now they are also considered promising substitutes to regular diesel fuel, because they are renewable and environmentally friendly. There are a large number of oil crops grown worldwide. Of them, rapeseed, soybean, mustard, peanut, sunflower, safflower, and cotton get considerable attention. Total global oilseed production from oil crops was 503.12 million tons (t) from 258.53 million hectares (ha) during 2013–2014 (http://www.fas.usda.gov/psdonline). The leading countries in oilseed production are China, India, the United States, Brazil, and Argentina. Of the several renewable energy and other byproduct sources, oilseed crops are progressively being considered as materials to fulfill the rising demand. In recent years the production of oilseed has increased somewhat due to heightened interest in vegetable oils and to facilitate the increasing demands of humankind (Zubr and Matthaus, 2002; Pilgeram et al., 2007). The lack of suitable conservation methods and other environment-related problems like salinity, drought, and metal stress limit the production of oilseed crops. Heavy metal contamination in the soil is a serious problem that not only affects the growth and yield of crops but also causes health hazardous effects. Heavy metals in excessive amounts cause disturbances in the nutrition uptake and carbohydrate metabolism of plants, and strongly reduces plant biomass (John et al., 2009). Heavy metals not only inhibit metabolic processes such as nitrogen assimilation and photosynthesis, but also are responsible for oxidative stress (Boussama et al., 1999). In spite of metal toxicity, some crops were reported to survive at higher concentration of metals (Ma et al., 2001). Therefore, significant interest has developed lately in the use the crop plants as green technology for the remediation of surface soils contaminated with toxic heavy metals (Vassilev et al., 2004). Based on genetic and functional studies, tolerance improvement has

Breeding Oilseed Crops for Sustainable Production. http://dx.doi.org/10.1016/B978-0-12-801309-0.00002-1

been made in different crops and can be used in phytoremediation. The use of such crops for remediation of polluted soil is important, so that the production of plant biomass could be high in response to establish the suitable agricultural management.

Trace Elements

Trace elements have been extensively studied in recent decades and have wide roles in biochemical mechanisms both as essential and toxic elements (Guengerich, 2009). Trace elements are normally found at low concentrations (mg kg^{-1} or less) in living organisms. It is well known that trace elements are essential for life and necessary for numerous metal-dependent enzyme and protein activities (Kramer et al., 2007). Rates of accumulation are necessarily governed by physiological requirements rather than toxicity (Di Toppi and Gabbrielli, 1999). Of the many naturally occurring elements of the Earth's crust, N, P, K, Ca, Mg, and S are macronutrients absolutely necessary for plant growth. There are a total of 53 heavy metals, but few of them (Co, Cr, Cu, Fe, Mg, Mn, Mo, Ni, V, Zn) have a direct bearing on living systems, being classified as essential elements.

Sources and Occurrence

There are 92 naturally occurring elements present in the environment in minute concentrations. In other words, there is widespread contamination of soil, water, air, and biota by all the toxic elements. Enhanced accumulation of metals in the environment mainly depends upon natural and anthropogenic activity. Natural sources accumulate trace elements in soil as a result of weathering of mineral rocks and atmospheric emissions from volcanic activity, but may be enhanced significantly in soil by commercial and anthropogenic activities (Kabata-Pendias, 2000). Inputs of trace elements through anthropogenic activities include the use of organic manure, fertilizers, pesticides, and as constituents of raw materials for different industrial byproducts. These processes add variable quantities of trace elements to the agroecosystem.

Trace Element Contamination

Small percentages of trace elements are bioavailable in soil and minute quantities are required for plant growth. These elements function as micronutrients and are essential for different plant metabolic processes such as energy metabolism, cell protection, hormone perception, gene regulation, and signal transduction (Hebbern et al., 2009). In spite of their essentiality for plant growth, the accumulation of trace elements, especially heavy metals from the soil, have the potential of restricting the soil's function, cause toxicity to plants, and contaminate the food chain. When the concentrations of these metal elements exceed permissible limits, they pose serious threats to both flora and fauna, and due to their high reactivity can directly influence plant growth. When the concentrations of toxic elements exceed permissible limits, they cause oxidative stress through free radicals and/or interrupting enzyme functions by replacing essential metals and nutrients (Henry, 2000). Moreover, the effect of metal accumulation is now considered one of the most disturbing factors to have arisen in the late nineteenth and early twentieth centuries (Azevedo et al., 2012).

WHY DO PLANTS TAKE UP TOXIC METALS?

Plant's life cycle duration requires not only macronutrients (N, P, K, Ca, S, and Mg), but also essential micronutrients such as Zn, Cu, Fe, Mn, Ni, and Mo. Plants have evolved highly specific and very efficient mechanisms to uptake, translocate in the vascular system, and store these nutrients in plant tissue. Toxic elements having chemical properties similar to those of essential elements also follow the same mechanism of movement in plant tissues. Plants have specialized root surfaces developed for the uptake of elemental nutrients from soil. Some metals accumulating in plants produce chelating agents that help in solubilization and uptake of metals like Al (Pellet et al., 1995), iron (Ma and Nomoto, 1996), and possibly zinc (Cakmak et al., 1996). Under nutrient deficiency, plants secrete siderophores (mugenic and avenic acids) which enhance metal bioavailability to plants that are bound tightly to the soil. Changes in pH and redox reactions in the rhizosphere also play a role in mobilizing plant nutrients (Marschner, 1995). Heavy metals are translocated apoplectically into plant tissue. The translocation mechanism is governed by specialized proteins, embedded in the plasma membrane, which involve cation–proton pumps where ATPases generate electrochemical gradients, or by ion channels that enable the transport of ions into the cell (Roberts and Tester, 1997).

TOXIC EFFECTS OF METALS ON OILSEED CROPS

Both essential and nonessential elements in elevated concentrations induce toxicity and inhibit plant growth. All metals are toxic at the source level of solubility. The overproduction and rapid accumulation of reactive oxygen species (ROS), also called "oxidative burst," is one of the earliest responses under excess of metal ions that mediates a series of alterations in biochemical and physiological processes. There are three different processes of metal induced–toxicity that can be found at the cellular level for biochemical changes:

1. Generation of free radicals and reactive oxygen species.
2. Blocking of essential functional groups in biomolecules.
3. Displacement of essential metal ions from biomolecules.

We now turn to the impacts of heavy metals on the morphological, physiological, and biochemical aspects of oilseed crops.

Retarded growth under metal toxicity occurs as a result of metabolic disorder. Toxic ions interact with polysaccharides and decrease the plasticity of cell walls. As roots are the first to contact toxic ions, root growth is more sensitive than shoot growth due to declines in cell division and rates of elongation in the root. Elongation in roots is inhibited due to the binding of toxic ions with homogalacturonan (HGA), which causes rigidity of the root cell wall. Generally, metal toxicity causes the decline in growth, reduces the yield (Raziuddin et al., 2011), and damages cell division, which leads to plant death. Recently, Ali et al. (2014) observed that oilseed growth decreased with exposure to different Cd concentrations. It is observed that root elongation stops after higher Cd concentration treatment in *Brassica napus.* The presence of higher concentrations of heavy metals in the soil can reduce the biological activity and fertility of soil, which leads to yield reduction (McGrath et al., 1996). Cr treatments decreased the absorption and accumulation of minerals like K^+, Ca^{2+}, and Mg^{2+} in sunflowers and interfered with their distribution into different plant parts as well, in accordance with previous findings (Shanker et al., 2004). Plant growth reduction among *Brassica* cultivars has been attributed to Pb-induced reduction in nutrient uptake and changed metabolic processes of cells (Ashraf et al., 2011). The transport of various important essential ions like N, Ca, Mg, P, K, and Zn was reported to get altered in both roots and shoots of all the cultivars of *Brassica* due to Pb toxicity (Gopal and Rizvi, 2008), thereby obstructing various physiological processes. Reduction in the absorption of nutrients may result from strong interaction of K^+ ions with Pb as both have similar radii and compete with each other for entrance into the plant through the same potassium channels. Shoot growth, leaf length, and leaf area also revealed the same results under Cr stress (Gill et al., 2015). The struggle of different elements such as Zn^{2+}, Cr^2, Ni^{2+}, and Cu^{2+} with Fe^{3+} for uptake from polluted soil creates the Fe deficiency in plants. Such uptake competition occurs between different elements because of the similar size of radii and uptake mechanisms. Metal toxicity changes the nature of membranes and shifts the ionic homeostasis in cells. A similar pattern in K deficiency was observed in crops as Mg^{2+} ions compete with K^+ for uptake. The acidic nature of soil induced by metal accumulation also contributes to nutrient deficiency. Cd^{2+} inhibits the transport of electrons and protons in the mitochondria, disorganizes the electron transport chain, and decreases adenosine triphosphate (ATP) formation (Heyno et al., 2008).

Crop productivity depends on its intrinsic capacity for photosynthesis and the availability of photosynthetic radiation within the canopy. Photosynthesis declines under metal stress and decreases the biomass of plant (Piotrowska et al., 2009). While a few studies reported yield enhancement with different metals (Breckle, 1991), a significant decrease in biomass after exposure to metal stress was observed by Imtiyaz et al. (2014) in soybean plants. Similar findings have also been reported recently (Gill et al., 2014; Ali et al., 2013a). Moreover, Ali et al. (2013b) and Raziuddin et al. (2011) reported that *B. napus* grown under metal stress was unable to take up essential elements for vegetative growth, which ultimately resulted in yield reduction.

Metal toxicity inhibits the photosynthetic rate by interrupting the constituents of photosynthetic parameters in plants, as documented by many scientists (Cenkci et al., 2010; Farooq et al., 2013; Touiserkani and Haddad, 2012). Due to the influence of metal ion or iron deficiency, the number of chloroplasts per cell decreased, which results in disturbing the composition of thylakoid membranes and the electron transport system (Bertrand and Poirier, 2005). The substitution of essential divalent ions in chloroplasts produces an ion imbalance that affects the functioning of chloroplasts (Solymosi et al., 2004), which is believed to be the cause of photosystem II (PSII) disassembly (Geiken et al., 1998). Heavy metal ions also interrupt the electron transport chain that disturbs the formation of ATP and nicotinamide adenine dinucleotide phosphate/oxidase (NADPH) in the Calvin cycle and breaks the influence of ribulose-1,5-biphosphate carboxylase/oxygenase (RUBISCO) and phosphoenolpyruvate carboxylase (PEPcase) activity (Chaffai et al., 2007). Essential metal ions play important roles in the functioning of various enzymes. Replacement of these ions with toxic ones leads to reduction in enzyme activity. For example, elements like Ni, Zn, and Co remove the Mg^{2+} from RUBISCO leading to loss of its activity. Similarly, Cd seeks out Ca^{2+} in calmodulin and results in inhibition of calmodulin-dependent phosphodiestrase activity (Garnier et al., 2006).

EFFECT OF HEAVY METAL STRESS ON OIL QUALITY

Mostly oilseed crops are grown for the production of oil. The oil content of small grains (wheat) is only 1–2%, while that of oilseed ranges from approximately 20% for soybeans to over 40% for sunflower and rapeseed (canola). The major world sources of edible seed oils are soybeans, sunflower, rapeseed, cotton, and peanut. Fats and oils are essential nutrients, comprising about 40% of the calories in diets. In recent years, oilseed has made remarkable progress throughout the world, especially in China, Canada, and India. However, due to overuse of pesticides and herbicides, large quantities of heavy metals have accumulated in the soil which, along with the contribution of industrial and sewage sludge to heavy metal accumulation, not only affect plant growth but also desaturate the composition of fatty acids. Fatty acids differ by length in commonly occurring 16–18 carbon atoms. Fatty acid components comprise phospholipids, glycolipids, and intracellular messengers, which contribute to the synthesis of hormones in cytosols. Reactive oxygen species formation under metal toxicity also induces peroxidation of membrane lipids (Askwith and Kaplan, 1998). In their study the decrease in glycolipid content in plants was attributed to metal stress. It has been observed that two molecules of glycolipids – monogalactosyldiacylglycerol (MGDG) and digalactosyldiacylglycerol (DGDG) – significantly decreased under Cd stress. This decrease in the MGDG/DGDG ratio can reduce the stability of the PSII complex (Quartacci et al., 2000), as MGDG is considered a lipid indispensable for PSII activity (Murata et al., 1990). Several reports indicate that environmental stresses have changed fatty acid composition, mainly in the content of linolenic acid (C18:3). Heavy metal stresses break down the citrate carrying an acetyl group from the mitochondria to the cytoplasm and for this process NADPH is required, which is produced by the pentose phosphate pathway. Fatty acids are triacylglycerols (neutral fats), which can be activated to acyl CoA by the hydroxylic action of lipases, translocated to the inner mitochondrial membrane by carnitinine, and simultaneously destroyed in the mitochondrial matrix. Similarly, Krupa and Baszynski (1989) found that loss of thylakoid activity under stress conditions lowered the contents of glycolipids and phospholipids. The removal of thylakoid acyl lipids decreased the amount of *trans*-δ-3-hexadecanoic acid, and excluded the linolenic acid that tends to occur in all lipids. So, heavy metals not only affect plant yield but also alter the composition of the fatty acids of membrane lipids, which reduces the oil quality of crops.

HYPERACCUMULATION AND OILSEED CROPS

The use of plants for remediation of metal-polluted soil was first reported by Baker and Brooks (1989) from the discovery of "hyperaccumulators." They are defined as plants that contain more than or up to 0.1% (i.e., more than 1 mg g^{-1}) of Cr, Cd, Cu, Pb, Co, and Ni or 1% (>10 mg g^{-1}) of Zn or Mg in dry matter, which can accumulate metals in their aerial parts at a level higher than the soil. Hyperaccumulators are plants that can accumulate high concentrations of metal contents in their roots, shoots, and/or leaves (Raskin et al., 1994; Cunningham and Ow, 1996). The extent of metal accumulation can vary in different species and ecotypes of the same species (Chen et al., 2004). Researchers have identified over 400 plant species as natural metal hyperaccumulators. Brassicaceae is a large family containing a large number of hyperaccumulating species (87 from 11 genera). Due to the slow growth rate of these plants, biomass production is limited. According to Szczygłowska et al. (2011), the most often cited species in phytoremediation was *B. juncea* Czern. (148 citations), followed by *Helianthus annuus* (57 citations), and *B. napus* and *Zea mays* (39 citations each). The potential of these crops for phytoextraction has been studied in recent years (Baker et al., 1994; Ebbs and Kochian, 1998), especially within Brassicaceae, in view of the large number of hyperaccumulators belonging to this family. Metal accumulation at higher levels has been observed in *Brassica* spp. like *B. juncea, B. napus,* and *B. oleracea* (Ali et al., 2014; Gill et al., 2015; Nanda-Kumar et al., 1995). Of the Brassicaceae, *B. juncea* has received a lot of attention regarding the desired accumulation of metal concentrations. It has been described as a hyperaccumulator for various metals such as Cd, Cr, Cu, Mn, Pb, Se, and Zn. Metal accumulation at higher levels has been observed in *Brassica* spp. like *B. juncea, B. napus,* and *B. oleracea* (Ali et al., 2013c; Gill et al., 2015; Nanda-Kumar et al., 1995). *B. juncea* showed great resistance against Cd toxicity and accumulated Cd at a level of 1450 μg Cd g^{-1} dry wt as compared to *B. napus* (555 μg g^{-1} dry wt) (Nouairi et al., 2006). In addition, this plant proved to be more effective at removing other metals such as Se (reduced between 13% and 48%) and Pb (28% reduction) (Salt et al., 1998). *B. napus* also has great capacity to accumulate Cd in its roots under hydroponic conditions (Ali et al., 2013d). The different metal accumulation potential in sunflowers has been reported by several authors (Fellet et al., 2007; Tassi et al., 2008). Due to their long and fibrous root systems, sunflowers are considered the most promising plants for metal removal in water. Their roots remove Pb (Dushenkov et al., 1995), U (Dushenkov et al., 1997a),^{137}Cs and^{90} Sr (Dushenkov et al., 1997b) from hydroponic solutions. Maize has been found to accumulate significant percentages of Cd and other metals such as Ni, Cu, Pb, and Zn (Murakami and Ae, 2009) with the help of amendments, chelators, bacteria, etc.

MOLECULAR ASPECTS OF METAL HYPERACCUMULATION

Metal Transporters

ATPases not only play critical roles in ion homeostasis but also in the overall strategy for heavy metal tolerance in plants (Williams et al., 2000). CPx-ATPases play key roles in the operation of ethylene signaling in plants through Responsive-to-antagonist 1 (RAN1). The suggested role of RAN1 is to deliver Cu reported from yeast complementation studies to the secretory system through this CPx-ATPase, which is required in the production of functional hormone receptors (Hirayama et al., 1999). So, these ATPase protein types not only contribute to the transport of sufficient heavy metal ions for essential cell functions but also prevent the accumulation of toxic levels. Natural resistance-associated macrophage proteins (NRAMPs) define a highly novel family of related proteins that are involved in the metal ion transport of divalent metal ions. Rice was the first plant in which this family was identified as having three NRAMPs (OsNramp 1–3) (Belouchi et al., 1995). NRAMPs play a role in Fe and Mn uptake in transport-deficient yeast, and disruption of the *AtNramps3* gene increased their Cd accumulation and sensitivity in *Arabidopsis* (Thomine et al., 2000). In barley, in the presence of Cd, the NRAMP transcript downregulated in adequate nitrogen supply but under N-deficiency strongly upregulated (Finkemeier et al., 2003).

Cation diffusion facilitator (CDF) proteins are families that transport heavy metals including Cd, Cr, Cu, and Zn. Heavy metal uptake depends upon certain members of the CDF protein family. Other members are located in the plasma membrane and in intracellular membranes. Van der Zaal et al. (1999) reported that ZAT protein is involved in the vesicular/vacuolar sequestration of Zn, thereby contributing to Zn tolerance in plants. Similarly, CDF family member BjCET2 proteins function as Zn transporters as reported in *B. juncea* on the basis of its sequence and general structural features. Overexpressing BjCET2 proteins possess an efflux mechanism that significantly accumulates more Cd and Zn in the shoots, indicating that the production of BjCET2 proteins enhances long-distance transport and shows great potential for phytoremediation in Cd-contaminated soil. Persans et al. (2001) have characterized a CDF transporter (TgMTP1) in *Thlaspi goesingense* that is thought to account for the accumulation of metal ions within shoot vacuoles. In yeast, TgMTP1t1 tolerates Co, Cd, and Zn accumulation, while TgMTP1t2 tolerates Ni accumulation. Both are highly expressed in *T. goesingense* compared with nonaccumulator species (Persans et al., 2001).

Phytochelatins

The synthesis of phytochelatins (PCs) is essential to cope with the higher levels of heavy metals in plants by binding them in complexes and sequestering the complexes inside their cells. PCs are metal-binding peptides, enzymatically synthesized in plants from glutathione (GSH) in the presence of heavy metal stress. More recently, PCs were found to be involved in long-distance transport of Cd in the phloem (Mendoza-Cózatl et al., 2008). Mutant lines of *Arabidopsis* that are deficient in the synthesis of PCs showed sensitivity against Cd and As tolerance (Clemens et al., 1999; Vatamaniuk et al., 1999). In *B. juncea*, expression of the GSH biosynthetic pathway was increased which showed that PC biosynthesis resulted in Cd tolerance (Yong et al., 1999). The conversion of GSH to phytochelatin by binding metal ions to the γ-glutamylcysteine dipeptidyltranspeptidase (PC synthase) enzyme hints at GSH biosynthesis being a probable pathway by which PC expression might be modulated.

Metallothioneins

Metallothioneins (MTs) are gene-encoded low-molecular-weight cysteine-rich metal-binding proteins, first identified in mammalian tissues as Cd-binding peptides (Maestri et al., 2010). In fungi and animals Cu induces heavy metal toxicity, providing evidence for a role in metal tolerance. It has also been reported that, in *Arabidopsis* seedlings, MT2 mRNA was induced by Cu which was shown to exploit high tolerance – less was noticed for Zn and Cd. Van Vliet et al. (1995) reported that MT genes can be induced by Cu and that MT2 RNA expression was enhanced in *Arabidopsis* Cu-sensitive mutants that accumulate higher concentrations of Cu. On the basis of cysteine protein content and structure, they can be divided into two different types (Vasak, 2005). The Cys-x-Cys motifs belong to type I, and Cys-Cys and Cys-x-x-Cys pairs belong to type II. Of the type I motifs reported, three were found in the Brassicaceae. Type I has 7 amino acid links in between 1 and 2 domains, while type II contains 27 genes in which domains 1 and 2 contain 8 and 6 cysteine residues, respectively. Introduction of the mammalian MT gene in *B. napus*, *B. campestris*, and *Nicotiana tabacum* confirms Cd tolerance by enhancing the chelation of metal with MTs (Karenlampi et al., 2000) and revealing that the MT gene plays a potential role in metal stress. Nevertheless, there is still a lack of information about MTs and their precise function in plants (Hall, 2002).

INTERACTING FACTORS IN OILSEED CROP BREEDING AND HEAVY METAL TOLERANCE

In recent years there has been marked improvement in the number of oilseed crops due to huge investments in biotechnological research by public and private research institutes as well as by commercial plant-breeding companies. Such techniques as tissue culturing, transformation response, embryogenesis, and DNA-based technologies have been used to study oilseed crops. Before the advances in hybrid technology, pedigree selection was widely used to develop new cultivars of oilseed crops. The reason for the huge investment in oilseed crops (especially in canola) was possible development of commercial hybrids. There has been remarkable heterosis in oilseed crops. The price of seeding is quite low, which enables farmers to purchase the requisite number of hybrid seeds. Basically, breeding has three objectives when it comes to oilseed crops: (i) improved seed yield, (ii) improved oil quality, and (iii) improvement biotic and abiotic stress tolerance.

However, there is little information regarding the relationship between oilseed crop breeding and heavy metals. Researchers are interested in plants that have the potential to transport heavy metals from roots to shoots. Similarly, there is demand for crops that reduce the accumulation of nontoxic or nonessential metals and metalloids in their edible parts for better quality of food. Oilseed crops show significant potential for phytoremediation of heavy metals due to higher biomass production and potential commercial value (rapeseed oil for biodiesel) with enhanced rates of metal accumulation in both roots and shoots (especially *Brassica* spp.) (Purakayastha et al., 2008). Veerle et al. (2006) investigated phytoremediation of Cd and Zn by different *B. napus* accessions and found little difference in metal accumulation patterns in both lab and field trials. Then, they crossed one accession having a high metal uptake rate with another having the highest metal translocation rate by hypothesizing that it would increase the metal uptake rate as well as its mobility from roots to shoots. However, data from filial generation 2 and parent accessions showed nonsignificant increase in metal uptake contents, suggesting that there is some potential to improve metal accumulation through crossing and selection. They suggested that selection based on this parameter would more likely result in higher biomass rather than in higher metal uptake and translocation. Chaney et al. (2007) claimed that the idea to breed and select oil crops, like *B. napus* varieties, with considerable enhanced metal phytoremediator properties is limited and that genetic engineering could play significant role as mercury is effectively remediated with transgenic crops (Heaton et al., 2003; Meagher and Heaton, 2005). However, to develop efficient phytoremdiators, much more knowledge of molecular and physiological uptake and tolerance of metals for a particular plant is required. For example, in Indian mustard the higher Cd accumulation brought about by transforming an *Escherichia coli gshI* gene that codes for γ-glutamylcysteine synthetase and an *E. coli gshII* gene that codes for ECS results in overexpression and increased Cd tolerance and accumulation (Zhu et al., 1999). The rhizosphere bacterium *Pseudomonas putida* 06909 is engineered to produce MBP-EC20, a metal-binding peptide that has high affinity for Cd. Due to the efflux pump in the metalloregulatory *cad* operon, bacterium *P. putida* 06909 showed moderate resistance to Cd (Lee et al., 2006). Increased microbial biosorption and plant–bacterium symbiosis not only protect the engineered strain but also the colonized sunflower plants against the toxic effects of Cd. Transgenic *B. napus* plants, which produce the 1-aminocyclopropane-1-carboxylate (ACC deaminase) enzyme to break down the ethylene precursor associated with plant growth-promoting bacteria, have also been used for remediation of Ni-contaminated soils (Farwell et al., 2007). It has been reported that gene *SAL13-2* binds to metal ions and plays a role in metal ion conjugation and resistance in this overexpression system (Tang et al., 2014).

CONCLUSIONS AND FUTURE PERSPECTIVES

Brassica spp. are widely cultivated crops throughout the world. Since time immemorial, oilseed crops have been used only for their seeds (as edible oils), but nowadays they provide other products in the form of food (vegetables) and fodder. The key strategy of plant breeding and genetics is to develop hybrids or cultivars of great economic importance. As a result of scientific advances, there has been significant growth in the commercial sectors in providing inputs to agriculture. Heavy metal toxicity has become a major global concern. Numerous studies have revealed the hazardous impact metal toxicity has on plant life due to increased urbanization, industrialization, and modern agricultural development. Injuries brought about by metal stress can eventually be traced back to changes in plants that adversely affect plant growth and productivity. Plants play an essential role in the removal of metals from soils. Coping mechanisms in oilseed crops, such as Indian mustard and canola cultivars, have the ability to tolerate and accumulate different heavy metals and develop sustainable ecosystems in polluted soil. Much progress has been made in oilseed crops as a result of transgenic technology. This is especially the case in *B. juncea* and *B. napus*, as they provide tolerance against metal stress. The other oil crops need further biotechnological intervention to improve specific traits like high yield and tolerance mechanisms. An overarching study is now needed to understand the genetic and molecular basis of oilseed crops as well as the mechanisms plants use to limit metal toxicity.

Acknowledgments

The authors acknowledge support from the National High Technology Research and Development Program of China (2013AA103007), Special Fund for Agro-scientific Research in the Public Interest (201303022), the Jiangsu Collaborative Innovation Center for Modern Crop Production (Nanjing, China), and the Science and Technology Department of Zhejiang Province (2012C12902-1, 2011R50026-5).

References

Ali, S., Farooq, M.A., Yasmeen, T., Hussain, S., Arif, M.S., Abbas, F., Bharwana, S.A., Guoping, Z., 2013a. The influence of silicon on barley growth, photosynthesis and ultra-structure under chromium stress. Ecotox. Environ. Safe. 89, 66–72.

Ali, B., Huang, C.R., Qi, Z.Y., Ali, S., Daud, M.K., Geng, X.X., Liu, H.B., Zhou, W.J., 2013b. 5-Aminolevulinic acid ameliorates cadmium-induced morphological, biochemical and ultrastructural changes in seedlings of oilseed rape. Environ. Sci. Pollut. Res. 20, 7256–7267.

Ali, B., Tao, Q.J., Zhou, Y.F., Gill, R.A., Ali, S., Rafiq, M.T., Xu, L., Zhou, W.J., 2013c. 5-Aminolevolinic acid mitigates the cadmium-induced changes in *Brassica napus* as revealed by the biochemical and ultra-structural evaluation of roots. Ecotox. Environ. Safety 92, 271–280.

Ali, B., Wang, B., Ali, S., Ghani, M.A., Hayat, M.T., Yang, C., Xu, L., Zhou, W.J., 2013d. 5-Aminolevulinic acid ameliorates the growth, photosynthetic gas exchange capacity and ultrastructural changes under cadmium stress in *Brassica napus* L. J. Plant Growth Regul. 32, 604–614.

Ali, B., Qian, P., Jin, R., Ali, S., Khan, M., Aziz, R., Tian, T., Zhou, W., 2014. Physiological and ultra-structural changes in *Brassica napus* seedlings induced by cadmium stress. Biol. Plant 58, 131–138.

Ashraf, M.Y., Azhar, N., Ashraf, M., Hussain, M., 2011. Influence of lead on growth and nutrient accumulation in canola (*Brassica napus* L.) cultivars. J. Environ. Biol. 32, 659–666.

Askwith, C., Kaplan, J., 1998. Iron and copper transport in yeast and its relevance to human disease. Trends Biochem. Sci. 23, 135–138.

Azevedo, R.A., Gratao, P.L., Monteiro, C.C., Carvalho, R.F., 2012. What is new in the research on cadmium-induced stress in plants? Food Energy Secur. 1, 133–140.

Baker, A.J.M., Brooks, R.R., 1989. Terrestrial higher plants which hyperaccumulate metallic elements – a review of their distribution, ecology and phytochemistry. Biorecovery 1, 1–7.

Baker, A.J.M., McGrath, S.P., Sidoli, C.M.D., Reeves, R.D., 1994. The possibility of *in situ* heavy metal decontamination of polluted soils using crops of metal accumulating plants. Resour. Conserv. Recy. 11, 42–49.

Belouchi, A., Cellier, M., Kwan, T., Saini, H.S., Leroux, G., Gros, P., 1995. The macrophage-specific membrane protein Nramp controlling natural resistance to infections in mice has homologues expressed in the root system of plants. Plant Mol. Biol. 29, 1181–1196.

Bertrand, M., Poirier, I., 2005. Photosynthetic organisms and excess of metals. Photosynthetica 43, 345–353.

Boussama, N., Ouariti, O., Suzuki, A., Ghorbal, M.H., 1999. Cadmium stress on nitrogen assimilation. J. Plant Physiol. 155, 310–317.

Breckle, S.W., 1991. Growth under stress: heavy metals. In: Waiser, Y., Eshel, A., Kafkafi, U. (Eds.), Plant Roots: The Hidden Half. Marcel Dekker, New York, pp. 351–373.

Cakmak, I., Ozturk, L., Karanlik, S., Marschner, H., Ekiz, H., 1996. Zinc-efficient wild grasses enhance release of phytosiderophores under Zn deficiency. J. Plant Nutr. 19, 551–563.

Cenkci, S., Cigerci, I.H., Yildiz, M., Özay, C., Bozdag, A., Terzi, H., 2010. Lead contamination reduces chlorophyll biosynthesis and genomic template stability in *Brassica rapa* L. Environ. Exp. Bot. 67, 467–473.

Chaffai, R., Ellhammadi, M.A., Seybou, T.N., Tekitek, A., Marzouk, B., Ferjani, E.E.I., 2007. Altered fatty acid profile of polar lipids in maize seedling in response to excess copper. J. Agron. Crop Sci. 193, 207–217.

Chaney, R.L., Angle, J.S., Broadhurst, C.L., Peters, C.A., Tappero, R.V., Sparks, D.L., 2007. Improved understanding of hyperaccumulation yields commercial phytoextraction and phytomining technologies. J. Environ. Qual. 36, 1429–1443.

Chen, Y., Li, X., Shen, Z., 2004. Leaching and uptake of heavy metals by ten different species of plants during an EDTA-assisted phytoextraction process. Chemosphere 57, 187–196.

Clemens, S., Kim, E.J., Neumann, D., Schroeder, J.I., 1999. Tolerance to toxic metals by a gene family of phytochelatin synthases from plants and yeast. EMBO J. 18, 3325–3333.

Cunningham, S.D., Ow, D.W., 1996. Promises and prospects of phytoremediation. Plant Physiol. 110, 715–719.

Di Toppi, L., Gabbrielli, R., 1999. Responses to cadmium in higher plants. Environ. Exp. Bot. 41, 105–130.

Dushenkov, V., Kumar, P.B.A.N., Motto, H., Raskin, I., 1995. Rhizofiltration: the use of plants to remove heavy metals from aqueous streams. Environ. Sci. Technol. 29, 1239–1245.

Dushenkov, S., Kapulnik, Y., Blaylock, M., Sorochisky, B., Raskin, I., Ensley, B., 1997a. Phytoremediation: a novel approach to an old problem. In: DLWise (Ed.), Global Environmental Biotechnology. Elsevier Science BV, Amsterdam, pp. 563–572.

Dushenkov, S., Vasudev, D., Kapulnik, Y., Gleba, D., Fleisher, D., Ting, K.C., Ensley, B., 1997b. Removal of uranium from water using terrestrial plants. Environ. Sci. Technol. 31, 3468–3474.

Ebbs, S.D., Kochian, L.V., 1998. Phytoextraction of zinc by oat (*Avena sativa*), barley (*Hordeum vulgare*), and Indian mustard (*Brassica juncea*). Environ. Sci. Technol. 32, 802–806.

Farooq, M.A., Ali, S., Hameed, A., Ishaque, W., Mahmood, K., Iqbal, Z., 2013. Alleviation of cadmium toxicity by silicon is related to elevated photosynthesis, antioxidant enzymes, suppressed cadmium uptake and oxidative stress in cotton. Ecotox. Environ. Safe. 96, 242–249.

Farwell, A.J., Vesely, S., Nero, V., Rodriguez, H., McCormack, K., Shah, S., Dixon, D.G., Glick, B.R., 2006. Tolerance of transgenic canola plants (*Brassica napus)* amended with plant growth-promoting bacteria to flooding stress at a metal-contaminated field site. Environ. Pollut. 147, 540–545.

Fellet, G., Marchiol, L., Perosa, D., Zerbi, G., 2007. The application of phytoremediation technology in a soil contaminated by pyrite cinders. Ecol. Eng. 31, 207–214.
Finkemeier, I., Kluge, C., Metwally, A., Georgi, M., Grotjohann, N., Dietz, K.-J., 2003. Alterations in Cd-induced gene expression under nitrogen deficiency in *Hordeum vulgare*. Plant Cell Environ. 26, 821–833.
Garnier, L., Simon-Plas, F., Thuleau, P., Agnel, J.P., Blein, J.P., Ranjeva, R., Montillet, J.L., 2006. Cadmium affects tobacco cells by a series of three waves of reactive oxygen species that contribute to cytotoxicity. Plant Cell Environ. 29, 1956–1969.
Geiken, B., Masojid, E.J., Rizzuto, M., Pompili, M.L., Giardi, M.T., 1998. Incorporation of [S35] methionine in higher plants reveals that stimulation of the D1 reaction center protein turnover accompanied tolerance to heavy stress. Plant Cell Environ. 21, 1265–1273.
Gill, R.A., Hu, X.Q., Ali, B., Yang, C., Shou, J.Y., Wu, Y.Y., Zhou, W.J., 2014. Genotypic variation of the responses to chromium toxicity in four oilseed rape cultivars. Biol. Plant 58, 539–550.
Gill, R.A., Zang, L., Ali, B., Farooq, M.A., Cui, P., Yang, S., Ali, S., Zhou, W., 2015. Chromium-induced physio-chemical and ultrastructural changes in four cultivars of *Brassica napus* L. Chemosphere 120, 154–164.
Gopal, R., Rizvi, A.H., 2008. Excess lead alters growth, metabolism and translocation of certain nutrients in radish. Chemosphere 70, 1539–1544.
Guengerich, F.P., 2009. Thematic mini review series: metals in biology. J. Biol. Chem. 284, 18557.
Hall, J.L., 2002. Cellular mechanisms for heavy metal detoxification and tolerance. J. Exp. Bot. 53, 1–11.
Heaton, A.C.P., Rugh, C.L., Kim, T., Wang, N.J., Meagher, R.B., 2003. Toward detoxifying mercury-polluted aquatic sediments with rice genetically engineered for mercury resistance. Environ. Toxicol. Chem. 22, 2940–2947.
Hebbern, C.A., Holst Laursen, K., Ladegaard, A.H., Schmidt, S.B., Pedas, P., Bruhn, D., Schjoerring, J.K., Wulfsohn, D., Husted, S., 2009. Latent manganese deficiency increases transpiration in barley (*Hordeum vulgare* L.). Physiol. Plant 135, 307–316.
Henry, J.R., 2000. An Overview of Phytoremediation of Lead and Mercury. NNEMS Report, National Network for Environmental Management Studies, Washington, D.C., pp. 3–9.
Heyno, E., Klose, C., Krieger-Liszkay, A., 2008. Origin of cadmium-induced reactive oxygen species production: mitochondrial electron transfer versus plasma membrane NADPH oxidase. New Phytol. 179, 687–699.
Hirayama, T., Kieber, J.J., Hirayama, N., et al.,1999. Responsive-to-antagonist 1, a Menkes/Wilson disease-related copper transporter, is required for ethylene signaling in *Arabidopsis*. Cell 97, 383–393.
Imtiyaz, S., Agnihotri, R.K., Ahmad, S., Sharma, R., 2014. Effect of cobalt and lead induced heavy metal stress on some physiological parameters in *Glycine max*. Int. J. Agri. Crop Sci. 7, 26–34.
John, R., Ahmad, P., Gadgil, K., Sharma, S., 2009. Heavy metal toxicity: effect on plant growth, biochemical parameters and metal accumulation by *Brassica juncea* L. Int. J. Plant Prod. 3, 65–75.
Kabata-Pendias, A., 2000. Trace Elements in Soils and Plants. CRC Press, USA.
Karenlampi, S., Schat, H., Vangronsveld, J., Verkleij, J.A.C., van der Lelie, D., Mergeay, M., Tervahauta, A.I., 2000. Genetic engineering in the improvement of plants for phytoremediation of metal polluted soils. Environ. Pollut. 107, 225–231.
Kramer, U., Talke, I.N., Hanikenne, M., 2007. Transition metal transport. FEBS Lett. 581, 2263–2272.
Krupa, Z., Baszynski T, 1989. Acyl lipid composition of thylakoid membranes of cadmium treated tomato plants. Acta Physiol. Plant 11 (2), 111–116.
Lee, W., Wood, T.K., Chen, W., 2006. Engineering TCE-degrading rhizobacteria for heavy metal accumulation and enhanced TCE degradation. Biotechnol. Bioeng. 95, 399–403.
Ma, J.F., Nomoto, K., 1996. Effective regulation of iron acquisition in graminaceous plants. The role of mugineic acids as phytosiderophores. Physiol. Plant 97, 609–617.
Ma, J.F., Ryan, P.R., Delhaize, E., 2001. Aluminum tolerance in plants and the complexing role of organic acids. Trends Plant Sci. 6, 273–278.
Maestri, E., Marmiroli, M., Visioli, G., Marmiroli, N., 2010. Metal tolerance and hyperaccumulation: costs and trade-offs between traits and environment. Environ. Exp. Bot. 68, 1–13.
Marschner, H., 1995. Mineral Nutrition of Higher Plants, second ed. Academic Press, San Diego, pp. 889.
McGrath, S.P., Zhao, F.J., Withers, P.J.A., Sinclair, A.H., Evans, E.J., 1996. Development of sulphur deficiency in crops and its treatments. Proceedings of the Fertiliser Society No. 379, The Fertiliser Society, Peterborough, UK.
Meagher, R.B., Heaton, A.C.P., 2005. Strategies for the engineered phytoremediation of toxic element pollution: mercury and arsenic. J. Ind. Microbiol. Biotechnol. 32, 502–513.
Mendoza-Cózatl, D.G., Butko, E., Springer, F., Torpey, J.W., Komives, E.A., Kehr, J., Schroeder, J.I., 2008. Identification of high levels of phytochelatins, glutathione and cadmium in the phloem sap of *Brassica napus*. A role

for thiol-peptides in the long-distance transport of cadmium and the effect of cadmium on iron translocation. Plant Cell 54, 249–259.

Murakami, M., Ae, N., 2009. Potential for phytoextraction of copper, lead, and zinc by rice (*Oryza sativa* L.), soybean (*Glycine max* [L.] Merr.), and maize (*Zea mays* L.). J. Hazard. Mater. 162, 1185–1192.

Murata, N., Higashi, S-I., Fujimura, Y., 1990. Glycerolipids in various preparations of photosystem II from spinach chloroplasts. Biochem. Biophys. Acta 1019, 261–268.

Nanda-Kumar, P.B.A.N., Dushenkov, V., Motto, H., Raskin, I., 1995. Phytoextraction: the use of plants to remove heavy metals from soils. Environ. Sci. Technol 29, 1232–1238.

Nouairi, I., Wided Ben Ammar, W., Youssef, N., Douja Ben Miled Daoud, D.B., Habib Ghorbal, M., Zarrouk, M., 2006. Comparative study of cadmium effects on membrane lipid composition of *Brassica juncea* and *Brassica napus* leaves. Plant Sci. 170, 511–519.

Pellet, M.D., Grunes, D.L., Kochian, L.V., 1995. Organic acid exudation as an aluminum tolerance mechanism in maize (*Zea mays* L.). Planta 196, 788–795.

Persans, M.W., Nieman, K., Salt, D.E., 2001. Functional activity and role of cation-efflux family members in Ni hyperaccumulation in *Thlaspi goesingense*. Proc. Natl. Acad. Sci. USA 98, 9995–10000.

Pilgeram, A., Sands, D., Boss, D., Dale, N., Wichman, D., Lamb, P., Lu, C., Barrows, R., Kirkpatrick, M., Thompson, B., 2007. *Camelina sativa*, a Montana omega-3 and fuel crop. In: Janick, J., Whipey, A. (Eds.), Issues in New Crops and New Uses. ASHA Press, Alexandria, VA, pp. 129–131.

Piotrowska, A., Bajguz, A., Godlewska, B., Czerpak, R., Kaminska, M., 2009. Jasmonic acid as modulator of lead toxicity in aquatic plant *Wolffia arrhiza* (Lamnaceae). Environ. Exp. Bot. 66, 507–513.

Purakayastha, T.J., Viswanath, T., Bhadraray, S., Chhonkar, P.K., Adhikari, P.P., Suribabu, K., 2008. Phytoextraction of zinc, copper, nickel and lead from a contaminated soil by different species of *Brassica*. Int. J. Phytoremed. 10, 61–72.

Quartacci, M.F., Pinzino, C., Sgherri, C.L.M., Dalla, F., Vecchia, F., Navari-Izzo, F., 2000. Growth in excess copper induces changes in the lipid composition and fluidity of PSII-enriched membranes in wheat. Physiol. Plant 108, 87–93.

Raskin, I., Kumar, P.B.A.N., Dushenkov, S., Salt, D., 1994. Bioconcentration of heavy metals by plants. Curr. Opin. Biotechnol. 5, 285–290.

Raziuddin, Farhatullah, Hassan, G., Akmal, M., Shah, S.S., Mohammad, F., Shafi, M., Bakht, J., Zhou, W., 2011. Effects of cadmium and salinity on growth and photosynthesis parameters of *Brassica* species. Pak. J. Bot. 43 (1), 333–340.

Roberts, S.K., Tester, M., 1997. Permeation of Ca^{2+} and monovalent cations through and outwardly rectifying channel in maize root stelar cells. J. Exp. Bot. 48, 839–846.

Salt, D.E., Smith, R.D., Raskin, I., 1998. Phytoremediation. Ann. Rev. Plant Physiol. Plant Mol. Biol. 49, 643–668.

Shanker, A.K., Djanaguiraman, M., Sudhagar, R., Jayaram, K., Pathmanabhan, G., 2004. Expression of metallothioneins3 like protein mRNA in sorghum cultivars under chromium (VI) stress. Curr. Sci. 86, 901–902.

Solymosi, K., Martinez, K., Kristóf, Z., Sundqvist, C., Böddi, B., 2004. Plastid differentiation and chlorophyll biosynthesis in different leaf layers of white cabbage (*Brassica oleracea* cv. *capitata*). Physiol. Plant 121, 520–529.

Szczygłowska, M., Piekarska, A., Konieczka, P., Namiesnik, J., 2011. Use of *Brassica* plants in the phytoremediation and biofumigation processes. Int. J. Mol. Sci. 12, 7760–7771.

Tang, Y., Cao, Y., Qiu, J., Gao, Z., Ou, Z., Wang, Y., Zheng, Y., 2014. Expression of a vacuole localized BURP-domain protein from soybean (SALI3-2) enhances tolerance to cadmium and copper stresses. PLOS One, DOI: 10.1371.

Tassi, E., Pouget, J., Petruzzelli, G., Barbafieri, M., 2008. The effects of exogenous plant growth regulators in the phytoextraction of heavy metals. Chemosphere 71, 66–73.

Thomine, S., Wang, R., Ward, J.M., Crawford, N.M., Schroeder, J.I., 2000. Cadmium and iron transport by members of a plant metal transporter family in *Arabidopsis* with homology to Nramp genes. Proc. Natl. Acad. Sci. USA 97, 4991–4996.

Touiserkani, T., Haddad, R., 2012. Cadmium-induced stress and antioxidative responses in different *Brassica napus* cultivars. J. Agric. Sci. Technol. 14, 929–937.

Van der Zaal, B.J., Neuteboom, L.W., Pinas, J.E., Chardonnens, A.N., Schat, H., Verkleij, J.A.C., Hooykaas, P.J.J., 1999. Over-expression of a novel *Arabidopsis* gene related to putative zinc-transporter genes from animals can lead to enhanced zinc resistance and accumulation. Plant Physiol. 119, 1047–1055.

Van Vliet, C., Anderson, C.R., Cobbett, C.S., 1995. Copper-sensitive mutant of *Arabidopsis thaliana*. Plant Physiol. 109, 871–878.

Vasak, M., 2005. Advances in metallothione in structure and functions. J. Trace Elem. Med. Biol. 19, 13–17.
Vassilev, A., Schwitzguebel, J.P., Thewys, T., van der Lelie, D., Vangronsveld, J., 2004. The use of plants for remediation of metal-contaminated soils. Scientific World J. 4, 9–34.
Vatamaniuk, O.K., Mari, S., Lu, Y.-P., Rea, P.A., 1999. AtPCS1, a phytochelatin synthase from *Arabidopsis:* isolation and *in vitro* reconstitution. Proc. Natl. Acad. Sci. USA 96, 7110–7115.
Veerle, G.M.J., Hans, N.J.M., Jos, V.A.C., 2006. Phytoextraction with *Brassica napus* L. A tool for sustainable management of heavy metal contaminated soil. Environ. Pollut. 144, 77–83.
Williams, L.E., Pittman, J.K., Hall, J.L., 2000. Emerging mechanisms for heavy metal transport in plants. Biochim. Biophys. Acta Biomembr. 1465, 104–126.
Yong, L.Z., Pilon-Smits, E.A.H., Tarun, A.S., Weber, S.U., Jouanin, L., Terry, N., 1999. Cadmium tolerance and accumulation in Indian mustard is enhanced by overexpressing γ-glutamylcysteine synthetase. Plant Physiol. 121, 1169–1177.
Zhu, Y.L., Zayed, A., Qian, C.H., de Souza, M., Terry, N., 1999. Phytoaccumulation of trace elements by wetland plants: II. Water hyacinth (*Eichhornia crassipes*). J. Environ. Qual. 28, 339–344.
Zubr, J., Matthaus, B., 2002. Effects of growth conditions on fatty acids and tocopherols in *Camelina sativa* oil. Ind. Crops Prod. 15, 155–162.

CHAPTER

3

Brassicas

Surinder Kumar Gupta

Division of Plant Breeding & Genetics, Faculty of Agriculture, Sher-e-Kashmir University of Agricultural Sciences & Technology Chatha, Jammu (J&K), India

INTRODUCTION

Brassica is the second largest oilseed crop after soybean (*Glycine max* (L.) Merr.) in world oilseed production (FAO, 2010; Raymer, 2002). Of the 37 species in the Brassica genus, the 4 most widely cultivated species for oilseed and vegetables are *Brassica rapa* L., *B. juncea* (L.) Czern. & Cosson, *B. napus* L., and *B. carinata* A. Braun. (Raymer, 2002; Rakow, 2004; Sovero, 1993). Oleiferous brassicas are generally derived from two species, *B. napus* L. and *B. campestris* L. (syn. *B. rapa* L.) *B. campestris* is also referred to by such names as toria, sarson, summer turnip rape, and Polish rape. Similarly, different names are also given to *B. napus* such as Argentine rape, Swede rape, and colza (Gupta and Pratap, 2007; Kalia and Gupta, 1997). All rapeseed-contributing cultivated *Brassica* spp. are highly polymorphic including oilseed crops, root crops, and vegetables such as Chinese cabbage, broccoli, and Brussels sprouts. However, a few of them are cultivated as salad, vegetable, and condiment crops as well.

B. juncea is of much importance in Asia, and *B. napus* in Europe and Canada. Under European and Canadian conditions, both winter and summer (spring-planted) forms of *B. campestris* (syn. *B. rapa*) and *B. napus* are being grown, but only the spring form of *B. juncea* has evolved. Winter types of *B. napus* are largely grown under north European, Chinese, and Canadian conditions (Rai et al., 2007). However, spring types of *B. campestris* are usually preferred and are largely grown in Sweden, Finland, some parts of Canada and northwest China. In the Indian subcontinent genetic improvement of seed yield is the prime-breeding objective, while in the Western world breeding for quality receives greater attention (Jonsson, 1973).

The two species *B. napus* and *B. rapa* are of commercial value in Canada and Australia (Rakow, 2004; Raymer, 2002). They are the third leading source of vegetable oil in the world after soy and palm and the world's second leading source of protein meal (Gupta, 2009).

Rapeseed is cultivated over an area of 28.23 million hectares with production of about 58.21 Mt, making it the third most important oil plant in the world after palm oil and soybean (FAO, 2010). Canada, Poland, the United Kingdom, and Australia contributed about 77% of total world rapeseed mustard production during 2008, with Canada the largest producer contributing 22%

Breeding Oilseed Crops for Sustainable Production. http://dx.doi.org/10.1016/B978-0-12-801309-0.00003-3

(FAO, 2010). Germany has highest productivity of rapeseed (37.60 q ha^{-1}) followed by the United Kingdom (32.98 q ha^{-1}) and the Czech Republic (29.38 q ha^{-1}). Because of its high yields, the European Community was the leading producer of rapeseed oil in 2008 (Gupta, 2012).

The development of low erucic acid rapeseed and simultaneous rapid growth in global rapeseed production started in Canada in 1968, with the commercial release of the single-low cultivar "Oro" followed by several other single-low cultivars and the first canola cultivar "Tower" in 1974 (Gupta and Pratap, 2008a). In Europe the transition started later with the release of the first single-low cultivars in 1974. The introduction of low erucic acid rapeseed is now under way in China and India. This change in crop quality has created the need for specialized production of industrial rapeseed. Oil cake is a better feed for cattle and poultry due to fewer glucosinolates (<30 μmol g^{-1} oil-free meal). It has been found to be at par with soybean meal with good potential for developing high-value protein food and feed.

The Cruciferae contains a number of species and a diversity of crop plants that have an amalgam of breeding systems ranging from complete crosspollination to a high level of selfpollination (Rai, 1997; Rai et al., 2007). Therefore, there is sufficient diversity in this crop. Some species are cultivated as salad, vegetable, and condiment crops as well. The important crucifers are propagated from seed, but a few minor crops such as horseradish (*Armoracia rusticana* Gaertnia, Meyer & Scherb (syn. *Cochlearia armoracia* L.)), seakale (*Crambe maritima*), and watercress (*Nasturtium* spp.) are vegetatively propagated. Since the mode of reproduction and the breeding objectives differ in different species, breeding methods may be quite different within each of them.

BREEDING OBJECTIVES

Genetic improvement for seed yield is the prime objective in Asia, whereas in Europe and Canada breeding for oil and meal quality receives greater attention in *Brassica* breeding. Seed yield is mainly determined by three important components (i.e., number of siliquae per unit area, number of seeds per siliqua, and seed weight. Further improvement in productivity can be achieved by manipulating yield-determining components as well as agronomic and morphological characters. Development of high-yielding and early-maturing varieties is a major objective in central China and western Canada where frost days in growing season are usually fewer than 100, because early-maturing varieties complete their life cycles during this period and escape frost injury. In the Indian subcontinent, early-maturing varieties (80–90 days) are required for fitting in relay, multiple, and intercropping systems.

Breeding for resistance to disease and insect pests is also a major objective in brassicas. In the Indian subcontinent, *Alternaria* blight, white rust, and downy and powdery mildews are the major diseases, while in Western countries blackleg (*Leptosphaeria maculans* Desm.) is important in Canada and Australia (Rai et al., 2007; Gupta and Pratap, 2008b). Besides these, club root (*Plasmodiophora brassicae*) and root rot (*Rhizoctinina solani*) are other important diseases. Among the major insects, mustard aphid (*Lipaphis erysimi* Kalt.), mustard sawfly (*Athalia proxmia*), and leaf miner (*Bagrada cruciferarum*) are the important insect pests that cause considerable economic losses. *B. juncea* selections are reported to possess better tolerance to mustard aphid than *B. campestris* selection (Rai and Sehgal, 1975; Rai et al., 1987).

In recent years, specific stress is being laid on the enhancement of the quantity and quality of oil. In this direction many efforts have been taken in different countries of the world. The composition of *Brassica* oil varies from species to species and further depends on various genetic and environmental factors. In Europe and Canada breeding for oil and cake better suited to human nutrition and livestock feeding has received higher research priority than in the Asian countries. Erucic acid free types of both *B. napus* (Stefansson et al., 1961) and *B. campestris* (Downey, 1964, 1966) have been found in Canada in fodder forms of these species and this trait has been incorporated into both annual and biannual oilseed forms in Canada and Europe (Jonsson, 1973). Currently, breeding for double-low quality is also gaining importance in China. The identification of naturally occurring zero erucic acid mutants in *B. napus* and *B. rapa* was the first discovery of many that have led to the evolution of zero erucic acid varieties. To improve oil quality further, the major objective is to raise the level of stearic acid in order to reduce the need for hydrogenation in the manufacture of margarine and frying oil. Moreover, reduced level of palmitic acid and development of commercial source of high oleic canola with more than 80% oleic acid to improve the stability of frying oils as well as suitability for industrial applications remains an important objective.

A number of cytoplasmic male sterility (CMS) systems are available in crucifers, which are being utilized for the production of hybrids. Moreover, both selfcompatible and selfincompatible forms occur in the two main species *B. campestris* var. "Toria" and *B. juncea*, which strongly suggests that hybrid cultivars based on selfincompatible material should be produced.

GENETIC RESOURCES

To strengthen the genetic resources of crucifers, germplasm is being continuously augmented by plant exploration and introductions worldwide. The germplasm collected are maintained in gene banks in countries like China, India, United Kingdom, United States, and Germany. These countries together hold more than 60% of total world rapeseed and mustard germplasm (Singh and Sharma, 2007). Genetic stock and wild crucifers are being utilized in interspecific crossing programs in India and other parts of the world to create new genetic variability. Some of these are also being utilized as a base population for breeding work. At the international level, Cruciferae germplasm is being maintained by the International Board for Plant Genetic Resources (IBPGR, now the International Plant Genetic Resources Institute, IPGRI) in Rome (Italy); Universidad Politécnica de Madrid (Spain); Tohoku University in Sendai (Japan); the Banco Nazionale di Germoplasma in Bari (Italy); Kew Gardens (United Kingdom) and the Horticulture Research Institute in Willesbourne (United Kingdom); and the Nordic Gene Bank (Sweden) (Gupta, 2012). In Australia, cultivated and wild crucifers are being maintained by the Australian Temperate Field Crops Collection, and the Victoria Institute for Dryland Agriculture in Horsham. In India, crucifer genetic resources are being maintained by the National Bureau of Plant Genetic Resources in New Delhi, which during the last three decades has introduced over 3950 accessions of rapeseed and mustard from more than 25 countries (Singh and Sharma, 2007).

CREATION OF GENETIC VARIABILITY

In crucifers, enough variability is available because of the crosspollinating nature of its primary species. The biological and chemical diversity available in this crop is well explained by Vaughan et al. (1976) and Tsunoda et al. (1980). If there is no variability in the germplasm, then breeders have to resort to purposeful intervarietal or distant hybridization. However, the problem is the low success rate of interspecific hybridization, which depends a lot upon the genetic relationship and genomic constitution of the parents used (Siemens, 2002).

In general, interspecific hybridization is more successful if an amphidiploid species is used as the female parent, which has one genome in common with the pollen parent (Zhang et al., 2003, 2006a,b). However, hybrids between monogenomic primary species are rather more difficult with a success rate of 0.002 hybrids per pollinated flower (Downey et al., 1980; Mahapatra and Bajaj, 1987; Quazi, 1988). To rescue these interspecific hybrids, ovaries containing embryos are cultured using various culture media including plant growth regulators, resulting in successful plant regeneration from interspecific hybridization between *B. rapa* and *B. oleracea* through ovary culture (Zhang et al., 2004; Zhang and Zhou, 2006).

Many of the wild relatives of oilseed rape are abundant in cultivated fields and are potential donors to oilseed rape, as explained by Gupta (2009) in his monograph and other workers (Chèvre et al., 2004; Scheffler and Dale, 1994). Four wild species belonging to the Brassicaceae are reported to hybridize spontaneously with *B. napus*. These are *B. rapa* (syn. *B. campestris*), *Raphanus raphanistrum, Sinapis arvensis*, and *Hirschfeldia incana* (Jørgensen, 2007). Successful artificial hybridization with *B. napus* has been achieved using *B. elongata, B. fruticulosa, B. souliei, Diplotaxis tenuifolia, H. incana, Coincya monensis*, and *S. arvensis* (Plumper, 1995), and *Sinapis alba* (Gupta, 1993). All these hybrids showed resistance to *Leptosphaeria maculans* in cotyledon tests, but resistance to *Alternaria* was lost in *B. napus* × *B. souliei* hybrids and reduced in *B. napus* × *B. elongata* hybrids compared with the wild species involved. Hybrids between *B. napus* and wild species, *B. napus* × *Hirschfeldia incana, B. napus* × *S. arvensis*, and *B. napus* × *C. monensis* have been successfully backcrossed to *B. napus* (Plumper, 1995). Successful artificial hybridization has also been achieved in crosses involving *D. cretacea, D. harra* ssp. *crassifolia* and *D. muralis* with *B. campestris* var. "Toria" (RSPT-1), when aided by postfertilization *in vivo* treatment of pollinated buds with 0.01% gibberellic acid (GA), followed by embryo rescue on modified Murashige and Skoog (MS) and Gamborg media after 18–20 days of pollination (Pratap et al., 2008).

Unusual floral structures such as apetalous genotypes are reported to have advantages because of increased photosynthetic efficiency and less severe disease infection spread by petals (Jiang and Becker, 2001). Bijral et al. (2004), in F_5 generation of 53 intergeneric crosses between *B. napus* and *Eruca sativa*, observed a plant with abnormal floral characteristics. On the advancement of generations of this plant, Pratap and Gupta (2007) also observed multipetalous and apetalous flowers with 7–8 stamens in many of its progeny.

It has now been possible to transfer the blackleg-resistant gene from *B. juncea* to *B. napus* because of possible recombination between the A and C genomes in *B. juncea* crosses and the A and B genomes in *B. carinata* crosses (Sacristan and Gerdemann, 1986). Iintrageneric hybridization with *B. napus* has been used to transfer the resistance gene to *Leptosphaeria maculans* into the gene pool of oilseed rape (Winter et al., 2002; Roy, 1984; Sacristan and Gerdemann, 1986; Zhu et al., 1993; Snowdon et al., 2000), which has been widely used by breeders. The triazine resistance has been transferred from *B. napus* to *B. oleracea* (Ayotte et al., 1986, 1987).

CMS has been transferred from radish to *B. oleracea* (Bannerot et al., 1974; McCollum, 1988). Various wild species – namely, *B. oxyrrhina* (Prakash and Chopra, 1988); *B. tournefortii* (Pradhan et al., 1991); *D. catholica; D. erucoides* (Malik et al., 1999); *D. berthautii* (Mallik et al., 1999); *D. harra; D. muralis* (Hinata and Konno, 1979); *D. siifolia* (Rao et al., 1994; Rao and Shivanna, 1996); *Hirschfeldia incana; Moricandia arvensis* (Prakash et al. 1998); and *Raphanus sativus* (Ogura, 1968) – have been used as donor species for the development of new CMS lines. Similarly, genes for earliness have been introgressed from *Erucastrum gallicum* into cultivated species. The genes for high-linoleic acid have been transferred from *B. juncea* to *B. napus* through selection in F_2 generation (Roy and Tarr, 1985; Roy and Tarr, 1986).

There are good possibilities of incorporating resistance genes from *B. juncea* and *B. carinata* to *B. napus* cultivars (Prakash and Chopra, 1988; Rao, 1990). *B. macrocarpa*; *B. juncea* (Prakash and Chopra, 1988); *B. tournefortii* (Salisbury, 1989); *Hirschfeldia incana* (Salisbury, 1989); and *Raphanus* spp. (Agnihotri et al., 1991) have been identified as potential donors for the development of shatter-resistant varieties in rapeseed. Moreover, wide hybridization has been reported with some degree of success in the crosses of *B. spinescens* ($2n = 16$) × *B. campestris* ($2n = 20$), *Eruca sativa* ($2n = 22$ EE) × *B. campestris* ($2n = 20$ AA) and for the production of *B. napus* × *Raphanobrassica* hybrids by embryo rescue and ovary culture techniques (Agnihotri et al., 1990; Agnihotri et al., 1990b,c). Four F_1 hybrids of *E. sativa* × *B. campestris* ($2n = 21$ EA) showed a maximum of 12 bivalents, of which 5 were attributed to allosyndetic pairing between *Eruca* and *Brassica* genomes. This suggests that the possibility exists for a flow of useful genes from *Eruca* to rapeseed for resistance to aphids and drought. Wild allies of *Brassica* were evaluated under natural drought conditions and identified as potential donors for drought resistance (Gupta et al., 1995). Wild crucifers – namely, *B. tournefortii* (Salisbury, 1989); *D. acris*; and *D. harra* – have also been identified as potential donors for drought-resistant genes and are being used in the interspecific hybridization program for the transfer of resistance genes into cultivated crucifers.

BREEDING METHODS

The Cruciferae includes a number of cultivated crops and wild species that have a breeding system ranging from complete crosspollination to a high level of selfpollination. Therefore, these are quite interesting material from the breeding point of view. The selection procedures in crosspollinated species vary from mass selection to recurrent selection, and in predominantly selfpollinated species the desirable plants are usually selected from broad-based populations such as land races, segregated populations, germplasm complexes, gene pools. and bulked seed. Bulked seed is repeatedly grown cycle after cycle. One cycle of mass selection in "Toria" is reported to have given a yield improvement of 8.2% (Chaubey, 1979).

Segregating populations or the progeny from crosses could also be used as a good base population for initiating a recurrent selection program. In this method, desirable individual open-pollinated plants (around 3000) are harvested and threshed separately. Some of this seed is saved and some is planted in progeny rows, evaluated visually, and superior rows are selected, tagged, and harvested separately. After harvesting and threshing, the seeds are analyzed for their 1000 seed weight, oil content, glucosinolates, and protein content. Thereafter, equal quantities of reserved seed from selected plants are composited. In this way, the first cycle in the recurrent selection program is completed, and this composited seed is grown again in fields in isolation, where intercrossing takes place among plants within composited populations. The second cycle of recurrent selection starts with the harvesting of single plants (around 1000) from this population. A sample of bulked seed is harvested from the remaining plants of the population for use in replicated yield trials to determine the response to selection in each recurrent cycle for characters under improvement such as oil content, seed yield, or tolerance to a disease. Recurrent cycle selection is continued until a reasonable level of improvement is achieved.

In selfpollinated crucifers, pure line selection is the procedure usually followed in India, which involves the isolation of superior performing lines from a genetically broad-based population based upon their progeny performance. Various improved varieties like "Varuna," "Krishna," "Kranti," "Shekhar," "Sita," "RH-30," and "Durgamani" have all been developed from such simple breeding efforts (Rai, 1983a, b).

PEDIGREE METHOD

Since *B. napus* and *B. juncea* are selfpollinated crops, the majority of cultivars derived are pure lines from pedigrees or pedigree modifications. In India as well as other parts of the world, various high-yielding varieties have been developed following this method. In this method 5–10 F_1 plants are grown to obtain F_2 seed and 1000–3000 F_2 plants are grown and harvested individually, from which F_3 progeny is secured. In F_4 generation, selection is practiced. The variation among F_4 families is a good indication of the effectiveness of further selection. This method has been utilized to develop a low erucic acid, high-yielding, and winter-hardy *B. napus* variety from a cross between high erucic acid winter *B. napus* variety "Rapol" and low erucic acid spring *B. napus* variety "Oro."

BACKCROSS BREEDING

When a desirable gene is available from an unadapted or wild population, backcrossing would be the right procedure, but if the favorable gene is available in adapted or cultivated material then the pedigree method of selection would be more appropriate. The spring *B. napus* variety "Wester" was developed by a combination of backcross and pedigree breeding. Backcross breeding has been used to transfer the low glucosinolate content of *B. napus* variety "Bronowski" into a number of commercial cultivars of gobhi sarson (*B. napus*) in various parts of the world. This method is also used to transfer new traits such as fatty acid composition, seed color, and herbicide and insect pest resistance.

DEVELOPMENT OF SYNTHETICS AND COMPOSITES

The development of composite and synthetic varieties is being viewed as a possible way to increase the average yield of brassicas. Although synthetic *B. napus* cultivars have also been marketed in Europe, they were often nonuniform. Therefore, this method of breeding is no longer used for *B. napus*. Equal quantity of seed from varieties or recurrent lines that arise from the widely different sources or from different gene pools mixed and sown in isolation plot. The seed harvested from Syn-0 constitutes Syn-1 seeds. This method has also been used for *B. napus* (Becker et al., 1991) but, after the discovery of viable CMS systems, breeding program are now directed toward hybrid varieties. The development of composite varieties involves the production of a number of intervarietal hybrids or blends of them. This is followed by evaluation of inbreeding depression in seed yield from F_2 hybrids followed by generation and evaluation of the performance of experimental checks against ruling checks (Rai, 1982). Whereas the development of synthetic varieties requires the development of inbred lines, testing them for general combining ability (GCA) by making all possible crosscombinations, predicting the F_2 hybrid performance of a number of experimental synthetic varieties, testing yield levels in gene trials across locations, and finally releasing those that exceed standards.

DEVELOPMENT OF HYBRIDS

A significant level of yield heterosis in brassicas generally has been reported by various workers (Schuster and Michael, 1976; Lefort-Buson and Datte, 1982; Brandle and McVetty, 1990), in *B. rapa* (Sernyk and Stefansson, 1983; Schuler et al., 1992), and in *B. juncea* (Singh, 1973; Larik and Hussain, 1990; Pradhan et al., 1993). In India, 11–82% yield heterosis has been reported in mustard (*B. juncea*), 10–72% in gobhi sarson, and 20–107% in *B. campestris* (Das and Rai, 1972; Labana et al., 1975; Yadava et al., 1974; Doloi, 1977; Srivastava and Rai, 1993), which is sufficiently high for its exploitation in hybrid cultivars. A range of 14–30% natural outcrossing is usually observed in these crops. Hence, this is sufficient to justify the efforts to develop CMS lines and search for usable fertility restorer lines to produce hybrids.

In oilseed brassicas more than 17 male sterile forms have been investigated (Stiewe and Robbelen, 1994; Prakash et al., 1995). Only a few have been developed to the commercial stage, but breeders worldwide are rapidly using CMS systems for the production of hybrids. A number of CMS sources such as *B. carinata* CMS, *B. juncea* CMS, *B. oryrhina* CMS, *B. tournefortii* CMS, *Raphanus*-based ogura CMS, *Brassica napus*–based Polima CMS, *Siettiana* CMS, and *Siifolia* CMS are now well known and some are being worked with intensively. Of these CMS sources, fertility restoration has been identified in *Raphanus*-based Ogura CMS, in Polima CMS in Western countries, and in CMS-based crosses in *B. tournefortii, B. juncea* CMS, Polima CMS, and *Siifolia* CMS in India. Fortunately, the Punjab Agricultural University in Ludhiana (India) has recommended release of the first CMS-based gobhi sarson hybrid PGSH-51 for cultivation in Punjab (India).

DOUBLED-HAPLOID BREEDING AND *IN VITRO* MUTAGENESIS

The doubled-haploid (DH) breeding technique is now widely used in *B. napus* and *B. juncea* breeding programs (Ferrie and Keller, 2002). This breeding tool not only eliminates the several generations needed to attain genetic stability and uniformity in breeding lines (Zhang et al., 2006b), but also significantly reduces the size of populations needed to find a desired genotype. DH breeding through microspore culture is very well developed in brassicas (Maluszynski et al., 2003; Xu et al., 2007). It is possible to obtain haploid and subsequently DH plants through this technology. The DH technology could be further used in mutation breeding (Zhu et al., 1993), genetic engineering, *in vitro* screening for complex traits like drought, cold, and salinity tolerance, and for developing mapping populations for linkage maps using molecular markers (Pratap et al., 2007). Several methods are available for DH production in brassicas such as microspore culture, anther culture, and ovary/ovule culture. Also microspore cultures provide the best material for mutation induction in haploid cells (Szarejko and Forster, 2007). The possibility of producing haploids in *B. napus* from anther culture (Keller and Armstrong, 1978) and microspore culture (Lichter, 1982) has provided breeders with a new tool for breeding improved cultivars of rapeseed and mustard (Zhou et al., 2002a, b).

The refinement of DH technology to its present form in brassicas has been undertaken by various workers (Charne and Beversdorf, 1988; Yu and Liu, 1995; Wang et al., 1999, 2002; Shi et al., 2002). Details on the microspore culture technique and various factors affecting its efficiency can be found in Gupta (2007, 2008, 2009). Microspore embryogenesis is affected by such factors as donor plant genotype and conditions, pretreatment, growth stage of the anther/microspore to be cultured, culture media and environment, and the diploidization process (Dunwell, 1996; Gu et al., 2003b, 2004c; Zhang et al., 2006a; Pratap et al., 2007). DHs also provide an efficient screening material for desired mutants and other material for complex traits. Since we can obtain a very large number of synchronously developing embryos through microspore-derived DHs, we can modify the system to screen them *in vitro* for various desirable traits. For example, for development of herbicide-resistant brassicas, the active chemical is introduced in the culture medium after mutation treatment (Beversdorf and Kott, 1987), and, after chromosome doubling, the surviving plants could be raised under controlled conditions and later screened for this trait. Similarly, effective selection could also be done for drought, cold, and salinity tolerance. By using this technique, several herbicide-resistant mutants have been developed in rapeseed (Kott, 1995, 1998; Swanson et al., 1988, 1989).

Though embryogenic microspores are the prime targets for mutagenic treatment, other haploid tissues and cells have also been treated with mutagens in brassicas. In *B. napus*, isolated microspores have been treated with chemicals such as ethyl methanesulfonate (EMS) (Beversdorf and Kott, 1987), sodium azide (NaN_3) (Polsoni et al., 1988), *N*-methyl-*N*-nitrosourea (MNU) (Cegielska-Taras et al., 1999), and *N*-ethyl-*N*-nitrosourea (ENU) (Swanson et al., 1988; Swanson et al., 1989). They have also been treated with physical mutagens such as gamma rays (Beversdorf and Kott, 1987; McDonald et al., 1991), X-rays (McDonald et al., 1991), and UV rays (Ahmad et al., 1991; McDonald et al., 1991). *B. napus* anthers have also been treated with gamma rays and fast neutrons by Jedrzejaszek et al. (1997). Similarly, microspores of *B. carinata* have been treated with EMS and UV rays (Barro et al., 2001, 2002) and microspores of *B. campestris* with UV rays (Zhang and Takahata, 1999; Ferrie and Keller, 2002). Isolated microspores and haploid embryos have been treated with chemical mutagens in *B. juncea* as well.

Despite great promise, the use of DH technology as a routine breeding tool for *B.* improvement has yet to materialize, mainly due to problems associated with anther/microspore culture (Pratap et al., 2007). These include low regeneration rate, high genotype-specific response, high frequency of callogenesis, and low recovery of DH plants. The focus of rapeseed breeders has lately shifted toward more specific and practical goals such as development of herbicide-tolerant varieties, development of male sterile lines for hybrid seed production, oil and meal quality improvement, and drug production (Gupta and Pratap, 2008). This entails DH breeding being adopted in conjugation with newer ideas such as directed *in vitro* mutagenesis, *in vitro* screening for desirable traits, and incorporation of molecular markers.

GENETIC TRANSFORMATION

With the development of genetic transformation techniques, it has become possible to maximize oil yield and to manipulate the quality to make the oil suitable for various applications. Much emphasis is now placed on transgenic technology, which seeks to improve cultivated brassicas. As a result, the global area of biotech canola had reached an estimated 5.5 million hectares in 2007 (James, 2007), the majority of it being herbicide-resistant canola.

Successful genetic transformation systems have been developed in such economically important brassicas as *B. napus* (Moloney et al., 1989), *B. oleracea* (De Block et al., 1989), *B. juncea* (Barfield and Pua, 1991), *B. carinata* (Narasimhulu et al., 1992), *B. rapa* (Radke et al., 1992), and *B. nigra* (Gupta et al., 1993). However, of all the systems developed, *Agrobacterium tumefaciens*–mediated gene transfer is most widely used for *Brassica*. Moreover, it is also efficient and practical in most species in the genus (Cardoza and Stewart, 2004).

Rapeseed cultivars tolerant to herbicides such as imidazoline, glyphosate, and glufosinate are available commercially in the United States and Canada (Cardoza and Stewart, 2004). For insect resistance, a gene from *Bacillus thuriengiensis* has been introduced in canola cultivars (Stewart et al., 1996; Halfhill et al., 2001), which leads to overproduction of δ-endotoxins by insects feeding on transgenic canola. This crystalline prototoxin gets inserted into the midgut plasma membrane of the insect, leading to lesion formulation and production of pores that disturb osmotic balance. These cause swelling and lysis of cells and, as a result, the larvae stop feeding and die (Hofte and Whiteley, 1989; Schnepf et al., 1998; Shelton et al., 2002).

Canola varieties with increased linolenic acid (Liu et al., 2001), stearate (Hawkins and Kridl, 1998), laurate (Knutzon et al., 1999), and increased enzyme activity (Facciotti et al., 1999) have been developed through genetic transformation. Furthermore, brassicas have been transformed to develop various industrial and pharmacological products like hirudin, a blood anticoagulant protein (Chaudhary et al., 1998). For example, *B. carinata* has been transformed for the production of hirudin, a blood anticoagulant protein (Chaudhary et al., 1998) while *B. napus* has been used for the production of carotenoids (Shewmaker et al., 1999). Development of male sterile lines and fertility restoration systems has also been achieved through genetic transformation in *B. napus* (Jagannath et al., 2001, 2002), which could have tremendous potential for the development of commercial hybrid cultivars. Salt and cold-tolerant lines have also been developed in *B. juncea* by engineering the bacterial *codA* gene (Prasad et al., 2000). Hitz et al. (1995) have also reported transgenic lines of *B. napus* var. "Wester" having high palmitic and stearic acids. High oleic acid containing *B. napus* and *B. juncea* lines with better shelf life have also been obtained through the transgenic technology (Stoutjesdijk et al., 2000).

DEVELOPMENT OF HERBICIDE-TOLERANT CULTIVARS

Herbicides provide an inexpensive and effective means of controlling weeds in crop *Brassica*. Development of herbicide-resistant cultivars in *Brassica* was started in 1960. Tolerance was cytoplasmically controlled and effective against the triazine family of herbicides. Identification of a triazine-tolerant biotype of birdsrape mustard led to the development of triazine-tolerant *B. napus* oilseed cultivars through introgression of the weed-tolerant biotype cytoplasm in oilseed rape. Beversdorf et al. (1987) developed the first triazine-tolerant cultivar "OAC Trinton" by means of an interspecific cross and backcross program. Triazine-tolerant *B. napus* cultivars are very useful and indeed essential in fields where highly competitive weeds such as wild mustard (*S. arvensis* L.), stickweed (*Thlaspi arvense* L.), and quack grass (*Agropyron repens* L.) are found.

QUALITY IMPROVEMENT

Oilseed *Brassica* is mainly processed to produce two main products: oil and meal. Oil is used for human consumption, as salad oil or margarine. It is also used for technical purposes such as lubricants and hydraulic oils, as base chemicals for oleochemistry or biodiesel fuel (Shahidi, 1990; Kimber and McGregor, 1995). *Brassica* oil contains few harmful saturated fatty acids and many mono and polyunsaturated fatty acids, which makes it nutritionally superior to most other edible oils (Agnihotri et al., 2007). However, its use is limited due to the presence of two major antinutritional substances: erucic acid (a long carbon chain unsaturated fatty acid) and glucosinolate (a sulfur-containing compound).

Fatty acid composition determines the oil quality of seed. High levels of erucic and eicosenoic acids in *Brassica* oil make it inferior in quality to other oilseeds (Gupta and Pratap, 2007). Therefore, the most important breeding objective in *Brassica* breeding has been genetic modification of seed quality by changing the proportion of fatty acids suitable for nutritional and industrial purposes. Modifications in the compositions in fatty acids have been achieved in the past by coupling various conventional breeding methods with biotechnological techniques such as induced mutation, *in vitro* embryo rescue, DH techniques, and genetic engineering, especially posttranscriptional gene-silencing (Agnihotri et al., 2007).

Gas liquid chromatography (GLC) (Craig and Murphy, 1959) together with a technique in which only half the cotyledon was used to test erucic acid content provide a quick means to screen very large populations necessary to identify genetically changed *Brassica* strains with low or zero erucic acid (Stefansson et al., 1961; Downey and Harvey, 1963, and *B. campestris* Downey, 1964) had been developed in the early 1960s. Later, such strains were developed in *B. juncea* (Kirk and Oram, 1981) and *B. carinata* (Alonso et al., 1991). Gupta et al. (1994, 1998) identified low erucic acid genetic stocks among the Indian accessions of *B. juncea.* Several low erucic acid *B. juncea* genotypes have been developed in India through interspecific hybridization (Khalatkar et al., 1991; Malode et al., 1995; Banga et al., 1998), and transgressive segregation through interspecific/intergeneric hybridization, followed by the pedigree method (Agnihotri et al., 1995; Agnihotri and Kaushik, 1998, 1999a,b). Similarly, other fatty acids have also been modified in oleiferous brassicas, and high oleic acid and low linolenic acid *B. juncea* genotypes have been developed (Oram et al., 1999; Potts et al., 1999).The meal that remains after oil extraction is nutritionally high in first-class proteins. However, because of the crude traditional oil-milling process, defatted meal or oilcakes are largely only used today in animal feed mixtures or in fertilizers. Despite it having been an important source of nutrition for animals, there are undesirable components in the meal like glucosinolates, which render them unfit for animal and human consumption. In high concentrations, in nonruminants like swine and poultry, it hydrolyzes to form thiocynates, isothiocynates, or nitriles and can adversely affect iodine uptake by the thyroid gland and can result in weight loss (Fenwick et al., 1983). The high-performance quantitative GLC technique (McGregor et al., 1983; Spinks et al., 1984; Brazezinski et al., 1986) has made it possible to obtain the profiles of glucosinolates and also measure their absolute levels. To improve the quality of

Brassica seed meal, glucosinolates should either be decreased or eliminated altogether from the meal through appropriate breeding techniques. Unfortunately, however, the genes controlling the glucosinolate content in rapeseed are either pleiotropic or linked to the seed-filling stage and have a positive correlation with 1000 seed weight (Oliveri and Parrini, 1986). This renders strict selection difficult for quality traits in early segregation, especially in light of genotypes for high seed yield possibly being lost. Therefore, it is advocated to keep the population heterozygous for quality characters and select plants for these characters in advanced generation. To date (2015), the "Bronowski" cultivar is the only known source with a low glucosinolate content and no natural germplasm source for stable low-glucosinolate genes has been reported (Agnihotri et al., 2007). "Bronowski" is a Polish *B. napus* cultivar, which has a glucosinolate content of about 12 μmol g^{-1} oil-free meal and 7–10% of erucic acid in the oil. Considerable success has been achieved in Australia in the development of low-glucosinolate genotypes using mutagenesis, interspecific hybridization, and tissue culture coupled with pedigree selection (Oram et al., 1999).

Breeding of canola-type cultivars in Brassica with less than 2% erucic acid in the oil and less than 30 μmol g^{-1} of glucosinolate in defatted meal (commonly known as "00") has been carried out and several varieties such as "Cyclone" (Denmark), "Shiralee" (Australia), and "AC Excel" (Canada) are now available (Rakow, 1995). In Australia, several double-low cultivars of *B. juncea* have shown promising yield potential (Burton et al., 2003a, b). In India, Agnihotri and Kaushik (2003a, b) have reported successful introgression of double-low traits in *B. juncea* "Varuna" using low erucic acid donors TERI (OE) M21 and Zem-1 and the low-glucosinolate line BJ-1058. The double-low *B. napus* varieties "GSC-865" and "TERI-Uttam-Jawahar" have been released for commercial cultivation in Punjab and Madhya Pradesh, respectively (Agnihotri et al., 2007).

Experimental work toward development and improvement of low erucic acid germplasm for other species is being pursued at global level (Rakow and Raney, 2003). At present, breeding efforts in the development of canola-quality double-low *B. napus* cultivars in improving oil composition and enhancing vitamin levels are under way in many countries of the world including Germany (Luhs et al., 2003), Canada (Raney et al., 2003a,b), United States (Corbett and Sernyk, 2003), Australia (Gororo et al., 2003), France (Carre et al., 2003), and Poland (Spasibionek et al., 2003). Yellow seed coat color also adds to high oil content, and therefore could be another breeding objective for improved brassicas.

FUTURE DEVELOPMENTS

Marker-assisted selection and chromosome mapping came into use with publication of the first linkage maps for *B. oleracea* (Slocum et al., 1990), *B. rapa* (Song et al., 1991), and *B. napus* (Landry et al., 1992).

SUSTAINABILITY

Plant breeders are basically genetic engineers and their main objective is to improve crops by utilizing different breeding techniques. Most breeding objectives will differ little from those in earlier days. However, with the remarkable expansion of breeding tools, in particular, derived from biotechnology and molecular genetics, breeders will be able to optimize the classical procedures mentioned in this publication. Seed yield, production stability, and product quality will remain pillars of the rapeseed industry (Gomez Campo, 1999).

Herbicide tolerance as well as disease and pest resistance produced by alien gene transfer will have an economic impact on all *Brassica* crops including vegetable cultivars. Tailoring seed oil composition to new technical and oleochemical uses will be as attractive as "molecular farming" of high-value materials such as polyhydroxybutyrate, drugs, or hormones (Gomez Campo, 1999).

NEW EMERGING CROPS AND POSSIBLE RESEARCH DEVELOPMENTS

Crucifer breeders all over the world have explored and domesticated a few of the oleiferous brassicas that are now emerging and that hold tremendous potential for oil and meal in the future. However, owing to the lack of suitability of these crops to local agronomic practices as well as the antinutritional compounds present in their oil and meal make their commercialization a difficult task. Still, a few species have proved their worth and could be potential industrial and domestic oil producers:

1. *B. carinata* is grown on a small scale in Ethopia. The seed is large and predominantly dark, although some yellow-seeded forms are available. Apart from being resistant to various major diseases (*Alternaria* blight and white rust) and pests, it also revealed high-yield potential in research and adapted trials (Gupta et al., 1999). *B. carinata* has an edge over other domesticated species especially under rain-fed and natural aphid infestation conditions (Anand and Rawat, 1984). Pure line selection is being used to develop high-yielding varieties in this crop. Since the available germplasm is late maturing, efforts have been made to induce earliness through mutation breeding or interspecific/intergeneric hybridization., At the Saskatoon Research Centre in Canada attempts are being made to develop high-yielding early-maturing varieties.
2. *Camelina sativa* is a major oilseed crop in Siberia (Francis and Campbell, 2003). Its crude oil is used in Canada and many European countries for industrial purposes. Camelina oil has very good potential for industrial use (Vollmann et al., 1996; Pilgeram et al., 2007). Vollmann et al. (2005) reported that there was enough genetic variation in the material studied by him for various traits such as oil content, seed yield, maturity, and drought resistance to be exploited. The variability inherent in *Camelina* has yet to be fully exploited. Therefore, conventional breeding techniques may be among the options utilized for the development of suitable ideotypes.
3. *Lesquerella* spp. are mainly native to the south central states of the United States and northern Mexico. Therefore, they provide the advantage of being suitable for dryland farming (Princen, 1983). They are highly crosspollinated crops with 86–90% outcrossing. Breeding efforts for *Lesquerella* are still at a relatively early stage, although it is already in commercial production. There are several registered cultivars in the USA, including a high oil cultivar of *Lesquerella* fendleri (WCL-LO3) registered in 2006 by Dierig et al. (2006a) and a mutant with a cream flower color which was also registered that year by Dierig et al. (2006b). Inter and intrapopulation improvement can be used to develop

high-yielding genotypes in this crop. Hybrid cultivar development may be another possibility because of the resistance of the male sterility system to *Lesquerella fendleri* (Dierig et al., 1996). Seventeen species have been described in terms of chemical, botanical, and habitat descriptors (Mikolajczak et al., 1962; Barclay et al., 1962).

4. *Crambe* spp. have been evaluated as a potential oilseed crop in Canada, Denmark, Germany, Poland, Russia, and Sweden since 1932 (White and Higgins, 1966). Due to high erucic acid content, high smoke point, and high viscosity, *Crambe* oil is potentially useful in the lubrication, emulsification, and refrigeration industries (Lazzeri et al., 1994). The main objective of *Crambe* is to increase seed yield, oil production, and protein meal quality. Mutagenesis and interspecific hybridization have been used as breeding approaches to produce new cultivars with improved agronomic and seed quality traits (Lessman and Meir, 1972).
5. *Eruca sativa* is mainly cultivated in Iran, Pakistan, and India. It is highly crosspollinated with 50–70% outcrossing and grown in the drier parts of India (Punjab, Haryana, Uttar Pradesh, and Rajasthan) and some parts of Pakistan. *Eruca* oil is characterized by a high content of sulfur and nitrogen compounds. It finds its main use in industry and pharmaceuticals. It is also highly resistant to aphids and thrives well under rain-fed and drought conditions. As a result, this species is used for the transfer of drought/aphid-resistant genes to cultivated species by using interspecific and intergeneric hybridization. Very little work has been done on genetic improvement of this crop. In India, variety type-27 has been developed by following mass selection either from local land races or from genetically variable germplasm entries. This variety has been released for commercial cultivation in India. Crucifer breeders are using new breeding methods, polycross methods, as well as synthetic and composite varieties to develop high-yielding varieties in this crop.

Acknowledgment

The authors acknowledge Prof. Wei Jun Zhou, Institute of Crop Science, College of Agriculture & Biotechnology, Zhejiang University, China for a critical review of this manuscript.

References

Agnihotri, A., Kaushik, N., 1998. Transgressive segregation and selection of zero erucic acid strains from intergeneric crosses of *Brassica*. Indian J. Plant Genet. Resour. 11 (2), 251–255.

Agnihotri, A., Kaushik, N., 1999a. Genetic enhancement for double low characteristics in Indian rapeseed mustard. Proceedings of Tenth International Rapeseed Congress, 26–29 September, Canberra, Australia.

Agnihotri, A., Kaushik, N., 1999b. Transfer of double low characteristics in Indian *B. napus*. J. Oilseeds Res. 16, 227–229.

Agnihotri, A., Kaushik, N., 2003a. Towards nutritional quality improvement in Indian mustard (*Brassica juncea* [L]. Czern and Coss) var. Pusa Bold. Proceedings Eleventh International Rapeseed Congress, Royal Veterinary and Agricultural University, Copenhagen, Denmark, 6–10 July, pp. 501–503.

Agnihotri, A., Kaushik, N., 2003b. Combining canola quality, early maturity and shattering tolerance in *B. napus* for Indian growing conditions. Proceedings Eleventh International Rapeseed Congress, Royal Veterinary and Agricultural University, Copenhagen, Denmark, 6–10 July, pp. 436–439.

Agnihotri, A., Lakshmikumaran, M., Shivanna, K.R., Jagannathan, V., 1990. Embryo rescue of interspecific hybrids of *Brassica spinescens* × *B. campestris* and DNA analysis: current plant science and biotechnology agriculture. Progr. Plant Cell. Molec. Biol. 9, 270–274.

Agnihotri, A., Gupta, V., Lakhmikumaran, M., Shivanna, K.R., Prakash, S., Jagannathan, V., 1990b. Production of *Eruca–Brassica* hybrids by embryo rescue. Plant Breed. 104, 281–289.

Agnihotri, A., Shivanna, K.R., Raina, S.N., Lakshmikumaran, M., Prakash, S., Jagannathan, V., 1990c. Production of *Brassica napus* × *Raphanobrassica* hybrids by embryo rescue. Plant Breed. 105, 292–299.

Agnihotri, A., Shivanni, K.R., Lakshmikumaran, M.S., Jagannathan, V., 1991. Micropropagation and DNA analysis of wide hybrids of cultivated Brassica. Proc. GCIRC Eighth Rapeseed Congr., Abstr., 151.

Agnihotri, A., Raney, J.P., Kaushik, N., Singh, N.K., Downey, R.K., 1995. Selection for better agronomical and nutritional characteristics in Indian rapeseed-mustard. Proceedings Ninth International Rapeseed Congress, 4–7 July, Cambridge, UK, pp. 425–427.

Agnihotri, A., Prem, D., Gupta, K., 2007. The chronicles of oil and meal quality improvement in rapeseed. In: Gupta, S.K. (Ed.), Advances in Botanical Research–Rapeseed Breeding. Academic Press/Elsevier Ltd, San Diego, CA, pp. 50–99, p. 554.

Ahmad, I., Day, J.P., MacDonald, M.V., Ingram, D.S., 1991. Haploid culture and UV mutagenesis in rapid cycling *Brassica napus* for the generation of resistance to chlorsulfuron and *Alternaria brassicola*. Ann. Bot. 67, 521–525.

Alonso, L.C., Fernandez-Serrano, O., Fernandez-Escobar, J., 1991. The outset of a new oilseed crop: *Brassica carinata* with low erucic acid. Proceeding of the Eighth International Rapeseed Conference, Saskatoon, Canada, pp. 170–176.

Anand, I.J., Rawat, D.S., 1984. Recent plant breeding efforts towards productivity breakthrough in rapeseed–mustard oilseed. Proceedings of the Symposium on Oilseed Production and Utilization –Constraints and Opportunities, New Delhi, India.

Ayotte, R., Harney, P.M., Machado, V.S., 1986. The transfer of triazine resistance from *B. napus* to *B. oleracea*. Cruciferae Newsl. 11, 95–96.

Ayotte, R., Harney, P.M., Machado, V.S., 1987. Transfer of triazine resistance from *Brassica napus* to *B. oleracea*. I. Production of F1 hybrids through embryo rescue. Euphytica 36, 615–624.

Banga, S.S., Banga, S.K., Gupta, M.L., Sandha, G.S., 1998. Synthesis of genotypes with specialized fatty acid and composition in Indian mustard (*Brassica juncea* L). Crop Improv. 25, 21–25.

Bannerot, H., Boulidar, L., Canderon, Y., Tompe, J., 1974. Transfer of cytoplasmic male sterility from *Raphanus sativus* to *B. oleracea*. Proc. Eucarpia Meeting Cruciferae Crop. Sci. 25, 52–54.

Barclay, A.S., Gentry, H.S., Jones, Q., 1962. The search for new industrial oils. II: *Lesquerella* (Cruciferae) as a source of new oilseeds. Econ. Bot. 16, 95–100.

Barfield, D.G., Pua, E.C., 1991. Gene transfer in plants of *Brassica juncea* using *Agrobacterium tumefaciens* mediated transformation. Plant Cell Rep. 10, 308–314.

Barro, F., Fernandez-Escobar, J., De La Vega, M., Martin, A., 2001. Doubled haploid lines of *Brassica carinata* with modified erucic acid content through mutagenesis by EMS treatment of isolated microspores. Plant Breed. 120, 262–264.

Barro, F., Fernandez-Escobar, J., De La Vega, M., Martin, A., 2002. Modification of glucosinolate and erucic in doubled haploid lines of *Brassica carinata* by UV treatment of isolated microspores. Euphytica 129, 1–6.

Becker, H.C., Karlsson, B., Damgaard, C., 1991. McGregor, D.I. (Ed.), Genotypic and environmental variation for outcrossing rate in rapeseed, 5, Proc. Eighth Int. Rapeseed Congress, Saskatoon, Canada, pp. 1454–1459.

Beversdorf, W.D., Kott, L.S., 1987. An *in vitro* mutagenesis/selection system for *Brassica napus*. Iowa State J. Res. 61, 435–443.

Bijral, J.S., Gupta, S.K., Dey, T., 2004. Unusual floral morphology in an advanced generation of an intergeneric cross of *Brassica napus* × *Eruca sativa*. J. Res. SKUAST-J 3, 61–64.

Brandle, J.E., McVetty, P.B.E., 1990. Geographical diversity, parental selection and heterosis in oilseeds. Can. J. Plant Sci. 70, 935–940.

Brazezinski, W., Mendelewski, P., Musse, B.G., 1986. Comparative study on determination of glucosinolates in rapeseed. Cruciferae Newsl. 11, 128–129.

Burton, W., Salisbury, P., Potts, D., 2003. Inheritance of allyl glucosinolate in the development of canola quality Brassica juncea for Australia. In: Proc. Eleventh Int. Rapeseed CongRoyal Veterinary and Agricultural University. Copenhagen, Denmark. p. 280, 6-10 July.

Burton, W., Salisbury, P., Potts, D., 2003b. The potential of canola quality *Brassica juncea* as an oilseed crop for Australia. Proceedings Eleventh International Rapeseed Congress, Royal Veterinary and Agricultural University, Copenhagen, Denmark, 6–10 July, pp. 5–7.

Cardoza, V., Stewart, Jr., N.C., 2004. *Brassica* biotechnology: progress in cellular and molecular biology. *In Vitro* Cell Dev. Biol. Plant 40, 542–551.

Carre, P., Dartenuc, C., Evrard, J., Judde, A., Labalette, F., Raoux, R., Renard, M., 2003. Frying stability of rapeseed oils with modified fatty acid composition. Proceedings Eleventh International Rapeseed Congress, Royal Veterinary and Agricultural University, Copenhagen, Denmark, 6–10 July, pp. 540–543.

Cegielska-Taras, T., Szala, L., Krzymanski, J., 1999. An *in vitro* mutagenesis-selection system for *Brassica napus* L. New horizons for an old crop. Proc.Tenth Int. Rapeseed Congress, Canberra, Australia, pp. 1–4.

Charne, D.G., Beversdorf, W.D., 1988. Improving microspore culture as a rapeseed breeding tool: The use of auxins and cytokinins in an induction medium. Can. J. Bot. 66, 1671–1675.

Chaubey, C.N., 1979. Mass selection in toria. Indian J. Genet. 39, 194–201.

Chaudhary, S., Parmenter, D.L., Moloney, M.M., 1998. Transgenic *Brassica carinata* as a vehicle for the production of recombinant proteins in seeds. Plant Cell Rep. 17, 195–200.

Chèvre, A.M., Ammitzbøll, H., Breckling, B., Dietz-Pfeilstetter, A., Frédérique Eber, F., Fargue, A., Gomez-Campo, C., Jenczewsk, E., Jørgensen, R.B., Lavigne, C., Meier, M.S., Nijs, H., Pascher, K., Seguin-Swartz, G., Sweet, J., Steward, C.N., Warwick, S., 2004. A review on interspecific gene flow from oilseed rape to wild relatives. In: Nijs, H., Bartsch, D., Sweet, J. (Eds.), Introgression from Genetically Modified Plants into Wild Relatives and Its Consequences. CAB Publishing, Wallingford, UK, pp. 235–251.

Corbett, P., Sernyk, L., 2003. Global opportunities for naturally stable canola/rapeseed oils. Proceedings Eleventh International Rapeseed Congress, Royal Veterinary and Agricultural University, Copenhagen, Denmark, 6–10 July, pp. 524–527.

Craig, B.M., Murfy, N.L., 1959. Quantitative fatty acids analysis of vegetable oils by gas chromatography. J. Am. Oil Chem. Soc. 36, 549–552.

Das, B., Rai, B., 1972. Heterosis in intervarietal crosses of toria. Indian J. Genet. 32, 197–202.

De Block, M., De Brower, D., Tenning, P., 1989. Transformation of *Brassica napus* and *Brassica oleracea* using *Agrobacterium tumefaciens* and the expression of bar and neo genes in the transgenic plants. Plant Physiol. 91, 694–701.

Dierig, D.A., Coffelt, T.A., Nakayama, F.S., Thompson, A.E., 1996. *Lesquerella* and *Vernonia*: oilseeds for arid land. In: Janick, J. (Ed.), Progress in New Crops. ASHS Press, Alexandria, VA, pp. 347–354.

Dierig, D.A., Dahlquist, G.H., Tomasi, P.M., 2006a. Registration of WCL-LO3 high oil *Lesquerella fendleri* germplasm. Crop Sci. 46, 1832–1833.

Dierig, D.A., Salywon, A.M., De Rodriquez, D.J., 2006b. Registration of a mutant *Lesquerella* genetic stock with cream flower color. Crop Sci. 46, 1836–1837.

Doloi, P.C., 1977. Levels of self-incompatibility, heterosis inbreeding depression in *Brassica campestris*. Unpublished PhD Thesis. Govind Ballabh Pant University of Agriculture and Technology, Pantnagar (Nainital), Uttar Pradesh, India.

Downey, R.K., 1964. A selection of *Brassica campestris* L. containing no erucic acid in its seed oil. Can. J. Plant Sci. 44, 295.

Downey, R.K., 1966. Breeding for fatty acid composition in oils of *Brassica napus* L. and *B. campestris* L. Qual. Plant. Mater. Veget. 13, 171–180.

Downey, R.K., Harvey, B.L., 1963. Method of breeding for oil quality in rape. Can. J. Plant Sci. 43, 271–275.

Downey, R.K., Klassen, A.J., Stringan, G.S., 1980. Rapeseed-Mustard. In: Fehr, W.R., Hadley, H.H. (Eds.), Hybridization of Crop Plant. Aust. Soc. Agron. Inc, Madison, WI, pp. 495–509.

Dunwell, J.M., 1996. Microspore culture. Jain, S.M., Sopory, S.K., Veilleux, R.E. (Eds.), In Vitro Haploid Production in Higher Plants, 1, Kluwer Academic Publishers, Dordrecht, pp. 205–216.

Facciotti, M.T., Bertain, P.B., Yuan, L., 1999. Improved stearate phenotype in transgenic canola expressing a modified acyl–acyl carrier protein thioesterase. Nat. Biotechnol. 17, 593–597.

FAO, 2010. FAO Agricultural Outlook 2010–19.

Fenwick, G.R., Heaney, R.K., Mullin, W.J., 1983. Glucosinolates and their breakdown products in food and food plants. Crit. Rev. Food Nutr. 18, 123–201.

Ferrie, A.M.R., Keller, W.A., 2002. Application of double haploidy and mutagenesis in *Brassica* 13th Cruciferae Genetics Workshop. University of California, Davis, CA, March 23–26.

Francis, C.M., Campbell, M.C., 2003. New High Quality Oil Seed Crops for Temperate and Tropical Australia, RIRDC Publication No. 03/045, RIRDC Project No. UWA-47A. Rural Industries Research and Development Corporation, Canberra, Australia.

Gomez-Campo, C., Prakash, S., 1999. Origin and domestication. In: Gomez-Campo, C. (Ed.), Biology of Brassica coenospecies. Elsevier Press, Amsterdam, pp. 3–32.

Gororo, N., Salisbury, P., Rebetzke, G., Burton, W., Bell, C., 2003. Genotypic variation for saturated fatty acid content of Victorian canola. Proceedings Eleventh International Rapeseed Congress, Royal Veterinary and Agricultural University, Copenhagen, Denmark, 6–10 July, pp. 215–217.

Gu, H.H., Hagberg, P., Zhou, W.J., 2004c. Cold pretreatment enhances microspore embryogenesis in oilseed rape (*Brassica napus* L.). Plant Growth Regul. 42, 137–143, 2004.

Gu, H.H., Zhou, W.J., Hagberg, P., 2003b. High frequency spontaneous production of doubled haploid plants in microspore cultures of Brassica rapa ssp. chinensis. Euphytica 134, 239–245.

Gupta, S.K., 1993. Wild hybridization in *Brassica*. Nucleus 36, 149–151.

Gupta, S.K., 2009. Biology and breeding of crucifers. CRC Press, Boca Raton, FL.

Gupta, S.K., 2012. Techonological Innovations in Major World Oil Crops, Vol. 1, Springer Science+Business Media B.V.

Gupta, S.K., Pratap, A., 2007. History, origin and evolution. Gupta, S.K. (Ed.), Advances in Botanical Research–Rapeseed Breeding, 45, Academic Press, London, UK, pp. 1–20, 554 pp.

Gupta, S.K., Pratap, A., 2008a. Evaluation of wild crucifers under sub-tropical conditions of Jammu & Kashmir (India). Cruciferae Newsletter 27, 97–98.

Gupta, S.K., Pratap, A., 2008b. Recent trends in oilseed brassicas. In: Nayyar, H. (Ed.), Crop Improvement: Challenges and Strategies. I. K. International, New Delhi, India, pp. 284–299.

Gupta, V., Sita, G.L., Shaila, M.S., Jagannathan, V., 1993. Genetic transformation in *Brassica nigra* by *Agrobacterium* based vector and direct plasmid uptake. Plant Cell Rep. 12, 418–421.

Gupta, M.L., Banga, S.K., Banga, S.S., Sandha, G.S., Ahuja, K.L., Raheja, R.K., 1994. A new genetic stock for low erucic acid in Indian mustard. Cruciferae Newsl. 16, 104–105.

Gupta, S.K., Sharma, D.R., Chib, H.S., 1995. Evaluation of wild allies of *Brassica* under natural conditions. Cruciferae Newsl. 17, 10–11.

Gupta, M.L., Ahuja, K.L., Raheja, R.K., Labana, K.S., 1998. Variation for biochemical quality traits in promising genotypes of Indian mustard. J. Res. 25, 1–5.

Gupta, S.K., Bhagat, K.L., Khanna, Y.P., Gupta, S.C., 1999. Performance of Abyssinian mustard under rainfed conditions of Jammu. J. Oilseed Res. 16 (2), 285–288.

Halfhill, M.D., Richards, H.A., Mabon, S.A., Stewart, Jr., N.C., 2001. Expression of GFP and Bt transgenes in *Brassica napus* and hybridization with *Brassica rapa*. Theor. Appl. Genet. 103, 151–156.

Hawkins, D., Kridl, L., 1998. Characterization of acyl-ACP thioesterase of mangosteen (*Garcinia mangosteena*) seed and high levels of state production in transgenic canola. Plant J. 13, 743–752.

Hinata, K., Konno, N., 1979. Studies on a male sterile strain having the *Brassica campestris* nucleus and the *Diplotaxis muralis* cytoplasm. I. On the breeding procedure and some characteristics of the male sterile strain. Jpn. J. Breed. 29, 305–311.

Hitz, W.D., Mauvis, C.J., Ripp, K.G., Reiter, R.J., DeBonte, L., Chen, Z., 1995. The use of cloned rapeseed genes for the cytoplasmic fatty acid desaturases and the plastid acyl-ACP thioesterases to alter relative levels of polyunsaturated and saturated fatty acids in rapeseed, D5: breeding oil quality. GCIRC, Cambridge, UK, pp. 470–478.

Hofte, H., Whiteley, H.R., 1989. Insecticidal crystal proteins of *Bacillus thuringiensis*. Microbiol. Rev. 53, 242–255.

Jagannath, A., Bandhopadhyay, P., Arumugam, N., Burma, P.K., Pental, D., 2001. The use of a spacer DNA fragment insulates the tissue specific expression of a cytotoxic gene (*barnase*) and allows high frequency generation of transgenic male sterile lines in *Brassica juncea* L. Mol. Breed. 8, 11–23.

Jagannath, A., Arumugam, N., Guipta, V., Pradhan, A., Burma, P.K., Pental, D., 2002. Development of transgenic *barstar* lines and identification of a male sterile (*barnase*/restorer *barstar*) combination for heterosis breeding in Indian oilseed mustard (*Brassica juncea*). Curr. Sci. 82, 46–52.

James, C., 2007. Global Status of Commercialized Biotech/GM Crops: 2007. ISAAA Brief No. 37. ISAAA, Ithaca, New York.

Jedrzejaszek, K., Kruczkowska, H., Pawlowska, H., Skucinska, B., 1997. Simulating effect of mutagens on *in vitro* plant regeneration. MBNL 43, 10–11.

Jiang, L., Becker, H.C., 2001. Effect of apetalous flowers on crop physiology in winter oilseed rape (*Brassica napus*). Pflanzenbauwissenchaft 5, 58–63.

Jonsson, 1973. Breeding for low erucic acid contents in summer turnip rape (*Brassica campestris* L. var. *annua*). Z. Pflanzenzücht 69, 1–18.

Jørgensen, R.B., 2007. Gene flow from wild species into rapeseed. Gupta, S.K. (Ed.), Advances in Botanical Research–Rapeseed Breeding, 45, Academic Press/Elsevier Ltd, San Diego, CA, pp. 452–483, 554 pp.

Kalia, H.R., Gupta, S.K., 1997. Recent advances in oilseed brassicas. In: Kalia, H.R., Gupta, S.K. (Eds.), Breeding Methods. Kalyani Publishers, p. 347.

Keller, W.A., Armstrong, K.C., 1978. High frequency production of microspore derived plants from *Brassica napus* anther cultures. Z. Pflanzenzücht 80, 100–108.

Khalatkar, A.S., Rakow, G., Downey, R.K., 1991. Selection for quality and disease resistance in *Brassica juncea* cv. Pusa Bold. Proceedings Eighth International Rapeseed Congress, Saskatoon, Canada, 9–11 July, 198 pp.

Kimber, D., McGregor, D.I., 1995. Brassica Oilseeds. Production and Utilization. CABI, Wallingford, UK, 394 pp.

Kirk, J.T.O., Oram, R.N., 1981. Isolation of erucic acid free lines of *Brassica juncea*: Indian mustard now a potential oilseed crop in Australia. J. Austral. Inst. Agric. Sci. 47, 51–52.

Knutzon, D.S., Hayes, T.R., Wyrick, A., Xiong, H., Davies, H.M., Voelker, T.A., 1999. Lysophosphatidic acid acyltransferase from coconut endosperm mediates the insertion of laurate at the sn-2 position of triaglycerols in lauric rapeseed oil and can increase total laurate levels. Plant Physiol. 120, 739–746.

Kott, L., 1995. Production of mutants using the rapeseed doubled haploid system. Induced Mutations and Molecular Techniques for Crop Improvement. IAEA, Vienna, pp. 505-515.

Kott, L., 1998. Application of doubled haploid technology in breeding of oilseed *Brassica napus*. Agric. Biotechnol. News Inform. 10, 69–74.

Labana, K.S., Badwal, S.S., Chaurasia, B.D., 1975. Heterosis and combining ability in *B. juncea*. Crop Improv. 2, 46–51.

Landry, B.S., Hubert, N., Crete, R., Chiang, M.S., Lincoln, S.E., Etoh, T., 1992. A genetic map of *Brassica oleracea* based on RFLP markers detected with expressed DNA sequences and mapping of resistance genes to race 2 of *Plasmodiophora brassicae* (woronin). Genome 35, 409–420.

Larik, A.S., Hussain, M., 1990. Heterosis in Indian mustard *Brassica juncea* (L.) Coss. Pak. J. Bot. 22 (2), 168–171.

Lazzeri, L., Leoni, O., Conte, L.S., Palmieri, S., 1994. Some technological characteristics and potential uses of *Crambe abyssinica* products. Indust. Crops Prod. 3, 103–112.

Lefort-Buson, M., Datte, Y., 1982. Genetic study of some agronomic characters in winter oilseed rape (*Brassica napus*). I: heterosis. Agronomie 2, 315–332.

Lessman, K.J., Meir, V.D., 1972. Agronomic evaluation of *Crambe* as a source of oil. Crop Sci. 12, 224–227.

Lichter, R., 1982. Induction of haploid plants from isolated pollen of *Brassica napus*. Z. Pflanzenphysiol. 103, 229–237.

Liu, J.W., DeMichele, S., Bergana, M., Bobik, E., Hastilow, C., Chuong, L.T., Mukerji, P., Huang, Y.S., 2001. Characterization of oil exhibiting high gamma-linolenic acid from a genetically transformed canola strain. J. Am. Oil Chem. Soc. 78, 489–493.

Luhs, W., Weier, D., Marwede, V., Frauen, M., Lekband, G., Becker, H.C., Frentzen, M., Friedt, W., 2003. Breeding of oilseed rape (*Brassica napus* L.) for modified tocopherol composition – synergy of conventional and modern approaches. Proceedings Eleventh International Rapeseed Congress, Royal Veterinary and Agricultural University, Copenhagen, Denmark, 6–10 July, pp. 194–197.

McDonald, M.V., Ahgmad, I., Menten, J.O.M., Ingram, D.S., 1991. Haploid culture and in vitro mutagenesis (UV light, X-rays, and gamma rays) of rapid cycling Brassica napus for improved resistance to disease. Plant Mutation Breeding for Crop Improvement, Vol. 2, IAEA, Vienna, pp. 129–138.

Mahapatra, D., Bajaj, Y.P.S., 1987. Interspecific hybridization of *B. juncea* × *B. hirta* using embryo rescue. Euphytica 36, 321–326.

Malik, M., Vyas, P., Rangaswamy, N.S., Shivanna, K.R., 1999. Development of two new cytoplasmic male-sterile lines of *Brassica juncea* through wide hybridization. Plant Breed. 118, 75–78.

Malode, S.N., Swamy, R.V., Khalatkar, A.S., 1995. Introgression of 'OO' quality characters in *Brassica juncea* cv. Pusa bold. Proceedings Ninth International Rapeseed Congress, Cambridge, UK, 4–7 July, pp. 431–438.

Maluszynski, M., Kasha, K.J., Forster, B.P., Szarejko, I., 2003. Doubled Haploid Production in Crop Plants: A Manual. Kluwer Academic Publisher, Dordrecht, The Netherlands, 428 pp.

McCollum, G.D., 1988. CMS (ESG-508) and CMS (FSG-512) cytoplasmic male sterile cabbage germplasm with radish cytoplasm. Hort. Sci. 23, 227–228.

McGregor, D.I., Mullim, W.J., Fenwick, G.R., 1983. Review of analysis of glucosinolate analytical methodology for determining glucosinolate composition and content. J. Assoc. Official Anal. Chem. 66, 825–849.

Mikolajczak, K.L., Earle, F.R., Wolff, I.A., 1962. Search for new industrial oils, VI: seed oils of the genus *Lesquerella*. J. Am. Oil Chem. Soc. 39, 78–80.

Moloney, M.M., Walker, J.M., Sharma, K.K., 1989. High efficiency transformation of *Brassica napus* using *Agrobacterium* vectors. Plant Cell Rep. 8, 238–242.

Narasimhulu, S.B., Kirti, P.B., Mohapatra, T., Prakash, S., Chopra, V.L., 1992. Shoot regeneration in stem explants and its amenability to *Agrobacterium tumefaciens* mediated gene transfer in *Brassica carinata*. Plant Cell Rep. 11, 359–362.

Ogura, H., 1968. Studies on a new male-sterility in Japanese radish, with special reference to utilization of this sterility towards the practical raising of hybrid seeds. Mem. Fac. Agr. Kogoshima Univ. 6, 39–78.

Oliveri, A.M., Parrini, P., 1986. Relationship between glucosinolate content and yield component in rapeseed. Cruciferae Newsl. 11, 126–127.

Oram, R.N., Salisbury. P.A., Krick, J.T.O., Burton, W.A., 1999. Development of early flowering, canola grade *Brassica juncea* germplasm. Proceedings of Tenth International Rapeseed Congress, Canberra, Australia, 26–29 September.

Pilgeram, A.L., Sanda, D.C., Boss, D., Dale, N., Wichman, D., Lamb, P., Lu, C., Barrows, R., Kirkpatrick, M., Thompson, B., Johnson, D.L., 2007. *Camelina sativa*, a Montana omega-3 and fuel crop. In: Janick, J., Whipkey, A. (Eds.), Issues in New Crops and New Uses. ASHS Press, Alexandria, VA, pp. 129–131.

Plumper, B., 1995. Somatische und sexuelle Hybridisierung für den Transfer von Krankheitsresistenzen und *Brassica napus*. PhD Thesis, Free Univ. of Berlin, Berlin.

Polsoni, L., Kott, L.S., Beversdorf, W.D., 1988. Large-scale microspore culture technique for mutation–selection studies in *Brassica napus*. Can. J. Bot. 66, 1681–1685.

Potts, D.A., Rakow, G.W., Males, D.R., 1999. Canola-quality *Brassica juncea*, a new oilseed crop for the Canadian prairies. Proceedings of Tenth GCIRC International Rapeseed Congress, Canberra, Australia, 26–29 September.

Pradhan, A.K., Mukhopadhyay, A., Pental, D., 1991. Identification of putative cytoplasmic donor of CMS system in *Brassica juncea*. Plant Breed. 106, 204–208.

Pradhan, A.K., Sodhi, Y.S., Mukhopadhyay, A., Pental, D., 1993. Heterosis breeding in Indian mustard (*Brassica juncea* L. Czern & Coss): analysis of component characters contributing to heterosis for yield. Euphytica 69, 219–229.

Prakash, S., Chopra, V.L., 1988. Introgression of resistance to shattering in *Brassica napus* from *Brassica juncea* through non-homologus recombination. Plant Breed. 101, 167–168.

Prakash, S., Kirti, P., Chopra, V., 1995. Cytoplasmic male sterility (CMS) other than ogura and polima in Brassiceae: current status. Ninth Int rapeseed congr., Cambridge, vol. 1, pp. 44–48.

Prakash, S., Kirti, P.B., Bhat, S.R., Gaikwad, K.V., Kumar, Dinesh, Chopra, V.L., 1998. A *Moricandia arvensis* based cytoplasmic male sterility and fertility restoration system in *Brassica juncea*. Theor. Appl. Genet. 97, 488–492.

Prasad, K.V.S.K., Sharmila, P., Kumar, P.A., Saradhi, P.P., 2000. Transformation of *Brassica juncea* (L.) Czern with bacterial coda gene enhances its tolerance to salt and cold stress. Mol. Breed. 6, 489–499.

Pratap, A., Gupta, S.K., 2007. Unusual floral morphology in advanced generations of intergeneric hybrids between *Brassica napus* and *Eruca sativa*. Extended summaries: National Seminar on Changing Vegetable Oils Scenario: Issue and Challenges Before India, Indian Society of Oilseeds Research, Hyderabad, India, pp. 50–51.

Pratap, A., Gupta, S.K., Vikas, 2007. Advances in doubled haploid technology of oilseed rape. Indian J. Crop Sci. 2, 267–271.

Pratap, A., Gupta, S.K., Sharma, M., 2008. Genetic amelioration of crop brassicas through intergeneric hybridization using Diplotaxis. National Seminar on Physiological and Biotechnological Approaches to Improve Plant Productivity, Center for Plant Biotechnology, Department of Science & Technology, Haryana, Hisar, India, 15–17 March, pp. 85–86.

Princen, L.H., 1983. New oilseed crops on the horizon. Econ. Bot. 37 (4), 478–492.

Quazi, M.H., 1988. Interspecific hybrids between *B. napus* and *B. oleracea* developed by embryo rescue. Theoret. Appl. Genet. 75, 309–318.

Radke, S.E., Turner, J.C., Facciotti, D., 1992. Transformation and regeneration of *Brassica rapa* using *Agrobacterium tumefaciens*. Plant Cell Rep. 11, 499–505.

Rai, B., 1976. Considerations in the genetic improvement of oil quality in rapeseed. Oilseed J. 6, 13–15.

Rai, B., 1979. Heterosis Breeding. Agro-Biological Publications, Azad Nagar, New Delhi, India, 183 pp.

Rai, B., 1982. Breeding strategy for developing high yielding varieties of toria (*Brassica campestris* var. toria). Research and Development Strategies for Oilseeds Production in IndiaICAR, New Delhi, pp. 131–135.

Rai, B., 1983a. Genetic improvement of seed yield and disease resistance in rapeseed and mustard oilcrops. Oilseeds J. 13, 6–13.

Rai, B., 1983b. Advances in rapeseed and mustard breeding research. Indian Farming 33, 3–8.

Rai, B., 1987. PT 303 the first national variety of toria. Indian Farming 37, 16–17.

Rai, B., 1997. Breeding methods. In: Kalia, H.R., Gupta, S.K. (Eds.), Recent advances in oilseed brassicas. Kalyani publishers, India, p. 2552.

Rai, B., Kumar, A., 1978. Rapeseed and mustard production programme. Indian Farming 28, 27–30.

Rai, B., Sehgal, V.K., 1975. Field resistance of *Brassica* germplasm to mustard aphids (*Lipaphis erysimi* Kalt.). Sci. Cult. 41, 444–445.

Rai, B., Singh, A., 1976. Commercial seed production in rapeseed. Indian Farming 26, 15–17.

Rai, B., Gupta, S.K., Pratap, A., 2007. Breeding methods. Gupta, S.K. (Ed.), Advances in Botanical Research–Rapeseed Breeding, 45, Academic Press/Elsevier Ltd, San Diego, CA, pp. 21–48, 554 pp.

Rakow, G., 1995. Developments in the breeding of oil in other *Brassica* species. Proceedings Ninth International Rapeseed Congress, Cambridge, UK, 4–7 July, pp. 401–406.

Rakow, G., 2004. Species Origin and Economic Importance of Brassica. Biotechnol. Agric. Forestr. 54, 3–11.

Rakow, G., Raney, J.P., 2003. Present status and future perspectives of breeding for seed quality in *Brassica* oilseed crops. Proceedings Eleventh International Rapeseed Congress, Royal Veterinary and Agricultural University, Copenhagen, Denmark, 6–10 July, pp. 181–185.

Raney, J.P., Olson, T.V., Rakow, G., Ripley, V.L., 2003a. *Brassica juncea* with a canola fatty acid composition from an interspecific cross with *Brassica napus*. Proceedings Eleventh International Rapeseed Congress, Royal Veterinary and Agricultural University, Copenhagen, Denmark, 6–10 July, pp. 281–283.

Raney, J.P., Olson, T.V., Rakow, G., Ripley, V.L., 2003b. Selection of near zero aliphatic glucosinolate *Brassica juncea* from an interspecific cross with *B. napus*. Proceedings Eleventh International Rapeseed Congress, Royal Veterinary and Agricultural University, Copenhagen, Denmark, 6–10 July, pp. 284–286.

Rao, M.V.B., 1990. Widening variability in cultivated digenomic *Brassica* through interspecific hybridization. PhD Thesis, Indian Agricultural Research Institute, New Delhi, India.

Rao, G.U., Shivanna, K.R., 1996. Development of a new alloplasmic CMS *Brassica napus* in the cytoplasmic background of *Diplotaxis siifolia*. Cruciferae Newsl. 18, 68–69.

Rao, G.U., Batra, V.S., Prakash, S., Shivanna, K.R., 1994. Development of a new cytoplasmic male sterile system in *Brassica juncea* through wide hybridization. Plant Breed. 112, 171–174.

Raymer, P.L., 2002. Canola: an emerging oilseed crop. In: Janick, J., Whipkey, A. (Eds.), Trends in New Crops and New Uses. ASHS Press, Alexandria, VA, pp. 122–126.

Roy, N.N., 1984. Interspecific transfer of *Brassica juncea*-type high blackleg resistance to *Brassica napus*. Euphytica 33, 295–303.

Roy, N.N., Tarr, A.W., 1985. IXLIN – an interspecific source for high linoleic acid content in rapeseed. Plant Breed. 95, 201–209.

Roy, N.N., Tarr, A.W., 1986. Development of new zero linoleic acid (18:3) lines of rapeseed (*Brassica napus* L.). Z. Planzenzücht 96, 218–233.

Sacristan, N.D., Gerdemann, M., 1986. Different behavior of *Brassica juncea* and *B. carinata* as sources of interspecific transfer to *B. napus*. Plant Breed. 97, 304–314.

Salisbury, P.A., 1989. Potential utilization of wild crucifer germplasm in oilseed Brassica breeding. Proc. ARAB Seventh Workshop Toowoombu, Queensland, Australia, pp. 51–53.

Scheffler, J.A., Dale, P.J., 1994. Opportunities for gene tarsnfer from transgeneic oilseed rape (*Brassica napus*) to related species. Transgen. Res. 3, 263–278.

Schnepf, E., Crickmore, N., Van Rie, J., Lereclus, D., Baum, J., Feitelson, J., Ziegler, D.R., Dean, D.H., 1998. *Bacillus thuringiensis* and its pesticidal crystal proteins. Microbiol. Molec. Biol. Rev. 62, 775–806

Schuler, T.J., Hutcheson, D.S., Downey, R.K., 1992. Heterosis in intervarietal hybrids of summer turnip rape in western Canada. Can. J. Plant Sci. 72, 127–136.

Schuster, W., Michael, J., 1976. Untersuchungen über Inzuchtdepressionen und Heterosis effekte bei Raps (*Brassica napus oleifera*). Z. Pflanzenzücht 77, 56–66.

Sernyk, J.L., Stefansson, B.R., 1983. Heterosis in summer rape (*Brassica napus* L.). Can. J. Plant Sci. 63, 407–413.

Shahidi, F., 1990. Canola and Rapeseed: Production, Chemistry, Nutrition, and Processing Technology. Van Nostrand Reinhold, New York.

Shelton, A.M., Zhao, J.Z., Roush, R.T., 2002. Economic, ecological, food safety and social consequences of the deployment of Bt transgenic plants. Ann. Rev. Entomol. 47, 845–881.

Shewmaker, C.K., Sheehy, J.A., Daley, M., Colburn, S., Ke, D.Y., 1999. Seed-specific overexpression of phytoene synthase: increase in carotenoids and other metabolic effects. Plant J. 20, 401–412.

Shi, S.W., Wu, J.S., Zhou, Y.M., Liu, H.L., 2002. Diploidization techniques of haploids from *in vitro* culture microspores of rapeseed (*Brassica napus* L.). Chin. J. Oil Crops Sci. 24, 1–5.

Siemens, J., 2002. Interspecific hybridization between wild relatives and *Brassica napus* to introduce new resistance traits into the oilseed rape gene pool.

Singh, S.P., 1973. Heterosis and combining ability estimates in Indian mustard, *Brassica juncea* (L.) Czern & Coss. Crop Sci. 13, 497–499.

Singh, R., Sharma, S.K., 2007. Evaluation, maintenance and conservation of germplasm. Gupta, S.K. (Ed.), Advances in Botanical Research–Rapeseed Breeding, 45, Academic Press, London, UK, pp. 465–481, 554 pp.

Slocum, M.K., Figdore, S.S., Kenard, W.C., Suzuki, J.Y., Osborn, T.C., 1990. Linkage arrangement of restriction fragmentlength polymorphic loci in *Brassica oleracea*. Theor. Appl. Genet. 80, 54–64.

Snowdon, R.J., Winter, H., Diestel, A., Sacristán, M.D., 2000. Development and characterization of *Brassica napus–Sinapis arvensis* addition lines exhibiting resistance to *Leptosphaeria maculans*. Theor. Appl. Genet. 101, 1008–1014.

Song, K.M., Suzuki, J.Y., Slocum, M.K., Williams, P.H., Osborn, T.C., 1991. A linkage map of *Brassica rapa* (syn *campestris*) based on restriction fragment length polymorphism loci. Theor. Appl. Genet. 82, 296–304.

Sovero, M., 1993. Rapeseed, a new oilseed crop for the United States. In: Janick, J., Simon, J.E. (Eds.), New Crops. New York, John Wiley & Sons, pp. 302–307.

Spasibionek, S., Krzymanski, J., Bartkowiak-Broda, I., 2003. Mutants of *Brassica napus* with changed fatty acid composition. Proceedings Eleventh International Rapeseed Congress, Royal Veterinary and Agricultural University, Copenhagen, Denmark, 6–10 July, pp. 221–224.

Spinks, E.A., Sones, K., Fenwick, G.R., 1984. The quantitative analysis of glucosinolates in cruciferous vegetables, oilseeds and forage crops using high performance liquid chromatography. Fette Seifen Anstrichmittel 86, 228–231.

Srivastava, K., Rai, B., 1993. Expression of heterosis for yield and its attributes in rapeseed. Indian J. Agric. Sci. 63, 243–245.

Stefansson, B.R., Hougen, F.W., Downey, R.K., 1961. Note on the isolation of rape plants with seed oil free from erucic acid. Can. J. Plant Sci. 41, 218–219.

Stewart, C.N., Adang, M.J., All, J.N., Raymer, P.L., Ramachandran, S., Parrott, W.A., 1996. Insect control and dosage effects in transgenic canola containing a synthetic *Bacillus thuringiensis* cry1Ac gene. Plant Physiol. 112, 115–120.

Stiewe, G., Robbelen, G., 1994. Estabilishing cytoplasmic male sterility in *Brassica napus* by mitochondrial recombination with *Brassica tournefortii*. Plant Breed 113, 294–304.

Stoutjesdijk, P.A., Hurlestone, C., Singh, S.P., Green, A.G., 2000. High oleic-acid Australian *Brassica napus* and *B. juncea* varieties produced by co-suppression of endogenous delta 12-desaturases. Biochem. Soc. Trans. 28, 938–940.

Swanson, E.B., Coumans, M.P., Brown, G.L., Patel, J.D., Beversdorf, W.D., 1988. The characterization of herbicide tolerant plants in *Brassica napus* L. after *in vitro* selection of microspores and protoplasts. Plant Cell Rep. 7, 83–87.

Swanson, E.B., Herrgesell, M.J., Arnoldo, M., Sippell, D.W., Wong, R.S.C., 1989. Microspore mutagenesis and selection: canola plants with field tolerance to the imidazolinones. Theor. Appl. Genet. 78, 525–530.

Szarejko, I., Forster, B.P., 2007. Doubled haploidy and induced mutation. Euphytica 158, 359–370.

Tsunoda, S., Hinata, K., Gomez-Campo, C., 1980. Brassica crops and wild allies. Japan Scientific Society Press, Tokyo.

Vaughan, I.G., Macleod, A.J., Jones, B.M.G., 1976. The Biology and Chemistry of Cruciferae. Academic Press, New York.

Vollmann, J., Damboeck, A., Eckl, A., Schrems, H., Ruckenbauer, P., 1996. Improvement of *Camelina sativa*, an underexploited oilseed. In: Janick (Ed.), Progress in New Crops. ASHS Press, Alexandria, VA, pp. 357–362.

Vollmann, J., Grausgrubber, H., Stift, G., Dryzhruk, V., Lelley, T., 2005. Genetic diversity in *Camelina* germplasm as revealed by seed quality characteristics and RAPD polymorphism. Plant Breed. 124, 446–453.

Wang, M., Farnham, M.W., Nannes, J.S.P., 1999. Ploidy of broccoli regenerated from microspore culture versus anther culture. Plant Breed. 118, 249–252.

Wang, H.Z., Liu, G.H., Zheng, Y.B., Wang, X.F., Yang, Q., 2002. Breeding of *Brassica napus* cultivar Zhongshuang No. 9 with resistance to *Sclerotinia sclerotiorum*. Chin. J. Oil Crops Sci. 24, 71–73.

White, G.A., Higgins, J.J., 1966. Culture of Crambe: A New Industrial Oilseed Crop, ARS Production Research Report 95. Agricultural Research Service, US Department of Agriculture, Washington, DC.

Winter, H., Diestel, A., Gärtig, S., Krone, N., Sterenberg, K., Sacristán, M.D., 2002. Transfer von Resistenzen gegen *Leptosphaeria maculans* aus Wilderuciferen in den Raps. Vortr. Pflanzenzücht.

Xing, G.M., Long, M.H., Tanaka, S., Fujieda, K., 1988. Clubroot resistance in *Brassica–Raphanus*. J. Fac. Agric. Kyushu Univ. 33, 189–194.

Xu, L., Najeeb, U., Tang, G.X., Gu, H.H., Zhang, G.Q., He, Y., Zhou, W.J., 2007. Haploid and doubled haploid technology. In: S.K. Gupta (Ed.), Rapeseed Breeding. Advances in Botanical Research, vol. 45. Academic Press/Elsevier, San Diego, CA, pp. 181–216.

Yadava, T.P., Singh, H., Gupta, V.P., Rana, R.K., 1974. Heterosis and combining ability in raya for yield and its component. Indian J. Genet. 34A, 648–695.

Yu, F.Q., Liu, H.L., 1995. Effects of donor materials and media on microspore embryoid yield of *Brassica napus*. J. Huazhong Agric. Univ. 14, 327–332.

Zhang, F., Takahata, Y., 1999. Microspore mutagenesis and *in vitro* selection for resistance to soft rot disease in Chinese cabbage (*Brassica campestris* L. ssp. *pekinensis*). Breed. Sci. 49, 161–166.

Zhang, G.Q., Zhou, W.J., 2006. Genetic analyses of agronomic and seed quality traits of synthetic oilseed *Brassica napus* produced from interspecific hybridization of *B. campestris* and *B. oleracea*. J. Genet. 85, 45–51.

Zhang, G.Q., Zhou, W.J., Gu, H.H., Song, W.J., Momoh, E.J.J., 2003. Plant regeneration from the hybridization of *Brassica juncea* and *B. napus* through embryo culture. J. Agron. Crop Sci. 189, 347–350.

Zhang, G.Q., Tang, G.X., Song, W.J., Zhou, W.J., 2004. Resynthesizing *Brassica napus* from interspecific hybridization between *Brassica rapa* and *B. oleracea* through ovary culture. Euphytica 140, 181–187.

Zhang, G.Q., Zhang, D.Q., Tang, G.X., He, Y., Zhou, W.J., 2006a. Plant development from microspore-derived embryos in oilseed rape as affected by chilling, desiccation and cotyledon excision. Biol. Plantarum 50, 180–186.

Zhang, G.Q., He, Y., Xu, L., Tang, G.X., Zhou, W.J., 2006b. Genetic analyses of agronomic and seed quality traits of doubled haploid population in *Brassica napus* through microspore culture. Euphytica 149, 169–177.

Zhou, W.J., Hagberg, P., Tang, G.X., 2002a. Increasing embryogenesis and doubling efficiency by immediate colchicine treatment of isolated microspores in spring *Brassica napus*. Euphytica 128, 27–34.

Zhou, W.J., Tang, G.X., Hagberg, P., 2002b. Efficient production of doubled haploid plants by immediate colchicine treatment of isolated microspores in winter *Brassica napus*. Plant Growth Regul. 37, 185–192.

Zhu, J., Struss, D., Röbbelen, G., 1993. Studies on resistance to *Phoma lingam* in *Brassica napus–Brassica nigra* addition lines. Plant Breed. 111, 192–197.

CHAPTER

4

Sunflower

Yalcin Kaya

Engineering Faculty, Department of Genetic and Bioengineering, Trakya University, Edirne, Turkey

INTRODUCTION

Sunflowers (*Helianthus annuus* L.) are one of the higher adaptation crops because their deep roots prevent erosion and allow drought and other unsuitable conditions to be survived. However, there is a need to develop new varieties requiring less water, energy, and fertilizer to improve and increase its adaption capability under different environmental conditions and get improved and sustainable yields.

Helianthus is a large genus with 51 species and offers great opportunities to transfer valuable genes from wild types with the most desirable traits, such as disease resistance, more seeds, and easy harvesting. In addition to classical breeding methods, new molecular techniques commonly used in recent years such as next generation genotyping and sequencing are needed to sequence, assemble, and annotate the sunflower genome and to locate genes that are responsible for agriculturally important traits. Since the *Helianthus* family is one the largest plant families on Earth its genome is 3.5 billion letters long, slightly longer than the human genome (Kaya et al., 2012).

BREEDING OPPORTUNITIES FOR SUSTAINABLE PRODUCTION OF SUNFLOWERS

To achieve desired goals in plant breeding, it is necessary to have adequate genetic variability and suitable genetic resources. Typically used sunflower genetic resources include both cultivated and wild species.

Available Genetic Sources

There are several important sunflower collections in the world that provide breeders with sunflower germplasm, supplying enough genetic variability for successful contribution to

Breeding Oilseed Crops for Sustainable Production. http://dx.doi.org/10.1016/B978-0-12-801309-0.00004-5

their breeding programs. The largest collection in the world in terms of genotypes of cultivated sunflowers is maintained at the All-Russian Research Institute of Plant Industry, N.I. Vavilov (VIR), St. Petersburg, (Russia). The second largest collection exists at the US Department of Agriculture (USDA) National Plant Germplasm Plant Introduction Station, Ames, IA (USA) containing the largest and the most important wild species collection of the genus *Helianthus* in the world. There are also notable collections of wild sunflower species maintained at the USDA Sunflower Research Unit, Fargo, ND (USA), Institute of Field and Vegetable Crops (IFVC), Novi Sad (Serbia); Institut National de la Recherche Agronomique (INRA), Montpellier (France); Instituto Nacional de Tecnología Agropecuaria (INTA), Pergamino (Argentina); INTA, Cordoba (Spain), and VIR, St. Petersburg (Russia). There are also some major collections regarding the germplasm of cultivated sunflowers at IFVC, Novi Sad (Serbia); Yurijev Institute, Kharkov (Ukraine); VNIIMK, Krasnodar (Russia); National Institute of Field Crops, Fundulea (Romania); Dobroudja Institute, General Toshevo (Bulgaria) (Fernandez et al., 2009; Kaya, 2010; Škorić, 2012; Kaya et al., 2012).

Sunflower biomass and yield performance increased rapidly in the second half of the twentieth century, mostly due to the development of productive Russian varieties and hybrids based on cytoplasmic male sterility (CMS). Nevertheless, due to extensive use and the negative consequent results of certain human activities, there was a reduction in genetic variability in both cultivated sunflowers and wild species at the same time. The construction of dams, roads, railroads, industrial facilities, and residential areas, and the expansion of arable land, especially in North America, have caused genetic erosion of wild *Helianthus* species. Breeders often discard a large number of genotypes that at a particular moment do not have the desired agronomic value. They are also discarded by the hybridization process. Cultural practices tend to reduce the genetic variability of cultivated sunflowers and lead to genetic erosion. It often happens that genotypes (varieties, lines, hybrids) collected in one region adapt poorly to the local conditions of the region where they are tested. In that way the variability of collected and multiplied samples lead to genetic drift and selection pressure. For these reasons sunflower breeders should not devote their attention entirely to immediate breeding success but should try to simultaneously work on the protection, maintenance, and improvement of genetic sunflower resources (Fick and Miller, 1997; Fernandez et al., 2009; Škorić, 2012; Kaya et al., 2012).

Landraces, Local Populations, Public Lines, etc

Land races and local populations have huge genetic variation and are well adapted to local soil and climatic conditions as well as other environmental factors. However, they are the source of many desirable genes, especially those addressing higher adaptability to environmental conditions and resistance to certain diseases. The idea behind developing composite lines is to obtain new populations with open-pollinated (OP) plants by intercrossing a number of genotypes selected for specified characters (resistance to certain diseases, agronomic traits, etc.) and developing new genetic variability with phenotypic recurrent selection. Some previously developed genotypes can be used as valuable initial germplasm in breeding programs. For instance, the composite CM 303 was the source of several well-known inbred lines, including HA-89. Breeding lines and populations developed in breeding programs, including homozygous lines, derived by induced mutation, interspecific hybridization, or by new biotechnology methods and transgenic lines are also useful sources for sunflower-breeding programs (Fick and Miller, 1997; Fernandez et al., 2009; Škorić, 2012; Kaya et al., 2012).

The USDA breeding program is widely known for its public lines and the many inbred lines (like restorer, CMS, and B male) developed in this program such as HA-89, HA-821, HA-446, RHA 274, and RHA 427 which contain genes determining different traits from being high oleic to being resistant to some important diseases. These lines have been used commonly in almost all sunflower-breeding programs up to today both as tester and initial breeding materials (Fick and Miller, 1997; Fernandez et al., 2009; Škorić, 2012).

OP Cultivars

OP lines have certain desirable genes and sometimes very useful at solving certain problems in sunflower production through breeding. They are a good source of genes that confer such traits as high yield, high quality, and disease resistance. The first OP varieties, which were especially resistant to broomrape, began to be developed in 1915 by researchers at the Saratov experimental station in the early breeding programs of the former USSR via methods of individual selection (Gorbachenko et al., 2011). Between 1918 and 1925, Saratovsky 169, 206, 1915, 420, Kruglik A-41, Zelenka 76, and Fuksinka 10 were developed by Russian breeders and were widely used by farmers. Saratovsky 169 was grown for several years on a plot that exceeded 1 million hectares (Škorić, 2012). Zhdanovsky 6432, 8281, 6393, 8884, 8885, and Stepnyak were widely cultivated in subsequent years in the former USSR (Škorić et al., 2010). Soon after, in Krasnodar (Russia), Pustovoit combined resistance to race B and high productivity and developed, among other, cultivars VNIIMK 1646 and 6540. This was followed by cultivars with great oil content, such as VNIIMK 8883, 8931, Peredovik, and Smena. OP varieties were developed and started to be used worldwide by sunflower producers (Fick and Miller, 1997; Fernandez et al., 2009; Škorić, 2012; Kaya et al., 2012).

Synthetic Varieties

The collections maintained at breeding centers in different parts of the world include developed synthetics, defined as populations derived by intercrossing a number of individual plants selected for their superior combining ability, determined by crossing them with a tester, which can be used as initial germplasm in the development of new inbred lines. Improved populations are made by means of genotypic recurrent selection. The benefits of synthetics maintained by open pollination are the development of inbred lines showing high values of combining ability as a result of hybrid combinations (Fick and Miller, 1997; Fernandez et al., 2009; Škorić, 2012; Kaya et al., 2012).

Wild Species from the Genus *Helianthus*

A large number of wild sunflower species have genes for resistance to diseases and insects, tolerance to abiotic stresses such as drought, cold, salinity, and herbicides, and high quality of proteins and oil. They have been successfully used for the discovery of CMS and Rf genes through interspecific hybridization. In addition, breeders use them to increase genetic variability so that the effect of heterosis in seed yield is manifested to a higher degree (Škorić, 2012; Kaya et al., 2012).

Interspecific and Intergenetic Sources

Interspecific hybridization is an essential means in sunflower breeding of obtaining and broadening genetic variability. However, there are higher barriers to distant hybridization and the introduction of new genes from wild germplasm into sunflower cultivars because of different ploidy levels ($2x$, $4x$, $6x$) and genetic incompatibilities. The *Helianthus* genus has a basic chromosome number of $n = 17$ and contains diploid ($2n = 2x = 34$), tetraploid ($2n = 4x = 68$), and hexaploid ($2n = 6x = 102$) species. The genus has 14 annual species, which are all diploid, and 38 perennial species, which include 25 diploid, 3 tetraploid, 7 hexaploid, and 3 mixaploid species. Successful interspecific hybrids have been developed that have many useful genes with traits such as resistance to abiotic stresses, higher oil content, and higher protein quality. There are also new CMS and pollen fertility sources (Vear, 2011; Škorić, 2012; Seiler, 2012; Kaya, 2014) (Table 4.1).

There are few examples of developed intergeneric hybrids due to higher crossing barriers. Successful hybridization was performed to enrich the sunflower gene pool by introgressing new genes with traits like early flowering: *Helianthus* × *Verbesina*; the HA-89 line with *V. encelioides*; and the Peredovik and HA-89 lines used as female parents with *Tithonia rotundifolia* to develop intergenetic hybrids resistant to mildew, *Phomopsis*, and *Sclerotinia* (Kaya, 2014).

Developing the Ideal Plant for Successful and Profitable Sunflower Production

Significant results have been achieved worldwide in sunflower breeding in recent decades; this is particularly so for the application of marker genes in breeding, which expands knowledge of genetic variability at the molecular level. To develop an optimal ideotype of sunflower hybrids for specific agroecological conditions, breeders should have a thorough knowledge of genetics and breeding and should be familiar with the main characteristics of the target environment and the response of sunflowers to current conditions there. Implementation of biometrical genetics may greatly accelerate the development of previously defined hybrid models, by providing effective breeding strategies for sunflower yield improvement. In the case of breeding programs that have already achieved significant results in yield improvement, an ideotype can be used to improve missing traits of promising cultivars; for example, increasing the resistance to biotic and abiotic stresses, improving oil quality, and modifying oil composition (Fick and Miller, 1997; Fernandez et al., 2009; Škorić, 2012; Kaya et al., 2012).

Breeders should concentrate on main characteristics, which is a more profitable approach to improving and stabilizing the yield of sunflowers, instead of focusing on improvement of a large number of traits or genes in a single breeding cycle. Furthermore, sunflower breeders should be familiar with the available germplasm in their nurseries and should be able to understand which genes exist there to define the right hybrid model for their breeding programs. Additionally, breeders should focus on and be capable of noticing the physiological responses of plants and the formation of plants such as optimal density and the rate at which plant roots and leaves grow under different environmental conditions (Fick and Miller, 1997; Fernandez et al., 2009; Škorić, 2012; Kaya et al., 2012).

TABLE 4.1 Sections, Series, and Subspecies for the *Helianthus* with Chromosome Numbers

Section	Series	Species	Subspecies	(2*n* =)
1. Agrestis		*Agrestis*		34
		Niveus	*niveus, tephrodes, canescens*	34
		Debilis	*debilis, vestitus, tardilorus, silvestris, cucumerifolius*	34
2. Annuui		*Praecox*	*praecox, runyonii, hirtus*	34
		petiolaris	*petiolaris, fallax*	34
		neglectus, annuus, argophyllus, bolanderi, anomalus, paradoxus		34
3. Ciliares	1. *Pumili*	*gracilenthus, pumilus, cusickii*		34
	2. *Ciliares*	*arizonensis, laciniatus*		34
		ciliaris		68
4. Atrorubentes	1. *Divaricati*	*mollis, divaricatus, decapetalus*		34
		occidentalis	*occidentalis, plantagineus*	34
		hirsutus, strumosus		68
		eggertii, tuberosus, strumosus		102
		rigidus	*rigidus, subrhomboideus*	102
		giganteus, grosseserratus		34
		nuttallii	*nuttallii, parishii, rydbergii*	34
	2. *Gigantei*	*maximiliani, salicifolius*		34
		resinosus, schweinitzii, californicus		102
	3. *Microcephali*	*microcephalus, glaucophyllus, smithii, longifolius*		34
		laevigatus		68
	4. *Angustifolius*	*angustifolius, simulans, floridanus*		34
	5. *Atrorubentes*	*silphioides, atrorubens, heterophyllus, radula, carnosus*		34

Fernandez et al (2009), Breton et al (2010), Vear (2011), Kaya et al (2013).

Seed yield is the most important characteristic for sunflower breeders formulating a model of selection schemes for efficient photosynthesis. The effect of heterosis in sunflower hybrids is based on using inbred lines such as CMS, female, and restorer male lines to obtain high-yielding hybrids. This is mainly achieved using single crosses since heterosis decreases frequently using two-way or more crosses. Hybrids are genetically narrower than varietal populations so it is necessary to develop hybrids that are highly adaptable to each agroecological region (Škorić, 2012). In addition to high seed yield and high oil content in seeds, sunflower hybrids should be integrated with resistance to some important diseases as well as high tolerance to broomrape. The discoveries of the first source of CMS by Leclercq (1969) and the distribution of genes by Kinman (1970) enabled practical use of the phenomenon for heterosis in sunflowers (Fick and Miller, 1997; Fernandez et al., 2009; Kaya et al., 2012). Shortly thereafter, work on the development of sunflower hybrids and models of sunflower hybrids were started in France (Leclercq), the United States (Kinman, Fick, Zimmer, and Miller), Romania (Vranceanu and Stonescu), Yugoslavia (Škorić), and other countries (Škorić, 2012).

To develop ideotypical sunflower plants for agroecological conditions in different regions, there needs to be a hybrid model showing the genes that should be incorporated in the current genotype to optimize direct seed yield, oil yield, and environmental requirements. Škorić (2012) indicated that the following characteristics count among direct yield components in addition to high seed yield and high oil yield per unit area: number of plants per unit area (ha) (55,000–75,000/ha), number of seeds per plant (1,500–2,000), weight of 1,000 seeds (up to 80 g for oil type, 120–150 g for confectionery type), hectoliter weight (50–55 kg hL^{-1} for oil type, over 90 kg hL^{-1} for confectionery type), low husk percentage (below 25% for oil type, below 35% for confectionery type), seed oil content (50–55% for oil type, below 35% for confectionery type). He also mentioned that the final objective for oilseed sunflower genotypes was to obtain seed yield above 3000 kg ha^{-1} and for confectionery genotypes above 4000 kg ha^{-1}.

However, to develop an ideal model of sunflower hybrids, sunflower breeders have to understand the main characteristics of the environment for developing the hybrids: soil type; potential growing season length; mean, minimum, and maximum temperatures (daily, monthly); and the amount and distribution of rainfall during the year. Furthermore, the sunflower breeder should in particular understand optimal sunflower crop management for that environment, determine the limiting factors, learn how to reduce their effects accurately, and then implement the current hybrid model (Fick and Miller, 1997; Fernandez et al., 2009; Škorić, 2012; Kaya et al., 2012).

Breeding for the Most Important Sunflower Traits

Sunflowers or any other crop for that matter are strongly influenced by the cultivation environment and respond to conditions adaptively according to the way their epigenetic systems deal with the dynamics of change. Therefore, sunflower breeders should fully understand the condition and type of gene inheritance to plan accurately for the breeding process and choose genotypes that contain characteristics suitable to the environment to develop an ideal cultivar or hybrid and design a breeding program. However, it is not easy to collect many of the genes responsible for these characteristics, so it is necessary to put them in order based on importance and then apply them in a breeding procedure that involves as many of the desirable traits as possible (Fick and Miller, 1997; Fernandez et al., 2009; Škorić, 2012; Kaya et al., 2012).

Seed and Oil Yield

Seed yield is a complex and quantitative trait that is not only controlled by many genes but also influenced highly by environmental conditions, so both additive and nonadditive genetic effects play an important role in the inheritance of seed yield. Seed yield consists of three main components: number of plants per hectare, seeds per plant, and 1000 seed weight. Test weight is another important component and is closely related to low husk levels and high oil content in seeds (interestingly, there is a diverse correlation between the two traits). Selection for higher seed yield, higher oil yield, and other traits should start during inbred line creation by defining the effects of heterosis and by analyzing and evaluating the correlations among them to develop a productive hybrid with the desired traits (Škorić et al., 2007; Škorić, 2012).

The genetic variability of cultivated sunflowers presents a possibility to develop hybrids with a genetic potential for seed yields of over 6 t ha^{-1} and seed oil content of over 55% (Škorić et al., 2007; Škorić, 2012). However, there are limiting factors for current sunflower yields being in the production range 1.5–3.0 t ha^{-1}. Therefore, the sunflower breeder should focus on removing or reducing environmental factors so that sunflower yields of at least at 4 t ha^{-1} can be stabilized. Škorić (2012) indicated that only 10–20% of the yield potential of sunflower plants come from genetic capacity and that general phenotypic variance is determined by genetic factors such as leaves and roots and environmental conditions such as soil fertility and climate.

Self-Fertility and Self-Incompatibility

Sunflowers are normally OP plants, but the self-pollination rate of inbred lines and hybrids has increased with longer breeding process and greater efforts. However, some breeding lines, especially those from wild species, still have self-fertility problems. This is because the self-fertility of sunflower inbred lines has a positive correlation with their F_1 hybrids depending on gametogenesis, pollination, fertilization, and embryogenesis. All of these procedures are controlled by genetic and environmental factors, making self-fertility very complex and making it difficult to determine which other traits are affected most (Škorić, 1989; Fick and Miller, 1997; Fernandez et al., 2009; Škorić, 2012; Kaya et al., 2012).

The opposite of self-fertility is self-incompatibility, a sprophytic incompatible system that is controlled genetically and contributes to the physiological traits in sunflowers. Škorić (2012) indicated that the sterile central part of the sunflower head and smaller seeds are the effect of some nutritional deficiency that could be solved by eliminating minor rows on buds before flowering during the breeding process.

Number of Seeds per Head and Test Weight

The main components of seed yield are the number of plants per hectare, seed number per plant head, and seed weight. Any increase in seed yield depends on increasing these traits. The number of seeds per sunflower head is directly related to the number of florets, which depends on the sunflower genotype and the conditions of cultivation in the early period of plant development as well as before and during the formation of floret stems. The number of florets can reach 10,000 but is mostly 600–1,200 and less frequently 3,000–4,000. Škorić (2012) indicated that total floret number is related to daily integrals of the generator surface or by the number of initial flower lines and crop density.

The 1000 seed weight of oil-type sunflowers varies from 40 to 100 g but confectionery types can be over 100 g. Seed size, shape, and color are especially important for confectionery-type sunflowers depending on the market. For instance, consumers from China, Turkey, and some other countries require seeds that are at least 2 cm bigger, whereas in Eastern Europe such as the Balkans, Ukraine, and Russia consumers prefer big black seeds with big kernels and reduced husk content. Moreover, hulless sunflower kernels are used for confectionery, cakes, and other purposes in the food industry in North America (Fick and Miller, 1997; Fernandez et al., 2009; Škorić, 2012; Kaya et al., 2012).

SUNFLOWER BREEDING FOR DESIRABLE PLANT ARCHITECTURE

Plant height; head size, shape and position; leaf number; and vegetation period play very important roles in determining the optimum plant architecture for sunflowers (Breton et al., 2012; Škorić, 2012).

Plant Height

Plant height is not considered as important to seed yield as other traits and plant structure, but it is strictly linked to total leaf area and petiole length, which are very important for seed yield per plant. The majority of sunflower hybrids are 150–180 cm in height but shorter heights have some advantages such as sustaining more plants per hectare as well as tolerance to lodging and some diseases. On the other hand, many researchers have observed positive correlations between plant height and seed yield, so plant height should be considered carefully when developing parental lines during the breeding process. Similarly, many authors have mentioned the importance of plant height as a contributor to seed yield in many studies based on path and correlation analysis as well as general combining ability/specific combining ability (GCA/SCA) analysis (Fick and Miller, 1997; Fernandez et al., 2009; Škorić, 2012; Kaya et al., 2012).

Head Diameter, Size, Shape, and Inclination

Head diameter is one the most important characters contributing to high yield. Many path and correlation analyses indicate that head diameter contributes as much as 55.56% toward sunflower seed yield. Normally, sunflowers provide about 1500 seeds per head when plant density is high. Head size and shape are also important in achieving high seed yield. Sunflower head size is usually between 20 and 30 cm in diameter. There is much variance in head shape, which can be flat, concave, or convex according to breeder preference. The most important point is that the head center should be pollinated and have filled seeds. Head inclination and position are strongly related to bird damage and susceptibility to head rot diseases. A head position at an angle of 135–180° and firmly attached to the stem is the most preferred configuration by breeders (Fick and Miller, 1997; Fernandez et al., 2009; Škorić, 2012; Kaya et al., 2012).

Plant Architecture

Sunflowers need both, new plant architecture and design to improve via selection plant number per area – one of the three main characteristics in seed yield – and to increase seed weight. There have recently been some increases in plant number per hectare by using shorter sunflower hybrids and more intensive cultivating technologies such as planting sunflowers in narrow rows less than 70 cm apart to cram in 50,000 plants per hectare. However, to reach the final goal of 70,000 plants per hectare, sunflower plants will need to undergo some changes. A principal one is increasing leaf area by selection during the breeding process. To achieve maximum leaf area, sunflower plants should be designed with new leaf arrays that grow both vertically and horizontally to maximize sunlight. Škorić (2012) indicated that it is the top leaves that are mainly involved in photosynthesis, use CO_2, and absorb as many sun rays as possible in the morning and evening, while the middle leaves perform well when the sun is overhead during the day. He also mentioned that sunflower leaves should be wrinkled and point downward and that maximum leaf area should be 6000–7000 cm^2 per plant for efficient plant assimilation as well as for high intensity and productivity of photosynthesis. Breeders should be very aware of the inheritance of leaf number and size per plant in determining total leaf and assimilation area of F_1 generation plants.

To increase test weight – the third most important trait determining sunflower seed yield – breeders should focus on developing bigger kernels which need to assimilate more during grain filling and should consider limiting factors for grain growth development during the flowering period (Pereira et al., 2000; Škorić, 2012). The effect of additive genes play an important role in determining the inheritance of total leaf number per plant and plant height, but nonadditive genes contribute more to total leaf area per plant, especially when combining F_1 and F_2 generations. Therefore, all leaf characteristics such as petiole angle, petiole length, total leaf number per plant, total leaf area per plant, plant height, and head diameter contribute greatly to determining sunflower seed yield . Similarly, leaf number, total area of assimilation per plant, and leaf area index (LAI) are other important parameters, so breeders should take these traits into consideration and be aware of the inheritance of these traits during the selection process. Foliar diseases have a serious effect too – especially on the active leaf area – so they should also be considered carefully for increasing seed yield (Hladni et al., 2003; Hladni et al., 2004; Hladni et al., 2008; Breton et al., 2012; Škorić, 2012).

Increasing the harvest index by 0.40–0.50 as a result of stem shortening leads to lodging tolerance, higher plant density, and resulting higher yield, especially with application of more fertilizers (Škorić and Marinkovic, 1981). However, if the harvest index is followed by petiole length shortening, there should be an increase in yield as a result of allowing more plants per hectare. It could be possible to transfer the gene for this trait from *H. maximiliani* or induced mutations. However, such changes in the new plant architecture and the harvest index have lower heredity (0.20–0.30) for existing cultivars and hybrids. Škorić (2012) mentioned that better harvest index was also related to improving and changing hormonal regulation in all of a plant's generative and vegetative organs resulting in tolerance to various damages. He also indicated that yield increase could be possible without more nitrogen by improving the capability to use the limited sources available more effectively. Yield increase may be possible by accumulating more nonnitrogen storage substances in seeds, alternating the relation between storage and active constitutional proteins, changing important traits in relation to biomass yield, seed protein, as well as relative oil content and other substances. Although the harvest index could be predictable by the correlative effect of selection, it could be hard to improve during the selection process due to it being a complex trait that is influenced by many factors. Developing sunflower plants as having short petiole could be another way to increase the number of plants per area (Fick and Miller, 1997; Fernandez et al., 2009; Škorić, 2012; Kaya et al., 2012).

Developing Improved Oil Yield and Content in Sunflower Seeds

The main goal of sunflower production is to produce more vegetable oil. Thus, the development of improved oil-yielding lines and hybrids is the most important target for sunflower breeders. However, this trait is largely influenced by environmental factors and other yield traits. Average daily temperatures and moisture levels during the grain-filling period play important roles in determining this trait. Favorable conditions for improved oil yield are average daily temperatures lower than 25°C, good soil moisture, and the absence of disease (Škorić, 1988; Škorić, 1989). However, oil content is the main factor determining sunflower oil yield. Genetic contribution also plays a major part in constituting the oil content of sunflower seeds. The first breeding efforts were started by the Russian breeders Pustovoit (Krasnodar) and Zhdanov (Rostov-on-Don) who developed higher OP oil varieties in the first half of the twentieth century (Škorić, 2012).

Once inbred lines with improved oil content have been developed, they should be tested for GCA and SCA regarding oil yield productivity and heterosis. Special attention should be given to ensuring both parent lines have high oil content due to significant positive correlations between parental lines and their hybrids. Škorić (1989) mentioned that lower husk percentages could lead to negative effects on oil quality during prolonged seed storage and exposure to insects if seed oil content is over 55%. On the other hand, many authors indicate a significant negative correlation between husk percentage and oil content in sunflower seeds. To bring about improved oil content, breeding programs could make use of inbred lines and hybrids, current high oil–yielding lines, varietal populations, hybrids, mutant lines, synthetics, and other sources. Recurrent selection –used successfully in many cases – is one of the best breeding methods for increasing oil content by high oil line intercrossing coupled with simultaneous selection for some other desirable traits. Škorić (2012) indicated that Pustovoit's Reserve Method could still be used effectively in selection for high oil content because many OP varieties such as Peredovik and VNIIMK 8931 have oil content in excess of 50%.

Developing Improved Oil Quality in Sunflower Seeds

Normal sunflowers produce linoleic-type oil and mainly have linoleic acid (C18:2) in their seeds. Linoleic-type sunflower oil is used directly in the food industry and as a valuable raw material in nonfood industries such as soaps, cosmetics, pharmaceuticals, emulsifiers, lubricants, drying oils in paints, varnishes, polymers, synthetic rubber, livestock feed, and biodiesel. In addition to linoleic acid, sunflower seeds also have a high oleic (HO) acid (C18:1) content, a very favorable and healthy vegetable oil especially for frying (Vick et al., 2002; Ebrahimi et al. 2008; Velasco et al. 2008; Škorić, 2012).

Fatty acid composition can be expressed as the inverse rate of biosynthesis of oleic and linoleic acids during the seed-filling period. It is significantly affected by environmental factors, especially night temperature in that period, and genetic factors (Kaya et al., 2012; Škorić, 2012). The inheritance of oil quality and fatty acid composition is determined by joint action of the embryo and the female parent genome in sunflower seeds. It is primarily determined by the genotype of the embryo rather than that of the female plants. This is the reason Škorić (2012) advised better genetic analysis of the specific origin of seeds than of average seed samples because there were significant differences in the composition of fatty acids among seeds due to environmental factors resulting in variation in the expression of fatty acids. However, some inbred lines are less affected by environmental changes, so there is a need to analyze inbred lines from different genetic sources for fatty acid composition.

The main contribution to sunflower oil quality was made by Soldatov (1976) who used a dimethyl sulfate solution to bring about modifications of induced and spontaneous mutations and then obtained mutants with a high content of oleic acid (80%). Then, these mutants and HO Pervenets populations were used widely for HO acid content as the source of genes by sunflower breeders worldwide. HO exceeding 95% is controlled in breeding programs by one, two, or three dominant genes. Midoleic hybrids (NuSun) having oleic acid content of 65–75% were developed and used widely by the food industry in the United States (Vick et al., 2002). Many molecular studies – especially those utilizing molecular markers – were performed for appropriate selection for developing HO inbred lines and hybrids (Perez-Vich et al., 2002, 2004). Fernandez-Martinez et al. (2007) indicated that it is possible to make new types of oil on the basis of mutant lines, such as a low-saturation oil (17%), a high-palmitic oil (>25%), a high-stearic oil (>25%), a high-oleic oil (>85%), a high-linoleic oil (>75%), as well as other combinations differing in fatty acid composition, in addition to the standard type of sunflower oil. However, breeders and geneticists should continue making efforts to create genetic variability for oil quality (e.g., induced mutations used to develop HO hybrids).

Genetic and biochemical studies indicate that sunflower oil quality is not only determined by fatty acids, but also by the tocopherol content in the seed – an important indicator of the antioxidative imbalance of sunflowers. So sunflower breeders need to study the inheritance of tocopherols to increase the oxidative stability of sunflower oil. Standard sunflower oil comprises the compounds a-tocopherol (95%), 1-tocopherol (3%), and y-tocopherol (2%), which accumulate in sunflower seeds through a biosynthetic pathway determining the buildup of triacylglycerols, natural antioxidants used for forming vitamin E (Demurin et al., 1996; Velasco and Fernandez-Martinez, 2003; Velasco et al., 2004; Škorić, 2009, 2012). Velasco et al. (2004) also found that some lines having a higher content of γ-tocopherol, δ-tocopherol (>65%), and β-tocopherol (>75%). Furthermore, incorporating genes for HO acid and various tocopherols (01 + tphl: 01 tphI tph1) in breeding lines allowed the development of new hybrids with different oil quality types (Vera-Ruiz and Fernandez-Martinez, 2006; Škorić et al., 2006, 2007, 2008). New developments in sunflower production have led to new uses in pharmaceutical, cosmetic, and other industries (biodiesel), in addition to their use as food products. As a result, sunflower breeders and food specialists should define new future parameters and novel uses of sunflower oil for high-quality final products widely such as high oleic, tocopherol (E vitamine) content, etc.

Škorić et al. (2006, 2008) proposed the following oil quality profiles as goals in future sunflower breeding programs: common oil hybrids, high oleic (01 genes), >90% oleic acid, high oleic (01 + tphl genes), high oleic (01 + tph2 genes), high oleic (01 + tphl tph2 genes), high linoleic (>80% linoleic acid), high palmitic (>30% palmitic acid), and finally intermediate hybrids containing different combinations of the above genes. Sunflower breeding programs for fatty acids could be set up in three directions: high oleic in combination with low-saturated fatty acid content, high linoleic in combination with low-saturated fatty acid content (<6%), and high-saturated genotypes (Škorić, 2012).

Developing Improved Protein Yield and Quality in Sunflower Seeds

Sunflowers are protein crops – producing some 16–19% of protein – as well as being mainly oil-producing crops. Protein content in sunflower seed is a variable trait determined by many genes and influenced by the environment and by a mixture of genetic and environmental factors (GxE) that interact with these genes. On the other hand, there is a negative correlation between the content of protein and oil in seed. Therefore, selection efforts to increase the protein content in sunflower seeds are focused in two directions: confectionery types and oil types. Furthermore, the genetic variability of confectionary types is significantly less than that of oil types, so it should be taken into account during the selection process. However, when selecting for protein, genotypes should be evaluated on increased protein yield per hectare – not on primary increase in the content of proteins. In confectionary genotypes, the ideal protein and oil content should be around 30%. So, when choosing the initial materials for selection, they should have enough variability before the application of adequate plant-breeding methods takes place. Recurrent selection is one of the most appropriate methods to increase sunflower seed protein content and seed yield (Fick and Miller, 1997; Fernandez et al., 2009; Škorić, 2012; Kaya et al., 2012).

Consumer demand varies widely, especially for seed color. However, most customers prefer tasty, high-quality, and longer seeds (at least 2 cm long) for confectionery types. It is not easy to develop bigger kernels in relation to husks in breeding programs. To develop high-quality and tasty confectionery hybrids, different traits need to be added such as higher oleic acid content as well as β-tocopherols and δ-tocopherols for longer shelf life and different protein quality like albumin, globulin, and gluten. To increase protein content and quality during the selection process, there needs to be a higher content of essential amino acids, including lysine but excluding chlorogenic acid. The introduction of molecular methods, especially the use of molecular markers, in addition to classical breeding efforts should facilitate proper and efficient selection to increase protein yield per hectare and quality (Fick and Miller, 1997; Fernandez et al., 2009; Škorić, 2012; Kaya et al., 2012).

Breeding for Earliness in Sunflower

Vegetation period is another important sunflower trait and could be a limiting factor in sunflower production in some regions where earliness (combined with higher yield) is the preferred option of most producers. Sunflower genotypes could be classified as too early, early, midearly, medium, midlate, and late based on their vegetation periods. Earliness could be used to escape biotic or abiotic stress conditions. So, sunflower breeders should consider the inheritance patterns of earliness during the selection process as one of the most important yield traits, especially regarding yield productivity. When selecting for earliness, the period from the beginning of budding to the beginning of flowering is one of the important parameters. However, reductions in biological and agronomical traits such as height, leaf number, and use of suitable initial genetic material are factors involved in many other appropriate methods. As a final goal regarding oil content and yield, sunflower breeders should ignore the negative effects of other agronomic traits and concentrate instead on developing inbred lines or productive hybrids which have vegetation periods shorter than 100 days and which are not influenced by biotic and abiotic stresses (Škorić, 2012).

Breeding for Bee Attractiveness in Sunflowers

Pollinators, mostly bees, play an important role in the pollination of sunflowers, especially in hybrid seed production. Domestic bees (*Apis mellifera*), bumble bees (*Bombus* spp.), and blue orchard bees (*Osmia lignaria*) are common pollinators of sunflowers. Meteorological factors – especially wind, temperature, rain, humidity, and clouds – influence the number of visits and behavior of these bees. Moreover, genetic factors such as yellow-colored ray flowers and disk florets, aromatic substances, corolla length longer than 1.2 mm, less sticky substances, and larger nectar amount also affect and increase bee and insect visits and attractiveness. Researchers further indicate that stomata number and location affect bee attractiveness. Sunflower nectar amount, especially, is a highly heritable trait that could be improved by selection at early generations of inbred line development (Škorić, 2012).

SUNFLOWER BREEDING STRATEGIES FOR CONSTRAINTS

Breeding for Resistance to Biotic Stresses

Diseases, broomrape parasitic plants, viruses, and bacteria are, respectively, the most serious biotic factors limiting sunflower production.

Sunflower Diseases

Diseases are serious limiting factors to sunflower production in almost all parts of the world, where various diseases appear in different growing areas depending on environmental conditions. More than 30 different pathogens attack sunflowers and some cause severe economic costs to growers in the sunflower production areas of the world. The origin of the cultivated sunflower has a very narrow base and, hence, there is a lack of resistant genes. Nevertheless, sunflower breeders have achieved significant successes in finding genes for resistance or high tolerance to certain diseases in wild species and incorporating them into today's cultivated sunflower genotypes. Sunflower breeders have a number of factors to take into consideration, such as general principles of resistance breeding, major approaches to managing resistance genes, knowledge about the inheritance of certain pathogens, monitoring interactions between sunflowers, pathogens, and the environment, and selecting better breeding methods to develop resistance genotypes vertically and horizontally (Vear 2004; Škorić, 2012; Talukder et al., 2014).

Current results in sunflower breeding for disease resistance could be classified into four groups: genetic resistance to sunflower diseases such as *Plasmopara halstedii, Puccinia helianthi, Verticilium*, and *Erysiphe cichoracearum*; high level of tolerance to sunflower diseases such as *Phomopsis, Macrophomina, Albugo*, and *Alternaria* spp.; satisfactory tolerance of diseases such as *Phoma macdonaldii* and *Sclerotinia sclerotiorum*; and partial disease tolerance against *Rhizopus* spp., *Botrytis cinerea*, and other pathogenic fungi (Škorić et al. 2006; Shobha Rani and Ravikumar, 2006; Vear et al., 2007; Skoric, 2012; Talukder et al., 2014). However, for efficient disease control in future breeding programs, breeders should combine vertical and horizontal resistance (if available) and utilize a marker-assisted selection process combined with quantitative trait locus (QTL) mechanisms (Mouzeyar, 2000; Liu et al., 2012; Liu and Jan, 2012; Škorić, 2012). The major diseases of sunflowers are now given based on respective importance.

Downy Mildew (**Plasmopara halstedii** *(Farl.) Berl. & de Toni)*

Downy mildew occurs in sunflower-growing areas in all parts of the world and becomes especially epidemic during wet spring years. Downy mildew is constantly developing new races. At least 18 downy mildew races have been observed up to the moment, but resistant genes have been successfully controlled by transferring PI genes from wild sunflower species (Škorić et al. 2006; Bán et al., 2014; Sakr, 2014). Using metalaxyl as a chemical control for seed treatment is an efficient method. However, new races cannot be controlled by metalaxyl treatment, so sunflower breeders use the US Department of Agriculture (USDA) public lines having PI genes such as RHA-340 and RHA-419 to develop durable resistance. Furthermore, to get better downy mildew resistance, sunflower breeders need to develop new resistant genotypes with genes that are more resistant by making use of different resistance sources to achieve vertical and horizontal resistance against downy mildew in breeding programs. In addition to classical breeding efforts, different molecular studies especially for marker-assisted selection (MAS) have been performed to speed up selection cycles and also to characterize dominant genes in the genotypes (Liu et al., 2012). Breeding materials have been tested by inoculation methods which has brought about continuous improvements in determining downy mildew resistance (Vear et al. 2000; Molinaro-Demilly et al., 2004; Tourvieille de Labrouhe et al., 2004; Škorić, 2012; Sakr, 2014).

Head, Root, and Stem Rot (**Sclerotinia sclerotiorum** *(Lib.) de Bary)*

Sclerotinia is a major problem in production areas that have a humid climate and extremely wet summer seasons. As a result of being heavily influenced by environmental factors, high variability, and having three different types of fungus controlled by different mechanisms of resistance affecting plant roots, stems, and heads, selection for resistance is not easy (Škorić, 2004). *Sclerotinia* tolerance is controlled by many genes with different behaviors and needs SCA during the breeding process (two tolerant inbred lines could produce susceptible hybrids or two susceptible lines could produce a tolerant F_1 hybrid) because of using different genetic materials and different inoculation methods (Baldini et al., 2002; Vear 2004; Dnoseva et al., 2008; Škorić, 2012; Wurschum, 2014).

Stem rot tolerance is related to sowing dates and to the relationship between mean temperature and rainfall during flowering. Genotypic effects and $G \times E$ interactions play important roles in determining tolerance to fungus (Hahn 2002; Henson and Gulya, 2004; Gulya et al., 2008; Paricsi et al., 2008). An early-stage evaluation technique was developed for *Sclerotinia* root tolerance. It was performed in the greenhouse using mycelial inoculation at the four leaf pair stage by placing barley grains infected by fungus in contact with plant roots. Genotypes were evaluated, as symptoms developed, 25–45 days after inoculation and the percentage of destroyed plants and results were relatively well correlated (Peres et al., 2000). Moreover, millet seeds were also used for *Sclerotinia* head rot tolerance instead of barley grain.

During selection for *Sclerotinia* tolerance, the GCA effects of female lines were relatively larger than those of male lines; hence, female lines had greater influence in the development of resistant hybrids (Castano et al., 2001, 2009; Vear et al., 2007). During sunflower-breeding processes aimed at getting resistance to all three forms of *Sclerotinia* attack, it is necessary to combine two or three different types (Castano et al., 2009). *H. maximiliani* exhibited the highest level of resistance among the wild types. Promising results, providing considerable assistance for sunflower breeders, were found in *Sclerotinia* resistance study at the molecular level, mainly in MAS to make it appropriate (Baldini et al., 2002; Bert et al., 2002; Schnabl et al., 2002; Vear, 2004; Vear et al., 2008; Liu and Jan, 2012; Škorić, 2012; Wurschum, 2014; Talukder et al., 2014).

Rust (Puccinia helianthi *Schw.*)

Rust is one of the most important and widely distributed diseases in sunflower-producing areas causing wide-ranging economic losses – from North America to South America and from Australia to Africa. Rust is not a severe disease in Europe. Five races have been determined in North America, and serious yield loss has been observed in some years in the sunflower-producing areas of Argentina. Ph genes have been found to offer protection against rust and standard differential lines have been constituted to determine them. The most appropriate method for rust inoculation was found to be a mixture of resistant genotypes held in spatial isolation coupled with selection of resistant plants. Molecular markers were also identified for some sunflower rust genes by combining the R genes with character pyramiding (Sendall et al., 2006). Based on recent research, 25 lines have been determined with putative resistance genes utilizing molecular markers in Australia (Sendall et al., 2006) and resistant genotypes and germplasm in Hungary by Viranyi (2008) for future sunflower breeding for rust resistance. Rust resistance genes have also been determined and transferred into elite lines from wild species, because the wild *Helianthus* genus has a rich gene pool for further identification of resistance genes and for possible new rust races (Vear et al., 2008; Škorić, 2012; Talukder et al., 2014).

Stem Canker (Phomopsis/Diaporthe helianthi)

Phomopsis is a severe disease in almost all sunflower-growing countries of Europe as well as in the United States and Canada. The first sunflower-resistant genotypes were determined by Škorić in the 1980s derived from interspecific hybrids (cultivated sunflower × *H. tuberosus*), a cross between *H. argophyllus* × Armavirski 9345 and the restorer line SNRF-69 after a long breeding process in Serbia (Škorić, 2012). Sunflower resistance to *Phomopsis* is horizontal and is positively correlated with the stay-green trait in stems at ripening, with *Macrophomina* and *Phoma* resistance, and with drought tolerance (Masirevic, 2000).

As inoculation methods to evaluate breeding materials, leaf and petiole infection were found to be successful methods. *In vitro* screening of sunflower calli to determine the reaction to a filtrate of *Phomopsis* is possible too (Quaglia and Zazzcrini, 2007). It is advised to eliminate susceptible plants from early breeding generations before observations are made about F_3 or F_4 families, because of the difficulty of distinguishing intermediate and resistant genotypes at the mycelium. A rich germplasm has been developed for *Phomopsis* resistance, especially from wild *Helianthus* species, to increase genetic variability and to develop resistant hybrids. Based on recent results, leaf resistance was found to be controlled by one major gene and at least two complex factors were found to influence resistance in petiole and stem tissues (Masirevic, 2000). The recent use of molecular markers in breeding has led to recombinant inbred lines derived by crossing LR4-17 (resistant) with HA89 (susceptible) major QTLs for different leaf and stem resistance components being determined and pyramided (Langar et al., 2004, Liu and Jan, 2012; Qi et al., 2012; Škorić, 2012).

Verticillium *Wilt* (Verticillium dahliae *Kleb.*)

Verticillium is one the most harmful and widespread diseases responsible for serious economic damage in sunflower-producing areas in many parts of the world. *Verticillium* wilt is controlled by a single dominant gene called *VI*, and resistance sources have been determined in wild *H. annuus*, *H. petiolaris* populations, and in interspecific hybrids (cultivated sunflower × *H. tuberosus*) and then transferred to cultivated ones (Gulya, 2007). OP Russian cultivars VNIIMK 1646, VNIIMK 6540, VNIIMK 8883, VNIIMK 8931, and Armavirskiy 9343, as well as four new USDA lines (HA-132, HA-313, HA-314, and HA-315) have also exhibited resistance to *Verticillium* wilt. As an inoculation method of breeding material, the injection technique is successful at screening breeding materials. *Verticillium* wilt–resistant genotypes have also exhibited high tolerance to *Sclerotinia* (Escande et al. 2006; Gonzales et al., 2007; 2008; Škorić, 2012; Talukder et al., 2014).

Charcoal Rot (Macrophomina phaseolina *(Tassi) Goid*)

Macrophomina rot is responsible for economic damage to sunflower production and is especially widespread in arid regions. Earlier sunflower genotypes were significantly more sensitive to the pathogen than later genotypes. Resistance to *Macrophomina* is horizontal and controlled by polygenes. Some resistant genotypes have been derived from varietal populations (VNIIMK, Krasnodar) developed by interspecific hybridization with *H. tuberosus*. The HAR-1 and HAR-2 inbred lines used to be tolerant to almost all charcoal rot isolates. However, recent isolates have broken the resistance of these inbred USDA lines (Talukder et al., 2014). A radical inoculation method, which involves injecting fungus suspension into the head tissue and inoculation by the toothpick method, has been successfully used to evaluate sunflower genotypes (De Oliveira et al., 2004; Škorić, 2012).

Alternaria *Blight* (Alternaria helianthi *(Hansf.) Tubaki & Nishih.*)

Alternaria blight is one of the most widespread diseases in sunflowers. An attack, especially under humid conditions, results in its toxins limiting pollen germination and reducing leaf area for photosynthesis (Regina et al., 2008). Some resistant lines have been developed from wild *H. tuberosus* and *H. occidentalis* populations. These genes were then transferred into selected genotypes of cultivated sunflowers (Škorić et al, 2006; Madhavi et al., 2005; Sujatha and Prakabaran, 2006). A successful inoculation method involved applying a pathogen culture filtrate to the stigma and style an hour before pollination, and then scoring disease responses at flowering, 15 days after flowering, and at physiological maturity. However, gametophytic selection seems a more promising method than pollen selection allowing favorable allelic combinations to be distinguished that could not be detected at the sporophytic level (Shaik and Ravikumar, 2003; Shobha-Rani and Ravikumar, 2006). Therefore, Škorić (2012) indicated that combining gametophytic selection with conventional sporophytic selection could be a more effective way for breeding programs to achieve higher levels of resistance quickly.

Phoma *Black Stem* (Phoma macdonaldii *Boerema*)

Phoma black stem is a widely observed sunflower disease in many countries responsible for economic damage to early-drying sunflower plants (Quiroz et al., 2014). Higher tolerance to *Phoma macdonaldii* has been observed in wild species such as *H. maximiliani*, *H. argophyllus*, *H. tuberosus*, and *H. pauciflorus*. This led to resistant genotypes being determined and developed by sunflower breeders (Škorić, 2012). Significant differences have recently been observed among genotypes, isolates, and their interactions for tolerance to *Phoma* black. Some genotypes have shown wide-ranging partial resistance , something that was also confirmed with QTL analysis at the molecular level (Alignan et al., 2006; Darvishzadeh et al., 2007).

Other Fungal Diseases

There are many other fungal diseases of sunflowers that limit sunflower production in some years and some parts of the world. Many species of *Fusarium* attack sunflowers, but little research has been carried out (Gontcharov et al., 2006). Recently, Gontcharov et al. (2006) developed tolerant lines based on laboratory tests and individual selection of plants. *Botrytis cinerea* Pers. causes major losses of yield in sunflower production in some regions. Some sunflower genotypes have been found to be tolerant to *Botrytis* under natural and artificial inoculation conditions. Dry rot (*Rhyzopus*) often causes major losses in yield, particularly as it reduces the oil content in seeds, and typically appears in sunflower-producing areas with dry climates and high temperatures. *Rhizopus* resistance or tolerance has been determined in both wild species and some OP Russian cultivars such as Armavirec, Armavirskiy 3497, EC 50277, and K-2217 (Škorić, 2012). White rust *(Albugo tragopogonis* Schr.) has been observed in many countries but does not cause such severe losses in yield in sunflowers as other diseases. Powdery mildew (*Erysiphe cichoracearum* DC.) is widespread in almost all sunflower-growing countries. White rot resistance was first identified in wild *H. debilis,* then in *H. tuberosus*, and finally got transferred to cultivated plants. Powdery mildew resistance was found to be partially dominant (Jan and Seiler, 2007; Castano and Giussani, 2009; Škorić, 2012; Reddy et al., 2013; Talukder et al., 2014).

Viral Diseases

Several viruses that are problematic in other crops are known to attack sunflowers. However, they mostly use sunflowers as a secondary crop. The most important include aster yellow virus, cucumber mosaic virus, sunflower mosaic virus, sunflower ringspot virus, sunflower yellow blotch virus, leaf crinkle virus, tobacco ringspot virus, tobacco streak virus, tomato spotted wilt virus, and potyvirus. Cucumber mosaic virus, tobacco ringspot virus, and sunflower mosaic virus are viral diseases that have been found in some sunflower-growing countries but have caused less economic loss (Gulya et al., 2002; Škorić, 2012).

Sunflower chlorotic mottle virus (SuCMoV) is one of the sunflower viruses that has been studied most intensively, has occurred most frequently, and caused much economic damage (especially in Argentina). After virus infection, plant height and seed yield were significantly reduced and other yield traits such as stem length and diameter, seed width and length, as well as 1000 seed weight were also severely influenced (Lenardon, 2008). Some resistant and partially tolerant genotypes have also been determined using an inoculation method in the greenhouse controlled by a single dominant gene (Rcmo-1) (Lenardon et al., 2001, 2005; Bazzalo et al., 2008). Recent molecular studies have been performed to characterize SuCMoV resistance, RNA isolation, and Rcmo-1 gene mapping (Duhovny et al., 2000; Lenardon et al., 2005; Arias et al., 2003, 2005; Mailo et al., 2008; Škorić, 2012). Furthermore, Bejerman et al. (2008, 2010) have pointed out that potyvirus resistance is also linked to SuCMoV after conducting a molecular study in Argentina and Brazil.

Bacterial Diseases

Bacterial diseases are rarely seen in sunflowers; they typically use sunflowers as a secondary host. The most widespread bacterial disease is *Erwinia caratovora*, which was first observed and reported in Hungary. Some resistant genotypes have been determined under natural conditions. Furthermore, *Agrobacterium tumefaciens*, *Pseudomonas syringae*, and *Xanthomonas campestris* are other pathogenic bacterial diseases found in sunflowers (Škorić, 2012).

Sunflower Broomrape (*Orobanche cumana* Wallr.)

Sunflower broomrape is a holoparasitic angiosperm – lacking chlorophyll or photosynthetic activity – which penetrates the vascular tissues of sunflower roots, getting both water and nutrients, and results in 100% yield reductions in sunflower-producing areas, especially in Eastern European countries, Spain, and some Asian countries such as Iran, Israel, China, and Kazakhstan. Broomrape seeds are small and the 1000 seed weight is about 0.001 g, so it spreads very easily over longer distances. *Orobanche* seeds are known to persist in the soil up to 20 years, but they need to germinate in wet soils with a soil temperature of about 20–25°C. *Orobanche* is a self-pollinated plant due to its flower morphology – with curved tubular corollas having small lower lips making it hard for pollinators to land – and low intrapopulation genetic diversity. However, it is not strictly self-pollinating and some extent of crosspollination should be expected (Pacueranu-Joita et al. 2008, 2009; Škorić, 2012; Rodríguez-Ojeda et al. 2013; Pineda-Martos et al., 2013).

Sunflower resistance to *Orobanche* involves different mechanisms – such as genetic, physiological, biochemical, and mechanical –but the most important is the genetic system used by sunflower breeders which is characterized by the existence of certain genes controlling resistance. Sunflower root cells contain substances such as phytoalexins and benzothiadiazoles that activate broomrape seed germination. Broomrapes attach to the roots of susceptible sunflower genotypes, then an appressorium develops into a wedge-shaped haustorium penetrating the skin and parenchyma of the sunflower root surface and entering the central cambium cylinder (Sauerborn et al., 2002; Wegmann 2004; Matucova et al., 2004; Honiges et al., 2008; Fernandez-Martinez et al., 2008, 2010, 2012; Antonova et al., 2009; Škorić, 2012).

Genetic resistance is related to the existence of a layer of lignin between the parenchyma and cambium in sunflower root cells which stops broomrape haustoria from penetrating. There are also some mechanical barriers that reduce broomrape seed germination or reduce the occurrence of necrosis (before blooming). A low pH value in the sunflower root system is a sign of susceptibility to broomrape, a reaction of invertase and peroxidase enzymes that play important roles in determining resistance to broomrape (Labrousse et al., 2004).

The fact that pathogens react quickly by creating new virulent races in sunflowers was first observed in Saratov (an oblast in Russia) in the 1890s. Then, early in the twentieth century, they spread throughout Russia and severely threatened sunflower production. Broomrape resistance is controlled by *Or* genes. Genetic resistance breeding efforts against broomrape first started at the Saratov Breeding Station in Russia. The first resistant cultivar was Saratovski 169. Later, Kruglik A/41, Zelenka, and Fuksinka were developed to be resistant to race A controlled by Or_1. Subsequently, a new race called B appeared in sunflower areas, spread quickly, and became resistant to sunflower cultivars Zhdanovsky 6432, Zhdanovsky 8281, and Stepnyak developed by L.A. Žhdanov in 1926 which were widely cultivated in the following years (Antonova et al., 2009; Gontcharov, 2009; Gorbachenko et al., 2011). Žhdanov developed other resistant cultivars with the resistance gene Or_2 and used wild species of *Helianthus* spp. for the first time in the world – particularly, *H. tuberosus*, *H. maximiliani* Schrad., and *H. mollis* Lam. – for sunflower breeding for broomrape resistance (Škorić et al., 2010). Once resistance genes had been identified in the germplasm of wild species, good agronomic characteristics were introgressed into these new resistance genes. Hence, several cultivars resistant to race B were developed between the 1930s and1950s which had both high yield and with high oil content – such as VNIIMK 8883, VNIIMK 8931, Armavirsky 3497, Armavirsky 9343, Armavirsky 9345, Peredovik, and Smena – and were distributed worldwide. A new race was announced in Russia, Romania, and Bulgaria in the 1960s and 1970s. Subsequently, in the late 1970s resistant sunflower cultivars to race C were developed such as Progress, Yuvileyniy 60, Oktobar, and Novinka. Moreover, there were Romanian cultivars derived from resistant germplasm developed in the former USSR, and populations derived from interspecific crosses between *H. annuus* and *H. tuberosus* (Vrânceanu et al., 1980; Dozet et al. 2000; Antonova et al., 2009; Gontcharov, 2009; Pacureanu-Joita et al., 2009; Škorić, 2010, 2012; Jan et al., 2014). Later on, the D and E races appeared in Eastern Europe.

Vranceanu et al. (1980) established a differential set to distinguish between five races, which were controlled by dominant genes: A (Or_1), B (Or_2), C (Or_3), D (Or_4), and E (Or_5). These differential lines were Kruglik A41 for race A, Jdanov-8281 for race B, the Romanian cultivar Record for race C, line S-1358 for race D, and line P-1380-2 for race E. Pacueranu-Joita et al. (2008) included line LC-1093 as a tester for race F controlled by the Or_6 gene. An accession of *H. debilis* ssp. *tardiflorus* (PI 468691) identified by Velasco et al. (2012) exhibited complete resistance and was classified as race G because it parasitized plants of several race F–resistant lines (Fernández-Martínez et al., 2004, 2008; Pineda-Martos et al., 2013; Jan et al., 2014). Hence, the number of broomrape races reached eight and they were observed in many sunflower-growing countries such as Russia, Ukraine, Moldova, Romania, Turkey, Bulgaria, and Spain (Rodriguez-Ojeda et al. 2001; Akhtouch et al., 2002; Perez-Vich et al. 2002, 2004; Velasco et al., 2006, 2007; Fernandez-Martinez et al., 2007, 2008; Pacueranu-Joita et al. 2008, 2009; Kaya et al., 2004, 2009; Gontcharov, 2009; Antonova et al., 2009; Pacureanu-Joita et al., 2009; Škorić and Pacureanu-Joita, 2011; Jan et al., 2014).

Dynamic changes in the composition of races calls for new permanent sources of broomrape resistance to be found. The sources of the A, B, C, and D races are the Russian, Ukrainian, and Romanian cultivars. The sources for races F and G are wild species. New broomrape-resistant sources have recently been discovered in wild *Helianthus* species. Perennial *Helianthus* spp., in particular, have new resistant genes but experience higher crossability difficulties due to different numbers of chromosomes and genetic divergence. Introgressing broomrape resistance genes from wild annual *Helianthus* species into an elite germplasm can easily be accomplished by conventional crossing and backcrossing (Fernández-Martínez et al., 2008, 2009), by using the embryo culture technique, and subsequent chromosome doubling of the F_1 (Velasco et al., 2012). A resistant germplasm incorporating genes from over 20 wild sources has been developed from both annual (Jan and Fernandez-Martinez, 2002; Fernandez-Martinez et al., 2000, 2007, 2008; Hladni et al., 2009; Velasco et al., 2012; Jan et al., 2014) and perennial *Helianthus* species (Christov et al., 2009; Cvejic et al., 2012; Jan et al., 2000, 2002; Jan and Seiler, 2007; Velasco et al., 2007; Petcu and Pacureanu, 2011; Jan et al., 2014).

Sunflower breeders must adopt a breeding strategy to develop resistant inbred lines and use well-known conventional breeding methods and hybrids utilizing current resistant germplasm. After discovering resistance genes, Škorić (2012) suggested the best method for breeders would be to cross their own elite B and Rf lines with a source of Or genes, then incorporate them into their breeding material by means of recurrent crossbreeding, together with screening for resistance in all B and C generations, creating genetic variability for broomrape resistance, and other desired goals. Breeders must first determine the current races in the region and test their selection materials for broomrape resistance in infested fields. However, naturally infested field-testing is not reliable enough because the spread of infestations is not equal and can bring about experimental errors. A more reliable way is to test genetic material against resistance by mixing broomrape seed artificially in a greenhouse mostly during the winter period. Kaya et al. (2014) discussed a current artificial method in which the resistance of genetic material to broomrape was tested in soil artificially infested with mixed broomrape seeds collected from infested areas. In growth chamber tests, plastic cups were filled with 1–2 g of broomrape seeds (half a teaspoon in each cup, mixed in the lower half of the cup) and mixed soil, turf, and sand (in equal parts). Plants were removed from the cups after 35 days from planting, then their roots were washed and any attached broomrape tubers were counted. Labrousse et al. (2004) used hydroponic culture successfully for broomrape tests.

However, the most reliable and practical method of screening breeding materials for broomrape resistance is the use of molecular markers such as QTL, RFLP (restriction fragment length polymorphism), RAPD (random amplified polymorphic DNA), and SSR (simple sequence repeats) (Tang et al., 2003; Iouras et al., 2004; Marquez-Lema et al., 2008; Liu and Jan, 2012; Škorić, 2012; Pineda-Martos et al., 2013, 2014). Sunflower breeders should utilize genetic transformations for long-lasting solutions for broomrape resistance via intensive use of MAS. Fernandez-Martinez et al. (2008) indicated that while resistance is controlled by major genes, quantitative resistance and epistatic effects were also present. However, molecular analysis has proved that, while phenotypic variance for race E resistance can mainly be explained by a major QTL, the existence of F race resistance can be explained by several QTLs coupled with a small moderate effect mainly associated with the number of broomrapes per plant.

BREEDING FOR RESISTANCE TO ABIOTIC STRESSES

Sunflowers are more adaptable and more tolerant of various abiotic stresses than other crops, growing as they do under semiarid conditions in many areas of the world (Škorić, 2009). Abiotic stresses such as drought and mineral toxicities/deficiencies are limiting factors for the production of sunflowers and other crops and recent climate change will cause more problems in the future.

Sunflower Breeding for Resistance to Drought

Of the abiotic stresses, drought is the most restraining factor in sunflower production worldwide, severely reducing yield, oil volume, oil quality, and other important yield traits (Ebrahimi et al., 2008). Sunflowers are considered moderately tolerant to drought, growing as they do in hot semiarid climatic regions. The most important stage affecting sunflower growth and yield (due to drought) is the period from flowering to physiological maturity. Drought influences physiological and biochemical processes such as reduced water uptake, strongly decreased transpiration intensity, increased respiration intensity, and reduced energy efficiency of respiration (Fick and Miller, 1997; Chimenti et al., 2004; Breton et al., 2010). Škorić (2012) indicated that most physiological traits such as leaf water potential, leaf rolling, root growth, root xylem diameter, osmotic adjustment, stomatal conductance, abscisic acid (ABA) accumulation, canopy temperature, seedling establishment and growth, seedling recovery after stress, growth stress, and proline accumulation play important roles in drought tolerance after studying over 30 different parameters.

To develop drought-tolerant lines and hybrids in sunflower breeding, breeders need to be aware of the relationship between drought resistance traits and yield and apply effective screening methods for these traits. They then need to incorporate into the breeding material those characteristics contributing to physiological drought tolerance such as depth-efficient roots for more water uptake, higher osmotic pressure tolerance, lower transpiration rates, and higher ability to recover after wilting under heat stress. However, currently there have not been enough genetics studies into drought resistance because of drought resistance being controlled by many genes and, of course, the recent availability of molecular markers helping to determine drought tolerance in sunflowers properly (Fick and Miller, 1997; Rauf, 2008; Breton et al., 2010; Liu and Jan, 2012; Škorić, 2009, 2012).

To develop high-yielding and drought-resistant cultivars, genotypes should have such advantages as enhanced leaf area, earliness, and earlier stomatal closure. This will deliver the highest yields under dry conditions. Drought tolerance could also be achieved by avoiding drought by developing earlier or very early hybrids or arranging sowing that depends on starting dry period in summer season. However, earlier sunflower hybrids generally have a lower LAI, total evapotranspiration, and yield potential than later hybrids (Fick and Miller, 1997). Generally, later genotypes consume more water under drought, produce more dry matter, have greater water use efficiency, better drought tolerance, and higher yields than earlier genotypes (used to escape from drought). For drought tolerance selection in breeding programs, genotypes need to be developed with such attributes as reduced transpiration or increased water uptake from the soil (powerful roots). There are huge sources of wild *Helianthus* species and breeders have mainly used *H. argophyllus* as a source of donor genes for novel tolerant germplasm. Nitrate reductase activity, free proline accumulation rate, stress susceptibility index (SSI), stress tolerance index (STI), ABA, crop water stress index, single-leaf temperature, LAI, etc. could be used as parameters for evaluation of sunflower genotypes for drought tolerance (Fick and Miller, 1997; Orta et al., 2002; Chimenti et al., 2004; Farrokhi and Daneshian, 2004; Fulda et al., 2008; Rauf, 2008; Fernandez et al., 2009; Hussain et al., 2012; Škorić, 2009, 2012). Petcu et al. (2008) indicated that water stress caused reductions in leaf petiole length, shoot size, chlorophyll content, and seed yield.

Sunflower genetic material could be classified on performance under different stress conditions into four groups: genotypes with similar good performance in both stressed and nonstressed environments, genotypes with good performance only in nonstressed environments, and genotypes with weak performance in both environments (Fernandez et al., 2009). Ghaffari et al. (2012) indicated that STI and SSI were better at identifying drought-tolerant and sensitive lines. Moreover, plant height, stem and head diameter, leaf area, leaf temperature,

Fv/Fm, and SPAD value could all be used as physiological indicators to screen sunflower genotypes for drought under stressed conditions.

Sunflower breeders have to decide and plan proper parameters as well as breeding method for the desired objectives based on previous drought studies of morphological and physiological parameters. In addition to the above parameters, the stay-green trait – which is also a phenomenon of *Macrophomina* and *Phomopsis* resistance – is the most practical criterium for selection of lines with higher self-fertility. Sunflower breeders could use USDA inbred lines having stay-green traits such as HA-48, HA-22, CMS-1-50, PH-BC-2-91, PR-ST-3, RHA-SES, and RHA-583 for their breeding programs to develop drought-tolerant lines and hybrids (Škorić, 2009). Stay-green sunflower hybrids had such attributes as higher rate and quantum yield of photosynthesis in leaves under drought conditions and increased content of total soluble proteins (especially RuBisCo). This phenomenon provides improved adaptability and higher tolerance to drought, making it suitable for use for drought resistance in sunflower breeding. Moreover, molecular markers gave promising results recently in sunflower breeding for drought tolerance, but their focus should be recognizing the chromosomal segments carrying genes contributing to determination and providing the possibility to screen the tolerance character (Rauf, 2008; Škorić, 2009, 2012).

Sunflower Breeding for Tolerance to Salinity, Mineral Deficiency, and Toxicity

As a result of failing to compete with other more profitable crops such as wheat, barley, and canola in rainfed areas, and corn, sugar beet, and some legumes in irrigated areas, sunflower production has pushed into marginal areas in recent years. However, world vegetable prices, rapidly changing cultural practices such as no-till planting or weed management, and increasing the frequency and severity of abiotic stress factors due to recent global climate change also play important roles here. Therefore, based on recent developments, sunflower breeders should efficiently combine plant physiology with new tools and models to understand yield, stress traits, and yield relationships and in this way improve selection efficiency. Furthermore, they should utilize highly molecular genetic tools such as MAS and QTL to discover the many loci and genes responsible for yield in term of stress tolerance–related traits limiting yield (Castano and Giussani, 2009; Liu and Jan, 2012; Sala et al., 2012; Škorić, 2012; Kaya, 2014).

Tolerance to Salinity

Higher levels of soil salinity are serious limiting factors to sunflower production. However, sunflowers can tolerate salinity and grow successfully in low to medium saline soils in many countries. Soil salinity can be improved by the application of nitrogen and potassium fertilizers, but sunflower plants need to adopt different strategies to reduce the effect of salinity, such as the accumulation of different types of organic and inorganic osmolytes. Many wild *Helianthus* species naturally grow in saline soils. Therefore, they could be used by breeders as sources of genes for resistance to salinity by applying effective screening methods to identify effective breeding methods and transfer these genes into cultivated sunflower genotypes (Fernandez et al., 2009; Seiler, 2012; Škorić, 2012).

Based on previous studies, proline accumulation had a greater impact on tolerance to salinity associated with callus development, seed germination, and vigor and could be used as a more reliable index of salt tolerance (Prakash et al., 1996). Tests at the young-leaf stages of sunflowers for Na^+ and for the K/Na^+ ratio could be a good way of measuring salt tolerance for selecting sunflower genotypes against salinity tolerance because the increases in sodium chloride enhanced sodium and potassium uptake leading to successful maintenance of the K/Na ratio. Emergence percentage and index, shoot length, and shoot fresh weight could also be used as selection criteria for salt tolerance at the seedling stage. For rapid salt tolerance screening, radical length elongation during Petri dish trials could also be used to evaluate genotypes. Additionally, inbred lines with fresher, drier, and longer shoots and greater weight under nonsaline conditions exhibited higher tolerance; so, they could also be used as selection criteria for developing high-yielding lines for saline soils (Hussain et al., 1996; Fernandez et al., 2009; Škorić, 2012).

Hebbara et al. (2003) indicated that leaf temperature increased while osmotic potential, stomatal conductance, and transpiration rate decreased with increases in soil salinity. Khan et al. (2014) suggested that the application of proline spray to leaves noticeably alleviated the adverse effects of salt stress on sunflower growth by producing sugar and at the same time maintained ion and water availability by protecting sunflower photosynthesis against salt-induced oxidative stress.

Of the wild *Helianthus* species, *H. paradoxus* – which thrives in salt marshes – has the highest salt tolerance at 1300 mM, while normal sunflowers perished at 250–400 mM NaCl concentrations. Therefore, sunflower breeders could choose to transfer salinity-resistant genes from *H. paradoxus* directly or from interspecific sources originating from the pseudoautosomal region (PAR) code. The ARG-1575, GIG-1616, TUB-1709, PAR-1673, HA-429, and HA-430 inbred lines from USDA sources could of course also use other wild *Helianthus* spp. in the breeding programs. Cell survival, seed germination, dry matter accumulation, leaf death or senescence, leaf ion content, leaf necrosis, root growth, osmotic regulations, etc. could also be used as other selection criteria for salinity studies (Singh, 2000; Lexer et al., 2004; Fernandez et al., 2009; Seiler, 2012; Škorić, 2012).

Tolerance to Mineral Deficiency and Mineral Toxicity

For better growth and plant development, sunflowers only need 10 macro elements (C, O, H, N, P, K, S, Ca, Fe, and Mg) and six trace elements (B, Mn, Cu, Zn, Mo, and Co); all other elements are taken up from the soil or from fertilizer applications. Unfortunately, no breeding program has been organized anywhere in the world that focuses on sunflower resistance to mineral deficiency and/or mineral toxicity (Fernandez et al., 2009; Seiler, 2012; Škorić, 2012).

Sunflowers are highly sensitive field crops to low levels of boron (B), which plays an important role in transporting nutrients from the roots to upper parts of the plant. B deficiency symptoms are first noticed in younger leaves taking on a bronze color and become hardened, malformed, and necrotic; another symptom is a malformed capitulum (Fick and Miller, 1997; Ceyhan et al., 2008). Boron sensitivity is subject to change by genotypes; so, sunflower breeders should look for these symptoms in their genetic material and evaluate them especially at times of water deficit and under poor soil conditions.

Sunflowers are very sensitive to soil acidity and to higher levels of aluminum, which inhibits growth by blocking the mechanisms involved in water and nutrient absorption and transport in sunflowers. In their greenhouse study in Brazil, Castro et al. (2012) indicated that the dry weight of roots of sunflower genotypes was significantly affected by the level of aluminum in solution and observed the different responses of sunflower genotypes to root development under different concentrations of aluminum.

Tolerance to Hot Temperatures

Sunflowers and other crops differ in their responses to temperature change for normal plant growth and development, depending on the growth stage and genotypes. Sunflowers demonstrate high adaptability to higher temperatures by increasing transpiration, which keeps the leaves relatively cool. However, the transpiration rate can only be increased if enough water is supplied by a deep and well-developed root system. Therefore, such a root type, tolerance to intensive transpiration, increased capacity of a plant to produce more pollen, seed filling rate, and rapid synthesis of oil under hot conditions are important criteria for selecting sunflower tolerance to high temperatures (Fernandez et al., 2009; Seiler, 2012; Škorić, 2012).

Seiler (2012) indicated that *H. anomalus* Blake and *H. deserticola* Heiser (wild annual species) are highly adapted to desert and sandy areas and harsh habitats in the United States. So, these two and some other wild species such as *H. nuttallii* and *H. petiolaris* could be used as germplasm for heat stress studies. Sunflower breeders should choose a strategy, parameters, and germplasm that are suitable for selecting resistance to high temperatures as well as air and soil drought in their breeding programs to develop tolerant lines and hybrids (Fick and Miller, 1997; Rauf, 2008; Fernandez et al., 2009; Škorić, 2009, 2012).

Tolerance to Cold Temperatures

Low temperatures significantly increase protoplasm viscosity and inhibit the intensity of the metabolism and decomposition of proteins in sunflowers. Sunflowers mainly encounter cold temperatures at early stages of development such as germination, emergence, and the 2–3 leaf stage, especially in early plantings and during the maturation period of sunflower production at higher altitudes. Chlorophyll content and specific leaf area have been genetically associated with cold tolerance; so, they could be used as selection criteria in sunflower breeding programs. QTL analyses have also illustrated that several putative genomic regions are involved in the variation of physiological traits studied under cold conditions (Fernandez et al., 2009; Škorić, 2012; Sala et al., 2012).

Kalaidzhyan et al. (2007, 2009) applied induced mutations using chemical mutagens (dimethyl sulfate, DMS) to develop cold-resistant genotypes, and some mutagenic progenies survived harsh winter temperatures of –20°C. Sunflower breeders could avail themselves of such a unique germplasm as well as *Helianthus* wild species, especially those that grow in the mountains where winters are harsh and springs are cold (Škorić, 2009).

SUNFLOWER BREEDING FOR HERBICIDE TOLERANCE

Significant results havee recently been achieved in sunflower breeding for tolerance to imidazolinone (IMI) and sulfonylurea (SU) herbicides. Herbicide tolerance is brought about by inhibiting acetolactate synthase (ALS), also called acetohydroxyacid synthase (AHAS), an enzyme that biosynthesizes branched chain amino acids and causes the herbicidal effect that metabolizes herbicides in their inactive form. Using herbicide-resistant cultivars gave sunflower farmers the opportunity to control broadleaf weeds, a major problem that has greatly increased recently in sunflower production. IMI-type sunflower hybrids are more common in such countries as Turkey and Russia due to the effective control of broomrape and key weeds such as *Xanthium* and *Cirsium*. However, SU-type hybrids are preferred in such countries as Hungary and Romania, because they are less expensive than IMI herbicides and broaden the control spectrum of weeds, especially in nonbroomrape and nonproblematic *Xanthium* areas. However, extensive use and the genetic mutability trait bring some risks such as the escape of resistant genes to wild types, crossresistance, and residue problems (Kaya and Evci, 2007; Sala et al., 2012; Škorić, 2012; Kaya et al., 2013).

TOLERANCE TO IMIDAZOLINONES

The ability of imidazolinone herbicides including imazapyr, imazapic, imazethapyr, and imazamox to induce resistance in the wild *H. annuus* population was first identified in a soybean field treated with the herbicide for seven consecutive years in Kansas in 1996, and then in South Dakota (Al-Khatib et al., 1998; Al-Khatib and Miller, 2000; Evci et al., 2012). These resistance genes were transferred into cultivated sunflowers by a USDA research team via backcrossing and embryo culture. This was followed by the release of public IMISUN lines in 1998 (Al-Khatib and Miller, 2000). At the moment, all public institutes and several private companies in the world have incorporated IMI resistance into their elite lines and developed IMI-resistant hybrids (Al-Khatib 1998; Miller and Al-Khatib, 2000; Sala et al., 2012; Škorić, 2012; Kaya et al., 2013).

IMI resistance is controlled by two genes: a major gene with a semidominant type of gene action (Imr1), and a second gene (Imr2) with a modifier or enhancer effect. The development of resistant sunflower hybrids depends on both inbred lines having IMI-resistant genes (Bruniard and Miller, 2001; Sala et al., 2008a, b, 2012; Škorić, 2012). Škorić (2012) advised that, to develop IMI-type inbred lines, after breeding backcross (BC) generations breeders should use a female plant as the recurrent parent, the male as the backcross, and then the pollen donor plants (BC) should be selfed by applying a double-dose of IMI herbicide at the 4–6 leaf stage. Fifteen days after treatment, while screening and selecting genotypes phenotypically, injured or killed plants were found not to be resistant, heterozygous plants were more yellow or less injured, and resistant plants demonstrated a reduced yellow flash or green flash. The introduction of genes for resistance to IMI from resistant genotypes into elite B or Rf lines should be done by backcrossing accompanied with repeated resistance screening and elimination of sensitive and yellow-flash plants (Malidza et al., 2000, 2002; Bruniard and Miller, 2001; Kaya and Evci, 2007; Sala et al., 2012; Škorić, 2012; Kaya et al., 2013).

Sala et al. (2008a, b) reported obtaining a new source of IMI resistance, CLHA-PLUS, developed by induced mutations (ethyl methanesulfonate mutagenesis) and selection with imazapyr herbicide. CLHA-PLUS is different from Imrl at the molecular level and is controlled by expression of the partially dominant nuclear allele AHAS1–3. The CL-Plus gene has higher IMI resistance than the Imr1 and Imr2 genes and higher oil content in hybrids due to the absence of *Ahasl1–3* genes from a wild source, which determines the oil content in sunflowers. Nowadays, both IMI-type hybrids are selling in the market (widely known as the Clearfield

System, a trademark of BASF Co.) (Kaya and Evci, 2007; Sala et al., 2012; Škorić, 2012; Kaya et al., 2013).

Imidazolinone herbicides control a broad spectrum of grass and broadleaf weeds such as *Xanthium strumarium*, *Sinapis arvensis*, *Chenopodium album*, *Cirsium arvense*, *Convolvulus arvensis*, *Avena* spp., *Datura stramonium*, and *Amaranthus* spp. successfully by postemergence application in sunflower production in the world. IMI herbicides Imazamox + Imazapyr (33 + 15 g L^{-1}) were registered by BASF Co. to control both *Orobanche* and key weeds in sunflower production in Turkey (Kaya et al., 2004). However, Imazapyr was recently banned by the EU as it led to residue problems for crops following it in the rotation; so, only Imazamox (40 g L^{-1}) was selling for that purpose. Since it controls broomrape and major problem weeds, the planting areas of IMI-resistant hybrids increased rapidly in Turkey and, in recent years, in all Balkan countries, Russia, and the Ukraine. However, sunflower hybrids with IMI Plus genetical broomrape-resistant genes have recently been developed. Hence its takeup rate in the market has increased because it gives farmers the option to apply herbicides (or not) based on the weed population in their fields (Kaya et al., 2013, 2014).

TOLERANCE TO SULFONYLUREAS

The first SU-resistant genes were discovered in the same wild Kansas population. They were transferred into cultural types via the classical backcrossing method in much the same way as done for IMI resistance. Then, the first SU-resistant lines were developed and released by USDA as SURES in 2001 (Miller and Al-Khatib, 2004). On the other hand, SU-resistant genetic material was also found in native *H. annuus* and *H. petiolaris* populations collected in Colorado, Kansas, Nebraska, South Dakota, and North Dakota. Of the accessions tested, 57% exhibited resistance to tribenuron. More resistance to tribenuron (belonging to SU herbicide group) was found in populations collected in Colorado, Kansas, and Nebraska, and later in North and South Dakota (Olson et al., 2004). These results clearly indicate that tribenuron resistance in wild sunflowers is not a new trait. This resistance was widespread among accessions collected across Canada, but it occurred in low frequencies in those accessions. Resistance occurred in the environment before releasing SU-tolerant sunflower hybrids for commercial production in Canada. Jocic et al. (2008) mentioned that sunflower resistance to tribenuron-methyl is controlled by a single dominant gene from the SURES-1 and SURES-2 populations.

Another SU-resistant gene was developed by Dupont using chemical mutagenesis and commercial sunflower hybrids released by Pioneer Seed Co. in the early 1990s. To develop SU-resistant inbred lines and hybrids, breeders should introduce resistant genes into elite lines via backcrossing and screen resistance in all BC generations, similar to the IMI procedure. Although SU resistance is controlled by a single dominant gene, both parents need to have the SU gene applied to escape practical problems and for better results. The general preference for SU herbicides in sunflower production could be due to their being less expensive than IMI herbicides and their ability to broaden the control spectrum of weeds, especially in nonbroomrape problem areas. Therefore, all SU-resistant hybrids in the market were instilled with genetical broomrape resistance (Kaya and Evci, 2007; Fernandez et al., 2009; Seiler, 2012; Sala et al., 2012; Škorić, 2012; Kaya et al., 2013).

SUNFLOWER BREEDING FOR SUSTAINABLE PRODUCTION

Sunflowers are one of the most important oil crops in the world and mostly grow in dry areas. As a spring crop, sunflowers are more susceptible to climatic change and yield is so variable as a result of critical vegetation periods starting under hot summer conditions. Sunflowers are more profitable when grown under irrigated conditions (Semerci et al., 2011). However, when grown under rain-fed conditions they also need to be tolerant to seasonal droughts to compete with other crops in the rotation. Nevertheless, sunflower production is being pushed year on year into marginal areas because it is less profitable than other crops (mainly wheat) and cannot compete with them on price or productivity. Sunflowers are highly adaptable crops and grow in different environments. Hence the expectation that new developed hybrids will exhibit improved performance overtime, under different conditions (climate, soils, etc.), and at various locations. With this in mind, sunflower breeders should first familiarize themselves with the adaptation capabilities of the current genetic material held at their nurseries as well as the adaptation mechanisms involved, and then get more information about the environmental factors of the major regions of sunflower cultivation. To develop adaptability strategies for future hybrids, sunflower breeders should decide in advance the types of hybrids they will develop (i.e., adapted to poor or rich conditions, or to all environments). Therefore, sunflower breeders should test their hybrids in yield trials conducted in different locations and years, analyze the many yield traits using a proper statistical method, and then interpret the results correctly. This should be done by considering the types of G × E interactions, which will help them choose the most productive and most adaptable hybrids (Fick and Miller, 1997; De la Vega, 2004; Škorić and Jocic 2004; De la Vega et al., 2007; Škorić, 2012).

Triboi et al. (2004) pointed out that total leaf area, residual leaf area, and leaf nitrogen content are heritable traits, and that residual leaf area could be a criterion for evaluating sunflower hybrid adaptations. De la Vega et al. (2004, 2007) examined many hybrids in many locations in Argentina and concluded that oil content was highly positively correlated with oil yield and that harvest index was a major factor affecting seed yield. Therefore, harvest index could also be used as an evaluation criterion for adaptation of sunflower hybrids. They also indicated that large environments should be divided into subregions to find the most adaptable hybrids and the female lines that play the major roles in determining oil yield traits. Moreover, stability parameters could be used as evaluation tools for breeders determining seed yield and comparing candidate sunflower hybrids (Balalic, 2010; Škorić, 2012).

Sunflower yield increased drastically after the discovery of CMS and restorer lines and the appearance of hybrids in oil production mainly in the 1980s. However, sunflower yield has remained stable in the last 20 years despite many new cultivars being developed and introduced to market in this period. At the same time, corn yield has doubled since 1990 as a result of new hybrids being developed. Sunflower breeders should investigate this lack of heterosis in the genetic background and reanalyze plant physiology and anatomy. This difference between corn and sunflower yield could be the result of using the same CMS sources to obtain sunflower hybrids. Therefore, there needs to be a new design for sunflower plants, one that utilizes land races, interspecific hybrids, and wild species.

Sunflowers have greater potential than corn because the *Helianthus* genus is very large and has 51 subspecies that hold very valuable traits for breeders. Seed yield is mostly related to three main attributes: plant number per hectare, seeds per plant, and seed weight. Although, physiologically, these traits are mostly related to higher photosynthesis, nutrient uptake, water use efficiency, etc., anatomically these traits include larger leaf area, larger flower heads, increased plant height, and longer seeds. Based on these parameters, with the exception of plant number, plant biomass can be increased by developing plants with larger leaf areas, increased height, leaves with fewer petioles, and short plant angles (thus reducing the diameter of plants). Then, the number of plants per hectare could be increased without decreasing leaf area. *H. divaricatus*, *H. californicus*, *H. floridanusi*, and *H. maximiliani* meet these criteria in having very suitable, narrow, longer leaves and shorter petioles.

Apart from seeds per plant, by having larger heads cultivated hybrids are less affected by stress at critical periods, so drought and stress tolerance genes could be transferred from *H. deserticola*, *H. anomalus*, etc. When it comes to seed weight, much like the previous trait, drought and stress-tolerant lines are less affected during the seed-filling period and yield increased seed weight. Moreover, by utilizing wild types and land races seed number per head should not decrease and seed length should increase (with reduced husk content). Of course, higher oil content is another factor in increased seed weight. Some OP varieties and wild species such as *H. niveus* and *H. salicifolius* are very valuable sources for this. As a result, armed with a new plant design and increased leaf area, new hybrids could achieve higher heterosis in seed yield.

References

Akhtouch, B., Munoz-Ruiz, J., Melero-Vara, J., Fernandez-Martinez, J.M., Dominguez, J., 2002. Inheritance of resistance to race F of broomrape (*Orobanche cumana* Wallr.) in sunflower lines of different origin. Plant Breed. 121 (3), 266–268.

Alignan, M., Hewezi, T., Petitprez, M., Dechamp-Guillaume, G., Gentzbittel, L., 2006. A cDNA microarray approach to decipher sunflower (*Helianthus annuus* L.) responses to the necrotrophic fungus *Phoma macdonaldii*. New Phytol. 170, 523–536.

Al-Khatib, K., Miller, J.F., 2000. Registration of four genetic stocks of sunflower resistant to imidazolinone herbicides. Crop Sci. 40, 869–870.

Al-Khatib, K., Beungurtner, J.R., Peterson, D.E., Currie, R.S., 1998. Imazethapyr resistance in common sunflower (*Helianthus annuus* L.). Weed Sci. 46, 403–407.

Antonova, T.S., Araslanova, N.M., Guchetl, S.Z., Tchelustnikova, T.A., Ramazanova, S.A., Trembak, E.N., 2009. Virulence of sunflower broomrape (*Orobanche cumana* Wallr.) in some regions of Northern Caucasus. Helia 32 (51), 101–110.

Arias, M.C., Lenardon, S., Taleisnik, E., 2003. Carbon metabolism alterations in sunflower plants infected with sunflower chlorotic mottle virus. J. Plant Pathol. 151, 267–273.

Arias, M.C., Luna, C., Rodriguez, M., Lenardon, S., Taleisnik, E., 2005. Sunflower chlorotic mottle virus in compatible interactions with sunflower: ROS generation and antioxidant response. Eur. J. Plant Pathol. 113, 223–232.

Balalić, I., Crnobarac, J., Dušanić, N., 2010. Interaction hybrid planting date for oil yield in sunflower. J. Agric. Sci. 55 (1), 9–16.

Baldini, M., Turi, M., Vischi, M., Vannozzi, G.R., Olivieri, A.M., 2002. Evaluation of genetic variability for *Sclerotinia sclerotiorum* (Lib.) de Bary resistance in sunflower and utilization of associated molecular markers. Helia 25 (36), 177–190.

Bán, R., Kovács, A., Perczel, M., Körösi, K., Turóczi, G., 2014. First report on the increased distribution of Pathotype 704 of *Plasmopara halstedii* in Hungary. Plant Dis. 98 (6), 844–1844.

Bazzalo, M.E., Giolitti, F., Galella, M.T., Leon, A., Lenardon, S., 2008. Pathological and morphological evaluation of sunflower isohybrids carrying or not the Rem-1 gene for sunflower chlorotic mottle virus resistance. In: Proceedings 17th International Sunflower Conference. Cordoba, Spain. pp. 205–210, Vol. 1.

Bejerman, N., Giolitti, F., Lenardon, S., 2008. Molecular characterization of a novel sunflower chlorotic mottle virus (SuCMoV) strain. Proceedings 17th International Sunflower Conference, Cordoba, Spain, Vol. 1, pp. 137–142.

Bejerman, N., Giolitti, F., de Breuil, S., Lenardon, S., 2010. Molecular characterization of Sunflower chlorotic mottle virus: a member of a distinct species in the genus *Potyvirus*. Arch. Virol. 155, 1331–1335.

Bert, P.F., Jouan, I., Tourvieille de Labrouhe, D., Serre, F., Nicolas, P., Vear, F., 2002. Comparative genetic analysis of quantitative traits in sunflower (*Helianthus annuus* L.). 1. Characterisation of QTL involved in resistance to *Sclerotinia sclerotiorum* and *Diaporthe helianthi*. Theor. Appl. Genet. 105, 985–993.

Breton, C., Serieys, H., Bervillé, A., 2010. Gene transfer from wild *Helianthus* to sunflower: topicalities and limits. Oleagineux 17, 104–114.

Breton, C., Gil, A., Wargnier, J., Serieys, H., Bervillé, A., 2012. Transfer of architectural traits from perennial *Helianthus mollis* Lam. to sunflower (*H. annuus* L.) and localisation of introgression. Euphytica 186, 552–557.

Bruniard, J.M., Miller, J.F., 2001. Inheritance of imidazolinone-herbicide resistance in sunflower. Helia 24 (35), 11–16.

Castano, F., Giussani, M.A., 2009. Effectiveness of components of partial resistance in assessing white rot of sunflower head. Helia 32 (50), 59–68.

Castano, F., Vear, F., Tourvieille de Labrouhe, D., 2001. The genetics of resistance in sunflower capitula to *Sclerotinia sclerotiorum* measured by mycelium infections combined with ascospore tests. Euphytica 122, 373–380.

Castro, C.D., Morales, L.A.C., Oliveira Junior, A.D., Leite, R.M.V.B.C., Oliveira, F.A.D., Rodak, B.W., 2012. Sunflower Tolerance to Aluminum. Proceedings of International Sunflower Conference, February 26–29 Mar del Plata. 612–615.

Ceyhan, E., Onder, M., Oztürk, O., Harmankaya, M., Hamurcu, M., Gezgin, S., 2008. Effects of application boron on yields, yield component and oil content of sunflower in boron-deficient calcareous soils. Afr. J. Biotechnol. 7 (16), 2854–2861.

Chimenti, C., Giuliano, J., Hall, A., 2004. Osmotic adjustment, its effects on yield maintenance under drought in sunflower. In: Proceedings 16th International Sunflower Conference. Fargo, ND. pp. 261–267, August 29–September 2, Vol. 1.

Christov, M., Batchvarova, R., Hristova-Cherbadzhi, M., 2009. Wild species of *Helianthus* L. Sources of resistance to the parasite *Orobanche cumana* Wallr. Helia 51, 65–74.

Cvejic S., Dedić, B., Jocić, S., Miladinović, D., Miklič, V., 2012. Broomrape resistance in newly developed sunflower inbred lines. In: Proceedings of International Sunflower Conference, Mar del Plata, Argentina. International Sunflower Association, Paris, France, 1037–1042.

Darvishzadeh, R., Dechamp-Guillaume, G., Hewezi, T., Sarre, A., 2007. Genotype–isolate interaction for resistance to black stem in sunflower (*Helianthus annuus*). Plant Pathol. 56, 654–660.

De la Vega, A.J., 2004. Does subdivision of the sunflower growing region of Argentina increase the response to selection. In: Proceedings Sixteenth International Sunflower Conference. Fargo, ND. pp. 459–464, August 29–September 2, Vol. 2.

De la Vega, A., Delacy, I.H., Chapman, S.C., 2007. Changes in agronomic traits of sunflower hybrids over 20 years of breeding in central Argentina. Field Crops Res. 100 (1), 73–81.

De Oliveira, M.E., Neto, A.T., Leite, R.M.V.B.C., Castiglioni, V.B.R., Arias, C.A.A., 2004. Mutation breeding in sunflower for resistance to *Alternaria* leaf spot. Helia 27 (41), 41–50.

Demurin, Y., Škorić, D., Karlovic, D., 1996. Genetic variability of tocopherol composition in sunflower seeds as basic of breeding for improved oil quality. Plant Breed. 115, 33–36.

Dnoseva, M., Nenov, N., Kiryakov, I., 2008. Study on an in vitro screening test for resistance to *Sclerotinia sclerotiorum* in sunflower. In: Proceedings Seventeenth International Sunflower Conference. Cordoba, Spain. pp. 181–186, June 8-12, Vol. 1.

Dozet, B., Škorić, D., Jovanovic, D., 2000. Sunflower breeding for resistance to broomrape (*Orobanche cumana* Wallr.). In: Proceedings Fifteenth International Sunflower Conference. Toulouse, France. pp. 320–325, June 12-15, Vol. 2.

Dujovny, G., Sasaya, T., Koganesawa, H., Usugi, T., Shahara, K., Leanordan, S., 2000. Molecular characterization of a new potyvirus innfecting potyvirus. Arch. Virol. 145, 2249–2258.

Ebrahimi, A., Maury, P., Berger, M., Shariati, F., Grieu, P., Sarrafi, A., 2008. Genetic improvement of oil quality in sunflower mutants under water stressed conditions. In: Proceedings Seventeenth International Sunflower Conference. Cordoba, Spain. pp. 509–512, June 8-12, Vol. 2.

Escande, A., Pereyra, V., Glares, K., 2006. Stability of sunflower resistance to *Verticilium wilt*. In: Proceedings Fifteenth International Sunflower Conference. Toulouse, France. pp. 102–107, June 12-15, Vol. 2.

Evci, G., Pekcan, V., Yılmaz, M.I., Kaya, Y., 2012. The resistance breeding to IMI and SU herbicides in sunflower. Plant Sci. 49, 6–11.

Farrokhi, E., Daneshian, J., 2004. Evaluation of single and three-way cross sunflower hybrids for drought tolerance. In: Proceedings Sixteenth International Sunflower Conference. Fargo, ND. pp. 481–487, August 29–September 2, Vol. 2.

Fernandez, M.J., Perez-Vich, B., Velasco, L., 2009. Sunflower. Vollmann, J., Rajcan, I. (Eds.), Oil Crops, Handbook of Plant Breeding, 4, Springer, New York, pp. 155–232.

Fernandez, J.M., Perez-Vich, B., Velasco, L., Dominguez, J., 2007. Breeding for speciality oil types in sunflower. Helia 30 (46), 75–84.

Fernandez, J.M., Dominguez, J., Perez-Vich, B., Velasco, L., 2008. Update on breeding for resistance to sunflower broomrape. Helia 31 (48), 73–84.

Fernández, J.M., Domínguez, J., Pérez-Vich, B., 2010. Update on breeding for resistance to sunflower broomrape. Helia 33 (52), 1–12.

Fernández, J.M., Melero-Vara, J.J., Muñoz-Ruz, J., Ruso, J., Domínguez, J., 2000. Selection of wild and cultivated sunflower for resistance to a new broomrape race that overcomes resistance to *Or5* gene. Crop Sci 40, 550–555.

Fernández, J.M., Velasco, L., Pérez-Vich, B., 2012. Progress in research on breeding for resistance to sunflower broomrape. Helia 35 (57), 47–56.

Fick, G.N., Miller, J.F., 1997. Sunflower breeding. In: Schneiter, A.A. (Ed.), Sunflower technology and production, vol. 35, Agronomy monograph. ASA, CSSA, SSSA, Madison, WI, pp. 395–439.

Fulda, S., Stegmann, H., Horn, R., 2008. Influence of drought stress on growth, protein expression and osmolyte accumulation in sunflower *Helianthus* annuus L. c.v. Peredovik. In: Proceedings of Seventeenth International Sunflower Conference. Cordoba, Spain. pp. 357–362, June 8-12, Vol. 1.

Ghaffari, M., Toorchi, M., Valizadeh, M., Shakiba, M.R., 2012. Morpho-physiological screening of sunflower inbred lines under drought stress condition. Turk. J. Field Crops 17 (2), 185–190.

Goncharov, S., 2009. Sunflower breeding for resistance to the new broomrape race in the Krasnodar region of Russia. Helia 32 (51), 75–80.

Gontcharov, S.V., Antonova, T.S., Saukova, S.L., 2006. Sunflower breeding for resistance to *Fusarium*. Helia 29 (45), 49–54.

Gorbachenko, F.I., Usatenko, T.V., Gorbachenko, O.F., 2011. Results of Sunflower Breeding in Resistance to Broomrape on Don. Helia 34, 9–18.

Gulya, T., 2007. New strain of *Verticilium dahlia* in North America. Helia 30 (47), 115–120.

Gulya, T.J., Shiel, R.J., Freeman, T., Isakeit, T., Berger, P.H., 2002. Host range and characterization of sunflower mosaic virus. Phytopathology 92, 694–702.

Gulya, T., Radi, S., Balbyshev, N., 2008. Large scale field evaluation for Sclerotinia stalk rot resistance in cultivated sunflower. In: Proceedings Seventeenth International Sunflower Conference. Cordoba, Spain. pp. 175–180, June 8-12, Vol. 1.

Hahn, V., 2002. Genetic variation for resistance to *Sclerotinia* head rot in sunflower inbred lines. Field Crops Res. 77, 153–159.

Hladni, N., Jocić, S., Miklič, V., Mijić, A., Saftić-Panković, D., Kraljević-Balalić, M., 2008. Direct and indirect effects of morphophysiological traits on oil yield of sunflower (*Helianthus annuus* L.). In: Kobiljski, B. (ed.), Proceedings of First International Conference on Conventional and Breeding of Field and Vegetable Crops, Novi Sad, Serbia, 1, 491–494.

Hladni, N., Jocić, S., Miklič, V., Saftić-Panković, D., Škorić, D., 2009. Using new R inbred lines originating from an interspecific population with *H. deserticola* for development of sunflower hybrids resistant to broomrape. Helia 32 (51), 81–90.

Hladni, N., Škorić, D., Kraljević-Balalić, M., 2003. Components of phenotypic variability for head diameter in sunflower (*Helianthus annuus* L.). Genetika 35 (2), 67–75.

Hladni, N., Skoric, D., Kraljevic-Balalic, M., Ivanovic, M., Sakac, Z., Jovanovic, D., 2004. Correlation of yield components and seed yield per plant in sunflower (*Helianthus annuus* L.). Proceedings of Sixteenth International Sunflower Conference, 491–496.

Henson, B., Gulya, T., 2004. Screening sunflower for Sclerotinia head rot. In: Proceedings Sixteenth International Sunflower Conference. Fargo, ND. pp. 141–146, August 29–September 2, Vol. 1.

Höniges, A., Wegmann, K., Ardelean, A., 2008. Orobanche resistance in sunflower. Helia 31 (49), 1–12.

Hussain, M.K., Rehman, O.U., Rakha, A., 1996. Breeding sunflower for salt tolerance: interrelationship of morpho-physiological parameters in sunflower (*Helianthus annuus* L.) for salt tolerance. Helia 19 (25), 119–132.

Hussain, S., Ali, A., Ibrahim, M., Saleem, M.F., Alias-Haji, M.A., Bukhsh, A., 2012. Exogenous application of abscisic acid for drought tolerance in sunflower (*Helianthus annuus* L.): a review. J. Animal Plant Sci. 22 (3), 806–826.

Iouras, M., Stancin, D., Cinca, M., Nastase, D., Geronzi, F., 2004. Preliminary studies related to the use of marker assisted selection for resistance to *Orobanche cumana* Wallr. in sunflower. Romanian Agric. Res. 21, 33–37.

Jan, C.C., Fernandez-Martinez, J.M., 2002. Interspecific hybridization, gene transfer, and the development of resistance to the broomrape race F in Spain. Helia 25 (36), 123–136.

Jan, C.C., Seiler, G.J., 2007. Sunflower. Singh, R.J. (Ed.), Genetic Resources, Chromosome, Engineering, and Crop Improvement, Vol. 4, Oilseed Crops. CRC Press, Boca Raton, FL, pp. 103–165, Vol. 4(5).

Jan, C.C., Liu, Z., Seiler, G.J., Velasco, L., Perez-Vich, B., Fernandez-Martinez, J., 2014. Broomrape (*Orobanche cumana* Wallr.) resistance breeding utilizing wild *Helianthus* species. Helia 37 (61), 141–150.

Jocic, S., Miklic, V., Malidza, G., Hladni, N., Gvozdenovic, S., 2008. New sunflower hybrids tolerant of tribenuron-methyl. Proceedings Seventeenth International Sunflower Conference, Cordoba, Spain, June 8–12, Vol. 2, pp. 505–508.

Kaya, Y., 2010. Public sunflower breeding in Turkey and future directions. Plant Sci. (Rastenievudni Naouki) 47 (1), 7–13.

Kaya, Y., 2014. Sunflower. Pratap, A. (Ed.), Alien Gene Transfer in Crop Plants, 2, Springer Press, pp. 281–315.

Kaya, Y., Evci, G., 2007. Herbicide resistance in sunflower (*Helianthus annuus* L.). International Research Conference, Plant Genetic Stocks – The Basis of Agriculture of Today, June 13–14, Plovdiv, Bulgaria, Vol. 2, pp. 45–47.

Kaya, Y., Evci, G., Peckan, V., Gucer, T., 2004. Determining new broomrape-infested areas, resistant lines and hybrids in Trakya region of Turkey. Helia 27 (40), 211–218.

Kaya, Y., Evci, G., Peckan, V., Gucer, T., Yilmaz, M.I., 2009. Evaluation of broomrape resistance in sunflower hybrids. Helia 32 (51), 161–169.

Kaya, Y., Jocic, S., Miladinovic, D., 2012. Sunflower. Gupta, S.K. (Ed.), Technological Innovations in Major World Oil Crops, 1, Springer Press, New York, pp. 85–129.

Kaya, Y., Evci, G., Pekcan, V., Yilmaz, I.M., 2013. Clearfield technology in sunflower and developing herbicide resistance sunflower hybrids. Soil–Water J. 2 (2), 1713–1720.

Kaya, Y., Evci, G., Pekcan, V., Yilmaz, M.I., 2014. Broomrape resistance breeding in sunflower: a case study in Turkey. Proceedings of the Third International Symposium on Broomrape (*Orobanche* spp.) in Sunflower, June 3–6, Cordoba, New York, pp. 194–199.

Khan, A., Iqbal, I., Ahmad, I., Nawaz, H., Nawaz, M., 2014. Role of proline to induce salinity tolerance in sunflower (*Helianthus annuus* L.). Sci. Tech. Dev. 33 (2), 88–93.

Kinman, M.L., 1970. New developments in the USDA and state experiment station sunflower breeding programs. Proceedings of Fourth International Sunflower Conference, 181–183.

Labrousse, P., Arnaud, M.C., Griveau, Y., Fer, A., Thalouarn, P., 2004. Analysis of resistance criteria of sunflower recombined inbred lines against *Orobanche cumana* Wallr. Crop Prot. 23, 407–413.

Langar, K., Griveau, Y., Serieys, H., Kaan, F., Berville, A., 2004. Mapping components of resistance to phomopsis (*Diaporthe helianthi*) in a population of sunflower recombinant inbred lines. Proceedings Sixteenth International Sunflower Conference, Fargo, ND, August 29–September, Vol. 2, pp. 643–649.

Leclercq, P., 1969. Une sterilité male cytoplasmique chez le tournesol. Ann. Amelior. Plantes 19 (2), 99–106.

Lenardon, S., 2008. Rao, Khurana, S.M.P., Lenardon, S.L. (Eds.), Sunflower Chlorotic Mottle Virus. Characterization, Diagnosis and Management of Plant Viruses. Industrial Crops, 1, Studium Press LLC, Houston, TX, pp. 349–359.

Lenardon, S.L., Giolitti, F., León, A., Bazzalo, M.E., Grondona, M., 2001. Effects of sunflower chlorotic mottle virus infections on sunflower yield parameters. Helia 2435, 55–66.

Lenardon, S.L., Bazzalo, M.E., Abratti, G., Cimino, C., Galella, M.T., Grondona, M., Giolitti, F., León, A.J., 2005. Screening sunflower for resistance to sunflower chlorotic mottle virus and mapping the Rcmo-1 resistance gene. Crop Sci. 45, 735–739.

Lexer, C., Lai, Z., Rieseberg, L.H., 2004. Candidate gene polymorphisms associated with salt tolerance in wild sunflower hybrids: implications for the origin of *Helianthus paradoxus*, a diploid hybrid species. New Phytol. 161, 225–233.

Liu, Z., Jan, C.C., 2012. In: Kovacevic, Z., Škorić, D., Sakac, Z. (Eds.), Molecular techniques for sunflower breeding. Sunflower Genetics and Breeding: International Monogram. Serbian Academy Science, Beograd, Serbia, pp. 355–441.

Liu, Z., Gulya, T.J., Seiler, G.J., Vick, B.A., Jan, C.C., 2012. Molecular mapping of the *Pl16* downy mildew resistance gene from HA-R4 to facilitate marker-assisted selection in sunflower. Theor. Appl. Genet. 125, 121–131.

Madhavi, K.J., Sujatha, M., Raja Ram Reddy, D., Chander Rao, S., 2005. Culture characteristics and histological changes in leaf tissues of cultivated and wild sunflowers infected with *Alternaria helianthi*. Helia 28, 1–12.

Mailo, D., Poverene, M., Giolitti, F., Lenardon, S., 2008. Differential gene expression in SuCMoV – tolerant and susceptible sunflower lines. Proceedings Seventeenth International Sunflower Conference, Cordoba, Spain, June 8–12, Vol. 2, pp. 635–640.

Malidza, G., Škorić, D., Jocic, S., 2000. Imidazolinone resistant sunflower (Helianthus annuus L.): inheritance of resistance and response towards selected sulfonyl urea herbicides. Proceedings Fifteenth International Sunflower Conference, Toulouse, France, June 12–15, Vol. 2, pp. 42–47.

Malidza, G., Jocić, S., Škorić, D., Orbović, B., 2002. Control of weeds and *Orobanchecernua* in imidazolinone-resistant sunflower. Twelvth Symposium on Plant Protection and Herbicide Application, Zlatibor, Sebia. 84.

Marquez-Lema, A., Delavault, Ph., Letousey, P., Hu, J., Perez-Vich, B., 2008. Candidate gene analysis and identification or TRAP and SSR markers linked to the Or5 gene, which confers sunflower resistance to race E of broomrape (*Orobanche cumana* Wallr.). Proceedings Seventeenth International Sunflower Conference, Cordoba, Spain, June 8–12, Vol. 2, pp. 661–666.

Masirevic, S., 2000. Evolution of sunflower germplasm for resistance to phomopsis stem canker. Proceedings Fifteenth International Sunflower Conference, Toulouse, France, June 12–15, Vol. 2, pp. 84–89.

Matusova, R., van Mourik, T., Bouwmeester, H.J., 2004. Changes in the sensitivity of parasitic weed seeds to germination stimulants. Seed Sci. Res. 14 (4), 335–344.

Miller, J.F., Al-Khatib, K., 2004. Registration of two oilseed still flower genetic stocks, SURES-1 and SURES-2, resistant to tribenuron herbicide. Crop Sci. 39, 301–302.

Miller, J.F., Al–Khatib, K., 2000. Development of herbicide resistant germplasm in sunflower. Proceedings of Fifteenth International Sunflower Conference. Toulouse (France). O39–O41.

Olson, B.L.S., Al-Khatib, K., Aiken, R.M., 2004. Distribution of resistance to imazamox and tribenuron-methyl in native sunflower. www.sunflowernsa.com/research/research-workshop/documents/158.pdf

Orta, A.H., Erdem, T., Erdem, Y., 2002. Determination of water stress index in sunflower. Helia 25 (37), 27–38.

Pacureanu-Joita, M., Raranciue, S., Procopovici, E., Sava, E., Nastase, D., 2008. The impact of the new races of broomrape (*Orobanche cumana* Wallr.) parasite in sunflower crop in Romania. Proceedings Seventeenth International Sunflower Conference, Cordoba, Spain, June 8–12, Vol. 1, pp. 225–231.

Pacureanu-Joita, M., Raranciue, S., Sava, E., Stancin, D., Nastase, D., 2009. Virulence and aggressiveness of sunflower broomrape (*Orobanche cumana* Wallr.) populations in Romania. Helia 32 (51), 119–126.

Paricsi, S., Tar, M., Nagyne-Kutni, R., 2008. Study of resistance to Sclerotinia head disease in *Helianthus annuus* sunflower genotypes. Proceedings Seventeenth International Sunflower Conference, Cordoba, Spain, June 8–12, Vol. 1, pp. 211–214.

Pereira, M.L., Sadras, V.O., Trápani, N., 1995. Genetic improvement of sunflower in Argentina between 1930 and 1995. III. Dry matter partitioning and grain composition. Field Crops Res 67, 215–221.

Peres, A., Poisson-Bamme, B., Drolon, G., Le syndrome 'pieds secs' du tournesol: etude des causes et approche de la nuisibilite, Proceedings of Fifteenth International Sunflower Conference, Toulouse, France, 2000, 331–338.

Perez-Vich, B., Garces, R., Fernandez-Martinez, J.M., 2002. Inheritance of high palmitic and its relationship with high oleic acid content in the sunflower mutant AS 12. Plant Breed. 121, 49–56.

Perez-Vich, B., Munoz-Ruiz, J., Fernandez-Martinez, J.M., 2004. Developing mid-stearic acid sunflower lines from a high stearic acid mutant. Crop Sci. 44, 70–75.

Petcu, E., Pacureanu, J.M., 2011. Developing drought and broomrape resistant sunflower germplasm utilizing wild *Helianthus* species. Helia 34 (54), 1–8.

Petcu, E., Stanciu, M., Stancio, F., Raducanu, D., 2008. Physiological traits for quantification of drought tolerance in sunflower. Proceedings Seventeenth International Sunflower Conference, Cordoba, Spain, June 8–12, Vol. 1, pp. 345–350.

Pineda-Martos, R., Velasco, L., Fernandez-Escobar, J., Fernandez-Martinez, J.M., Perez-Vich, B., 2013. Genetic diversity of *Orobanche cumana* populations from Spain assessed using SSR markers. Weed Res. 53, 279–289.

Pineda-Martos, R., Velasco, L., Pérez-Vich, B., 2014. Identification, characterization and discriminatory power of microsatellite markers in the parasitic weed *Orobanche cumana*. Weed Res. 54 (2), 120–132.

Qi, L.L., Seiler, G.J., Vick, B.A., Gulya, T.J., 2012. Genetics and mapping of the R*11* gene conferring resistance to recently emerged rust races, tightly linked to male fertility restoration, in sunflower (*Helianthus annuus* L.). Theor. Appl. Genet. 125, 921–932.

Quaglia, M., Zazzerini, A., 2007. *In vitro* screening for sunflower (L.) resistant to *Diaphorte helianthi* fungal culture filtrate. Eur. J. Plant Pathol. 118, 340–393.

Quiroz, F.J., Molina, J.P.E., Dosio, G.A.A., 2014. Black stem by *Phoma macdonaldii* affected ecophysiological components that determine grain yield in sunflower (*Helianthus annuus* L.). Field Crops Res. 160, 31–40.

Rauf, S., 2008. Breeding sunflower (*Helianthus annuus* L.) for drought tolerance. Commun. Biomet. Crop Sci. 3 (1), 29–44.

Reddy, K.P., Sankaraneni, C.R., Pulugurtha, B.K., Mulpuri, S., 2013. Development of a scoring scale for powdery mildew *G. cichoracearum* disease and identification of resistance sources in cultivated and wild sunflowers. Euphytica 190, 385–399.

Regina, M.V., Leite, B.C., Amorirn, L., Filho, A.B., Oliveira, M.C., De Castro, C. (2008) Effect of sowing date and initial inoculums of *Alternaria* on sunflower in the south region of Brazil. Proceedings Seventeenth International Sunflower Conference, Cordoba, Spain, June 8–12, Vol. 1, pp. 93–198.

Rodríguez-Ojeda, M.I., Pineda-Martos, R., Alonso, L.C., Fernández-Escobar, J., Fernández-Martínez, J.M., Pérez-Vich, B., Velasco, L., 2013. A dominant avirulence gene in *Orobanche cumana* triggers *Or5* resistance in sunflower. Weed Res 53, 322–327.

Sakr, N., 2014. Pathogenicity in *Plasmopara halstedii* (sunflower downy mildew) races 710, 714 and 704. Acta Phytopathol. Hun. 49 (1), 11–24.

Sala C., Bulos M., Echarte M., Whitt S., Budziszewski G., Howie W., Singh B., Weston B. (2008a) Development of CLHA-PLUS: a novel herbicide tolerance trait in sunflower conferring superior imidazolinone tolerance and ease of breeding. Proceedings of Seventeenth International Sunflower Conference. Vol. Cordoba, Spain. June 8–12, 2: 489–494.

Sala, C.A., Bulos, M., Echarte, A.M., 2008b. Genetic analysis of an induced mutation conferring imidazolinone resistance in sunflower. Crop Sci. 48, 1817–1822.

Sala, C.A., Bulos, M., Altieri, E., Ramos, M.L., 2012. Sunflower: improving crop productivity and abiotic stress tolerance. In: Tuteja, N., Gill, S., Tubercio, A.F., Tuteja, R. (Eds.), Improving Crop Resistance to Abiotic Stress. Wiley-Blackwell Co, Germany, pp. 1205–1250.

Sauerborn, J., Buschman, H., Ghiasvand-Ghiasi, K., Kogel, H., 2002. Benzothiadia-zole activities resistance in sunflower (*Helianthus annuus* L.) to the root-parasitic weed orobanche cumana. Phytophatology 92 (1), 59–64.

Schnabl, H., Binsfeld, P.C., Cerboncini, C., Dresen, B., Peisker, H., Wingender, R., Henn, A., 2002. Biotechnological methods applied to produce *Sclerotinia sclerotiorum* resistant sunflower. Helia 25 (36), 191–198.

Seiler, G.J., 2012. Utilization of wild *Helianthus* species in sunflower breeding. In: Kovacevic, Z., Škorić, D., Sakac, Z. (Eds.), Sunflower Genetics and Breeding-International monogram. Serbian Academy of Science, Serbia, pp. 355–413.

Semerci, A., Kaya, Y., Peker, K., Sahin, I., Citak, N., 2011. The analysis of sunflower yield and water productivity in Trakya region. Bulgarian J. Agric. Sci. 17, 207–217.

Sendall, B.C., Goulter, K.C., Aitken, E.A.B., Goulter, K.C., Thompson, S.M., Mitchell, J.H.M., Kochman, J.K., Walters, W., Shatte, T., Gulya, T.J., 2006. Review of research on the sunflower: *Puccinia helianthi* pathosystem in Australia. Austr. Plant Pathol. 35, 657–670.

Shaik, S.M., Ravikumar, R.L., 2003. Association between field and in vitro reaction to *Alternaria* leaf blight in sunflower genotypes. Helia 26 (38), 109–114.

Shobha Rani, T., Ravikumar, R.L., 2006. Sporophytic and gametohytic recurrent selection for improvement of partial resistance to *Altemaria* leaf blight in sunflower (*Helianthus annuus* L.). Euphytica 147, 421–431.

Singh, B.D., 2000. Plant Breeding, Principles and Methods. Kalyani Publishers, New Delhi, India, 896.

Škorić, D., 2009. Sunflower breeding for resistance to abiotic stresses. Helia 32 (50), 1–16.

Škorić, D., 2012. The Genetics of Sunflower. In: Kovacevic, Z., Škorić, D., Sakac, Z. (Eds.), Sunflower Genetics and Breeding-International Monogram. Serbian Academy of Sciences, Serbia, pp. 1–125.

Škorić, D., Jocic, S. (2004) Achievements of sunflower breeding at the IFVC in Novi Sad. Proceedings of Sixteenth International Sunflower Conference, Fargo, ND, USA, 29 August–4 September, 2: 451–458.

Škorić, D., Pacureanu-Joita, M., 2011. Possibilities for increasing sunflower resistance to broomrape (*Orobanche cumana* Wallr.). J. Agric. Sci. Technol. B1, 151–162.

Škorić, D., Jocic, S., Lecic, N., Sakac, Z., 2007. Development of sunflower hybrids with different oil quality. Helia 30 (47), 205–212.

Skoric, D., 1988. Sunflower breeding. J. Edible Oil Ind. 25 (1), 1–90.

Skoric, D., Jocic, S., Sakac, Z., Lecic, N., 2008. Genetic possibilities for altering sunflower oil quality to obtain novel oils. Can. J. Physiol. Pharmacol. 86 (4), 215–221.

Soldatov, K.I., 1976. Chemical mutagenesis in sunflower breeding. In: Proceedings of Seventh International Sunflower Conference. Krasnodar, Russia. pp. 352–357, June 27–July 3.

Sujatha, M., Prakabaran, A.J., 2006. Ploidy manipulation and introgression of resistance to *Alternaria helianthi* from wild hexaploid *Helianthus* species to cultivated sunflower (*Helianthus* annuus L.) aided by another culture. Euphytica 152, 201–2015.

Talukder, Z.I., Hulke, B.S., Marek, L.F., Gulya, T., 2014. Sources of resistance to sunflower diseases in a global collection of domesticated USDA plant introductions. Crop Sci. 54, 694–705.

Tang, S., Heesacker, A., Kishore, V.K., Fernandez, A., Sadik, E.S., Cole, G., Knnap, S.J., 2003. Genetic mapping of the Or_5 gene for resistance to *Orobanche* race E in sunflower. Crop Sci. 43, 1021–1028.

Tourvieille de Labrouche, D., Serre, F., Walser, P., Philippon, J., Vear, F., Tardin, M.C., Andre, T., Castellanet, R., Chatre, S., Costes, M., Jouve, P., Madeuf, J.L., Mezzaroba, A., Plegades, J., Pauchet, I., Mestries, E., Penaud, A., Pinochet, X., Serieys, H., Griveau, Y. (2004) Partial, non-race-specific resistance to downy mildew in cultivated sunflower lines. Proc. 16th Int. Sunflower Conf. Fargo, ND, USA. August 29–September 2. 1: 111–116.

Triboi, A.M., Messaoud, J., Debaeke, P., Lecoeur, J., Vear, F. (2004) Heredity of sunflower leaf characters as yield predictors. Proceedings of Sixteenth International Sunflower Conference, Fargo, ND, USA. August 29–September 2. 2: 517–524.

Vear F (2004) Breeding for durable resistance to the main diseases of sunflower. Proceedings of Sixteenth International Sunflower Conference, Fargo, ND, USA. August 29–September 2. 1: 15–28.

Vear, F., Serre, F., Roche, S., Walser, P., Tourvieille de Labrouhe, D., 2007. Improvement of *Sclerotinia sclerotiorum* head rot resistance in sunflower by recurrent selection of a restorer population. Helia 30 (46), 1–12.

Vear F, Jouan-Dufournet I, Bert PF, Serre F, Cambon F, Pont C, Walser P, Roche S, Tourvieille de Labrouhe D, Vincourt P (2008) QTL for capitulum resistance to *Sclerotinia sclerotiorum* in sunflower. Proceedings of Seventeenth International Sunflower Conference, 2: 605–610, Cordoba, Spain, June 8–12.

Vear, F., 2011. Helianthus. In: Kole, C. (Ed.), Wild Crop Relatives: Genomic and Breeding Resources: Oilseeds. Springer, New York, pp. 161–170.

Vear, F., Philippon, J., Roche, S., Walser, P., Tourvieille de Labrouhe, D., Mouzeyar, S., Nicolas, P., 2000. Genetical analyses of the sunflower downy mildew resistance gene *Pl5*. Proceedings of Fifteenth International Sunflower Conference, Toulouse, France, June 12–16, 2000, J31–J36.

Velasco, L., Fernandez-Martinez, J.M., 2003. Identification and genetic characterization of new sources of beta and gamma tocopherol in sunflower germoplasm. Helia 26 (38), 17–24.

Velasco, L., Perez-Vich, B., Fernandez-Martinez, J.M., 2004. Development of sunflower germplasm with high delta-tocopherol content. Helia 27 (40), 99–106.

Velasco, L., Perez-Vich, B., Jan, C.C., Fernandez-Martinez, J.M., 2007. Inheritance of resistance to broomrape (*Orobanche cumana* Wallr.) race F in a sunflower line derived from wild sunflower species. Plant Breed. 126, 67–71.

Velasco, L., Perez-Vich, B., Fernandez-Martinez, J.M., 2008. A new sunflower mutant with increased levels of palmitic acid in seed oil. Helia 31 (48), 55–60.

Velasco, L., Pérez-Vich, B., Yassein, A.A.M., Jan, C.C., Fernández-Martínez, J.M., 2012. Inheritance of resistance to sunflower broomrape (*Orobanche cumana* Wallr.) in a interspecific cross between *Helianthus annuus* and *Helianthus debilis* subsp. tardiflorus. Plant Breed. 131, 220–221.

Vera-Ruiz, E.M., Fernandez-Martinez, J.M., 2006. Genetic mapping Tphl gene controlling beta-tocopherol accumulation in sunflower seeds. Mol. Breed. 17, 291–296.

Vick, B.A., Jan, C.C., Miller, J., 2002. Inheritance of reduced saturated fatty acid content in sunflower oil. Helia 25, 113–122.

Viranyi F (2008) Research progress in sunflower diseases and their management. In: Proceedings of Seventeenth International Sunflower Conference, Vol.1: 1–13. Cordoba, Spain, June 8–12.

Vranceanu, A.V., Tudor, V.A., Stonescu, F.M., Parvu, V. 1980. Virulence groups of broomrape (*Orobanche cumana Wallr.)* differential hosts and resistance sources and genes in sunflower. In Proceedings of the Ninth International Sunflower Conference. Torremolinos, Spain. June, 8–13: 74–81.

Würschum, T., Anyanga, W.O., Hahn, V., 2014. Inheritance of *Sclerotinia* midstalk rot resistance in elite sunflower breeding germplasm. Helia 37 (61), 193–203.

CHAPTER

5

Groundnut

Mothilal Alagirisamy

All India Co-ordinated Research Project on Groundnut, Division of Plant Breeding and Genetics, Regional Research Station, Tamil Nadu Agricultural University, Vridhachalam, India

INTRODUCTION

General Description and Crop Use

Groundnut (*A. hypogaea* L.), also called peanut, is the third most important oilseed crop in the world. It is cultivated in tropical, subtropical, and temperate countries between 40°N and 40°S. It is an excellent source of edible oil and protein while the main byproducts of the crop, namely oilcake and haulms, serve as important animal feed. Groundnut derives its name from a Greek word *Arachos* meaning a weed and *hypogaea* meaning an underground chamber, i.e., a weedy plant having fruits below the soil. Due to its subterranean fruiting habit, groundnut is popularly known as the "unpredictable legume." Groundnut is an herbaceous annual legume plant belonging to the family Fabaceae (Leguminosae). The plant produces a typical tubular hypanthium or flower stalk. After fertilization, the peg or gynophore carrying the fertilized ovary grows toward the soil (positively geotrophic) where it develops into a pod bearing two or three kernels.

A groundnut kernel contains approximately 40–54% oil, 22–36% protein, and 10–20% carbohydrate. It is also a good source of B vitamins, namely thiamine (B_1), riboflavin (B_2), niacin (B_3), and tocopherol but is deficient in fat-soluble vitamins A and D and almost lacking in vitamin C (Nagaraj, 1995). The groundnut shells are used for making particleboards or used as a fuel or a filler in fertilizers and the feed industry. Groundnut haulms serve as nutritious fodder for livestock. They contain protein (8–15%), lipids (1–3%), minerals (9–17%), and carbohydrate (38–45%). The digestibility of nutrients in groundnut haulm is around 53% and that of crude protein 88% when fed to cattle (Jambunathan, 1991). Being a legume crop, groundnut helps in improving soil health and fertility by leaving behind N_2 and organic matter in the soil.

World Area and Production

The world groundnut (in shell) area was 24.59 million hectares with a total production of 40.47 million metric tons in 2012 (FAOSTAT, 2012). The world's average productivity

Breeding Oilseed Crops for Sustainable Production. http://dx.doi.org/10.1016/B978-0-12-801309-0.00005-7

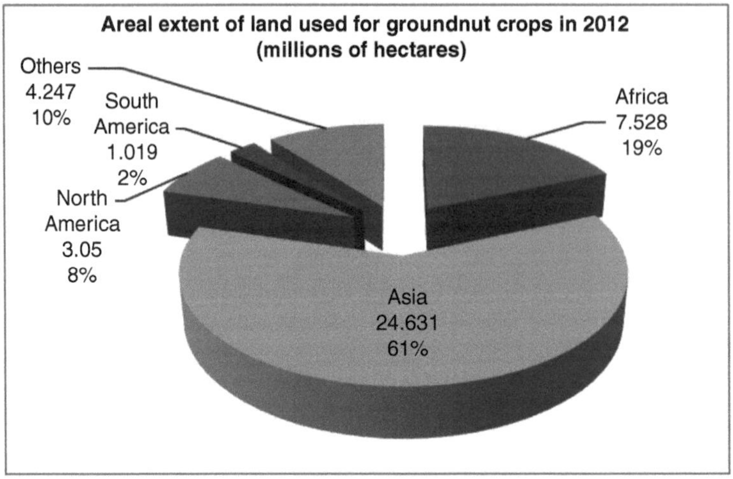

FIGURE 5.1 **Area under groundnut cultivation (in million hectares) at the global level.**

of groundnut was 1645 kg ha^{-1}. As compared to 2010, the area and production of groundnut declined slightly by 0.89 million hectares and 2.25 million metric tons. Groundnut is a major oilseed crop in Asian and African countries. Together they contribute 80% of the total groundnut area and production (Figs 5.1 and 5.2). The most important groundnut-producing countries are China, India, United States, Nigeria, Indonesia, Myanmar, Senegal, Sudan, Argentina, and Vietnam.

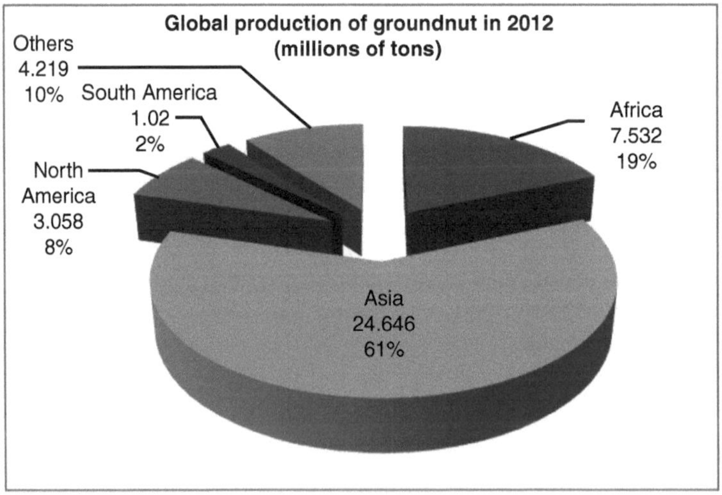

FIGURE 5.2 **Groundnut production (in million tons) at the global level.**

Utilization and Consumption Patterns

Approximately 60% of the world's groundnut production is marketed for direct consumption or oil production, while 32% is utilized for food, and 8% for seed (Birthal et al., 2010). Consumption patterns vary from country to country. Groundnuts are consumed in many diverse ways. They can be eaten as raw, boiled, salted, or dry roasted such as honey roasted, hickory smoked, or chili flavored. Also, they are used for confectionery and chocolate-based products. Groundnut can be used in soups and as toppings on meat and rice based dishes and in sauces (satay). Groundnut butter is also made from either whole groundnuts or from the cake left after oil extraction. Groundnut oil is an excellent cooking medium because of its high smoking point (Singh and Diwakar, 1993). Each of the botanical varieties has its own preferential use. Larger sized Virginia kernels are preferred for cocktail and salted nuts, while medium-sized Runner and small Spanish varieties are suitable for peanut butter, oil, and candies. Valencia bunch varieties are suitable for roasting in their shells.

BOTANY

Origin and Distribution of Wild and Cultivated Species

Wild Species

The genus *Arachis* originated from South America and its distribution is confined to Argentina, Bolivia, Brazil, Paraguay, and Uruguay (Krapovickas and Rigoni, 1957). However, the greatest number of wild species occur in western central Brazil, followed by Bolivia (Valls et al., 1985). The genus *Arachis* belongs to the family Leguminosae, tribe Aeschynomeneae, and subtribe Stylosanthenae. Based on morphology, geographical distribution, chromosomal features, cross compatibility reactions, and pollen stainability of interspecific hybrids, the genus *Arachis* is subdivided into nine sections – namely, Arachis, Erectoides, Heteranthae, Caulorrhizae, Rhizomatosae, Extranervosae, Triseminatae, Procumbentes, and Trierectoides (Krapovickas and Gregory, 1994). Nearly 80 annual and perennial species were reported in all nine sections. The key for identifying different sections was proposed by Krapovickas and Gregory (2007) and is given below.

The section Trierectoides consists of perennial plant species possessing thick spindle shaped hypocotyls; erect plants with trifoliolate leaves; stipules with margins which are fused, forming a tube that encloses one or two internodes; flowers and pods clustered at the base of the plant; a well-developed hypanthium; standard orange petals with reddish lines on the dorsal side; peg very long, extended up to 1 m, subterranean pods, horizontal, biarticulate, well-developed isthmus; single-seeded articles; smooth pericarp; and a species chromosome number of $2n = 20$. The two species belonging to this section are *A. guaranitica* Chodat & Hassl and *A. tuberosa* Bong. ex Benth.

The section Erectoides consists of pernennial plants with roots having thickened branches; branches which are erect or decumbent with tetrafoliolate leaves; flowers which are clustered at the base of the plant; a hypanthium that is well developed; standard orange petals with reddish lines on the dorsal side; fruit which is subterranean and biarticulate; a very long peg, horizontal, not deep; an isthmus which is well developed; articles which are single seeded; a pericarp which is smooth; and a species chromosome number of $2n = 20$. Fourteen species were described under this section – namely, *A. martii* Handro, *A. brevipetiolata* Krapov. & W.C. Greg., *A. oteroi* Krapov. & W.C. Greg., *A. hatschbachii* Krapov. & W.C. Greg., *A. cryptopotamica* Krapov. & W.C. Greg., *A. major* Krapov. & W.C. Greg., *A. benthamii* Handro, *A. douradiana* Krapov. & W.C. Greg., *A. gracilis* Krapov. & W.C. Greg., *A. hermannii* Krapov. & W.C. Greg., *A. archeri* Krapov. & W.C. Greg., *A. stenophylla* Krapov. & W.C. Greg., *A. paraguariensis* Chodat & Hassl. ssp. *paraguariensis*, *A. paraguariensis* ssp. *capibarensis* Krapov. & W.C. Greg., and *A. porphyrocalyx* Valls & C.E. Simpson.

The section Extranervosae consists of perennial plants with roots with tuberiform thickenings linked together in a chain; a mainstem which is erect with branches which are procumbent and stipules with margins free to the base; tetrafoliate leaves with their upper surface bright green and shiny; flowers which are small to medium sized with a hypanthium, which is 4 cm long, and are standard orange or yellow with red lines on the ventral surface; fruit which is subterranean and biarticulate; a peg with adventitious roots which is short and almost vertical, 20 cm long horizontally; an isthmus which is short; articles which are smooth, covered with a dense layer of hairs; and a species chromosome number of $2n = 20$. The section consists of 10 species, namely, *A. setinervosa* Krapov. & W.C. Greg., *A. macedoi* Krapov. & W.C. Greg., *A. marginata* Gardner, *A. prostrata* Benth., *A. lutescens* Krapov. & Rigoni, *A. retusa* Krapov., W.C. Greg. & Valls., *A. burchellii* Krapov. & W.C. Greg., *A. pietrarellii* Krapov. & W.C. Greg., *A. villosulicarpa* Hoehne., and *A. submarginata* Valls & Krapov. & C.E. Simp.

The section Triseminatae consists of perennial plants with a taproot, without thickenings; a mainstem which is erect and has branches which are decumbent; stipules which are somewhat fused at the base forming a very short tube and with the apices which are subulate; leaves which are tetrafoliolate; flowers which are small, along the length of the branches, with a hypanthium which is 4.5 cm long, standard orange with a conspicuous purple spot toward the base on the dorsal surface and with fainter red lines on the ventral surface; fruit which is subterranean, frequently with three segments; a peg and isthmi which are elongated, horizontal, and shallow; articles which are single seeded; a pericarp which is smooth and covered with a dense layer of hairs; and a species chromosome number of $2n = 20$. The lone species belonging to this section is *A. triseminata* Krapov. & W.C. Greg.

The section Heteranthae consists of annual or biennial herbs with a taproot with slender branches; a mainstem which is erect with branches which are procumbent; stipules with the margins free to the base and with acicular apices; leaves which are tetrafoliolate; flowers which are dimorphic, distributed along the length of the lateral branches which are very small, with the hypanthium from 0.3 cm to 2.0 cm long and the calyx and corolla closed, 2–3 mm long, in addition to normal flowers with hypanthium extended, a larger calyx, and an expanded corolla, standard orange or yellow, with reddish lines on the outside surface or on both dorsal and ventral surfaces; fruit which is subterranean and biarticulate; a peg and isthmus which is horizontal, not deep; articles which are single seeded; a pericarp which is smooth and covered with a dense layer of hair; and a species chromosome number of $2n = 20$. The six species described under this section are *A. giacomettii* Krapov., W.C. Greg. Valls & C.E. Simp., *A. sylvestris* (A. Chev.), *A. pusilla* Benth., *A. dardani* Krapov. & W.C. Greg., *A. interrupta* Valls & C.E. Simp., and *A. seridoensis* Valls & C.E. Simp., Krapov. & R. Veiga.

The section Caulorrhizae consists of perennial plants with a taproot, without thickenings; branches which are spreading, procumbent, and rooting at the nodes; stems which are hollow; stipules with the margins free to the base; leaves which are tetrafoliolate; flowers positioned along the length of the branches, well-developed hypanthium, up to 10 cm long or more, standard, commonly yellow, expanded; fruit which is subterranean and biarticulate; pegs which are short, vertical, or inclined; an isthmus which is short, up to 5 cm long; articles which are single seeded; and a species chromosome number of $2n = 20$. The two species belonging to this section are *A. repens* Handro and *A. pintoi* Krapov. & W.C. Greg.

The section Procumbentes consists of perennial plants with a taproot, without thickenings; branches which are procumbent and rarely decumbent; leaves which are tetrafoliolate, generally glabrous or subglabrous; stipules with free margins; flowers positioned along the length of the branches, a well-developed hypanthium; standard expanded, yellow or orange, with reddish lines on the dorsal side; fruit which is subterranean and biarticulate; a peg and isthmus which is very extended, horizontal, and shallow; articles which are smooth; and a species chromosome number of $2n = 20$. The 10 species belonging to this section are *A. lignosa* (Chodat & Hassl.) Krapov. & W.C. Greg., *A. kretschmeri* Krapov. & W.C. Greg., *A. rigonii* Krapov. & W.C. Greg., *A. chiquitana* Krapov., W.C. Greg. & C.E. Simp., *A. matiensis* Krapov., W.C. Greg. & C.E. Simp., *A. appressipila* Krapov. & W.C. Greg., *A. vallsii* Krapov. & W.C. Greg., *A. subcoriacea* Krapov. & W.C. Greg., *A. pflugeae* C.E.Simp. Krapov. & Valls, and *A. hassleri* Krapov., Valls & C.E. Simp.

The section Rhizomatosae consists of perennial plants which are rhizomatous; have a taproot, without thickenings; branches which are procumbent; leaves which are tetrafoliolate; flowers positioned along the length of the branches, a hypanthium that is well developed, standard expanded, orange, rarely yellow; fruit which is subterranean and biarticulate; a peg and isthmus which is short and more or less vertically growing. The section Rhizomatosae are classified into two series, namely, Prorhizomatosae (leaflets which are coriaceous, with the margin outstanding on both surfaces; standard orange flowers with red lines on both surfaces; and diploids with $2n = 20$ chromosomes) and Rhizomatosae (leaflets which are more or less soft, with the margin slightly marked, standard orange or yellow flowers, with red lines only on the front; and tetraploids with $2n = 4x = 40$ chromosomes). The series Prorhizomatosae consists of a lone species, namely, *A. burkartii* Handro. The series Rhizomatosae consists of three species: *A. pseudovillosa* (Chodat & Hassl.) Krapov. & W.C. Greg., *A. glabrata* Benth. var. *glabrata*, *A. glabrata* Benth. var. *hagenbeckii* (Harms ex Kuntze) F.J. Herm, and *A. nitida* Valls, Krapov. & C.E. Simp.

The section Arachis consists of annual or perennial plants without rhizomes or stolons; with a taproot, without thickenings; a main stem which is erect, with branches which are procumbent and spreading; stipules with open margins; leaves which are tetrafoliolate; flowers distributed along the length of the branches, a hypanthium which is well developed, standard always expanded, orange or yellow, without reddish lines on the dorsal surface or with a pinkish purple tinge (*A. batizocoi*); fruit which is subterranean, generally with two single-seeded articles or one multi-seeded article (*A. hypogaea*); a peg which is vertical or inclined, not horizontal; a pericarp which is smooth or reticulate; and species chromosome numbers of $2n = 20$ (diploids) and $2n = 4x = 40$ (tetraploids). The Arachis section is the largest section among the nine sections in which 31 species were described: namely, *A. glandulifera* Stalker, *A. cruziana* Krapov., W.C. Greg. & C.E. Simp., *A. monticola* Krapov. & Rigoni, *A. magna* Krapov., W.C. Greg. & C.E. Simp., *A. ipaënsis* Krapov. & W.C. Greg., *A. valida* Krapov. & W.C. Greg., *A. williamsii* Krapov. & W.C. Greg., *A. batizocoi* Krapov. & W.C. Greg., *A. duranensis* Krapov. & W.C. Greg., *A. hoehnei* Krapov. & W.C. Greg., *A. stenosperma* Krapov. & W.C. Greg., *A. praecox*

Krapov., W.C. Greg. & Valls, *A. palustris* Krapov., W.C. Greg. & Valls, *A. benensis* Krapov., W.C. Greg. & C.E. Simp., *A. trinitensis* Krapov. & W.C. Greg., *A. decora* Krapov., W.C. Greg. & Valls, *A. herzogii* Krapov., W.C. Greg. & C.E. Simp., *A. microsperma* Krapov., W.C. Greg. & Valls, *A. villosa* Benth., *A. helodes* Martius ex Krapov. & Rigoni, *A. correntina* (Burkart) Krapov. & W.C. Greg., *A. simpsonii* Krapov. & W.C. Greg., *A. cardenasii* Krapov. & W.C. Greg., *A. kempff-mercadoi* Krapov., W.C. Greg. & C.E. Simp., *A. diogoi* Hoehne, *A. kuhlmannii* Krapov. & W.C. Greg., *A. hypogaea* L., *A. linearifolia* Valls, Krapov. & C.E. Simp., *A. schininii* Krapov. Valls & C.E. Simp., *A. gregoryi* C.E. Simp. Krapov. & Valls, and *A. krapovickasii* C.E. Simp., D.E. Williams, Valls & I.G. Vargas.

Cultivated Species

Anthropologists discovered the earliest known evidence of domesticated groundnut in the Ñanchoc Valley which dates back 7600 years. This is located approximately 500 m above mean sea level (msl) on the lower western slope of the Andes in northern Peru (Dillehay et al., 2007). The cultivated groundnut species *A. hypogaea* L. is a segmental allopolyploid with a chromosome constitution of $2n = 4x = 40$. The cultivated species is made up of two genomes A and B. The Agenome is characterized by a small pair of chromosomes with a lower level of euchromatin condensation whereas the B genome is characterized by a chromosome pair with a secondary constriction and a satellite (Smartt et al., 1978). It is believed that the cultivated tetraploid groundnut may be originated most probably from hybridization of two diploid wild species each with the A and B genome followed by natural chromosome doubling (Halward et al., 1991; Young et al., 1996; Seijo et al., 2004, 2007). There are numerous reports available in the literature regarding the donors of the A and B genomes to the cultivated species. The A and B genome might be contributed by *A. cardenasii* and *A. batizocoi*, respectively (Smartt et al., 1978). Mallikarjuna et al. (2006) proposed *A. hoehnei* as the probable B genome donor of the cultivated groundnut. Halward et al. (1991), Kochert et al. (1996), Seijo et al., (2004, Seijo et al., 2007), Ramos et al. (2006), Robledo and Seijo (2010), and Moretzsohn et al. (2013) identified two diploid wild species *A. duranensis* and *A. ipaensis* as the contributor of the A and B genomes to the cultivated groundnut.

Classification of Cultivated Species

The cultivated groundnut species *A. hypogaea* L. has been classified based on its growth habit, branching pattern, presence or absence of reproductive nodes on the main stem, position of vegetative and reproductive axes on the primary branches, pod and seed characteristics, and seed dormancy. The cultivated groundnut is broadly classified as follows:

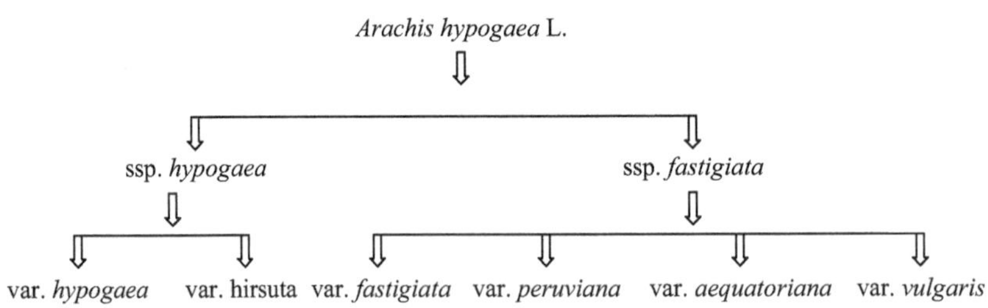

A brief description of the subspecies and botanical varieties of *A. hypogaea* L. as indicated by Krapovickas and Gregory (2007) is presented below:

1. *Subspecies hypogaea.* Habit procumbent or decumbent or erect; main stem without flowers and $n + 1$ branches which alternate regularly with two vegetative and two reproductive (alternate branching) nodes. The first branch on the cotyledonary lateral is always vegetative. Inflorescences are simple. There are 2 or 2–4 seeds per pod; the pod beak can be either prominent, moderate, or absent; the pod constriction can be prominent, moderate, or absent; the pod size can be very large (>20 mm) or small (<10 mm). Its testa color is commonly tan but can be red, white, or purple with variegated forms also present. Its seed dormancy is usually present. Its foliage is dark green.
   a. The *hypogaea* variety. Commonly known as Virginia, its habit is procumbent, decumbent, or erect, with the main axis of procumbent forms being short (not exceeding 40–50 cm) – stem is usually not very hairy. Its leaflets have undersides which are glabrous or with a few hairs apearing on the midvein. It usually produces two seeds and is medium to late to mature.
   b. The *hirsute* variety. Its habit is procumbent and its main axis may exceed 1.0 m. Its stem is fairly hairy. Its leaflets have hairs, which are 1–2 mm long on the underside, and are dispersed across entire surface. Its pods are strongly beaked, with 2–4 seeds and it is very late to mature.
2. *Subspecies fastigiata*. Its habit is erect to decumbent; its main stem holds flowers and lateral branches on which the reproductive and vegetative branches present themselves in no specific order (sequential branching); can be inflorescence, simple, or compound; its first branches on cotyledonary laterals are reproductive; seed dormancy usually absent; and foliage is usually more light green in color than in ssp. *hypogaea*.
   a. The *fastigiata* variety. Commonly known as Valencia. Vegetative branches on primaries are absent or regularly placed at the distal nodes. Its reproductive branches are usually short and slender. Its inflorescence is usually simple. Its fruits show subtle or slightly marked reticulation, without pronounced longitudinal ribs. Its pods have 2 or 2–4 (rarely 5) seeds. A beak can be absent, slight, or prominent. It is medium to small in size. Its testa being tan, red, white, yellow, purple, or variegated.
   b. The *peruviana* viariety. It has a main stem with vegetative branches toward the base and reproductive branches toward the apex. Its lateral branches are decumbent, with either very few or no vegetative branches. Its reproductive branches are 5–10 cm long, thick, multiflowered, sometimes with leaves toward the apex, and abundant on the main stem as well as on the lateral branches. Its leaves are large and somewhat thick. Its leaflets are glabrous on both sides, with hairs only on the margin and on the underside of the midvein. Its fruit usually has three seeds, in some landraces this increases to four, however; its pericarp is strongly reticulated having outstanding longitudinal ribs. Its fruits always demonstrate sharply marked reticulation and outstanding longitudinal ribs.
   c. The *aequatoriana* variety. Its main stem is erect, more or less branched, with reproductive branches usually short, less than 5 cm long. Its lateral branches are

decumbent, with reproductive branches up to 20 cm long. It has large leaves. Its leaflets have glabrous upper surfaces and undersides showing hairs 1–2 mm long dispersed across their entire surface. Its fruit has 3–4 seeds; its pericarp is very reticulated with outstanding longitudinal ribs.

d. The *vulgaris* variety. This variety is commonly known as Spanish. Its vegetative branches are occasional and irregularly placed. Its inflorescence is compound. Its pods usually have 2 seeds. Its fructification is clustered around the base of the plant. A beak may be present or absent. It is medium to small in size. Its testa being tan, red, white, or purple.

Flowering and Floral Biology

The inflorescence of groundnut is either solitary (simple inflorescence bearing a single flower) or raceme (compound inflorescence bearing three to several flowers). The inflorescences occur in the axils of foliage leaves or cataphylls. The *hypogaea* variety bears simple inflorescence while *vulgaris* possesses compound inflorescence. Groundnut flowers are papilionaceous and sessile. An elongated pedicellate-like structure called a "hypanthium" or "calyx tube" is inserted in the flowering branch from which cataphyll bracts appear. The hypanthium is usually pubescent and bears five calyx lobes, four of which become fused except for their tips and appear back of the standard, while the fifth lobe lies under the keel. The standard petals are usually lemon yellow in color. However, orange to dark orange, garnet, and occasionally white or creamy white may also be observed in some accessions. Dark color crescent lines radiating from the base of the standard is observed in all the genotypes. The wing petals wrap around the keel petal. The keel petal completely wraps the stamens and pistils. The stamenal column is composed of 10 monodelphous stamens, of which two are sterile with nonfunctional anthers. The remaining eight functional stamens are dimorphic, i.e., there are four oblong, adnate, introrse anthers alternated with four globose, and dorsifixed anthers. Among the four oblong stamens, three are biloculate and one is uniloculate. The filaments are fused two thirds of the length basally. The tortuous style is long and filiform ending with club-shaped stigma above the anther column. The style has profuse, slanting, tiny upward hairs on the distal end. The ovary is located at the base of the hypanthium, usually superior, and is about 1.5 mm long having 2–4 ovules – occasionally 5 or 6 (Ramanatha Rao, 1988). Flowering commences 20–30 days after emergence depending upon the botanical variety and environment. Although groundnut produces enormous flowers, less than 15–20% produce mature pods (Lim and Hamdan, 1984; Ramanatha Rao and Murty, 1994). Usually one flower out of seven produces a mature fruit. The flowering period consists of two peaks. The first peak occurs during the first 3 weeks of flowering, while the second peak occurs during the latter period of flowering. Flowering declines 60–70 days after emergence. In the Spanish bunch varieties, flowering lasts 3–6 weeks, while in Virginia types, the flowering may last 6–8 weeks. The Virginia type produces more flowers and the total number of flowers produced may vary from 18–142 (Ishag, 2000). The groundnut flower opens early in the morning between 6.00 A.M. and 9.00 A.M. and the anther dehisces 1–2 h prior to the flower opening, which ensures complete self-pollination. Occasionally, cross-pollination occurs, but only up to 2.5%.

CYTOGENETICS

Chromosome Description and Numbers

The genus *Arachis* consists of diploids ($2n = 2x = 20$), tetraploids ($2n = 4x = 40$), and aneuploids ($2n = 2x = 18$). The cytogenetics of *A. hypogaea* L. was first reported by Husted (1933, 1936). He confirmed the somatic chromosome number as $2n = 40$. Chromosome pairing was normal with 20 bivalents and occasional multivalents also noticed in metaphase I of the pollen mother cells. Intersubspecific hybrids between var. *vulgaris*, var. *fastigiata*, and var. *hypogaea* showed mostly bivalents with rare univalents and multivalents (Stalker, 1980). Prevalence of "sticky chromosome" in two diploid species (*A. villosa* and *A. spegazzinii*), due to quadrivalents, was also observed (Singh and Moss, 1982). Since all the chromosomes are metacentric, differences could not be distinguished between them. However, one pair of chromosomes appeared contrastingly smaller than the others and one pair demonstrated a secondary constriction. This helped to distinguish the chromosomes of *A. hypogaea* L. (Husted 1933).

Interspecific Hybridization

The first interspecific hybrid was obtained by crossing *A. hypogaea* L. ($2n = 40$) with *A. correntina* ($2n = 20$). The F_1 ($2n = 30$) hybrid was completely sterile due to irregular meiosis (Krapovickas and Rigoni, 1949, 1951). Another sterile triploid was obtained by crossing *A. hypogaea* L. ($2n = 40$) with *A. villosa* ($2n = 20$) (Krapovickas and Rigoni, 1957). The first fertile interspecific hybrid was obtained by crossing *A. hypogaea* L. ($2n = 40$) with *A. monticola* ($2n = 40$) (Krapovickas and Rigoni, 1954, 1957) demonstrating *A. monticola* to be a closely related wild species.

The chromosomes of section Arachis are small, ranging from 1.51 μm to 3.82 μm (Stalker and Dalmacio, 1981). In another study, Lavia et al. (2009) observed a mean chromosome length of 1.85–2.23 μm within the section Arachis. The mean chromosome length of subspecies *hypogaea* ranged from 1.69 μm to 2.21 μm; while for subspecies *fastigiata* the range was between 1.81 μm in var. *acquatoriana*, *peruviana*, and *vulgaris* to 1.97 μm in var. *fastigiata*. SAT chromosome type 3 was observed in subspecies *fastigiata* while SAT chromosome type 6 was found in subspecies *hypogaea*. In general, all the chromosomes of cultivated groundnut are metacentric (Lavia and Fernandez, 2004).

Species with the smallest chromosomes and asymmetrical karyotypes are considered primitive types from which advanced sections, with longer chromosome and symmetrical karyotypes, are evolved. The extra length might be accompanied by an unequal distribution of DNA in the chromosome arms (Lavia et al., 2009). Like diploids and tetraploids, the chromosomes of aneuploids (*A. praecox* and *A. palustris*) are small, metacentric, and with satellites. However, the basic chromosome number is $x = 9$. These two species differ with respect to size of satellite, interchromosomal asymmetry index, type of chromosome that bears the satellite, and its position in the ideogram (Lavia, 1998).

Genomic Relationship between Section Arachis and Other Sections

The genetic diversity of the genus *Arachis* is classified into four gene pools (Singh and Simpson, 1994). The primary gene pool consisting of *A. hypogaea* and *A. monticola*;

the secondary gene pool consisting of diploid species from section Arachis that are cross-compatible with *A. hypogaea*; the tertiary gene pool consisting of species of the section Procumbentes that are weakly cross-compatible with *A. hypogaea*; and the fourth gene pool consisting of the remaining wild *Arachis* species classified into seven other sections. The perennial species of section Arachis possessed very small stigmas guarded by a ring of hairs which were very difficult to pollinate while most of the annuals had larger stigmas, were not surrounded by dense hairs, and were easy to pollinate (Lu et al., 1990). To overcome cross incompatibility barriers, mentor pollen or application of growth regulators (Stalker et al., 1987) may be adopted to enhance pollen germination and fertilization. Bud pollination produced normal pollen tube growth in pistils of the wild annual *A. spegazzinii* (Lu et al., 1990). Though various *in vitro* techniques are used to recover the hybrid embryos, the intersectional hybrids are either male or female sterile; occasionally they did not attain a reproductive stage (Stalker and Simpson, 1995). Hull and Carver (1938) first attempted intersectional hybridization between *A. hypogaea* and *A. glabrata* with unsuccessful results.

Section Arachis × *section Rhizomatosae*

Mallikarjuna and Sastri (2002) reported a successful hybrid between annual *A. hypogaea* L. and perennial *A. glabrata* var. *glabrata*. The hybrids were perennial without underground rhizomes. Morphological and fertility differences were observed in the hybrids. The chromosome associations in the tetraploid hybrids were highly variable. The number of bivalents and univalents ranged from 9–19 and 0–12, respectively. Occasional trivalents and tetravalents were also observed. Higher mean ring bivalents observed in the hybrid indicated the high degree of homeology between the genomes of *A. hypogaea* and *A. glabrata*. The backcross hybrid [(*A. hypogaea* × *A. glabrata*) × *A. hypogaea*] was more stable having a mean of 10–20 bivalents with more ring bivalents (11), indicating greater homology as more *A. hypogaea* chromosomes had the opportunity to pair with the genome of *A. hypogaea* × *A. glabrata*. More univalents in the BC_1 hybrid indicated the preferential pairing of the A genome of triploids and tetraploids (*A. hypogaea*) and nonhomology observed between the B genome of triploids and the R genome of *A. glabrata* (Mallikarjuna, 2004).

The triploid hybrid between *A. duranensis* and *A. glabrata* recorded a mean of 10 bivalents and 9 univalents. Trivalents are rarely observed. The presence of univalents may be due to the nonhomologous R chromosomes of *A. glabrata*. The triploids showed an erect growth habit with a short hypanthium and pale yellow flowers. However, higher pollen fertility could be obtained (58.0%) by backcrossing the triploid hybrid with *A. hypogaea* (Mallikarjuna, 2004). Another triploid hybrid was obtained by crossing *A. diogoi* with *A. glabrata*. At meiosis, a mean of 9 bivalents and 12 univalents were observed. However, a restriction in chromosome pairing was observed as the meiocytes showed more rod bivalents instead of ring bivalents (Mallikarjuna, 2004). According to Stalker and Simpson (1995) and Smartt et al. (1978), one of the genomes in *A. hypogaea* is closely related to the R genome of *A. glabrata*.

Section Arachis × *Section Procumbentes*

Mallikarjuna (2005) obtained a triploid hybrid by crossing *A. hypogaea* and *A. chiquitana*. The hybrids showed intermediate morphology. Their pollen fertility ranged from 5–24%. The hybrids were backcrossed using *A. hypogaea* as the male parent. Some of the BC_1 plants showed normal vegetative growth but did not reach the flowering stage, while the remaining hybrids flowered normally. There was significant improvement in pollen fertility in BC_1 hybrids. Mallikarjuna and Hoisington (2009) also reported intersectional hybrids of the cross *A. hypogaea* × *A. kretschmeri*. Hybrids were successfully obtained by applying growth hormones to the pollinated pistils followed by embryo rescue. Although fertile F_1 was obtained, the hybrids were slow growing. The triploid hybrids were backcrossed to *A. hypogaea*. All the BC_1F_1 plants produced single-seeded pods with prominent beaks. Another backcross was made with *A. hypogaea*. The resultant BC_2F_1 plants produced two-seeded pods, which resembled *A. hypogaea*.

Section Arachis* × *Section Erectoides

Stalker (1981) made crosses between diploid *A. rigonii* and *A.* spp. (PI 262278). Stem cuttings obtained from the resulting hybrids were treated with colchicine for doubling the chromosomes. The seeds obtained from the induced polyploidy plants were sown and crossed with two diploid species, *A. stenosperma* and *A. duranensis*. The result showed that introgression of genes from *Erectoides* is possible by using diploids as bridging species. Bera *et al.* (2002) attempted to produce interspecific hybrids between annual tetraploid *A. hypogaea* and perennial diploid *A. paraguariensis*. The hybrids were in general poor in vigor and produced 95–98% fertile pollen. Meiotic analysis indicated the presence of 19 bivalents and the plant was categorized as nullisomic. Meiosis was normal and equal distribution of chromosomes at anaphase I was observed.

Genomic Relationship within Section Arachis

To understand genomic homology, Bera et al. (2002) attempted hybridization between annual tetraploid *A. hypogaea* and perennial runner *A. cardenasii*. Hybridity was confirmed by the presence of $2n = 30$ chromosomes. At meiosis, the mean chromosomal association was 3.45 univalents, 10.42 bivalents, 1.14 trivalents, 0.26 quadrivalents, and 0.04 pentavalents. The triploid was found to be sterile and it was induced to produce fertile hexaploids. The mean chromosomal configurations in the hexaploids were 26.29 univalents, 14.49 bivalents, 0.02 trivalents, and 0.06 quadrivalents. A higher number of univalents observed in hexaploids might be due to preferential pairing (Riley and Law, 1965) and mechanical adaptation (Manton, 1950). Bera et al. (2002) observed autosyndetic tetravalents in the triploids. He suggested that the basic chromosome number of *Arachis* might be $x = 9$, and that during the course of evolution, species with the karyotype $x = 10$ originated by selective duplication of chromosomes.

Vindhiyavarman et al. (2000) reported a triploid hybrid between *A. hypogaea* and *A. cadenasii*. He backcrossed the triploid with *A. hypogaea*. The BC_1F_1 hybrids were profuse in growth habit, had dark green broader leaves, profuse flowering, and rhizomatous roots. The BC_1F_1 hybrid was amenable to multiple cutting and has been recommended as a forage legume. Mallikarjuna et al. (2004) produced triploids from cross combinations of *A. hypogaea* and *A. kempff-mercadoi*. The triploids showed a mean pollen fertility of 0–5%. Backcrosses were made with *A. hypogaea* to get BC_1F_1 seeds. A normal seed set was observed only when the triploids were used as females with *A. hypogaea* as the male parent while the reciprocal crosses failed. This may be due to a discrepancy in the chromosome number or loss of a few segments of a chromosome in the females. Pompeu (1983) attempted to cross tetraploid (*A. hypogaea*) and diploid wild species (*A. diogoi*). The triploid F_1 hybrid recorded pollen fertility of around 10%. Vindhiyavarman et al. (2000) observed low pollen fertility (5.0%) when *A. hypogaea* and an amphidiploid (*A. stenosperma* × *A. cardenasii*) were crossed.

Direct [*A. hypogaea* × (*A. cardenasii* × *A. villosa*)] and reciprocal [(*A. cardenasii* × *A. villosa*) × *A. hypogaea*] cross combinations recorded a mean pollen fertility of 30.5–38.5%. Although the reciprocal cross [(*A. stenosperma* × *A. cardenasii*) × *A. hypogaea*] showed high mean pollen fertility (38.0%), the direct cross [*A. hypogaea* × (*A. stenosperma* × *A. cardenasii*)] registered a lower mean fertility of 5.0% (Vindhiyavarman et al., 2000).

GERMPLASM RESOURCES FOR SUSTAINABLE PRODUCTION

Germplasm not only provides broad genetic diversity in groundnut breeding but also serves as a resource for identification of genes and functional genomics for sustainable groundnut production. Genetic resources comprising local land races or primitive varieties, modern and obsolete varieties, genetic stocks, breeding lines, and wild species will serve as the basis of world food security. The International Crop Research Institute for Semi-Arid Tropics (ICRISAT) is maintaining the world's largest collection of groundnut germplasm, which includes 14,968 cultivated types and 478 accessions of 48 wild species. The collections are represented by six botanical varieties, namely var. *hypogaea* (6838 accessions, 45.8%), var. *vulgaris* (5493 accessions, 36.6%), var. *fastigiata* (2351 accessions, 15.7%), var. *aequitoriana* (14 accessions, 0.10%), var. *peruviana* (251 accessions, 1.7%), and var. *hirsuta* (19 accessions, 0.13%). The US Department of Agriculture (USDA) Southern Regional Plant Introduction Station, Griffin, GA, also maintains 9027 groundnut accessions of different botanical varieties.

Core and Minicore Collections

The core collection approach was introduced to increase the efficiency of characterization and utilization of collection in gene banks. Frankel (1984) defined a core collection as a limited set of accessions representing, with minimum repetitiveness, the genetic diversity of a crop species and its wild relatives. The number of genotypes in the core collection should be a manageable number, i.e., 10% of the total collection, which allows the breeder to access important traits or combination of traits. Minicore collections consist of 10% of the core or 1% of the entire collection. Minicore collections were developed for various traits such as resistance to various diseases such as tomato spotted wilt virus (Anderson et al., 1996a), *Rhizoctonia* limb rot (Franke et al., 1999), late leaf spot (Holbrook and Anderson, 1995), *Cylindrocladium* black rot *Sclerotinia* blight and pepper spot (Damicone et al., 2010), *Aspergillus* resistance (Jiang Hui-Fang et al., 2010); drought tolerance (Upadhyaya, 2005), cold tolerance (Upadhyaya et al., 2001a), seed dormancy (Wang et al., 2012), multienvironmental conditions (Upadhyaya et al., 2005), early maturity (Upadhyaya et al., 2006) and high oil content (Holbrook et al., 1998). Genotypes with proven trait of interest may be selected carefully from minicore collections for evolving genetically superior genotypes for sustainable groundnut production. When the diversity of the parents involved in the breeding program increases, dominance and epistatic gene action play significant roles in the inheritance of quantitative traits (Halward and

Wynne, 1991). Less diverged parents showed additive mode of inheritance for quantitative traits (Isleib and Wynne, 1983).

Role of Land Races in Sustainable Groundnut Production

Land races are dynamic populations of a cultivated plant that has a historical origin, distinct identity, and lacks formal crop improvement, as well as often being genetically diverse, locally adapted, and associated with a traditional farming system (Camacho Villa et al., 2005). Land races play a vital role in sustainable agriculture. Land races offer rich sources of genes for resistance to pests, diseases, and adverse environmental conditions. They are traditional gifts to resource-poor farmers. Land races of groundnut are extremely popular in the traditional food systems of Brazilian communities. For several centuries the cultivation of land races has been carried out directly by farmers with no inputs (namely, fertilizers, pesticides, and water). Native farmers used the seeds of land races passed down from generation to generation and exchanged them between themselves (Rosa Lía Barbieri et al., 2014).

Karen A. Williams made explorations during 1995 and 1996 and collected more than 200 land races belonging to 6 botanical varieties from 3 areas of Ecuator, namely, Quito, Guayaquil, and Loja. Most of the *hirsuta* types possessed hairy stem and leaves, which may deter insects from feeding and egg laying. Also, they are tolerant to cold and drought conditions. *Hirsuta* land races were also collected from Puebla and Guanajuato, two states in the central highlands of Mexico. Local Mexican people preferred land race PI 576616 for their children due to its reduced oil content. Barrientos-Priego et al. (2002) reported the lowest oil content in *hirsuta* land races. They observed greater genetic variation for oil content within Mexican land races. Land races serve as a new source for developing a genotype with low oil content. Peruvian *hirsuta* land races are good sources for high oil content, which could be better used in breeding programs for evolving groundnut cultivar with high oil content. Sanchez-Dominguez and Williams (1993) reported superior flavor and quality of Mexican *hirsuta* land races (PI 576633, PI 576634, PI 576635, PI 576636, PI 576637, and PI 576638). Pattee et al. (1995) noticed significantly higher intensities of sweet attributes in *hirsuta* land races, which might be the reason for their preference by local Mexicans. Krapovickas et al. (2009) collected 62 land races of Bolivian origin which include 42 *hypogaea*, 17 *fastigiata*, 1 *vulgaris*, and 2 *peruviana* types. The most important Bolivian land races – namely, Crema, Colorado San Simón, Bayo Americano, Overo, and Overo Carenado – are widely cultivated throughout the country. Eleven *hypogaea* land races from the Yungas region of La Paz, 18 *hypogaea* land races from Santa Cruz and Cochabamba, and 14 landraces from the Department of Tarija were collected. Similarly, huge genetic diversity has been identified in Río Beni where 10 *fastigiata* landraces were collected. The genetic diversity available could be better used in sustainable breeding.

BREEDING FOR SUSTAINABLE PRODUCTION

Breeding for Sustainable Yield

No doubt, the Green Revolution has significantly reduced world famine and malnutrition by adopting a high-input system that involves growing high-yielding varieties and hybrids, using synthetic inorganic fertilizers and applying pesticides to boost crop production and productivity. However, with declining resource availability and a growing population, high-input agriculture has become less sustainable and unrealistic (Fess et al., 2011). Sustainable agriculture entails crop production for the human population going hand in hand with conservation of environment and natural resources needed for future generations.

Sustainable breeding programs are more focused to exploit land races or old traditional varieties possessing rich genetic diversity as they thrive well under low-input conditions and are more adapted to unpredictable environmental conditions. Modern high-yielding varieties developed under high-input conditions may not perform well under low-input or stress conditions due to their narrow genetic base. The most efficient method of enhancing the yieldability of a genotype is to screen it under low-input or stress conditions. Such varieties not only give moderate yields but they cope well in abnormal situations. For several decades, the objectives of breeding have been development of high-yielding cultivars for high-input conditions, which has resulted in loss of important alleles needed for low-input conditions (Fess et al., 2011). Sustainable breeding encompasses both evolutionary and participatory plant breeding. Evolutionary breeding is a technique in which mass selection is practiced, which is favored by natural selection. Participatory plant breeding (PPB) involves plant breeders and farmer–producers developing new crop varieties more suited to low-input and adverse environmental conditions. According to the needs of the farmer-producers, plant breeders identify the trait they consider important and attempt several potential crosses to generate adequate genetic variability. The selection in the segregating populations will be exercised by the breeder, keeping the trait of interest in mind (Witcombe et al., 1996). The final evaluation, selection, and seed multiplication will be carried out in farmers' fields. The objective of the program is to develop varieties that are more suited to local environmental conditions and production practices, ultimately increasing the sustainability and profitability of low-input systems. In most cases, the involvement of farmer–producers is restricted to selection of advanced breeding lines prior to release. This is known as participatory varietal selection (PVS). When PPB and PVS are combined, this is called the participatory crop improvement (PCI) approach. Farmer-participatory varietal selections have been made to identify ideal groundnut varieties (Nigam et al., 2005; Adu-Dapaah et al., 2007; Tulole Lugendo Bucheyeki et al., 2008; Vindhiyavarman et al., 2010).

Breeding for Biotic Stress Resistance

Breeding for Early and Late Leaf Spot Resistance

Early leaf spot caused by *Cercospora arachidicola* Hori. and late leaf spot caused by *Phaeoisariopsis personata* (Berk. & Curt.) V. Arx. [= *Cercosporidium personatum* (Berk. & Curt.) Deighton] are commonly prevalent serious diseases in groundnut worldwide. The two leaf spot diseases between them cause yield losses from 10% to 60% (Ghuge et al., 1981; McDonald et al., 1985; Waliyar, 1991). Consequently, market quality is also significantly reduced (Knauft et al., 1988). However, late leaf spot causes severe defoliation and is more destructive than early leaf spot, which further decreases yield. Cultural and management practices such as crop rotation and burial of affected crop residues reduce the inoculum load of both soil-borne fungi. Chemical control measures are not economical as they involve huge cost and the possibility of new races of pathogens evolving. The most economic method of sustainable disease control is the development of resistant cultivars.

COMPONENTS OF DISEASE RESISTANCE

Resistance to early leaf spot has been associated with resistance components: namely, longer latent period, reduced sporulation, smaller lesion diameter, and lower infection frequencies (Waliyar et al., 1993). A combination of larger incubation and latent periods, fewer lesions per leaf, smaller lesion diameter, lower sporulation index, reduced leaf area damage and disease score, which primarily depends on defoliation percentage, decides the resistance reaction of groundnut to late leaf spot disease (Dwivedi et al., 2002).

GENETICS OF RESISTANCE

Resistance components are controlled by dominant gene action while resistance is governed by 3–4 loci with recessive genes (Nevill, 1980). Resistance to early leaf spot is controlled by nonadditive gene action. In particular, epistasis plays a major role in the inheritance of resistance components (latent period) (Green and Wynne, 1986). Anderson et al. (1986) reported high broad-sense heritability for all resistance components. Resistance to late leaf spot is governed by additive gene action (Jogloy et al., 1987). Anderson et al. (1991) observed low to high broad-sense heritability for resistance components and significant narrow-sense realized heritability for late leaf spot resistance. Jogloy et al. (1999a) also noticed an additive mode of inheritance. Janila et al. (2013a) observed a combination of both nuclear and maternal gene effects for the control of late leaf spot. Epistatic effects – namely, additive × dominant, additive–maternal effects, and dominant–maternal– are involved in inheritance. High heritability and genetic advance for late leaf spot resistance have also been reported by Ashish et al. (2014).

1. *Early leaf spot*: Genotypes NC 3033, NcAc 3139, and NC 5 have been reported as resistant parents to early leaf spot (Hassan and Beute, 1977). Kornegay et al. (1980) and Johnson et al. (1986) also noticed a higher level of resistance in NC 3033 and GP-NC 343. Fávero et al., 2009 reported 22 accessions of wild species, including A genome species [*A. kuhlmannii* (11 accessions), *A. stenosperma* (5), *A. duranensis* (1), *A. helodes* (1), *A. kempff-mercadoi* (1), and non–A genome species [*A. magna* (2) and *A. gregoryi* (1)] as highly resistant to early leaf spot disease.
2. *Late leaf spot*: Mehan et al. (1996) screened 979 germplasm accessions belonging to *vulgaris*, *fastigiata*, and *hypogaea*. Of these ICG 6843, ICG 10890, ICG 11567, ICG 12112, BPZ 69, and 259-2 (red) were found to show a higher level of resistance to late leaf spot diseases. Interspecific genotypes – namely, ICGV 99010, ICGV 99005, ICGV 99006, and ICGV 99013 – which have different resistance components were also highly resistant (Dwivedi et al., 2002). Izge et al. (2007) identified ICGV-SM-93531, ICGV-IS-96802, ICGV-IS-96827, and ICGV-IS-96808 as resistant parents as they recorded the lowest levels of leaf spot incidence. Groundnut cultivars screened for resistance under natural conditions revealed a resistant reaction of FDRS 10 and a moderately resistant reaction of K-134, TCGS-156, and TPT-3 (Jyosthna et al., 2004). Lukanda et al., 2012 reported two varieties – namely, A65 and ICGM 281 – as having field-level resistance to late leaf spot. Janila et al. (2013a) noticed a higher level of resistance in ICG 11337 and suggested using the female as parent to tap into the maternal effects of resistance to late leaf spot.

In addition to the cultivated species, some wild species – namely *A. chacoense* and *A. stenosperma* – have been found to possess a higher level of resistance to both leaf spot diseases (Subrahmanyam et al., 1980). Of the 91 accessions of wild species, 54 species of *Arachis* from

A genome species [*A. stenosperma* (24 accessions), *A. kuhlmannii* (6), *A. helodes* (4), *A. simpsonii* (3), *A. diogoi* (1), *A.* aff. *diogoi* (1), *A. microsperma* (2), *A. linearifolia* (1), *A. cardenasii* (1)] and non–A genome species [*A. cruziana* (1), *A. hoehnei* (3), *A. magna* (3), *A. valida* (2), *A. batizocoi* (1), *A. williamsii* (1)] were found to be highly resistant (Fávero et al., 2009).

BREEDING STRATEGY

Since the narrow-sense heritability of resistance components is low, selection should necessarily be postponed to later generations instead of the F_2 generation. In addition, a positive correlation between disease resistance and pod length/seed size has been observed. Hence, selection for these traits will simultaneously improve disease resistance (Jogloy et al., 1987). Anderson et al. (1986) suggested early-generation selection for early and late leaf spot resistance. They also suggested the possibility of significant genetic gain when selection is based on family means instead of on individual plants within F_2 families. Jogloy et al. (1999b) also suggested early-generation selection for the components of disease resistance (disease score). Janila et al. (2013a) suggested adopting a disease scoring system as a selection criterion in the field for screening segregating populations because of its high heritability and ease of measurement. Ashish et al. (2014) also indicated the success of direct selection for resistance to late leaf spot.

Breeding for Rust Resistance

Groundnut rust caused by the fungus *Puccinia arachidis* Speg. is another economically important disease which causes yield losses often in excess of 50% (Subrahmanyam et al., 1979). Rust is equally highly destructive disease and limits commercial production in South America, Central America, and the Caribbean. In the United States, the losses due to rust at two locations of Texas were reported to be 50 and 70%, respectively (Subrahmanyam et al., 1985). In India, rust is more prevalent in rainy season crops and causes considerable yield losses of 40–55% (Ghuge et al., 1981; Mayee, 1987). In China also, yield losses due to rust range from 15% to 59% depending upon disease intensity and crop stage (Zhou and Liang-gao, 1987). The disease also affects pod, kernel, and haulm quality.

COMPONENTS OF DISEASE RESISTANCE

The ability of uredospores to germinate and the capacity of the mycelium to penetrate the leaf through the substomatal cavity and invade the leaf tissue determine the resistance reaction. Resistance is associated with failure of the fungus to invade host tissues at all infection sites resulting in low infection frequency (Subrahmanyam et al., 1983). Resistance to rust is governed by a number of components – namely, longer incubation and latent periods, very few pustules per leaf, relatively smaller pustule diameter, lower sporulation index, and reduced leaf area damage and disease score (Dwivedi et al., 2002).

GENETICS OF RESISTANCE

Bromfield and Bailey (1972) reported digenic inheritance with resistance being recessive. Nigam et al. (1980) observed the involvement of two genes for the control of rust disease. However, both digenic inheritance and trigenic inheritance were recorded from F_2 populations of rust-resistant and rust-susceptible genotypes (Kishore, 1981). Inheritance of rust disease is governed by two recessive duplicate genes (Knauft and Norden, 1983) based on studies of F_2 and F_3 generations of crosses between resistant and susceptible cultivars. Jinks and Jones (1958) reported the importance of epistasis – particularly, additive, additive × additive, and additive × dominant gene effects – for the inheritance of rust resistance. Vasanthi and Reddy (1997) noticed the involvement of 2–3 genes acting in a duplicate complementary manner in rust resistance. Motagi et al. (2013) indicated the role of independent duplicate recessive genes to control rust resistance in mutational and interspecific sources of resistance. High heritability and genetic advances indicated the success of direct selection for rust resistance (Ashish et al., 2014).

SOURCES OF RESISTANCE

Fourteen rust-resistant germplasm lines – namely, ICG 7881, ICG 7882, ICG 7883, ICG 7886, ICG 7887, ICG 7888, ICG 7889, ICG 7890, ICG 7891, ICG 7893, ICG 7894, ICG 7895, ICG 7896, and ICG 7898 – have been jointly released by the USDA and ICRISAT (Subrahmanyam et al., 1985). Mehan et al. (1996) identified four genotypes – ICG 6843, ICG 10890, ICG 11567, and ICG 12112 – as resistant to rust disease. The accessions M-5, NcAc-17090, 255/88, 259/88, and 262/88 showed resistance to rust disease (Hossain et al., 2007).

Species belonging to section Arachis (*A. batizocoi, A. duranensis, A. spegazzinii, A. correntina, A. stenosperma, A. cardenasii, A. chacoense, A. villosa*), Erectoides (*A. apressipila, A. paraguariensis*), Triseminale (*A. pusilla*), Extranervosae (*A. villosulicarpa*), Rhizomatosae (*A. hagenbeckii, A. glabrata*) are highly immune/resistant to rust (Subrahmanyam et al., 1983). A genome species (*A. khulmannii, A. helodes, A. stenosperma, A. duranensis, A. cardenasii, A. diogoi, A. kempff-mercadoi, A. linearifolia, A. microsperma, A. schininii, A. simpsonii, A. villosa*) and non–A genome species (*A. batizocoi, A. cruziana, A. hoehnei, A. magna*) showed hypersensitivity to rust (Fávero et al., 2009).

BREEDING STRATEGY

Segregating populations that show increased incubation period, reduced infection frequency, reduced lesion diameter, and low sporulation index should be selected to improve rust resistance (Subrahmanyam et al., 1983; Mehan et al., 1994).

PATHWAYS FOR INTROGRESSION OF RESISTANCE

The introgression of alien genes for resistance to both leaf spot and rust from highly resistant cross-compatible wild species could be achieved by following the four different pathways suggested by Simpson (2001) – namely, (i) triploid pathway, (ii) hexaploid pathway, (iii) amphidiploid pathway, and (iv) autotetraploid pathway.

1. *Triploid pathway*
   The cultivated species *A. hypogaea* L. ($2n = 4x = 40$) with an AABB genome is crossed with diploid wild species ($2n = 2x = 20$) with an AA or BB genome. The triploids obtained are mostly sterile but occasionally set seeds due to unreduced gametes ($2n$). The resultant hexaploids are selfed or backcrossed with cultivated species. Segregating populations show different ploidy level (2–6x). Select stable tetraploid erect plants with ($2n = 4x = 40$) in F_9 or later generations.
2. *Hexaploid pathway*
   The cultivated species *A. hypogaea* L. ($2n = 4x = 40$) with an AABB genome is crossed with diploid wild species ($2n = 2x = 20$) with an AA or BB genome. The resultant F_1 hybrid is a triploid ($2n = 3x = 30$) with an AAB or ABB genome. Treat the triploids with

colchicines to double the chromosome number. The hexaploid ($2n = 6x = 60$) with an AAAABB or AABBBB genome is backcrossed with *A. hypogaea* L ($2n = 4x = 40$). The hybrids may be selfed or backcrossed with *A. hypogaea* L. until recovery of a stable tetraploid after eliminating the excess chromosome during segregation.

3. *Amphidiploid pathway*

   The diploid wild species ($2n = 2x = 20$) with an AA genome is crossed with another diploid wild species ($2n = 2x = 20$) with an AA genome. The resultant F_1 hybrid ($2n = 2x = 20$) with an AA genome is used as male and backcrossed with a B genome species ($2n = 2x = 20$). The three-way hybrids so produced ($2n = 2x = 20$) are treated with colchicines to double the chromosome number ($2n = 2x = 40$). The resultant amphidiploids ($2n = 2x = 40$) are crossed with the cultivated species *A. hypogaea* L. ($2n = 2x = 40$). Repeat the backcrosses until stabilized tetraploids ($2n = 4x = 40$) are recovered.

4. *Autotetraploid pathway*

   The diploid wild species with an AA genome or BB genome is treated with colchicines to produce autotetraploids. The autotetraploids with an AAAA or BBBB genome are crossed with cultivated species *A. hypogaea* L. ($2n = 4x = 40$) with an AABB genome. The hybrids will have a genomic constitution of AAAB or ABBB with a normal chromosome ($2n = 4x = 40$) complement. The hybrids are backcrossed with cultivated species *A. hypogaea* L. until stabilized tetraploids ($2n = 4x = 40$) are recovered.

Breeding for aspergillus resistance

Groundnut is highly prone to infestation by two species of fungi – namely, *Aspergillus flavus* and *Aspergillus parasiticus*. *A. flavus* is a facultative, soil-borne, and air-borne saprophytic fungus capable of producing a secondary metabolite called aflatoxin in seeds (Diener et al., 1987). Germinating spores may sometimes invade groundnut flowers and pegs and be retained in developing seeds (Styer et al., 1983). Groundnut pods are in direct contact with the soil, which is a primary source of inoculum for *A. flavus* and *A. parasiticus* (Horn et al., 1995; Horn and Dorner, 1998). Infection by *A. flavus* predominantly occurs at preharvest and at postharvest during drying and storage. Preharvest contamination by *A. flavus* could be achieved either through pods which are predisposed to dry weather/drought or damage caused by white grubs, termites, millipedes, nematodes, and even sometimes natural cracks in the soil may expose seeds to fungal invasion (Schroeder and Ashworth, 1965). Larvae of the lesser corn stalk borer damage pods and kernels which in turn favors fungal invasion (Lynch, 1984; Lynch and Mack, 1995; Lynch and Wilson, 1991). Of the mycotoxins, aflatoxins are acutely toxic, carcinogenic, and immunosuppressive. This class of mycotoxin poses a serious health hazard to humans and animals that consume contaminated groundnuts. Aflatoxins are associated with various diseases such as aflatoxicosis in livestock, domestic animals, and humans. Aflatoxin acting synergistically with hepatitis virus causes human liver cancer in Africa and South East Asia (Wild and Turner, 2002). *A. flavus* produces four types of aflatoxins – namely, B1, B2, G1, and G2. Of the four, B1 is the most toxic form followed by G1, B2, and G2. About 25% of the world kernel production gets contaminated with mycotoxins every year leading to the serious and complex problems of food safety and food that is unfit for human consumption. Groundnut industries in the southeastern United States alone have invested nearly $26 million for the management of aflatoxin (Lamb and Sternitzke, 2001).

TYPES OF RESISTANCE

In general, the mechanism of resistance to *A flavus* in groundnuts is of three types: (1) resistance to *in vitro* seed colonization by *A. flavus*, (2) preharvest resistance to *A. flavus* contamination, and (3) resistance to aflatoxin production.

1. *Resistance to* in vitro *seed colonization by A.* flavus

   Since the groundnut shell is in direct contact with the soil fungal population, this acts as an effective natural barrier to prevent fungal invasion. Resistance to fungal invasion is influenced by shell structure (Zambettakis, 1975), thickness of shell (Janpengphat et al., 2014), and the presence of antagonistic microflora in the shell (Kushalappa et al., 1979; Mixon, 1980).

   Both the physical and biochemical characteristics of the testa are correlated with the resistance mechanism. Seed coat thickness or permeability (LaPrade et al., 1973), compact palisade layer, thicker and coarser waxy surface (Liang et al. 2006), and small hila (Taber et al., 1973) are associated with resistance to invasion and colonization by *A. flavus*.

   Groundnut pods that have invisible mycelia and are least affected by kernel infection can be considered tolerant to *A. flavus* colonization and infection (Olwari et al., 2013). The calcium content in groundnut shells, seed coats, and seed may be considered a good indicator for the selection of genotypes resistant to *A. flavus* infection, especially as genotype 511CC which had a higher calcium concentration reported lower incidence of *A. flavus* infection (Janpengphat et al., 2014).
2. *Preharvest resistance to A. flavus contamination*

   Temperature, soil moisture, and inoculum load of *A. flavus* in the soil during pod maturity influence resistance to preharvest aflatoxin contamination. Genotypes resistant to *in vitro* seed colonization by *A. flavus* observed the lowest level of aflatoxin (Mehan, 1989). Holbrook et al. (1994) and Anderson et al. (1996b) developed a technique to screen genotypes for resistance to preharvest aflatoxin contamination, which helps identify reliable resistant lines.
3. *Resistance to aflatoxin production*

   The failure of *A. flavus* to produce an adequate level of aflatoxin in seeds when inoculated with *Aspergillus* is another approach to screening genotypes for resistance. No germplasm has been found to be resistant to all three types of resistance. Each type of resistance is governed by different genes, which are expressed differently in different environments (Utomo et al., 1990). However, breeders are making continuous attempts to combine all three types of resistance into a single genotype.

GENETICS OF RESISTANCE

Utomo et al. (1990) observed low heritability for seed infection. However, moderate to high narrow-sense heritability estimates observed for percent severity kernel infection with *A. flavus* and aflatoxin accumulation indicated the importance of additive genetic variation for *A. flavus* resistance (Ozimati et al., 2014).

SOURCES OF RESISTANCE

The germplasm available has been screened for resistance to *A. flavus*. Singh et al. (1997) identified resistance in ICG 1122, ICG 1173, ICG 1323, ICG 1326, ICG 1859, ICG 1994, ICG 3263, ICG 3267, ICG 3336, ICG 3700, ICG 4589, ICG 4749, ICG 4888, ICG 7412, ICG 7633,

ICG 8666, ICG 9407, ICG 9610, ICG 10020, ICG 10094, and ICG 10933. Zambettakis et al. (1981), Mehan (1989), and Mehan et al. (1991) also observed resistance in ICG 1326 and ICG 4749. Mehan (1989) also noticed resistance in PI 337409, AH 78223, VAR 27, Faizpur, and Monir 240-30.

BREEDING STRATEGY

The genetic mechanism behind *Aspergillus* colonization and aflatoxin accumulation are independent (Utomo et al., 1990; Upadhyaya et al., 1997). The genes governing these two traits can be deployed in a single genetic background for evolving a resistant cultivar.

Breeding for Sclerotinia blight resistance

Sclerotinia blight disease of groundnuts is caused by the soil-borne fungi *Sclerotinia minor* Jagger and *Sclerotinia sclerotiorum* (Lib.) de Bary. *Sclerotinia* blight is also an economically important destructive disease that causes substantial yield losses and affects the quality of kernels. Yield loss due to disease incidence has been estimated to be 10%, but may exceed 50% in severe cases (Porter and Melouk, 1997). The control of *sclerotinia* blight is rather difficult and expensive.

GENETICS OF RESISTANCE

Inheritance of resistance to *Sclerotinia* blight might be controlled by several genes (Wildman et al., 1992).

SOURCES OF RESISTANCE

Growth habit is considered an avoidance mechanism for *Sclerotinia* blight (Coffelt and Porter, 1982; Chappell et al., 1995). In general, Spanish and Valencia genotypes exhibited a lower incidence of *Sclerotinia* blight than Virginia genotypes (Akem et al. 1992) as the former two possessed upright growth habits. Both growth habit and resistance to stem colonization contribute to the resistance mechanism to *Sclerotinia* blight (Chappell et al., 1995). Most late-maturing and some medium-maturing genotypes manifested a good level of resistance to *Sclerotinia* blight (Gorbet et al., 2004). The genotypes Chico, VGP 1, and NC 3033 are desirable resistant donors in breeding for resistance to *Sclerotinia* blight, while VA 71-347 could be used as a resistant variety (Coffelt and Porter, 1982). Cultivars Andru II and GP 1 showed better resistance (Gorbet et al., 2004).

BREEDING STRATEGY

Making use of highly resistant donors for hybridization and backcrossing followed by family selection could be a viable approach to transferring resistance to agronomically desirable cultivars (Wildman et al., 1992).

Breeding for Resistance to Cylindrocladium Black Rot (CBR)

Cylindrocladium black rot (CBR) or black root rot is a soil-borne fungus caused by *Cylindrocladium parasiticum* Crous, M.J. Wingfield & Alfenas (syn. *C. crotalariae* (Loos) Bell & Sobers). The disease was first noticed in the United States in southwest Georgia in 1965 (Bell and Sobers, 1966), in Japan in 1970, and in China in 2008 (Pan et al., 2009). The pathogen causes necrosis in peg, pod, and root. In severe cases, both developing and developed pods are affected (Bell and Sober, 1966; Johnston and Beute, 1975). Though it is not as severe as other soil-borne fungi, economically it is an important limiting factor reducing pod yield up to 50% (Bell et al., 1973).

GENETICS OF RESISTANCE

Resistance to CBR is governed by both morphological and physiological traits. Physiological resistance is partially controlled by cytoplasmic factors (Coffelt and Porter, 1982). Hadley et al. (1979) reported a moderate to high heritability estimate for CBR resistance. However, Green et al. (1983) observed a moderate heritability estimate for CBR resistance.

SOURCES OF RESISTANCE

Early studies suggested that Spanish bunch genotypes were generally the most resistant to CBR followed by Valencia genotypes. However, Virginia genotypes are moderately resistant (Coffelt and Garren, 1982). The first CBR-resistant germplasm – NC 3033 – was released in 1976 (Beute et al., 1976). Branch and Brenneman (2012) also identified five resistant lines – namely, Georgia Greener, Georgia-06G, Georgia-07W, Georgia-02C, and Carver. Coffelt and Garren (1982) identified Tifton-8 as the most resistant genotype of the 22 Virginia germplasm lines that showed reduced susceptibility to CBR.

BREEDING STRATEGY

Green et al. (1983) suggested early-generation selection for CBR resistance. However, Pataky et al. (1983) approved late-generation selection to get a desirable genotype with CBR resistance. Often, resistance to CBR is linked to other undesirable traits. Hence, it is necessary to break unfavorable linkages to recover high-yielding, acceptable pod and kernel traits with CBR resistance. This could be achieved by intermating desirable transgressive segregants in the F_2 population. Recurrent selection can also be adopted to obtain a resistant line. However, if the line is resistant to CBR and yield is governed by pleiotropy, it may not be possible to get a desirable resistant line (Green et al., 1983).

Breeding for Resistance to Nematodes

The root-knot nematode is the most important sedentary endoplant parasitic nematode, causing as it does severe damage to groundnuts by affecting groundnut pods. Four species of nematodes have been recognized as economically important species – namely, *Meloidogyne arenaria* (Neal) Chitwood race 1 (Koenning and Barker, 1992; McSorley et al., 1992), *M. javanica* (Tomaszewski et al., 1994), *M. haplanaria* (Eisenback et al., 2003), and *M. hapla* Chitwood (Culbreath et al., 1992; Schmitt and Barker, 1988). *Meloidogyne* species are distributed all over the world, from North, Central, and South America to Asia, Europe, Africa, and Australia (Sasser, 1977; Song et al., 1992). It is usually second-stage juveniles (J2s) that penetrate the roots and migrate intercellularly to reach vascular bundles where they establish feeding sites (Wyss et al., 1992; Williamson and Hussey, 1996). The nematode causes yield losses to the tune of 20–30%. However, under extreme cases, losses may reach 70–100% (Song et al., 1995). Root-knot nematode and CBR fungus are always found together in the soil. Infestation of root knot nematode increases the severity of CBR (Culbreath et al., 1992; Diomande and Beute 1981a, b and Diomande et al., 1981). Host plant resistance is the right approach to minimize damage achieved through integrated pest management strategies.

GENETICS OF RESISTANCE

Garcia et al. (1996) reported the role played by two dominant genes – namely, Mag and Mae. The former gene inhibits root galling and the latter inhibits egg production. However, Choi et al. (1999) observed the involvement of a single gene, two dominant genes that were heterozygous (AaBb), and two dominant genes one of which was homozygous and the other heterozygous (AABb) in the resistance mechanism. In another experiment, one dominant and one recessive gene govern nematode resistance (Church et al., 2005). However, several dominant genes are involved in the expression of resistance reaction in *A. cardenasii* (Starr and Simpson, 1991). Muitia et al. (2006) reported the role played by a single dominant gene for root-knot nematode resistance.

SOURCES OF RESISTANCE

On screening the groundnut germplasm for resistance to root-knot nematode, several resistant genotypes were identified. The donors – namely, COAN (Simpson and Starr, 2001), TxAG 6, TxAG 7 (Simpson et al., 1993), GP-NC WS 5, GP-NC WS 6 (Stalker et al., 2002), NemaTAM (Simpson et al., 2003) – were reported to be highly resistant. Resistance to *Meloidogyne arenaria* was observed in 17 genotypes – namely, Ah 25, Ah 3328, Ah 4515, Ah 6902, Ah 6719, Ah 7188, Ah 7299, AK 10-2, 55-437, C 83, C 149, EC 24118, Kigung, Khargaon 3, NcAc 50, A 3, Abuarbac (Patel et al., 1996). Osei et al. (2013) also reported resistance in three genotypes (ICGV97040, ICGV97051, and ICGV97058). A higher level of resistance was observed in NR 0812 and NR 0817 (Anderson et al., 2006), ICG 5341, ICG 6330 (Patel et al., 1996), COAN (Simpson and Starr, 2001), NemaTAM (Simpson et al., 2003) for *M. javanica*, while resistance was noticed in 26 genotypes – namely, ICG 411, ICG852, ICG 859, ICG 1268, ICG 2248, ICG 2496, ICG 3053, ICG 3104, ICG 6323, ICG 6826, ICG 10047, Apexy, B 1, C 162, C 166, Dangi, EC 85994, JH 223, KG 61-22, PI 268594, PI 270787, S 7-2-1, S -24-3, U 4-7-3, No. 923, No. 523 – for *M. javanica* (Patel et al., 1996).

Wild species serve as potential resistant donors for the transfer of nematode resistance. Baltensperger et al. (1986) identified *A. glabrata*, an incompatible wild species, as a resistant source for *M. hapla*. Diploid species – namely, *A. batizocoi* Krapov. & W.C. Greg., *A. cardenasii* Krapov. & W.C. Greg., and *A. diogoi* are resistant to root-knot nematode (Nelson et al., 1989). *A. cardenasii* Krapov. & W.C. Greg. showed a hypersensitive reaction by developing necrosis as a result of penetration but there was no postinfectional development. Resistance is governed by a few genes and is easy to manipulate in a breeding program. However, *A. batizocoi* Krapov. & W.C. Greg. did not show a hypersensitive reaction. However, nematode development was slow. Resistance is governed by a larger number of genes and difficult to manipulate in a breeding program (Nelson et al., 1990). Another diploid species, *A. stenosperma* Krapov. & W.C. Greg., exhibited an early hypersensitive response (Morgante et al., 2013).

BREEDING STRATEGY

For root-knot nematode resistance, reduction in root galling (gall index based on percentage of galled roots), egg count per root system, egg count per gram of root, and numbers of galls in the segregating populations may be identified as factors important to developing resistant lines.

Breeding for Resistance to Insect Pests

Nearly 400 species of insect pests threaten preharvest groundnuts (Smith and Barfield, 1982) and postharvest groundnuts are threatened by more than 80 insect pests (Redlinger and

TABLE 5.1 Insect Pests Causing Damage to Groundnuts

Category	Type of insects	Examples
1. Sucking pests (cellular feeders)	(a) Intracellular feeders	Aphids (*Aphis craccivora* Koch.) and jassids (*Empoasca fabae* [Harris], *Empoasca kerri* Pruthi, groundnut hopper, *Hilda patruelis* Stal.)
	(b) Intercellular feeders	Thrips (*Calliothrips* spp., *Frankliniella fusca* Hinds, *Frankliniella schultzei* Trybom, *Scirtothrips dorsalis* (Hood), *Caliothrips indicus* (Bagnall), two-spotted spider mite (*Tetranychus urlicae* Koch.)
2. Defoliators	(a) Foliage feeders/leaf eaters	Red hairy caterpillar (*Amsacta albistriga*), tobacco army worm (*Spodoptera litura* Fabricius), cotton leafworm *Spodoptera littoralis* (Biosd.), fall army worm *Spodoptera frugiperda* (J.E. Smith); granulate cutworm *Agrotis subterranea* (F.); corn earworm *Helicoverpa zea* (Boddie), leaf miner (*Aproaerema modicella* Deventer), velvet bean caterpillar *Anticarsia gemmatalis* (Hubner).
	(b) Soil pests/root, peg or pod feeders	Wireworm (*Conoderus scissus* Schaffer), corn rootworm (*Diabrotica undecimpunctata howardi* Barber), lesser cornstalk borer *Elasmopalpus lignosellus* (Zeller), pod borer (*Euborellia stali* Dohn.), white grub *Lachnosternaconsanguinea* (Blanchard), termites (*Odontotermes* spp., *Microtermes* spp.), millipedes (*Peridontopyge* spp.), earwigs, ants (*Dorylus orientalis* Westwood), coleopterous and lepidopterous larvae).
3. Transmitters of virus		Aphids – groundnut rosette and peanut stripe virus, tobacco thrips and the western flower thrips *Frankliniella occidentalis* (Pergande), tomato spotted wilt virus, and peanut yellow spot virus.
4. Storage pests		Bruchid beetle *Caryedon serratus* (Ol.), red flour beetle *Tribolium castaneum* (Herbst), rice moth *Corcyra cephalonica* (Stainton), and pod sucking bug *Elasmolomus sordidus* (Fabricius).

Davis, 1982). However, a few insects are considered economically important as they cause considerable yield loss. Changing cropping patterns, multiple cropping seasons, and climatic change all aggravate the insect pest problem. In general, the insect pests causing damage to groundnuts can broadly be grouped into four categories (Table 5.1).

MECHANISM OF RESISTANCE

Long and dense trichomes on leaves are linked to higher levels of jassid resistance. A host suitability index has been extensively used to classify the resistant and susceptible genotypes of fall armyworms (Lynch et al., 1981). A severity index can also be used to discriminate resistant and susceptible individuals. Vegetative traits such as main stem thickness and hairiness, leaflet shape and length, leaf hairiness, basal leaflet width, and stipule adnation length and width as well as reproductive traits such as hypanthium length, standard petal length and petal markings, and peg length can be used as markers for resistance to *Heliothis armigera* and *Spodoptera litura* (Sharma et al., 2003).

GENETICS OF RESISTANCE

Dwivedi et al. (1986) reported additive genetic variance for long trichomes in genotypes for jassid resistance and a high general combining ability for the presence of trichomes.

SOURCES OF RESISTANCE

1. *Jassids.* Resistance to jassids has been found in three breeding lines – namely, NcAc 2214, NcAc 2240, NcAc 2230 (Amin et al., 1985). Stalker and Campbell (1983) noticed resistance in wild species of groundnuts – namely, *A. correntina*, *A. cardenasii*, *A. duranensis* and *A. villosa*. Similarly, a higher level of resistance has been reported in certain wild species – namely, *A. cardenasii*, *A. duranensis*, *A. kempff-mercadoi*, *A. monticola*, *A. stenosperma*, *A. paraguariensis*, *A. pusilla* and *A. triseminata* (Sharma et al., 2003).
2. *Thrips.* Amin et al. (1985) identified six genotypes as resistant sources (NC Ac 2242, NC Ac 2214, NC Ac 2243, NC Ac 2240, NC Ac 2232, and NC Ac 2230). Wild species also showed a considerable level of resistance to this insect pest, especially *A. batizocoi*, *A. pusilla*, *A. paraguariensis*, *A. repens*, *A villosa* (Stalker and Campbell, 1983). Amin (1985) also observed resistance in two wild species – namely, *A. chacoense* and *A. duranensis*.
3. *Aphids.* Genes for aphid resistance are harbored in a number of wild species – namely, *A. chacoense*, *A. villosa*, *A. correntina*, and *A. glabrata* (Amin, 1985).
4. *Tobacco army worm.* Sharma et al. (2002) identified ICGV 86031 as a resistant genotype for tobacco army worm. Prasad and Gowda (2006) also reported resistance in three genotypes – namely, NcAc 343, Mutant 28-2, and R 9227. Resistance to tobacco army worm has been reported in several wild species including *A. kemph-mercadoi*, *A. paraguariensis*, *A. major*, *A. appresipila*, and *A. ipaensis*, (Stevenson et al., 1993; Sharma et al., 2003).
5. *Fall army worm.* Genotypes NC 6, Tifton 8 (Todd et al., 1991) and IAC 22, IAC 886 (Campos et al., 2010) have been reported as resistant to fall army worm. Lynch et al (1981) observed resistance in *A. villosa*, *A. burkartii* Handro, and *A. correntina*.
6. *Leaf miner.* Groundnut lines ICG 5040, ICGS 2741, ICG (FDRS) 10, NC AC 17090, and GBPRS 312 (ICGV 86031) have been found to be highly resistant to leaf miner (Rao et al., 1998). Rao (2000) observed resistance in NCAc 2575. Resistance to leaf miner has also been reported in some wild species – namely, *A. cardenasii*, *A. duranensis*, *A. kempff-mercadoi*, *A. monticola*, *A. stenosperma*, *A. paraguariensis*, *A. pusilla* and *A. triseminata* (Sharma et al., 2003).
7. *Helicoverpa* spp. Sharma et al. (2003) identified resistance in some wild species of groundnut – namely, *A. cardenasii*, *A. duranensis*, *A. kempff-mercadoi*, A. *monticola*, *A. stenosperma*, *A. paraguariensis*, *A. pusilla*, and *A. triseminata*.
8. *Termites.* Some genotypes – namely, NC Ac 2243, NC Ac 2240, and NC 343 – have been reported to have durable resistance to termites (Amin et al., 1985).

BREEDING STRATEGY

Cultivars/genotypes identified as resistant sources to insect pests have expressed partial resistance to insects, but this is often insufficient for introgression into adapted cultivars. Moreover, when crosses were made between resistant and susceptible genotypes, the level of resistance became diluted in the breeding program (Linch and Stalker, 1997). Resistant genotypes in which different genes have various mechanism of resistance could be crossed to enhance the level and effectiveness of resistance (Prasad and Gowda, 2006).

Breeding for Abiotic Stress Resistance

Breeding for Drought Tolerance

Drought is a condition or stress experienced by crops during their life cycles which is due to insufficient rainfall or when water availability is limited. Drought may be of different duration or intensity. Drought may occur during any of the different phonological phases of yields. Drought contributes to losses in excess of 6.7 million metric tons in annual world groundnut production (Subbarao et al., 1995). Drought resistance/tolerance is not governed by a single trait; it is a cumulative effect of various physiological, biochemical, and anatomical traits.

Classification of Drought Pattern

Drought patterns can be broadly grouped into three types – namely, (i) early-season drought, (ii) mid-season drought, and (iii) end-season drought.

1. *Early-season drought*

   Moisture stress in the early stages of crop growth (i.e., 20–25 days of stress) is also called preflowering drought. Early-season drought followed by copious irrigation results in higher productivity (Puangbut et al., 2009, 2010; Nageswara Rao et al., 1988). Increased productivity might be due to profuse and uniform flowering which in turn leads to more pegs and pods resulting in higher pod yield at harvest (Nautiyal et al., 1999; Nageswara Rao et al., 1988). Preflowering drought has enhanced yield levels to the tune of 13–57% in some genotypes (Nageswara Rao et al., 1985; Nautiyal et al., 1999; Puangbut et al., 2010). Increased yield is due to increases in vegetative growth, crop growth rate (CGR), pod growth rate (PGR) (Nageswara Rao et al., 1985; Nautiyal et al., 1999), chlorophyll concentration, stomatal conductance, photosynthesis, and relative growth rate (RGR) (Awal and Ikeda, 2002). Imposition of early-season drought followed by rewatering results in increases in SPAD chlorophyll meter reading (SCMR), leaf area index (LAI), and biomass production. Increase in yield has been associated with high biomass production after recovery combined with increased green leaf area and concentration of leaf chlorophyll (Puangbut et al., 2009).
2. *Mid-season drought*

   Mid-season drought significantly affects the pegging, pod, and kernel development stages which in turn lead to lower productivity. Late flowering and pod formation are the most sensitive stages to moisture stress. Drought during late flowering, pod formation, and pod-filling periods has reduced yields significantly (Roy et al., 1988). The relationship between root length density percentage (%RLD) in the deeper soil layer (60–90 cm) and yield has been found to be significant and positive indicating that %RLD in the lower layer is an important trait affecting pod yield under mid-season drought conditions (Jongrungklang et al., 2012).
3. *End-season drought*

   End-season drought is also called terminal drought. It affects kernel development which in turn reduces crop productivity: the content of total oil, linoleic, and behenic fatty acid gets reduced. However, total protein, stearic, and oleic fatty acid increases as a result of end-season drought (Dwivedi et al., 1996). End-season drought predisposes aflatoxin contamination.

Genetics of Drought Resistance/Tolerance

HERITABILITY OF DROUGHT RESISTANCE TRAITS

Heritability estimates in the broad sense were low to intermediate for drought resistance traits (Painawadee et al., 2009). However, Songsri et al. (2008) observed high heritability for biomass, pod yield, DTI (drought tolerance index) (pod yield), DTI (biomass), harvest index, specific leaf area (SLA), and SCMR. Puangbut et al. (2011) observed higher estimates for harvest index (HI), SLA, and SCMR. Higher estimates of heritability suggested that selection for drought tolerance traits might be more advantageous than selecting for yield per se because of higher heritability and lower G × E interaction. The heritability estimate is higher for SCMR (Upadhyaya, 2005).

GENOTYPIC CORRELATION AMONG DROUGHT RESISTANCE TRAITS

The HI had a high genetic correlation with pod yield and number of mature pods under end-season conditions. The SCMR had a moderate genetic correlation with numbers of mature pods indicating that these traits were most important to yield under early-season drought (ESD) conditions (Puangbut et al., 2011). ESD increased the SCMR, transpiration efficiency (TE), and root dry weight (RDW), but it reduced SLA. Genotypes having high TE also had larger root systems under drought conditions. RDW had a significant correlation with TE (Puangbut et al., 2009). Upadhyaya (2005) observed a negative correlation between the SCMR and SLA. The SCMR also had a stronger association than SLA with economic traits like pod yield and 100-seed weight. This will have implications in breeding programs, as selection for the SCMR would result in greater correlated response to yield and other economic traits than selection for SLA. Further, the SCMR showed a reduced genotype × season interaction. Songsri et al. (2008) observed a strong and negative genotypic correlation between SLA and SCMR under drought conditions. Further, they observed a strong positive correlation between the DTI for pod yield and the DTI for biomass. TE was strongly and positively correlated with N_2 fixation. High N_2 fixation might contribute to high TE under ESD conditions. Enhanced N_2 fixation also increased pod yield (Puangbut et al., 2011). SLA, relative water content (RWC), chlorophyll density, and drought stress ratings were positively correlated with lower preharvest aflatoxin contamination (PAC). Breeding for drought tolerance using these traits as selection criteria may help to accelerate progress in developing resistance to PAC (Girdthai et al., 2010).

GENOTYPIC CORRELATION BETWEEN DROUGHT RESISTANCE TRAITS, YIELD, AND YIELD COMPONENTS

The SCMR had higher genotypic correlations with pod yield, biomass, and other agronomic traits under stress conditions than did SLA (Songsri et al., 2008). The SCMR is an indicator of the photosynthetically active light transmittance characteristics of leaves and is positively correlated with chlorophyll content (Akkasaeng et al., 2003), chlorophyll density (Arunyanark et al., 2008), and water use efficiency (WUE) (Sheshshayee et al., 2006).

GENOTYPIC CORRELATION BETWEEN YIELD AND YIELD COMPONENTS

The correlation coefficient revealed that 100-seed weight, total number of kernels per plant, total number of pods per plant, and biomass had a high positive correlation with seed yield under drought stress conditions. Path analysis suggested that 100-seed weight, total number of kernels per plant, total number of pods per plant, and biomass had the greatest positive direct effects on seed yield (Sadeghi and Niyaki, 2012). Sadeghi and Niyaki (2012) also observed the high positive correlation of seed yield, total number of kernels per plant, plant height, and 100-seed weight with oil yield. Path analysis revealed that 100-seed weight, seed yield, and biomass had the greatest positive direct effects on seed yield.

GENE EFFECTS OF DROUGHT TOLERANCE TRAITS

Nigam et al. (2001) observed the importance of additive gene effects in the expression of SLA and HI. The additive × additive type of epistasis, which can be a fixable component, was also noticed. Jayalakshmi et al. (1999) also reported the predominance of additive gene action for SLA. Similarly, the importance of additive gene action for SLA and SCMR has also been recognized by Chuni et al. (2005). Upadhyaya et al. (2010) observed an additive effect for the SCMR at 80 days after seeding (DAS) while the predominance of dominant effects with duplicate epistasis was observed for the SCMR at 60 DAS. The predominance of additive effects for the SCMR at 80 DAS suggested effective selection could be practiced even in an early generation, whereas for the SCMR or SLA at 60 DAS it would be better to defer selection to a later generation.

Sources of Drought Resistance

Upadhyaya (2005) identified 18 diverse accessions for drought tolerance which were superior or similar to the best control cultivar for SCMR and SLA. The selected accessions also had good pod yield (ICG 862, ICG 8285, ICG 11855), shelling percentage (ICG 118, ICG 2106, ICG 5827), 100-seed weight (ICG 11219, ICG 11855), protein content (ICG 5236, ICG 6654), and oil content (ICG 118, ICG 862, ICG 11219). Nigam et al. (2002) selected several lines with superior performance under different kinds of drought (ICG 3086, ICG 3141, ICG 2738, ICGV 91151, ICGV 94127, ICGV 92209, and ICGV 91109 for mid-season drought; ICG 2213, ICGS 76, ICGV 90226, ICGV 91074, ICGV 91185, ICGV 91192, ICGV 92004, ICGV 92022, ICGV 92023, ICGV 92028, ICGV 92029, and ICGV 92033 among others for end-season drought). They are available for inclusion in a breeding program. Jharna et al. (2001) identified three entries – namely, 3203, ICGV 93269, and ICGV 93232 – as drought-tolerant genotypes based on biochemical screening. The released cultivar KKU 60 was identified as drought tolerant by SCMR, SLW, and pod yield (Boontang et al. 2010).

Breeding Strategy

The conventional method of breeding for drought tolerance is based on yield *per se* in a given drought environment. Direct selection for yield involves huge resource investment and poor repeatability of results owing to the high G × E interaction. Parents belonging to a similar age group (110–120 days) are used in hybridization programs. Such parents produce progeny whose maturity is relatively uniform, a good basis on which selection can be exercised. At ICRISAT, maturity is calculated according to growing degree days (GDD). GDD varies slightly between locations. Both trait-based selection and empirical-based selection approaches are followed extensively to identify genotypes that have drought tolerance.

Breeding for Earliness

Earliness is one of the key objectives of groundnut improvement programs. Earliness in maturity is very desirable to escape end-season drought or to avoid *in situ* germination by escaping rain or, sometimes, escaping the onset of diseases/pests. Short-growing-season groundnuts can be accommodated in any cropping sequence or raised in residual moisture situations after harvesting the rice crop in South East Asia. More often, these early-maturity cultivars are raised as an intercrop with sugarcane, which gives some remuneration to farmers. Yield is often positively correlated with crop duration. Hence, the yield of early-maturing genotypes cannot be compared with normal-maturing genotypes. Earliness imposes a substantial yield penalty (Evans, 1993). A certain quantity of yield must be compromised when earliness is considered. Of the botanical types, Valencia and Spanish bunch varieties are classified as early varieties, while Virginia bunch or Runner are late maturers. Jogloy et al. (2011) suggested the possibility of reducing the reproductive duration and maturity duration by improving the CGR, PGR, and partitioning efficiency (PE). Selection for higher CGR, PGR, and PE results in shorter reproductive duration and, ultimately, shorter crop duration. Groundnut maturity is assessed based on a predetermined cumulative thermal time (CTT), GDD, or accumulated heat units expressed as °Cd. At ICRISAT, CTT is followed to screen larger segregating populations. The CTT of 1240, 1470, 1710, and 1950 °Cd is equivalent to 75, 90, 105, and 120 days after planting during the rainy season and 105, 120, 133, and 145 days after planting in the postrainy season at ICRISAT (Vasudeva Rao et al., 1992).

Genetics of Earliness

Earliness has been found to be recessive to late maturity (Badami, 1923, 1928). Incomplete dominance for late maturity over earliness has been reported by Patel et al. (1936) and Hassan (1964). Holbrook et al. (1989) observed the role of 4 or 5 genes with complete dominance for late maturity over earliness. Days to first flowering was conditioned by a single gene with additive gene action. Days to accumulation of 25 flowers from the appearance of the first flower were governed by 3 independent genes with complete dominance. The segregation of 13:3 for late and early maturity indicated the components were governed by dominant–recessive epistasis (Upadhyaya and Nigam, 1994). Varman and Raveendran (1996) noticed the involvement of two recessive genes that act in an additive manner to influence days to first flowering. Jogloy et al. (2011) reported a negative correlation of maturity between pod yield and harvest index, which suggested the possibility of selecting early-maturity cultivars without affecting pod yield or harvest index.

Sources of Earliness

The genotype Chico is an early-maturing Spanish bunch type which takes 75–80 days to mature. Although, it is early maturing, its extremely small pods are not preferred commercially. The cultivar Gangapuri is a Valencia bunch that matures in 80–90 days and the cultivar JL 24 is a Spanish bunch variety that matures in 85–90 days. They are important early-maturing cultivars in India. Early-maturity cultivars – namely, ICGV IS-96808, ICGV IS-96855, ICGV 96891, and ICGV IS 96894 – with reasonable pod yield and rosette resistance are highly preferred in West Africa. Upadhyaya et al. (2006) observed early maturity in 17 *vulgaris* – namely, ICG 3200, ICG 3540, ICG 3631, ICG 4729, ICG 5512, ICG 5560, ICG 5881, ICG 9427, ICG 9968, ICG 11914, ICG 13585, ICG 13606, ICG 13647, ICG 14390, ICG 14788, ICG 14814, ICG 14815 – and 4 *fastigiata* – namely, ICG 4558, ICG 4890, ICG 9930, ICG 11605 – land races from 13 countries. Such land races could be used effectively in breeding programs to develop early-maturing genotypes.

Breeding Strategies

Nigam and Aruna (2008) suggested the following strategies to evolve early-maturing cultivars: (i) selection for low base temperature (Tb) and CTT for various phenological stages, (ii) selection for tolerance to high temperature, (iii) selection for photoperiod-insensitive genotypes, (iv) selection for high CGR, (v) selection for high WUE, (vi) diversification of sources of earliness in maturity, and (vii) evaluation in target environments/cropping systems.

Breeding for Improved Quality Characteristics

Groundnut quality is decided by the way in which groundnuts are consumed and utilized. Groundnuts are consumed in many ways. As a food crop, they are directly consumed as a wholesome food. As a feed supplement, groundnuts affect quality as a result of having less oil but a high protein, energy, and mineral composition. As an edible oilseed crop, the high oil content of groundnuts is desirable. For enhanced shelf life, high oleic to linoleic (O/L) acid ratios are important quality parameters. As a confectionery food, kernel color, size, shape, flavor, texture, greater proportion of sound mature kernel (SMK), higher 100-kernel weight (>55 g), more sugar (>11%), more protein (>24%), higher blanchability (>60%), and low oil (<45%) are considered important quality traits (Janila et al., 2013b). Quality attributes can be broadly classified into (i) morphological and physical quality and (ii) biochemical quality.

Morphological and Physical Quality

These traits are related to the physical appearance of the kernel and play a vital role in consumer preference/acceptance. A high proportion of SMK is considered most important because SMK is an attribute that appeals to the eyes.

1. *Kernel color*. Basically, five testa colors are common to groundnuts – namely, white, tan, pink, and red, as well as purple in nonvariegated types. However, variegated testa with red and white stripes, purple and white stripes, purple stripes on tan, and white spots on red are also observed. Kernel color is controlled by qualitative genes. Solid colors are governed by 12 genes (F_1, F_2, D_1, D_2, R_1, R_2, R_3, P, W_1, W_2, Wh_1, and Wh_2) and variegated colors are controlled by five genes (V, Vsp_1, Vsp_2, Wsp_1, and Wsp_2) (Branch, 2011). Light brown to reddish brown is the color of preference and each type has its own specific use.
2. *Kernel size*. Uniformity in size and ease of processing are major criteria. Virginia kernels (large red skinned) are suitable for cocktail and salted nuts; Runner market-type kernels (medium sized) and Spanish kernels (small) are suitable for snacks, peanut butter, oil, and confectionery purposes; Valencia kernels (long shelled) are mostly preferred for in-shell roasting. Seed size has been reported as being governed qualitatively by a single gene (Balaiah et al., 1977), 3 genes (Pattanashetti et al., 2008), and 5 genes (Martin, 1967). Kernel size is inherited quantitatively, and both additive and nonadditive gene action were involved in the expression of this trait (Parmar et al., 2000). Godoy and Norden, (1981) and Parameshwarappa et al. (2005) reported additive genetic variance for kernel size. Nonadditive gene action has also been reported for this trait (Hariprasanna et al., 2008). This character is highly amenable for selection. Selection could be easily made to improve kernel size through a conventional breeding approach. However,

selection may be postponed to a later generation until homozygosity reaches all the loci governing kernel size.

3. *Kernel shape and texture.* Round or elongated kernels with tapering ends, which are easy for blanching, are mostly preferred. Kernels should be firm and crispy.
4. Absence of extraneous material or immature/shriveled kernels.
5. *Moisture content.* This determines the extent of splitting and the growth of mould.

Biochemical Quality

OIL QUALITY

Genotypes with high oil content are preferred for oil extraction. However, low oil content is desirable for confectionery groundnuts. Since groundnuts are predominantly considered oilseed crops, oil is deemed the main component followed by protein and carbohydrate. Groundnuts contain 40–54% oil by weight depending upon the type and variety (Nagaraj, 1995). The oil is rich in monounsaturated fatty acids (MUFA) and polyunsaturated fatty acids (PUFA). The presence of various desirable fatty acids determines the quality of oil. The most important unsaturated fatty acids are oleic (C18:1), linoleic (C18:2), and palmitic (C16:0) acids, which contribute nearly 80% of total fatty acids (Dwivedi et al. 1993). Along with the above fatty acids, behenic (C22:0), stearic (C18:0), arachidic (C20:0), eicosenoic (C20:1), lignoceric (C24:0), γ-linolenic (C18:3), erucic (C22:1), and myristic (C14:0) are also present in trace quantities. A high O/L ratio is associated with longer shelf life and lower rancidity (Braddock et al. 1995). High-oleic groundnuts lower low-density lipoprotein (LDL), improve high-density lipoprotein (HDL), and have a good flavor (O'Byrne et al., 1997). Due to having high levels of polyunsaturated fatty acids, groundnut oil is highly susceptible to rancidity and off-flavors (Ory et al., 1992; Braddock et al., 1995; Mugendi et al., 1998).

GENETICS OF OIL QUALITY. Both qualitative and quantitative modes of inheritance have been reported in the literature. The high-oleic trait was found to be conditioned by two recessive genes ol_1 and ol_2 in a natural mutant (Moore and Knauft, 1989). López et al. (2001) found a digenic mode of inheritance for the high-oleic trait, which was conditioned by two loci designated Ol_1 and Ol_2.

Epistasis also plays a major role in inheritance of the high-oleic trait. Partial dominant, additive, and additive × additive types of epistasis have been reported for this trait (Isleib et al., 2006; Mercer et al., 1990; Singkham et al., 2011; Upadhyaya and Nigam, 1999). Isleib et al. (2006) noticed the role the pleiotropic effect plays in the expression of oleic and linoleic acid. However, Aruna and Nigam (2009), Singkham et al. (2011), and Singkham et al. (2012) observed an additive mode of inheritance for this trait. High heritability for the high-oleic trait indicated the additive mode of inheritance (Singkham et al., 2010). They also suggested selection of the high-oleic trait for early generation. Since the association between oleic and linoleic acid is negative, selection for high-oleic acid will automatically result in a lower level of linoleic acid (Raheja et al., 1987; Singkham et al., 2010). Wang et al. (2011) also noticed a negative correlation between oleic acid and linoleic and palmitic acids, which suggested the possibility of improving the content of oleic acid.

SOURCES FOR HIGH-OLEIC FATTY ACID. In general, Virginia and Runner-type groundnuts have a higher oleic acid concentration and a lower linoleic acid concentration (Raheja et al., 1987). Sharma et al. (1985) reported higher oleic and lower linoleic content in the induced mutant TG-18 from its parent Spanish Improved. Norden et al. (1987) identified two high-oleic groundnut lines – namely, F 435-2-1 and F 435-2-2 – from a naturally occurring mutant with a high-oleic (80.0%) and low-linoleic (2.0%) level. In China, the first high-oleic groundnut variety Huayu 32 was developed for commercial cultivation. SunOleic 97R was evolved from a natural mutant (Gorbet and Knauft, 2000), while Georgia-02C was developed through induced mutation by gamma irradiation (Branch, 2003).

PROTEIN QUALITY

Groundnuts contain more high-quality protein than is available from any animal source. Protein content ranges from 22% to 36.0% (Nagaraj, 1995). Groundnut protein is composed of 90% globulins and 10% albumins. The two major globulins are arachin and conarchin/nonarachin which constitute 63% and 33%, respectively. Groundnut protein contains various essential amino acids. Acidic amino acids such as aspartic acid, glutamic acid, and arginine are relatively higher and account for 45% of total amino acids. However, lysine, methionine, and threonine are deficient. Hence, it is nutritively poor. Groundnut protein is easily digestible (97%) with a biological value of 57.9% (Nagaraj, 1995). Parameshwarappa et al. (2005) observed the predominant role played by additive gene action in protein content. Hence, selection might be possible to evolve genotypes with higher protein content.

CARBOHYDRATE QUALITY

Groundnuts contain 10–20% carbohydrate. Sugars such as sucrose, glucose, fructose, galactose, stachyose, and raffinose are present. However, sucrose is considered the most important sugar in imparting the sweet attribute to groundnuts. Groundnuts contain 2.86–6.35% sucrose (Nagaraj, 1995). The sweet attribute is a heritable trait (Pattee et al., 2000). Eating quality is highly correlated with sweetness. Luo et al. (2004) observed higher consumption of in-shell groundnuts containing >6% sucrose. Gocho (1992) recorded the highest sucrose content (9.0%) in groundnut seed. However, Wang et al. (2011) reported an exorbitantly higher sucrose content of 14.65% in J 81 and confirmed the results by seven repetitions.

VITAMINS AND MINERAL NUTRITION QUALITY

Groundnuts are excellent sources of vitamins B_1 (thiamine), B_2 (riboflavin), and B_3 (niacin). In addition, considerable amounts of vitamin E and minerals (Ca, Mg, Fe, and Zn) are present. The enhanced levels of iron and zinc in groundnuts could be of benefit in finding a cheaper and more sustainable approach to fighting malnutrition.

ANTINUTRITIONAL QUALITY

Plants have an inherent ability to produce substances (antinutritional factors) to protect themselves from being eaten. Antinutritional factors limit the availability of nutrients to the body and sometimes destroy essential vitamins. To get the full potential of food substances, antinutritional factors must be removed. Groundnuts are low in oxalic acid, phytic acid, cyanogenic glycoside, and tripsin inhibitors; they are also aflatoxin free. These are some of the key quality attributes.

ORGANOLEPTIC QUALITY

Flavor and aroma are other important sensory quality traits that influence the processing of groundnuts. Composition determines the flavor and texture of roasted groundnuts.

Certain amino acids such as aspartic acid, asparagines, glutamic acid, phenylalanine, and histidine were found to be the precursors of typical groundnut flavors, while threonine, tyrosine, and lysine contributed to atypical groundnut flavors (Newell et al., 1967). The genotype, environment, and their interaction also predominantly control groundnut flavor. Johnsen et al. (1988) proposed seven descriptors to identify flavor quality: roast peanutty, woody/skins/hulls, raw/green/beany, cardboardy, sweet, salty, and astringent. Groundnuts with a sweet and roast-peanutty flavor are considered ideal. Kernel maturity is directly correlated with ideal flavor (McNeill and Sanders, 1996; McNeill and Sanders, 1998, Mozingo et al., 1991; Pattee et al., 1995; Sanders et al., 1989 a,b; Sanders and Bett, 1995). Kernels obtained from well-developed and matured pods taste better than those from immature pods. Immature kernels are reported to have a fruity fermented flavor (McNeill and Sanders, 1996).

MOLECULAR BREEDING

Although conventional breeding is the most widely globally used approach to introgress the trait of interest, it has its own demerits in being labor intensive and time consuming. Occasionally, the targeted trait may have an undesirable linkage to other traits, which poses a difficulty in groundnut breeding. Modern biotechnological tools help precisely transfer the desirable gene without altering much of the genetic background of the donor cultivar/recipient variety. The two most important biotechnological approaches for groundnut improvement are (i) marker-assisted breeding and (ii) genetic transformation.

Marker-Assisted Breeding

Molecular markers are also called DNA markers. They are highly polymorphic and not influenced by environmental factors. With the advent of molecular markers, targeted traits can be deployed to a breeding line without altering much of the genetic background. This process is known as genomic-assisted breeding (Varshney et al., 2005).

Of the 8 quantitative trait loci (QTLs) observed for resistance to late leaf spot (LLS), 2 (*Qlf2* and *Qlf7*) explained 32.7% of total phenotypic variation for lesion frequency per square centimeter of leaf, 3 (*Qlc4*, *Qlc7*, and *Qlc8*) observed 40.3% of total phenotypic variation for lesion number per leaf, and 3 QTLs (*Qlad3*, *Qlad6*, and *Qlad8*) explained 31.2% of total phenotypic variation for leaf area damage (Cuc et al., 2008). Khedikar et al. (2010) reported that one major QTL linked with rust resistance explained 6.90–55.20% variation. They observed that linking the simple sequence repeat (SSR) marker IPAHM 103 to this QTL could be used for marker-assisted backcrossing. Shoba et al. (2012) reported 32–52% phenotypic variation for LLS resistance and suggested using the primer PM 384100 for marker-assisted breeding. Sujay et al. (2012) identified 28 QTLs for LLS and 15 QTLs for rust. A major QTL for LLS, namely QTLLLS01, explained 10.27–62.34% of phenotypic variance. Four new markers (GM2009, GM1536, GM2301, and GM2079) for rust resistance have been identified as explaining 82.96% of phenotypic variance. Rust and LLS-linked markers were found in linkage group 03 (8 SSRs) and linkage group 04 (3 SSRs) (Navinbhai Gajjar et al., 2014). On the other hand, Varshney et al. (2014) observed one dominant (IPAHM 103) and three codominant (GM 2079, GM 1536, and GM 2301) markers for rust resistance in one QTL region in linkage group

AhXV which explained phenotypic variation of 82.66%. Three markers – namely, RKN410, RKN440, and RKN229 – were linked to an identified resistance gene from *A. cardenasii* or *A. diogoi* for root-knot nematode (Burow et al., 1996). The dominant polymerase chain reaction (PCR)-based marker 197/909 has been reported to be linked to nematode resistance (Chu et al., 2007). Five SSR markers (PM375, PM36, PM45, pPGPseq8D9, and Ah-041) have been associated with seed length, pod length, number of pods per plant, 100-seed weight, maturity, and oil content (Selvaraj et al., 2009)

Genetic Transformation

Genetic transformation is another approach to transferring the gene of interest into the desired genotype. Attempts were made to transfer genes for resistance to leaf spot, rust, *Sclerotinia* blight, peanut stem necrosis disease, white grub, leaf-feeding insects, as well as drought and salinity tolerance. Asif et al. (2011) introduced *AtNHX1*, a vacuolar-type Na(+)/H(+) antiporter gene driven by a 35S promoter using *Agrobacterium*-mediated transformation. Transgenic plants were resistant to both salt and water stress conditions. They possessed higher salt and proline levels in their leaves than nontransformed controls. A barley oxalate oxidase gene conferring resistance to *Sclerotinia minor* was introduced to three groundnut cultivars – namely, Wilson, Perry, and NC-7. The transformed plants were regenerated from embryogenic cultures. Lesion size was reduced significantly (75–97%) in transgenic plants (Livingstone et al., 2005). Rana and Mohanty (2012) used the groundnut cultivar Smruti for *Agrobacterium*-mediated transformation. Using the strain GV3107 with the binary vector pCAMBIA2300 containing the gene that increases tolerance to moisture stress, *CBF3*, driven by a stress-inducible promoter rd29A. Of the different explants used, leaf explants performed the best. Transgenics showed increased tolerance to moisture stress (Rana and Mohanty, 2012). Vadawale et al. (2012) introduced the *cox* gene in the groundnut cultivar GG 20 through *Agrobacterium tumefaciens* LBA 4404 with binary vector pHS724. Transformed plants showed improved salinity tolerance. Geng et al. (2013) transformed groundnut plants with the synthetic gene *cry8Ea1* against white grubs. The gene was toxic to *Holotrichia parallela* larvae and was expressed in chimeric groundnut roots using the *Agrobacterium* rhizogene-mediated transformation system. The rice chitinase gene *Rchit* was introduced into three varieties of groundnuts. Transformed plants showed 2–14 fold greater chitinase activity in the leaves. The seeds of transgenics showed reduced *A. flavus* infection during *in vitro* seed inoculation. The leaves of transgenics showed longer incubation, longer latent period, and lower infection frequencies for LLS and rust disease (Prasad et al., 2013). A coat protein (CP) gene was introduced to the groundnut cultivar Kadiri 6 (K6) and cultivar Kadiri 134 (K134) via *Agrobacterium*-mediated transformation of deembryonated cotyledons and immature leaves of groundnuts. Transgenics showed traces or no accumulation of tobacco streak virus (TSV) (Mehta et al., 2013). The peroxisomal ascorbate peroxidase gene *SbpAPX* was transformed to a local variety of groundnut through *Agrobacterium*-mediated transformation. Under salt stress conditions, transgenics remained green while nontransgenics showed bleaching and yellowish leaves (Singh et al., 2014). Pandurangaiah et al. (2014) reported MuNAC4, a nitrogen assimilation control (NAC) transcription factor from horse gram that confers drought tolerance. Transgenics were tolerant to long-term desiccation stress by reducing damage to membrane structures and enhancing osmotic adjustment and antioxidative enzymes under stress. Plant β-1,3-glucanases are commonly involved in disease resistance. The gene was transformed into the groundnut variety Huayu 22 by *Agrobacterium*-mediated transformation. Transgenics were found to be more resistant to *Cercospora personata* than nontransgenics (Qiao et al., 2014). Transgenics harboring the chimeric Bt gene, *cry1X*, evolved from the cultivar TMV 2 were found to be highly resistant against the two major defoliators of groundnuts (*Helicoverpa armigera* and *Spodoptera litura*) (Entoori et al., 2008). Transgenic groundnut plants expressing the oxalate oxidase gene for resistance to *Sclerotinia* blight were developed. Transgenics showed a higher level of resistance and 78–90% lower area under the disease progress curve (AUDPC) (Telenko et al., 2011). Transgenics were evolved by *Agrobacterium*-mediated transformation using the gene *cry1AcF* against the defoliator *Spodoptera litura*. Highest mean larval mortalities of 80–85% with mean larval mortality of 16.25% were recorded in transgenic plants, showing the effectiveness of the introduce gene (Keshavareddy et al., 2013).

Acknowledgment

The author acknowledges his wife Mrs. A. Ezhil, Assistant Professor, TKG Government Arts College, Vridhachalam and his daughters M. Giritha and M. Megna for extending support in the preparation of manuscript. He also acknowledges Dr. V. Thiruvengadam, Assistant Professor (Plant Breeding and Genetics) for his timely help in fine-tuning the manuscript.

References

Adu-Dapaah, H.K., Asumadu, H., Lamptey, J.N.L., Haleegoah, J., Asafo-Adjei, B., 2007. Farmer participation in groundnut varietal selection. Proc. African Crop Sci. Conf. 8, 1435–1439.

Akem, C.N., Melouk, H.A., Smith, O.D., 1992. Field evaluation of peanut genotypes for resistance to Sclerotinia blight. Crop Prot. 11, 345–349.

Akkasaeng, C., Vorasoot, N., Jogloy, S., Patanothai, A., 2003. Relationship between SPAD readings and chlorophyll contents in leaves of peanut (*Arachis hypogaea* L.). Thai J. Agric. Sci. 36 (3), 279–284.

Amin, P.W., 1985. Resistance of wild species of groundnut to insect and mite pests. In: International Workshop on Cytogenetics of Arachis, 31 Oct–2 Nov 1983, Patancheru, Andhra Padesh, India.

Amin, P.W., Singh, K.N., Dwivedi, S.L., Rao, V.R., 1985. Sources of resistance to the jassid (*Empoasca kerri* Pruthi), thrips (*Frankliniella schultzei* [Trybom]) and termites (*Odontoterrnes* sp.) in groundnut (*Arachis hypogaea* L.). Peanut Sci. 12, 58–60.

Anderson, W.F., Wynne, J.C., Green, C.C., Beute, M.K., 1986. Combining ability and heritability of resistance to early and late leaf spot of peanut. Peanut Sci. 13, 13–14.

Anderson, W.F., Holbrook, C.C., Wynne, J.C., 1991. Heritability and early generation selection for resistance to early and late leaf spot in peanut. Crop Sci. 31, 588–593.

Anderson, W.F., Holbrook, C.C., Culbreath, A.K., 1996a. Screening the peanut core collection for resistance to tomato spotted wilt virus. Peanut Sci. 23 (1), 57–61.

Anderson, W.F., Holbrook, C.C., Wilson, D.M., 1996b. Development of greenhouse screening for resistance to *Aspergillus parasiticus* infection and preharvest aflatoxin contamination in peanut. Mycopathologia 135, 115–118.

Anderson, W.F., Timper, P., Holbrook, C.C., 2006. Registration of root-knot nematode resistant peanut germplasm lines NR 0812 and NR 0817. Crop Sci. 46, 481–482.

Aruna, R., Nigam, S.N., 2009. Inheritance of fatty acid content and related quality traits in groundnut, *Arachis hypogaea* L. J. Oilseeds Res. 26 (1), 10–17.

Arunyanark, A., Jogloy, S., Akkasaeng, C., Vorasoot, N., Kesmala, T., Nageswara Rao, R.C., Wright, G.C., Patanothai, A., 2008. Chlorophyll stability is an indicator of drought tolerance in peanut. J. Agron. Crop Sci. 194, 113–125.

Ashish, J., Nadaf, H.L., Gangadhara, K., 2014. Genetic Analysis of rust and late leaf spot in advanced generation recombinant Inbred Lines of groundnut (*Arachis hypogaea* L.). Int. J. Genet. Eng. Biotechnol. 5 (2), 109–114.

Asif, M.A., Zafar, Y., Iqbal, J., Iqbal, M.M., Rashid, U., Ali, G.M., Arif, A., Nazir, F., 2011. Enhanced expression of AtNHX1, in transgenic groundnut (*Arachis hypogaea* L.) improves salt and drought tolerence. Mol. Biotechnol. 49 (3), 250–256.

Awal, M.A., Ikeda, T., 2002. Recovery strategy following the imposition of episodic soil moisture deficit in stands of peanut (*Arachis hypogaea* L.). J. Agron. Crop Sci. 188, 185–192.

Badami, V.K. 1923. Hybridization work on groundnut. Agrlculture Department Report for 1922–23, Mysore, India, pp. 29–30.

Badami, V.K. 1928. *Arachis hypogaea* – (the groundnut). PhD thesis. Cambridge (Eng.), University Library. Inheritance studies, pp. 297–374.

Balaiah, C., Reddy, P.S., Reddi, M.V., 1977. Genetic analysis in groundnut: I. Inheritance studies on 18 morphological characters in crosses with gujarat narrow leaf mutant. Proc. Indian Acad. Sci. 85 B (5), 340–350.

Baltensperger, D.D., Prine, G.M., Dunn, R.A., 1986. Root-knot nematode resistance in *Arachis glabrata*. Peanut Sci. 13, 78–80.

Barrientos-Priego, L., Isleib, T.G., Pattee, H.E., 2002. Variation in oil content among Mexican and Peruvian runner *hirsuta* peanut landraces and virginia-type *hypogaea* lines. Peanut Sci. 29, 72–77.

Bell, D.K., Sobers, E.K., 1966. A peg, pod, and root necrosis of peanuts caused by a species of Calonectria. Phytopathology 56, 1361–1364.

Bell, D.K., Locke, B.J., Thompson, S.S., 1973. The status of Cylindrocladium black rot of peanut in Georgia since its discovery in 1965. Plant Dis. Rep. 57, 90–94.

Bera, S.K., Paria, P., Radhakrisnan, T., 2002. Wide hybridization in *Arachis hypogaea*: genetic improvement and species relationship. Int. Arachis Newsl. 22, 37–39.

Beute, M.K., Wynne, J.C., Emery, D.A., 1976. Registration of NC 3033 peanut germplasm. Crop Sci. 16, 887.

Birthal, P.S., Rao, P.P., Nigam, S.N., Bantilan, C.S., Bhagavatulu, S., 2010. Groundnut and Soybean Economies in Asia: Facts, Trends and Outlook. International Crops Research Institute for the Semi-Arid Tropics, Patancheru.

Boontang, S., Girdthai, T., Jogloy, S., Akkasaeng, C., Vorasoot, N., Patanothai, A., Tantissuwichwong, N., 2010. Responses of Released cultivars of peanut to terminal drought for traits related to drought tolerance. Asian J. Plant Sci. 9, 423–431.

Braddock, J.C., Sims, C.A., O'Keefe, S.F., 1995. Flavor and oxidative stability of roasted high oleic acid peanut. J. Food Sci. 60, 489–493.

Branch, W.D., 2003. Registration of 'Georgia-02C' peanut. Crop Sci. 43, 1883–1884.

Branch, W.D., 2011. First 100 Years – Inheritance of Testa Color in Peanut (*Arachis hypogaea* L.). Crop Sci. 51, 1–4.

Branch, W.D., Brenneman, T.B., 2012. New Sources of *Cylindrocladium* Black Rot Resistance among Runner-Type Peanut Cultivars. Peanut Sci. 39 (1), 38–42.

Bromfield, K.R., Bailey, W.K., 1972. Inheritance of resistance to *Puccinia arachidis* in peanut. Phytopathology 62, 748.

Burow, M.D., Simpson, C.E., Paterson, A.H., Starr, J.L., 1996. Identification of RAPD markers in peanut (*Arachis hypogaea* L.) associated with root-knot nematode (*Melaidogyne arenaria*) resistance, derived from *A. cardenasii*. Mol. Breed. 2, 307–309.

Camacho Villa, T.C., Maxted, N., Scholten, M.A., Ford-Lloyd, B.V., 2005. Defining and identifying crop landraces. Plant Genet. Res. 3 (3), 373–384.

Campos, A.P., Boiça Júnior, A.L., Ribeiro, Z.A., 2010. Não preferência para oviposição e alimentação de *Spodoptera frugiperda* (J. E. Smith, 1797) (Lepidoptera: Noctuidae) por cultivares de amendoim. Arquivos do Instituto Biológico 77, 251–258.

Chappell, G.F., Shew, B., Ferguson, B., Ferguson, J.M., Beute, M.K., 1995. Mechanisms of Resistance to *Sclerotinia* minor in Selected Peanut Genotypes. Crop Sci. 35, 362–696.

Choi, K., Burow, M.D., Church, G., Burow, G., Paterson, A.H., Simpson, C.E., Starr, J.L., 1999. Genetics and mechanism of resistance to *Meloidogyne arenaria* in peanut germplasm. J. Nematol. 31, 283–290.

Chu, Y., Holbrook, C.C., Timper, P., Ozias-Akins, P., 2007. Development of a PCR-based molecular marker to select for nematode resistance in peanut. Crop Sci. 47 (2), 841–845.

Chuni, Lal., Rathnakumar, A.L., Hariprasanna, K., Gor, H.K., Chikani, B.M., 2005. Promising parental lines for the development of high water-use efficient groundnut varieties. Int. Arachis Newsl. 25, 8–10.

Church, G.T., Starr, J.L., Simpson, C.E., 2005. A recessive gene for resistance to *Meloidogyne arenaria* in interspecific *Arachis* spp. hybrids. J. Nematol. 37, 178–184.

Coffelt, T.A., Garren, K.H., 1982. Screening for resistance to *Cylindrocladium* black rot in peanuts (*Arachis hypogaea* L.). Peanut Sci. 9 (1), 1–5.

Coffelt, T.A., Porter, D.M., 1982. Screening peanuts for resistance to *Sclerotinia* blight. Plant Dis. 66, 385–387.

Cuc, L.M., Mace, E.S., Crouch, J.H., Quang, V.D., Long, T.D., Varshney, R.K., 2008. Isolation and characterization of novel microsatellite markers and their application for diversity assessment in cultivated groundnut (*Arachis hypogaea* L.). BMC Plant Biol. 8, 55–65.

Culbreath, A.K., Beute, M.K., Shew, B.B., Barker, K.R., 1992. Effects of *Meloidogyne hapla* and *M. arenaria* on black rot severity in new *Cylindrocladium*-resistant peanut genotypes. Plant Dis. 76, 352–357.

Damicone, J.P., Holbrook, Jr., C.C., Smith, D.L., Melouk, H.A., Chamberlin, K.D., 2010. Reaction of the core collection of peanut germplasm to Sclerotinia blight and pepper spot. Peanut Sci. 37, 1–11.

Diener, U.L., Cole, R.J., Sanders, T.H., Payne, G.A., Lee, L.S., Klich, M.A., 1987. Epidemiology of aflatoxin formation by *Aspergillus flavus*. Ann. Rev. Phytopathol. 25, 249–270.

Dillehay, T.D., Rossen, J., Andres, T.C., Williams, D.E., 2007. Pre-ceramic adoption of Peanut, Squash, and Cotton in Northern Peru. Science, 1890–1893, 316, 5833.

Diomande, M., Beute, M.K., 1981a. Effects of *Meloidogyne hapla* and *Macroposthonia ornate* on *Cylindrocladium* black rot of peanuts. Phytopathology 71, 491–496.

Diomande, M., Beute, M.K., 1981b. Relation of *Meloidogyne hapla* and *Macroposthonig ornata* populations to *Cylindrocladium* black rot in peanuts. Plant Dis. 65, 339–342.

Diomande, M., Black, M.C., Beute, M.K., Barker, K.R., 1981. Enhancement of *Cylindrocladium crotalariae* root rot by *Meloidogyne arenaira* (race 2) on a peanut cultivar resistant to both pathogens. J. Nematol. 13, 321–327.

Dwivedi, S.L., Amin, P.W., Rasheedunisa, Nigam, S.N., Nagabhushanam, G.V.S., Rao, V.R., Gibbons, R.W., 1986. Genetic analysis of trichome characters associated with resistance to jassid (*Empoasca kerri* Pruthi) in peanut. Peanut Sci. 13, 15–18.

Dwivedi, S.L., Nigam, S.N., Jambunathan, R., Sahrawat, K.L., 1993. Effect of genotypes and environments on oil content and oil quality parameters and their correlation in peanut (*Arachis hypogaea* L.). Peanut Sci. 20, 84–89.

Dwivedi, S.L., Nigam, S.N., Nageswara Rao, R.C., Singh, U., Rao, K.V.S., 1996. Effect of drought on oil, fatty acids and protein contents of groundnut (*Arachis hypogaea* L.) seeds. Field Crops Res. 48, 125–133.

Dwivedi, S.L., Pande, S., Rao, J.N., Nigam, S.N., 2002. Components of resistance to late leaf spot and rust among interspecific derivatives and their significance in a foliar disease resistance breeding in groundnut (*Arachis hypogaea* L.). Euphytica 125, 81–88.

Eisenback, J.D., Bernard, E.C., Starr, J.L., Lee, T.A., Tomaszewski, E.K., 2003. *Meloidogyne haplanarian*.sp. (Nematoda: Meloidogynidae), a root-knot nematode parasitizing peanut in Texas. J. Nematol. 35, 395–403.

Entoori, K., Sreevathsa, R., Arthikala, M.K., Ananda Kumar, P., Vinoda Kumar, A.R., Madhusudhan, B., Makarla, U., 2008. A chimeric cry1X gene imparts resistance to *Spodoptera litura* and *Helicoverpa armigera* in the transgenic groundnut. EurAsian J. BioSci. 2, 53–65.

Evans, L.T., 1993. Crop Evolution, Adaptation and Yield. Cambridge University Press, Cambridge.

FAOSTAT, 2012. Groundnut world production. Food and Agricultural Organization, Rome, Italy. http://www.faostat.fao.org.

Fávero, A.P., Moraes, S.A., Garcia, A.A.F., Valls, J.F.M., Vello, N.A., 2009. Characterization of rust, early and late leaf spot resistance in wild and cultivated peanut germplasm. Scientia Agricola (Piracicaba, Braz.) 66 (1), 110–117.

Fess, T.L., Kotcon, J.B., Vagner, A., Benedito, V.A., 2011. Crop breeding for low input agriculture: a sustainable response to feed a growing world population. Sustainability 3, 1742–1772.

Franke, M.D., Brenneman, T.B., Holbrook, C.C., 1999. Use of a core collection to identify resistance to *Rhizoctonia* limb rot in peanut and the correlation of resistance with hypocotyle infections of seedlings. Plant Dis. 83, 944–948.

Frankel, O.H., 1984. Genetic perspectives of germplasm conservation. In: Arber, W.K., Llimensee, K., Peacock, W.J., Starlinger, P. (Eds.), Genetic Manipulation: Impact on Man and Society. Cambridge University Press, Cambridge, UK, pp. 161–170.

Garcia, G.M., Stalker, H.T., Shroeder, E., Kochert, G., 1996. Identification of RAPD, SCAR, and RFLP markers tightly linked to nematode resistance genes introgressed from *Arachis cardenasii* into *Arachis hypogaea*. Genome 39, 836–845.

Geng, L., Chi, J., Shu, C., Gresshoff, P.M., Song, F., Huang, D., 2013. A chimeric cry8Ea1 gene flanked by MARs efficiently controls *Holotrichia parallela*. Plant Cell Rep. 32, 1211–1218.

Ghuge, S.S., Mayee, C.D., Godbole, G.M., 1981. Assessment of losses in groundnut due to rust and tikka leaf spot. Indian Phytopathol. 34, 179–182.

Girdthai, T., Jogloy, S., Vorasoot, N., Akkasaeng, C., Wongkaew, S., Holbrook, C.C., Patanothai, A., 2010. Heritability of, and genotypic correlations between, aflatoxin traits and physiological traits for drought tolerance under end of season drought in peanut (*Arachis hypogaea* L.). Field Crop Res. 118, 169–176.

Gocho, H., 1992. Breeding for eating quality of groundnut in Japan. In: Nigam, S.N. (ed.). Groundnut – a global perspective: proceedings of an International Workshop, Nov 25–29, 1991, ICRISAT Center, India (Patancheru 502 324, Andhra Pradesh, India. pp. 367–372.

Godoy, I.J.G., Norden, A.J., 1981. Shell and seed size relationship in peanuts. Peanut Sci. 8, 21–24.

Gorbet, D.W., Knauft, D.A., 2000. Registration of 'SunOleic 97R' peanut. Crop Sci. 40, 1190–1191.

Gorbet, D.W., Kucharek, T.A., Shokes, E.M., Brenneman, T.B., 2004. Field evaluations of peanut germplasm for resistance to stem rot. Peanut Sci. 31, 91–95.

Green, C.C., Wynne, J.C., 1986. Diallel and generation mean analyses for the components of resistance to *Cercospora arachidicola* in peanut. Theor. Appl. Genet. 73, 228–235.

Green, C.C., Wynne, J.C., Beute, M.K., 1983. Genetic variability and heritability in early generation crosses of large-seeded virginia-type peanuts with lines resistant to *Cylindrocladium* black rot. Peanut Sci. 10, 47–51.

Hadley, B.A., Beute, M.K., Wynne, J.C., 1979. Heritability of *Cylindrocladium* black rot resistance in peanut. Peanut Sci. 6, 51–54.

Halward, T.M., Wynne, J.C., 1991. Generation means analysis for productivity in two diverse peanut crosses. Theor. Appl. Genet. 82, 784–792.

Halward, T., Stalker, T., LaRue, E., Kochert, G., 1991. Genetic variation detectable with molecular markers among unadapted germplasm resources of cultivated peanut and related wild species. Genome 34, 1013–1020.

Hariprasanna, K., Chuni, Lal., Radhakrishnan, T., Gor, H.K., Chikani, B.M., 2008. Analysis of diallel cross for some physical-quality traits in peanut (*Arachis hypogaea* L.). Euphytica 160, 49–57.

Hassan, M.A., 1964. Genetic, floral, biological and maturity studies in groundnut. MS thesis (unpublished). Ranchi Agriculture College, Ranchi University, Kanke, Ranchi, India.

Hassan, H.N., Beute, M.K., 1977. Evaluation of resistance to *Cercospora* leaf spot in peanut germplasm potentially useful in a breeding program. Peanut Sci. 4, 78–83.

Holbrook, C.C., Anderson, W.F., 1995. Evaluation of a core collection to identify resistance to late leaf spot in peanut. Crop Sci. 35, 1700–1702.

Holbrook, C.C., Kvien, C.S., Branch, W.D., 1989. Genetic control of peanut maturity as measured by the hull-scrape procedure. Oleagineux 44, 359–364.

Holbrook, C.C., Matheron, M.E., Wilson, D.W., Anderson, W.F., Will, M.E., Norden, A.J., 1994. Development of a large-scale field screening system for resistance to pre-harvest aflatoxin contamination. Peanut Sci. 21, 20–22.

Holbrook, C.C., Bruniard, J., Moore, K.M., Knauft, D.A., 1998. Evaluation of the peanut core collection for oil content. Agron. Abs., p. 159.

Horn, B.W., Dorner, J.W., 1998. Soil populations of *Aspergillus* species from section *Flavi* along a transect through peanut-growing regions of the United States. Mycologia 90, 767–776.

Horn, B.W., Greene, R.L., Dorner, J.W., 1995. Effect of corn and peanut cultivation on soil populations of *Aspergillus flavus* and *A. parasiticus* in southwestern Georgia. Appl. Environ. Microbiol. 61, 2472–2475.

Hossain, M.D., Rahman, M.Z., Abeda, K., Rahman, M.M., 2007. Screening of groundnut genotypes for leaf spots and rust resistance. Int. J. Sust. Crop Prod. 2, 7–10.

Hull, E.H., Carver, W.A., 1938. Peanut Improvement. Florida Agric. Exp. Station Ann. Rep., 39–40.

Husted, L., 1933. Cytological Studies on the Peanut, *Arachis*. I. Chromosome number and morphology. Cytologia 5, 109–117.

Husted, L., 1936. Cytological Studies on the Peanut, *Arachis*. II. Chromosome number, morphology and their application to the problem of the origin of the cultivated forms. Cytologia 7, 396–423.

Ishag, H.M., 2000. Phenotypic and yield response of irrigated groundnut cultivars in a hot environment. Exp. Agric. 36, 303–312.

Isleib, T.G., Wynne, J.C., 1983. Heterosis in testcrosses of 27 exotic peanut cultivars. Crop Sci. 23, 832–841.

Isleib, T.G., Wilson, R.F., Novitzky, W.P., 2006. Partial dominant, pleiotropism, and epistasis in the inheritance of the high-oleate trait in peanut. Crop Sci. 46, 1331–1335.

Izge, A.U., Mohammed, Z.H., Goni, A., 2007. Levels of variability in groundnut (*Arachis hypogaea* L.) to Cercospora leaf spot disease – Implication for selection. Afr. J. Agric. Res. 2 (4), 182–186.

Jambunathan, R., 1991. Groundnut quality characteristics, In: Uses of Tropical Grain Legumes: Proceedings of a Consultants Meeting, ICRISAT, Patancheru, Andhra Pradesh, India. March 27–30, 1989. pp. 267–275.

Janila, P., Ramaiah, V., Rathore, A., Upakula, A., Reddy, R.K., Waliyar, F., Nigam, S.N., 2013a. Genetic analysis of resistance to late leaf spot in interspecific groundnuts. Euphytica 193, 13–25.

Janila, P., Nigam, S.N., Pandey, M.K., Nagesh, P., Varshney, R.K., 2013b. Groundnut improvement: use of genetic and genomic tools. Plant Sci. 4, 1–16.

Janpengphat, P., Senthong, C., Meechoui, S., Ingram, K.T., 2014. Calcium in Shell, Seed Coat, and Seed in Relation to Resistance of Peanut Genotypes to *Aspergillus flavus* Infection. Chiang Mai J. Sci. 41 (3), 531–541.

Jayalakshmi, V., Rajareddy, C., Reddy, P.V., Nageswara Rao, R.C., 1999. Genetic analysis of carbon isotope discrimination and specific leaf area in groundnut (*Arachis hypogaea* L.). J. Oilseeds Res. 16, 1–5.

Jharna, D.E., Chowdhury, B.L.D., Haque, M.A., Bhuiyan, M.R.H., Husain, M.M., 2001. Biochemical screening of some groundnut (*Arachis hypogaea* L.) genotypes for drought tolerance. J. Biol. Sci. 1 (11), 1009–1011.

Jiang Hui-Fang, Ren Xiao-Ping, Wang Sheng-Yu, Zhang Xiao-Jie, Huang Jia-Quan, Liao Bo-Shou, Holbrook, C.C., Upadhyaya, H.D., 2010. Development and evaluation of peanut germplasm with resistance to *Aspergillus flavus* from core collection. Acta Agronomica Sinica 36 (3), 428–434.

Jinks, J.L., Jones, R.M., 1958. Estimation of the components of heterosis. Genetics 43, 223–234.

Jogloy, S., Wynne, J.C., Beute, M.K., 1987. Inheritance of late leaf spot resistance and agronomic traits in peanut. Peanut Sci. 14, 86–90.

Jogloy, S., Pensuk, V., Patanothai, A., Wongkaew, S., 1999a. Combining ability of resistance to late leafspot and rust of peanut. Thai J. Agric. Sci. 32, 281–287.

Jogloy, S., Trisuvanwat, W., Wongkaew, S., Patanothai, A., 1999b. Combining ability and heritability of rust resistance and agronomic traits in peanut (*Arachis hypogaea* L). Thai J. Agric. Sci. 32, 487–494.

Jogloy, C., Jaisil, P., Akkasaeng, C., Kesmala, T., Jogloy, S., 2011. Heritability and Correlation for Components of Crop Partitioning in Advanced Generations of Peanut Crosses. Asian J. Plant Sci. 10, 60–66.

Johnsen, P.B., Civille, G.V., Vercellotti, J.R., Sanders, T.H., Dus, C.A., 1988. Development of a lexicon for the description of peanut flavor. J. Sensory Studies 3, 9–17.

Johnson, C.S., Phips, P.M., Beute, M.K., 1986. *Cercospora* leaf spot management decisions: Uses of a correlation between rainfall and disease severity to evaluate the Virginia leaf spot advisory. Phytopathology 76, 860–863.

Johnston, S.A., Beute, M.K., 1975. Histopathology of *Cylindrocladium* black rot of peanut. Phytopathology 65, 649–653.

Jongrungklang, N., Toomsan, B., Vorasoot, N., Jogloy, S., Boote, K.J., Hoogenboom, G., 2012. Classification of root distribution patterns and their contributions to yield in peanut genotypes under mid-season drought stress. Field Crops Res. 127, 181–190.

Jyosthna, M.K., Eswara Reddy, N.P., Chalam, T.V., Reddy, G.L.K., 2004. Morphological and biochemical characterization of Phaeoisariopsis personata resistant and susceptible cultivars of groundnut (*Arachis hypogaea*). Plant Pathol. Bull. 13, 243–250.

Keshavareddy, G., Rohini, S., Ramu, S.V., Sundaresha, S., Kumar, A.R.V., Kumar, P.A., Udayakumar, M., 2013. Transgenics in groundnut (*Arachis hypogaea* L.) expressing cry1AcF gene for resistance to Spodoptera litura F. Physiol. Mol. Biol. Plants 19, 343–352.

Khedikar, Y.P., Gowda, M.V.C., Sarvamangala, C., Patgar, K.V., Upadhyaya, H.D., Varshney, R.K., 2010. A QTL study on late leaf spot and rust revealed one major QTL for molecular breeding for rust resistance in groundnut (*Arachis hypogaea* L.). Theor. Appl. Genet. 121, 71–984.

Kishore, B., 1981. Rust inheritance studies in groundnut (*Arachls hypogaea* L.). Unpublished thesis, Andhra Pradesh Agricultural University, India.

Knauft, D.A., Norden, A.J., 1983. Inheritance of rust resistance in peanuts. Proc. Am. Peanut Res. Edu. Soc. 15, 96, (abstract).

Knauft, D.A., Gorbet, D.W., Norden, A.J., 1988. Yield and market quality of seven peanut genotypes as affected by leaf spot disease and harvest date. Peanut Sci. 15, 9–13.

Kochert, G., Stalker, H.T., Gimenes, M., Galgaro, L., Lopes, C.R., Moore, K., 1996. RFLP and cytogenetic evidence on the origin and evolution of allotetraploid domesticated peanut *Arachis hypogaea* L. (Leguminosae). Am. J. Bot. 83, 1282–1291.

Koenning, S.R., Barker, K.R., 1992. Relative damage functions and reproductive potentials of *Meloidogyne arenaria* and *M. hapla* on peanut. J. Nematol. 24, 187–192.

Kornegay, J.L., Beute, M.K., Wynne, J.C., 1980. Inheritance of resistance to *Cercospora arachidicola* and *Cercosporidium personatum* in six Virginia-type peanut lines. Peanut Sci. 7, 4–9.

Krapovickas, A., Gregory, W.C., 1994. Taxonomía del género *Arachis* (Leguminosae). Bonplandia 8, 1–186.

Krapovickas, A., Gregory, W.C., 2007. Taxonomy of the genus *Arachis* (Leguminosae). Translated by Williams, D.E., Simpson, C.E., Bonplandia 16 (Suppl.), 1–205.

Krapovickas, A., Rigoni, V.A., 1949. Chromosomas de un a especie silvestre de *Arachis*. Idia Buenos Air es 2, 23–24, (abstract).

Krapovickas, A., Rigoni, V.A., 1951. Estudios citológicos en el género *Arachis*. Revista de Investigaciones Agricolas 5, 289–293.

Krapovickas, A., Rigoni, V.A., 1954. Cruzamientos interespecíficos en *Arachis* y consideraciones sobre el origen del maní cultivado. In Anais do 28 Congreso Panamericano de Agronomía, Piracicaba, Sao Paulo, Brazil, pp. 266–267.

Krapovickas, A., Rigoni, V.A., 1957. Nuevas especies de *Arachis* vinculadas al problema del origen del mani. Darwiniana 11, 431–455.

Krapovickas, A., Vanni, R.O., Pietrarelli, J.R., Williams, D.E., Simpson, C.E., 2009. The peanut landraces from Bolivia. Bonplandia 18 (2), 95–189.

Kushalappa, A.C., Bartz, J.A., Norden, A.J., 1979. Susceptibility of pods of different peanut genotypes to *Aspergillus flavus* group fungi. Phytopathology 69, 159–162.

Lamb, M.C., Sternitzke, D.A., 2001. Cost of aflatoxin to the farmer, buying point, and sheller segments of the Southeast United States peanut industry. Peanut Sci. 28, 59–63.

LaPrade, J.C., Bartz, J.A., Norden, A.J., Demuynk, T.J., 1973. Correlation of peanut seed-coat surface wax accumulations with tolerance to colonization by *Aspergillus flavus*. Proc. Am. Peanut Res. Edu. Soc. 5, 89–94, (abstract).

Lavia, G., 1998. Karyotypes of *Arachis palustris* and *A. praecox* (Section *Arachis*), two species with basic chromosme number x = 9. Cytologia 63, 177–181.

Lavia, G.I., Fernández, A., 2004. Karyotypic studies in *Arachis hypogaea* L. varieties. Caryologia. 57 (4), 353–359.

Lavia, G., Ortiz, A., Fernández, A., 2009. Karyotypic studies in wild germplasm of *Arachis* (Leguminosae). Genet. Res. Crop Evol. 56, 755–764.

Liang, X.Q., Luo, M., Guo, B.Z., 2006. Resistance mechanisms to *Aspergillus flavus* infection and aflatoxin contamination in peanut (*Arachis hypogaea*). Plant Pathol. 5, 115–124.

Lim, E.S., Hamdan, O., 1984. The reproductive characters of four varieties of groundnuts (*Arachis hypogaea* L.). Pertanica 7, 25–31.

Linch, R.E., Stalker, H.T., 1997. Evaluation of *Arachis hypogaea* × *A. cardenasii* interspecific lines for resistance to insect pests. Peanut Sci. 24, 89–96.

Livingstone, D.M., Hampton, J.L., Phipps, P.M., Grabau, E.A., 2005. Enhancing resistance to *Sclerotinia minor* in peanut by expressing a barley oxalate oxidase Gene. Plant Physiol. 137, 1354–1362.

Lu, J., Mayer, A., Pickersgill, B., 1990. Stigma morphology and pollination in *Arachis* L. (Leguminosae). Ann. Bot. 66, 73–82.

Lukanda, L.T., Nkongolo, K.K.C., Kalonji-Mbuyi, A., Kizungu, F R.V., 2012. Epidemiology of the groundnut (*Arachis hypogaea* L.) leaf spot disease: genetic analysis and developmental cycles. Am. J. Plant Sci. 3, 582–588.

Luo, H., Zhou, G.Y., Fang, H.B., Dong, G.J., 2004. Studies on biochemical characters relative to direct edible peanut. J. Peanut Sci. 4, 1–4.

Lynch, R.E., 1984. Damage and preference of lesser cornstalk borer (Lepidoptera: Pyralidae) larvae for peanut pods in different stages of maturity. J. Econ. Entomol. 77, 360–363.

Lynch, R.E., Mack, T.P., 1995. Biological and biotechnical advances for insect management in peanut. In: Pattee, H.E., Stalker, H.T. (Eds.), Advances in Peanut Science. American Peanut Research and Education Society, Stillwater, Oklahoma, pp. 95–159.

Lynch, R.E., Wilson, D.M., 1991. Enhanced infection of peanut, *Arachis hypogaea* L., seeds with *Aspergillus flavus* group fungi due to external scarification of peanut pods by the lesser cornstalk borer, *Elasmopalpus lignosellus* (Zeller). Peanut Sci. 18, 110–116.

Lynch, R.E., Branch, W.D., Garner, J.W., 1981. Resistance of *Arachis* species to the fall armyworm, *Spodoptera frugiperda*. Peanut Sci. 8, 106–109.

López, Y., Smith, O.D., Senseman, S.A., Rooney, W.L., 2001. Genetic factor influencing high oleic acid content in Spanish market-type peanut cultivars. Crop Sci. 41, 51–56.

Mallikarjuna, N., 2004. Meiotic study of intersectional hybrids between *Arachis hypogaea, A. duranensis and A. diogoi with A. glabrata*. Int. Arachis Newsl. 24, 7–8.

Mallikarjuna, N., 2005. Production of hybrids between *Arachis hypogaea* and *A. chiquitana* (section *procumbentes*). Peanut Sci. 32, 148–152.

Mallikarjuna, N., Hoisington, D., 2009. Peanut improvement: production of fertile hybrids and backcross progeny between *Arachis hypogaea* and *A. kretschmeri*. Food Sec. 1, 457–462.

Mallikarjuna, N., Sastri, D.C., 2002. Morphological, cytological and disease resistance studies of the intersectional hybrid between *Arachis hypogaea* L. and *A. glabrata* Benth. Euphytica 126 (2), 161–167.

Mallikarjuna, N., Pande, S., Jadhav, D.R., Sastri, D.C., Rao, J.N., 2004. Introgression of disease resistance genes from *Arachis kempffmercadoi* into cultivated groundnut. Plant Breed. 123, 573–576.

Mallikarjuna, N., Tandra, S.K., Jadhav, D.R., 2006. *Arachis hoehnei* the probable B genome donor of *Arachis hypogaea* based on crossability, cytogenetical and molecular studies. SAT EJ. 2 (1), 1–2.

Manton, I., 1950. Problems of cytology and evolution in the Pteridophyta. Cambridge University Press, Cambridge.

Martin, J.P., 1967. Contribution at l etude de certain characters d'importnace agronomique chez l'arachide etude de l'heredite de la richesse eb huile, du sendement eb decortigage et de la grosseur des grains dans le groups des varieties tardives. Oleagineux 22, 673–676.

Mayee, C.D., 1987. Rust disease of groundnut in Maharashtra state of India. In: Proceedings of the Discussion Group Meeting on Groundnut Rust Diseases, September 24–28, 1984, ICRISAT center, Patancheru, Andhra Pradesh, India. pp. 81–89.

Mcdonald, D., Subrahmanyan, P., Gibbon, R. W., Smith, D. H., 1985. Early and late leaf spots of groundnut. Information Bulletin no. 21. ICRISAT, Panthancheru, Andra Pradesh, India. pp. 24.

McNeill, K.L., Sanders, T.H., 1996. Pod and seed size relation to maturity and in-shell quality potential in Virginia-type peanuts. Peanut Sci. 23, 133–137.

McNeill, K.L., Sanders, T.H., 1998. Maturity effects on sensory and storage quality of roasted Virginia-type peanuts. J. Food Sci. 63 (2), 366–369.

McSorley, R., Dickson, D.W., Candeanedo-Lay, E.M., Hewlett, T.E., Frederick, J.J., 1992. Damage functions for *Meloidogyne arenaria* on peanut. J. Nematol. 4, 193–198.

Mehan, V.K., 1989. Screening groundnuts for resistance to seed invasion by Aspergillus flavus and to aflatoxin production. In: McDonald, D., Mehan, V.K., (eds.), Aflatoxin Contamination of Groundnut. Proceedings of International Workshop, 6–9 Oct. 1987, ICRISAT Center, India. Patancheru, A.P., India. pp. 323–334.

Mehan, V.K., Ba, A., McDonald, D., Renard, J.L., Rao, R.C.N., Jayanthi, S., 1991. Field screening of groundnuts for resistance to seed infection by *Aspergillus flavus*. Oléagineux 46, 109–118.

Mehan, V.K., Reddy, P.M., Rao, V.K., McDonald, D., 1994. Components of rust resistance in peanut genotypes. Phytopathology 84, 1421–1426.

Mehan, V.K., Reddy, P.M., Subrahmanyam, P., McDonald, D., Singh, A.K., 1996. Identification of new sources of resistance to rust and late leaf spot in peanut. Int. J. Pest Manag. 42, 267–271.

Mehta, R., Radhakrishnan, T., Kumar, A., Yadav, R., Dobaria, J.R., Palanisamy, P., Thirumalaisamy, P., Jain, R.K., Chigurupati, P., 2013. Coat protein-mediated transgenic resistance of peanut (*Arachis hypogaea* L.) to peanut stem necrosis disease through Agrobacterium-mediated genetic transformation. Indian J. Virol. 24 (2), 205–213.

Mercer, L.C., Wynne, J.C., Young, C.T., 1990. Inheritance of fatty acid content in peanut oil. Peanut Sci. 17, 17–21.

Mixon, A.C., 1980. Comparison of Pod and Seed Screening Methods on *Aspergillus* spp. Infection of Peanut Genotypes. Peanut Sci. 7, 13.

Moore, K.M., Knauft, D.A., 1989. The inheritance of high oleic acid in peanut. J. Hered. 80, 252–253.

Moretzsohn, M.C., Gouvea, E.G., Inglis, P.W., Leal-Bertioli, S.C.M., Valls, J.F.M., Bertioli, D.J., 2013. A study of the relationships of cultivated peanut (*Arachis hypogaea* L) and its most closely related wild species using intron sequences and microsatellite markers. Ann. Bot. 111, 113–126.

Morgante, C.V., Brasileiro, A.C.M., Roberts, P.A., Guimaraes, L.A., Araujo, A.C.G., Fonseca, L.N., 2013. A survey of genes involved in *Arachis stenosperma* resistance to *Meloidogyne arenaria* race 1. Funct. Plant Biol. 40 (12), 1298–1309.

Motagi, B.N., Angadi, C.C., Gowda, M.V.C., Naidu, G.K., 2013. Genetics of rust resistance in mutants and interspecific derivatives of groundnut (*Arachis hypogaea* L.). Karnataka J. Agric. Sci. 26 (1), 14–16.

Mozingo, R.W., Coffelt, T.A., Wright, F.S., 1991. The influence of planting and digging dates on yield, value, and grade of four Virginia-type peanut genotypes. Peanut Sci. 18, 55–62.

Mugendi, J.B., Sims, C.A., Gorbet, D.W., O'Keefe, S.F., 1998. Flavor stability of high-oleic peanut stored at low humidity. J. Am. Oil Chem. Soc. 75, 21–25.

Muitia, A., López, Y., Starr, J.L., Schubert, A.M., Burow, M.D., 2006. Introduction of resistance to root-knot nematode (*Meloidogyne arenaria* neal (chitwood)) into high-oleic peanut. Peanut Sci. 33, 97–103.

Nagaraj, C., 1995. Quality and utility of oilseeds, Directorate of Oilseeds Research, Rajendranagar, Hyderabad.

Nageswara Rao, R.C., Sardar, Singh, Sivakumar, M.V.K., Srivastava, K.L., Williams, J.H., 1985. Effect of water deficit at different growth phases of peanut I yield responses. Agron. J. 77, 782–786.

Nageswara Rao, R.C., Williams, J.H., Sivakumar, M.V.K., Wadia, K.D.R., 1988. Effect of water deficit at different growth phases of peanut. II. Response to drought during pre-flowering phase. Agron. J. 80, 431–438.

Nautiyal, P.C., Ravindra, V., Zala, P.V., Joshi, Y.C., 1999. Enhancement of yield in groundnut following the imposition of transient soil-moisture stress during the vegetative phase. Exp. Agric. 35, 371–385.

Navinbhai Gajjar, K., Mishra, G.P., Radhakrishnan, T., Meramanbhai Dodia, S., Rathnakumar, A.L., Narendra Kumar, Santosh Kumar, Dobaria, J.R., Kumar, Abhay, 2014. Validation of SSR markers linked to the rust and late leaf spot diseases resistance in diverse peanut genotypes. Australian J. Crop Sci. 8 (6), 927–936.

Nelson, S.C., Simpson, C.E., Starr, J.L., 1989. Resistance to *Meloidogyne arenaria* in *Arachis* spp. germplasm. J. Nematol. 21, 654–660.

Nelson, S.C., Starr, J.L., Simpson, C.E., 1990. Expression of resistance to *Meloidogyne arenaria* in *Arachis batizocoi* and *A. cardenasii*. J. Nematol. 22, 423–425.

Nevill, D.J., 1980. Studies of Resistance to Foliar Diseases of Groundnut. Proceedings of International Workshop on Groundnuts. ICRISAT, Hyderabad, India, pp. 199–202.

Newell, J.A., Mason, M.E., Matlock, R.S., 1967. Precursors of typical and atypical roasted peanut flavor. J. Agric. Food Chem. 15, 767–772.

Nigam, S.N., Aruna, R., 2008. Improving breeding efficiency for early maturity in peanut. In: Jules Janick (Eds.), Plant Breeding Reviews. Wiley, Hoboken, vol. 30, pp. 295–323.

Nigam, S.N., Dwivedi, S.L., Gibbons, R.W., 1980. Groundnut breeding at ICRISAT. In: Proceedings of the International Workshop on Groundnuts, 13–17 October-1980, ICRISAT Center, India. Patancheru, A.P., India: pp. 62–68.

Nigam, S.N., Upadhyaya, H.D., Chandra, S., Nageswara Rao, R.C., Wright, G.C., Reddy, A.G.S., 2001. Gene effects for specific leaf area and harvest index in three crosses of groundnut (*Arachis hypogaea* L.). Ann. Appl. Biol. 139, 301–306.

Nigam, S.N., Nageswara Rao, R.C., Wright, G.C., 2002. Field Screening for Drought Tolerance in Groundnut. In : Saxena, N.P., O'Toole, John, C., (eds.). 2002. Field Screening for Drought Tolerance in Crop Plants with Emphasis on Rice: Proceedings of an International Workshop on Field Screening for Drought Tolerance in Rice, Dec 11–14, 2000, ICRISAT, Patancheru, India. Patancheru, Andhra Pradesh, India, and the Rockefeller Foundation, New York, New York. pp. 147–151.

Nigam, S.N., Aruna, R., Yadagiri, D., Reddy, T.K., Subramanyam, K., Reddy, B.R.R., Kareem, K.A., 2005. Farmer participatory varietal selection in groundnut – a success story in Anantapur, Andhra Pradesh, India. SAT EJ. 1 (1), 1–3.

Norden, A.J., Gorbet, D.W., Knauft, D.A., Young, C.T., 1987. Variability in oil quality among peanut genotype in the Florida breeding programme. Peanut Sci. 14, 7–11.

O'Byrne, D.J., Knauft, D.A., Shireman, R.B., 1997. Low fat-monounsaturated rich diets containing high-oleic peanuts improve serum lipoprotein profiles. Lipids 32, 687–695.

Olwari, F., Bisikwa, J., Kaaya, A.N., Okello, D.K., 2013. Tolerance levels of peanut varieties against *Aspergillus flavus* infection. J. Plant Pathol. Microbiol. 4 (8.), 1–4.

Ory, R.L., Crippen, K.L., Lovegren, N.V., 1992. Off-flavors in peanuts and peanut products. In: Charalambous, G. (Ed.), Developments in Food Science v. 29: Off-Flavors in Foods and Beverages. Elsevier Science Publishers, Amsterdam, The Netherlands, pp. 57–75.

Osei, K., Asibuo, J.Y., Agyeman, A., Osei-Bonsu, P., Danso, Y., Adomako, J., 2013. Reactions of some confectionery groundnut accessions to plant parasitic nematodes infection. Agrosearch 13 (2), 1–11.

Ozimati, A.A., Rubaihayo, P.R., Gibson, P., Edema, R., Kayondo, I.S., Ntare, B.R., Okello, D.K., 2014. Inheritance of resistance to kernel infection by *Aspergillus flavus* and aflatoxin accumulation in groundnut. African J. Crop Sci. 2 (1), 51–59.

Painawadee, M., Jogloy, S., Kesmala, T., Akkasaeng, C., Patanothai, A., 2009. Heritability and correlation of drought resistance traits and agronomic traits in peanut (*Arachis hyogaea* L.). Asian J. Plant Sci. 8, 325–334.

Pan, R., Guan, M., Xu, D., Gao, X., Yan, X., Liao, H., 2009. Cylindrocladium black rot caused by*Cylindrocladium parasiticum* newly reported on peanut in China. Plant Pathol. 58, 1176.

Pandurangaiah, M., Loganatha Rao, G., Sudhakarbalu, O., Nareshkumar, A., Kiranmai, K., Lokesh, U., Thapa, G., Sudhakar, C., 2014. Over expression of horsegram (*Macrotyloma uniflorum* Lam. Verdc.) NAC transcriptional factors (MuNaC4) in groundnut confers enhanced drought tolerance. Mol. Biotechnol. 56, 758–769.

Parameshwarappa, K.G., Kenchanagoudar, P.V., Bentur, M.G., Patil, R.K., 2005. Genetic variability and correlation in adopted genotypes of Spanish bunch groundnut. Proceedings of the National Symposium on Enhancing Productivity of Groundnut for Sustaining Food and National Security, NRCG, Junagadh, Oct. 11–12, 2004. pp. 74–75.

Parmar, D.L., Kumar, A.L.R., Bharodia, P.S., 2000. Genetic analysis of pod and seed characters in crosses of large seeded Virginia genotypes of groundnut. Int. Arachis Newsl. 20, 10–11.

Pataky, J.K., Black, M.C., Beute, M.K., Wynne, J.C., 1983. Comparative analysis of *Cylindrocladium* black rot resistance in peanut: greenhouse, microplot and field testing procedures. Phytopathology 73, 1615–1620.

Patel, J.S., John, C.M., Seshadri, C.R., 1936. The inheritance of characters in the groundnut – *Arachis hypogaea*. Proc. Indian Acad. Sci. 3, 214–233.

Patel, B.A., Patel, D.J., Sharma, S.B., Patel, H.V., Patel, S.K., 1996. Nematode problems of groundnut and their management in Gujarat, India. Int. Arachis Newsl. 16, 38–39.

Pattanashetti, S.K., Gowda, M.V.C., Girija, 2008. Inheritance of morphological traits and pod features in groundnut (*Arachis hypogaea* L.). Indian J. Genet. Plant Breed. 68, 157–162.

Pattee, H.E., Williams, D.E., Sanchez-Dominguez, S., Ciesbrecht, F.G., 1995. Evaluation of Six Landrace Accessions of *Arachis hypogaea* ssp. *hypogaea* var. *hirsute* Kohler. I. Descriptive and Sensory'. Peanut Sci. 22, 18–22.

Pattee, H.E., Isleib, T.G., Giesbrech, F.G., Mc Feeters, R.F., 2000. Relationship of sweet, bitter and roaosted peanut sensory attributes with carbohydrate components in peanut. J. Agric. Food Chem. 48, 757–763.

Pompeu, A.S., 1983. Cruzamentos entre *Arachis hypogaea* e as especies *A. diogoi* e A. spp. (30006, 30005). Bragantia 42 (6), 261–265.

Porter, D.M., Melouk, H.M., 1997. Sclerotnia blight. In: Kokalis-Burelle, N., Porter, D.M., Rodriguez-Kabana, R., Smith, D.H., Subrahmanyam, P. (Eds.), The compendium of peanut diseases. second ed. American Phytopathological Society Press, St. Paul, MN, pp. 34–35.

Prasad, M.N.R., Gowda, M.V.C., 2006. Mechanisms of resistance to tobacco cutworm (*Spodoptera litura* F.) and their implications to screening for resistance in groundnut. Euphytica 149, 387–399.

Prasad, K., Bhatnagar-Mathur, P., Waliyar, F., Sharma, K.K., 2013. Overexpression of chitinase gene in transgenic peanut confers enhanced resistance to major soil borne and fungal pathogens. J. Plant Biochem. Biotechnol. 22, 222–233.

Puangbut, D., Jogloy, S., Vorasoot, N., Akkasaeng, C., Kesmala, T., Patanothai, A., 2009. Variability in yield responses of peanut (*Arachis hypogaea* L.) genotypes under early season drought. Asian J. Plant Sci. 8, 254–264.

Puangbut, D., Jogloy, S., Toomsan, B., Vorasoot, N., Akkasaeng, C., Kesmala, T., Rachaputi Rao, C.N., Wright, G.C., Patanothai, A., 2010. Physiological basis for genotypic variation in tolerance to and recovery from pre-flowering drought in peanut. J. Agron. Crop Sci. 196, 358–367.

Puangbut, D., Jogloy, S., Kesmala, T., Vorasoot, N., Akkasaeng, C., Patanothai, A., Puppala, N., 2011. Heritability of early season drought resistance traits and genotypic correlation of early season drought resistance and agronomic traits in peanut. SABRAO J. Breed. Genet. 43 (2), 165–187.

Qiao, L.X., Ding, X., Wang, H.C., Sui, J.M., Wang, J.S., 2014. Characterization of the β-1,3-glucanase gene in peanut (*Arachis hypogaea* L.) by cloning and genetic transformation. Genet. Mol. Res. 13 (1), 1893–1904.

Raheja, R.K., Batta, S.K., Ahuja, K.L., Labana, K.S., Singh, M., 1987. Comparison of oil content and fatty acid composition of peanut genotypes differing in growth habit. Plant Foods Hum. Nutr. 37, 103–108.

Ramanatha Rao, V., 1988. Botany. In: Reddy, P.S. (Ed.), Groundnut. Indian Council of Agricultural Research, New Delhi, India, pp. 24–64.

Ramanatha Rao, V., Murty, U.R., 1994. Genotypic variations in dry matter production, chemical compositions and calcium efficiency ratio of groundnut grown on acid sands. Field Crops Res. 4, 155–160.

Ramos, M.L., Fleming, G., Chu, Y., Akiyama, Y., Gallo, M., 2006. Chromosomal and phylo-genetic context for conglutin genes in *Arachis* based on genomic sequence. Mol. Genet. Genom. 275, 578–592.

Rana, K., Mohanty, I.C., 2012. *In vitro* regeneration and genetic transformation in groundnut (*Arachis hypogaea* L. cv. Smruti) for abiotic stress tolerance mediated by *Agrobacterium tumefaciens*. J. Today's Biol. Sci.: Res.Rev. 1 (1), 62–85.

Rao, R.V.S., 2000. Resistance to leafminer. Aproaerema modicella in groundnut *Arachis hypogaea*. Indian J. Entomol. 62 (3), 280–285.

Rao, R.V.S., Sridhar, R., Singh, U., Rao, G.V.R., 1998. Bio-chemical basis in groundnut (*Arachis hypogaea*) resistant to leafminer *(Aproaerema modicella)*. Indian J. Agric. Sci. 68 (2), 104–109.

Redlinger, L.M., Davis, R., 1982. Insect control in postharvest peanuts. In: Pattee, H.E., Young, C.T. (Eds.), Peanut Science and Technology. American Peanut Research and Education Society, Yoakum, Texas, USA, pp. 520–571.

Riley, R., Law, C.N., 1965. Genetic variation in chromosome pairing. Adv. Genet. 13, 57–114.

Robledo, G., Seijo, G., 2010. Species relationships among the wild B genome of *Arachis* species (section *Arachis*) based on FISH mapping of rDNA loci and heterochromatin detection: a new proposal for genome arrangement. Theor. Appl. Genet. 121, 1033–1046.

Rosa Lía, Barbieri Gomes, João Carlos Costa, Alercia, Adriana, Padulosi, Stefano, 2014. Agricultural biodiversity in southern Brazil: integrating efforts for conservation and use of neglected and underutilized species. Sustainability 6, 741–757.

Roy, R.C., Stonehouse, D.P., Francois, B., Brown, D.M., 1988. Peanut response to imposed drought conditions in southern Ontario. Peanut Sci. 15, 85–89.

Sadeghi, S.M., Niyaki, S.A.N., 2012. Genetic correlation and path-coefficient analysis of oil yield and its components in peanut (*Arachis hypogaea* L.) genotypes under drought and non-drought stress condition. J. Basic Appl. Sci. Res. 2 (7), 6561–6565.

Sanchez-Dominguez, S., Williams, D.E., 1993. Results of a recent plant exploration in Mexico to collect the *Hirsuta* peanut. Proc. Am. Peanut Res. Edu. Soc. 25, p. 35.

Sanders, T.H., Vercellotti, J.R., Crippen, K.L., Civille, G.V., 1989a. Effect of maturity on roast color and descriptive flavor of peanuts. J. Food Sci. 54, 475–477.

Sanders, T.H., Vercellotti, J.R., Crippen, K.L., Civille, G.V., 1989b. Interaction of maturity and curing temperature on descriptive flavor of peanuts. J. Food Sci. 54, 1066–1069.

Sanders, T.H., Bett, K.L., 1995. Effect of harvest date on maturity, maturity distribution, and flavor of florunner peanuts. Peanut Sci. 22, 124–129.

Sasser, J.N., 1977. Worldwide dissemination and importance of the root-knot nematode, *Meloidogyne* spp. J. Nematol. 9, 26–29.

Schmitt, D.P., Barker, K.R., 1988. Incidence of plant-parasitic nematodes in the coastal plains of North Carolina. Plant Dis. 72, 107–110.

Schroeder, H.W., Ashworth, Jr., L.J., 1965. Aflatoxins in Spanish peanuts in relation to pod and kernel conditions. Phytopathology 55, 464–465.

Seijo, J.G., Lavia, G.I., Fernandez, A., Krapovickas, A., Ducasse, D., Moscone, E.A., 2004. Physical mapping of the 5S and 18S-25S rRNA genes by FISH as evidence that *Arachis duranensis* and *A. ipaënsis* are the wild diploid progenitors of *A. hypogaea* (Leguminosae). Am. J. Bot. 91, 1294–1303.

Seijo, G., Lavia, G.I., Fernandez, A., 2007. Genomic relationships between the cultivated peanut (*Arachis hypogaea*, Leguminosae) and its close relatives revealed by double GISH. Am. J. Bot. 94, 1963–1971.

Selvaraj, M.G., Narayana, M., Schubert, A.M., Ayers, J.L., Baring, M.R., Burow, M.D., 2009. Identification of QTLs for pod and kernel traits in cultivated peanut by bulked segregant analysis. Elect. J. Biotechnol. 12, 1–10.

Sharma, N.D., Santha, I.M., Patil, S.H., Metha, S.L., 1985. Fatty acid and amino acid composition of groundnut mutants. Qualitas Plantarum Plant Foods Hum. Nutr. 35, 3–8.

Sharma, H.C., Pampapathy, G., Kumar, R., 2002. Techniqueto screenpeanuts for resistance tothe tobaccoarmyworm, *Spodoptera litura* (Lepidoptera: noctuidae) under no-choice cage conditions. PeanutSci. 29, 35–40.

Sharma, H.C., Pampapathy, G., Dwivedi, S.L., Reddy, L.J., 2003. Mechanisms and diversity of resistance to insect pests in wild relatives of groundnut. J. Econ. Entomol. 96 (6), 1886–1897.

Sheshshayee, M.S., Bindumadhava, H., Rachaputi, N.R., Prasad, T.G., Udayakumar, M., Wright, G.C., Nigam, S.N., 2006. Leaf chlorophyll concentration relates to transpiration efficiency in peanut. Ann. Appl. Biol. 148, 7–15.

Shoba, D., Manivannan, N., Vindhiyavarman, P., Nigam, S.N., 2012. SSR markers associated for late leaf spot disease resistance by bulked segregant analysis in groundnut (*Arachis hypogaea* L.). Euphytica 188, 265–272.

Simpson, C.E., 2001. Use of wild *Arachis* species/introgression of genes into *A. hypogaea* L. Peanut Sci. 28 (2), 114–116.

Simpson, C.E., Starr, J.L., 2001. Registration of 'COAN' peanut. Crop Sci. 41, 918.

Simpson, C.E., Nelson, S.C., Starr, J.L., Woodard, K.E., Smith, O.D., 1993. Registration of TxAG-6 and TxAG-7 peanut germplasm lines. Crop Sci. 33, 1418.

Simpson, C.E., Starr, J.L., Church, G.T., Burow, M.D., Paterson, A.H., 2003. Registration of 'NemaTAM' peanut. Crop Sci. 43, 1561.

Singh, F., Diwakar, B., 1993. Nutritive Value and Uses of Pigeonpea and Groundnut. Skill Development Series no. 14. Human Resource Development Program. International Crops Research Institute for the Semi-Arid Tropics, Patancheru, Andhra Pradesh, India.

Singh, A.K., Moss, J.P., 1982. Utilization of wild relatives in genetic improvement of *Arachis hypogaea* L. 2. Chromosome complement of species of section Arachis. Theor. Appl. Genet. 61, 305–314.

Singh, A.K., Simpson, C.E., 1994. Biosystematics and genetic resources. In: Smartt, J. (Ed.), The Groundnut Crop: a Scientific Basis for Improvement. Chapman & Hall, London, pp. 96–137.

Singh, A.K., Mehan, V.K., Nigam, S.N., 1997. Sources of resistance to groundnut fungal and bacterial diseases: an update and appraisal. (In En. Summaries in En., Fr.) Information Bulletin no. 50. Patancheru 502 234, Andhra Pradesh, India: International Crops Research Institute for the Semi-Arid Tropics, 48.

Singh, N., Mishra, A., Jha, B., 2014. Ectopic over-expression of peroxisomal ascorbate peroxidase (SbpAPX) gene confers salt stress tolerance in transgenic peanut (*Arachis hypogaea* L.). Gene 547, 119–125.

Singkham, N., Jogloy, S., Kesmala, T., Swatsitang, P., Jaisil, P., Puppala, N., Patanothai, A., 2010. Estimation of heritability by parent-offspring regression for high-oleic acid in peanut. Asian J. Plant Sci. 9 (6), 358–363.

Singkham, N., Jogloy, S., Kesmala, T., Swatsitang, P., Jaisil, P., Puppala, N., Patanothai, A., 2011. Combining ability for oleic acid in peanut (*Arachis hypogaea* L.). SABRAO J. Breed. Genet. 43 (1), 59–72.

Singkham, N., Jogloy, S., Suriharn, B., Kesmala, T., Swatsitang, P., Jaisil, P., Puppala, N., Patanothai, A., 2012. Types of gene effects governing the inheritance of oleic and linoleic acids in peanut (*Arachis hypogaea* L.). African J. Biotechnol. 11 (67), 13147–13152.

Smartt, J., Gregory, W.C., Gregory, M.P., 1978. The genomes of *Arachis hypogaea*. 1. Cytogenetic studies of putative genome donors. Euphytica 27, 665–675.

Smith, Jr., J.W., Barfield, C.S., 1982. Management of preharvest insects. In: Pattee, H.E., Young, C.T. (Eds.), Peanut Science and Technology. American Peanut Research Education and Society, Yoakum, Tex, pp. 250–325.

Song, X.S., Sharma, S.B., Hu, J.P., 1992. Occurrence and damage of root-knot nematode (*Meloidogyne hapla* Chitwood) on peanut in Shandong, China. Int. Arachis Newsl. 12, 34–36.

Song, X.S., Lan, W., Dong, W.B., Min, P., 1995. Identification of resistance to root-knot nematode (*Meloidogyne hapla* Chitwood) in peanut germplasm. Acta Phytopathologica-Sinica 25, 139–141.

Songsri, P., Jogloy, S., Kesmala, T., Vorasoot, N., Akkasaeng, C., Patanothai, A., Holbrook, C.C., 2008. Heritability of drought resistance traits and correlation of drought resistance and agronomic traits in peanut. Crop Sci. 48, 2245–2253.

Stalker, H.T., 1980. Utilization of wild species for crop improvement. Agron. J. 33, 111–147.

Stalker, H.T., 1981. Hybrids in the genus *Arachis* between sections *Erectoides* and *Arachis*. Crop Sci. 21, 359–362.

Stalker, H.T., Campbell, W.V., 1983. Resistance of wild species of peanut to an insect complex. Peanut Sci. 10, 30–33.

Stalker, H.T., Dalmacio, R.D., 1981. Chromosomes of *Arachis* species, section *Arachis* (Leguminosae). J. Hered. 72, 403–408.

Stalker, H.T., Simpson, C.E., 1995. Genetic resources in Arachis. In: Pattee, H.E., Stalker, H.T. (Eds.), Advances in Peanut Science. American Peanut Research Education and Society, Stillwater, OK, pp. 14–53.

Stalker, H.T., Seitz, M.H., Reece, P., 1987. Effect of gibberellic acid on pegging and seed set of *Arachis* species. Peanut Sci. 14, 17–21.

Stalker, H.T., Beute, M.K., Shew, B.B., Barker, K.R., 2002. Registration of two root-knot nematode-resistant peanut germplasm lines. Crop Sci. 42, 312–313.

Starr, J.L., Simpson, C.E., 1991. Segregation of resistance to *Meloidogyne arenaria* in progeny of interspecific hybrids. Proc. Am. Peanut Res. Edu. Soc. 23, 23, (Abstract).

Stevenson, P.C., Blaney, W.M., Simmonds, M.J.S., Wightman, J.A., 1993. The identification and characterization of resistance in wild species of *Arachis* to *Spodoptera litura* (Lepidoptera: Noctuidae). Bull. Entomol. Res. 83, 421–429.

Styer, C.H., Cloe, R.J., Hill, R.A., 1983. Inoculation and infection of peanut flowers by *Aspergillus flavus*. Proc. Am. Peanut Res. Edu. Soc. 15, 91, (abstract).

Subbarao, G.V., Johansen, C., Slinkard, A.E., Nageswara Rao, R.C., Saxena, N.P., Chauhan, Y.S., 1995. Strategies for improving drought resistance in grain legumes. Crit. Rev. Plant Sci. 14, 469–523.

Subrahmanyam, P., Reddy, D.V.R., Gibbons, R.W., Rao, V.R., Garren, K.H., 1979. Current distribution of groundnut rust in India. Pest Articles and News Summaries 25, 25–29.

Subrahmanyam, P., Mehan, V.K., Nevil, D.J., McDonald, D., 1980. Research on fungal diseases of groundnut at ICRISAT. In: Proceedings of International Workshop in Groundnuts, International Crop Research Institute for the Semi-Arid Tropics, Patancheru, India, pp 193–198.

Subrahmanyam, P., McDonald, D., Gibbons, R.W., Rao, P.V.S., 1983. Components of resistance to *Puccinia arachidis*. Phytopathology 73, 253–256.

Subrahmanyam, P., Reddy, L.R., McDonald, D., Gibbons, R.W., 1985. Peanut rust: a major threat to peanut production in the semiarid tropics. Plant Dis. 69, 813–819.

Sujay, V., Gowda, M.V.C., Pandey, M.K., Bhat, R.S., Khedikar, Y.P., Nadaf, H.L., Gautami, B., Sarvamangala, C., Lingaraju, S., Radhakrishan, T., Knapp, S.J., Varshney, R.K., 2012. Quantitative trait locus analysis and construction of consensus genetic map for foliar disease resistance based on two recombinant inbred line populations in cultivated groundnut (*Arachis hypogaea* L.). Mol. Breed. 30, 773–788.

Taber, R.A., Pettit, R.E., Benedict, C.R., Dieckert, J.W., Ketring, D.L., 1973. Comparison of *Aspergillus flavus* tolerant and susceptible peanut lines I. Light microscope investigation. Am. Peanut Res. Edu. Soc. 5, 206–207.

Telenko, P.D.E., Hu, J., Livingstone, D.M., Shew, B.B., Phipps, P.M., Grabau, E.A., 2011. Sclerotinia blight resistance in Virginia-type peanut transformed with a barley oxalate oxidase gene. Phytopathology 101, 786–793.
Todd, J.W., Beach, R.M., Branch, W.D., 1991. Resistance in eight peanut genotypes to foliar feeding of fall armyworm, velvetbean caterpillar, and com earworm. Peanut Sci. 18, 38–40.
Tomaszewski, E.K., Khalil, M.A.M., El-Deeb, A.A., Powers, T.O., Starr, J.L., 1994. *Meloidogyne javanica* parasitic on peanut. J. Nematol. 26, 436–441.
Tulole Lugendo, Bucheyeki, Shenkalwa, Erasto M., Mapunda, Theofora##X., Matata, Leah W., 2008. On-farm evaluation of promising groundnut varieties for adaptation and adoption in Tanzania. African J. Agric. Res. 3 (8), 531–536.
Upadhyaya, H.D., 2005. Variability for drought resistance related traits in the mini core collection of peanut. Crop Sci. 45, 1432–1440.
Upadhyaya, H.D., Nigam, S.N., 1994. Inheritance of two components of early maturity in groundnut (*Arachis hypogea* L.). Euphytica 78, 59–66.
Upadhyaya, H.D., Nigam, S.N., 1999. Detection of epistasis for protein and oil contents and oil quality parameters in peanut. Crop Sci. 39, 115–118.
Upadhyaya, H.D., Nigam, S.N., Mehan, V.K., Lenne, J.M., 1997. Aflatoxin contamination of groundnut: prospects for a genetic solution through conventional breeding. In: Mehan, V.K., Gowda, C.L.L. (Eds.), Aflatoxin Contamination Problems in Groundnut in Asia: Proceedings of the First Asia Working Group Meeting. ICRISAT, Patancheru, India, pp. 81–85.
Upadhyaya, H.D., Nigam, S.N., Singh, S., 2001. Evaluation of groundnut core collection to identify sources of tolerance to low temperature at germination. Indian J. Plant Genet. Res. 14 (2), 165–167.
Upadhyaya, H.D., Mallikarjuna Swamy, B.P., Goudar, P.V.K., Kullaiswamy, B.Y., Singh, S., 2005. Identification of diverse groundnut germplasm through multienvironment evaluation of a core collection for Asia. Field Crops Res. 93, 293–299.
Upadhyaya, H.D., Reddy, L.J., Gowda, C.L.L., Singh, S., 2006. Identification of diverse groundnut germplasm sources of early maturity in a core collection. Field Crops Res. 97, 261–267.
Upadhyaya, H.D., Sharma, S., Singh, S., Singh, M., 2010. Inheritance of drought resistance related traits in two crosses of groundnut (*Arachis hypogaea* L.). Euphytica 177, 55–66.
Utomo, S.D., Anderson, W.F., Wynne, J.C., Beute, M.K., Hagler, W.M.Jr., Payne, G.A., 1990. Estimates of heritability and correlation among three mechanisms of resistance to *Aspergillus parasiticus* in peanut. Proc. Am. Peanut Res. Edu. Soc. 22, 26, (Abstract.).
Vadawale, V., Mihani, R., Mathew, A., Robi, P., 2012. Transformation of groundnut – *Arachis hypogaea* L. var. GG 20 with the cox gene – an attempt to develop salinity tolerance. Int. J. Pharma Bio Sci. 3 (1), 591–599.
Valls, J.F.M., Ramanatha Rao, V., Simpson, C.E., Krapovickas, A., 1985. Current status of collection and conservation of South American groundnut germplasm with emphasis on wild species of Arachis. In: Moss, J.P. (Ed.), In: Proceedings of International Workshop on Cytogenetics of Arachis. International Crops Research Institute for Semi-Arid Tropics, Patancheru, A. P. India, pp. 15–35, October 31–November 2, 1983.
Varman, P.V., Raveendran, T.S., 1996. Genetic variability and causal relationship in groundnut. Madras Agric. J. 83 (12), 784–786.
Varshney, R.K., Graner, A., Sorrells, M.E., 2005. Genomics assisted breeding for crop improvement. Trends Plant Sci. 10, 621–630.
Varshney, R.K., Pandey, M.K., Janila, P., Nigam, S.N., Sudini, H., Gowda, M.V.C., Sriswathi, M., Radhakrishnan, T., Manohar, S.S., Nagesh, P., 2014. Marker-assisted introgression of a QTL region to improve rust resistance in three elite and popular varieties of peanut (*Arachis hypogaea* L.). Theor. Appl. Genet. 127 (8), 1771–1781.
Vasanthi, .R.P., Reddy, C.R., 1997. Inheritance of testa colour and resistance to late leaf spot and rust in groundnut (*Arachis hypogaea* L.). J. Oilseeds Res. 14 (2), 244–248.
Vasudeva Rao, M.J., Nigam, S.N., Huda, A.K.S., 1992. The thermal time concept as a selection criterion for earliness in peanut. Peanut Sci. 19, 7–10.
Vindhiyavarman, P., Mothilal, A., Ganesan, K.N., Naina Mohammed, S.E., 2000. An appraisal of triploids of amphidiploids in the genus *Arachis*. Int. Arachis Newsl. 20, 8–9.
Vindhiyavarman, P., Manivannan, N., Nigam, S.N., Muralidharan, V., 2010. Farmers participatory varietal selection in groundnut a case study from Tamil Nadu, India. Elect. J. Plant Breed. 1 (4), 878–881.
Waliyar, F., 1991. Evaluation of yield losses due to groundnut leaf diseases in West Africa. In: Proceedings of the Second ICRISAT Regional Groundnut Meeting for West Africa, September 11–14, 1991, Niamey, Niger. ICRISAT, Patancheru, India, pp. 32–33.

Waliyar, F., Bosc, J.P., Bonkoungou, S., 1993. Sources of resistance to foliar diseases of groundnut and their stability in West Africa. Oleagineux 48, 283–287.

Wang, C.T., Tang, Y.Y., Wang, X.Z., Chen, D.X., Cui, F.G., Chi, Y.C., Zhang, J.C., Yu, S.L., 2011. Evaluation of groundnut genotypes from China for quality traits. J. SAT Agric. Res. 9, 1–5.

Wang, M.L., Chen, C.Y., Pinnow, D.L., Barkley, N.A., Pittman, R.N., Lamb, M., Pederson, G.A., 2012. Seed Dormancy Variability in the U.S. Peanut Mini-Core Collection. Res. J. Seed Sci. 5, 84–95.

Wild, C.P., Turner, P.C., 2002. The toxicology of aflatoxins as a basis for public health decisions. Mutagenesis 17, 471–481.

Wildman, L.G., Smith, O.D., Simpson, C.E., Taber, R.A., 1992. Inheritance of Resistance to *Sclerotinia* minor in selected Spanish peanut crosses. Peanut Sci. 19, 31–34.

Williamson, V.M., Hussey, R.S., 1996. Nematode pathogenesis and resistance in plants. Plant Cell 8, 1735–1745.

Witcombe, J.R., Joshi, A., Joshi, K.D., Sthapit, B.R., 1996. Farmer participatory crop improvement I: varietal selection and breeding methods and their impact on biodiversity. Exp. Agric. 32, 445–460.

Wyss, U., Grundler, F.M.W., Múnch, A., 1992. The parasitic behaviour of second-stage juveniles of *Meloidogyne incognita* in roots of *Arabidopsis thaliana*. Nematologica 38, 98–111.

Young, N.D., Weeden, N.F., Kochert, G., 1996. Genome mapping in legumes (Fam. Fabaceae). In: Paterson, A., Austin, H. (Eds.), Genome Mapping in Plants. Landes Company, USA, pp. 211–227.

Zambettakis, C., 1975. Study of the contamination of several varieties of groundnut by *Aspergillus flavus*. Etude de la contamination de quelques variétés d'arachide par *Aspergillus flavus*. Oléagineux 30, 161–167.

Zambettakis, C., Waliyar, F., Bockelee-Morvan, A., de Pinds, O., 1981. Results of four years' research on resistance of groundnut varieties to *Aspergillus flavus*. Oleagineux 36, 377–385.

Zhou Liang-gao., 1987. The Groundnut Rust Situation in the People's Republic of China 1ICRISAT (International Crops Research Institute for the Semi-Arid Tropics). Groundnut rust disease. Proceedings of a Discussion Group Meeting, 24–28 September 1984, ICRISAT Center, India, Patancheru, A.P. 502 324, India.

CHAPTER

6

Sesame

Faisal Islam*, Rafaqat A. Gill*, Basharat Ali*, Muhammad A. Farooq*, Ling Xu**, Ullah Najeeb[†], Weijun Zhou*

*Institute of Crop Science and Zhejiang Key Laboratory of Crop Germplasm, Zhejiang University, Hangzhou, China

**Zhejiang Province Key Laboratory of Plant Secondary Metabolism and Regulation, College of Life Sciences, Zhejiang Sci-Tech University, Hangzhou, China

[†]Department of Plant and Food Sciences, Faculty of Agriculture and Environment, University of Sydney, Eveleigh, NSW, Australia

INTRODUCTION

History of Sesame Cultivation and Utilization

Sesame (*Sesamum indicum* L.), a member of the Pedaliaceae family, is an erect annual herb commonly known as sesamum, benniseed, or simsim. It is one of the oldest and most traditional oilseed crops, valued for its high-quality seed oil. According to recent archeological findings, sesame cultivation was derived from wild populations native to South Asia, and its cultivation was established in South Asia from the time of the Harappan civilization and spread west to Mesopotamia before 2000 B.C. (Fuller, 2003). Despite other claims, it was first cultivated in Africa and later taken to India at a very early date (Alegbejo et al., 2003; Purseglove, 1969). Tunde-Akintunde et al. (2012) suggested that sesame was the main oil crop grown by the Indus Valley Civilization and was likely transferred to Mesopotamia around 2500 B.C. The Assyrians used its oil for different purposes such as food, salves (ointments), and medicine, while Hindus believed it to be sacred. Sesame is also known as the "queen of oilseeds," but it is actually an orphan crop. Little research into sesame has been undertaken and, hence, it is not a crop mandated by any international crop research institute (Bedigian and Harlan, 1986; Bhat et al., 1999), despite being cultivated in both tropical and temperate zones of Africa, Asia, Latin America, and some parts of the southern United States (Bedigian, 2010d; IPGRI and NBPGR, 2004).

Sesame is adaptable to a range of soil types, although it performs well in well-drained, fertile soils of medium texture (typically sandy loam) at neutral pH. Generally, sesame is a short-day plant that may grow in long-day areas. Depending upon light intensity and day period in various regions, sesame has produced genotypes with different photoperiod requirements. Some cultivars

Breeding Oilseed Crops for Sustainable Production. http://dx.doi.org/10.1016/B978-0-12-801309-0.00006-9

like Venezuela 51 are day neutral. Depending upon the cultivar, the crop matures in 75–150 days after sowing (Ashri, 2007). Sesame seeds have both nutritional and medicinal value because they are rich in fat, protein, carbohydrates, fiber, and essential minerals. They are used in sweets such as sesame bars and halva (dessert), and in bakery products or milled to get high-grade edible oil (Bedigian, 2004). Seeds are chemically composed of 44–57% oil, 18–25% protein, 13–14% carbohydrates (Borchani et al., 2010). Sesame oil is famous for its stability as a result of its resistance to oxidative rancidity after long exposure to air (Global Agri Systems, 2010). Generally, the oil contains 35% monounsaturated fatty acids and 44% polyunsaturated fatty acids (Hansen, 2011).

Sesame oil has significant resistance against oxidation as a result of it containing endogenous antioxidants including lignins and tocopherols (Elleuch et al., 2007; Lee et al., 2008). There are two types of lignins, (i) sesamin and (ii) sesamolin, in sesame oil. Sesamolin is converted to sesamol after roasting. The molecular structure of sesamol consists of phenolic and benzodioxide groups. The phenolic group is responsible for antioxidant activities in a number of natural products, while the benzodioxide group is involved in anticancer and antioxidant activities. Recent research into sesame showed that it contains immunoglobulin E (IgE)–mediated food allergens (Agne et al., 2003; Dalal et al., 2002; Pastorello et al., 2001). The preponderance of allergy to sesame seed is associated with its wider use in baked and fast food products.

SESAME PRODUCTION AND TRENDS

Average global sesame yield in 2010 was 3.84 million metric tons grown on an area of 7.8 million hectares (Fig. 6.1). The largest producer of sesame seeds in 2013 was Burma. The world's largest exporter of sesame seeds was India, and Japan the largest importer (FAOSTAT, 2013). World total cultivation area under sesame was 9,398,770 ha, producing 4.76 million tons (FAO, 2013), which has risen from 1.12 million tons in the early 1961s (FAOSTAT, 2015). Major sesame-producing countries in 2007 were India, China, Burma (Myanmar), Sudan, Ethiopia, Uganda, and Nigeria, while in 2001 the largest producers of sesame were China and India followed by Burma (4.2 million tons) and Sudan (3 million tons). Asia and Africa grow 70% and 26% of world sesame, respectively (Hansen, 2011). At present, 2 million

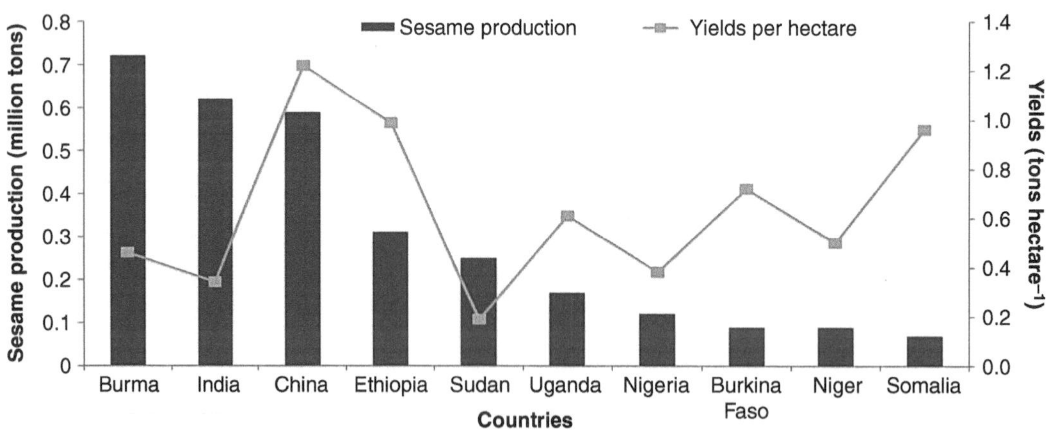

FIGURE 6.1 **Top 10 sesame producers in the world.** *Source: FAO (2012)*

TABLE 6.1 Production of Sesame Seed and Oil in Different Regions of the World

Region	Sesame seed (tons, %)	Sesame oil (tons, %)	Total (tons)
Africa	808,864	1,336,763	2,145,627
America	176,469	22,467	198,936
Asia	2,180,387	809,355	2,989,742
Australia	0	2,000	2,000
Europe	1,430	31,025	32,455

Source: FAO (2012).

tons of sesame seed is produced in Asia, which comprises more than 50% of world sesame seed production, while Africa produces 1.8 million tons of sesame seed (43% of world sesame seed production). Some 0.8 million tons of sesame oil is produced in Asia, while Africa produces 28% of world total sesame oil. Europe produces a small amount of sesame seeds and converts them into oil, showing that there is higher demand for oil than for the seeds themselves, whereas the opposite applies in both Asia and Africa (Hahm and Kuei, 2014). These data should be considered only a rough estimate of the situation as much of the harvest is consumed locally, and there is no record of domestic trade and local manufacturing. Only a small portion of the global sesame harvest enters international trade. In most cases, the oil is used locally, especially for frying.

The cultivable land used worldwide for sesame production has generally remained constant over the years, but in a number of countries the crop has become marginalized due to higher remuneration from other crops and labor shortages pushing sesame to less fertile areas. It looks likely that sesame production might decrease in the future (Bennett, 2008). Countries in the European Union are more productive in terms of average sesame yield per hectare; for example, Italy produces 7.2 metric tons per hectare (Table 6.1) (FAO, 2012). In contrast, some Asian and African countries have a relatively low sesame yield. For example, Pakistan and Kenya produce 1.2 metric tons and 400 kg per hectare of sesame, respectively (Agricultural Statistics of Pakistan, 2008–09; OngInjo and Ayiecho, 2009). The total area under sesame cultivation in Kenya has grown at a slow rate, from 20,000 ha in 1980 to 27,000 ha in 2010 (FAO, 2012). These huge yield gaps among different sesame growers are due to knowledge gaps, poor crop management techniques, and lack of advanced technologies. Countries such as Nigeria have a large market potential for the production of sesame seeds for domestic purposes and export markets, where production figures have increased steadily since 1980, reaching 67,000 metric tons by 1997 (Alegbejo et al., 2003) and have reached 149,685 metric tons in 2013 (FAOSTAT, 2015). There was a rise in sesame seed production from 98 million kg to 152 million kg from 2003 to 2007 worldwide (CBN, 2009).

MAJOR CHALLENGES TO SUSTAINABLE SESAME CROP PRODUCTION

For the sustainable production of sesame, there is a need for abiotic stress tolerance to be integrated in sesame. This requires identification of the important stress factors that reduce sesame productivity. Basic studies have been carried out on the consequences of various abiotic stresses on sesame, mostly regarding breeding. Cultivated sesame encounters a number of stress factors including salinity, drought, waterlogging, and chilling. For example, sesame is mostly grown on arid and semiarid land where its productivity is limited by drought and salinity. Sesame has been found to be sensitive to salinity (Yousif et al., 1972), mostly to excessive calcium/sodium chloride ions in soil solutes. According to FAO crop tolerance data published in 1985, sesame salt tolerance parameters such as threshold – electrical conductivity of the extract in decisiemens per meter (ECe–dS/m) – and slope (percent per decisiemens per meter) have yet to be determined (Maas and Grattan, 1999). Moreover, sesame cultivars show significant variation in the degree of salt tolerance (Yahya, 1998).

Koca et al. (2007) observed the negative effects of salinity (50 and 100 mM) on the biochemical and antioxidant defense system of sesame, and consequently on growth. They found that some sesame cultivars (i.e., Orhangazi and Cumhuriyet) showed considerable reduction in root and shoot lengths, while lipid peroxidation was increased in response to salinity. They further found that Cumhuriyet was relatively more tolerant than Orhangazi. The enhanced tolerance of Cumhuriyet to salt stress was linked to its high-proline level, which plays a crucial role in inducing salt tolerance in plants by stabilizing proteins, regulating cytosolic pH, and scavenging hydroxyl radicals (Matysik et al., 2002). A similar phenomenon is observed in *Beta vulgaris* (Gzik, 1996), *Brassica juncea* (Jain et al., 1991), and *Medicago sativa* (alfalfa) (Petrusa and Winicov, 1997), where proline plays a significant role in plant growth under saline stress. Activation of reactive oxygen species (ROS) scavenging enzymes also seemed to play a role in salt tolerance, as was evident from the high level of superoxide dismutase enzymes along with H_2O_2 scavenging enzymes – ascorbate peroxidise, catalase, peroxidise – in Cumhuriyet compared with Orhangazi (Koca et al., 2007). Moreover, it was observed that growth parameters, lipid peroxidation, and proline accumulation were positively correlated with the salt tolerance of sesame (Matysik et al., 2002).

Sesame is usually grown under rain-fed conditions where precipitation is irregular. It is regularly subjected to mild to severe water deficit stress. The crop is sensitive to drought, especially at the vegetative stage (Boureima et al., 2011). In semiarid regions across the world, the production potential of sesame is often limited by drought stress (Boureima et al., 2012). The sensitivity of sesame to drought is reflected in the changes that occur subsequently in plant metabolism, growth development, and yield. Variable behavior in response to drought stress has also been noticed among various sesame cultivars, with some cultivars being highly resistant and others more susceptible (Boureima et al., 2011). Sesame has the ability to overcome drought by developing an extensive rooting system, although it experiences substantial yield losses if drought occurs when it is cultivated on marginal and low rain–fed areas. The effect of drought is more pronounced on sesame seed yield than other morphological characters. Kim et al. (2007) investigated the effect of drought at the postharvest stage on yield and found that water stress significantly decreased sesame yield by decreasing the number of seeds, although seed size was unaffected. This shows that the postflowering response to drought in sesame is associated with the production of fewer seeds, instead of sacrificing seed size.

Sesame falls into the category of chilling-sensitive plants (0–15°C), along with crops such as rice, maize, soybean, cotton, and tomato. These plants are unable to enhance freezing tolerance when exposed to low temperature (Levitt, 1980). For commercial sesame varieties, the optimum temperature ranges from 25°C to 27°C, while it requires 90–120 frost-free days to achieve optimal yields in cold regions. Sesame showed a considerable growth reduction below 20°C, while seed germination and growth were completely inhibited below 10°C (Oplinger et al., 1990). Low temperature not only affects plant growth but also deteriorates oil quality by reducing the lignin content of seeds (Beroza et al., 1955). Moreover, Beroza et al. (1955) further stated that low temperature reduces the efficiency of ROS scavenging enzymes produced in response to chilling stress, which shows the seriousness of the situation. Chilling stress could be the main player in cellular injury and the senescence of sesame plants. Limited work has been carried out on improving chilling tolerance in sesame, and researchers are trying to decipher the pathway for cold acclimation and come up with molecular changes (Gong et al., 2005; Dong et al., 2006) that might help in inducing chilling tolerance in sesame.

Sesame crops experience a reduction in growth and yield after 2–3 days of waterlogging, which frequently occurs when they are grown on soils that are poorly drained. This results in immediate senescence and a decline in crop conditions. Crops may experience substantial yield losses under excessive irrigation (Ucan et al., 2007). At various stages of growth, waterlogging considerably reduces plant growth, leaf axils per plant, biomass, seed yield, and net photosynthesis (Sun et al., 2009). This can lead to fungal attack by *Fusarium oxysporum* and *Macrophomina phaseolina*, which induce two serious diseases *Fusarium* wilt and charcoal rot, respectively, in sesame (Liu et al., 1993). Even drought-resistant sesame cultivars are very sensitive to high moisture levels in fields (Langham and Wiemers, 2001). At the genomic level, the sesame response to hypoxia results in downregulation of a number of genes that are involved in energy metabolism and changes the expression of genes related to flavone and flavonol biosynthesis, steroid biosynthesis, photosynthesis, cysteine and methionine metabolism, and glutathione metabolism pathways (Wang et al., 2012). Comprehensive evaluation of the waterlogging tolerance of different sesame varieties and plant types has identified several germplasm lines that have waterlogging tolerance, providing important material for genetic improvement (Sun et al., 2010).

BREEDING EFFORTS IN SESAME CROP FOR SUSTAINABLE PRODUCTION

Breeding efforts regarding sesame meal quality are aimed at reducing antinutritive components. Increasing the oil content, quality, and yield also remains a major aim in oilseed crop breeding. Here we undertake a general overview of breeding aims for sustainable and high-yielding production of sesame. Although sesame is an important world oilseed crop, few breeding efforts have been made for crop improvement (Ashri, 1987). Commercial production and extension is limited by low yield, pests and diseases, mature seed capsule destruction, and nonuniform capsule maturity (Langham and Wiemers, 2002). Other breeding challenges include flavor improvement (i.e., removal of bittering) and market rules that limit whole-seed marketing (Oplinger et al., 1990) with the drought tolerance as sesame usually grows in areas with average rainfall less than 600 mm per year.

The main breeding objectives of sesame are: (i) improvement of seed retention in the leaf axil (capsule) during/after ripening, (ii) high seed yield, (iii) pest and disease resistance, (iv) uniformity in plant type and agronomic characters like uniform leaf axil maturation, (v) lowering the production rate of bitter seeds, (vi) formation of higher numbers of leaf axils, and (vii) adaptation to various stresses (Ashri, 1987). In the past, sesame breeding was based on the use of naturally occurring variation or by inducing variation through inter and intraspecific hybridization or mutations (Lee and Choi, 1985; Pathirana, 1992; Dharmalingam and Ramanathan, 1993; Prabakaran et al., 1995). Sesame has a number of natural phenotypic variants (i.e., plant height; days to flowering; leaf axil number; length, shape, hairiness and difference in seed shape, color, and length; resistance to stress (drought), pests, diseases; and oil content). The existence of a number of these contrasting traits may help identify crossed offspring and accelerate crop improvement (Weiss, 1983). Sesame is a selfpollinated crop, although, depending upon topical conditions and insect populations, some outcrossing (10–50%) may also take place in field-grown crops. This is the reason approximately 180–360 m distance is maintained between plants for seed purity (Pathirana, 1994, Van Rheenen, 1968; Uzo, 1977, Weiss, 1983). Cross control is simple and can produce up to 50 seeds, which can help in developing varieties with desired traits.

Sesame yield is influenced by such diseases as *Alternaria,* white and angular leaf spot, stem rot, bacterial diseases, powdery mildew, and wilts (Nyanapah et al., 1995; Ojiambo et al., 1999; Oplinger et al., 1990). Farmers mostly control these diseases through fungicide application and cultivational techniques. Some of these diseases can also be handled by resistance breeding, especially as natural genetic variation for disease resistance exists. In Kenyan sesame cultivars, a varying degree of resistance against white and angular leaf spot has been observed. SIK 031 and SIK 013 showed resistance to white leaf spot, whereas SIK 031 and SPS 045 showed resistance to angular leaf spot (Nyanapah et al., 1995). Breeding plans that include resistant germplasm such as this can assist in producing high-yield varieties for disease-infected areas.

As sesame breeds produced by natural variation and selection are partially resistant to diseases like bacterial leaf spot, *Fusarium* wilt, and phyllody (Weiss, 1983), no cultivar with absolute resistance to these diseases exists. In some cases, resistance to diseases (*Sclerotinia, Fusarium, Rhizoctonia)* exists but these traits are not incorporated in commercial sesame cultivars. Similarly, many wild sesame lines show resistance to pests but these traits are also not incorporated in commercial varieties. Since wild sesame species in India and Nigeria are resistant to root rot, bacterial diseases, and pests, a number of researchers crossed *Sesamum indicum* with wild *Sesamum* species and selected breeds resistant to root rot. To reduce diseases in sesame, mutation breeding is also practiced, in which a number of useful morphological and physiological mutants are cultivated in disease-borne areas (Ashri, 1981, 1985; Micke et al., 1987). Similarly, gamma rays have also been used to produce mutants resistant to *Phytophthora* blight (Pathirana, 1992).

In many countries, inland production of sesame is limited due to seed shattering that results in huge yield losses. Up to 50% of sesame seeds are lost due to shattering before harvesting (Langham and Wiemers, 2002). Indeterminate sesame varieties contain both mature and immature seeds and, at or before harvesting, early capsules dry, open, and shed seeds. Sesame crops are manually harvested (99%), and harvesting costs are about 70% of capital costs since seed shattering makes harvesting difficult and more expensive (Langham and Wiemers, 2002; Khidir, 1972). Recently, Sesaco has developed shattering-resistant varieties named SESACOND that can be mechanically harvested (http://sesaco.publishpath.com/growing-sesame). Shattering-resistant or indehiscent-type mutants in sesame have been developed using gamma rays (Cagirgan, 1996, 2001). These mutants have closed capsules that prevent seed loss during or at the time of harvest but they are difficult to thresh and carry many other unwanted characteristics such as sterility, cupped leaves, twisted stems, decreased size of capsules, and low yield production (Langham and Wiemers, 2002; Day, 2000).

Crop-genetic diversity can be determined with the aid of morphological, biochemical, and molecular markers (Stuber, 1992). Morphological traits underlie genetic diversity and divergence in sesame and, based on agromorphological traits, several studies have exploited high genetic diversity in populations of sesame (Bisht et al., 1998; Arriel et al., 2007). Polymerase chain reaction (PCR)-based techniques such as amplified fragment length polymorphism (AFLP), simple sequence repeats (SSRs), intersimple sequence repeats (ISSRs), and random amplified polymorphic DNA (RAPD) have also been widely used in studies of genetic diversity in sesame (Salazar et al., 2006; Pham et al., 2009). The germplasm can be characterized by combined use of morphological and molecular markers. This represents the best option as it provides relative analysis of phenotypes from field experiments with molecular phenotypes and genotypes from laboratory studies. By comparing various methods in genetic studies, useful information can be provided to scientists and plant breeders regarding screening and selection (Pham et al., 2011).

Using molecular techniques, selection for the breeding process entirely depends on genotypes rather than phenotypes to make breeding programs more effective and rapid, although little information is available on the use of markers in sesame (Bhat et al., 1999). Genetic maps and closely linked molecular markers (using the AFLP technique) were commonly used to study genetic diversity. For the closed-capsule mutant trait, a link molecular marker was developed by Uzun et al. (2003) that helped identify and integrate shattering resistance during the breeding program. Sesame plants containing the indehiscent trait are extensively used for breeding in the United States to produce high-yielding sesame varieties. Mechanical harvesting of these cultivars is suitable and increases profitability due to increased seed yield per acre and reduced labor cost (Langham and Wiemers, 2002).

The main constraints to sesame production are associated with its highly variable performance and low seed production. Average world sesame production is about 340 kg per hectare, but up to 2250 kg per hectare has been obtained using advanced crop practices (Brigham, 1985, 1987). The main reason for low yields has to do with a lack of both breeding practices and research specially related to yield structures that act as the basis for sesame-breeding progress (Baydar 2005; Pathirana, 1995). Exploiting heterosis in sesame hybrids may increase seed yield by up to 100–500% (Uzun et al., 2003; Jada and Layrisse, 1995). New hybrids have fast growth rates and a high leaf area index (LAI). The increased oil content in new hybrids might be the result of the combined heterosis effects of different physiological traits (Banerjee and Kole, 2011). Moreover, Lee et al. (2005) demonstrated that some morphological traits are linked to maize grain yield. However, they were unable to explain the underlying mechanism of grain yield formation. Banerjee and Kole (2006) also showed that increased LAI in sesame plays a crucial role in oil production. This demonstrates the great potential of hybrid sesame plants for yield. The International Atomic Energy Agency (IAEA) mutation-breeding program has also developed higher yielding lines (www.JointFAO-IAEA-mutation.htm). Pungsankkae is an example of a high-yield mutant variety, which was released in Korea by crossing a Korean variety with the Israeli determinate mutant dr-45.

Sesame oil demand is increasing due to its high quality. It is generally accepted that sesame seeds contain 44–57% oil (Borchani et al., 2010), but Ashri (1998) and Baydar et al. (1999)

claimed it contained 34–63% oil. Like its variability in agronomic traits, sesame also has variability in oil content. Estimation of oil content of African sesame showed 29–51% variation (Were et al., 2006). This indicates that it is feasible to improve the oil content of sesame. The fatty acid composition in sesame seed oil has been reported by different researchers. Azeez and Morakinyo (2011) found variation in the fatty acid composition in Nigeria's sesame accessions and suggested that high-quality seed oil accession should be utilized in sesame-breeding programs to increase seed oil quantity and quality. The high linoleic and linolenic acid content in seeds reduces oil stability and deteriorates its quality. The fatty acid composition of sesame has also been improved using genetically modified sesame having high oleic acid and low linoleic and linolenic acids content. Conventional breeding methods were used to increase oil content in sesame and, successively, produced sesame varieties with high oil content (Baydar et al., 1999).

STEPS TOWARD THE ENHANCEMENT OF SUSTAINABLE DEVELOPMENT

Sesame is both a commercial and important crop for small farmers. Thus, as is the case with other oilseed crops, significant attention must be given to sesame usage and the concerns of farmers. Efficient seed dissemination will be key to obtaining optimal effects by investing in breeding programs and biotechnology (Fig. 6.2). Faltering sesame production and productivity in cultivated areas around the world is a matter of concern. Sesame is the sixth largest oil-producing crop in the world (FAO, 2002). The progressive increase in sesame cultivation and production holds the potential to improve food security and opportunities for economic growth in rural areas of the world. The main weakness in sesame cultivation and production is susceptibility to biotic and abiotic stresses including soil fertility, drought,

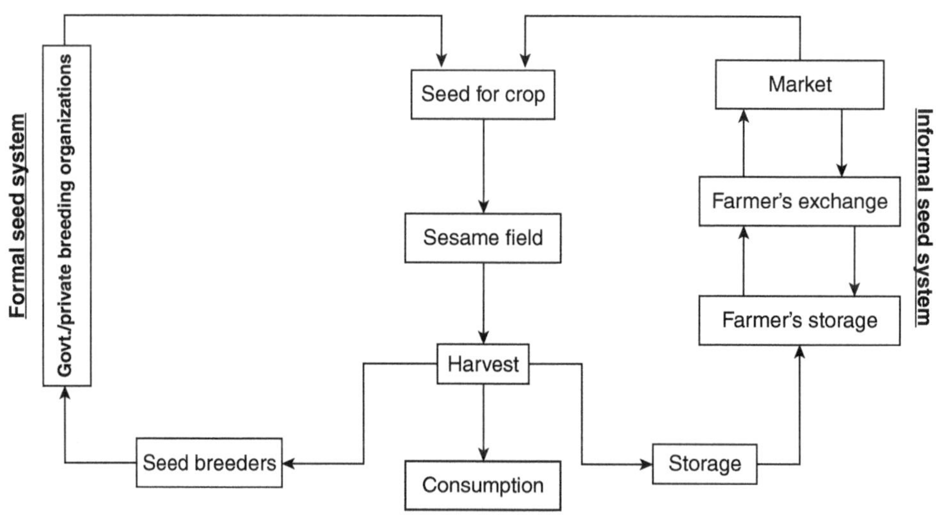

FIGURE 6.2 **Sesame seed system.**

pests, and diseases. Therefore, any scheme that enhances the stability of sesame crop production by developing resistant and tolerant plants will be effective for small-scale sesame-farming systems. Outbreaks of different diseases and pests in sesame crop in different regions of the world are well documented. These events may indicate a trend that could escalate in the future. In light of this trend, there needs to be a progressive, iterative process focused on limiting these restrictions.

For the sustainable development of sesame, the following recommendations are made:

- There is a need for strategies to promote development of sesame by enhancing key breeding programs both at the national and international level, so that open-pollinated and hybrid varieties can be developed.
- There is a need to enhance partnerships between national and international sesame breeding groups/organizations, so that effective and product-oriented breeding strategies can be implemented.
- There is a need to identify important diseases, pests, and other biotic stresses so that research can be intensified with the help of advanced techniques of biotechnology and breeding.
- There is little information in the literature on the sesame varietal preferences of farmers and the factors governing acceptance of new varieties. Variety substitution is known to occur between farmers but it is poorly understood. More knowledge is required on farmer varietal preferences and how they relate to adaptation in diverse sesame agroecologies.
- There is also a need to understand how sesame seeds are distributed among farmers in developing countries, even though a study of traditional seed collection and conservation has reported at local levels. This resulted in the construction of community seed banks such as those in India and Nepal. The lack of response by the private sector to seed distribution compels researchers to develop an informal seed system. However, to date, there is no clear plan for uncommercialized distribution of sesame seeds.

RECENT SUSTAINABLE DEVELOPMENTS IN SESAME

Development of Varieties Resistant to Biotic Stresses

Biotic stresses such as diseases, insects, and pests affect sesame crops adversely resulting in unpredicted losses in productivity and production. These losses are more significant where there is lack of resistant/tolerant varieties. Thus, the development of resistant cultivars will not only enhance sesame yield but also help cope with biotic stresses.

Drought-Tolerant Varieties with Enhanced Water Use Efficiency

Since sesame crops are generally cultivated in marginal areas where they face water stress, development of varieties with higher tolerance to moisture stress and water use efficiency will benefit both sesame cultivation and production.

Development of Hybrids

To improve sesame yield, research needs to target activities that are known to increase yield. Conventional breeding is unable to mobilize sufficient genetic variation, whereas hybrids extend the chance to mobilize greater genetic variation and heterosis. To maximize sesame yield there is a need to intensify hybrid development programs. With the encouraging results from sunflower hybrids in mind, it is likely a special network program may be launched to develop sesame and other oilseed crops.

Increase National Breeding Capacity

There have been few projects set up to breed new sesame varieties. A breeding program should be adopted such that sesame can be grown under a wider range of agroecological conditions, based on combining the resistance and tolerance traits for major constraints in each area. At the national and international level, a well-developed program should be provided to test and disseminate the varieties that come from such efforts. The breeding potential of germplasm accessions has scarcely been exploited to date. Recently, Korea, Kenya, India, and Israel have set up gene banks for efficient management and facilitation in sesame crop development (Mahajan et al., 2007).

Acknowledgments

The author acknowledges support from the National High Technology Research and Development Program of China (2013AA103007), the Special Fund for Agroscientific Research in the Public Interest (201303022), Zhejiang Provincial Top Key Discipline of Biology and its Open Foundation (2014C03, 2015D19), Jiangsu Collaborative Innovation Center for Modern Crop Production, and the Science and Technology Department of Zhejiang Province (2012C12902-1).

References

Agne, P.S.E., Rance, F., Bidat, E., 2003. Sesame Seed Allergy. Rev. Franc. Allergol. Immunol. Clin. 43, 507–516.

Alegbejo, M.D., Iwo, G.A., Abo, E.M., Idowu, A.A., 2003. Sesame: a potential industrial and export oilseed crop in Nigeria. J. Sus. Agri. 23, 59–76.

Arriel, N.H., Di-Mauro, A.O., Arriel, E.F., 2007. Genetic divergence in sesame based on morphological and agronomic traits. Crop Breed. Appl. Biotechnol. 7, 253–261.

Ashri, A., 1981. Increased genetic variability for sesame improvement by hybridization and induced mutations. In: Ashri, A. (ed.) Sesame: Status and Improvement. FAO Plant Production and Protection, Paper 29, Rome. pp. 141–145.

Ashri, A., 1985. Sesame and safflower: status and potentials. Proceedings of Expert Consultations. Food and Agriculture Organization of the United Nations, Rome, Italy.

Ashri, A., 1998. Sesame breeding. Plant Breeding Rev. 16, 179–228.

Ashri, A., 1987. Report on FAO/IAEA. Expert Consultation on Breeding Improved Sesame Cultivars. Hebrew University, Israel.

Ashri, A., 2007. Sesame (*Sesamum indicum* L.). In: Singh, R.J. (Ed.), Genetic Resources, Chromosome Engineering, and Crop Improvement: Oilseed crops, vol. 4, CRC Press, Boca Raton, FL, pp. 231–289.

Azeez, M.A., Morakinyo, J.A., 2011. Genetic diversity of fatty acids in sesame and its relatives in Nigeria. Eur. J. Lipid Sci. Technol. 113, 238–244.

Banerjee, P.P., Kole, P.C., 2006. Genetic variability for some physiological characters in sesame (*Sesamun indicum* L.). Sesame Safflower Newsl. 21, 20–24.

Banerjee, P.P., Kole, P.C., 2011. Heterosis, inbreeding depression and genotypic divergence for some physiological parameters in sesame (*Sesamun indicum* L.). J. Crop Improv. 25, 11–25.

Baydar, H., 2005. Breeding for the improvement of the ideal plant type of sesame. Plant Breeding 124 (3), 263–267.

Baydar, H., Marquard, R., Turgut, I., 1999. Pure line selection for improved yield, oil content and different fatty acid composition of sesame, *Sesamum indicum*. Plant Breed. 118, 462–464.

Bedigian, D., 2004. History and lore of sesame in Southwest Asia. Econ. Bot. 58, 329–353.

Bedigian, D., 2010. Introduction and early use of sesame in America: medicine, cookery and folkways. In: Bedigian, D. (Ed.), Sesame: The Genus *Sesamum*, Medicinal and aromatic plants–Industrial profiles series, Boca Raton, Florida, CRC Press, Taylor & Francis Group, pp. 389–421.

Bedigian, D., Harlan, J.R., 1986. Evidence for cultivation of sesame in the ancient world. Econ. Bot. 40, 137–154.

Bennett, B., 2008. Adapting to New Marketing Realities: Value Chain Analysis of Sesame in Tanzania. Natural Resources Institute, Kent, UK. http://www.sadc.int/fanr/agricresearch/icart/grantprojects/CRARF_SESAME/SESAME.

Beroza, M., Kinman, M.L., 1955. Sesamin, sesamolin, and sesamol content of the oil of sesame seed as affected by strain, location grown, ageing, and frost damage. J. Am. Oil Chem. Soc. 32 (6), 348–350.

Bhat, K.V., Babrekar, P.P., Lakhanpaul, S., 1999. Study of genetic diversity in Indian and exotic sesame (*Sesamum indicum* L.) germplasm using random amplified polymorphic DNA (RAPD) markers. Euphytica 110, 21–33.

Bisht, I.S., Mahajan, R.K., Loknathan, T.R., 1998. Diversity in Indian sesame collection and stratifi cation of germplasm accessions in different diversity groups. Genet. Res. Crop Evol. 45, 325–335.

Borchani, C., Besbes, S., Blecker, C.H., Attia, H., 2010. Chemical characteristics and oxidative stability of sesame seed, sesame paste, and olive oils. J. Agri. Sci. Tech. 12, 585–596.

Boureima, S., Eyletters, M., Diouf, M., Diop, T.A., Van Damme, P., 2011. Sensitivity of seed germination and seedling radicle growth to drought stress in sesame (*Sesamum indicum* L.). Res. J. Environ. Sci. 5, 557–564.

Boureima, S., Oukarroum, A., Diouf, M., Cisse, N., Van Damme, P., 2012. Screening for drought tolerance in mutant germplasm of sesame (*Sesamum indicum*) probing by chlorophyll *a* inflorescence. Environ. Exp. Bot. 81, 37–43.

Brigham, R.D., 1985. Status of sesame research and production in Texas and the USA. In: Ashri, A. (Ed.), Sesame and Sallower: Status and Potentials. Publ. 66, FAO, Rome, pp. 73–74.

Brigham, R.D., 1987. Status of sesame (Sesamum indicum L.) breeding in the USA. Agronomy abstracts. Annual meetings, p. 57. Am. Soc. Agron. Madison, WI (abstract).

Cagirgan, M.I., 1996. A preliminary report on the first induced mindehiscent capsule mutants in sesame. Eucarpia Meeting on Tropical Plants, Communications and Posters. CIRAD, Montpellier, France, p. 248.

Cagirgan, M.I., 2001. Mutation techniques in sesame (Sesamum indicum L.) for intensive management: confirmed mutants. Sesame improvement by induced mutations. IAEATE CDOC-1195, IAEA, Vienna, Austria, 31–40.

CBN, 2009. Annual report and statements of accounts for the year ended 31st December, 2008. Central Bank of Nigeria. pp. 142–145.

Dalal, I., Binson, I., Reifen, R., Amitai, Z., Shohat, T., Rahmani, S., Levine, A., Ballin, A., Somekh, E., 2002. Food allergy is a matter of geography after all: Sesame as a major cause of severe mediated food allergic reactions among infants and young children in Israel. Allergy 57, 362–365.

Day, J.S., 2000. Anatomy of capsule dehiscence in sesame varieties. J. Agri. Sci. 134, 45–53.

Dharmalingam, V., Ramanathan, T., 1993. Combining ability for yield and its components in sesame. Oleagineux 48, 421–424.

Dong, C.H., Hu, X., Tang, W., Zheng, X., Kim, Y.S., Lee, B.H., Zhu, J.K., 2006. A putative *Arabidopsis* nucleoporin, AtNUP160, is critical for RNA export and required for plant tolerance to cold stress. Mol. Cell Biol. 26, 9533–9543.

Elleuch, M., Besbes, S., Roiseux, O., Blecker, C., Attia, H., 2007. Quality characteristics of sesame seeds and by-products. Food Chem. 103, 641–650.

FAO, 2012. Food and Agriculture Organization. FAOSTAT Database. Available at http://faostat.fao.org/site/567/default.

FAO, 2002. Food and agriculture organization. FAOSTAT Database. Available from: http://faostat.fao.org/site/567/DesktopDefault.aspx?PageID=567#ancor.

FAOSTAT, 2015. Food and agriculture organization. FAOSTAT Database. Available from: http://faostat.fao.org/site/567/DesktopDefault.aspx?#ancor.

FAO, 2013. Food and agricultural commodities production: countries by commodity. FAOSTAT, Food and Agriculture Organization of the United Nations. Available from: http://faostat.fao.org/site/567/DesktopDefault.aspx?PageID=878#ancor.

Fuller, D.Q., 2003. Further evidence on the prehistory of sesame. Asian Agri-History 7, 127–137.

Global Agri Systems, 2010. Dehulled and roasted sesame seed oil processing unit. 18th August 2011. Available at http://mpstateagro.nic.in/Project%20Reports%20pdf/Dehulled%20and%20Roasted%20Sesame%20Seed%20Oil%20Processing%20Unit.pdf.

Gong, Z., Dong, C.H., Lee, H., Zhu, J., Xiong, L., Gong, D., Zhu, J.K., 2005. A dead box RNA helicase is essential for mRNA export and important for development and stress responses in *Arabidopsis*. Plant Cell 17, 256–267.

Gzik, A., 1996. Accumulation of proline and pattern of amino acids in sugar beet plants in response to osmotic, water and salt stress. Environ. Exp. Bot. 36, 29–38.

Hahm, T.S., Kuei, C.Y., 2014. Present and potential industrial applications of sesame: a mini review. J. Food Process. Pres., DOI: 10.1111/jfpp.12381 [accessed 26.08.2015].

Hansen, R., 2011. Sesame profile. 19th August 2011. Available at http://www.agmrc.org/commodities__products/grains__oilseeds/sesame_profile.

IPGRI and NBPGR (International Plant Genetic Resources Institute and National Bureau of Plant Genetic Resources), 2004. Descriptors for sesame (*Sesamum* spp.). National Bureau of Plant Genetic Resources, New Delhi, India.

Jada, Q., Layrisse, A., 1995. Heterosis and combining ability in hybrids among 12 commercial varieties of sesame (*Sesamum indicum* L.). Plant Breed. 114, 239–242.

Jain, S., Nainawatee, H.S., Jain, R.K., Chowdhury, J.B., 1991. Proline status of genetically stable salt-tolerant *Brassica juncea* L. somaclones and their parent cv 'Parkash'. Plant Cell Rep. 9, 684–687.

Khidir, M.O., 1972. Natural cross-fertilization in sesame under Sundan conditions. Exp. Agri. 8, 55–59.

Kim, K.S., Park, S.H., Jenks, M.A., 2007. Changes in leaf cuticular waxes of sesame (*Sesamum indicum* L.) plants exposed to water deficit. J. Plant Physiol. 164, 1134–1143.

Koca, H., Bor, M., Zdemir, F.O., Turkan, I., 2007. The effect of salt stress on lipid peroxidation, antioxidative enzymes and proline content of sesame cultivars. Environ. Exp. Bot. 60, 344–351.

Langham, D.R., Wiemers, T., 2001. Progress in mechanizing sesame in the US through breeding. In: Janick, J., Whipkey, A. (Eds.), Trends in New Crops and New Uses. ASHS Press, Alexandria, VA, pp. 157–173.

Langham, D.R., Wiemers, T., 2002. Progress in mechanizing sesame in the US through breeding. In: Janick, J., Whipkey, A. (Eds.), Trends in New Crops and New Uses. ASHS Press, Alexandria, pp. 157–173.

Lee, J.I., Choi, B.H., 1985. Progress and prospects of sesame breeding in Korea. In: Ashri, A. (Ed.), Sesame and Safflower: Status and Potentials. FAO Plant Production and Protection paper 66, FAO, Rome, pp. 137–214.

Lee, E.A., Ahmadzadeh, A., Tollenaar, M., 2005. Quantative genetic analysis of the physiological process underlying maize grain yield. Crop Sci. 45, 981–987.

Lee, J.Y., Lee, Y.S., Choe, E.O., 2008. Effects of sesamol, sesamin and sesamol in extracted effects of sesamol, sesamin, and sesamolin extracted from roasted sesame oil on the thermal oxidation of methyl linoleate 41, 1871–1875.

Levitt, J., 1980. Responses of Plants to Environmental Stress, in Chilling, Freezing, and High Temperature Stress, 1, Academic Press, New York.

Liu, J., Tu, L., Xu, R., Zheng, Y., 1993. The relationship between the waterlogging resistance and the genotypes and the vigor of root system in sesame (*Sesamum indicum* L.). Acta Agric. Bor. Sin. 8, 82–86, (in Chinese).

Maas, E.V., Grattan, S.R., 1999. Crop yields as affected by salinity. In: Skaggs, R.W., van Schilfgaarde, J. (Eds.), Agricultural Drainage, Argon. Monogr. No. 38. American Society of Agronomics, Madison, WI, pp. 55–108.

Mahajan, R.K., Bisht1, I.S., Dhillon, B.S., 2007. Establishment of a core collection of world sesame (*Sesamum indicum* l.) germplasm accessions. J. Breed. Genet. 39, 53–64.

Matysik, J., Bhalu, B., Mohanty, P., 2002. Molecular mechanisms of quenching of reactive oxygen species by proline under stress in plants. Curr. Sci. 82, 525–532.

Micke, A., Donini, B., Maluszynski, M., 1987. Induced mutations for crop improvement – a review. Trop. Agric. (Trinidad) 64, 259–278.

Nyanapah, J.O., Ayiecho, P.O., Nyabundi, J.O., 1995. Evaluation of sesame cultivars for resistance to *Cercospora* leaf spot. Afr. Agric. For. 60, 115–121.

Ojiambo, P.S., Ayiecho, P.O., Nyabundi, J.O., 1999. Effect of plant age on sesame infection by *Alternaria* leaf spot. Afr. Crop Sci. J. 7, 91–96.

OngInjo, E.O., Ayiecho, P.O., 2009. Genotypic variability in sesame mutant lines in Kenya. Afr. Crop Sci. J. 17, 101–107.

Oplinger, E.S., Putnam, D.H., Kaminski, A.R., Hanson, C.V., Oelke, E.A., Schulte, E.E., Doll, J.D., 1990. Sesame: alternative field crops manual. University of Wisconsin Extension, Madison, WI, USA, University of Minnesota Extension, St. Paul, USA. http://www.hort.purdue.edu/newcrop/afcm/sesame.html.

Pastorello, E.A., Pravettoni, V., Trambaioli, C., Pompei, C., Brenna, O., Farioli, L., Conti, A., 2001. Lipid transfer proteins and 2S albumins as allergens. Allergy 56, 45–47.

Pathirana, 1992. Gamma ray-induced field tolerance to *Phytophthora* blight in sesame. Plant Breed. 108, 314–319.

Pathirana, R., 1994. Natural cross-pollination in sesame (*Sesamum indicum* L.). Plant Breed. 112, 167–170.

Pathirana, R., 1995. Comparison of selection procedures in breeding for seed yield in segregating sesame populations. Euphytica 82, 73–78.

Petrusa, L., Winicov, I., 1997. Proline status in salt tolerant and salt sensitive *alfalfa* cell lines and plants in response to NaCl. Plant Physiol. Biochem. 35, 303–310.

Pham, D.T., Bui, M.T., Werlemark, G., 2009. A study of genetic diversity of sesame (*Sesamum indicum* L.) in Vietnam and Cambodia estimated by RAPD markers. Genet. Resour. Crop. Evol. 56, 679–690.

Pham, T., Geleta, M., Bui, T.M., Cach, B.T., Merker, A., Carlsson, A.S., 2011. Comparative analysis of genetic diversity of sesame (*Sesamum indicum* L.) from Vietnam and Cambodia using agro-morphological and molecular markers. Hereditas 148, 28–35.

Prabakaran, A.J., Rangasamy, S.R.S., Ramalingam, R.S., 1995. Identification of cytoplasm-induced male sterility in sesame through wide hybridization. Curr. Sci. 68, 1044–1047.

Purseglove, J.W., 1969. Tropical crops dicotyledons. Volume 1, Longman, London.

Salazar, B., Laurentin, H., Davila, M., 2006. Reliability of the RAPD technique for germplasm analysis of sesame (*Sesamum indicum* L.) from Venezuela. Interciencia 31, 456–460.

Sun, J., Zhang, X.R., Zhang, Y.X., Wang, L.H., Huang, B., 2009. Effects of waterlogging on leaf protective enzyme activities and seed yield of sesame at different growth stages. Chin. J. Appl. Environ. Biol. 15, 790–795.

Sun, J., Zhang, X.R., Zhang, Y.X., Wang, L.H., Li, D.H., 2010. Evaluation of yield characteristics and waterlogging tolerance of sesame germplasm with different plant types after waterlogging. J. Plant Genet. Res. 11, 139–146.

Stuber, C.W., 1992. Biochemical and molecular markers in plant breeding. Plant Breeding Rev. 9, 37–61.

Tunde-Akintunde, T.Y., Oke, M.O., Akintunde, B.O., 2012. Sesame Seed, Oilseeds, Uduak G. Akpan (Ed.), In Tech. Available from: http://www.intechopen.com/books/oilseeds/sesameseed.

Ucan, K., Killi, F., Gencoglan, C., Merdun, H., 2007. Effect of irrigation frequency and amount on water use efficiency and yield of sesame (*Sesamum indicum* L.) under field conditions. Field Crop Res. 101 (3), 249–258.

Uzo, J.O., 1977. Expression of hybrid vigour in sesame. Diss. Abs. Int. 8, 37–49.

Uzun, B., Lee, D., Donini, P., 2003. Identification of a molecular marker linked to the closed capsule mutant trait in sesame using AFLP. Plant Breed. 122, 95–97.

Van Rheenen, H.A., 1968. Natural cross-fertilization in sesame (Sesamum indicum L.). Samaru research bulletin/ Institute for agricultural research. Ahmadu Bello University (94).

Wang, L., Zhang, Y., Qi, X., Li, D., Wei, W., Zhang, X., 2012. Global gene expression responses to waterlogging in roots of sesame (*Sesamum indicum* L.). Acta Physiol. Plant 34, 2241–2249.

Weiss, E.A., 1983. Oilseed Crops. New York, Longman, Inc, pp. 282–340.

Were, B.A., Onkware, A.O., Gudu, S., Welander, M., Carlsson, A.S., 2006. Seed oil content and fatty acid composition in East African sesame (*Sesamum indicum* L.) accessions evaluated over 3 years. Field Crops Res. 97, 254–260.

Yahya, A., 1998. Responses to soil salinity of sesame (*Sesamum indicum* L.) and sugar beet (*Beta vulgaris* L.). Doctoral Thesis, Uppsala, Sweden.

Yousif, H.Y., Bingham, F.T., Yermason, D.M., 1972. Growth, mineral composition, and seed oil of sesame (*Sesamum indicum* L.) as affected by NaCl. Soil Sci. Soc. Am. Proc. 36, 450–453.

CHAPTER

7

Safflower

Vrijendra Singh, Nandini Nimbkar

Department of Genetics and Plant Breeding, Nimbkar Agricultural Research Institute, Phaltan, Maharashtra, India

INTRODUCTION

Safflower is humanity's oldest crop and has survived for over 4000 years producing quality oil rich in poly and monounsaturated fatty acids, flowers which have pharmaceutical properties which can cure many chronic diseases, and natural dyes. Safflower is a drought-resistant oilseed crop grown under rainfed conditions all over the world. Traditionally, safflower has been produced from China to the Mediterranean region and along the Nile Valley up to Ethiopia (Weiss, 1971). Currently, it is commercially cultivated in India, Mexico, the United States, Kazakhstan, Australia, China, Uzbekistan, Ethiopia, Spain, Turkey, Iran, Canada, the Russian Federation, and Pakistan. The cultivation of safflower, since time immemorial under diverse climatic conditions in different countries, suggests the sustainability of the crop is very high. Sustainability of a crop not only depends upon improvements made in terms of productivity and resistance to biotic and abiotic stresses, but also depends upon identifying new alternate uses for the crop in order to maintain high demand for it. Diversity in the crop for different traits plays an important role in making the crop sustainable. Similarly, identifying or developing new ideotypes also increases its suitability to diverse growing situations allowing it to fit into varied intercropping schemes, crop sequences, and population geometries, thus enlarging the scope for its sustainable production. In view of the immense importance of these attributes for making safflower sustainable, their role and scope is discussed in this chapter.

ECONOMIC IMPORTANCE

As mentioned above safflower is a multipurpose crop giving quality oil, rich in polyunsaturated fatty acids, which help reduce cholesterol levels in the blood, and producing brilliantly colored flowers which are used as a source of natural dyes for food and fabrics. In addition to meeting our energy and aesthetic demand for coloring food and fabrics, different parts of the safflower plant also have medicinal properties and are useful in treating many chronic diseases. The health benefits accrued from these are also discussed.

Breeding Oilseed Crops for Sustainable Production. http://dx.doi.org/10.1016/B978-0-12-801309-0.00007-0

PLANT AS A LEAFY VEGETABLE

Safflower plants at a tender age are consumed as a leafy vegetable in India. They are rich in vitamin A, iron, phosphorus, and calcium. Young seedlings are sold in the form of small bundles in rural markets in regions growing safflower, both in India and neighboring countries (Knowles, 1969).

SAFFLOWER SEED

Safflower seeds are diuretic and tonic (The Wealth of India, 1950). A study on the effects of safflower seed extract (SSE) supplementation on cardiovascular risk markers in healthy human beings revealed significant reductions in circulating oxidized LDL, autoantibody titres to malondialdehyde-modified LDL, the soluble form of vascular cell adhesion molecule-1 (sVCAM-1), and urinary 8-isoprostane. The study also indicated the index of arterial stiffness, brachial-ankle pulse wave velocity (baPWV), to be lower than the baseline in 11 of 20 subjects. This was also accompanied by a reduction in blood pressure. The study concluded that individuals with elevated oxidative stress, inflammation, and/or arterial stiffness may receive more benefits from SSE supplementation (Koyama et al., 2009). The investigation into the effect of *Carthamus tinctorius* extract (CTE) on bone morphogenetic proteins (BMPs), especially BMP-2 gene expression in human osteosarcoma MG63 cells as well as in rats, revealed no effect compared to a control group up to 8 h after incubation, but its effect was found to be significantly higher, 31% more than that of the control group, at 16 h incubation. BMP-2 gene expression by *in situ* hybridization was remarkably increased by a CTE-supplemented diet in a fracture group compared to the control group. Thus, it was concluded that *C. tinctorius* L. increased BMP-2 gene expression in both human osteosarcoma cells and fractured bone (Lee et al., 2006).

Studies carried out to test the hypolipidemic effect of powdered safflower seed (SSP) and safflower seed extracts prepared with ethanol (SSE) or hot water (SSW) on high fat and high cholesterol–fed rats for 5 weeks showed that all the safflower seed preparations significantly lowered the plasma cholesterol concentration while the plasma triglyceride concentration was only lowered by the supplementation of SSE and SSW. The hepatic total cholesterol contents were significantly lowered in the SSW group compared with the control group, whereas the hepatic triglyceride contents were significantly lower in both the SSE and SSW groups compared with the control group. The results suggested that supplementation of SSE or SSW is more effective than SSP in improving the artherogenic risk factors in high cholesterol–fed rats (Moon et al., 2001).

SAFFLOWER OIL

Safflower oil is the richest in polyunsaturated fatty acid (linoleic acid, 77%). The medicinal properties of linoleic acid were first reported in the United States in the 1960s indicating its usefulness in lowering serum cholesterol levels in laboratory tests on animals and humans and reducing the risk of heart attacks. This led to one of the first diet booms. This gave safflower oil recognition as healthy oil and it became the best-selling oil over a period of a few years (Smith, 1993). The high demand for polyunsatured fatty acid oil during this period sustained safflower production despite its prices doubling in order to compete with the high-yield, dwarf Mexican wheats which had just been introduced into the country (Smith, 1993). The development of safflower with high oleic content (which enhanced the shelf life of safflower oil and was suitable for deep and refrying with all the advantages of polyunsaturated fatty acid oil) in the United States further widened its utilization and hence sustainability.

A recent study from Ohio State University (Asp et al., 2011) reported that an 8 g daily dose of safflower oil, common cooking oil, for 16 weeks can improve health measures such as glycemia, inflammation, and blood lipids in obese, postmenopausal women who have Type 2 diabetes. The same group also discovered that daily supplementation with safflower oil for 16 weeks reduced abdominal fat and increased muscle tissue in the same group of women (Ohio State University, 2011). These findings suggested that a daily dose of 1 2/3 teaspoons of safflower oil in your diet is a safe way to reduce the risk of cardiovascular disease.

A study involving dietary supplementation with safflower oil for a period of 8 weeks discovered that platelet linoleic acid was increased significantly and this was also associated with an acute change in thromboxane B_2 levels. Moreover, an 8-week study of dietary modification with safflower oil revealed a reduced total serum cholesterol level of 9–15, low-density lipoprotein cholesterol by 12–20, and apolipoprotein B levels by 21–24 (all compared with baseline figures). It was also indicated that actual cholesterol synthesis is lower in diets rich in safflower oil compared to diets rich in butter (Cox et al., 1998). Safflower oil contains *N*-(*p*-coumaroyl) serotonin (CS), a potent antioxidant compound. An *in vitro* study revealed that this compound expresses unique growth-promoting activity for mouse and human fibroblasts. It is proposed that the growth-promoting effect of CS may not result from its antioxidant activity, since both antioxidative activity and an inhibitory effect of CS on proinflammatory cytokine production from human monocytes were noticed for similar doses (Takii et al., 1999).

SAFFLOWER FLOWERS

Safflower flowers are regarded as stimulants, sedatives, and emmenagogues. They act as a laxative if taken in large doses. They are used as a substitute and adulterant for saffron in the treatment of measles, scarlatina, and other exanthematus diseases (The Wealth of India, 1950). Safflower flowers dilate arteries, reduce blood pressure, increase blood flow, and thereby promote oxygenation of tissues. Safflower inhibits thrombus formation and over time, dissolves thrombi (Anonymous, 1972). Safflower usage three times a day for 4 weeks helped reduce heart arrythmia and hypertension (Wang et al., 1978). The products of safflower decoction can be used to treat male sterility (Qin, 1990) and infertility in women (Zhou, 1986). Safflower is useful in treating menstrual disorders (Zhong, 1992), arthritis (Yao, 1985), and respiratory ailments (Wang and Li, 1985).

Safflower has also been produced for its use as a dye, yielding red and yellow pigments with the crop commonly known as dyer's saffron. Safflower-dyed linen was used to wrap mummies in Egypt. India was the major trader of safflower flowers and dye and as noted by Watt (1908) India's export of dried flowers rose from 24,700 pounds in 1804 to 1,015,700 pounds in 1874–1875. Another report indicates that in the nineteenth century, India was exporting safflower dye, to the tune of 3289 cwt (164.45 t) valued at Rs 87,630, to Hong Kong, Java, Japan, and other countries in the East in 1941–1942. Most of the safflower dye exported was from Bengal (The Wealth of India, 1950). This clearly suggests that apart from the seed, safflower flowers were also a source of revenue for the farmers. With the advent of synthetic aniline dyes the safflower dye trade practically ceased from 1942 onwards, and there has been no revival of it so far despite of the release of a number of high-yielding, nonspiny cultivars recently in the country. In Japan, safflower dye used on silk was reserved for women of high rank. Safflower has great potential to be used for both freshly cut and dried flowers and its exploitation in European floriculture rapidly increased during the last two decades of the last century (Uher, 2008). Thus, the varied usage of safflower has helped to increase monetary returns from the crop, thereby imparting greater sustainability.

GENETIC RESOURCES

Safflower (*Carthamus tinctorius* L.) belongs to family Compositae or Asteraceae. The genus *Carthamus* comprises 25 species having 10, 11, 12, 22, and 32 pairs of chromosomes. Of the 25 species only *C. tinctorius*, having $2n = 24$, is cultivated and the rest are wild and weedy in habit. Among the wild relatives *C. oxyacantha* M. Bieb and *C. palaestinus* Eig have $2n = 24$ chromosomes and are cross compatible with cultivated safflower and also with each other (Ashri and Knowles 1960). Therefore, these two species also add variability to the gene pool of cultivated safflower. Safflower holds rich variability for different traits of economic importance. A germplasm collection of over 6000 accessions is maintained at the Directorate of Oilseeds Research, Rajendranagar, Hyderabad, India (Mukta, 2012). Similarly the US collection of 2383 accessions has been maintained at the Western Regional Plant Introduction Station (WRPIS) at Pullman, WA (Mukta, 2012). The National Crop Gene Bank, Institute of Crop Germplasm Resources, Chinese Academy of Sciences in Beijing collected 1100 accessions during the Eighth Five Year Plan (1996–2001) (Li et al., 1993). Other safflower-growing countries also maintain the germplasm.

PRESENT STATUS OF RESEARCH

Safflower research in general can be broadly categorized into four sections:

- Crop improvement.
- Development of agroproduction technologies for different agroclimatic conditions.
- Management of pests and diseases.
- Development of value-added products produced from safflower.

Research and development in safflower is being carried out in all four priority areas mentioned earlier. However, in the present chapter only the status of research in crop improvement for making safflower sustainable will be discussed in detail.

CROP IMPROVEMENT

The major objective in crop improvement in safflower has been to develop varieties for (i) high seed yield, (ii) high oil content, and (iii) greater resistance to diseases and pests, since the improvement in these traits makes the crop more remunerative and sustainable.

SEED YIELD

Since the beginning of safflower agriculture seed yield enhancement has been accorded the highest priority in safflower improvement. This was based upon understanding the interrelationship between seed yield and its components. This revealed that the number of branches per plant, seed weight per capitulum, and 100-seed weight either directly or indirectly made the largest contribution to seed yield (Nie et al., 1988; Patil et al., 1990; Singh et al., 2004). Since seed yield and oil content are complex traits their direct selection does not give the desired results due to the considerable genetic–environment interaction in safflower (Ranga Rao and Ramachandram, 1979). Similar observations for seed yield are reported by Nie et al. (1988).

Safflower shows a negative association between seed yield and oil content (Ranga Rao et al., 1977; Mandal, 1990; Patil et al., 1990). Similarly hull content also showed a negative association with oil content but had a positive association with seed yield (Ranga Rao et al., 1977; Sangale et al., 1982; Mandal, 1990). Seed weight and seed number, which contribute to seed weight per capitulum are considered to be direct components of seed yield (Ramachandram and Goud 1982; Ramachandram, 1985; Ghongade et al., 1993; Nie et al., 1993; Singh et al., 2004).

Thus, correlation studies illustrate the usefulness of the number of capitula per plant, yield per capitulum, and seed weight for seed yield enhancement in safflower. Similarly, to improve flower yield in safflower, the component traits, namely, number of primary branches per plant, number of capitula per plant, capitulum diameter, number of flowers per capitulum, petal area per flower, stigma length, and seed yield per plant should be considered for selection (Singh, 2004a).

The breeding methods used to develop cultivars in safflower have been summarized by Singh and Nimbkar (2007) and Mundel and Bergman (2008). The breeding methods commonly used to develop safflower cultivars are pureline and mass selection, the bulk population method, pedigree breeding, backcross breeding, recurrent selection, and the single-seed descent method (Singh and Nimbkar, 2007). Concerted efforts for varietal improvement in India have resulted in the development and release of 30 varieties for commercial production. Similarly, other major safflower-growing countries such as the United States, China, Mexico, Spain, and Canada have also made significant achievements in raising the productivity of the crop to make its cultivation more sustainable. The seed yield potential of safflower cultivars in different countries was further maximized with the determination of appropriate management approaches for achieving optimum yields. The combined crop improvement work for safflower in India has been undertaken while keeping in mind that the crop is to be grown under deep black cotton soils, having both high nutrient content and good water-holding capacity, both of which help the crop to attain its optimum growth achieving high seed yields. As a result of this, an ideotype producing a maximum number of branches and capitula, with high seed weight per capitulum and seed yield per plant was identified. As a consequence, a number of varieties having similar plant conformations, giving high seed yields, have been developed for different agroclimatic situations.

OIL CONTENT

Improvement in oil content results in an increase in oil yield per unit area of agriculture. Therefore, this trait has also received an equally high priority for improvement in almost all the oilseed crops grown for human consumption, as well as industrial use, the world over. The oil content improvement in safflower in the United States received high priority and the results have been spectacular. The industry in the United States in the mid-1970s paid a premium for seed containing greater than 34% oil. Over a period of 30 years, the oil content was increased from 15% to 48–50% (Nutrasaff developed by J. W. Bergman's Montana State University Safflower Breeding Program) thus achieving a 45% increase in oil levels (Mundel, 2008). The increase in oil content in the United States was attained incrementally by reducing hull content and using single genes such as those associated with striped hull, reduced hull, and partial hull (Mundel, 2008).

Genetic improvement of oil in India has not been given sufficient importance and as a consequence most of the cultivars developed so far have oil contents between 28% and 32%. However, the safflower improvement program at the Nimbkar Agricultural Research Institute (NARI), Phaltan, Maharashtra has paid sincere attention to oil improvement since the onset of the All India Coordinated Research Projects (AICRP) Center in 1980. The center's philosophy in this regard has been to enhance the oil content without compromising seed yield so that even without the premium for higher oil content, growers' interests would be secured. The efforts for improving oil content in safflower resulted in the development of a nonspiny safflower cultivar, NARI-6, containing 35% oil in its seeds. It was released for commercial production in 2001 and the development and release of a genetic male sterility based, nonspiny hybrid, NARI-NH-1, giving seed oil contents of 35%, was released in 2002. In addition, a thermosensitive genic male sterile (TGMS)-based hybrid NARI-H-23, having 35% oil in its seeds, has been developed and was released for commercial cultivation in 2014. Continued efforts to further enhance oil content have paid rich dividends and as a result NARI has developed a safflower variety, NARI-57, giving oil contents as high as 38–39%. This cultivar has been released for commercial production in India in 2015.

All these cultivars developed for higher oil content also recorded seed yield superiority over their counterparts, thus, giving simultaneous improvement for both seed and oil yields. In addition, the contribution of the AICRP (Safflower), which functioned up to 1992 at the Banaras Hindu University, deserves special attention since it too included improvement of oil content as one of its objectives. As a result, safflower variety HUS-305, possessing 35–36% oil content, was developed and released for nontraditional areas of safflower production in India during 1986–1987. However, the seed yield of HUS-305 in traditional areas of safflower cultivation was 10–15% lower than that of the national check A-1. Other AICRP centers have in general paid greater attention to the improvement of seed yield rather than oil content.

RESISTANCE TO DISEASES

Safflower is attacked by many diseases, which can be broadly categorized as (i) foliar and (ii) root diseases. The important diseases in each category are described below.

FOLIAR DISEASES

Resistance to foliar diseases is important in safflower since it enhances the sustainability of the crop. Safflower, under normal growing conditions, suffers a yield loss of 10–25% (Indi et al., 1988) and up to 50% under severe foliar disease infestation in India (Indi et al., 1986). Under a disease epidemic the entire crop can be wiped out – something that happened in India in 1997–1998 (Anonymous, 1998). Diseases cause direct yield losses of approximately 30% annually in Montana (USA), and in years when wet weather occurs as the crop matures losses in seed germination of 40% or more can occur (Irwin, 1976; Bramhankar et al., 2002; Bergman and Jacobsen, 2005). Among the foliar diseases the most prevalent are (i) *Alternaria carthami* Chowdhury, (ii) *Ramularia carthami* Zaprom., (iii) *Cercospora*, and (iv) rust.

Mundel and Huang (2003) reviewed disease resistance in safflower and indicated that major diseases of safflower can be controlled by breeding for disease resistance as well as by the use of cultural practices. A number of varieties developed in the United States were indicated to have resistance to *Alternaria* leaf blight. These varieties were developed by following mass selection of resistant plants from crossings of existing varieties in a disease nursery initiated in the early 1960s (Mundel, 2008). The varieties having field resistance to bacterial blight caused by *Pseudomonas syringae* van Hall were also developed by following the same method. The work for *Alternaria* resistance in safflower has also been successfully undertaken in Australia where the *Alternaria*-resistant cultivar Sironaria was developed by backcrossing. Gila was used as one of the parents in the backcrossing program used for the development of Sironaria. This variety occupied a sizable area in Australia. Espinoza et al. (2012) reported the development of *Ramularia*-tolerant, oleic-rich, elite safflower lines in Mexico. The oleic-rich lines were indicated to have a yielding ability as high as 3670 kg ha^{-1} in addition to possessing a high tolerance to *Ramularia*.

Safflower improvement, with specific reference to resistance to foliar diseases in India, is limited to the extent of screening genotypes for resistance to different foliar diseases. Though absolute resistance to foliar diseases has not been reported to date, moderate resistance for foliar diseases has been indicated. The genetics of inheritance of resistance to foliar diseases, studied by using moderately resistant genotypes, revealed the role of single dominant genes in controlling *A. carthami* Chowdhari, *Cercospora carthami* Sund. & Ramak, and *R. carthami* Zaprom. (Karve et al., 1981).

FUSARIUM OXYSPORUM AND ROOT DISEASES

Safflower wilt caused by *Fusarium oxysporum* Schlecht is the most important safflower disease in India (Sastry, 1996), Egypt (Zayad et al., 1980), and the United States (Klisiewicz and Houston 1962). The occurrence of *F. oxysporum* has been reported to be as high as 80% with average wilting of 53% under irrigated conditions causing heavy yield losses in the Indian Deccan

Plateau (17–20°E and 70–74°N) (Sastry, 1996). Similar observations in the region have also been indicated by Pedgaonkar et al., (1990). Resistance to root diseases (wilt) in general exists in the locally available safflower germplasm (Sastry and Ramachandram, 1992). Inheritance of resistance to *F. oxysporum* Schl. ex Fries, *Rhizoctonia bataticola* Bult., and *R. solani* Kuhn is reported to be controlled by single dominant genes (Karve et al., 1981). Further studies of inheritance of *F. oxysporum* resistance in safflower revealed the role of inhibitory gene action (Singh et al., 2001).

In order to develop wilt-resistant cultivars backcross breeding was carried out. These genotypes showed yield advantages of 31% over the national check A-1 (Singh et al., 2003). The screening of wilt-resistant BC_4F_6 genotypes developed through backcross breeding showed wilting ranged from 7.7% to 34% as against wilting of 92.4% in the wilt-susceptible check Nira under wilt-sick plot conditions. The seed yield from wilt-resistant genotypes was reported to be significantly higher than that of the wilt-susceptible check cultivar in the trial, thereby suggesting the usefulness of breeding for wilt resistance in safflower (Singh et al., 2008a). Wilt-resistant varieties, NARI-38 and NARI-57, were developed for commercial production in India during 2007 and 2015, respectively.

The development of wilt-resistant cultivars has made safflower production, in areas endemic to wilt, highly sustainable because of an enhanced yield due to a lack of wilt compared with wilting of up to 80% in wilt-susceptible cultivars.

PEST RESISTANCE

Safflower aphid (*Uroleucon compositae* Theobald) is the major safflower pest in India. The losses in yield of safflower due to aphids are reported to be 55–60% (Suryawanshi and Pawar, 1980), 55.9–67.9% (Basavan-Goud et al., 1981), and 24.2% (Shetgar et al., 1992). The source of tolerance to aphids is present in the locally available germplasm. Aphids are a serious problem and can be so devastating that the entire crop may dry up without giving any economic yield if left unprotected. Despite the serious threat to the crop, breeding for aphid resistance has not been attempted sincerely in India mainly due to the fact that the pest is easily manageable with the help of chemical sprays and manipulation of cultural practices especially advancing the time of sowing of the crop from the normal date for planting in the region.

Aphid resistance in safflower is reported to be under the control of both additive and nonadditive gene actions with a predominance demonstrated for nonadditive gene action (Singh and Nimbkar, 1993). Breeding for aphid resistance has been initiated recently in India since it is the most economical, time-tested, and ecofriendly method for controlling aphids. Aphid-tolerant safflower reduces the usage of harmful chemicals. This in turn helps to reduce the cost of crop production and keeps the environment safe by way of avoiding chemical usage. This all leads to a more sustainable production of safflower.

DEVELOPMENT OF HYBRIDS

The development of hybrids of safflower has been a major achievement in safflower research. In the United States presently this is being done with the use of cytoplasmic male sterility (Hill, 2005). India is the only country in the world which has released safflower hybrids based on genic male sterility (GMS), cytoplasmic genic male sterility (CMS), and thermosensitive genic male sterility (TGMS).

The hybrids, irrespective of the male sterility system they belong to, gave a 20–30% increase in seed and oil yields over the varieties. The success of hybrids based on GMS was limited due to seed production problems associated with the fact that the system warrants roguing of 50% fully grown fertile plants at the time of flowering, which is time consuming, labor intensive, and difficult to carry out on a commercial scale. Despite these seed production difficulties with GMS-based hybrids, a limited quantity of 1.5–2.0 quintals of seeds, especially of the nonspiny hybrid NARI-NH-1 since 2001 and spiny hybrid NARI-H-15 since 2005, have been produced every year and distributed to safflower growers across the country to assess their performance under different agroclimatic conditions. The evaluation of hybrids in different agroclimatic regions showed them to give excellent performance compared with varieties and thus demonstrated their superiority over them. This has made them popular in safflower-growing areas and established their reputation. The hybrid based on CMS developed by Mahyco and released in 2006 did not succeed however due to the low-yielding nature of the female parent.

Recently, the TGMS-based hybrid NARI-H-23 developed by NARI during 2012–2013 has been approved by the Central Variety Release Committee for commercial cultivation in India. The major advantage of the TGMS-based hybrid over GMS-based hybrids is that hybrid seed production problems which existed in the GMS system have been overcome. Since TGMS expresses 100% male sterility if grown in the winter, the roguing of 50% fertile plants, as is done in case of seed production of hybrids based on GMS system, is not required. The TGMS system gives 100% male sterility in winter when average minimum and maximum temperatures during capitulum development are less than 15 and 32°C, respectively (Singh et al., 2008b). The development of the TGMS system has made hybrid seed production commercially feasible in safflower. Thus, the popularization of the TGMS-based hybrid, NARI-H-23, is likely to enhance seed and oil yield output by 20–25% over the varieties which would help increase monetary returns and the areal extent of safflower in the country.

PROBLEMS CAUSING REDUCED SAFFLOWER AREA, PRODUCTION, AND PRODUCTIVITY

The higher profitability of pulses like gram, cereals like sorghum, and *Bacillus thuringiensis* (Bt) cotton due to higher yields and market prices has considerably increased the area in which they grow under rainfed cropping systems in India and of competing crops in other countries. The increase in area of these crops has affected less profitable crops like oilseed and particularly safflower. As a consequence there has been a reduction in the area used for oilseeds and a shifting of these crops to marginal soils which have poor nutrient levels and water holding capacities, thus, causing a further decline in their productivity and profitability. The oilseed situation has been further aggravated due to a change in climate – a rise in temperatures and the early or late withdrawal of the monsoon giving deficient or excess rain which adversely affects the sowing time and thus the production of safflower.

By moving some of the safflower crop from its preferred domain of black cotton soils to marginal soils, the productivity of this crop has declined drastically. This is not only because of the poor soil conditions, but also because the crop duration and plant type of the present cultivars was suited to black cotton soils with a sufficient moisture-holding capacity to sustain the crop for its entire duration of 4 months. Poor soils neither have adequate quantities of nutrients nor the ability to retain moisture. When current cultivars are sown on land with poor soil, the available moisture and nutrients can sustain them only up to 60–75 days, after which they produce branches and capitula, with flowering initiated in only a few of them. By the time flowering in 3–4 capitula of the plant is completed, the entire moisture within the soil is exhausted with none left for seed filling. Therefore, not only is the energy, used to produce the additional branches and the unflowered capitula, wasted but seed filling in the flowered capitula is badly affected and as a consequence a poor seed yield is obtained from these cultivars.

In view of the above, the major bottlenecks which need to be overcome to boost safflower productivity and production are:

- Lack of varieties suited to different growing conditions.
- Low harvest index.
- No screening of germplasm to identify suitable ideotypes meeting the needs of different growing conditions.
- Absence of agronomic evaluation of existing ideotypes to determine their usefulness for seed yield enhancement in safflower.
- No initiatives for tailoring plants to develop new, high-yielding ideotypes suited to mechanization for harvesting of petals and seeds.

OPPORTUNITIES TO OVERCOME THE BOTTLENECKS AFFECTING PRODUCTIVITY IN SAFFLOWER

Merits and Demerits of the Present Safflower Ideotypes

The modification of conventional ideotypes into more efficient forms with high harvest indices and greater responses to inputs and management has enabled a major breakthrough in the productivity of many crops. The productivity breakthrough in wheat and rice is the best example of ideotype modification in crop plants (Ramachandram and Ranga Rao, 1991).

As described earlier, improvement of the entire safflower crop since its inception has been completed to develop plant types (varieties) suited to deep black cotton soils, which have enough nutrients and water-holding capacity to sustain the crop for a duration of 4–4.5 months. This ideotype branches profusely giving primary, secondary, and tertiary branches producing 30–40 capitula, each having 20–25 seeds with a 100-seed weight around 5–6 g (Fig. 7.1). This ideotype works very well under normal seasons having average and timely rainfall capable of sustaining the crop for the entire period of its growth, but fails miserably to produce enough yield if the moisture in the soil is inadequate for sustaining the crop throughout its period of growth. This is because if moisture stress is experienced during the middle or at the end of the season, seed setting and seed filling are affected as the excessive vigor of the plant imposes a penalty on the plant in that more moisture is required to sustain it. As a result the available moisture is exhausted before seed setting and seed filling starts.

Presently, one ideotype is recommended for all safflower-growing areas irrespective of the agroclimate and soil types prevalent in different regions. This results in lower yields.

FIGURE 7.1 **Safflower with primary, secondary, and tertiary branches.**

Hence, to achieve an optimum potential there is a need to develop different ideotypes suited to particular soils and moisture regimes. Present safflower ideotypes developed for traditional areas of production need to be optimized in order to enhance harvest index to make the crop competitive with alternative crops. This is also needed to make safflower suitable for the efficient mechanized harvesting of flowers (petals), something that is difficult to achieve due to the presence of many small capitula on plants maturing at different times. Though a machine for petal harvesting has been developed at NARI, its efficiency is not comparable with manual (hand) picking of flowers of nonspiny cultivars (Rajvanshi, 2005). The gradual decline of manpower in agriculture and its working efficiency warrants the mechanized cultivation of crops, with safflower being no exception to this.

RESTRUCTURING OF THE SAFFLOWER IDEOTYPE

Knowles (1991) suggested that the ideal safflower plant (ideotype) should be investigated and cited an example of plants with appressed branching and further added that such genotypes when sown in denser populations may result in more heads per acre giving increased yields. The major drawbacks of the present safflower ideotype have already been discussed in this chapter. There are many opportunities to restructure the safflower ideotype depending upon growing conditions and the degree of mechanized production of the crop. The safflower germplasm is endowed with good genetic variability for different traits. Its utilization has resulted in a short-duration safflower developed at NARI and an extra early cultivar, JSI-99, developed at the AICRP Center at Indore. Plants producing only primary branches, single-headed unbranched plants, plants with dwarf statures, plants with large capitulum sizes, and plants with fasciated stems and capitula, etc. are some of the other important plant types. Each of these types has the potential to be used in combination with others to produce different ideotypes suited for different growing situations and hence might be used to increase the scope for a greater and wider sustainability of the crop. The different ideotypes envisioned for safflower are given below.

Development of Safflowers Producing Only Primary Branches

Studies carried out to examine the contribution of primary and secondary branch-based capitula to total yield of safflower grown under irrigated conditions at NARI during 2012–2013 revealed that primary branch-based capitula contribute 60% of the total seed yield as compared with 40% by the secondary branch-based capitula (Anonymous, 2012). The contribution of secondary-order capitula is likely to decline further when the crop is grown under rainfed conditions. Therefore, under rainfed conditions the secondary-order capitula, rather than contributing to seed yield, become a liability to the plant, adversely affecting seed setting and filling in the primary capitula and consequently decreasing its productivity. Similar studies were carried out by Ramachandram and Ranga Rao (1991) in safflower in which they studied intraplant variability for the yielding ability of capitula flowering over a period of 25–35 days. The study revealed that capitula which flowered during the first 15 days (mainly primary branch-based capitula) contributed more than 65% to the potential yield and the later-order capitula (secondary branch-based capitula) which flower 10–15 days after the first flowering are less productive.

A study reported that pruning of the side branches improved the number of seeds per capitulum, seed weight per capitulum, and weight per seed significantly compared with plants with branches, which were intact (Karve et al., 1976). These studies clearly suggest that restructuring the safflower plant to have only primary branches is likely to make it more productive, early maturing, and having an enhanced harvest index compared with the present ideotype which consists of secondary and tertiary branches. We have identified a genotype producing only primary branches in safflower (Fig. 7.2). The genetics of such genotypes needs to be studied in order to design a suitable breeding strategy to develop high-yielding cultivars with this ideotype. The population geometry may be suitably worked out to further enhance the potential of the new ideotype. The productivity thus enhanced would help to increase monetary returns from the crop to the grower.

Development of Short-Duration Safflowers

As elaborated earlier in this chapter, presently we have only one plant type recommended for all soil types irrespective of the nutrient status and water-holding ability of the soil. For a farmer, safflower in general is not a priority crop. It always receives secondary treatment and as a result it is often sown on poor soils with limited moisture. The present cultivars give very low yields in poor soils. Therefore, development of short-duration safflowers may be a boon to areas having poor soils. Poor soils can support the varieties maturing over 3 months

FIGURE 7.2 **Safflower with only primary branches.**

as compared with present varieties maturing in 4 months or more. Short-duration and short-statured safflowers, flowering at 40–45 days after sowing, producing 8–12 capitula, and maturing in 80–90 days after sowing, have been developed at NARI, and are appropriate for growing in relatively poor soils (Fig. 7.3).

A preliminary investigation to assess the potential of short-duration safflowers, at different plant spacings under rainfed conditions on poor soils as compared with regular safflowers, revealed that short-duration safflowers gave seed yields on par with the regular variety under normal spacing of 45×20 cm between rows and plants, respectively. The seed yield of the short-duration variety gradually increased over regular safflowers as the spacing between rows and plants was decreased. The short-duration genotype recorded a 20% increase in seed yield over regular-duration safflowers at a spacing of 30×10 cm between rows and plants, respectively. This suggests that on poor soils short-duration safflowers are more profitable than regular safflowers (Anonymous, 2014).

Rubis (2001) developed similar genotypes as described above. They were 3–4 weeks earlier than the normal plants and attained half their height. These types were characterized as daylength-neutral safflowers. Initially, the productivity of such genotypes was low and was improved by crossing them with very big, stiff stem plants and tall, late plants. Future safflower cultivars will no doubt be daylength neutral.

FIGURE 7.3 **Short-duration safflower (in flowering) compared with normal safflower.**

For Mediterranean environments a new safflower ideotype, which is characterized as a daylength-neutral plant, having a reduced rosette period, and maturing in 1800 growing degree days (GDD) as compared with 2500 GDD required for the regular safflower varieties cultivated in Southern Italy, has been suggested. The regular safflower in southern Italian conditions flowers and matures during the period when soil is facing moisture stress, which adversely affects the productivity of the crop. The shorter growing period of the new ideotype of safflower, requiring 1800 GDD, will enable the crop to flower and mature during the period when moisture is still present in the soil, helping the crop to attain its maximum productivity and making it economically remunerative to farmers (Corleto, 2008).

In India, the high suitability and profitability of the extra early variety, JSI-99, under nonconventional areas of safflower production have been demonstrated (Deshpande et al., 2005).

Ramamurthy (2013) indicated that short-duration varieties maturing in 70–100 days need to be developed in safflowers since this is the length of the growing period in most safflower-growing areas. The reduced length of growing period adversely affects the seed yield of presently grown varieties with 125–130 days to maturity.

In view of the above, short-duration safflowers developed at NARI are able to meet their requirements of nutrient and moisture from poor soils. Improvement of yielding ability of short-duration safflowers would further enhance their profitability and sustainability over other crops grown in such soils.

Development of Single-headed, Unbranched Safflowers

The occurrence of single-headed, unbranched safflowers is a regular feature in breeding nurseries. The unbranched plants in safflower, in general, show normal stems of regular height with dark-green, thick leaves, terminating in a single capitulum facing the sky. The capitula are compact having thick bracts producing either a few flowers or no flowers at all and giving either one or two seeds or no seeds. Contrary to the occasional occurrence of such plants in breeding nurseries, the safflower-breeding nursery at NARI during the rabi (dry season) 2013–2014, showed single-plant derivatives of a genotype segregating into a sizable number of single-headed, unbranched plants exhibiting a wide variability for number of flowers per capitulum, with some giving no seeds and others giving 50–60 seeds (Fig. 7.4). The occurrence of such plants has also been previously reported in safflowers (Sheelavanter et al., 1974; Deokar et al., 1978). The inheritance of the single-headed, unbranched plant was reported to be monogenic recessive (Deokar et al., 1978).

FIGURE 7.4 **Single-headed unbranched safflower.**

FIGURE 7.5 **Fasciated safflower.**

The development of the single-capitulum, unbranched ideotype in safflowers would be of specific significance in raising the productivity of the crop in terms of seed and flower yield and such plants would be highly suitable for mechanized harvesting of both flowers and seeds. Single-headed, unbranched plants in their present form cannot be used for commercial production, since they need to be improved to have a bigger head size containing 400–500 seeds or converted into the fasciated safflower (Singh et al., 2010) having a large capitulum with a 6–7 cm diameter (Fig. 7.5) and a short-duration background. The genotypes with big capitulum, fasciated stem, and short duration, as indicated previously, do exist in safflowers. The major advantages of developing single-headed, unbranched plants in normal and short-duration backgrounds will be the reduction in overall duration of the crop, which will not only help with proper seed filling but will also help avoid diseases and pests, thereby reducing losses. The change in ideotype would help in mechanized harvesting of flowers and seeds. This would further make the production of the crop more economical resulting in higher monetary returns to the farmers. The new ideotype would fit into many cropping systems in traditional and nontraditional areas and also into different intercropping systems with conventional and unconventional crops in different seasons, thus, raising overall safflower area and production in India.

References

Anonymous, 1972. Study of coronary heart disease prescription No. 2 on dilating blood artery and relieving smooth muscle spasms (in Chinese). Research Association of Coronary Heart Disease in Beijing, China.

Anonymous, 1998. Annual Progress Report on Safflower (1997–1998), Directorate of Oilseeds Research, Hyderabad, pp. 5.

Anonymous, 2012–2013. Annual Progress Report, All India Coordinated Project, Safflower, Phaltan, pp. 100.

Anonymous, 2014. Annual Progress Report of Nimbkar Agricultural Research Institute, p. 75.

Ashri, A., Knowles, P.F., 1960. Cytogenetics of Safflower (*Carthamus* L.) species and their hybrids. Agron. J. 52, 11–17.

Asp, M.L., Collene, A.L., Norris, L.E., et al., 2011. Time-dependent effects of safflower oil to improve glycemia, inflammation and blood lipids in obese, post-menopausal women with type 2 diabetes : a randomized, double-masked crossover study. Clin. Nutr. 30, 443–449.

Basavan-Goud, K., Kulkarni, R.A., Thontadarya, T.S., 1981. Estimation of crop loss in safflower due to the aphid *Dactynotus compositae Theobald* (Hemiptera Aphididae). Mysore J. Agric. Sci. 15, 279–282.

Bergman, J.W., Jacobsen, B.J., 2005. Control of Alternaria blight (*Alternaria carthami* Choud.) of safflower (*Carthamus tinctorius* L.) in the United States Northern Plains Region In: Esendal, E. (Ed.), Proceedings of Sixth International Safflower Conference, Istanbul-Turkey, June 6–10, 2005, pp. 215–221.

Bramhankar, S.B., Asalmol, M.N., Pardey, V.P., Ingle, Y.V., 2002. Effect of foliar spray of different fungicides on transmission of *Alternaria carthami* (Chowdhury) from plant to seed and seed to plant. Ann. Plant Physiol. 16, 112–114.

Corleto, A., 2008. A note on safflower plant ideotype suitable for Mediterranean environments. In: Kinghts, S.E., Pottar, T.D. (Eds.), Safflower unexploited potential and world adaptability. Proceedings of Seventh International Safflower Conference, Wagga Wagga, New South Wales, Australia, November 3–7, 2008.

Cox, C., Sutherland, W., Mann, J., De Jong, S., Chisholm, A., Skeaff, M., 1998. Effects of dietary coconut oil, butter and safflower oil on plasma lipids, lipoproteins and lathosterol levels. Eur. J. Clin. Nutr. 52, 650–654.

Deokar, A.B., Patil, F.B., Patil, G.D., 1978. Inheritance of monocapitulum character in safflower. J. Maharashtra Agric. Univ. 3, 257–258.

Deshpande, S.L., Saxena, M.K., Bharaj, G.S., Singh, J., 2005. Extra early safflower variety is the need of non-traditional region. In: Esendal, E. (Ed.), Proceedings of Sixth International Safflower Conference, Istanbul, Turkey, June 6–10, 2005, pp. 374.

Espinoza, X.M.O., Gracia, A.B., Casillas, E.A., Coronado, L.M., Molina, C.I.C.B.Y.N.A.A., 2012. Evaluation of elite oleic lines of safflower (*Carthamus tinctorius* L.) tolerant to *Ramularia carthami* in the north west of Mexico. J. Oilseeds Res. 29, 348–350, (Special issue).

Ghongade, R.A., Joshi, B.P., Navale, P.A., 1993. Correlation and path analysis of some yield components in safflower. J. Maharashtra Agric. Univ. 18, 240–243.

Hill, A.B., 2005. Hybrid safflower yield results. In: Esendal, E. (Ed.), Proceedings of Sixth International Safflower Conference, Istanbul, Turkey, June 6–10, 2005, pp. 352–359.

Indi, D.V., Lukade, G.M., Patil, P.S., 1986. Influence of *Alternaria* leaf spot (*Alternaria carthami* Chowdhury) on growth and yield of safflower. Curr. Res. Rep. 2, 137–139.

Indi, D.V., Lukade, G.M., Patil, P.S., Shambharkar, D.A., 1988. Estimation of yield losses due to *Alternaria* leaf spot in safflower (c.o. *Alternaria carthami* Chowdhury) under dryland conditions. Pesticides 22, 41–43.

Irwin, J.A.G., 1976. *Alternaria carthami*, a seed-borne pathogen of safflower. Aust. J. Exp. Agric. Anim. Husb. 16, 921–925.

Karve, A.D., Nagvekar, D.V., Nimbkar, N.B., 1976. Seed set on self-pollination in safflower. Indian J. Genet. 36, 108–110.

Karve, A.D., Deshmukh, A.K., Qadri, S.M. H., 1981. Breeding disease resistant safflower for cultivation in the Deccan peninsula of India. In: Proceedings of First International Safflower Conference, Davis, California, July 12–16, 1981, pp. 103–107.

Klisiewicz, J.M., Houston, B.R., 1962. Fusarium wilt of safflower. Plant Dis. Rep. 46, 748–749.

Knowles, P.F., 1969. Centers of plant diversity and conservation of crop germ plasm Safflower. Econ. Bot. 23, 324–329.

Knowles, P.F., 1991. Global Perspectives of Safflower. In: Ranga Rao, V., Ramachandram, M. (Eds.), Proceedings of Second International Safflower Conference, Hyderabad, India, 9–13 January 1989. ISOR, Directorate of Oilseeds Research, Hyderabad, pp. 13–16.

Koyama, N., Suzuki, K., Furukawa, Y., et al., 2009. Effects of safflower seed extract supplementation on oxidation and cardiovascular risk markers in healthy human volunteers. Br. J. Nutr. 101, 568–575.

Lee, K.-H., Sohn, O.-j., Ahn, J.-C., Kim, Y.-W., Park, S.-Y., Kim, J.-Y., 2006. *Carthamus tinctorius* L. increases BMP-2 gene expression during bone fracture healing in rats. Korean J. Physiol. Pharmacol. 10, 143–147.

Li, D., Zhou, M., Ramanatha Rao, V., 1993. Characterization and evaluation of safflower germplasm. Geological Publishing House, Beijing, China, pp. 260.

Mandal, A.B., 1990. Variability, correlation and path coefficient analysis of safflower (*Carthamus tinctorius* L.). Phytobreedon 6, 9–18.

Moon, K.D., Back, S.S., Kim, J.-H., et al., 2001. Safflower seed extract lowers plasma and hepatic lipids in rats fed high-cholesterol diet. Nutr. Res. 21, 895–904.

Mukta, N., 2012. Global strategies for safflower germplasm resource management. In: Murthy, I.Y.L.N., Basappa, H., Varaprasad, K.S., Padmavathy, P. (Eds.), Safflower Research and Development in the World: Status and Strategies. Indian Society of Oilseeds Research, Hyderabad, pp. 29–44.

Mundel, H.-H., 2008. Major achievements in safflower breeding and future challenges. In: Knights, S.E., Pottar, T.D. (Eds.), Safflower : Unexploited potential and world adaptability. Proceedings of Seventh International Safflower Conference, Wagga Wagga, New South Wales, Australia, November 3–7, 2008.

Mundel, H.H., Bergman, J.W., 2008. Safflower Breeding. In: Vollmann, J., Rajcan, I. (Eds.), Oil Crop Breeding Springer Series. Handbook of Plant Breeding. Springer, New York/Berlin, Chapter 13.

Mundel, H.-H., Huang, H.C., 2003. Control of major diseases of safflower by breeding for resistance and using cultural practices. In: Huang, H.C., Acharya, S.A. (Eds.), Advances in Plant Disease Management. Research Signpost, Trivandrum, Kerala, India, pp. 293–310.

Nie, Z., Chen, F., Shi, X.C., 1988. A study on selection indices of seed yield in safflower. Oil Crops China 1, 15–19.

Nie, Z., Chen, F., Shi, X.C., 1993. Path analysis of characters related to seed yield in safflower. Oil Crops China 3, 26–29.

Ohio State University, 2011. A dose of safflower oil each day might help keep heart disease at bay. Research News. Available from (researchnews.osu.edu/archive/saffoil.htm).

Patil, B.R., Deshmukh, S.G., Deshmukh, M.P., 1990. Studies on correlation and path analysis in safflower. Ann. Plant Physiol. 4, 86–91.

Pedgaonkar, S.M., Mehtre, G.A., Kurundkar, B.P., 1990. Incidence of safflower wilt in Marathwada region of Maharashtra. J. Maharashtra Agric. Univ. 15, 231–232.

Qin, Y., 1990. An analysis on the clinical treatment of male sterility of 300 cases by kidney-benefited and invigorating blood-circulation decoction (in Chinese). Jiangxi Trad. Chn. Med. 21, 21–22.

Rajvanshi, A.K., 2005. Development of safflower petal collector. In: Esendal, E. (Ed.) Proceedings of Sixth International Safflower Conference, Istanbul, Turkey, June 6–10, 2005, pp. 80–85.

Ramachandram, M., 1985. Genetic improvement of oil yield in safflower: problems and prospects. J. Oilseeds Res. 2, 1–9.

Ramachandram, M., Goud, J.V., 1982. Components of seed yield in safflower (*Carthamus tinctorius* L.). Genetica Agraria 36, 211–221.

Ramachandram, M., Ranga Rao, V., 1991. Some considerations for restructuring traditional plant type in safflower. In: Ranga Rao, V., Ramachandram, M. (Eds.), Proceedings of Second International Safflower Conference, Hyderabad, India, January 9–13, 1989. ISOR, Directorate of Oilseeds Research, Hyderabad, pp. 133–144.

Ramamurthy, V., 2013. Natural Resourcs management vis-à-vis safflower production. In Proceedings of Annual Group Meeting of Safflower. Nimbkar Agricultural Research Institute,August 29–31, pp. PB–2.

Ranga Rao, V., Ramachandram, M., 1979. Stability parameters for yield and its components in safflower. Mysore J. Agric. Sci. 13, 297–308.

Ranga Rao, V., Arunachalam, V., Ramachandram, M., 1977. An analysis of association of components of yield and oil in safflower (*Carthamus tinctorius* L.). Theor. Appl. Genet. 50, 185–191.

Rubis, D.D., 2001. Developing new characteristics during 50 years of safflower breeding. In: Bergman, J.W., Mundel, H.H. (Eds.), Proceedings of Fifth International Safflower Conference, Williston, N.D., U.S.A., 23–27 July 2001, pp. 109–111.

Sangale, P.B., Pagar, S.D., Daftardar, S.Y., 1982. Note on the association of seed oil content with hull and kernel and seed sizes of safflower. Indian J. Agric. Sci. 52, 613–614.

Sastry, K.R., 1996. Symptoms of wilt disease-clues for use in resistance breeding. In: Hegde, D.M., Raghavaiah, C.V., Pati, D. (Eds.), In: Proc. Training Programme on Breeding Approaches for Improving Productivity of Safflower and Group Meeting on Heterosis Breeding in Safflower. Directorate of Oilseeds Research, Hyderabad, pp. 25–32.

Sastry, K.R., Ramachandram, M., 1992. Differential genotypic response to progressive development of wilts of safflower. J. Oilseeds Res. 9, 297–305.

Sheelavanter, M.N., Krishnamurthy, K., Nalini, A.S., Shetty, A.N., 1974. A new monocapitulum non-branching safflower (*Carthamus tinctorius* L.). Sci. Cult. 40, 73–75.

Shetgar, S.S., Bilapte, G.G., Puri, S.N., Patil, V.V., Londhe, G.M., 1992. Assessment of yield loss in safflower due to aphids. J. Maharashtra Agric. Univ. 17, 303–304.

Singh, V., 2004. Annual Report of Adhoc Project on "Biometrical investigations of flower yield and its components and their maximization in safflower," Submitted to ICAR, New Delhi, pp. 35.

Singh, V., Nimbkar, N., 1993. Genetics of aphid resistance in safflower (*Carthamus tinctorius* L.). Sesame Safflower Newsl. 8, 101–106.

Singh, V., Nimbkar, N., 2007. Safflower (*Carthamus tinctorius* L.). In: Singh, R.J. (Ed.), Genetic Resources Chromosome Engineering and Crop Improvement Oilseed Crops, 4, CRC Press, Boca Raton, London, New York, pp. 167–194.

Singh, V., Galande, M.K., Deshpande, M.B., Nimbkar, N., 2001. Inheritance of wilt (*Fusarium oxysporum* f sp. Carthami) resistance in safflower. In: Bergman, J.W., Mundel, H.-H. (Eds.), Proceedings of Fifth International Safflower Conference Williston, ND and Sidney, MT, USA. 23–27 July 2001, pp. 127–131.

Singh, V. Rathod, D.R., Deshpande, M.B., Deshmukh, S.R., Nimbkar, N., 2003. Breeding for wilt resistance in safflower, In: Extended Summaries, National Seminar on stress management in oilseeds for attaining self-reliance in vegetable oils. Indian Society of Oilseeds Research, Hyderabad, India. 28–30 January 2003, pp. 368–370.

Singh, V., Deshpande, M.B., Choudhari, S.V., Nimbkar, N., 2004. Correlation and path coefficient analysis in safflower (*Carthamus tinctorius* L.). Sesame Safflower Newsl. 19, 77–81.

Singh, V., Ranaware, A.M., Nimbkar, N., 2008. Breeding for Fusarium wilt resistance in safflower. In: Knights, S.E., Pottar, T.D. (Eds.), Safflower: Unexploited potential and world adaptability. Proceedings of Seventh International Safflower Conference, Wagga Wagga, New South Wales, Australia. 37th November 2008.

Singh, V., Deshmukh, S.R., Deshpande, M.B., Nimbkar, N., 2008. Potential use of thermosensitive genetic male sterility for hybrid development in safflower. In: Knights, S.E., Pottar, T.D. (Eds.), Safflower: Unexploited potential and world adaptability. Proceedings of Seventh International Safflower Conference, Wagga Wagga, New South Wales, Australia. 3–7 November 2008.

Singh, V., Akade, J.H., Nimbkar, N., 2010. Inheritance of stem fasciation and twin/multiembryo seeds and genetic linkage between them in safflower. Indian J. Genet. 70, 281–287.

Smith, J.R., 1993. More than four decades of safflower development. In: Li, D., Han, Y. (Eds.), Proceedings of the Third International Safflower Conference, Beijing, China, 14–18 June 1993, pp. 861–867.

Suryawanshi, D., Pawar, V.M., 1980. Effect on the growth and yield due to aphid *Dactynotus sonchi* in safflower. Proc. Indian Sci. 89, 347–349.

Takii, T., Hayashi, M., Hiroma, H., et al., 1999. Serotonin derivative, *N*-(*p*-Coumaroyl) serotonin, isolated from safflower (*Carthamus tinctorius* L.) oil cake augments the proliferation of normal human and mouse fibroblasts in synergy with basic fibroblast growth factor (bFGF) or epidermal growth factor (EGF). J. Biochem. 125, 910–915.

The Wealth of India, 1950. A dictionary of Indian raw materials and industrial products. Raw Mater. 2, 83–88.

Uher, J., 2008. Safflower in European floriculture: a review, In: Knights, S.E., Potter, T.D. (Eds.), Safflower: Unexploited potential and world adaptability. Proceedings of Seventh International Safflower Conference, Wagga Wagga, Australia, 3–6 November 2008, pp. 1–5.

Wang, G., Li, Y., 1985. Clinical application of safflower (*Carthamus tinctorius* L.) (in Chinese). Zhejiang Trad. Chn. Med. Sci. J. 20, 42–43.

Wang, B., Yang, M., Pang, L., Yu, Z., 1978. The effects of safflower (*Carthamus tinctorius* L.) liquor on ischemic degree of cardiac muscle in different infarct regions of experimental myocardial infarction dog (in Chinese). Acta Pharmaceut. Sinica 14, 474–478.

Watt, G., 1908. The Commercial Products of India. John Murray, London, pp. 276–283.

Weiss, E.A., 1971. Castor Sesame and Safflower. Leonard Hill Books, London, pp. 901.

Yao, H., 1985. Treatment on bone arthritis with Shenjindan (muscle stretching powder) of 126 cases (in Chinese). Shandong Trad. Chn. Med. Sci. J. 1, 21–25.

Zayad, M.A., Yehia, A.H., El-Sabaey, M.A., Gowily, A.M., 1980. Studies on the host-parasite relationships of safflower root rot disease caused by *Fusarium oxysporum* Schlecht. Egypt J. Phytopathol. 12, 63–70.

Zhong, X., 1992. Clinical observation on the effect of Taohongsiwu decoction plus other herbs for treatment of vaginal bleeding of 128 cases (in Chinese). Zhejiang Trad. Chn. Med. Sci. J. 27, 300.

Zhou, W., 1986. Tian Ying's prescription was used for treatment on sterility of 77 cases (in Chinese). J. Trad. Chn. Med. 27, 31–32.

CHAPTER

8

Niger

A.R.G. Ranganatha, Anand Kumar Panday, Rajani Bisen, Surabhi Jain, Shikha Sharma

Project Coordinating Unit, All India Coordinated Research Project (Sesame and Niger), Indian Council of Agricultural Research (ICAR), JNKVV Campus, Jabalpur, India

INTRODUCTION AND ECONOMIC IMPORTANCE

Niger (*G. abyssinica*) is a little-known oilseed crop belonging to the Asteraceae. The major niger-growing countries of the world are Ethiopia and India. Niger is also reported to be grown in the West Indies, East Africa, and the United States (Kandel and Porter, 2002). In Ethiopia, 50–60% of the edible oil requirement of the country is met by niger seed oil, whereas only about 2% of the edible oil requirement is provided by niger seed in India (Riley and Belayneh, 1989; Dutta et al., 1994; Hiremath and Murthy, 1988). It is primarily grown on denuded soils in hilly and tribal pockets under input-starved conditions; however, it is a lifeline for tribal agriculture and the economy in India. India ranks first in areal extent, production, and export of niger in the world. Niger is less vulnerable to biotic and abiotic stresses than other oilseeds. The crop is capable of giving good seed yield under low soil fertility, moisture stress, and sensible crop management. It has the advantage of yielding oil and has a good degree of tolerance to insect pests, diseases, and wild animal attacks (Ranganatha et al., 2014). It has good potential for soil conservation, land rehabilitation, and as a biofertilizer. This means that crops following niger always demonstrate good growth. These attributes favor its cultivation on hilly areas, and marginal and submarginal land in and around forests (Ranganatha et al., 2009).

Niger seed is used as a human food that contains 35–45% oil with 18–20% protein. The oil is pale yellow with a nutty taste and a pleasant odor. The oil and seeds are free from any toxins and its taste is similar to desi ghee. The oil is readily subject to oxidative rancidation rendering its shelf life poor due to its high content of unsaturated fatty acid (oleic acid 38% and linoleic acid 50%). Niger oil is slow drying, used in food, paints, soaps, and as an illuminant. It is used as a substitute for olive oil, can be mixed with linseed oil, and is used as an adulterant for rapeseed oil, sesame oil, etc. It is also applied to treat burns. Niger oil is good at absorbing the fragrance of flowers and is therefore used as base oil by the perfume industry. Niger sprouts mixed with garlic and *tej* (a type of mead) are used to treat coughs. The seed is eaten fried, used as a condiment, or dried then grounded into a powder and mixed with flour

Breeding Oilseed Crops for Sustainable Production. http://dx.doi.org/10.1016/B978-0-12-801309-0.00008-2

etc. to make sweet cakes. Niger seeds, pressed with honey, are made into cakes in Ethiopia. Niger seed cake is a valuable cattle feed particularly for dairy cattle. Niger meal, with 30% protein and 17% crude fiber, in India can be used to produce feed cakes for calves. Niger plants are also used as a bee attractant plants for higher seed set.

PRODUCTIVITY SCENARIO

The AICRP on Niger, under the Indian Council of Agricultural Research, has refined the improved production technology to enhance the productivity of niger. The states contributing primarily to niger production in India are Madhya Pradesh, Odisha, Maharashtra, Karnataka, and Chhattisgarh. Besides these areas, the crop is also cultivated to some extent in the hilly areas of Andhra Pradesh, Bihar/Jharkhand, Gujarat, Uttar Pradesh, Rajasthan, Tamil Nadu, West Bengal, Assam, and Arunachal Pradesh in the northeastern hills region.

Productivity of niger is low, around 300–350 kg ha^{-1} in India. Niger production during 2012–2013 has increased by 10% and productivity by 86%, even after a reduction of 41% in the areal extent farmed since 1965–1966 (Ranganatha et al., 2014). Despite the fact that the largest area (86,900 ha) and production (29,800 t) is in Madhya Pradesh, Andhra Pradesh had the greatest seed yield of 750 kg ha^{-1} during 2012–2013 (Tables 8.1–8.3).

TABLE 8.1 Change in Area, Production, and Productivity of Niger From 1965–1966

	Area ('000 hectares)							
State	**1965–1966**	**1975–1976**	**1985–1986**	**1995–1996**	**2005–2006**	**2011–2012**	**2012–2013**	**% change between 1965–1966**
Madhya Pradesh	282.2	275.9	227.8	216.4	178.7	117.8	86.9	−69.2
Maharashtra	74.5	104.5	95.9	76.8	54.0	37.0	40.0	−46.3
Orissa	75.9	112.9	201.2	210.0	119.4	85.7	76.0	0.1
Andhra Pradesh	14.0	9.4	12.6	19.3	16.0	7.0	8.0	−42.8
Karnataka	22.5	47.4	54.8	44.7	30.0	21.0	12.0	−46.7
All India	525.4	615.3	635.0	600.6	414.4	364.4	310.4	−40.9

TABLE 8.2 Niger Production in the Main Growing States

	Production ('000 tons)							
State	**1965–1966**	**1975–1976**	**1985–1986**	**1995–1996**	**2005–2006**	**2011–2012**	**2012–2013**	**% change between 1965–1966**
Madhya Pradesh	42.6	52.6	49.2	44.6	36.0	20.7	29.8	−30.0
Maharashtra	10.5	13.4	20.7	15.2	14.0	12.0	12.0	14.3
Orissa	15.6	47.6	94.0	98.8	37.7	31.8	27.4	75.6
Andhra Pradesh	1.8	7.9	2.7	7.9	6.0	3.0	6.0	233.3
Karnataka	2.9	9.3	9.8	7.0	6.0	7.0	3.0	3.4
All India	91.9	150.5	192.3	190.3	108.0	98.1	100.8	9.68

TABLE 8.3 Niger Productivity in the Main Growing States

	Yield (kg ha^{-1})							
State	**1965–1966**	**1975–1976**	**1985–1986**	**1995–1996**	**2005–2006**	**2011–2012**	**2012–2013**	**% change between 1965–1966**
Madhya Pradesh	151.0	191.0	216.0	206.0	392.7	176.0	343	127.1
Maharashtra	141.0	128.0	216.0	198.0	259.3	324.0	300	112.8
Orissa	206.0	422.0	467.0	470.0	318.3	371.0	360	74.9
Andhra Pradesh	129.0	840.0	214.0	409.0	375.0	429.0	750	481.4
Karnataka	129.0	196.0	179.0	157.0	200.0	333.0	250	93.8
All India	175.0	245.0	303.0	317.0	261.0	269.0	325	85.7

Source: Directorate of Economics and Statistics, New Delhi.

It is a matter of concern for researchers and development personnel to ensure the real potential of new varieties is realized in the fields of farmers. In reality, a large gap always prevails between what is projected as the potential yield of a new variety at a research station and what is obtained in farm trials and further what is harvested by farmers. Technically, this is referred to as the yield gap for realizing the potential yield of new technologies. The yield gap is defined as the difference between the maximum attainable yield and the farm level yield. There are number of empirical pieces of evidence pertaining to yield gap analysis in agricultural crops.

ORIGIN AND DOMESTICATION

The evidence suggests that niger originated in the highlands of Ethiopia, north of 10°N latitude. The genus *Guizotia* belongs to the family Compositae, tribe Heliantheae, and subtribe Coreopsideae. Cultivated niger might have originated from the wild species *G. scabra* ssp. *schimperi* due to selection by Ethiopian farmers thousands of years ago. The genus *Guizotia* has six species. All species except *G. abyssinica* are wild and are endemic to East Africa especially Ethiopia. From Ethiopia, cultivated niger is believed to have spread to India during the third millennium B.C. along with other crops, such as finger millet. In India, variability in niger is concentrated in the central and eastern peninsular tracts. Bold-seeded late types are more prevalent in Orissa, whereas medium-seeded, high oil content types (40–43%) occur in Karnataka. Cold adaptable germplasm sporadically occurs in the eastern hills, especially in Sikkim. Early types are common in the Western Ghats and drought-tolerant types occur in the central peninsular region of India. Moderate salt-tolerant types of genetic resources have been reported in India. Such materials are valuable because there is a vast salt-affected area in India.

In Ethiopia, niger is grown mainly in midaltitude and highland areas (1600–2200 m above sea level). It is also cultivated in lower (500–1600 m) and higher (2500–2980 m) altitudes, which have enough rainfall. In Ethiopia, it is cultivated on waterlogged soils where most crops and all other oilseeds fail to grow and contributes a great deal to soil conservation and land rehabilitation.

In general, niger is a crop found in the cooler parts of the tropics. The major niger-producing areas in Ethiopia are characterized by a moderate temperature ranging between 15–23°C during the growing season. The Ethiopian types are short day and are suitable crops for hilly regions having high rainfall and humidity in India. Niger will grow on almost any soil as long as it is not coarse textured or extremely heavy. It is usually sown in areas with a rather poor soil or on clay soil under poor cultural conditions. It thrives well at pH levels between 5.2 and 7.3. Niger tolerates waterlogged soils, growing equally well on either drained soils or waterlogged clays. Niger is extraordinarily resistant to poor oxygen supply in the soil because of its ability to develop aerenchymas under these conditions. The aerenchymas develop only when niger plants are grown under highly waterlogged conditions and transport oxygen within the cormus into the root system (Prinz, 1976). Indian types are quantitative short-day types. Niger is adapted to areas where rainfall does not exceed 1000 mm per year. A higher precipitation of 1000–1200 mm and lower levels up to 350–500 mm are suitable, depending upon their variety and the distribution of rainfall. In India, rainfall of 600–1300 mm is optimum but a well-distributed rainfall of 500–800 mm can produce a reasonable yield (Sharma, 1990). Growth may be reduced with rainfall over 2000 mm, but plants can withstand high rainfall during the vegetative phase. For this reason, niger is the most suitable crop for hilly regions with high rainfall and humidity in India. Rainfall during seed setting and maturity leads to seed shattering and hence results in low yields. The microorganism involved in mycorrhiza association, *Glomus macrocarpus*, has been identified. Niger is salt tolerant (Abebe, 1975) but flowering is delayed and yield reduced with increasing salinity. It has been observed that crops planted after niger grow better and cultivation in soil in which niger was grown resulted in increased growth of the crop which followed niger (Yantasath, 1975).

TAXONOMY AND SPECIES RELATIONSHIPS

Niger is an annual dicotyledonous herb. Germination is epigeal and seedlings have pale green to brownish hypocotyls and cotyledons (Seegeler, 1983). The cotyledons remain on the plant for a long time. The first leaf is paired, small, and successive leaves are larger. The leaves are arranged on opposite sides of the stem, but at the top of the stem leaves are arranged in an alternate fashion. Leaves are 10–20 cm long and 3–5 cm wide. The leaf margin varies from pointed to smooth, leaf color varies from light green to dark green, and the leaf surface is smooth. The stem is smooth to slightly rough and the plant is usually moderate to well branched. The niger stem is hollow and easily breakable. The number of branches varies from 5 to 12 and in a very dense plant stand, fewer branches are formed. The color of the stem varies from dark purple to light green and the stem is about 1.5 cm in diameter at the base. In India, the plant height of niger is on average 1.4 m, but can vary considerably as a result of environmental influence and heights of up to 2 m have been reported in Ethiopia. The niger flower is yellow, white, and rarely slightly green. The heads are 15–50 mm in diameter with 5–20 mm long ray florets. The 2–3 capitulae grow together and have ray and disk florets. The receptacle has a semispherical shape and is 1–2 cm in diameter and 0.5–0.8 cm high. The receptacle is surrounded by two rows of involucral bracts. The capitulum consists of 6–8 fertile female ray florets with narrowly elliptic, obovate ovules. The stigma has two curled branches about 2 mm long. The hermaphrodite disk florets, usually 40–60 per capitulum, are arranged in three whorls. The disk florets are yellow to orange with yellow anthers, and a densely hairy stigma. The achene is club shaped, obovoid, and narrow (Seegeler, 1983). The head produces about 40 seeds. The achenes are black with white to yellow scars on the top and base and have a hard testa. The embryo is white.

Vavilov (1951) has described that niger belongs to the VIth Center of Origin of Cultivated Plants, the Abyssinian Center. The botanical name of niger is *G. abyssinica* (L.f.) Cass. The genus *Guizotia* was named in 1829 by Cassini for the French historian Guizot after karyomorphological studies. The ploidy level of niger is diploid with a somatic chromosome number of $2n = 2x = 30$. The genus *Guizotia* comprises annual (*G. abyssinica*, *G. scabra* ssp. *schimperi*, and *G. villosa*) and perennial (*G. scabra* ssp. *scabra*, *G. reptans*, *G. zavatarii*, and *G. arborescens*) types (Table 8.4). All except *G. abyssinica* are wild species, endemic to East Africa, especially Ethiopia. Based on morphological, phytogeographical, and cytological grounds, it is reported that *G. abyssinica* originated as a result of a large-grained mutant of *G. schimperi*, a weedy type of *G. abyssinica* in northern Ethiopia. *G. abyssinica* and *G. schimperi* cross easily, their F_1 meiosis

TABLE 8.4 Distribution and Characteristics of Niger Species

Distribution and characteristics of niger species	Distribution	Characteristics
Annuals *G. abyssinica* (L.f.) Cass	Cultivated in East Africa and Indian subcontinent.	—
G. scabra (Vis.) Chiov. ssp. *schimperi* (Schultz Bip. in Walp.) Baagoe	Native to Ethiopian highlands.	Moderately branched weed especially in niger cultivation, with outer involucral leaves being ovate, shorter than disk center.
G. villosa Sch. Bip.	Distributed in northern and south-western Ethiopian highlands.	Highly branched, weed found in open spaces.
Perennials *G. scabra* (Vis.) Chiov. ssp. *scabra*	Distributed widely from Ethiopia to Zimbabwe in the south and to the Nigerian highlands in the west.	Moderately branched, scabrous, suffrutescent herb, differentiated from annual ssp. *schimperi* by having outer involucral leaves that are lanceolate.
G. reptans Hutch	A rare species with a distribution restricted to Mount Kenya, the Aberdares, and Mount Elgon regions in East Africa. It is the only species not reported in Ethiopia.	Sparsely branched, a creeping, mat-forming herb.
G. zavattarii Lanza	Endemic in distribution around Mount Mega in southern Ethiopia and the Hari Hills of northern Kenya.	Erect, glandulous, with a predominantly shrub habit.
G. arborescens I. Friis.	Endemic to the southwest of Ethiopia and Mount Imontong on the border between Sudan and Uganda.	A rare arboreal species.

Source: Anonymous (2009).

is nearly complete (F_1's are highly fertile). Hence, *G. schimperi* may have an important role to play in the improvement of *G. abyssinica.*

G. villosa is an annual weed found in open spaces in northern parts of the Ethiopian highlands. Based on morphological, phytogeographical, and cytogenetical criteria, *G. villosa* likely evolved from *G. schimperi* or *G. schimperi*-like stock without involving any significant change in genome size. *G. abyssinica* and *G. scabra* are reported to be more closely related to *G. villosa* than to *G. scabra* ssp. *schimperi,* and that *G. zavattari* has relatively low affinity to the rest of the taxa. *G. reptans* is the only creeping mat-forming perennial species in *Guizotia* differing from the other species within the genus to such a degree that its habit and morphology are unique. These three taxa seem to be primitive. Karyotypes of *G. scabra* ssp. *scabra, G. reptans* and *G. zavattari* are asymmetrical suggesting these to be advanced taxa.

G. scabra contains two subspecies, namely *scabra* and *schimperi*. *G. scabra* ssp. *schimperi,* known as "mech" is a common annual weed in Ethiopia. *G. abyssinica, G. schimperi,* and *G. scabra* are diploid species characterized by 15 bivalents during prophase I/metaphase I of meiosis (Murthy et al., 1993). The former species is cultivated whereas the latter two are wild. Interspecific hybrids between these three species were generated and the F_1 hybrids were analyzed to assess cytogenetic relationships and crop evolution within the genus *Guizotia.* Meiotic chromosome configurations at diakenesis/metaphase I in pollen mother cells of hybrids averaged $0.25_{I} + 14.60_{II} + 0.15_{IV}$ for *G. abyssinica* × *G. schimperi;* $0.05_{I} + 13.6_{II} + 0.14_{III} + 0.58_{IV}$ for *G. abyssinica* × *G. scabra,* and $0.8_{I} + 12.7_{II} + 0.08_{III} + 0.88_{IV}$ for *G. schimperi* × *G. scabra* (I = univalent, II = bivalent, III = trivalent, and IV = quadrivalent). The results indicated that the genomes of *G. abyssinica* and *G. schimperi* are similar and homologous, whereas the *G. scabra* genome is only partially homologous to that of *G. abyssinica* and *G. schimperi.* Further, *G. abyssinica* might have originated from *G. schimperi* through selection and cultivation; chromosome translocations appear to have played a significant role in the divergence and differentiation of the three species.

ANTHESIS AND POLLINATION

Niger is essentially a cross pollinated crop. Niger has an exclusively entomophilous mode of pollination. The inflorescence of niger is called a capitulum or a head with two types of florets: ray florets and disk florets. The flowers develop from leaf axils in clusters of 2–5 capitula (heads). The receptacle has a semispherical shape, which is 1–2 cm in diameter and 0.5–0.8 cm high. It is surrounded by leafy involucral bracts arranged in numerous rows, the outer bracts being leaf-like, lanceolate, and up to 2 cm in length. The inner rows of bracts become progressively smaller and finally merge into flattened paleas of the receptacle, which is semispherical in shape and usually 1–3 cm in diameter. The niger flower has protandrous, sporophytic self-incompatibility due to which cross pollination by insects, especially bees, is the only alternative. The capitulum consists of 8–15 sterile ray florets on the periphery of the capitulum. They are sessile, zygomorphic, unisexual, pistillate, ligulate, epigynous, and yellow.

The ray florets are generally female, but often neuter. The ray florets form an outer whorl that is yellow to orange. The disk florets are hermaphrodite (bisexual florets), usually 45–65 per capitulum arranged in 3 whorls in centripetal order from the periphery. They are sessile, bisexual, complete, actinomorphic, and epigynous. The corolla is made of 5 petals (5 lobes fused together forming a tube-like structure) with a distinct upper section surrounding the style and stamens and a lower section surrounding the ovary, which is gamopetalous, pale yellow, and tubular. However, not all the florets end up forming seeds, which are generally 30 per head. The androecium has 5 stamens that are epipetalous, syngenesious, have long filaments, and basifixed anthers. The filaments are usually free. The points of dehiscence of the anther seem to be in the center between two pollen chambers in each. The gynecium is bicarpellary, syncarpous, inferior, and with a unilocular ovary with one ovule on the basal placenta. The stigma is bifid and densely hairy. The stigma lobes separate and curl backward in the evening. The stigma lobes rarely curl sufficiently to touch the style and the plant thus becomes selfincompatible. The ovary is surrounded by the lower section of the corolla tube. The stigma of both disk and ray florets are dry and the surface is unicellular, papillate, and appressed. Scanning electron microscope studies indicate that the stigma surface is the site of the incompatibility mechanism. Stigma receptivity studies showed that the stigma remain nonreceptive in the bud stage of a flower. High seed set was noticed when pollination was completed 24 h after anthesis. The opening of the flower starts in the outer whorl with the process continuing inward. A couple of hours after sunrise the florets start opening. Pollen grains are sticky and are liberated by anther sacs. At the same time the style elongates carrying the bifid stigma that extends well above the anthers, so that no contact is established between pollen grain and stigma of the same floret. Unfurling of the ray florets is visible with flowering proceeding concentrically over a period of 8 days.

Studies on controlled self pollination and cross pollination indicated niger is self incompatible. Two mechanisms operate to confer the cross-pollination nature of niger: (i) multiple oppositional alleles for self incompatibility and (ii) functional protandry. Niger is highly or entirely cross pollinated due to the fact that the flowers are very attractive to bees and other insects because of the bright color of the ray florets and the production of profuse pollen grains visible on the flower head. For complete flowering of the crop (capitulum) a minimum of 8–12 days are required. Niger seed is an achene 3–5 mm in length and 1.5 mm in width and has a lanceolate shape. The testa is hard, glossy, and normally black. The 1000-seed weight ranges from 1.6–6.6 g having a mean value of 3.8 g.

PLANT GENETIC RESOURCES

Genetic diversity of lines plays an important role in sustainable development (Esquinas-Alcazar, 2005) as it allows cultivation in the presence of various biotic and abiotic stresses. It is also important for selection of parents that can be used in breeding programs. Genetic diversity is studied by using morphological, biochemical, and molecular markers. Morphology has been a primary tool to estimate genetic differences among niger genotypes. However, morphological markers have a limited ability to estimate genetic diversity because of the strong influence of environmental factors, which make them highly dependent on cultivation conditions. Molecular markers overcome this limitation. Molecular marker techniques such as amplified fragment length polymorphism (AFLP), random amplified polymorphic DNA (RAPD), intersimple sequence repeats (ISSR), and simple sequence repeats (SSR) have been used in genetic diversity.

For niger to be competitive with other oilseed crops, its seed yield must be significantly improved. To achieve this objective, dwarf types must be developed which have uniform maturity resulting in reduced shattering losses. The Ethiopian germplasm collection contains short-stature plants that could be used for the development of dwarf types. There is genetic variation for the number of heads, something that could be utilized in breeding to select single-headed types. The presently used normal height niger material has many leaves and a low harvest index. Reducing plant height would decrease the number of leaves per plant and result in a better harvest index. The second important breeding objective for niger improvement is increasing the seed oil content. There exists great genetic variability for oil content in Ethiopian and Indian germplasm collections, which could be used, in a breeding program, to significantly increase oil content. An increase in oil content of 4–5% seems to be feasible. A genetic improvement program for niger must be based on its pollination behavior. Because of its self incompatibility, breeding procedures used in the improvement of cross pollinating crops are the methods of choice for niger breeding. The standard breeding procedure for cross-pollinating crops is recurrent selection. The resulting varieties are open-pollinated population varieties. The pollination behavior of niger is similar to that of sunflowers. Thus, niger is also a candidate for hybrid development. The identification of genetic male sterility in India, and recently in Ethiopia, has opened the way for the exploitation of heterosis in niger. Shattering losses in niger is a problem, therefore, the development of determinate types is the prime objective to reduce shattering losses as well as labor costs by promoting mechanical harvesting.

As modern, high-yielding, and genetically uniform cultivars are used, threats from diseases will increase. This will require increased emphasis on resistance breeding. Wild species of the genus *Guizotia* could serve as sources for disease resistance genes that could be introgressed into cultivated species. During the last few years, modern techniques of plant tissue culture, doubled haploid technology, and transformation are increasingly used by breeders. Protocols to regenerate plants from niger hypocotyl and cotyledon tissues and seedlings have been developed. Plant regeneration is dependent on genotype and media composition. If niger is susceptible to *Agrobacterium tumefaciens* infection, then it will be a good candidate for gene transfer. Dihaploid plants of niger have been produced by anther culture. Selfcompatible lines – such as dwarf and single-headed double-haploid plants – were obtained from anther cultures. Anther and microspore-derived dihaploids can be used to develop homozygous inbred lines in a short time. Recessive, simply inherited, and easily identifiable marker traits, which are important for niger seed production to ensure the genetic purity of varieties, could be obtained through microspore culture technology.

GENETIC DIVERSITY

Cultivated niger and its wild and weedy relatives have a pattern of distribution, and in areas where they occur naturally intercrossing takes place as cultivated and other species retain genomic homology and some interfertility. The wild and weedy relatives of niger are potential sources of genes which may be responsible for resistance to pests, diseases, and abiotic stresses. All the related species of cultivated niger are distributed in Africa. All the niger grown in India shows little variation with a narrow genetic base. However, some variability has been observed in the habit of the plant type, stem color, seed color, and time to maturity.

The niger accessions at the Bioversity International (formerly IPGRI, Rome) were collected from relatively secure areas of the country accessible by major roads. Germplasm-collecting missions deep into villages and remote areas are necessary. Niger germplasm apparently has only been collected from Ethiopia and India; there is no information regarding germplasm collection from other countries. The niger land races in Ethiopia, India, and possibly other countries have been geographically isolated for a very long time and therefore may carry valuable genes. Germplasm collection in other countries such as Uganda and Zimbabwe should also be carried out. The collection of germplasm should include wild relatives within the genus *Guizotia*. Characterization and evaluation of germplasm should be standardized and better descriptors developed. Valuable germplasm, particularly the Ethiopian gene pool, has not been characterized and needs thorough characterization and evaluation. Such germplasm evaluation could result in identification of valuable germplasm with high oil content, high seed yield, and male sterile and dwarf lines. Documentation of germplasm is equally important. Documentation of available germplasm, in addition to the present practice, should include a bar-coding system using desktop computers in a similar manner to the techniques used by large plant-breeding companies.

Germplasm exchange between Ethiopia and India, already a delicate issue, should be explored. However, the Indian materials are early maturing in comparison to long duration of Ethiopian lines. Therefore, elite lines (e.g., male-sterile and dwarf lines), rather than accessions, would be preferred for Ethiopia. Striking genetic differences exist between Ethiopian and Indian niger. These differences could be investigated using isozyme and molecular markers. It would be interesting to investigate which niger ecotype migrated to India. The variation among abat, bungne, and mesno niger ecotypes could be differentiated using isozyme and molecular markers. All species within the genus *Guizotia* are diploids with a chromosome number of $2n = 30$. Speciation within the genus *Guizotia* was not as a result of changes in chromosome number. The four species *G. abyssinica*, *G. scabra* ssp. *scabra*, *G. villosa*, and *G. scabra* ssp. *schimperi* are not reproductively isolated so hybrids among these species could be obtained with ease. It would be important to study the progenitor of niger using isozyme and molecular markers such as RAPDs. Solution to the problem of the phylogeny of the species could come from molecular techniques.

Niger originated in Africa and was only recently introduced as a crop to India; therefore, there is little possibility for exploration activities within the country. Thus, special attention must be given to the introduction of cultivated and wild germplasm. The available records indicate that a large number of accessions of niger germplasm have been introduced by the National Bureau of Plant Genetic Resources (NBPGR). Presently, 1600 accessions of cultivated germplasm are available for utilization in the country according to the All India Coordinated Research Project (Sesame and Niger). Introduced niger germplasm has been characterized and evaluated for important morphoagronomic traits.

EXPLORATION AND COLLECTION

Sincere efforts have been made for the collection and conservation of niger genetic resources through crop-specific and a few multicrop/regional-specific explorations. So far, a total of 1600 indigenous collections have been made in major niger-growing areas and the diversity is mainly represented in roadside fields. The collections exhibited variability in days to maturity, plant height (46–196 cm), number of primary branches per plant (3–17), number of capitula per plant (30–112), number of seeds per capitulum (13–47), 1000-seed weight (1.6–6.0 g), seed yield per plant (0.7–4.1 g), oil content (30–47%), seed color (light black, black, dark black, gray, and golden), seed shape (cylindrical, straight, curved, and sickle shaped), flower color (pale, yellow, and orange), stem color (green, purplish green, purple, and deep purple both with and without blotches), leaf color (pale green, light green, green, and dark green) and leaf margin (entire, serrate, and dentate).

Local native types of niger still constitute a reservoir of underutilized diversity. The collections maintained in the germplasm depositories in India do not represent wide genetic diversity. Many of these collections are either sporadic or confined to roadsides and, more so, the germplasm from some pockets of the country are not at all represented. Therefore, interior areas where maximum variability is expected need to be explored to broaden the existing base. In addition, interior areas in the tribal belts of Gujarat and Rajasthan, Dharmpuri in Tamil Nadu, the northeastern hills region of Assam, Arunachal Pradesh, Manipur, Mizoram, Nagaland, Meghalaya, and particularly Garo and Khasi Hills should be explored. The collections also need to be made from other countries of the Indian subcontinent where variability is expected: namely, the Surlahi and Palchauk areas of Nepal; and the hilly regions of Comilla, Jamalpur, Faridpur (winter niger), Tangail, Anikgonj, Dhaka, Chandpur, Lakshmipur, Barisa, and Sarapur of Bangladesh.

Fresh collections in remote areas, so far not explored, and recollections warranted by the loss of germplasm are imperative. Genetic resources such as released varieties, obsolete varieties, breeding lines with specific traits, prebreeding materials, advanced cultivars, mutants, and cytogenetic stocks/testers need to be conserved. It is essential to (i) identify the gaps in the existing collection on the basis of crop geography, biosystematics, and crop ecology; (ii) set up priorities with regard to types to be collected and areas to be covered; and (iii) formulate a sampling strategy and constitute a team for collection.

African germplasm, particularly from Ethiopia, is considered a good source of high-yield, bold-seeded material that shows resistance to waterlogging and drought, grows in lowland areas, and exhibits vigorous growth and late maturity. Genetic differences caused by a long history of geographical isolation are reported to exist between Indian and African gene pools. To broaden the genetic base, introductions need to be made from African countries like Ethiopia, Sudan, Uganda, Tanzania, Malawi, Zimbabwe, and Zaire and Asian countries like Nepal and Bangladesh.

CONSERVATION

In India, the facilities for *ex situ*, long-term conservation of germplasm has been developed at NBPGR. For safe conservation the base collection has been preserved for long-term storage at –20°C. In addition to long-term storage, medium-term storage at 4°C is also used. There is an urgent need to identify the fraction of diversity that has already been conserved and, therefore, the fraction that has not, and also to find out what can be exploited and how this can be achieved. Simultaneously, threatened resources, including wild species, should be accorded priority.

Niger, being a cross-pollinated crop due to selfincompatibility, has its germplasm stored at different centers, maintained by sibbing (bagging together a group of plants) to avoid intercrossing among accessions. Presently, a total of 1600 working collections of niger are being maintained and regenerated.

In India, genetic variability has been tapped through straight selection of locally adapted material. A total of 20 improved varieties have been developed for different niger-growing states. Improved plant-breeding methods have so far been least employed in niger. The genetic base of cultivars is narrow. Therefore, germplasm utilization needs to be intensified for genetic improvement of niger.

GENETIC IMPROVEMENT

Development of Inbred Lines and Hybrids

Pollination in niger occurs primarily by insects and only to a limited extent by the wind. Development of hybrids is one of the primary objectives of breeding, although improved open-pollinated and synthetic varieties also have importance, especially in tribal areas where hybrid varieties are not feasible for economic reasons. Development of inbred lines and hybrids of niger is difficult, and concrete efforts in this direction are required. The major reason for hybrid development is inbred development. For the development of inbred lines optimal individual plant selection (IPS) from the population is chosen on the basis of following characters:

1. Higher number of branches per plant.
2. Higher number of flowers per plant.
3. Higher number of capitula per plant.
4. Higher number of seeds per capitula.
5. Bold-seededness and high 1000-seed weight.
6. Higher seed yield.

The seeds of individual plants are stored separately after threshing. Plants and their progeny should be raised in isolation of location, space, time, and pollinator barrier. Again the earlier procedure is to be repeated until fixation of the inbred has occurred.

Factors Affecting Genetic Potential

From agroeco analysis of growing seasons it is evident that drought stress is an important factor limiting productivity in the major niger-growing states. In addition, other biotic stresses, especially disease, are causing significant recurring yield losses. Niger production in India continues to face challenges as the crop is often grown on marginal land, which is inherently deficient in moisture and nutrients. The limited genetic variability in the working collections maintained and used in breeding research has restricted genetic advances. The crop is often placed in the vulnerable position of being affected by many diseases and pests. Therefore, it is necessary to restructure crop improvement strategies with a focus on agroecology and agronomic traits in order to increase productivity and stability. Considering the adverse ecologies in which niger are cultivated, breeding research should have a geographical or regional focus. However, seed yield and oil content will continue to be important traits in niger crop improvement.

SELFINCOMPATIBILITY

The selfincompatible nature of niger and its exclusively entomophilous mode of pollination renders selfing a difficult process. High inbreeding depression, up to 91%, is associated with poor seed setting with generation advancement from S_3 onwards (Ramachandran and Menon, 1979). The outcrossing nature of *Guizotia* species can be inferred from cross compatibility between various taxa that produces viable and fertile hybrids (Murthy et al., 1993; Dagne, 1994, 2001). However, direct evidence regarding selfincompatibility is only available for *G. abyssinica* (Ramachandran and Menon, 1979; Riley and Belayneh, 1989; Nemomissa et al., 1998). Self incompatibility in niger is of the sporophytic type (Prasad, 1990; Nemomissa et al., 1998) which causes the inhibition of pollen germination or the twisting of the pollen tube over the surface of the papillae (Prasad, 1990). Selfcompatible genotypes were also reported in low frequencies (up to 5%) in the Ethiopian gene pool (Getinet and Sharma, 1996; Nemomissa et al., 1998).

Several aspects of self incompatibility have been compared in self and cross pollination experiments in which seed set was considered to be a measure of compatibility. Results indicate the presence of around 10 self incompatibility alleles and 1 self compatibility allele at the S locus in niger. Most of these alleles behave differently in pollen and pistil interactions. About two thirds of the allelic interactions in pollen and pistil were dominant/recessive and one third codominant. Reciprocal cross pollinations (RCPs) resulted in progeny with similar levels of compatibility within and among populations because of a wide distribution of S alleles in the populations and, consequently, low population differentiation at the S locus. An analysis of molecular variance revealed that only 2% of the total genetic variation at the S locus differentiates the populations.

During flowering initiation, selfing or sib mating is followed by covering two or more representative adjacent plants in a muslin cloth bag or nylon mosquito net bag in order to exclude insects. The shaking of bags or bagged plants is attended to on alternate days to ensure pollination, the seeds within the bag will be S_0 and will be used to raise the S_1 generation. Shrivastava and Shomwanshi (1974) have suggested a procedure for crossing and producing crossed seed, and indicated that flowers from two plants should be rubbed together. However, rubbing the florets may result in some selfing. Thus, it may be preferable to attempt specific crosses by removing all the disk florets, just as the bud begins to open, while leaving the female ray florets intact. By bagging the emasculated capitula and pollinating the ray florets the next morning, when the stigma arms have opened, several crossed seeds per capitulum can be obtained. In niger, a very high amount of cross pollination, almost reaching 100%, can be obtained. Selfing with parchment paper bags reduced seed setting drastically to only 0.2 seeds per capitulum as compared with 40–60 seeds in open pollination.

Anthesis takes place between 8.00 A.M. and 8.45 A.M. and disk florets open around 10.30 A.M. during sunny days and 11.30 A.M. during cooler days under Indian conditions. The number of capitula open ranged from 2 to 5 with a flowering period of about 6–7 days. The percentage of seeds set ranged from 0.10% to 2.11% when bagging a capitulum; 0–5.35% when bagging two capitula; 3.33–10.16% when bagging three capitula; 0–1.415% when bagging a single branch; and 2.43–4.68% when bagging a whole plant. This could be attributed to self incompatibility in niger. Under open-pollinated conditions the percentage of seeds set ranged from 26.32% to 56.57% with a mean of 42.59%.

Maintenance of line is a problem because of the self incompatibility of the crop. However, sibbing of the collections can be carried out by covering 10–50 plants within a muslin bag or mosquito net to exclude insects and then rubbing the flowering capitula together 2–3 times a week. Seed increase of advanced lines can be done in isolated plots, preferably at least 1 km distant from other niger crops.

Inheritance of Characteristics

As the nutritional quality of niger oil is good, efforts were first made to get genetic information on different yield attributes. Since academic efforts were being continued and applied efforts were meager little progress has been made. However, of late a few genetic studies on seed yield and yield component characters have been carried out (summarized in Table 8.5). Studies into yield component analysis have been carried out by different workers. High-heritability coupled with high genetic advances results in gene action being additive. However, high-heritability coupled with low genetic advances results in gene action being nonadditive in controlling that particular character. Accordingly, the reports of different workers are furnished in Table 8.5. Based on these reports, niger breeders should exploit additive gene action by using mass selection to improve high-yielding varieties. For long-term improvement of the polygenic characters of both additive and nonadditive gene action, population improvement and synthetic development should be followed.

BREEDING METHODS

Mass Selection

Mass selection is a procedure in which individual plants are selected for specific traits with high heritability of phenotypic appearance. Procedures are listed below:

- S_1 selection.
- Half-sib progeny selection.
- Full-sib progeny selection.

Recurrent Selection

Recurrent selection is reselection generation after generation, with intermating of selected plants to produce a population for the next cycle of the selection (Figure 8.1). The idea of this method is to ensure the isolation of superior inbreds from the population subject to recurrent selection. The isolation of an outstanding inbred line depends on two factors: (i) the proportion of superior genotypes present in the base population from which lines are isolated and (ii) the effectiveness of selection during the inbreeding of desirable genes. The process is as follows:

- A large number of plants (10,000–20,000) are selected from open-pollinated populations.
- Seeds from selected plants are analyzed in laboratories for oil content, husk percentage, etc. Finally, about 1,000 plants are selected.
- A portion of seeds from each selected plant are sown in the progeny-testing nursery, one row per plant, in two replications, every third row being a control.

TABLE 8.5 Gene Action for Different Traits Based on Heritability and Genetic Advance

Character	Gene action	References
Seed yield and yield attributes	Nonadditive	Bilaiya (1987)
Plant height	Additive	Channarayappa (1987)
Capitula per plant, 1000-seed weight, and days to flower	Additive	Bilaiya et al. (1992)
Seed yield and yield traits	Nonadditive	Upadhyaya and Reddy (1997); Patil et al. (2000); Patil et al. (2001)
Capitula per plant and plant height	Additive	Nema and Singh (1965)
Seed yield and capitula per plant	Additive	Payasi et al. (1987)
Days to maturity	Nonadditive	
Capitula per plant	Additive	Misra (1992)
Days to flower, maturity, and plant height	Nonadditive	
Seed yield and harvest index	Additive	Mathur and Gupta (1993)
Days to flower, maturity, and protein and oil content	Nonadditive	
Branches per plant and 1000-seed weight	Additive	Patel et al. (1993)
Days to flower, branches per plant, and seeds per capitulum	Additive	Mishra (1995)
Days to flower, plant height, and seeds per capitulum	Additive	Pradhan et al. (1995)
Secondary branches and seeds per capitulum	Additive	Borole and Patil (1997)
Days to flower and maturity	Nonadditive	
Plant height and capitula per plant	Additive	Patil (2000)
Seed yield, capsules per plant, and seeds per capitulum	Additive	Sreedhar et al. (2005)
Days to maturity and 1000-seed weight	Nonadditive	
Seed yield, branches per plant, capitula per plant, seeds per capitulum, and 1000-seed weight	Additive	Patil and Duhoon (2005, 2007), Patil (2003)
Days to flower and maturity	Nonadditive	
Plant height	Environment plays a major role	

- Progeny-testing of plants, whose families performed well during the first season (Season 1), is repeated during Season 2 using remnant seeds from selected plants.
- Remnant seeds of elite plants giving the best families are sown in spatially isolated multiplication fields during Season 3 for crossing *inter se.*
- Each entry in the controlled pollination plot is harvested and the identity of individual plants maintained. After lab analysis, the seeds from selected plants are utilized in two ways: (i) seeds from the best plants are retained for the next breeding cycle and (ii) the remaining seeds are bulked family-wise to test in preliminary trials.

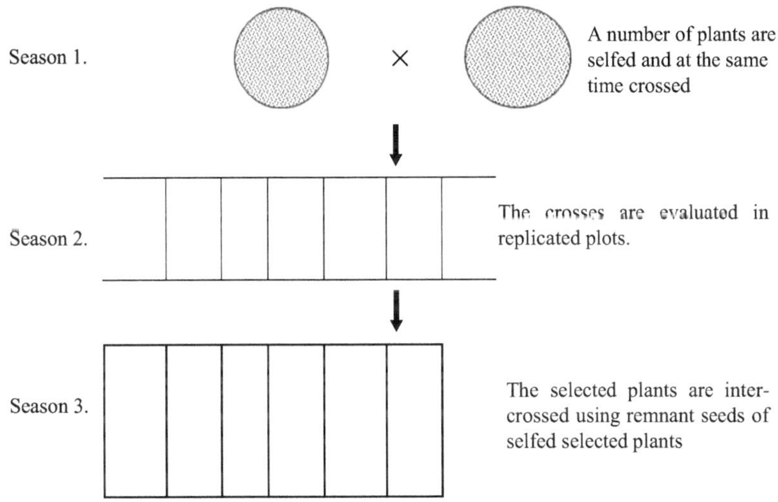

FIGURE 8.1 **Controlled pollination.**

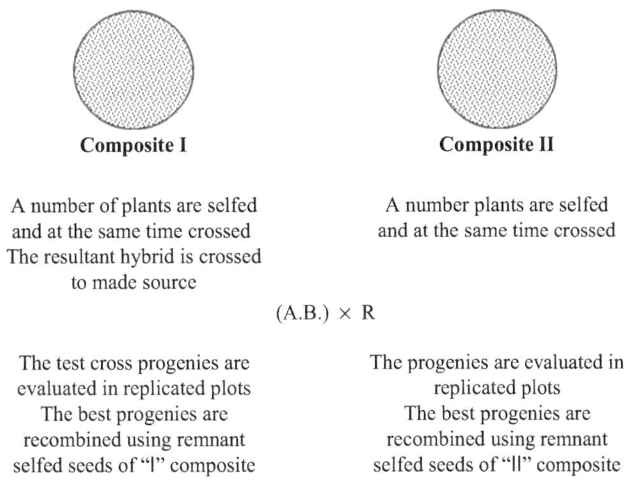

FIGURE 8.2 **Scheme of recurrent selection for SCA using homozygous testers.**

Second cycle of selection is initiated using crossed seeds obtained in season 3 (Figure 8.2).

Selfing and Crossing

The selfincompatible nature and exclusively entomophilous mode of pollination renders selfing a difficult process in niger. Very high inbreeding depression, up to 91% (Ramachandran and Menon, 1979) is associated with poor seed setting with generation advancement from S_3 onwards. The outcrossing nature of *Guizotia* species can be inferred from the cross compatibility between various taxa that produce viable and fertile hybrids

(Murthy et al., 1993; Dagne, 1994, 2001). However, direct evidence regarding self incompatibility is only available for *G. abyssinica* (Ramachandran and Menon, 1979; Riley and Belayneh, 1989; Nemomissa et al., 1999). Self incompatibility in niger is of the sporophytic type (Prasad, 1990; Nemomissa et al., 1999) which causes inhibition of pollen germination or twisting of the pollen tube over the surface of the papillae (Prasad, 1990). Self compatible niger genotypes were also reported in low frequencies in the Ethiopian gene pool, which can be as high as 5% in some populations (Getinet and Sharma, 1996; Nemomissa et al., 1999).

POPULATION IMPROVEMENT

Synthetic Variety

Synthetic variety is a population, or variety, produced by crossing all possible combinations of a number of genotypes that combine well with each other, which is maintained further by random open pollination in isolation. A synthetic variety can be developed from inbreds, sibbed lines, open-pollinated varieties, or other populations tested for their general combining ability.

PROCEDURES IN SYNTHETIC DEVELOPMENT

Development of inbred lines is a necessity for the development of synthetics. Inbred lines, with one generation selfing (short-term inbreds), can also be used for the development of synthetic varieties. The reconstitution of synthetic varieties will be possible when parental lines are inbreds. However, exact reconstitution is not possible in the case of short-term inbreds or open-populated varieties. Development of inbred lines in niger crops is a difficult task and sincere efforts in this direction are required. The selfincompatible nature and extensively entomophilous mode of pollination renders selfing a difficult process in niger. Very high inbreeding depression up to 91% (Ramchandran and Menon, 1979) is associated with poor seed set with generation advancement from first selfing (S_0) onward. When the initiation of flowering takes place, selfing or sib mating is done by covering two or more representative adjacent plants in a line covered by muslin cloth bags or nylon mosquito net bags to exclude insects. Shaking these bags, or bagged plants, is practiced on alternate days to ensure pollination. At maturity, the seeds within the bag will be S_0 and will be used to raise the S_1 generation.

Evaluation of Inbreds for General Combining Ability

Inbred lines are evaluated for their general combining ability (GCA). There are three methods used: top cross, poly cross, and single cross. In top cross, crosses are made between inbreds and either a tester or an open-pollinated variety. This is also known as an inbred–variety cross. In poly cross, selected inbreds are allowed to intermate by open pollination with a random sample of pollen from a number of selected lines. These lines are grown in the same nursery. The lines are grown at random in a poly cross-nursery with a minimum of 10 replications. Free outcrossing takes place and seed set on a line in replicates is pooled and the performance tested. In the third method, all possible crosses are made among selected inbreds, each in isolation. The crosses are evaluated for their GCA, yield, or other desirable traits in replicated trials compared with checks. The lines performing better on the basis of GCA are selected as constituents of the synthetic variety.

Production of a Synthetic Variety

Synthetic varieties may be produced in two ways. The first way requires an equal amount of seed from inbreds (Syn_0) to be mixed and planted in isolation. Open pollination is allowed and it is assumed that it is followed by random mating or crosses between all possible combinations. The seeds from this population are harvested in bulk. The population raised from this seed is known as the Syn_1 generation. The second way requires all possible crosses among the selected lines to be made in isolation. The parental lines constitute the Syn_0 generation. Equal amounts of seed from all the crosses are mixed to produce the synthetic variety. The population derived from this mixed seed is known as the Syn_1 generation.

Method 1. Equal seed from all the lines mixed and planted in isolation. Open-pollinated seed harvested as the synthetic variety (Syn_1).

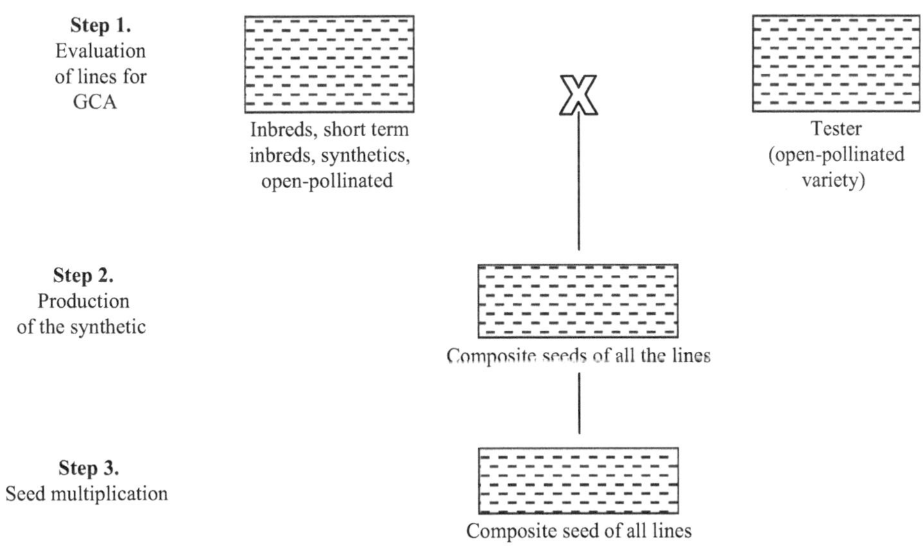

Method 2. The lines are planted in a crossing block. All possible intercrosses are made. Equal seed from all the crosses mixed to produce the synthetic variety (Syn_1).

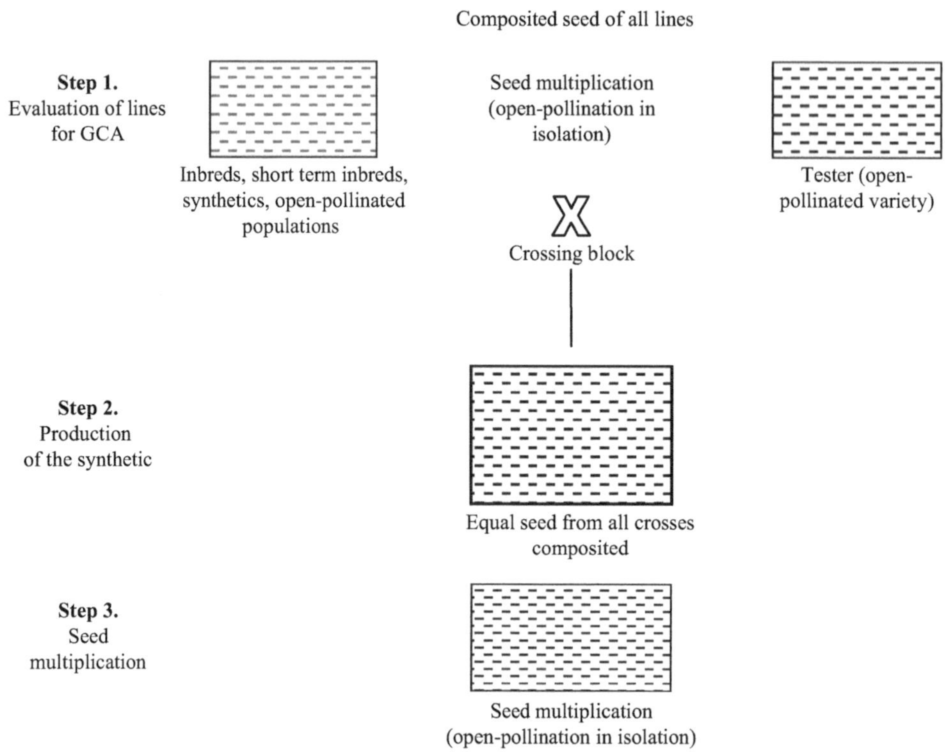

Until the crop was brought into the fold of the AICRP there was practically no research input either for crop improvement or agronomy. From 1972 onward the program strengthened by evolving varieties and developing new packages. Based on selection from native populations, and limited hybridization at different centers, 20 high-yielding varieties have been released, of which 13 are developed from selection and 5 being composites (Table 8.6). Under proper management, the varieties have the genetic potential to produce an average 8–10 q ha^{-1} seed yield. Niger production in Ethiopia is mainly based on local land race populations. Improved varieties such as Sendafa, Fogera-1, Esete-1, and Kuyu were released by the Institute of Agricultural Research, Holetta Research Center, Addis Ababa (Ethiopia) (Table 8.7). In India, both the niger-breeding program and niger seed production are much stronger than in Ethiopia, with improved varieties released. However, the yield of niger is higher in Ethiopia than India.

Being a minor crop, concerted breeding efforts in the past were not made. Of the 20 varieties released so far, 13 have been developed from secondary selection and 5 are composites. The narrow genetic base is the main constraint for genetic improvement and the extent of heterosis. The extent of heterosis is moderate (62.79%) for yield and low (7.23%) for oil content (Singh and Trivedi, 1993). However, by using diverse parents in large number of crosses, variability can be generated and the extent of heterosis can be exploited. For future improvement of niger, both hybrid and synthetic varieties are good options.

TABLE 8.6 Characteristics of Indian Niger Varieties

Cultivar name	Recommended regions/areas (area of adoption)	Days to maturity	Oil content (%)	Average yield (kg ha^{-1})	Salient features
MADHYA PRADESH					
Ootacmund	All MP	95–105	36–38	500–550	Black seeds
JNC-6	MP, Bihar, MS, Karnataka, and Rajasthan.	95–100	37–38	650–700	Shiny, dark black seeds
JNC-1	MP, MS, Bihar, Rajasthan, and Karnataka	90–100	38–40	650–700	Black seeds
JNS-9 (JNC-9)	All niger-growing states	95–100	38–40	650–700	Black seeds that are tolerant to moisture stress
MAHARASHTRA					
IGP-76	Niger-growing areas of MS, Orissa, Gujarat, Tripura, and Daman	95–105	36–39	600–650	Black seeds
N-5	Maharashtra and Bihar	95–100	36–39	600–650	Small black and sickle-shaped seeds
IGPN 2004-1, (Phule Karala-1)	Kharif season in MS and Karnataka (high-rainfall areas of Maharashtra)	95–105	39–41	650–700	Shiny black seeds that are tolerant to *Alternaria* leaf spot, powdery mildew, and root rot
KARNATAKA					
RCR-317	Karnataka	90–95	35–38	600–650	Black seeds
No. 71	KK, AP, WB, TN, Orissa, and NE Hills regions	95–105	36–38	500–600	Black bold seeds
KBN-1	Karnataka	85–95	36–38	600–650	Black seeds
RCR-18	Zones 1, 2, 3, and Karnataka	95–110	33–35	500–600	Shiny black seeds
ORISSA					
GA-10	Orissa (tribal areas)	110–115	39–41	600–650	Dark black seeds
Utkal Niger-150	Orissa	105–110	38–40	650–700	Black seeds that are tolerant to *Alternaria* and *Cercospora*
JHARKHAND					
Birsa Niger-1	Uplands of Bihar and MP	95–100	36–38	550–600	Gray seeds
Birsa Niger-2 (BNS-8)	All niger-growing states	95–100	35–38	600–650	Black seed
BNS-10 (Pooja-1)	All niger-growing states	95–100	36–38	650–700	Shiny black seeds

(Continued)

TABLE 8.6 Characteristics of Indian Niger Varieties *(cont.)*

Cultivar name	Recommended regions/areas (area of adoption)	Days to maturity	Oil content (%)	Average yield (kg ha^{-1})	Salient features
TAMIL NADU					
Paiyur-1	Tamil Nadu (hilly regions)	90–95	35–38	600–650	Black seeds
GUJARAT					
Gujarat Niger-1	Gujarat	95–100	35–38	600–650	Black seeds
NRS-96-1	All niger-growing states	90–95	35–38	650–700	Black seeds

TABLE 8.7 Characteristics of the Niger Varieties of Ethiopia

Variety	Maturity (days)	Plant height (cm)	Average yield (kg ha^{-1})	Oil (%)
Sendafa	145	133	780	40
Fogera-1	146	138	820	41
Esete-1	146	139	830	39
Kuyu	138	131	1060	41

Line Development

Line performance is determined by the parents involved in its development. Parental lines showing good vigor, high *per se* performance and combining ability, possession of multiple resistance and/or tolerance to major diseases and pests, and exhibiting high seed oil content will generally result in hybrids showing superior performance and yield. Therefore, it is of great importance to concentrate on breeding for line development. In niger, breeding for seed parents is more complex. The lack of diverse productive lines, especially seed parents, has limited yield potential. Therefore, in hybrid research, line development is a very important step.

The selection of female parents for good GCA, high seed density, and high oil content is one of the major factors to consider. An analysis of the extent to which the other germplasm accessions have been used in line development is needed for infusing newer materials. In terms of number of accessions available in India, there is certainly germplasm available. However, not much is known about the diversity present in different accessions collected from various other countries. Identification of heterotic accessions can enhance breeding efficiency. An evaluation strategy has to be designed in order to identify heterotic material from germplasm collections.

Trait-Based Selection

The potential for high achene yields in lines of both *per se* and hybrids (combining ability) is an important consideration in parental line development. High yield potential should be the first trait to be explored and its selection can be made visually. High achene yield must accompany other economic attributes if lines are to be successful. Some of the other traits that are to be considered are dark broad leaves, high leaf number, bold seeds, high seed number, test weight, and oil content. Other characters such as standability, plant height, and tolerance to pests and diseases are also important. Most of these characters have high heritability, and visual selection during generational advance is quite effective. Such trait-based selection is fairly effective in the development of lines showing a high level of performance. The traits present in these lines are expressed in synthetic varieties to varying levels.

Resistance Breeding

In India, niger is mainly grown as a *kharif* (monsoon) crop but in a few pockets it is also grown as a *rabi* (dry season) crop, mainly in the summer. At low temperature the chances of aphid infestation is more therefore for the development of low temperature tolerant varieties along with tolerance to aphid infestation is a point to be considered by the research workers. Salinity and alkalinity of the prevailing soils also represent issues to be tackled. At low temperature, the severity of powdery mildew is increased, therefore, the development of tolerant lines against powdery mildew is desirable. The development of a variety that is resistant to insect pests and diseases is also an important requirement. So far there has not been much work carried out in these areas since losses due to these factors are low. A number of insect pests have been recorded as being damaging to niger crops both in India and Ethopia. Among these pests are the capsule fly, niger caterpillar, semilooper, cutworm, and aphid which infest crops at several different growth stages. Niger crops are also attacked by a number of diseases of which *Alternaria* sp. and leaf spot are the most threatening. Among weed pests, the parasitic dodder (*Cuscuta campestris*) is the main threat. However, vigorous efforts in this direction are needed for the development of resistant varieties. Promising lines N-10, N-116, N-126, N-133, N-142, and N-144 were found to be tolerant to capsule fly; JNS-14 to aphids and flea beetle; and N-6 and N-11were found to be tolerant to white fly. These promising lines can be utilized in a resistance-breeding program to counter several insect pests and diseases that damage niger crops (Table 8.8).

Cercospora leaf spot (*Cercospora guizoticola*) appears as small, straw to brown colored spots with gray centers on the leaves. There spots may coalesce and cause defoliation. *Alternaria* (*Alternaria* sp.) spots are brown to black forming concentric rings. Powdery mildew (*Sphaerotheca* sp.) is another important disease of niger which appears as small powdery spots on the leaves

TABLE 8.8 Tolerant Lines Identified for Different Niger Diseases

Disease	Tolerant lines
Cercospora	JN-13, JN-106, JN-107, JN-112, JN-118, JN-128, JN-130, JNS-9, PCU-45, PCU-46, N-24, N-128, AJSR-48, ONS-157
Alternaria	RCR-328, JN-121, N-17, N-18, N-24, N-87, N-128, N-141, N-142, N-165, N-187, N-128, AJSR-47, AJSR-48
Powdery mildew	KEC-6, RCR-238 , RCR-290, JN-17,JN-19, JN-20, JN-21, JN-37, JN-68, JN-60, JN-72, JN-77, JN-78, JN-85, JN-86, JN-87

which gradually spread on the lamina and stem resulting in defoliation. In the case of stem/root rot (*Macrophomina phaseolina*) the infected roots turn light black to black in color, are covered with black sclerotia, and are brittle. The blackening extends from the ground up the stem giving it a black color.

During development of these lines they should be subjected to repeated evaluation for important diseases so that selection for both disease resistance and agronomic characteristics are considered together. This will ensure that the lines finally produced will possess an acceptable level of resistance to diseases. With such breeding lines it is necessary to ensure the diversity of plant type as well as other morphological characteristics such as maturity duration, varying oil content, and plant height. The diversity present in the lines would get expressed in synthetic varieties. In view of IPR matters becoming increasingly critical, incorporation of distinguishable traits during line development is also important. Niger seed populations in Ethiopia and India are very heterogeneous, indicating a great potential for yield increases through breeding. As such, breeding programs exist in both countries. Both large variation and high heritability were found for plant height and days to flowering; both variation and heritability were lower for number of branches, number of flower heads, 1000-seed weight, and seed yield. The breeding objectives for niger seed are to increase seed yield and oil content and reduce shattering.

Cuscuta weed (*Cuscuta chinensis/C. hyalina*) is a parasitic plant that infests niger and has stunted, pale yellow flowers. Removal of *Cuscuta* seed can be achieved by sieving before sowing. In addition niger seed can be steeped in brine solution before it is sown. Once developed, removal of *Cuscuta*-infested niger seedlings is recommended at an early crop growth stage. Presowing a soil application of Fluchloralin (1 kg active ingredient/ha) or preemergence application of Pendimethalin (1 kg a.i. ha^{-1}) is also affective.

QUALITY BREEDING

Niger oil is a rich source of energy and nutrition. Its oils and fats are useful as food fats and industrial raw materials. The proteins present in the seed and cake are edible and useful as animal feed. Vegetable oils are useful as lubricants, surface coatings, cosmetics, and as a raw material for various industrial products. The oil is good-quality edible oil with a pleasant nutty, sweet taste. It is pale yellow in color and may have a bluish tinge. The raw oil has low acidity and can be used directly for cooking. Even now, in some pockets of tribal hilly areas, niger oil is used for illumination purposes. The oil normally has poor shelf life and becomes rancid when stored. The oil contains 0.5–1.0% unsaponifiables and 0.4–3.0% free fatty acids. Niger cake is dark in color and contains 20–30% protein, 4–14% oil, 8–24% crude fiber, 20–28% sugars, and 8–12% ash. Niger cake is mostly utilized as animal feed. Low-grade cake can be used as manure. Blending of niger cake is suggested for improving the nutritional quality of other cakes. Niger seed fried in ghee can accompany other foods to improve their palatability and nutritional quality. Additionally, they are parched and mixed with pulses as a snack food. Chutney, a spicy preparation prepared by mixing chilli powder with roasted and pounded niger seed, is commonly prepared in some tribal and rural areas of India. Inferior quality oil is used as a fuel for illumination and in the preparation of soaps and detergents for industrial use. It is also used as body massage oil for smooth skin and supple joints. Dagne and Jonsson (1997) have analyzed the seed of *G. abyssinica* and wild *Guizotia* for oil content and fatty acid composition. They have inferred that, when gene transfer is desired, crossing between wild and cultivated taxa may not affect later oil quality and the oils of wild taxa will possibly remain safe for human consumption. The utilization of wild species for the genetic enhancement of cultivated niger is small. However, there have been considerable efforts to augment collections at the NBPGR, New Delhi. The collection of wild species has increased the availability of wild *Guizotia* for use in breeding programs. Lines of Ethiopian origin were found to be taller and later to maturity but more tolerant of cool winters than the lines from Nepal.

The oil extracted from the seed of niger is the oil of choice in Ethiopia. The major fatty acid in the oil produced in Ethiopia is linoleic acid and usually comprises around 75–80% of total fatty acids in the oil (Table 8.9). Cultivars that were grown in a uniform environment in a screen house at Shenton Park Research Station in Western Australia showed genetic differences. The seed originating from Nepal had much more variability in fatty acid composition. The percentage of linoleic acid in the oil can vary from 30–77%. Results showed that it is the more saturated fatty acids that increase with any decrease in the level of linoleic acid. The range of differences in the levels of palmitic acid and stearic acid were 22.6% and 19.4%, which largely accounted for the 47.1% difference in the concentration of linoleic acid found in the Nepal lines. The oleic acid level was more stable, with its difference being 7.5%. Vles and Gottenbos (1989) found Ethiopian niger oil rich in both linoleic acid, which prevents cardiovascular disorders such as coronary heart disease, sclerosis, and high blood pressure, and linoleic acid derivatives which serve as structural components of the plasma membrane and as a precursor of some metabolic regulatory compounds. The high-linoleic acid content makes the niger oil nutritionally valuable and good for human consumption. Fujimato et al. (1990) also reported the medicinal properties of niger oil as being very beneficial to human beings. Niger seed is a good bird feed, which generates exports. Nagaraj and Patil (2004) have reviewed the quality and composition of niger. Their findings are given in Table 8.10.

The oil content of niger is variously reported as 29–39% (Dutta et al., 1994), 30–35% (Kandel and Porter, 2002), and 42–44% (Dagne and Jonsson, 1997). Dutta et al. (1994) reported that the Ethiopian niger seed oil contains more than 70% linoleic acid, whereas Dagne and Jonsson (1997) reported 66–69% linoleic acid. In all the works so far undertaken on the fatty acid composition of niger, linoleic acid is unequivocally the dominant fatty acid present in seed oil, followed by palmitic, oleic, and stearic acids (Dutta et al., 1994; Dagne and Jonsson, 1997; Ramadan and Morsel, 2003). The percentage of oleic acid in the Ethiopian niger seed oil

TABLE 8.9 Variations in the Fatty Acid Profiles of Niger

Origin	No. of lines	% Palmitic acid	% Stearic acid	% Oleic acid	% Linoleic acid
Ethiopia	4	8.8–9.5	6.5–6.7	7.1–9.5	73.1–75.4
Nepal	37	8.1–30.7	5.1–24.5	3.8–11.3	30–77.1
Poland	1	8.3	8.4	18.2	63.3
Portugal	1	8.8	7.6	8.6	73.4
Spain	1	8.3	6.7	14.9	68.7

TABLE 8.10 Niger Seed Composition and Oil Quality

	Composition	
Constituents	**Nasirullah et al. (1992)**	**Nagaraj (1999)**
Oil (%)	30–43.2	32.0–32.4
Protein (%)	10–30	26–30
Crude fiber	7–10	—
Soluble sugars (%)	6.9–18	
Color of oil	2.9 Y + 0.5 R	2.0 Y + 0.6 R
Iodine value	138.2	112.8–129.0
Refractive index 250C	1.471	1.469–1.472
Unsaponifiables	–	0.2–1.7
1000-seed weight (g)	1.6–6.0	—
Moisture (%)	5.5–6.0	

was reported to be in the range of 6–11% (Dutta et al., 1994) and 5.4–7.5% (Dagne and Jonsson, 1997). It has been indicated that the oil content and the fatty acid profile may vary depending on the origin of the material and the maturity level of the seed (Riley and Belayneh, 1989; Stymne and Appelqvist, 1980). The quality of oil and its suitability for a particular purpose, be it for human consumption or industrial use, depends on the proportion of the different fatty acids it contains. Oils where linoleic acid is the predominant fatty acid are reported to have a poor shelf life, whereas those with high oleic acid content are more stable. For cooking oils, it becomes imperative that oils be suitable for the kind of cooking they are intended for (Mugendi et al., 1998; Warner and Knowlton, 1997; Fehr, 2007). It has also been reported that the presence of high proportions of linoleic acid in oils positively contributes to frying flavor intensity, a fact that also renders its shelf life poor (Warner et al., 1997). Thus, in deep-fat frying, which has now become a common way of preparing food, the oil should be optimized for frying flavor and frying time (Mugendi et al., 1998; Warner and Knowlton, 1997; Fehr, 2007).

MAINTENANCE BREEDING AND NUCLEUS SEED PRODUCTION

Commercial crop–growing areas may not necessarily be suitable for seed production due to a parallel buildup of diseases and pests. Seed production needs to be undertaken in areas where the environment allows full expression of diagnostic characters, facilities for protective irrigation, productivity and seed quality to be higher. *Cuscuta* weed is a major menace to seed production. Areas for seed production have to be essentially free from *Cuscuta*. Suitable areas for seed production are Vishakhapatnam and Rangareddy in Andhra Pradesh; Singhbhun, Dumka, and Ranchi in Bihar; Chitradurga, Tumkur, and Bangalore Rural in Karnataka; Sidhi, Narsingpur, Vidisha, Khandwa, and Shivpuri in Madhya Pradesh; Bilaspur in Chattisgrah; Solapur, Ratanagiri, Nasik, and Latur in Maharashtra; and Malakangiri, Koraput, Kalahandi, and Navrangpur in Orissa. The estimated area required for production of quality seed of different categories are listed in (Table 8.11).

TABLE 8.11 Estimated Area Required for Production of Quality Seed in India

Category	Seed (kg)	Area (ha)
Nucleus seed	15	0.03
Breeder seed	1050	2.0
Foundation seed	37500	80 .0
Certified seed	2000000	4000.0

SEED PRODUCTION SYSTEMS

Presently two seed production systems are operating in the country.

Formal System

This system is being operated through public sector agencies such as the National Seed Corporation, State Farms Corporation of India, State Seed Corporations, State Agricultural Universities, and the oil federations. The seed multiplication ratio in this system is extremely poor. The main advantage of this system is that the identity, genetic purity, quality, and source of the seed is known to farmers.

Informal System

This system includes multiplication of varieties by private growers or individual farmers with sharing of the seed between farmers. The seeds of most of niger varieties under cultivation are being produced and supplied through this system. The main disadvantage of this system is that the identity, genetic purity, quality, and source of the seed is not authenticated. However, the seed produced through this system is less expensive and readily available to farmers.

Alternative Systems

The existing formal system of seed production is hardly able to cope with the demand for seed. Minor crops like niger represent a low priority to seed-producing agencies, and therefore the production of quality seed for farmers of niger is poor. The possibility of improving the supply of quality seed for crops like niger, through the formal system, in the near future appears unlikely. Therefore, in this crop, alternative systems of seed supply may prove worthy.

Direct Supply

Both formal and informal systems of seed supply have their limitations. To overcome these limitations, seed production could be undertaken by research institutes and distributed via farmers' fairs, field days, and sales counters. Direct supply becomes more feasible and effective in view of its timely supply of quality seed to the farmers. This system has been successful to a certain extent in that it has supplied a large area with quality seeds of improved varieties.

Seed Village

Another option to augment the seed supply of niger is a seed village. Institutes can choose a single variety and produce sufficient seed to cover the demand from an entire village. The idea is that a village grows only one variety. Local or other varieties should not be grown in a seed village. Combining together demonstrations and seed village will prove better for the improvement of seed replacement rate.

There is need to integrate niger-breeding programs with potential seed growers so that farmers can gain the benefit of improved varieties over a shorter period of time. Hence, seed grower associations should be established on a regional basis with the following objectives:

1. To develop an effective local supply system for quality niger seed.
2. To increase the availability of quality seed, of improved varieties, to farmers along with the transfer of technology.
3. To rapidly multiply recently developed varieties through conscious seed production.

Seed Production Agronomy

Niger is a cross-pollinated crop; therefore, it is difficult to maintain genetic purity without adopting appropriate isolation distances. The identification of improved varieties of niger through diagnostic morphological characteristics is a very difficult task for seed production and certification personnel. In this crop the breeder seed production is always surplus but the quantity of certified seed is always lower to meet the demand of farmers. The seed demands of niger are erratic due to low management levels. Delay in the onset of the monsoon and low-rainfall distribution are the two most important factors determining seed demand in niger-growing areas. It is very difficult to meet such sudden seed demands unless the Government of India and state governments have a buffer stock of niger seed as a contingency. Seed multiplication is relatively easy due to the very high rate of seed multiplication. Niger is, however, adapted to a wide range of soil types from clay loam to sandy loam; however, it thrives best on well-drained, loamy soils of good depth and texture. Heavy clay and black-cotton soils are not suitable for producing high yields (Ranganatha et al., 2014). Two ploughings followed by harrowing and planking are recommended to obtain optimum soil tilth to ensure an even depth of seed placement and subsequent emergence.

Niger is mainly grown in kharif seasons. The optimum sowing time is the middle of July to early August for kharif crops with September being ideal for semirabi crops. The appropriate sowing periods for different states are: Madhya Pradesh and Chhattisgarh, the third week of July to second week of August; Maharashtra, July continuing to early September; Orissa, second fortnight of August to the first week of September; Bihar/Jharkhand, second fortnight of August to the first week of September; Andhra Pradesh, second week of August; and Gujarat and Karnataka, July–August (Ranganatha et al., 2014). Generally 5 kg ha^{-1} of seed is required for a sole crop. To protect the crop from seed and soil-borne pathogens, seed should be treated with Thiram or Captan at 3.0 g kg^{-1} seed before sowing. Line sowing has been recommended. Seeds are mixed with sand, powdered Farm Yard Manure (FYM), or ash to increase its bulk. This should be repeated 20 times to ensure an even distribution of seed. Planking is done to cover the seeds. Seeds should be sown 2–3 cm deep depending upon soil type and moisture. A seed bed temperature of 15–22°C is the optimum. Temperatures below 10°C and above 35°C impair germination. Spacing depends upon soil type and variety. To obtain the optimum yield, an appropriate spacing is 30–45 cm between rows and 10–30 cm within lines.

The recommended seed rate gives a higher plant stand under optimum soil moisture. To maintain an optimum plant population, thinning is recommended to remove the extra plants after 2 weeks of sowing or when the seedlings attain a height of 8–10 cm. Niger has shown a good response to fertilizer applications in trials, on the basis of which the following fertilizer doses are recommended. An application of N via urea + seed treatment with phosphate solubilizing biofertilizer (PSB) of 10 g kg^{-1} of seed enhances yield significantly. An application of sulfur (20–30 kg ha^{-1}) increases seed yield and oil content. Regionally, the following is recommended: Madhya Pradesh, 10 kg N + 20 kg P_2O_5 ha^{-1} at sowing and 10 kg N ha^{-1} 35 days after sowing; Maharashtra, 4 tons of farmyard manure (FYM) and 20 kg N ha^{-1} at sowing; Orissa, 20 kg N + 40 kg P_2O_5 ha^{-1} at sowing and then 20 kg N ha^{-1} 30 days after sowing; Bihar/Jharkhand, 20 kg N + 20 kg P_2O_5 + 20 kg K_2O + 15 kg $ZnSO_4$ as a basal application; Andhra Pradesh, 5 tons of FYM and 10 kg N ha^{-1} at sowing; Karnataka, 20 kg N + 20–40 kg P_2O_5 + 10 kg K_2O ha^{-1} at sowing. Since seed production is planned with assured moisture/supplemental irrigation, a 50% higher fertilizer dose is recommended. Further organic biofertilizers and micronutrients are suggested to encourage better seed set, higher yield, and superior seed quality (Deshmukh et al., 2009).

Weeding is done 15–20 days after sowing and is recommended at the time the crop undergoes thinning. Second weeding should be undertaken a month after sowing before top dressing with nitrogenous fertilizer. In Orissa, *Cuscuta* (*Cuscuta hyalina*/ *C. chinensis*) infestation has become a major problem. Seeds should ideally be obtained from *Cuscuta*-free areas, and if *Cuscuta* seed is found mixed with niger seed, sowing should be done after separation *via* sieving with a 1 mm sieve. Niger is generally produced in the rainy season. Prolonged moisture stress adversely affects the plant stand and plant growth. In such situations, protective irrigation helps to establish the plant stand and gives good yields. For semirabi seed crops it is important to provide two times of irrigation, one at flowering and other at the seed-filling stage. This will help seed production yield.

Isolation Distance

Since niger is a cross pollinated crop with a selfincompatibility mechanism, it makes it very difficult for it to maintain its genetic purity without adopting an appropriate isolation distance (Ranganatha et al., 2012). An isolation distance of 1000 m is recommended for nucleus, breeder, and the foundation stages of seed production, whereas 500 m is recommended for certified stage production and should be rigorously followed to produce genetically pure, quality seed.

Rouging

Rouging should be done strictly to remove all off-type plants, which exhibit variation from the parental variety. Any plants infested by diseases and pests, especially by *Cuscuta* weed, should be removed. Fields should be inspected thoroughly at seedling, vegetative, flowering, and maturity stages by a monitoring team consisting of experts.

Bee Keeping

Selfincompatibility in niger coupled with sticky pollen grains that are not amenable to wind pollination leads to entomophilous pollination. A large number of attractive, colored flowers and a well-distributed, long-flowering period of 45–80 days makes niger an ideal crop for bee keeping and therefore insect pollination. The productivity and profitability of niger is substantially increased as a consequence of bee keeping, with additional yields (10–20%) and honey as a byproduct of open-pollinated niger crops.

Postharvest Processing

Threshed niger seed is cleaned by winnowing. The produce is dried, reducing the moisture content by 8%, and then stored properly. The quality of the produce should be upgraded through sieving, because clean produce (i.e., bold lustrous seeds that are free from trash, pests, and discolored unfilled seeds) has greater demand and will fetch a higher price in the market.

Niger seed in Ethiopia is widely grown in smallholdings on fragmented land. It is the dominant oil crop under production and has an areal extent larger than other oilseed crops. Niger seed in Ethiopia is cultivated mainly for the production of edible oil but is also directly consumed after being fried and mixed with sunflower seeds. Pressed cake from oil extraction is used for livestock feed especially in and around cities and large dairy farms. The country's requirement for edible oil has been met through domestic production and imports.

Great variation exists in niger productivity from state to state ranging from 150 kg ha^{-1} in Madhya Pradesh to 750 kg ha^{-1} in Jharkhand. There is an urgent need to identify clusters of villages/districts, demonstrating high yield potentials in major niger-growing states, as export zones. These zones should grow niger exclusively for the production of export-quality seed, attaining international export standards. Contract farming for niger exports, based on the model of sesame oil in Sudan, should be encouraged. Identification of progressive/responsive farmers happy to take up niger cultivation exclusively for export, alongside the development of specialized marketing and other related facilities, will boost export performance.

A critical understanding of the export requirements of a crop is a prerequisite to formulating an export promotion strategy. Uniform, lustrous, well-filled, bold, heavy seeds of a shiny black color are preferred and demand a premium in international markets. Clean niger seeds, free from foreign objects, particularly from *Cuscuta* and other weeds, are the standard required for international export.

SWOT ANALYSIS FOR NIGER

A strengths, weaknesses, opportunities, threats (SWOT) analysis for niger reveals:

- *Strengths.* A life line for tribal agriculture and tribal economies; at maximum production it is the largest areal crop in world; major export in the international market; high export potential; best bird feed; low vulnerability to pests and diseases; low input requirement; grows successfully without costly chemicals.

- *Weaknesses*. Low yield levels; low harvest index; poor transfer and adoption of improved technology; high susceptibility to *Cuscuta*.
- *Opportunities*. Best crop for waste lands; honeybee rearing for higher productivity and profitability; contributes to soil conservation and land rehabilitation; weed suppression due to allelopathic effect; oil good, taste akin to desi ghee; good for health ($<70\%$ unsaturated fatty acids); organic production of edible oil; a commodity for the twenty-first century.
- *Threats*. Exploitation of farmers by traders; Nepal as a potential competitor in the international market; *Cuscuta*.

FUTURE STRATEGIES

Niger populations in Ethiopia and India are very heterogeneous, indicating significant potential for yield increases through breeding, and breeding programs, which exist in both countries. Large variation and high heritability were found for plant height and days to flowering; both variation and heritability were lower for number of branches, number of flower heads, 1000-seed weight, and seed yield per plant. The breeding objectives for niger seed are to increase seed yield and oil content and to reduce shattering. Following the developments made with sunflowers and safflowers, it has been postulated that reduced capitula, dwarf types, with uniform maturity should also be developed to achieve breeding objectives. An increase in oil content appears feasible because of existing genetic variability, something, which can be used in breeding. Screening for self-compatible lines with little or no inbreeding depression should be carried out to minimize the problem of low seed set due to self-incompatibility. Systems should be developed to monitor genetic erosion in specific areas. Gene pools need to constitute mass reservoirs or composite populations of large number of diverse parents as a means of long-term conservation. To meet future needs, conservation of dwindling germplasm, especially the wild species, needs to be addressed. Germplasm should be stored at more than one site, so as to provide additional security and economize conservation of genetic resources. The strengths of niger include high yield potential and input responsiveness; resistance to pests such as caterpillars, semiloopers, cutworms; and resistance to *Alternaria*, powdery mildew, root rot, and *Cercospora* leaf spot. As niger seed is selfincompatible, breeders in India and Ethiopia have adopted means of population improvement such as mass selection and sibbing. Recently, a protocol for *Agrobacterium tumefaciens*–mediated genetic modification has been developed; hence, marker-assisted selection (MAS) can be exploited as a long-term program for quantum jumps in niger productivity. With male-sterile line development and isolation of inbreds, exploitation of heterosis in niger would be possible. Repeated male sterility is genetic in nature and governed by a recessive gene, which is closely linked to ligulate-homomorphic florets in capitula; while capitula consisting of heteromorphic ray florets and disk florets are fertile. This has opened up a new dimension in heterosis breeding in niger improvement, in addition to exploiting selfincompatibility.

References

Abebe, M. 1975. Ecophysiology of noog (*Guizotia abyssinica* Cass). PhD Thesis. University of California, Riverside.

Ahmed, E., Ram, S., Kerketta, V., Singh, D.N., Singh, S.K., 2002. Combining ability analysis in niger (*Guizotia abyssinica Cass*). J. Res. 14 (2), 203–209.

Anonymous, 2009. Annual Progress Report Sesame and Niger 2008–2009. All India Coordinated Research Project (Sesame and Niger), J.N. Agricultural University Campus, Jabalpur, M.P. pp.249–251.

Bilaiya, S.K. 1987. Genetic analysis for yield and it's components in niger (*Guizotia abyssinica* Cass). PhD Thesis, Jawaharlal Nehru Krishi Vishwa Vidyalaya, College of Agriculture, Jabalpur.

Bilaiya, S.K., Singh, C.B., Yadav, H.S., 1992. Diallel analysis for seed yield and oil content in niger. Ann. Agric. Res 13 (1), 89–92.

Borole, R.D., Patil, H.S., 1997. Genetic variability, correlation and path analysis in niger. J. Oilseeds Res. 14 (1), 40–43.

Channarayappa, 1987. Response to selection, correlation and path analysis in Niger (*Guizotia abyssinica* Cass). Mysore J. Agric. Sci. 21, 92–93.

Dagne, K., 1994. Meiosis in interspecific hybrids and genomic interrelationships in *Guizotia* Cass. (Compositae). Hereditas 121, 119–129.

Dagne, K., 2001. Cytogenetics of new *Guizotia* Cass. (Compositae), interspecific hybrids pertaining to genomic and phylogenetic affinities. Plant Syst. Evol. 230, 1–11.

Dagne, K., Jonsson, A., 1997. Oil content and fatty acid composition of seed of *Guizotia* Cass (Compositeae). J. Sci. Food Agric. 73, 274–278.

Deshmukh, M.R., Jyotishi, Alok, Ranganatha, A.R.G., 2009. Efficient nutrient management in niger on different soils of India JNKVV Research 43 (1), 23–26.

Dutta, P.C., Helmersson, S., Kebedu, E., Appelqvist, L.A., 1994. Variation in lipid composition of niger seed (*Guizotia abyssinica* Cass.) Samples collected from different regions in Ethiopia. J. Am. Oil Chem. Soc. 71, 839–843.

Esquinas-Alcazar, J., 2005. Protecting crop genetic diversity for food security: political, ethical and technical challenges. Nat. Rev. Genet. 6 (12), 946–953.

Fehr, R.W., 2007. Breeding for modified fatty acid composition in soybean. Crop Sci. 47, S-72.

Fujimato, Y., Kakinumay, K., Eguchi, T., Ikekawa, N., Hirayama, N., Mbarushimana, A., Nathomvukiye, D., 1990. 12, 15-Dehydroxylabed-8 (17), 13-Dien-19-oic acid from Guizotia scabra. Phytochemistry 29 (1), 319–321.

Getinet, A., Sharma, S.M., 1996. Niger, *Guizotia abyssinica* (L.f.) Cass. Promoting the conservation and use of under utilized and neglected crops. S. Institute of Plant Genetics and Crop Plant Research. Gatersleben/International Plant Genetic Resources Institute, Rome.

Hiremath, S.C., Murthy, H.N., 1988. Domestication of niger (*Guizotia abyssinica*). Euphytica 37, 225–228.

Kandel, H., Porter, P. (Eds.), 2002. Niger: *Guizotia abyssinica* (L.f.) Cass. *Production in northwest Minnesota*. University of Minnesota extension service.

Mathur, R.K., Gupta, S.C., 1993. Yield component analysis and it's implication for selection in niger (*Guizotia abyssinica* Cass). J. Oilseeds Res. 10 (1), 6–10.

Mishra, D., 1995. Variability and correlation studies in niger (*Guizotia abyssinica Cass*). Envir. Eco. 13 (2), 439–442.

Misra, R.C., 1992. Genetic parameters and character association in niger (*Guizotia abyssinica*). Oil Crops Newsl. 9, 23–24.

Mugendi, J.B., Sims, C.A., Gorbet, D.W., O'Keefe, S.F., 1998. Flavor stability of high oleic peanuts stored at low humidity. J. Am. Oil Chem. Soc. 75, 21–25.

Murthy, H.N., Hiremath, S.C., Salimath, S.S., 1993. Origin, evolution and genome differentiation in *Guizotia abyssinica* and its wild species. Theor. Appl. Genet. 87, 587–592.

Nagaraj, G., 1999. Quality and production characteristics of some speciality accessions of niger. J. Oil Tech. Assoc. India 31 (4), 187–188.

Nagaraj, G., Patil, H.S., 2004. Quality and composition of niger, *Guizotia abyssinica* (L) Cass – a review. J. Oilseeds Res. 21 (2), 224–229.

Nasirullah, K., Mallika, T., Rajlaxmi, S., Pasupathi, K.S., Ankaiah, N.N., Vibhakar, S., Krishna, M.N., Murtthy, K., Nagaraja, V., Kapoor, O.P., 1992. Niger (*Guizotia abyssinica*) seed oil. J. Food Sci. Tech. (India) 19, 147–149.

Nema, N.P., Singh, L., 1965. Estimates of genotypic and environmental variability in niger. Indian Oilseed J. 9, 192–194.

Nemomissa, Sileshi, Endashaw, Bekele, Kifle, Dagne, 1999. On some novel genotypes of Ethiopian populations of *Guizotia abyssinica* (L.f.) Cass [Noug]. *Sinet*. Ethiopian J. Sci. 21 (2), 297–303.

Patel, O.P., Mishra, R.K., Yadav, R.K., 1993. Variability and correlation in niger (*Guizotia abyssinica* Cass). Adv.Plant Sci. 6 (2), 186–190.

Patil, H.S., 2000. Yield component analysis in niger (*Guizotia abyssinica* Cass). Madras Agric. J. 87 (4–6), 260–263.

Patil, H.S., 2003. Nature of variability, correlation and path analysis in niger [*Guizotia abyssinica (L. f.)* Cass]. Crop Improv. 30 (2), 212–215.

Patil, H.S., Duhoon, S.S., 2005. Nature of variability, association and path analysis in niger, *Guizotia abyssinica (L. f.) Cass*. J. Oilseeds Res. 22 (2), 438–440.

Patil, H.S., Duhoon, S.S., 2007. Component analysis in niger (*Guizotia abyssinica* Cass). J. Oilseeds Res. 24 (2), 317–318.

Patil, H.S., Purkar, J.K., Deshmukh, R.B., 2000. Line × Tester analysis in niger (*Guizotia abyssinica* Cass). National seminar: Oilseeds and oils – Research and Development Needs in the Millennium Directorate of oilseeds Research, Hyderabad, Extended Summary No.62, 97.

Patil, H.S., Purkar, J.K., Deshmukh, R.B., 2001. Line × Tester analysis in niger (*Guizotia abyssinica* Cass). J. Oilseeds Res. 18 (1), 105–107.

Payasi, S.K., Mishra, Y., Kottu, B.K., Bilaiya, S.K., Yadav, L.N., 1987. Character association and component analysis in niger. J. Oilseeds Res. 4 (1), 70–74.

Pradhan, K., Mishra, R.C., Paikaray, R.K., 1995. Genetic variability and character association in niger. Indian J. Genet. Pl. Breed. 55 (4), 457–459.

Prasad, V., 1990. Pollen tube growth and site of incompatibility reaction in niger (*Guizotia abyssinica* Cass.). Curr. Sci. 59, 466–468.

Prinz, K.D., 1976. Unterschungen zur Ökophysiologie von Nigersaat (*Guizotia abyssinica* athiopischer und indischer Herkunft. PhD thesis. University of Göttingen, Germany.

Ramachandran, T.K., Menon, P.M., 1979. Pollination mechanism and inbreeding depression in niger (*Guizotia abyssinica* Cass). Madras Agric. J. 66 (7), 449–454.

Ramadan, M.F., Morsel, J.T., 2003. Determination of the lipid classes and fatty acid profile of niger (*Guizotia abyssinica* Cass.). seed oil. Phytochem. Ana. 14, 366–370.

Ramchandran, T.K., Menon, P.M., 1979. Pollination mechanism and inbreeding depression in niger (*Guizotia abyssinica* Cass) JNKVV. Res. J. 8 (2), 11–12.

Ranganatha, A.R.G., Tripathi, A., Jyotishi, A., Paroha, S., Deshmukh, M.R., Shrivastava, N., 2009. Strategies to enhance the productivity of sesame, linseed and niger. In: Proceedings of Platinum Jubilee Celebrations. UAS, Raichur.

Ranganatha, A.R.G., Mishra, Rajdeep, Tabassum, Aasfa, 2012. Seed production of sesame and niger. Oilseeds Seed Production, Training Programme. DOR, Hyderabad.

Ranganatha, A.R.G., Panse, R.K., Panday, A.K., Deshmukh, M.R., 2014. Strategies for Maximizing Sesame and Niger Production. In Recent Advances in Weed Management organized at Directorate of Weed Science Research (ICAR), Jabalpur.

Riley, K.W., Belayneh, H., 1989. Niger. In: Robbelen, G., Downey, R.K., Ashri, A. (Eds.), Oil Crops of the World. McGraw Hill Publishing Company, New York, pp. 394–403.

Seegeler, C.J.P. 1983. Oil plants in Ethiopia. Their taxonomy and agricultural significance, 2.9 *Guizotia abyssinica* (L. f.) Cassini (Compositae). Centre for Agril. Publi Docu, PUDOC, Wageningen, The Netherlands. pp. 122–146.

Sharma, S.M., 1990. New potential areas of niger in India. in Proceedings of the meetings held at Pantnagar and Hyderabad India, 4–7 January 1989. (A. Omran, ed.). IDRC-MR 252e. pp. 169-170.

Shrivastava, P.S., Shomwanshi, K.P.S., 1974. Investigation on the extent of cross pollination and selfing and crossing techniques in niger (*Guizotia abyssinica* Cass). JNKV Res. J. 8, 110–112.

Singh, P.K., Trivedi, H.B.P., 1993. Exploitation of male sterility for heterosis in niger. Oil Crops Newslett. 10, 75–77.

Sreedhar, R.V., Gangaprasad, S., Salimath, P.M., Ravikumar, S.R., 2005. Assessment of genetic variability and character association in niger, *Guizotia abyssinica* Cass. J. Oilseeds Res. 22 (2), 441–443.

Stymne, S., Appelqvist, L.A., 1980. The biosynthesis of linoleate and linolenate in homogenates from developing soya bean cotyledons. Plant Science Letters, 17, pp. 287–294.

Upadhyaya, P.C., Reddy, R.K., 1997. Combining ability analysis for seed yield and it's attributes in niger under rainfed condition. J. Oilseeds Res. 14 (1), 36–39.

Vavilov, N.I., 1951. Origin, variation, immunity and breeding of cultivated plants. Chronica Botanica 13, 1–364.

Vles, R.O., Gottenbos, J.J., 1989. Nutritional characteristics and food uses of vegetable oils: Niger. In: Robbelen, G., Downy, R.K., Ashri, A. (Eds.), Oilcrops of the World. McGraw Hill, New York, pp. 63–86.

Warner, K., Knowlton, S., 1997. Frying quality and oxidative stability of high oleic corn oil. J. Am. Oil Chem. Soc. 74, 1317–1322.

Warner, K., Orr, P., Glynn, M., 1997. Effects of fatty acid composition of oils on flavor and stability of fried foods. J. Am. Oil Chem. Soc. 74, 347–356.

Yantasath, K., 1975. Influence of nitrogen on growth and nutrient uptake of niger seed (*Guizotia abyssinica*) of Ethiopian and Indian origin. PhD thesis Georg-August University, Göttingen, Germany.

CHAPTER

9

Coconut

Suriya A.C.N. Perera

Division of Genetics and Plant Breeding, Coconut Research Institute, Lunuwila, Sri Lanka

INTRODUCTION

Coconuts are major tropical oil crops. The coconut palm, *Cocos nucifera*, is a member of the family Arecaceae and is the sole species in the genus *Cocos*. Coconuts are grown and used by millions of people in many tropical countries mainly as an oil source but also for a number of other uses.

Coconuts are grown in humid areas of the intertropical zone. They are found in coastal areas between latitudes 20°N and 20°S of the equator and up to an altitude of about 1200 m above sea level. Environments with high humidity, temperatures of 27–30°C, and moderate to well-aerated deep soils are the best for coconut palms.

Coconuts are both smallholder crops and major perennial plantation crops that contribute enormously to rural and urban industries. They have high potential for poverty alleviation in rural communities in countries where they are grown. Total world area under coconut production was estimated to be 12,442,000 ha in 2012 producing a total of 71.29 billion nuts (Anonymous, 2012). This area is mainly centered in Asia and distributed between the major coconut-growing countries. Indonesia, the Philippines, and India provide in excess of 75% of world cultivated land area. Brazil and Sri Lanka also have extensive areas dedicated to coconut production, although in Brazil the main use is not oil and fats.

Coconuts are predominantly oil crops. Relatively high melting point, narrow melting range, absence of unpleasant odor, and a certain resistance to oxidation and rancidity (Anonymous, 1979) are some of the main characteristics of coconut oil. They are widely used for domestic consumption as cooking and frying oil in coconut-growing countries. Coconut oil is rich in short-chain fatty acids and exhibits very good digestibility (Banzon, 1977). Coconuts are high-quality fat sources in infant milk, in ice cream fat and coatings, in cream fillings, and in confectionery oil. It can be converted to hard butter by hydrogenation and interesterification for use in a variety of confectionery formulations (Marcus and Puri, 1978).

The main constituent in coconut oil is lauric acid, which exceeds 40% (ranging from 44% to 49%) (Jones, 1991) of total fatty acid content. It is followed by myristic acid, which ranges from 13% to 19%. In addition, caprylic acid, palmitic acid, oleic acid, and capric acid are other

Breeding Oilseed Crops for Sustainable Production. http://dx.doi.org/10.1016/B978-0-12-801309-0.00009-4

fatty acids found to be present in coconut oil at percentages varying from 5% to 8%. Furthermore, fatty acids such as stearic, linoleic, and caproic acid have also been found to occur in lower percentages not exceeding 3% (Anonymous, 1979).

Coconut oil is used as an industrial vegetable oil and as a raw material in soap, glycerine, and margarine. Lauric acid from coconut oil is used to manufacture detergents, cosmetics, and pharmaceuticals. Virgin coconut oil (Bawalan and Chapman, 2006) – which is expelled under low heat from fresh coconut meat to preserve its natural vitamins and enzymes (Marikkar et al., 2007) – has gained much attention in recent times for its merits. It has now secured a significant international market in the pharmaceutical industry as well.

Coconut palms have a multitude of uses besides oil. This is the reason coconuts are known as "one of nature's gifts to mankind," "the tree of life," "the tree of a thousand uses," and "the milk bottle on the doorstep of mankind" (Bourdeix et al., 2005). In addition to oil, the coconut industry makes use of coconut kernels for a variety of products such as desiccated coconut, coconut cream, coconut milk powder, and defatted coconut. Coconut sap tapped from the inflorescence can be fermented to produce coconut vinegar, alcoholic beverages like toddy and arrack, and treacle and jaggery, which are produced using unfermented coconut sap. There are nonfood products such as coconut fiber, fiber products, shell charcoal, and activated carbon which are obtained from different components of the coconut fruit. Coconut water, which is rich in vitamins and minerals, has gained much recognition throughout the world in recent times as a natural nutritional and energy drink. There are international markets for coconut water; countries such as Brazil and India have expansive markets for tender coconut water. Coconut wood is heavily used in the furniture, handicraft, and building material industries. In addition, coconut palms have gained recognition as an integral part of the tropical landscape forming the iconic picture of a desert island composed of the sea, beaches, the sun, and coconut palms. Considering the many uses of coconut, it is very important to sustain plantations, stabilize production, and maintain the supply of raw material to safeguard the different industries and livelihoods that depend on this palm.

GENETICS AND BREEDING OF COCONUTS

Coconut Varieties

Coconuts are monocots, possessing a diploid genome with 16 pairs ($2n = 2x = 32$) of chromosomes. Global classifications divide coconuts into two groups/varieties as tall (*typica*) and dwarf (*nana*). These two groups of coconuts differ in stature and genetic behavior. Morphologically, tall coconuts are tall in habit, produce a root bole (a swollen base of the stem), and possess comparatively larger crowns and nuts. They are naturally crosspollinating resulting in heterozygous individuals and heterogeneous populations. The vegetative phase of tall coconuts is longer, taking about 6–7 years for flower initiation. They live longer, over 60 years under favorable conditions, than the dwarf varieties. Once they reach the reproductive stage, inflorescence emission is continuous resulting in nonseasonality in nut production.

Dwarf coconuts, in contrast, are short in stature and do not produce a root bole. Dwarf coconuts produce smaller crowns and bear smaller nuts than tall coconuts. They are naturally selfpollinating due to male and female phases of the same inflorescence overlapping and thus result in homozygous individuals and pureline populations. Compared with tall coconuts, the vegetative phase of dwarf coconuts is shorter, taking only about 3–4 years for initial flowering. However, the life span of dwarf coconuts is shorter and they display seasonality in nut production due to delayed inflorescence emission especially during unfavorable environmental conditions.

Challenges for Genetic Improvement of Coconut

Coconuts are difficult crops to breed. The long generation interval, the cross-pollinating breeding behavior of tall coconuts, the lack of a viable vegetative propagation method, the low number of seeds produced per palm, and the massive stature of the palm are the most important constraints in coconut breeding. Due to these limitations, genetic improvement in coconuts has been limited to mass selection and hybridization mainly between tall and dwarf coconuts (Perera, 2010a).

Conventional Coconut-Breeding Methods

Cocos is a monospecific genus and thus coconut breeding by conventional methods is limited to the intraspecies level. Mass selection was scientifically recognized as the most fundamental method for coconut breeding (Liyanage, 1955). Consequently, the majority of coconut plantations in the world are derived from mass selection. Yield of copra per palm or one of its components – such as number of fruits produced or per nut copra content – have been used as the main selection criteria for mass selection since the mid twentieth century. In many selection programs, seed nuts were selected mainly based on the size, weight, and shape of the nuts. Populations were subjected to successive selection for fruit characteristics. Depending on the reproduction system used for selection, the methods used to produce genetically improved cultivars for desirable traits were open pollination, selfing, or intercrossing between parents (Liyanage et al., 1988).

Controlled cross-pollination is feasible in coconuts since it facilitates hybridization between different varieties to extract hybrid vigor (Liyanage, 1972). Hybridization of coconuts so far has mainly been between tall and dwarf types. This method has been successfully practiced in coconut breeding to extract hybrid vigor for economically important traits (Wright, 1980). The resultant hybrids have been observed to be improved for precocity, nut yield, and fruit components.

So far in coconut breeding, single crosses between tall and dwarf coconuts have been the most common type of hybrids. Two parents are selected to combine the desirable features of one parent and remove weaknesses from the second parent. When developing hybrid coconuts using a tall coconut as a parent, intercrossing for several generations to derive nearly homozygous lines is not practiced due to practical difficulties. Parental populations and individual parents are usually selected based on the observed ability of such populations or genotypes to realize the desired traits as per the objectives of a particular breeding program. However, the resultant hybrids cannot be considered "true hybrids" because tall parents are not homozygous purelines as per the strict definition of hybrids. However, the resultant hybrids have successfully demonstrated their improved performance for the characters intended (Dissanayaka et al., 2008; Perera, 2010a).

Normally, promising hybrids are identified at the F_1 stage in coconut breeding rather than proceeding to advanced breeding lines, mainly because of the longer periods involved in crossing several generations. However, termination at the F_1 stage limits full exploration of possible phylogenic recombinations of a cross. Nevertheless, a few countries – such as Thailand and Ivory Coast – are reported to be testing multiple crosses to develop varieties with multiple desirable traits or simply continuing the selection process through further generations.

Conventional breeding methods such as these have mainly been targeted at increasing production, productivity, and precocity, which were considered important traits for sustainable coconut production in the early days of coconut production. However, at the present time, breeding for other biotic and abiotic stresses are also being addressed in conventional coconut-breeding programs (Anonymous, 1988).

Use of Molecular Techniques in Coconut Breeding

Conventional coconut-breeding methods have been successful at genetic improvement of coconuts for certain important traits. Yet, coconut breeding through conventional methods is not only a very long–term venture, it also has the difficulty of distinguishing the actual genetic effect from the very high environmental effects on important traits. Molecular markers in different crops have helped immensely in shortening the time period of a breeding program and in targeting the direct gene/quantitative trait locus (QTL) effect. For about two decades molecular techniques have been used to enhance the ability of coconut-breeding programs in several countries to assess the genetic diversity of local and global germplasm collections and to develop genetic linkage maps.

The first studies using molecular markers in coconuts aimed at assessing coconut genetic diversity. These studies used universal marker techniques such as random amplified polymorphic DNA (RAPD) (Ashburner et al., 1997; Duran et al., 1997; Everard, 1996; Dasanayake et al., 2003), restriction fragment length polymorphism (RFLP) (Lebrun et al., 1999), amplified fragment length polymorphism (AFLP) (Perera et al., 1998; Teulat et al., 2000), and inverse sequence-tagged repeat (ISTR) (Duran et al., 1997; Rohde et al., 2000). Later, coconut-specific microsatellite markers were developed (Perera et al., 1999; Rivera et al., 1999). Microsatellites markers are codominant markers and, therefore, they have been highly useful in analyzing heterozygous coconut populations for studies on genetic diversity and genetic relatedness, on germplasm characterization (Perera et al., 2000; Teulat et al., 2000; Dasanayake et al., 2003; Meerow et al., 2003), and on detecting somoclonal variation in tissue-cultured coconut plants. Moreover, they have been used for construction of genetic linkage maps (Herran et al., 2000; Lebrun et al., 2001; Baudouin et al., 2006). A microsatellite kit comprising 14 primers and associated software for data analysis has also been developed (Baudouin and Lebrun, 2002) standardizing the techniques across laboratories for comparable results and for efficient detection of diversity and identification of varieties. At a later stage, high-throughput microarray-based diversity array technology (DArT) was validated for coconuts and used in coconut diversity studies (Perera, 2005). Of all these marker systems, microsatellites have so far been the most extensively used in analyzing the coconut genome.

Microsatallite marker information on genetic diversity has revealed a high level of genetic diversity in a worldwide collection of 130 tall-coconut individuals and 49 dwarf-coconut individuals (Perera et al., 2000). A reduction was observed in the amount of genetic diversity in dwarf coconuts than in tall coconuts with a similar reduction in the number of alleles in dwarf coconuts. Rivera et al. (1999) and Teulat et al. (2000) reported similar results by analyzing coconut varieties from the south coast of Asia and the Pacific. These studies further revealed nonuniformity in the genetic diversity between the tall and dwarf groups of coconut. Accordingly, the tall group reported comparably higher genetic variation and higher heterozygosity (30%) levels. The naturally self-pollinating dwarf coconuts revealed a low heterozygosity percentage of 2.5%.

The development of linkage maps in coconuts is a highly challenging process (Perera, 2010b). Construction of a mapping population is one of the most critical decisions in linkage mapping using DNA markers. It is hindered by the practical difficulty of developing suitable mapping populations due to the perennial nature and massive stature of the coconut palm. A mapping population of a crop plant should comprise parents and their segregating progeny. Such segregating populations – such as F_2s and backcrosses – or advanced segregating populations – such as recombinant inbred lines (RILs) or doubled haploid lines (DHLs) – are commonly used mapping populations for linkage mapping in selfpollinated crops (Kuittinen et al., 1997; Marquez-Cedillo et al., 2001). Yet the development of such mapping populations is hindered by the cross-pollinating nature and perenniality of the coconut palm. In contrast, an alternative approach has been adopted in coconuts to produce an F_1 segregating population. In view of the heterozygosity of tall coconuts and the homozygosity of dwarf coconuts a segregating population could be developed by crossing the two coconut varieties, tall and dwarf (Bandaranayake and Kearsey, 2003). However, a critical consideration here is the selection of tall parents possessing sufficient levels of heterozygosity to result in a considerable level of polymorphism in the subsequent progeny (Perera, 2010b).

Yet another constraint in developing a sufficiently large mapping population in coconuts is the limited number of seeds produced by a particular mother palm and the low rate of success of artificial pollination. As a result, developing a mapping progeny with an adequate number of individuals takes a long time and, hence, will introduce an age gap among individuals of the resulting population. Combining several separate half-sib families to produce a single mapping population is an alternative way of increasing progeny size within a shorter time span. To achieve this, several dwarf palms should be chosen from a population as female parents and a single highly heterozygous male tall coconut should be chosen as the pollen donor to construct the mapping population (Perera, 2010b).

Despite these difficulties, several linkage maps of coconuts have been derived. The first genetic linkage map for coconuts was developed by using an F_1 population obtained by crossing East African Tall × Laguna Tall and by genotyping using ISTR markers (Rohde et al., 1999). A second mapping population was developed in the Philippines by crossing Malayan Yellow Dwarf × Laguna Tall. The resultant linkage map was developed by genotyping using such markers as AFLP, ISTR, RAPD, and inter simple sequence repeat (ISSR). The framework map of this population included 382 markers in 16 linkage groups. QTL analysis using this population identified 6 QTLs governing early germination (Herran et al., 2000). With the previously established correlations among early germination, flowering, and high nut yield this QTL map provides the opportunity for marker-assisted selection in coconuts. The same QTL map was later enriched by a QTL for growth parameters such as leaf production, girth at collar, and height (Ritter et al., 2000). A third mapping population was also developed in the

Ivory Coast by crossing Cameroon Red Dwarf ×Rennel Island Tall. By placing 280 markers in 16 linkage groups, the resultant framework map allowed identification of a QTL for yield parameters such as nut number, number of bunches, and fruit components (Lebrun et al., 2001; Baudouin et al., 2006).

Physical mapping is a molecular method that has been tested for coconuts. It provides the likely order of defined genomic DNA segments along chromosomes and, if contiguous regions of overlapping cloned DNA exist, physical mapping provides an estimate of the true distance in base pairs between distinct locations on any specific chromosome. Physical mapping can be used for studies on genome-wide gene discovery, expressed sequence tag (EST) mapping (functional genomics), or comparative genomics (synteny studies). With the availability of sequence information for oil palm (another type of palm), such studies may well be possible for coconuts.

Objectives of Coconut Breeding

The objectives of coconut breeding in the early days mainly related to increasing copra yields and reducing vegetative phases. Presently, coconut-breeding objectives in many countries have expanded to include breeding for resistance to and tolerance of biotic stresses. Efforts are also being made to genetically improve coconuts against abiotic stresses, mainly moisture stress or drought. These factors are important for coconut production to be sustained. Hence, breeding for sustainable coconut production has gradually been included as an objective in many coconut-growing countries.

BREEDING COCONUTS FOR SUSTAINABLE PRODUCTION

Factors Affecting Sustainability in Coconut Production

Coconuts are perennial crops. The sustainability of coconut production depends on both inherent and environmental factors. The comparatively long juvenile phase of about 4–7 years, the long economically productive life span (generally exceeding 50 years), the genetic potential of different cultivars, germplasm diversity, and the multitude of uses offered by the palm are the main inherent factors affecting sustainable production in coconuts. Sustainability of production in such a crop is greatly affected by environmental factors in addition to the inherent traits of cultivars. Biotic stresses (i.e., pests and diseases) and abiotic stresses (i.e., drought) also pose serious limitations to sustainable coconut production. Lowlands that have high water tables and experience strong winds (cyclones) result to a lesser degree in the death of coconut seedlings and occasional uprooting of palms in large numbers, respectively. Moreover, general limiting factors such as soil degradation (resulting in low soil fertility), old age of palms (resulting in senility of plantations), and more recently global climate change are important factors to be considered in ensuring the sustainability in coconut production. Coconuts, partly because of being a smallholder crop and partly because of their diverse uses, are subject to both positive and negative effects on the sustainability of production.

Diversity in its uses is an important characteristic of the coconut palm, something that provides coconut producers with many options when marketing their products. Yield from coconuts extends beyond the kernel, which is the traditional economically important yield parameter. At present, the fruit, husk, water, shell, and kernel are all used during processing into economically valued goods, which have gone beyond the boundaries of the producing countries and entered international markets. In addition to the fruit components, the fronds and the stem also are used in various industries. This unique feature of the coconut palm provides coconut farmers with opportunities for economic sustainability of the coconut palm even when there is a downturn in the market for a specific item.

Availability of Genetic Resources

Diversity in the germplasm of a crop species is key to breeding for important characters. Genetic diversity is referred to as the amount of genetic variability among individuals of a variety or population of a species (Brown, 1983). It results from the many genetic differences between individuals and may be manifested in differences in DNA sequences, in biochemical characteristics (as in protein structure or isoenzyme properties), in physiological properties (abiotic stress resistance, growth rate, etc.), or in morphological characters such as flower color or plant form (Rao and Hodgkin, 2001). Sustainability in production depends on several basic factors, as has already been stated. However, breeding for enhancement of such factors depends on the availability of positive genes governing the trait of concern, which in turn translates into germplasm diversity. In the early domestication of crops, the main selection criterion was yield. As a result, the majority of cultivated crops are now lacking genes governing tolerance or resistance, especially for biotic and abiotic stresses that limit the sustainability of production. Such genes are usually found in the wild relatives of respective crop plants. This means that wild plants can be used to introgress these genes into cultivated material by conventional methods – such as backcross breeding or marker-assisted backcross breeding – whenever the need arises.

A drawback coconuts suffer from in germplasm diversity is the extinction of wild relatives. Furthermore, since *Cocos* is a monospecific genus, there are no close relatives to act as gene donors for genetic improvement *via* gene introgression. Therefore, the best option available to breeders for diversity are land races and indigenous varieties that are present in countries where coconuts have been grown for longer periods. Despite this drawback, coconuts do benefit from genetic diversity because much of the global coconut germplasm has been conserved in *ex situ* field gene banks in an effort made by the International Coconut Genetic Resources Network (COGENT), which was established by Bioversity International. There are five regional *ex situ* field gene banks of coconuts established in India, Papua New Guinea, Ivory Coast, Brazil, and Indonesia in addition to the national germplasm repositories established in each of these countries. These *ex situ* field gene banks house a comprehensive collection of global coconut germplasm, which can be utilized in breeding for sustainability based on characterization and evaluation for important traits.

Coconut Breeding for Biotic Stresses: Pests

Coconut palms are attacked by several major pests that attack the leaves, crown, or nuts of the plant. Of them, the coconut leaf beetle (*Brontispa longissima*) and coconut caterpillar (*Opisina arenosella*) are the main pests attacking coconut leaves. Two beetles, the black beetle (*Oryctes rhinoceros*) and red weevil (*Rhynchophorus ferrugineus*) damage the growing point of

FIGURE 9.1 Eryophyid mites (*Aceria guerreronis*) cause substantial damage to developing nuts, especially in plantations in dry areas in many coconut-growing countries including Oman, India, and Sri Lanka.

the coconut palm at the seedling stage and young bearing palm stage, respectively, which often results in death of the palm especially if the latter is not timely treated. Eryophyid mites (*Aceria guerreronis*) are known to cause substantial damage to developing nuts, especially in plantations in dry areas in many coconut-growing countries including Oman, India, and Sri Lanka (Fig. 9.1). Integrated pest management strategies have been developed for all these pests by blending chemical control methods and biological control methods.

Genetic improvement for tolerance/resistance is being tested for *Aceria* mite infestation in India and Sri Lanka. Screening coconut hybrids for tolerance to *Aceria* mite infestation has been tested and has revealed variations among different hybrids that could lead to genetic governance for the trait (Perera et al., 2013). In addition, Shalini et al. (2006) reported molecular markers associated with *Aceria* mite tolerance in a molecular study using a collection of palms selected on a standard scale that varied from resistance to susceptibility as well as using genotyping with RAPD and simple sequence repeat (SSR) markers. Both these studies give positive indications for selecting and breeding coconuts for tolerance/resistance to *Aceria* mite infestation. An extensive evaluation is under way to screen indigenous varieties of coconut in Sri Lanka for *Aceria* mite infestation and a hybridization program is also in progress to combine tolerance in economically productive tall cultivars (Perera, 2011).

Coconut Breeding for Biotic Stresses: Diseases

Coconut palms are affected by bacterial pathogens causing various rots, fungal pathogens causing such symptoms as stem bleeding and leaf damage, and more importantly intracellular pathogens such as phytoplasma and viroids causing lethal or debilitating diseases. However, practical breeding programs for resistance to coconut diseases are limited to diseases caused by intracellular pathogens. Intracellular pathogens are obligatory parasites and depend on the palm for the resources they need for life and multiplication. When they multiply rapidly, they manifest themselves as diseases in the palms. The control of diseases caused by intracellular pathogens by chemical or other methods has so far been found practically impossible. Therefore, breeding for resistance/tolerance has been identified as the best way of managing diseases caused by intracellular pathogens.

The most prominent group of intracellular pathogens affecting coconut palms is phytoplasma. These organisms are referred to as cell wall–less bacteria. They inhibit phloem tissues in the plant (McCoy et al., 1983). There are several phytoplasma causing coconut diseases, of which lethal yellowing is the most devastating. Premature fall of coconuts is the first symptom shown by plants that have contracted lethal yellowing. The symptoms then progress to yellowing of fronds, blackening of inflorescences, rotting, and ultimately the entire crown falls off resulting in death of the palm in as short a period of time as 3 months (Dollet et al., 2009). Lethal yellowing first appeared in the Caribbean and then spread to Central America and Africa, devastating coconut plantations in a number of countries.

Breeding coconut palms for resistance to lethal yellowing first began in Jamaica and, later, in other disease-plagued countries. Accordingly, the first screening trials began in 1961 when the coconut variety Malayan Yellow Dwarf was identified as highly tolerant and was recommended for planting. Later, the F_1 hybrid Maypan, a cross between Malayan Dwarf × Panama Tall, was recommended as a fairly tolerant hybrid to lethal yellowing disease. Baudouin et al. (2009) reviewed the results of these screening trials and reported high levels of resistance in Malayan Dwarf and Sri Lanka Green Dwarf and high levels of susceptibility in most commercially cultivated tall coconut varieties. In addition to conventional screening and breeding trials, some efforts have been made to identify resistant genotypes to lethal yellowing through molecular methods and some positive results have been reported (Omamor et al., 2010; Zizumbo-Villarreal et al., 2006).

Cape St. Paul Wilt disease (CSPWD), which was first reported in Ghana, is another lethal yellowing–type disease caused by phytoplasma (Nkansah-Poku et al., 2009). Both screening and breeding trials have been conducted for CSPWD and once again Sri Lanka Green Dwarf and Malayan Yellow Dwarf were found to show high levels of resistance/tolerance to this disease (Quaicoe et al., 2009).

In addition to lethal diseases caused by phytoplasma as just described, coconut palms are also subject to debilitating diseases with phytoplasma as the causal organism. Kerala Root Wilt disease, which was first detected in India in the 1880s, and Weligama Coconut Leaf Wilt disease (WCLWD), which was reported in Sri Lanka in 2006, are symptomatically similar debilitating coconut diseases . Screening and breeding programs have been in place in both countries (Nair et al., 1996, 2006; Perera and Dissanayake, 2013) resulting in the identification of high disease resistance levels in Chowghat Green Dwarf and Malayan Green Dwarf in India (Nair et al., 2000) and Sri Lanka Green Dwarf in Sri Lanka (Perera et al., 2014). As far as screening resistant genotypes is concerned, studies analyzing the population structure and genetic relatedness among palms affected by root wilt diseases have been reported (Devakumar et al., 2011).

Coconuts are subject to a disease called Cadang-Cadang, which is caused by the lethal viroid CCCVd. This disease is also responsible for substantial palm deaths in the Philippines,

FIGURE 9.2 Drought reduces nut yield in coconuts and is capable of killing palms when prolonged.

which holds the second largest extent of coconuts in the world. Every year one million coconut palms are reported to be killed by CCCVd and over 30 million coconut palms have succumbed since the discovery of Cadang-Cadang in the Philippines in 1927/1928, highlighting the threat posed by this disease for coconut sustainability . Hence, breeding and selection of varieties/genotypes with resistance/tolerance is vital for the revival of coconut palms in affected areas.

Coconut Breeding for Abiotic Stresses: Drought

Drought is a multidimensional factor of stress, which affects plants at various levels of growth, development, and reproduction (Lambers et al., 1998). It is a natural climatic phenomenon in which less than expected precipitation occurs within a specified period of time (Feng and Zhang, 2005). Drought stress induces various biochemical and physiological responses in plants and represents one of the most adverse environmental factors of plant growth and production (Gomes and Prado, 2007). It causes significant nut yield reduction in coconuts and even the death of palms in the prolonged droughts that occur in dry areas (Fig. 9.2).

Coconut palms grow well in areas with mean annual rainfall of 1500 mm that is adequately distributed throughout the year. Both the amount and distribution of rain is especially important for fruit set and coconut growth. Monthly rainfall of 150 mm is considered ideal (Rajagopal et al., 1996). However, coconuts are frequently exposed to soil and atmospheric water deficit as a result of being a perennial palm with a long productive life span (Rajagopal and Kasturi Bai, 2002). Since coconuts are perennials, it is not possible to avoid drought by adjusting the timing of planting, as can be done with some annual crops. Hence, breeding for tolerance to drought stress or moisture deficit is the best way of mitigating the problem and ensuring sustainable coconut production.

In a study conducted in Sri Lanka some accessions in the conserved coconut germplasm were evaluated during a prolonged drought for their response to drought (Nainanayake and Morison 2007). Although no significant differences between genotypes were observed during the 80-day drought, the susceptibility index that was first used by Fischer and Maurer (1978) for spring wheat cultivars was identified as a reliable physiological indicator for screening coconut palms for drought tolerance. Gomes and Prado (2007) suggested that genotypic differences in physiological and morphological traits could largely explain the agronomic performance of field-grown coconut trees under water stress. This indicates the possibility of breeding for moisture stress expecting only tolerance for water stress, because complete resistance for drought cannot be expected in a live plant. Physiological studies aimed at screening and breeding coconuts for tolerance to drought stress are currently in progress. However, actual gene identification and introgressing breeding programs remain scarce.

Coconut Breeding for Abiotic Stresses: Climate Change

Earth's climate has always been changing mainly due to natural causes such as volcanic activity, although it has been relatively stable for about the last 300 years. However, since 1900 more rapid changes have taken place. They are considered by many to be mainly due to human activities. Global warming has resulted in increased air and ocean temperatures, widespread melting of snow and ice, and rising global sea levels. Such effects are expected to substantially affect crop production and sustainability. Climate change is expected to reduce crop yields, increase disease incidence due to faster evolution of pathogens and the emergence of new pathogenic life forms, increase pest populations, etc. However, by adopting suitable strategies as a result of careful scientific consideration, it is believed such effects can be mitigated. Perennial crops like coconuts will be more prone to such effects. Breeding for tolerance is seen as the best way of dealing with the expected ill effects of global climate change.

Some positive aspects of climate change on crop production have also been reported. It is said that elevated CO_2 levels will lead to higher biomass production. However, while higher CO_2 levels increase water use efficiency in plantation crops, they also increase disease incidence risks as a result of the reduced phenolic content in plant tissues. These factors should be carefully considered when developing coconut varieties that can accumulate more biomass and act as carbon sinks under elevated CO_2 levels.

An important element in breeding coconuts for sustainable production against the backdrop of climate change is the breeding of varieties with the genetic potential to perform well in varying environments. Other attempts aimed at coconut sustainability are the traditional coconut-breeding programs addressing this issue not just to increase productivity but also to expand coconuts into new environments (Dissanayaka et al, 2008).

Opportunities and Constraints

Increased nut production and productivity, despite not being mentioned above as factors limiting the sustainability of coconut production, need to be seriously considered for coconut profitability to be increased and sustainability to be insured. In addition, efforts need to be made to consider the wide range of coconut products when breeding coconuts for sustainability. Conventional breeding methods – such as selecting for tolerant genes and developing hybrids – are further useful tools in breeding coconuts for increased productivity and production sustainability. Major constraints to such a conventional approach have already been mentioned in this chapter.

The vast majority of germplasm has yet to be identified, let alone conserved. Germplasm that has been conserved in regional or national gene banks has yet to be characterized for optimizing its use in breeding for stress tolerance. There is both a high priority and high potential for obtaining raw material for breeding once underexploited materials are conserved and characterized, especially for breeding for tolerance to stresses such as *Aceria* mite and drought. Consequently, the importance of characterizing lesser-evaluated genetic material cannot be overstated. The major limitation of coconut germplasm for breeding for resistance/tolerance for production sustainability is the extinction of wild relatives of coconuts and the lack of close relatives for techniques such as alien gene transfer via wide hybridization to be applied. This is a major constraint because in most crops genes (especially for resistance/tolerance) are known to be present in their own wild relatives or sometimes in the wild relatives of related species.

As mentioned earlier in this chapter, many molecular biological applications have been given to coconuts to enhance the conventional breeding techniques applied in coconut breeding. Advances in molecular breeding techniques in coconuts will provide opportunities for coconut-breeding programs that are more targeted at overcoming barriers to sustainability in coconut production. For example, molecular breeding will pave the way to identifying genotypes possessing specific traits that are important for tolerance to various stresses through QTL mapping and marker-assisted selection. To facilitate marker-assisted selection efficiently and effectively it is essential to increase the density of markers in the currently available coconut genome map. To achieve this, additional polymorphic and high-throughput molecular markers and marker systems need be identified for coconuts. Furthermore, there is a need to map a greater number of specific genes/QTLs that govern important traits. There is a need for future genome-mapping projects to use different mapping populations that segregate for the trait of interest. In addition to linkage mapping, strategies like linkage disequilibrium or association mapping will provide greater opportunities to overcome the limitations in gene discovery by linkage mapping of coconuts, the major constraint being the difficulty of developing segregating populations for linkage mapping due to coconuts being perennials. On the positive side, the basic background and scientific methods necessary for molecular breeding have already been developed and are progressing further. Studies such as transcriptome profiling of stress-tolerant genotypes can be used for candidate gene identification for tolerance to important stress conditions in coconuts. Advanced high-throughput genotyping facilities are emerging and more sequencing facilities are being introduced, even for orphan crops such as coconuts. Hence, coconuts can benefit from these novel molecular breeding methods as a result of the emergence of additional genomic resources for less-studied crops (Varshney et al., 2010).

Resistant and tolerant varieties of coconuts have already been bred against lethal yellowing. Moreover, research is being conducted at present using quantitative polymerase chain reaction (qPCR) and terminal restriction fragment length polymorphism (T-RFLP) techniques to quantify the levels of tolerance in different cultivars. Such efforts demonstrate the higher potential for combating other diseases caused by intracellular pathogens (such as palm death), which adversely affect coconut sustainability in many coconut-growing countries.

Coconuts face the same general difficulties as other perennial crops. However, coconut palms face an additional difficulty due to the lack of a viable vegetative propagation method. This has always been a major drawback to coconut genetic improvement because of the inability of getting superior genotypes fixed by vegetative propagation. However, there are now reports shedding light on the feasibility of coconut tissue culture via various explants including unfertilized ovaries (Perera et al., 2007). Such a major step toward creating a vegetative propagation method will raise expectations of greater achievements in breeding coconuts for sustainable production. In recent years, considerable progress has been made in the development of tolerance to and isolation of stress, mainly in disease and salt-tolerant cell/callus lines using *in vitro* techniques. *In vitro* selection will save the time required to develop disease-resistant and abiotic stress-tolerant lines of important plant species (Rai et al., 2011). Despite certain limitations, many reports are available on successful regeneration of plants from selected cell/callus lines showing stress tolerance to be stable in the whole plant (Nabors et al., 1980; Singh et al., 2003; Hossain et al., 2007). The feasibility of tissue-cultured coconuts opens up avenues to *in vitro* screening and developing of stress-tolerant genotypes. Transformation techniques of coconut have so far been hindered by the lack of viable protocols for coconut tissue culture. The hope for tissue-cultured coconuts will increase the potential for subsequent genetic modification of this palm via transformation to drive stress resistant or tolerant palms. Although further advances are needed in this respect, genetic improvement through transformation gives the opportunity to negate the lack of resistant genes from wild coconuts and the possibility of gene transfer from completely unrelated species.

Acknowledgments

The author acknowledges Dr Nayani Aratchige and Mrs Nilmini Jayalath for providing Figs 9.1 and 9.2.

References

Anonymous, 1979. Technical Data Handbook on the Coconut, its Products and By Products, A Compilation. Philippines Coconut Authority, Quezon City, Philippines, 123.

Anonymous, 1988. Rapport scientifique cocotier. Station Cocotier Marc Delorme. Institut de Recherches pour les huiles et Oleagineux, Cote d' Ivoire. pp. 217.

Anonymous, 2012. Asia Pacific Statistical Year Book. Asia Pacific Coconut Community Jakartha, Indonesia.

Ashburner, G.R., Thompson, W.K., Halloran, G.M., 1997. RAPD analysis of South Pacific coconut palm populations. Crop Sci. 37, 992–997.

Bandaranayake, C.K., Kearsey, M.J., 2003. Genome mapping, QTL analysis and MAS: importance, principle, constraints and application in coconut. Int. Plant Genet. Res. Newsl 142, 47–54.

Banzon, J.A., 1977. Industrial coconut research and development. Philippines J. Coconut Studies 2, 13–17.

Baudouin, L., Lebrun, P., 2002. The development of a microsatellite kit and dedicated software use with coconuts. Burotrop Bull. 17, 16–20.

Baudouin, L., Lebrun, P., Konan, J.L., Ritter, E., Berger, A., Billottee, N., 2006. QTL analysis of fruit components in the progeny of a Rennell Island Tall coconut (Cocos nucifera L.) individual. Theor. Appl. Genet. 112, 258–268.

Baudouin, L., Philippe, R., Quaicoe, R., Dery, S., Dollet, M., 2009. General overview of genetic research and experimentation on coconut varieties tolerant/resistant to Lethal Yellowing. OCL vol. 16 No. 2 MARS-AVRIL.

Bawalan, D.D., Chapman, K.R., 2006. Virgin coconut oil; Production Manual for Macro- and Village-Scale Processing, FAO. Thammada Press, Bangkok, Thailand.

Bourdeix, R., Santos, G., Labouisse, J.P., Baudouin, L., 2005. Useful definition of terms and nomenclature. In: Batugal, P., Ramanatha Rao, V., Jeffrey, Oliver (Eds.), Coconut Genetic Resources. IPGRI, Rome, pp. 9–10.

Brown, W.L., 1983. Genetic diversity and genetic vulnerability: an appraisal. Econ. Bot. 37 (1), 4–12.

Dasanayake, P.N., Everard, J.M.D.T., Karunanayake, E.H., Nandadasa, H.G., 2003. Characterization of coconut germplasm by microsatellite markers. Trop. Agric. Res. 15, 51–61.

Devakumar, K., Thomas, R.J., Nair, R.V., Jerard, B.A., Rajesh, M.K., Jacob, P.M., Jayadev, K., Parthasarathy, V.A., 2011. Analysis of population structure and genetic relatedness among root (wilt) disease-resistant and susceptible west coast tall coconut palms (*Cocos nucifera* L.) using microsatellite markers. Indian J. Agric. Sci. 81, 487–493.

Dissanayaka H.D.M.A.C., Perera, S.A.C.N., Fernando, W.B.S., Attanayake, R.B., Meegahakumbura, M.G.M.K., Perera, L., 2008. Evaluationof the comparative performance of five commercial coconut cultivars under two different agro-ecological zones in Sri Lanka. In: Ninanayake, A. and Jayamanne, E. (Eds.), Proceedings of the Plantation Second Plantation Crop Research Symposium, BMICH, Colombo, Sri Lanka. , Samayawardena printers, Colombo. pp. 126–138.

Dollet, M., Quaicoe, R., Pilet, F., 2009. Review of coconut lethal yellowing diseases. Diversity, variability and diagnosis. OCL vol. 16, No. 2, MARS-AVRIL. pp. 1–5.

Duran, Y., Rohde, W., Kullaya, A., Goikoetxea, P., Ritter, E., 1997. Molecular analysis of east African Tall coconut genotypes by DNA marker technology. J. Genet. Breed. 51, 279–288.

Everard, J.M.D.T., 1996. Use of molecular markers for breeding of the coconut palm (*Cocos nucifera* L.), MSc Thesis, University of New England, Armidale, Australia.

Feng, L.H., Zhang, X.C., 2005. Quantitative expression on drought magnitude and disaster intensity. Nat. Haz. Earth Syst. Sci 5, 495–498.

Fischer, R.A., Maurer, R., 1978. Drought tolerance in spring wheat cultivars I. Grain yield responses. Aust. J. Agric. Res. 3, 191–198.

Gomes, F.P., Prado, C.H.B.A., 2007. Ecophysiology of coconut palm under water stress. Braz. J. Plant Physiol. 19 (4), 377–391.

Herran, A., Estioko, L., Becker, D., Rodriguez, M.J.B., Rohde, W., Ritter, E., 2000. Linkage mapping and QTL analysis in coconut (*Cocos nucifera* L.). Theor. Appl. Genet. 101, 292–300.

Hossain, Z., Mandal, A.K.A., Datta, S.K., Biswas, A.K., 2007. Development of NaCltolerant, line in *Chrysanthemum morifolium* Ramat through shoot organogenesis of selected callus line. J. Biotechnol. 129, 658–667.

Jones, L.H., 1991. Perennial vegetable oil crop. In: Persely, G.J. (Ed.), Agricultural Biotechnology: Opportunities for International Developments. Commonwealth Agricultural Bureaux International, Wallingford, UK, pp. 213–224.

Kuittinen, H., Mattila, A., Savolainen, O., 1997. Genetic variation at marker loci and in quantitative traits in natural populations of *Arabidopsis thaliana*. Heredity 79, 144–152.

Lambers, H., Chapin, III, F.S., Pons, T.L., 1998. Plant Physiological Ecology. Springer-Verlag, New York.

Lebrun, P., N'Cho, Y.P., Bourdeix, R., Baudouin, L., 1999. Le cocotier. In: Hamon, P., Seguin, M., Perrier, X., Glaszmann, J.C. (Eds.), Diversité génétique des plantes cultivées. CIRAD, Montpellier, pp. 219–240.

Lebrun, P., Baudouin, L., Bourdeix, R., Louis Konan, J., Barker, J.H.A., Aldam, C., Herran, A., Ritter, E., 2001. Construction of a linkage map of the Rennell Island Tall coconut type (*Cocos nucifera* L.) and QTL analysis for yield characters. Genome 44, 962–970.

Liyanage, D.V., 1955. Planting materials for coconut. Ceylon Coconut Quarterly 6, 75–80.

Liyanage, D.V., 1972. Production of improved coconut seeds by hybridization. Oleagineux 27 (12), 597–599.

Liyanage, D.V., Wickramaratne, M.R.T., Jayasekera, C., 1988. Coconut breeding in Sri Lanka. Cocos 6, 1–26.

Marcus, J.G., Puri, P.S., 1978. Manufactured hard butters. J. Am. Oil Chem. Soc. 55, 610A–612A.

Marikkar, J.M.N., Jayasundara, J.M.M.A., Prasadika, S.A.H., Jayasingha, C.V.L., Premakumara, G.A.S., 2007. Assessment of stability of vergin coconut oil during deep frying. CORD 23 (1), 62–70.

Marquez-Cedillo, L.A., Hayes, P.M., Kleinhofs, A., Legge, W.G., Rossnagel, B.G., Sato, K., Ullrich, S.E., Wesenberg, D.M., 2001. QTL analysis of agronomic traits in barley based on the doubled haploid progeny of two elite North American varieties representing different germplasm groups. Theor. Appl. Genet. 103, 625–637.

McCoy, R.E., Howard, F.W., Tsai, J.H., Donselman, H.M., Thomas, D.L., Basham, H.G., Atilano, R.A., Escafi, F.M., Britt, L., Collins, M.E., 1983. Lethal Yellowing of Palms. University of Florida, Gainesville, Florida.

Meerow, A.W., Wisser, R.J., Brown, J.S., Kuhn, D.N., Schnell, R.J., Broschat, T.K., 2003. Analysis of genetic diversity and population structure within Florida coconut (*Cocos nucifera* L.) germplasm using microsatelite DNA, with special emphasis on the Fiji Dwarf cultivar. Theor. Appl. Genet. 106, 715–726.

Nabors, N.W., Gibbs, G.E., Bernstein, C.S., Meis, M.E., 1980. NaCl tolerant tobacco plants from cultured cells. Z. Pflanzen Physiol. 97, 13–17.

Nainanayake, N.P.A.D., Morison, J.I.L., 2007. A case study on physiology-based drought screening of coconut with selected accessions. Cocos 18, 1–20.

Nair, M.K., Koshy, P.K., Nair, R.V., Rao, E.V.V.B., Nampoothiri, K.U.K., Iyer, R.D., 1996. A root (wilt) disease resistant coconut hybrid and strategy for resistance breeding. Indian Coconut J. 27 (1), 2–5.

Nair, R.V., Rajesh, M.G., Jacob, P.M., 2000. Chowghat green dwarf – a suitable variety for the coconut root (wilt) diseased tract. Indian Coconut J. 31 (3), 55.

Nair, R.V., Thomas R.J., Jacob, P.M., Nair, S.N., 2006. Breeding for resistance to coconut root (wilt) disease. In: Nair, R. V., Krishnakumar, V., Thomas, R.J., Muralidharan, K., and Thomas, G.V. (Eds.), Breeding for Resistance to Diseases and Insect Pests in Plantation Crops . pp. 63–69.

Nkansah-Poku, J., Philippe, R., Quacoe, N., Dery, S., Ransford, A. 2009. Cape Saint Paul Wilt Disease of Coconut in Ghana: Surveillance and Management of Disease Spread. OCL vol. 16 No. 2, MARS-AVRIL.

Omamor, I.B., Eziashi, E.I., Asemota, O.A., Aisagbonhi, C.I., Ogunkanmi, L.A., 2010. DNA molecular analysis between lethal yellowing disease and non-lethal yellowing disease of coconut palms (*Cocos nucifera* L.) in Nigeria. Trends Mol. Sci., 1–8.

Perera, S.A.C.N., 2005. Report of the Genetics and Plant Breeding Division, Annual Report of the Coconut Research Institute, Sri Lanka.

Perera, S.A.C.N., 2010a. Technological Innovations in Major World Oil Crops'. In: Gupta, S.H. (Ed.), Springer, USA, pp. 201–218.

Perera S.A.C.N, 2010b. QTL analysis in coconut via genome mapping; Principles, requirements and prospects, Cocos, J. Coconut Res. Inst. Sri Lanka. 20: 1–9.

Perera, S. A. C. N., 2011. Development of coconut hybrids for tolerance to Acria mite infestation. Annual Report of the Coconut Research Institute. Coconut Research Institute, Sri Lanka, 36–37.

Perera, S.A.C.N., Dissanayake, H.D.M.A.C., 2013. Management of the Weligama Coconut Leaf Wilt Disease Screening and Breeding Coconuts for Resistance/Tolerance to WCLWD. Weligama Coconut Leaf Wilt Disease. Coconut Research Institute, Sri Lanka, 96–106.

Perera, L., Russell, J.R., Provan, J., McNicol, J.W., Powell, W., 1998. Evaluating genetic relationships between indigenous coconut (*Cocos nucifera* L.) accessions from Sri Lanka by means of AFLP profiling. Theor. Appl. Genet. 96, 545–550.

Perera, L., Russell, J.R., Provan, J., Powell, W., 1999. Identification and characterization of microsatellites in coconut (*Cocos nucifera* L.) and the analysis of coconut populations in Sri Lanka. Molec. Ecol. 8, 344–346.

Perera, L., Russell, J.R., Provan, J., Powell, W., 2000. Use of Microsatellite DNA markers to investigate the level of genetic diversity and population genetic structure of coconut (*Cocos nucifera* L.). Genome 43, 15–21.

Perera, P.I.P., Hocher, V., Verdeil, J., Doulbeau, S., Yakandawala, D.M.D, Weerakoon, L.K, 2007. Unfertilized ovary: a novel explant for coconut (*Cocos nucifera* L.) somatic embryogenesis. Plant Cell Rep. 26, 1–8.

Perera, L., Sarathchandra, S.R., Wickramananda, I.R., 2013. Screening Coconut cultivars for tolerance to infestation by the Coconut Mite, *Aceria guerreronis* (Keifer) in Sri Lanka. CORD 29 (1), 46–51.

Perera, S.A.C.N., Herath, H.M.N.B., Herath, Wijesekera, H.T.R., Subhathma, W.G.R., Subhathma, Weerakkody, W.A.T.L., 2014. Evaluation of Coconut Germplasm in Weligama and Matara area of the Southern Province of Sri Lanka for Resistance to Weligama Coconut Leaf Wilt Disease. Cocos 21, 7–14.

Quaicoe, R., Dery, S., Philippe, R., Baudouin, L., Nipah, J., Nkansah Poku, J., Arthur, R., Dare, D., Yankey, E., Pilet, F., Dollet, M., 2009. Resistance screening trials on coconut varieties to Cape Saint Paul Wilt Disease in Ghana. OCL vol. 16, No. 2, MARS-AVRIL.

Rai, M.K., Kalia, R.K., Singh, R., Gangola, M.P., Dhawan, A.K., 2011. Developing stress tolerant plants through in vitro selection – an overview of the recent progress. Environ. Exp. Bot. 71, 89–98.

Rajagopal, V., Kasturi Bai, K.V., 2002. Drought tolerance mechanism in coconut. Burot. Bull. 17, 21–22.

Rajagopal, V., Shivashankar, S., Mathew, J., 1996. Impact of dry spells on the ontogeny of coconut fruits and its relation to yield. Plant. Rech. Develop. 3, 251–255.

Rao, V.R., Hodgkin, 2001. Genetic diversity and conservation and utilization of plant genetic resources. Plant Cell Tissue Organ Cult. 68, 1–19.

Ritter, E., Rodriguez, M.J.B., Herran, A., Estioko, L., Becker, D., Rohde, W., 2000. Analysis of quantitative trait loci (QTL) based on linkage maps in coconut (*Cocos nucifera* L.). In: Arencibia, A. (Ed.), Plant Genetic Engineering Towards the Third Millennium. Elsevier Science B.V., Amsterdam, The Netherlands, pp. 42–48.

Rivera, R., Edwards, K.J., Barker, J.H.A., Arnold, G.M., Ayad, G., Hodgkin, T., Karp, A., 1999. Isolation and characterization of polymorphic microsatellites in *Cocos nucifera* L. Genome 42, 668–675.

Rohde, W., Becker, D., Kullaya, A., Rodriguez, M.J.B., Herran, A., Ritter, E., 1999. analysis of coconut germplasm biodiversity by DNA marker technologies and construction of a first genetic linkage map. In: Oropeza, C., Verdeil, J.L., ashburner, G.R., Cardega, R., Santamaria, J.M. (Eds.), Current Advances in Coconut Biotechnology. Kluwer Academic Publishers, Dordrecht, The Netherlands, pp. 99–120.

Rohde, W., Herran, A., Estioko, L., Sinje, S., Becker, D., Kullaya, A., Rodriguez, M.J.B., and Ritter, E., 2000. Mapping of DNA markers, homeotic genes and QTLs in coconut (*Cocos nucifera* L.) and synteny studies with oil palm. International Symposium on oil palm genetic resources and Utilization, pp. AC1–AC2.

Shalini, K.V., Manjunatha, S., Lebrun, P., Berger, L., Baudouin, N., Pirany, N., Ranganath, R.M., Prasad, D.T., 2006. Identification of molecular markers associated with mite resistance in coconut, *Cocos nucifera* L. Genome 50, 35–42.

Singh, M., Jaiswal, U., Jaiswal, V.S., 2003. *In vitro* selection of NaCl-tolerant callus line and regeneration of plantlets in a bamboo (*Dendrocalamus strictus* Nees). *In vitro* Cell Dev. Bio. Plant 39, 229–233.

Teulat, B., Aldam, C., Trehin, R., Lebrun, L., Barker, G.M., Karp, A., Baudouin, L., Rognon, F., 2000. An analysis of genetic diversity in coconut (*Cococs nucifera*) populations from across the geographic range using sequence-tagged microsatellites (SSRs) and AFLPs. Theor. Appl. Genet. 100, 764–771.

Varshney, R.K., Glaszmann, J., Leung, H., Ribaut, J., 2010. More genomic resources for less-studied crops. Trends Biotechnol. 28, 452–460.

Wright, H., 1980. Commercial hybrid production. In: Fehr, W.R., Hadley, H.H. (Eds.), Hybridization of Crop Plants. The American Society of Agronomy, Wisconsin, USA, p. 162.

Zizumbo-Villarreal, D., Ruiz-Rodriguez, M., Harries, H., Colunga-GarciaMartin, C., 2006. Population genetics, lethal yellowingdisease, and relationships among Mexican and imported coconut ecotypes. Crop Sci. 46, 2509–2516.

CHAPTER

10

Oil Palm

Ntsomboh-Ntsefong Godswill,**, Ngando-Ebongue G. Frank*, Ajambang-Nchu Walter*,†, Maho-Yalen J. Edson‡, Tabi-Mbi Kingsley*,**, Vincent Arondel§, Bell J. Martin**, Youmbi Emmanuel**,¶*

*Selection and Genetic Improvement Section, Specialized Centre for Oil Palm Research of La Dibamba, Institute of Agricultural Research for Development (IRAD), Douala, Cameroon
**Department of Plant Biology, University of Yaounde 1, Yaounde, Cameroon
†Plant Molecular Biology Laboratory, Department of Plant Breeding and Biotechnology, Bogor Agricultural University, Bogor, Indonesia
‡Department of Biological Sciences, Higher Teachers' Training College, University of Yaounde 1, Yaounde, Cameroon
§Membrane Biogenesis Laboratory, UMR5200 CNRS, University of Bordeaux, France
¶Tissue Culture Laboratory, Centre Africain de Recherche sur Bananiers et Plantains (CARBAP), Njombé, Cameroun

INTRODUCTION

The issue of sustainability is the greatest challenge facing the current and coming decades, which will witness further global population growth and increased consumption of edible oils and fats. Simultaneously, fuel prices and weather patterns are becoming increasingly volatile. Within this context, the agricultural sector is facing scrutiny from governments, regulators, NGOs, investors, and consumers regarding how its business practices impact the environment and the wider world. Oil palms (*Elaeis guineensis* Jacq.), in particular, have been at the forefront of these environmental and reputational challenges (Cochard et al., 2005; Durand-Gasselin et al., 2011; WWF Report, 2012; Feintrenie, 2012). The high and increasing yields of oil palms from the onset have led to a rapidly expanding world industry based in the tropical areas of Asia, Africa, and America (Corley and Tinker, 2003). Despite incurring considerable expenses, especially during the first 4 years of investment, palm cultivation overall is profitable for small, medium, and large-scale farmers and is considered to be one of the most viable livelihood opportunities currently available in oil crop cultivation (Okwuagwu et al., 2011; Beggs and Moore, 2013).

Breeding Oilseed Crops for Sustainable Production. http://dx.doi.org/10.1016/B978-0-12-801309-0.00010-0

The oil palm is the highest yielding oil crop per hectare (Ariffin et al., 2000; Corley and Tinker, 2003; Singh et al., 2007) with oil yield per hectare about 10 times greater than soy oil (Hartley, 1988; Durand-Gasselin et al., 2010; Skurtis et al., 2010; Mohd et al., 2011; Jacquemard, 2011). This crop produces about 4–6 Mt ha^{-1} $year^{-1}$ of crude palm oil (CPO) (Corley, 2009; Sunilkumar and Sparjan Babu, 2013) as against 0.4 t ha^{-1} $year^{-1}$ of soybean oil and 0.6 t ha^{-1} $year^{-1}$ of rapeseed oil. Moreover, the crop presents huge potential for growth as present day best genotypes can produce over 10 t ha^{-1} $year^{-1}$. Oil palm's share of the market keeps increasing with time. Due to its high productivity, it is likely to be the most sustainable oil crop worldwide (Oil World, 2008; RSPO, 2008).

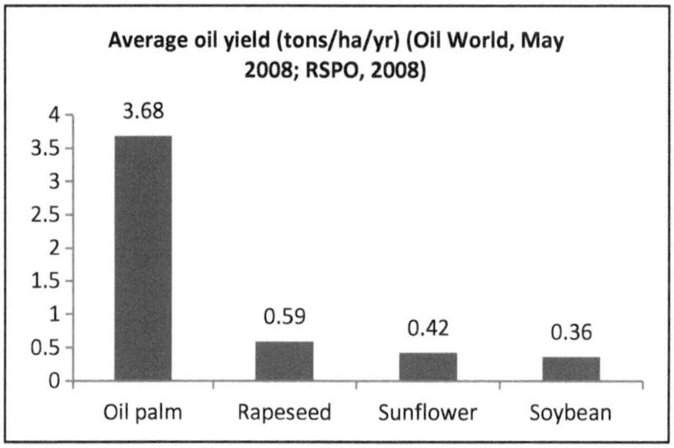

Contributing 39% to world production of vegetable oils, palm oil is currently the most important vegetable oil commodity in the world (Ngando-Ebongue et al., 2013). Several products are obtained from this crop in plantations and mills. Fresh fruit bunch (FFB) processing produces CPO from the mesocarp, palm kernel oil (PKO) from the seed, and organic byproducts and residues (Redshaw, 2003; Maria and Helena, 2007; Schlegel, 2010). Kernel and sludge cakes and wastes from palm and kernel oil extraction processes are used as animal feed. Fruit bunch stalk and fiber waste can be used as organic mulches or processed into organic compost. About 80% of CPO is used in the food industry while PKO has more uses in oleochemical industries (Soh et al., 2003). Refined CPO is used solely or blended with other oils such as cooking oil, salad oil, margarines, and spreads to make soaps and livestock feeds. The refined oil is fractionated to give olein and stearin, and further fractionation gives fatty acids and alcohols as well as intermediate commodities traded and used in the food and oleochemical industries (Schlegel, 2010; Soh et al., 2003). It can also be used directly (i.e., without hydrogenation) as margarine without the production of unhealthy transfatty acids. Secondary and waste products from the palm oil industry include palm wine tapped from the apex of felled trees, carotene from the CPO, which can be processed into vitamin A supplement and a natural dye for snack foods. Tocopherols and tocotrienols are extracted from palm oil for industrial vitamin E production. CPO also contains antioxidants in its carotene and tocopherol/tocotrienol contents, which appear to have anticancer properties. Palm oil and its products are now being certified independently for sustainable production compliance (Mohd et al., 2011).

Like other vegetable oils, palm oil can also be used as a biofuel (Choo and Cheah, 2000; Maria and Helena, 2007; Corley, 2009). Although there is comparative advantage in some developing countries for biofuel feedstock production (Enongene-Ndobe, 2014), there are concurrent constraints such as the fear that a rise in biofuel feedstocks might contribute to increased food insecurity and/or prove to be environmentally disruptive, especially in those countries displacing natural forest for palm plantations (Maria and Helena, 2007). Since palm oil seems to be the most efficient feedstock in terms of biodiesel production, there is utmost need to analyze other alternatives which will also support socioeconomic development, notably higuerilla (*Ricinus communis* L.), *Jatropha curcas*, and *Acrocomia aculeata*. This notwithstanding, the promotion of more sustainable development actions especially in developing countries is crucial through the creation of a legal framework for biofuels that not only fosters biodiesel and bioethanol production, but also encourages sustainability throughout the production and commercialization processes (Maria and Helena, 2007).

Given the current growth rate of the global population (Godfray et al., 2010), particularly in Third World countries, oil palm cultivation will continue to expand in order to meet increasing global demands for vegetable oil (Soh et al., 2003; Kushairi et al., 2010). However, this growth could be impeded by several constraints and challenges such as insufficient improved planting material, low-quality oil-producing germplasm, slow natural seed germination that limits the establishment of planting materials (Rees, 1963d), pests and diseases, and soil health management (Petronella et al., 2009) in addition to controversies related to environmental and health concerns. The solution to feeding the world population sustainably is by crop improvement through classical and modern breeding approaches in addition to sustainable agricultural practices. Sustainable agriculture is a systematic approach to agriculture that focuses on ensuring the long-term productivity of human and natural resources for meeting food and industrial needs. It is an integrated system of plant and animal production practices having a site-specific application that over the long term will satisfy human food or industrial needs; enhance environmental quality and the natural resource base upon which the agricultural economy depends; make the most efficient use of nonrenewable resources and on-farm resources and integrate, where appropriate, natural biological cycles and controls; sustain the economic viability of farm operations; and enhance the quality of life for farmers and society as a whole (Schlegel, 2010; Black, 2006). In fact, production can be intensified while reducing inputs and lowering the environmental footprint of agriculture (Biodiversity International, 2010). Plant breeders need to enhance food security in a sustainable manner.

Hence, the production of high value–added products (in terms of yield, ease of harvest, mill performance, olein content, etc.) (Durand-Gasselin et al., 2011) might be strategic for sustaining the oil palm industry (Zamzuri, 2011). In this context, the approach of enhancing production by increasing the availability of land for cultivation has limited options. Precision agriculture (Fairhurst et al., 2003) and long-term ecological sustainability is needed through appropriate breeding and management techniques. A viable alternative is to increase productivity per unit land area sustainably by yield intensification (Fairhurst and Hardter, 2003; Kushairi et al., 2010) through genetic improvement programs that generate superior individuals with improved productivity and stability. Genetic diversity study is a prerequisite to such breeding programs (Khush, 2002), and the key to success of any genetic improvement program lies in the availability of genetic variability of desired traits (Heller, 1993; Ghafoor et al., 2001; Cleber da Silva et al., 2014). This is important for the development of elite planting materials as potential genetic resources for growers and for further use in breeding programs (Okoye et al., 2009).

Most breeding programs aim at deriving a variety that demonstrates a significant improvement in quantitative (Demol, 2002) and qualitative yield and greater adaptability to environmental constraints. In oil palm breeding, these qualities are obtained in hybrids demonstrating early production, high quality and quantity of CPO and PKO as well as tolerance to diseases (Rajanaidu et al., 2000; Kushairi et al., 2011) and environmental stress. Such goals can be attained by crossing parents that have demonstrated consistent expression of interesting traits for at least 10 years (Hartley, 1988). FFB yield is a major economic trait that determines the productivity of oil palms. For breeders and agronomists, it is crucial to know the heritability estimates of bunch production traits, the relationship between these components, and their contribution toward oil yield (Okoye et al., 2009). High heritability estimates indicate the presence of large numbers of fixable variable additive factors or traits, which may be improved by selection (Manju and Sreelathakumary, 2002; Okoye et al., 2009). Plants respond and adapt differently to the selection pressures and environmental conditions imposed by distinct agroclimatic zones (Singh et al., 1998). This is why morphological and agronomic features cannot be relied upon alone (Muthamia et al., 2013). Hence, grouping accessions by multivariate methods is of practical value in a breeding program (Ghafoor et al., 2001). Accessions are obtained from prospection and collection of germplasm especially from centers of origin and diversity in order to conserve the wide range of oil palm resources for posterity, broaden the genetic base of breeding material in order to raise yields, improve nutritional qualities, and tackle future challenges facing the industry (Marhalil and Rajanaidu, 2011a).

For a better understanding of techniques geared toward sustainable oil palm breeding, it is important to master the reproductive biology of the plant as well as developments in traditional and modern breeding approaches and innovative crop production concepts like organic farming. Traditional improvement programs for oil palm such as reciprocal recurrent selection (RRS) employ different breeding strategies and generate superior lines or improved varieties. However, these approaches are costly and usually take more time as the breeding cycle can take up to 10 years (i.e., 1 year of pollination, 2–3 months of seed preparation and germination, 1 year in the nursery, 3 years in the field before harvesting and 4–6 years of yield evaluation (Soh, 1986; Rajanaidu et al., 2000; Kushairi et al., 2011). Adding these to progeny testing, it takes almost 20 years to develop new planting material (Kushairi et al., 2011). Sometimes such breeding is referred to as "chance breeding" due to uncertainty in predicted success.

These shortcomings have brought about interest in vegetative propagation (tissue culture) of oil palm (Low et al., 2008). Tissue culture or micropropagation has been studied and applied to oil palm breeding (Kushairi et al., 2010; Zamzuri, 2011; Taufiq et al., 2011; Musa et al., 2011; Mohaimi, 2011) with a view to commercial distribution of elite clones capable of increasing oil production (de Touchet et al., 1991). In addition to the application of tissue culture techniques to oil palm breeding are technologies such as genetic engineering and genomics. The use of molecular markers and genetic engineering is already playing an important role in oil palm breeding (Jalani et al., 1998). Marker-assisted selection is likely to bring tremendous improvements as it should allow selection of a given trait in a plantlet, thus saving time and space and removing the need to grow the tree for 10 years to assess the trait. For this to happen, detailed quantitative trait locus (QTL) analyses (Singh et al., 2009; Billotte et al., 2010; Chandusingh et al., 2013) are required to genetically map the trait close to molecular markers to be able to select for the latter rather than for the agronomic trait itself.

Another aspect of food crop production applied to oil palm breeding is integrated pest and disease management (IPDM), which is directly linked to the general concept of sustainability (Durand-Gasselin et al., 2011). Through IPDM and other approaches, breeders are able to make strategic choices by selecting for durable resistance to diseases (Diabate et al., 2010; Durand-Gasselin et al., 2011) and managing the dynamics of pest populations and their natural enemies. Sustainability means production with environmental friendliness, economic viability, and social welfare. For sustainability to happen, oil palm breeding and production activities must thus be carried out without harmful effects to society and the environment and without jeopardizing the potential for future generations to meet their own needs (Ekins et al., 2008). Sustainable breeding is defined as crossing different plants to produce offspring with desirable characteristics (Black, 2006).

This chapter focuses on technical aspects or choices that influence the sustainability of oil palm production (Durand-Gasselin et al., 2011). It presents the origin and biology of the oil palm, crop-breeding techniques used to improve the crop, and highlights in the history of breeding for better production. Germplasm conservation and IPDM approaches applied to oil palms for sustainable production are mentioned. Two case studies – the South East Asian and Cameroonian breeding experiences toward sustainable oil palm production – are presented. On the whole, sustainability is highlighted with major related constraints, strengths, and opportunities. The chapter closes with future challenges to sustainable oil palm production.

Origin of the Oil Palm

Elaeis guineensis grows in the wild, semiwild, and cultivated parts of the tropics, within ±10° latitude of the equator, in Africa, South East Asia, and South and Central America (Sapey et al., 2012). Evidence of an African origin of the oil palm has accumulated over time by dating pollen in sediments (Zeven, 1964; Hartley, 1988; Corley and Tinker, 2003). The center of origin and diversity appears to be concentrated in the tropical rainforests of west and central Africa where wild and semiwild palm groves are found on the coastal belt, starting with northernmost occurrences along the West African coast. The first concentration of palms is in the highlands of the Fouta Djallon district of Guinea, at 10–11°N. The palm belt of Africa then runs through Sierra Leone, Liberia, the Ivory Coast, Ghana, Togoland, Benin (previously Dahomey), Nigeria, Cameroon, the People's Republic of Congo, and the Democratic Republic of Congo (DRC, formerly Zaire).

In central Africa palms grow through much of the Congo region and even into Angola. Most grow between 3°N and 7°S, but a few can be found as far south as 15°S. The eastern part of Nigeria is the center for highest diversity of oil palm germplasm (Hartley, 1977; Maizura et al., 2001; Okwuagwu et al., 2011). It is endemic to the tropical lowlands of West and Central Africa (Okwuagwu et al., 2011) stretching from the Atlantic coast (Guinea) in the west to Zanzibar and the island of Madagascar in the east and from the Sub-Sahara in the north (Senegal, 16°N) to the south of Angola (15°S) bordering Namibia (Hartley, 1977; Hartley, 1988; Soh et al., 2003; Sapey et al., 2012). Oil palms are also found on the island of Madagascar and, remarkably, reach as far south as 21°S.

World Distribution of the Oil Palm Industry

Oil palms are grown industrially in tropical Asia, Africa, and America where there has been a considerable expansion of its acreage during recent years. The establishment of oil palm plantations started in South East Asia, with little or no direct connection between existing West African groves and this new industry.

The Asian oil palm industry has developed at an astonishing rate, and now leads the world. In 1935, Indonesia became the global leader in palm oil export, with a plantation area of 75,000 ha. The introduction and expansion of oil palms in Indonesia reflects their economic attractiveness and benefits, but the social interactions between companies and local communities have had a mixed track record that requires attention. Ambiguous and contested land tenure in local communities, the state issuing concession licenses, the debate about potential negative environmental consequences of forest conversion to oil palm plantations (Feintrenie, 2012), development impacts (negative in situations dominated by conflict, potentially positive elsewhere) have all been at the heart of controversy (Budidarsono et al., 2012). In addition, Basal Stem Rot (BSR) of oil palms caused by *Ganoderma boninense* is a major disease in oil palm plantations in South East Asia (Idris et al., 2004; Susanto, 2009). BSR incidence of 0–1.5% in palms below 6 years of age and 87% in palms above 16 years of age have been reported in plantations located in Sumatra (Indonesia) (Virdiana et al., 2011). Future growth of the industry is expected in the remaining parts of equatorial South East Asia (Philippines, Thailand, Papua New Guinea) (Corley and Tinker, 2003).

The American oil palm industry has grown rather slowly, but has become established in many Central and South American countries. Exports are on the small side and a large part of production is used locally. South America has several serious diseases that affect oil palms, which are not found elsewhere. The industry has developed through small to medium-sized plantations, avoiding both the very large plantations of Asia and the wild palm culture of Africa (Corley and Tinker, 2003). South American production will probably grow more rapidly in future.

The African oil palm industry still supplies a large part of the domestic demand for oil, and the large surplus of kernels has allowed substantial but declining export trade in them to continue (Corley and Tinker, 2003). Hence, the industry in Africa has taken a few backward steps in most countries, but there are exceptions like the Ivory Coast. A major constraint in Africa (especially in Western and Central Africa) is vascular wilt disease caused by a soil-borne phytopathogenic fungus *Fusarium oxysporum* f. sp. *elaeidis* (Aderungboye, 1982; Ho et al., 1985; Ho and Varghese, 1986; Diabate et al., 1992; Corley and Tinker, 2003; de Franqueville and Diabate, 2004; Ntsomboh-Ntsefong et al., 2012; Ngando-Ebongue et al., 2013). This disease was observed at the Ndian Estate of PAMOL Plantations Plc in Cameroon where it spread rapidly from the neighboring Cowan Estate in Nigeria as far back as 1940 (Prendergast, 1957). It has been reported in Ivory Coast, Ghana, Benin, DRC, and Liberia (Flood et al., 1989; de Franqueville et al., 2011). The complexity of the disease calls for concerted research in breeding and selection for resistant progeny (Prendergast, 1957; Guldentops, 1962).

Appropriate Growing Conditions

Oil palms do not tolerate saline conditions or fresh but stagnant water above the soil surface, but they do accept fluctuating freshwater tables, water tables 50 cm or more below the soil surface, and moving soil water in light-textured soils and sediments. Palms are resistant to relatively high water tables in hydromorphic soils. The main limiting factor for oil palm growth and production is rainfall. Palms require minimum rainfall of 2000 mm $year^{-1}$ that is well distributed all year round and at least 1800 h of sunshine per year with average solar radiation of 12 MJ m^{-2} day^{-1}. Apart from rainfall, altitude is the greatest barrier to oil palm establishment, and normally palms are not grown above 300 m or so. However, where rainfall is sufficient, palms grow at up to 1300 m in the Cameroon Mountains, at up to 1000 m in the Fouta Djallon in Guinea, in East Cameroons and in many localities in East Africa, although yields tend to be low. The development of palms that are tolerant of higher altitudes is now proceeding (Corley and Tinker, 2003). The best planting density for this is 143 plants ha^{-1} arranged in equilateral triangles with 9 m sides and 7.80 m line spacing. Monthly temperature minima and maxima should be above 20°C and between 28°C and 34°C, respectively (Jacquemard, 1995).

Classification of the Oil Palm

The oil palm received its botanical name from Jacquin (1763) as *Elaeis guineensis* Jacq. *Elaeis* is derived from the Greek word *elaion*, meaning oil, while the specific name *guineensis* shows that Jacquin attributed its origin to the Guinea coast. *Elaeis guineensis* Jacq. is an allogamous species with a diploid genome comprised of 16 pairs of homologous chromosomes (Diana et al., 2012); i.e its chromosome number is $2n = 2x = 32$ (Schlegel, 2010; Ting et al., 2010). It belongs to the family Arecaceae, which forms a distinct group of plants among the monocotyledons. The Arecaceae are placed in the order Arecales (Corley and Tinker, 2003). The oil palm is grouped with *Cocos* and other genera in the subfamily Cocosoideae and tribe Cocoeae (Ngando-Ebongue, 2009). There are three accepted species of *Elaeis*. The first two, *E. guineensis* and *E. melanococca* (also named *E. oleifera)* are, respectively, the African and American oil palms which were presumed to have diverged when the American and African continents drifted apart in prehistoric times (Zeven, 1965; Soh et al., 2003). The third species, *Elaeis odora* (previously *Barcella odora*), is not cultivated and little is known about it even though molecular markers indicated that its inclusion within the genus *Elaeis* is justified (Corley and Tinker, 2003). Another lesser known species of *Elaeis* is *E. madagascariensis* (Soh et al., 2003).

Morphology and Growth of *Elaeis guineensis*

The oil palm seed is a nut covered by a soft oily mesocarp. It consists of a shell, or endocarp, and one, two, or three kernels. In the majority of cases the seed contains only one kernel, since two of the three ovules in the tricarpellate ovary usually abort. Abnormal ovaries sometimes occur and four or five-seeded nuts may, very rarely, arise from these. Inside the shell lies the kernel surrounded by a dark-brown testa covered with a network of fibers. Embedded in the endosperm and opposite one of the germ pores lies the embryo. During seed germination, the emerging embryo forms a hypocotyl. The plumule and radicle both emerge through a cylindrical, persistent ligule close to the seed. During the first few weeks of growth, the seedling is entirely dependent on the supply of stored material from the endosperm. The plumule does not emerge from the plumular projection until the radicle has reached 1 cm in length (Fig. 10.1a).

FIGURE 10.1 (A) Germinated oil palm seeds. (B) Oil palm seedling ready for field planting.

The first adventitious roots are produced in a ring just above the radical–hypocotyl junction, and they give rise to secondary roots before the first foliage leaf emerges. The radicle continues to grow for about 6 months, after which numerous primary roots develop in its place. A green leaf emerges about 1 month after germination. One leaf per month is produced until the seedling is 6 months old. After 3–4 months the base of the stem becomes a swollen "bulb" and the first true primary roots emerge from it. Secondary roots grow out in all directions while the leaves become successively larger and change in shape (Fig. 10.1b) (Corley and Tinker, 2003).

The adult *E. guineensis is* a large, pinnate-leaved palm having a solitary columnar stem with short internodes. There are short spines on leaf petioles and within the fruit bunch. The palm (Fig.10.2a) is normally monoecious with male or female, but sometimes mixed, inflorescences developing in the axils of the leaves. The fruits are borne on a large, compact bunch. The fruit pulp, which provides palm oil, surrounds a nut, the shell of which encloses the palm kernel (Hartley, 1988; Corley and Tinker, 2003). The crown of a mature palm (Fig. 10.2b) consists of between 30–50 leaves. The leaf bases adhere to the stem until the palm is at least 12 years old, sometimes much longer. In the crown of an adult palm a continuous succession of

FIGURE 10.2 (A) Oil palm trees in a plantation. (B) Oil palm crown with fresh fruit bunches.

leaf buds or primordia separate laterally from the apical meristem. The apical bud develops into a central "spear" which develops into a mature leaf bearing linear leaflets or pinnae on each side of the leaf stalk. The leaflets are produced by the splitting of an entire leaf during elongation of the leaf axis. After the leaf opens, it is progressively displaced outward and downward as younger leaves emerge (Corley and Tinker, 2003). Some leaflet variations have been noted in oil palms such as in the *idolatrica* palm where the leaflets do not separate normally and an entire or semientire leaf is formed. The midribs of all unseparated leaflets are in a single plane (Hartley, 1988).

The oil palm has a columnar stem with one terminal growing point though occasionally branched palms develop with two or more growing points. The rate of extension of the stem is variable and depends on both environmental and hereditary factors. The diameter of the stem, excluding the attached leaf bases, varies from 20–75 cm. A wide central cylinder is separated from a narrow cortex through which the leaf traces pass. The cylinder has a peripheral zone of congested vascular bundles with fibrous phloem sheaths, and the intervening parenchyma cells are sclerotic; thus, this zone provides the main mechanical support of the stem. Vascular bundles are much less densely packed in the central zone, where most of the storage tissue is located (Corley and Tinker, 2003).

In mature palms, primary roots spread from the stem base (bole), with new primaries continually replacing dead ones. The primary roots extend either downward from the base of the palm or outward in a more or less horizontal direction. Roots sometimes develop on the stem up to 1 m above ground, but these normally dry out before reaching the soil. Secondary roots grow both downward and upward, and ascending secondaries generally reach the surface of the soil. The secondaries give rise to tertiaries from which are developed quaternaries. Roots of all classes show preferential growth under optimal conditions of water and nutrient supply and when there is ample rotting felled vegetation or heaps of palm leaves. When there is good *Pueraria* cover, this may lead to a high density of quaternaries in the center of the interline (Hartley, 1988; Corley and Tinker, 2003).

Sex Differentiation

Oil palms are monoecious producing functional unisexual separate male and female inflorescences (Fig. 10.3) in an alternating cycle on the same plant (Corley and Tinker, 2003),

FIGURE 10.3 (A) Male inflorescence. (B) Female inflorescence.

resulting in an allogamous mode of reproduction. The sex ratio of an oil palm stand is influenced by both genetic and environmental factors. In particular, the enhancement of male inflorescence production is possible under water stress (Helene et al., 2011). Inflorescences are borne from leaf axils, with either male or female inflorescences per axil but never both. Each flower primordium of oil palms is a potential producer of both male and female organs (Corley and Tinker, 2003) though in very rare cases both the androecium and gynecium develop fully to give a hermaphrodite flower (Hartley, 1988). Mixed inflorescences, with both male and female spikelets, are more common in young palms (Corley and Tinker, 2003). Often a proportion of the inflorescence aborts prematurely before emergence such that some leaf axils appear not to possess any inflorescence.

The female inflorescence reaches a length of 30 cm or more before anthesis. The flowers are arranged spirally around the rachis of the spikelet, each in a shallow cavity and subtended by a bract that develops into sharp spines. The number of flowers in an inflorescence varies depending mainly on palm age. The male inflorescence is borne on a longer peduncle than that of the female inflorescence and contains long, finger-like, cylindrical spikelets without spines. All flowers on the spikelet usually open within 2 days, though during rainy weather the opening may prolong to 4 days. Before opening, the sessile flower is completely enclosed by a triangular bract. The anthers of male flowers dehisce through vertical slits (Corley and Tinker, 2003). Male, female, and mixed spikelets may appear on the same bunch in oil palms. Spikelets can contain all three types of flowers, male flowers in singles and pairs, or they may have the superficial appearance of a female spikelet but containing pairs of male flowers (Williams and Thomas, 1970; Corley and Tinker, 2003).

Pollination, Fruit Set, and Bunch Development

Oil palms are both wind and insect pollinated, although the latter is prevalent and more effective. Numerous species are involved in the pollination of oil palms, but the main pollinating agents are from a subfamily of weevils, the Derelominae, among which species of *Elaeidobius* predominate (Corley and Tinker, 2003). Weevils are attracted by the strong fennel-like fragrance released by the male inflorescence at anthesis. Female inflorescences have the same aniseed scent as males, but weevils apparently visit them only by accident in the search for male inflorescences (Syed, 1979; Corley and Tinker, 2003). Their introduction in a plantation generally brings about an increase in fruit set and bunch weight. In Africa, *Elaeidobius* spp. account for 99% of oil palm pollination, with *Elaeidobius kamerunicus* and *E. plagiatus* being the most efficient species (Ponnemma et al., 2004; Yalamoussa et al., 2011).

In Malaysia, the main pollinator is *Thrips hawaiiensis* (Syed, 1979; Corley and Tinker, 2003), although it is not a very efficient pollinator. *E. kamerunicus* was released in two plantations in Malaysia in February 1981 (Syed, 1982; Syed et al., 1982; Kang, 1999; Corley and Tinker, 2003). Later, there were releases in Indonesia, Papua New Guinea, and Colombia (Corley and Tinker, 2003). In Sabah, it was found that some pollen was dispersed by a *Pyroderces* sp. moth (Syed, 1979; Tan and Basri-Wahid, 1985; Corley and Tinker, 2003). Meanwhile, in Latin America two pollinating beetles, the sap beetle, *Mystrops costaricensis* reportedly the commonest and most efficient oil palm pollinator (Syed, 1982), and the weevil, *E. subvittatus*, a native of Africa, have been found on both *E. guineensis* and *E. oleifera* (Chinchilla and Richardson, 1990; Corley and Tinker, 2003). Fruit set in America was not always satisfactory until *E. kamerunicus* was introduced (Syed, 1984; Corley and Tinker, 2003).

Anthesis in a normal inflorescence usually lasts 36–48 h but may extend to a week (Tandon et al., 2001; Corley and Tinker, 2003). During anthesis, male flowers open between 08:00 am and 09:00 am while female flowers open from 09:00 am and 10:00 am. Visitations by weevils start at about 07:00 am and continue until 15:00 pm with maximum activity between 11:00 am and 12:00 am (Tandon et al., 2001). Most pollen is shed within the first 2–3 days following the start of anthesis and production ceases within 5 days. Inflorescences produce between 25 g and 300 g of fresh pollen. Pollen can remain viable at room temperature for up to 6 days after release (Hardon and Turner, 1967; Tan and Rao, 1979; Corley and Tinker, 2003). It can remain viable for decades in a freezer, but the viability of pollen produced toward the end of the anthesis period may be low (Corley and Tinker, 2003).

After pollination, between 30% and 60% of flowers normally develop into fruits, depending on the efficiency of pollination. The fruit develops steadily in size and weight from pollination until 100 days or more after pollination (DAP). The kernel is at first liquid, then gelatinous, and solidifies at about 100 DAP; the embryo matures by 70–80 DAP and the shell becomes hard and lignified by about 120 DAP. Oil formation in the kernel commences at about 70 DAP and is probably complete by 120 DAP. Mesocarp oil synthesis starts at about 120 DAP and continues until the fruit detaches from the bunch (Corley and Tinker, 2003; Ngalle-Bille et al., 2013). Oil palm fruit is a sessile drupe varying in shape from nearly spherical to ovoid or elongated and bulging somewhat at the top. Different fruit forms exist. All are grouped in three naturally occurring fruit types of oil palm: Dura has a relatively thick shell (Fig. 10.4a), Tenera has a relatively thin shell (Figure 10.4b), and the rare more or less female-sterile form Pisifera (Figure 10.4c) has no shell (Corley and Tinker, 2003).

Tenera is a thin-shelled (sh^+sh^-) hybrid between Pisifera (sh^-sh^-) × Dura (sh^+sh^+) (Beirnaert and Vanderweyen, 1941; Corley and Tinker, 2003). There is a rare abnormal fruit type, variously known as Poissoni, mantled (Durand-Gasselin et al., 1993), or *diwakkawakka*, in which fleshy outgrowths or supplementary carpels surround the main part of the fruit. The mesocarp of all fruit types contains fibers that run longitudinally through the oil-bearing tissue. The oil content of the mesocarp of ripe fruit varies from under 40% to over 60% (Corley and Tinker, 2003). Ripeness takes between 4.5 months and 6 months, and by the time the bunch is ripe it is subtended by about leaf 30–32. Well-set bunches (Fig. 10.5) carry from 500 fruits to 4000 fruits depending on bunch size.

At maturity, fruits start to detach from the bunch apex downward and loose fruits are a signal that the bunch is ripe. Bunch maturity can also be determined from fruit color (Sunilkumar and Sparjan Babu, 2013) by expression of the VIRESCENS (VIR) gene (Singh

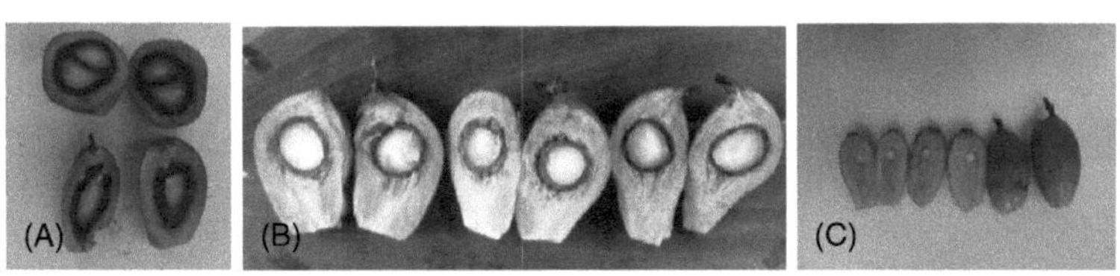

FIGURE 10.4 **Oil palm fruits.** (A) *Dura*; (B) *Tenera*; (C) *Pisifera*.

FIGURE 10.5 **Dura oil palm fresh fruit bunch.**

et al., 2014). Ripe bunches are ovoid and may reach 50 cm or more in length and 35 cm in breadth. Bunch number produced per palm per year declines with age while bunch weight increases with palm age, with mean values less than 5 kg in 3-year-old palms and over 25 kg in 15-year-old palms. Individual bunches weighing up to 100 kg have been recorded. Peak production of oil palms is reached at 6–10 years, thereafter yield remains more or less constant for many years, despite steadily increasing mean bunch weight (Corley and Tinker, 2003).

Elaeis oleifera (H.B.K.) Cortés

Elaeis oleifera is endemic to tropical Latin America stretching from Mexico in the north to the Amazonas of Bolivia, Brazil, Colombia, and Peru in the south and straddling the Pacific and Atlantic coasts. It occurs naturally in groves in open grasslands and riverine areas and its distribution appears to be associated with indigenous Indian migratory movements (Santos et al., 1986; Soh et al., 2003; Corley and Tinker, 2003). The American oil palm produces oil in smaller quantities and is of interest in breeding because of its desirable attributes such as good oil quality (higher oleic acid content and iodine value), slow growth rate (Vallejo and Cassalett, 1975; Corley and Tinker, 2003), short-spined bunches, and resistance to fatal yellowing; although it has a low fruit-to-bunch ratio.

The spathe of the female inflorescence is more persistent than in *E. guineensis*, so that even ripe bunches are still partially covered by fibrous material. The period of anthesis in the female inflorescence is much more erratic than in *E. guineensis* and may last for 3–4 weeks, or have two peaks, while other palms complete anthesis in 3–4 days (Rao and Chang, 1982; Corley and Tinker, 2003). Bunches usually weigh between 8–12 kg but occasionally reach 30 kg. The fruits are small and numerous and may number more than 5000 (Ooi et al., 1981). Parthenocarpic fruit are often numerous, and may constitute up to 90% of the total. There is no evidence that fruit forms comparable with Dura, Tenera, and Pisifera exist in populations of *E. oleifera*. After about 15 years, the palm adopts a procumbent habit with the crown remaining in an erect position and the distal 1.5–2 m of the trunk being erect (Corley and Tinker, 2003).

Elaeis guineensis × *Elaeis oleifera* Hybrid

These two species have frequently been hybridized on an experimental scale. The first commercial plantation of the hybrid was established following the discovery that *E. oleifera* was disease resistant (Turner, 1981). The hybrid has leaves that are larger than those of either parent, but retains the leaflet arrangement of *E. oleifera*. The characteristics of the latter palm regarding height increment, falling leaf bases, persistent spathes, parthenocarpy, and fruit shape and color are also retained in the hybrid. Hybrid pollen viability and germination percentage are low (Hardon and Tan, 1969). This, coupled with the fact that hybrid inflorescences appear to be less attractive to *E. kamerunicus*, results in poor fruit set in the hybrid (Tan, 1985; Corley and Tinker, 2003). The internal fruit characters of the hybrid depend on the fruit form (Dura, Tenera, or Pisifera) of the *E. guineensis* parent. Crosses between *E. oleifera* × Pisifera have thin shells, but lack the fiber ring seen in *E. guineensis* Tenera. Oil to mesocarp is intermediate between that of the parent species, and the distribution of fatty acids also appears to be intermediate (Hardon, 1969; Corley and Tinker, 2003).

OBJECTIVES AND DEVELOPMENTS IN SUSTAINABLE OIL PALM BREEDING

Objectives of Sustainable Oil Palm Breeding

The oil palm product and transformation industry is a sector that is globalized, complex, and multistakeholder (producers, agroindustrialists, smallholders, consumers, scientists, distributors, transformers, NGOs, governments, media). The sector evokes several antagonistic interests in public opinion. Certain governments and NGOs have played a major role in controversial oil palm debates in catalyzing the evolution of thought toward sustainability. Oil palm controversy stems from concerns about conservation of natural species against development (Rival and Levang, 2013). This industry is land intensive and, hence, is linked to deforestation (Maria and Helena, 2007); climate change and instability; concern for the health of consumers; the use of inorganic pesticides, herbicides, fertilizers, and their effects on the environment; increasing urbanization and reduction in labor; and economic and social issues. Unplanned oil palm development will lead to degradation of forest systems, loss of plant and animal habitats, extreme land degradation, and pollution.

An interesting development in oil palm breeding for sustainable production was the establishment of the Roundtable for Sustainable Palm Oil (RSPO). The RSPO, which was set up in 2004, consists of palm oil producers, processors, traders, consumers, manufacturers, retailers, and NGOs, and is tasked with developing the principles and criteria to make the palm oil industry sustainable, and facilitate the development of sustainable palm oil production. The RSPO guidelines include commitment to transparency; compliance with all applicable local, national, and ratified international regulations; adoption of sustainable cultivation practices such as water management, pesticide control, and soil erosion mitigation (Durand-Gasselin et al., 2011); conservation of resources and biodiversity (managing and monitoring high conservation values, HCVs); and community development (UNDP, 2007; Abdullah and Sulaiman, 2013; RSPO, 2013). Sustainable breeding has thus been adopted by design (since some actors strive to be part of the current sustainability trend) and by pressure (usually from environmentalist NGO advocacies) because of the challenges facing agriculture today.

Sustainability is something breeders always strive to achieve and has to take into account interactions with the world beyond plantations (Durand-Gasselin et al., 2011). The objectives of oil palm breeding for sustainability include breeding for high yield (intensification) (Abdullah and Sulaiman, 2013), resistance to diseases, oil quality (high levels of unsaturated fatty acids) through the use of *E. oleifera*, selection for palms naturally rich in vitamin A and E for health reasons, fertilizer efficiency of planting material through agronomic and physiological approaches, selection for low-lipase palm lines yielding oil with fewer free fatty acids (Morcillo et al., 2013), and indehiscent fruits to introduce flexibility in organizing harvests (Durand-Gasselin et al., 2011). The aims of sustainable breeding in oil palm are to satisfy food and industrial needs; reduce the cultivated area while improving yields; reduce the use of environment-degrading substances such as inorganic fertilizers and pesticides, including nonrenewable resources such as fossil fuels; reduce variable costs; and reduce the use of manpower by breeding technology-friendly oil palm varieties. Achieving sustainability requires the availability of planting material of high production potential grown using environmentally friendly (sustainable) cultural practices. Such material can only be obtained through appropriate breeding techniques.

Startup and Evolution of Sustainable Oil Palm Breeding

Interest in the oil palm grew once it was realized that it could be used as a substitute for animal fat in the manufacture of soap, candle wax, and margarine (Jalani et al., 1998). There have so far been three major periods of genetic improvement for oil palm yield: mass selection, the discovery of heterosis in oil palm and Dura × Pisifera (D × P) development after the Second World War, and the modern breeding schemes widely used now (Rosenquist, 1986; Hartley, 1988; Durand-Gasselin et al., 2002). A fourth will be the routine implementation of marker-assisted selection (MAS) and clonal propagation (Fernanda et al., 2014; Durand-Gasselin et al., 2011).

Two Dura seedlings were taken from the botanic gardens of Amsterdam and another two from Mauritius (derived presumably from the same palm in West Africa) and grown in the Bogor Botanic Gardens, Java (Indonesia) in 1848 (Jalani et al., 1998). The progeny from these four Bogor palms were distributed to the Deli province in Sumatra and Malaysia (then Malaya); hence they become known as the Deli Dura population (Rosenquist, 1986; Soh et al., 2003). The Deli Dura was used as commercial planting material for the rapidly expanding plantations in the Far East from 1911 until the early 1960s.

In the 1920s meanwhile, mass selection resulting in breeding populations of restricted origin (BPRO) was carried out in South East Asia, in the Belgian Congo by INEAC (Institut National pour l'Etude Agronomique du Congo Belge), and in French West Africa by a few growers (Durand-Gasselin et al., 2011). The revelation of monogenic inheritance of the shell (Sh) gene (Beirnaert and Vanderweyen, 1941) and the fact that hybrid progeny of D × P are 100% thin-shelled Teneras with thick mesocarps and higher oil yield, led to rapid change to Tenera hybrid planting material, marking the end of the mass selection era, the time the first effective controlled pollination techniques were developed (Durand-Gasselin et al., 2011). Teneras from West Africa were then imported and bred by researchers in the Far East as the source of Pisiferas (Hartley, 1988; Soh et al., 2003). Breeding progress in Deli Dura planting material was about 15% per generation; a breeding generation in oil palm requires 8–10 years

(Soh et al., 2003). Nearly 90% of the world's oil palm seed production is based on Deli Dura, which originated from the four Bogor palms. Since the switchover to Tenera hybrid planting material several generations of this material have been developed and oil yields have improved (Hardon et al., 1987). However, this could not be attributed entirely to breeding as materials were planted at different times and locations and perhaps with different agronomic inputs (Lee and Toh, 1992; Soh et al., 2003).

Most oil palm breeding populations (Deli, AVROS, Yangambi, La Me, Ekona, and Calabar) were descended from one or a few palms (narrow genetic base) and were termed BPROs by Rosenquist (1986). Inbreeding depression in selfs was reported at the end of the 1940s leading to the International Experiment initiated by the IRHO (Institut de Recherches pour les Huiles et Oléagineux). Expeditions and collections were made by researchers at various centers in West Africa, especially after the Second World War during the "International Experiment," which involved exchanges of material and crosses for breeding purposes among various BPROs from several research institutes (Durand-Gasselin et al., 2011). These formed the genetic base and diversity of respective oil palm breeding programs (Rosenquist, 1986; Hartley, 1988). The International Experiment led to three major findings (Gascon and De Berchoux, 1964; Ngando-Ebongue et al., 2013):

- The existence of substantial variability between the different BPROs studied.
- The additive nature of heredity of the number and average weight of bunches of the different BPROs.
- The superiority of between-origin and generally complementary crosses over within-origin crosses. These results were used for seed production (Durand-Gasselin et al., 2011).

Later on, two modern breeding schemes were developed: family/individual palm selection (FIPS) and reciprocal recurrent selection (RRS). FIPS which is widely used in Malaysia and Papua, as well as by PAMOL (palm oil producing company in Cameroon) and Pamol, Unipalm in Africa (Rosenquist, 1989; Durand-Gasselin et al., 2011) takes advantage of selected parents based on their phenotypic value and general combining ability (GCA) and that of their families (Durand-Gasselin et al., 2011). RRS (Fig. 10.6) was adopted in the IRHO/CIRAD (Centre de Coopération Internationale en Recherche Agronomique pour le Développement) network in 1958 within which the different BPROs were divided into two distinct groups based on the complementarities of their traits: group A (Dura female including Deli and Angola populations) comprising origins with a small number of large bunches and group B (Pisifera male parents of African populations: La Me, Yangambi-Sibiti, NIFOR, Cameroon) including origins with a large number of smaller bunches (Meunier and Gascon, 1972).

The first selection cycle was launched in the late 1950s and several trials were undertaken at IRHO network research stations resulting in overall yield improvement of about 18% compared with the starting population. Various interesting combinations were found among crosses between Deli × La Me and Deli × Yangambi populations. During the second selection cycle, emphasis was laid on Deli × La Me crosses due to their generally high degree of wilt tolerance and slow growth rate alongside traits relating to yield improvement (Gascon et al., 1988; Nouy et al., 1991). The average yield of this Deli × La Me second-cycle planting material was about 4.5 tons oil ha^{-1} $year^{-1}$ under moderate water-deficit conditions in La Dibamba (Cameroon) (Ngando-Ebongue et al., 2013). The oil palm planting material currently used worldwide is mainly Tenera hybrid obtained from D × P crosses. D × P hybrid

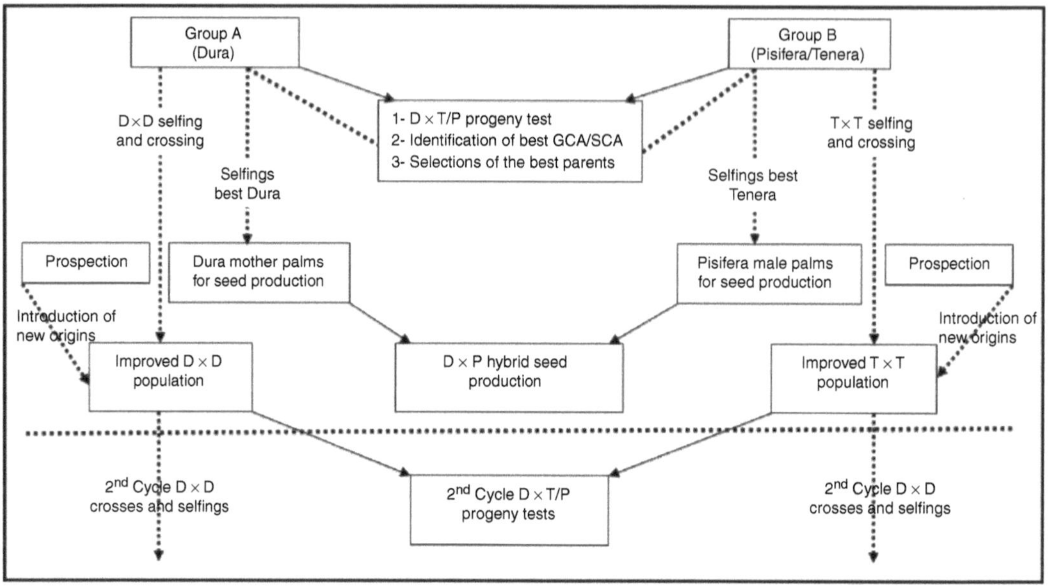

FIGURE 10.6 **The reciprocal recurrent selection scheme (RRS) applied to the oil palm (Ngando-Ebongue et al., 2012).**

seeds are largely based on Deli Dura selection at research centers such as the Malaysian Palm Oil Board (MPOB) and others at Chemara, Banting, Dami, Socfindo, and Dabou. The main sources of Pisifera are Algemeene Vereniging voor Rubber Ondememingen ter Oostkust van Sumatra (AVROS), NIFOR (Calabar), Ekona, Yangambi, and La Me (Kushairi et al., 2010).

DEVELOPMENTS IN OIL PALM BREEDING

The establishment of plantations on a large scale commenced in the 1920s, both in the Congo and in the Far East, and work on improvement of the crop by selection and breeding quickly followed (Corley and Tinker, 2003). In Asia, research work was undertaken by plantation companies in Indonesia and Malaysia and by AVROS, while in Africa work first began in the French colonies under Service de l'Agriculture, which was taken over in 1946 by IRHO (Corley and Tinker, 2003; Ngando-Ebongue et al., 2012).

Breeding Objectives of Improvement Programs

Efforts have been devoted to oil palm breeding and seed production such as developing technologies for semiclonals, biclonals and dihaploids, in addition to constant monitoring of the quality of commercial planting materials and protection of new varieties to ensure sustainability (Kushairi et al., 2011). The promotion of sustainable palm oil is thus an opportunity for palm oil seed distributors (Durand-Gasselin et al., 2010, 2011). Oil palm breeding objectives include increased oil yield, short palms, improved oil quality, disease resistance, physiological traits (bunch index, total dry matter, and bunch dry matter), and exploitation of G × E interaction (Rajanaidu et al., 2000; Kushairi et al., 2011). Research conducted on oil palms in most improvement programs covers the following main areas:

- Agronomy, which includes oil palm mineral nutrition, water management, soil fertility, ecophysiology studies, and the development of sustainable cultivation practices.
- Breeding, which involves continual improvement of the oil palm through conventional selection programs, assisted molecular breeding, and tissue culture.
- Crop protection, predominantly through an IPDM approach, which includes pest, disease, and weed control.
- Sustainability, which involves developing methodologies and tools to assess the impact of field practices on the environment, and subsequently to develop and test more sustainable agricultural practices for palm oil production.

Prospection for Oil Palm Germplasm Collection

Prospection for germplasm collection constitutes a major first step in breeding. As in any breeding program, oil palms need genetic variability as a prerequisite for improvement (Ting et al., 2010; Kushairi et al., 2011). The role of germplasm in the improvement of cultivated plants has been well recognized (Frankel and Hawkes, 1975; Hawkes, 1981; Holden and Williams, 1984; Gill, 1984; Jalani et al., 1998; Sapey et al., 2012). It is generally recognized that the narrowness of effective gene pools has been a major obstacle to rapid selection progress (Ting et al., 2010). Despite the vast wild or semiwild populations existing in the palm belt across west and central Africa, most breeding programs have a narrow genetic base; hence, the need for new material of sufficient genetic variability to provide scope for future breeding progress (Corley and Tinker, 2003). Deli Dura and a limited number of other Teneras and Pisiferas form the basis of oil palm breeding (Kushairi et al., 2011). In this respect, the collection of germplasm from different genetic resources is essential to broaden the genetic base of oil palm breeding material for a number of purposes such as increasing yield potential (Fadila et al., 2011). It was concern about this situation that provided the initial impetus for prospection for new oil palm genetic materials (Hardon, 1974; Jalani et al., 1998).

Prospection for germplasm collection and conservation by many research institutes has been accelerated in recent decades for several reasons: to broaden the genetic base of oil palm breeding material; in response to population pressure and rapid urbanization of the ecologies of oil palm groves; and to safeguard against crop vulnerability and prevent the extinction of land races and wild relatives (Reddy et al., 2006; Okwuagwu et al., 2011; Sapey et al., 2012).

Collection of **Elaeis oleifera** *Germplasm*

Major interest in interspecific hybrids between *E. oleifera* and *E. guineensis* led oil palm breeders to put much more effort into collection of *E. oleifera*, at least until MPOB started work on *E. guineensis*. Collections have been made by the MPOB prospection program for *E. oleifera* in South America (Colombia, Panama, Costa Rica, Honduras, Nicaragua, Brazil, and Surinam) as well as for other oil-bearing palm species (*Jessenia* spp., *Oenocarpus* spp., *Bactris gasipaes*) (Rajanaidu and Jalani, 1994), intended to be used in future genetic modification of palm oil (Rajanaidu et al., 1986; Soh et al., 2003). Collections have taken place in Malaysia,

Ivory Coast, Costa Rica, Brazil, and elsewhere (de Blank, 1952; Meunier, 1975; Escobar, 1982; Ooi et al., 1981; Barcelos et al., 1986; Miranda Santos et al., 1986; Richardson, 1995). Despite these collections, many of the hybrids studied in Malaysia have been derived from a single palm of uncertain origin, possibly from Brazil, known as the KLM (Kuala Lumpur *Melanococca*) palm.

Collection of **Elaeis guineensis** *Germplasm*

In the Far East, the palm oil industry was satisfied with Deli Dura material for more than a third of a century. In spite of the good quality of the Deli material, interest later arose in importing germplasm from Africa and Latin America (Rajanaidu et al., 1999; Ting et al., 2010) in order to widen the genetic base of oil palms in South East Asia (Ting et al., 2010). The earliest important prospection was that carried out in Congo in the 1920s (Corley and Tinker, 2003). The La Me (Ivory Coast) population resulted from a survey made between 1924 and 1930 in the Bingerville region of the Ivory Coast (Cochard et al., 2000; Corley and Tinker, 2003). The Pobe population (Benin) is descended from 38 Tenera palms selected in 1927 from wild groves in the Porto-Novo and Pobe regions of Benin (Gascon and De Berchoux, 1964; Corley and Tinker, 2003). Systematic prospection was also carried out by Meunier (1969) in the Ivory Coast (Corley and Tinker, 2003).

Meanwhile, in Nigeria there have been a number of collections from various parts of the country including the 1954 Ufuma and Aba collections; the 1960 coastal, inland, and marginal region collections; the 1973 NIFOR/MARDI major prospection exercise jointly carried out by the Nigerian Institute for Oil Palm Research (NIFOR) and the Malaysian Agricultural Research and Development Institute (MARDI). The oil palm work carried out by MARDI was later taken over by the Palm Oil Research Institute of Malaysia (PORIM), now known as the MPOB. The 1973 NIFOR/MARDI collections came from 45 Nigerian locations and 1991 collections from the marginal zone on the northern edge of the oil palm belt of the old Nsukka province (Okwuagwu, 1985; Corley and Tinker, 2003; Okwuagwu et al., 2011). Large quantities of seeds issued within West Africa by NIFOR owe their origin to the Aba and Calabar programs and to later generations from them (Corley and Tinker, 2003). The germplasm accessions in NIFOR are not limited to Nigerian grove accessions and their progeny. Due to the liberal genetic exchange policy adopted by oil palm–growing countries (Okwuagwu et al., 2011), the introduction of various African and Deli origins has been attributed to them (Okwuagwu, 1985). Evaluation of these collections has been carried out both in Nigeria and elsewhere (Ataga and Fatokun, 1989; Rajanaidu et al., 2001; Seng et al., 2011). NIFOR has intensified its efforts to conserve oil palm genotypes with specific adaptations in various ecologies.

In Cameroon, experimental plots were planted in the botanic gardens at Victoria in 1903, which included thin-shelled palms, apparently from a village called Lisombe (Corley and Tinker, 2003). The first plantations in Cameroon such as those at the Ikassa Estate were established before 1913, probably with seeds from the Ekona district (Rosenquist, 1986; Rao et al., 1999). The United Africa Company planted the Ndian Estate from 1928 onward, using seeds from Ikassa and from Calabar plots, together with a small number of seeds from Congo, and Deli Dura from Malaysia (Corley and Tinker, 2003). Blaak collected wild material in Cameroon in the 1960s from the Bamenda Highlands (Rosenquist, 1986; Corley and Tinker, 2003). The IRHO also undertook prospection in Cameroon in the 1970s. Populations in seven different areas were observed, and open-pollinated seeds from three palms and pollen from five others were collected from Widikum (Cochard et al., 2000; Corley and Tinker, 2003). In May 2007, prospection was jointly undertaken in Cameroon by the Institute of Agricultural Research for Development (IRAD) and the Colombian Institute for Research on Oil Palm (CENIPALMA). In May 2008, another prospection was jointly carried out in Cameroon by IRAD and the Indonesian Palm Oil Board (IPOB). Accessions have been planted in the countries of parties involved in these organizations and are being evaluated for subsequent expoitation in ongoing improvement schemes.

The MPOB (formerly PORIM) has undertaken a series of explorations to collect germplasm of *E. guineensis*, *E. oleifera*, and other economic palms (Rajanaidu, 1994; Rajanaidu and Rao, 1988; Jalani et al., 1998; Kushairi et al., 2011). Post–Second World War prospection in Congo was done on estates planted with Yangambi material, among palms of local origin on estates, and in grove areas (Corley and Tinker, 2003). Another prospection was made in the early 1970s jointly with NIFOR in Nigeria. This collection was planted in Malaysia, and at NIFOR (Corley and Tinker, 2003). Since 1973, collections of oil palm germplasm have been carried out with numerous plant explorations in African countries such as that undertaken in Angola in 1991 and 2010 by MPOB (Marhalil and Rajanaidu, 2011). Information gleaned from analyses of prospection collections was used to plan the sampling for subsequent prospections, the aim being to maximize the amount of diversity represented per sample collected (Corley and Tinker, 2003). In 1984, collections were made in Cameroon and Congo, jointly by PORIM and plantation companies owned by Unilever in those countries (Rajanaidu, 1986; Corley and Tinker, 2003). Material from these collections was planted in Malaysia and Colombia, as well as in the countries of origin. Collection in Tanzania and Madagascar (Rajanaidu, 1987) and in coastal parts of Angola (Rajanaidu et al., 1991) followed in the late 1980s, and in Senegal, Sierra Leone, Guinea, Gambia, and Ghana in the 1990s (Rajanaidu and Jalani, 1994; Rajanaidu et al., 2000; Corley and Tinker, 2003). The MPOB has established a collection of material that comprehensively covers the entire oil palm belt of West and Central Africa (Corley and Tinker, 2003).

The MPOB currently has the largest collection of palm germplasm from various parts of the world, which have been field-planted at the MPOB Research Station, Kluang, Johore (Malaysia) (Fadila et al., 2011; Norziha et al., 2011). Assessment and selection (Rajanaidu et al., 2011b; Sujadi et al., 2011; Beyegue et al., 2011) for traits of interest have been carried out on these germplasm collections resulting in regular transfer of new technologies in the form of breeding and planting materials to the industry termed PS Series palms. Some 13 PS Series palms have so far been transferred: PS1 of slow height increment; PS2 of high iodine value; PS3 of high kernel; PS4 of high carotene; PS5 of thin shelled Teneras; PS6 of large fruit Duras; PS7 of high bunch index; PS8 of high vitamin E; PS9 of *Bactris gasipaes* which is bred for palm heart production and recommended for diversification, agroforestry, and intercropping; PS10 of long stalk palms; PS11 of high carotene; PS12 of high oleic palm oil; and PS13 of low lipase (Kushairi et al., 2011). The MPOB has also built capability and capacity to carry out genetic engineering and introduce desirable genes into oil palms for the development of novel planting materials (Jalani et al., 1998).

Conservation, Evaluation, and Use of Collected Germplasm

Conservation of oil palm genetic resources provides a reservoir of genes for the development of novel traits, a prerequisite for the improvement of profitability of the crop (Hayati et al., 2004; Sapey et al., 2012). A collection has little practical use until it has been properly evaluated and its attributes become known to breeders (Sapey et al., 2012). Thus, the assessment of oil palm genetic diversity is an important step toward effective utility in breeding programs (Ting et al., 2010). Germplasm evaluation, in the broad sense and in the context of genetic resources, is the description of the material in a collection. It covers the whole range of activities starting from the receipt of new samples by the curator and growing these for seed increase, characterization, and preliminary evaluation and for further or detailed evaluation and documentation (Sapey et al., 2012). Germplasm evaluation involves definition of selection criteria with long-term validity, precision of selection tools, assessment of available genetic diversity, study of the heritability of interesting characters and their possible impact on breeding methods as well as measurement of the expected investment/benefit ratio (Durand-Gasselin et al., 2011).

Two different approaches to prospection have been adopted (Corley and Tinker, 2003). The IRHO – now Centre de Coopération Internationale en Recherche Agronomique pour le Développement (CIRAD-CP) – has concentrated on introducing new sources of variation (Kushairi et al., 1999) into existing breeding populations. The MPOB, in contrast, has aimed to collect representative samples from all across the oil palm belt, sampling being done at random, to try and capture the full range of genetic variation present.

The need for further collection to broaden the genetic base of oil palm breeding material and as a safeguard against crop vulnerability inherent in growing uniform and closely related cultivars over wide areas remains imperative (Okwuagwu et al., 2011).

Ex situ *and Laboratory Gene Bank Management*

Conserving resources for present exploitation and future generations is a pillar of sustainability. Germplasm collections constituting life gene banks planted at various research centers serve as an available source of breeding material, even though wild oil palm groves have been lost. These collections have been evaluated (Fadila et al., 2011) both in the field and laboratory for traits such as disease tolerance and growth rate alongside traits relating to oil quality and yield performance (Blaak et al., 1963; Gascon et al., 1988; Nouy et al., 1991; Isa et al., 2011; Junaidah et al., 2011; Wong and Mariam, 2011) in order to identify individuals or accessions with the best traits for eventual exploitation in breeding programs. However, *ex situ* conservation of genetic material incurs the high cost of maintenance, needs large areas of land, and palms are exposed to diseases and changing weather conditions (Norziha et al., 2011).

Cryopreservation (storage at the ultralow temperature of –196°C) in liquid nitrogen is a cost-effective alternative to field gene banks. It needs less space and protects genetic material from diseases (Norziha et al., 2011). It is possible to revive and germinate oil palm zygotic embryos after freezing in liquid nitrogen (Grout et al., 1983). Embryoids formed after culture on a medium with high sucrose content can survive freezing in liquid nitrogen (Engelmann and Duval, 1986). Moreover, introduction of a desiccation stage by drying the embryoids to between 19% and 35% moisture content leads to freezing survival by a proportion of any type of embryoid (Dumet et al., 1993a). Storage at –80°C is as good as at –196°C for desiccated embryoids (Dumet et al., 1994a) or pollen.

However, a cheap alternative to cryopreservation is storage under conditions allowing little or no growth. Engelmann (1990) showed that the interval between transferring fresh medium could be extended to 4 months for cultures below 1% oxygen. Cultures maintained at 15 °C, instead of the usual 25–30 °C, could be kept for 6–9 months without transfer (Paranjothy et al., 1990; Tarmizi and Marziah, 1995).

BREEDING TECHNIQUES FOR SUSTAINABLE PRODUCTION

Due to competition from other vegetable oils and fats, additional gains in oil palm productivity are needed at an ever faster pace (Singh et al., 2007). The traditional method of oil palm propagation is via seeds. Current oil palm planting materials are mostly produced through the hybridization of D × P by RRS. In nature, germination of the oil palm is very slow (between 1 year and 3 years) making it difficult to establish planting materials (Rees, 1963d). The traditional breeding technique for the crop enhances yield but is however impeded by the long selection cycle of 10–12 years and the enormous resources involved in land, labor, and field management of oil palm breeding programs. In order to make today's agriculture sustainable, it is necessary for plant breeders to adopt innovative technologies, particularly through advances in modern genetics and genetic engineering (Schlegel, 2010), which can increase the efficiency of selection with more precision (Tester and Langridge, 2010). Molecular approaches (genomics and genetic engineering technologies) are powerful tools to assure rapid and precise selection for the traits of interest. Recent advances in high-throughput genomics technologies including next-generation sequencing and high-throughput genotyping have helped immensely in understanding the functions and regulation of genes in crop plants (Varshney et al., 2009c). These techniques are already being used in oil palms to understand oil formation (Bourgis et al., 2011; Tranbarger et al., 2011; Dussert et al., 2013). The oil palm genome has also been sequenced (Singh et al., 2013a). *E. guineensis* is an exceptional example where a quantum leap in yield improvement has been obtained through a single gene (Singh et al., 2013b).

To accelerate and capture the maximum yield potential of a selected genotype, oil palms need to be vegetatively propagated. High-yielding genetic material is available in the oil palm industry (Kushairi et al., 2010). It has been reported that the best experimental plot produced an oil yield of 8.6 t^{-1} ha^{-1} $year^{-1}$, selected progeny produced 12.2 t^{-1} ha^{-1} $year^{-1}$, individual palms yielded 13.6 t^{-1} ha^{-1} $year^{-1}$, and maximum theoretical oil yield was 18.2 t^{-1} ha^{-1} $year^{-1}$. Cloning such palms could provide high-yielding planting material for the industry, which would constitute a second wave in oil palm yield improvement (Kushairi et al., 2010). Integrated breeding approaches such as MAS (Billotte et al., 2010; Montoya et al., 2014), marker-assisted recurrent selection (MARS), and genome-wide selection (GS) offer the opportunity of precision breeding with great potential as against chance breeding. For instance, genetics and genomics studies involving the fatty acid composition in the *Elaeis* genus have given rise to results that clarify information about genes, enzymes, and regulatory mechanisms for fatty acid composition traits (Siti Nor Akmar et al., 2003; Tranbarger et al., 2011; Xu et al., 2011; Bourgis et al., 2011; Al-Shanfari et al., 2011; Montoya et al., 2013). Moreover, useful information provided by genetic maps, QTLs and marker-type single-nucleotide polymorphisms (SNPs) involved in fatty acid composition (Singh et al., 2009; Billotte et al., 2010; Montoya et al., 2013) facilitate the implementation of classical breeding strategies aimed at modifying the fatty acid composition of *E. guineensis* (Montoya et al., 2014).

Reciprocal Recurrent Selection

As a breeding technique for the genetic improvement of quantitative traits (Comstock et al., 1949), RRS has been applied to a variety of economic animal and plant species. The RRS (Figure 10.6) scheme has certain theoretical advantages over classical pure-bred methods (Meunier and Gascon, 1972; Meunier and Jacquemard, 1988). In this scheme, the basic population of the selection program is made up of all ecotypes divided into two complementary groups, A and B, of diversified populations (Adlin and Kiswito, 1975) with respect to bunch yield components and other heritable characters (Meunier and Jacquemard, 1988). The two populations are improved through selection combining a phenotypic choice of individuals based on high-heritability characters and selection based on the value of their progeny for low-heritability characters (Meunier and Jacquemard, 1988). Recombining selected individuals and creating new improved basic populations serve as the basis for new improvement cycles (Meunier and Jacquemard, 1988; Gascon et al., 1987).

Strict division between ecotypes makes it possible to avoid the risk of inbreeding (e.g., roll-up leaflet morphology and albinism) in subsequent cycles (Meunier and Jacquemard, 1988; Heri et al., 2011). RRS thus helps to improve populations for their general combining ability, with specific combining ability exploited directly for seed production or cloning. It is easy to introduce new individuals with interesting characters into this scheme at any moment. To gain time in the RRS breeding process, trials were set up to correspond with Dura and Tenera progeny originating from the selfing or recombination of parents tested in the trials. As such, from the trial results, seed production can be effected by reproduction of the best crosses when selfs of the two parents are available and by utilization of parents whose good GCA has been demonstrated or which come from the best populations. The strength of this scheme is the reasonable amount of work and land area required, even though the workload increases as more individuals are selected (Meunier and Jacquemard, 1988).

The basic RRS scheme can be modified depending on the material and results available (Meunier and Jacquemard, 1988). Two basic selection schemes have been reported in oil palm breeding programs: the modified recurrent selection scheme practiced in the Far East and the modified reciprocal recurrent selection scheme practiced in Africa (Hartley, 1988).

Modified Recurrent Selection Scheme

The modified recurrent selection scheme is highly practiced by top breeding programs in the Far East (Hartley, 1988). It evolved from initial use of Deli Dura as the commercial planting material to its subsequent use as a maternal parent in commercial Tenera (T) hybrid seed production. In this scheme, Pisiferas (P), introduced or bred (and being female sterile), are initially selected based on their Tenera–sib performance in the T × T/P cross or based on proven lineage. They are then progeny-tested (top-crossed) with a sample of phenotypically selected Deli Dura mother palms. Proven Pisiferas and phenotypically selected Duras are then used in commercial mixed Tenera hybrid seed production. Parents used for further breeding in D × D and T × T/P crosses are also phenotypically selected although progeny-tested parents are also sometimes included. As parent selection often involves family/individual palm selection, this scheme is sometimes referred to as FIPS (Rosenquist, 1990). However, as selected parents are interbred in each cycle, it is preferably referred to as a form of recurrent selection (Allard, 1960; Soh et al., 1990). This scheme emphasizes and exploits GCA effects.

The main strengths of this scheme are that more recombinant crosses can be made and tested while saving time, space, and effort from extensive progeny testings. The main disadvantage is that parents chosen for further breeding and Dura mother palms for hybrid seed production usually have not been hybrid progeny-tested. The key assumption that additive or general combining effects expressed within population crosses (determined by family and individual means or combining ability analyses) will be reflected in interpopulation hybrid crosses may be untenable (Soh, 1987). The scheme may also lead to too great a dependence on Deli as the source of Dura mother palms.

Modified Reciprocal Recurrent Selection Scheme

This scheme is mostly practiced in West Africa (e.g., Ivory Coast, Nigeria) (Meunier and Gascon, 1972) and of late in Indonesia (Lubis, 1988; Lubis et al., 1990). It was adapted from the RRS method developed in maize breeding (Comstock et al., 1949). It exploits general combining ability effects and specific combining ability effects. The choice of this scheme presumably arose because heterosis effects with respect to yields were noticed in Deli Duras × African Teneras or Pisiferas crosses (Hartley, 1988).

The scheme has three main strengths. First, selfed parents selected for commercial hybrid production and for further breeding are both based on progeny tests (from D and T grandparents) of prospective commercial interpopulation hybrids. Second, the scheme has two distinct phases: the within-hybrid improvement phase and the recombinant hybrid improvement phase. The former phase permits short-term commercial exploitation of the best test cross and its improvement by recurrent selection within the selfs/sibs of selected parents. The latter phase, in contrast, involves outcrosses that are run in parallel in a similar way. It allows accumulation of favorable additive and nonadditive alleles and maintenance of genetic variability to sustained longer term improvement. Third, commercial hybrids can be reproduced using the D and P from the selfs/sibs of the parents (which have been planted concurrently as progeny tests) once the interpopulation hybrid test results are known (Hartley, 1988). Its main disadvantage is the requirement for a large program size. To produce 3–4 million hybrid seeds from reproduction of the top 15% of crosses, about 500 crosses and 180 parental selfs have to be planted over 600 ha and evaluated over 15–25 years (Soh, 1999).

Backcross Breeding

Backcross breeding is practiced in introgression to improve a particular genotype or population; for instance, the dumpy trait has been introgressed into the AVROS population to reduce height (Lee and Toh, 1992). The backcross approach is also used to introgress better oil quality traits from Nigerian *guineensis* and American *oleifera* materials and to restore the yielding ability and fertility of recurrent advanced breeding parents, respectively (Soh, 1999).

Genomic *In Situ* Hybridization

Genomic *in situ* hybridization (GISH), in which total genomic DNA is used as a probe to hybridize chromosome spreads, has been used to separate parental chromosomes in *oleifera* × *guineensis* hybrids and could be useful in backcross-breeding programs (Soh et al., 2003). Genomics interventions have the potential, especially when used in an integrated manner, to improve the breeding efficiency of oil palms for sustainable production. These include increasing crop productivity, diversification, enhancing the nutritional value of oil (biofortification), and reducing the environmental impacts of agricultural production.

Recombinant Inbred Lines

The development of recombinant inbred lines has been proposed and attempted in oil palms. However, production of Dura inbred line cultivars (since Tenera is heterozygotic and Pisifera homozygotic and female sterile) with oil yields superior to the best thin-shelled Tenera hybrid is unlikely through this approach. Other disadvantages would be the possible loss of desirable genes linked to yield in infertile gametes/zygotes in the inbreeding process, and longer time to produce a commercial variety. However, the recombinant inbred line approach combined with single-seed descent may be considered for development of inbred parents for hybrid production. This will put oil palm hybrid seed production on a similar footing to hybrid maize production for the generation of hybrid combinations for testing (Soh, 1987).

Marker-assisted Selection

In marker-assisted selection (MAS), molecular markers can allow a trait based on DNA to be selected. This can be carried out on plantlets rather than on mature trees allowing time and space savings. The basic requirements of MAS are the availability of markers (molecular, morphological, and biochemical). The use of molecular markers is based on naturally occurring DNA polymorphism, which forms the basis for designing strategies that can be exploited for application purposes. Linkage maps in turn serve many useful purposes, such as improved selection in breeding programs, studying the evolutionary relationship between related species, and starting points for map-based cloning (Moretzsohn et al., 2000; Singh et al., 2007; Singh et al., 2013b; Singh et al., 2014). The use of molecular markers and genetic engineering is useful in oil palm breeding (Jalani et al., 1998). In Malaysia, the use of molecular markers to screen the germplasm of the oil palm included isozymes and restriction fragment length polymorphism (RFLP) (Hayati et al., 2004; Maizura et al., 2006). However, the application of other marker systems is necessary to further exploit and understand their genetic structure (Ting et al., 2010). Other MAS techniques tested include simple sequence repeat (SSR) (Powell et al., 1996; Rival et al., 1998; Brondani et al., 1998; Castilho et al., 2000; Perera et al., 2000, 2001; Billotte et al., 2001; Adin et al., 2004; Zehdi et al., 2004, 2005; Singh et al., 2007, 2008; Ting et al., 2010; Diana et al., 2012), flow cytometry (Rival et al., 1997), amplified fragment length polymorphism (AFLP) (Ulrich and Wolfenbarger, 1999; Xu et al., 2000), and random amplified polymorphic DNA (RAPD) (Morentzsohn et al., 2000; Corley and Tinker, 2003). AFLP and microsatellite maps for oil palms are currently available (Chua et al., 2001; Billotte et al., 2001).

Simple Sequence Repeat

High information content and ease with which polymerase chain reaction (PCR) assays can be automated for identifying microsatellites or SSR polymorphisms make SSRs ideal genetic markers (Cheah and Wooi, 1995; Peakall et al., 1998; Singh et al., 2007). They are preferred for high-throughput mapping, genetic analyses, and marker-assisted plant improvement programs (Powell et al., 1996; Singh et al., 2007). As codominant and locus-specific markers, SSRs have been widely used as tools in genotype identification and population genetic studies in plants including oil palms (Provan et al., 1996; Brondani et al., 1998; Korzun et al., 1999; Ramsay et al., 2000; Castilho et al., 2000; Eujayl et al., 2001; Billotte et al., 2001; Singh et al., 2007; Bakoumé et al., 2009a,b; Ting et al., 2010; Diana et al., 2012; Noorhariza et al., 2012; Ajambang et al., 2012). The highly polymorphic nature of SSRs is of particular value for oil palms, as breeding programs often involve narrowly adapted gene pools. SSR markers that are PCR based also have the advantage of reliability, reproducibility, discriminative ability, and cost-effectiveness when compared with other marker systems. The method is also effective at fingerprinting tissue culture clones as a means of quality control with the added advantage that it can be applied at any stage of the tissue culture process (Singh et al., 2007).

The SSR technique is not without constraints. The construction of large-scale enriched libraries using SSR technology requires considerable technical skills, time, and money for successful execution (Singh et al., 2007). In order to circumvent technical difficulties, several modifications, especially using the PCR technique with degenerate primer(s) that contain repeat sequences to produce enriched genomic libraries, have been introduced (Fisher et al., 1996).

Oil Palm Tissue Culture

Stages of Tissue Culture in Oil Palm Breeding

Tissue culture (i.e., *in vitro* or clonal propagation) (Ahmad et al., 2011) has the potential to develop high-yielding collections while circumventing the long generation time required with traditional breeding techniques (Zamzuri, 2011). Tissue culture has been progressively improved over time to take advantage of potential genetic progress (Soh et al., 2001; Potier et al., 2006; Soh, 2010; Durand-Gasselin et al., 2011). The ability to propagate oil palms from both young leaves and zygotic embryos through somatic embryogenesis is a sustainable (spatiotemporal gain) approach to its propagation and breeding. Plant cell technologies, especially the protoplast technique, can be used as tools for producing protoclones and somatic hybrids in oil palms. Tissue culture generally involves five stages (Debergh and Maene, 1981; George, 1993; Ahmad et al., 2011):

- mother palm selection and preparation (ortet selection and cabbage sampling);
- establishing an aseptic culture (explant inoculation and callus culture);
- multiplication of suitable propagules (embryoid culture, polyembryoid multiplication);
- formation of shoots and roots (shoot elongation and root initiation);
- transfer to natural environment or weaning.

Being a mixture of hybrids from nonfully inbred parents, commercial oil palm planting material has considerable between-palm genetic variability arising from between-family and within-family genetic variability and depending on the relatedness and inbred status between Dura × Pisifera parents (Hardon et al., 1987; Meunier et al., 1989). Although the differences might not be entirely genetic, they did however provide the rationale and impetus in the 1960s to develop *in vitro* propagation of oil palms. From the outset, the oil palm – being a monocot and a perennial tree – was considered a recalcitrant species for *in vitro* propagation. However, with concerted research efforts, the first *in vitro* oil palms were achieved by the mid-1970s (Jones, 1974; Soh et al., 2003; Raberchault et al., 1970; Soh et al., 2003).

The success in propagating oil palms through tissue culture techniques using various plant parts has been widely reported (Staritsky, 1970; Smith and Thomas, 1973). The early success of plantlet production in the 1970s and increased yield potentials inspired many oil palm organizations to exploit *in vitro* propagation techniques. In 1986, the problem of clonal mantled fruits (i.e., floral and vegetative abnormalities of clonal palms) (Jalani et al., 1998; Corley and Tinker, 2003; Morcillo et al., 2006) planted at United Plantations Berhad (Malaysia) was brought to the attention of the scientific community. This boosted interest among oil palm tissue culturists. As more information and understanding on the tissue culture process and the problems arising from it accumulated, there was renewed interest to proceed with large-scale propagation of oil palm clones to increase productivity. However, scaling up the process revealed several constraints that limit commercial development of the technique.

Constraints of Tissue Culture in Oil Palm Breeding

The main difficulties encountered at the pilot scale are the production costs and the occurrence of variant palms in embryogenically derived plant material (Corley et al., 1986). Somaclonal variation is manifested as abnormal flower and fruit development. In the pistillate flower the vestigial androecium develops into supernumery carpels forming a mantle over the developing fruit similar to the rare genetic mantled or Poissoni fruit variant found in nature (Hartley, 1988). This abnormality is accompanied by parthenocarpic fruit set, severe bunch failure, and palm sterility (Hartley, 1988; Corley et al., 1986; Paranjothy et al., 1990; Corley and Tinker, 2003). However, palms bearing mantled bunches can recover to become normal bunch bearing once again (Durand-Gasselin et al., 1995; Duval et al., 1997). Susceptibility to the abnormality varies between and within clones and the risk appears to increase with extended culture. The earliest batches of ramets (clonal plants) tended to be mantle free (Paranjothy et al., 1995). The mantled fruit phenomenon was experienced by all laboratories and confidence in mass propagation was seriously eroded. Speculations were made on the physiological, epigenetic (gene expression), and genetic bases of the mantle somaclonal variant (Corley et al., 1986; Soh, 1987). Mantling was apparently found to be sexually transmissible (Rao and Donough, 1990) but this still did not rule out the epigenetic basis. Use of a fast-growing instead of a nodular callus (Soh et al., 2003) and/or phytohormones such as cytokinins (Jones et al., 1995) in the proliferation stage was suggested as the cause but this soon proved untenable as laboratories that did not use either still produced mantled ramets.

Another inconsistency with tissue culture is the inefficiency of the cloning process. Although callogenesis rates and percent embryogenic palms appear not to be process limiting, embryogenesis rates on callus cultures varied from 0.2% to 36%, averaging 4% per ortet. The capability for active proliferation and thus mass propagation of the embryos obtained is more important. Generally, about half are capable (Soh et al., 2001), while the remainder will either persist in producing shoots to the detriment of embryo proliferation or fail to develop further.

Yet another constraint to tissue culture is ortet selection. In the breeding of many crops and in clonally propagated crops (Soh et al., 2003), single-plant selection especially for yield from a mixed seedling–derived population has been inefficient if not futile. Soh (1986) using data from trials of an advanced Tenera hybrid (Deli × AVROS) population estimated that selecting the top 5% of palms for cloning would give clones yielding 10–15% more than the mean of hybrids and that clones yielding more than 30% were unlikely because of low heritability and genetic variability. Moreover, better clones could only be identified after clonal testing. Elucidation of low heritability for oil yield in these advanced Tenera hybrid populations implied that the same experiences in other crops could apply in oil palms (Soh, 1998). This has been further supported by other clonal trial results (Donough and Lee, 1995; Cochard et al., 1999; Soh et al., 2001) confirming the inefficiency of ortet selection based on its phenotype and the dependence of good clonal tests to pick out the infrequent outstanding clones, especially clones tending to exhibit G × E effects (Lee and Donough, 1993; Corley et al., 1995b; Soh et al., 1995). There is also a low rate of embryogenesis for large-scale propagation. In addition, agronomic inputs such as fertile land and tailored fertilizer regimes (site and planting material specific) (Chin et al., 2011) are necessary for clones to express their genetic potential. In sum, to obtain large-scale commercial propagation of proven Tenera clones, a number of critical issues need to be resolved including somaclonal variation (Hartley, 1988), cloning efficiency, ortet (parent palm of clones) selection efficiency, feasibility of recloning and the liquid suspension system, clonal field evaluation requirements, and expected genetic progress. Tissue culture requires sophisticated facilities that have greater skill in handling and maintenance than conventional techniques. Moreover, shoot tip–derived plants may show genetic instability. Micropropagation also leads to decrease in genetic variability.

Strengths of Micropropagation in Oil Palm Breeding

Countering the constraints of micropropagation are opportunities or strengths such as the low rate of embryogenesis for large-scale propagation being addressed *via* liquid culture, innovative shake flask, and bioreactor systems; production of clonal seeds from clonal parents with good specific combining ability; and biomarker approaches that lead to the development of diagnostic tools for screening clonal amenity and conformity (Kushairi et al., 2010). Moreover, micropropagation requires relatively little growing space and the technique enhances the rate of multiplication; over a million plants can be grown from a small, even microscopic, piece of plant tissue within 12 months. Shoot multiplication usually has a short cycle (2–6 weeks) and each cycle results in logarithmic increase in the numbers of shoots. Tissue culture provides propagules such as minitubers or microcorms for plant multiplication throughout the year irrespective of the season. The small size of propagules and their ability to proliferate in a soil-free environment facilitate their large-scale storage and large-scale dissemination by suitable means of transport across international boundaries. Furthermore, stocks of germplasm can be maintained for many years while pathogen-free plants can be raised and maintained economically.

Tissue Culture as a Prospective Alternative to Seedling Production

Prior to the advent of tissue culture, there was no known reliable method for vegetative propagation of oil palms (Wong et al., 1997). This made tissue culture an important factor in the development of the oil palm industry (Gorret et al., 2004), especially in the generation of superior and uniform oil palm planting material (Durand-Gasselin et al., 1990). The *in vitro* technique allows oil palm seedlings to be sustainably generated in two ways: direct regeneration of plantlets from immature and mature zygotic embryos and somatic embryogenesis. Zygotic embryo culture can be used for a variety of purposes including bypassing seed dormancy (Bürün and Çoban, 2002; Das et al., 1999) and inducing a faster growth rate (Chang and Yang, 1996). Pioneering work in oil palm tissue culture was carried out by Raberchault et al. (1970). The use of tissue culture was predicted to improve oil production and this was later confirmed by significant increases in yield (up to 30%) compared with commercial D × P seedlings in large-scale field trials (Soh, 1986; Wahid et al., 2005).

The first regenerable oil palm embryogenic cell suspension cultures were obtained from oil palm leaf–derived calli (de Touchet et al., 1991). Later, mature zygotic embryos excised from fruits were used to generate friable embryogenic tissue and to establish embryogenic suspensions (Teixeira et al., 1995). Although these two protocols were successful at establishing single somatic embryos from embryogenic suspension cultures, the regeneration rate into plantlets was poor. This issue was overcome by a protocol developed to improve the quality of shoot formation (Aberlenc-Bertossi et al., 1999). The first results obtained from suspension culture–derived clones were encouraging (Rival et al., 2001). An innovation in clonal propagation was the successful culture of oil palm suspension cells in a bioreactor. The advantage of using bioreactors in culture processes is that parameters such as pH, temperature, dissolved oxygen concentration, and agitation can be monitored and controlled (Gorrett et al., 2004).

However, large-scale commercial propagation and planting of proven Tenera clones have yet to take off despite long-standing research and development work involving the setting up of more than 20 tissue culture laboratories worldwide (Soh, 1986) and concomitant high expenditure in clonal propagation for trial and pilot commercial test plantings. Most of the laboratories involved in the oil palm industry now have well-established tissue culture media and protocols. The large number of variables in the culture process has meant that different laboratories often obtain different results from superficially similar treatments. Even in the same laboratory, results are not always reproducible, so there remains some uncertainty as to the best procedure (Corley and Tinker, 2003).

BREEDING FOR SUSTAINABLE PRODUCTION

Oil palm breeding and hybrid seed production actually began with the revelation of simple Mendelian inheritance of the shell thickness trait in the palm fruit (Beirnaert and Vanderweyen, 1941). Since oil palms are cross-pollinated crops, they borrow a number of breeding methodologies developed in maize breeding such as recurrent selection (Comstock et al., 1949) and top-cross testing. The breeding techniques being practiced for sustainable oil palm improvement include reciprocal recurrent selection, tissue culture, and genetic engineering. The D × P hybrid obtained through breeding is the thin-shelled Tenera heterozygote (the commercial variety). The Duras used as female parents are regenerated from D × D crosses while the Pisifera male parents are regenerated from T × T (25% Pisifera segregants) and T × P (50% Pisifera segregants) crosses. D × T crosses used in progeny tests segregate 1:1 for Dura and Tenera palms. Fertile Pisiferas do occur occasionally and fertile P × fertile P crosses produce 100% Pisifera progeny (Soh et al., 2003). Tenera hybrid seed production in oil palm breeding is done by controlled pollination (Ntsomboh-Ntsefong, 2007). The resulting seeds need to undergo heat treatment under controlled conditions before they can germinate. Germination of shell-less Pisifera and thin-shelled Tenera seeds is erratic, and *in vitro* embryo rescue germination is sometimes needed (Arasu, 1970).

Breeding for Yield Improvement

Since oil yield has been the major concern in oil palm cultivation (Fadila et al., 2011; Kushairi et al., 2011), one possible way of enhancing it is by increasing the bunch index (BI) (Fadila et al., 2011). Genetic variance and the heritability of traits of interest are essential for yield improvement. Estimations of genetic variance and heritability have been attempted toward increased breeding efficiency involving information on the existence of sufficient genetic variability, choice of the population to be improved, and adoption of the appropriate breeding procedure to obtain the best selection response. For selection decisions or prediction of selection responses to a particular population and environment, a specific heritability estimate obtained from the population concerned is used. North Carolina Model 1 (hierarchical) and Model 2 (factorial) mating designs and the parent–offspring method are commonly used to estimate genetic variance and heritability in oil palms (Soh et al., 2003).

Yield improvement in oil palm plantations over the past 50–60 years has been attributed to a 70% improvement in cultivar practices and a 30% improvement in agronomic practices (Davidson, 1993; Jacquemard et al., 1997). The first wave in yield improvement was the result of switching from Dura to Tenera (D × P) planting material. The switchover brought about an improvement of at least 30% in oil yield mainly through better oil content from the thicker mesocarp (Hardon et al., 1987). Meanwhile, high unsaturated fatty acid content, large kernel, and dwarf high-yielding lines have been developed from the Nigerian collection and distributed to private industry breeders (Soh et al., 2003), who are reluctant to exploit them directly in commercial hybrid seed production (Rajanaidu and Jalani, 1999; Rajanaidu et al., 2000). They prefer to introgress the desirable features into their advanced breeding parents to maintain the integrity, uniformity, and consistency of their commercial hybrids (Sharma et al., 1999; Soh, 1999).

Estimated potential annual oil palm yield of 10 t ha^{-1} is 2–5 times that of annual oilseed crops such as rapeseed. The theoretical yield potential based on crop physiological computations is 17 t oil ha^{-1} from 45 tons of FFBs. Oil yield exceeding 12 t ha^{-1}, FFB yield of 45 t/ha, and oil content in bunch of 35% have been reported in experimental plantings, although not necessarily jointly. With these estimated yield potentials, current world petroleum demand could be met by cultivating an area of roughly 4 million km^2 (40% of the area of the United States or over half the land mass of Brazil) (Abdullah and Sulaiman, 2013). Palm oil production witnessed a rapid increase from 3 million tons in 1970 to over 30 million tons in 2005. In 2009 the Malaysian palm oil industry achieved an average oil yield of 3.93 t ha^{-1}, FFB yield of 19.20 t^{-1} ha^{-1} yr^{-1}, and PKO yield of 1.01 t^{-1} ha^{-1} yr^{-1} (Kushairi et al., 2010). Malaysia hopes to increase FFB yield to 35 t^{-1} ha^{-1} yr^{-1} and oil extraction rate (OER) to 25% by the year 2020 (Kushairi et al., 2010).

Planting material derived from mixed hybrids thus possesses high-yield potential. For larger quantum yield improvement, there is a need for drastic changes to crop architectural design (at the plant form and planting pattern level) and to cultivation and harvesting methods.

The ability to select early will have a major impact on reducing the time and resources required for variety improvement in oil palms (Singh et al., 2007). Efforts are under way to breed high-efficiency palms with such traits as more efficient light capture, higher photosynthetic rate, higher leaf expansion ratio, higher leaf area ratio, better conversion efficiency of energy captured to dry matter, reduced respiratory loss, and improved harvest index (Breure and Corley, 1983; Breure, 1985, 1986; Smith, 1993a,b; Soh et al., 2003). Dwarf palms with longer bunch stalks will facilitate harvesting (Soh et al., 1994b). Similarly, palms with ripe fruits exhibiting a distinct color change will ensure harvesting at optimum ripeness (Singh et al., 2014) while those with a nonabscising fruit habit will minimize loose ripe fruit loss (Corley, 2006; Morcillo et al., 2013). Such programs have been proposed using breeding or genetic engineering approaches (Osborne et al., 1992; Rao et al., 1998; Jalani et al., 1998; Soh et al., 2003) as these traits are found in less advanced breeding or germplasm material (Soh et al., 2003).

Breeding for Oil Quality

The oil produced by oil palms is multipurpose in that it can be used in the food, biodiesel, and oleochemical industries. The high-saturated fatty acid content of CPO, which has a melting point of 36°C, makes the oil solidify at colder temperatures (Yusoff, 2000), thereby limiting its use. Moreover, dietary oils are considered unfit for human consumption above the limit of 5% (w/w) free fatty acids (Codex Alimentarius/FAO/OMS, 2005; Ngando-Ebongue et al., 2012). Thus, the presence of free fatty acid moieties due to lipase activity is an indication of the impairment of oil quality (Sambanthamurthi et al., 1991, 2000; Ngando-Ebongue et al., 2006, 2008). Consumers therefore commonly classify CPO with coconut oil as a saturated fat unhealthy for human consumption. This is despite the fact that palm oil contains about 50% unsaturated fatty acids and that the saturated component is mainly palmitic acid which Khosla and Sundram (1996) contended to be neutral in its cholesterolemic behavior. Brett (2009) has defined the fatty acid composition of the ideal palm oil based on the American Heart Association's recommended fat dietary intake (Soh et al., 2003). In line with the above, end user and consumer bias have pressured breeders to breed a more liquid palm oil with low-lipase (Morcillo et al., 2013) high-oleic acid, high-vitamin E, high-carotene, and high-unsaturated fatty acid content (Gascon and Wuidart, 1975; Basri, 2003; Mohd Din et al., 2005b; Singh et al., 2009; Billotte et al., 2010; Kushairi et al., 2011). Advanced oil palm–breeding populations have low genetic variability for unsaturated fatty acid contents, generally expressed as the iodine value. The situation is better in West African semi-wild prospected material, but the highest values can be found in American oil palm, *E. oleifera* collections. The Malaysian Palm Oil Board recommended that to put palm oil on a competitive basis with olive oil, breeding programs should be focused on progeny with an iodine value of about 70. To do this without sacrificing current progress in yield and agronomic attributes, selecting from collected *E. guineensis* material would bring 4–5 generations of backcross breeding to advanced breeding material. With *E. oleifera* material, more backcross generations are envisaged because of interspecific hybrid infertility (Sharma, 2000). Genetic engineering provides an opportunity to avert this shortcoming (Ngando-Ebongue et al., 2012). With rapid advances in genetic engineering, especially transgenic technology, it is possible to introduce alien desirable genes into oil palms (Jalani et al., 1998).

Breeding for Adaptability

Much of the reported improvement in oil palm planting material is likely to have been through adaptability to site limitations. This involves moving site yield potential toward genetic yield potential (Tinker, 1984) by introducing planting material that is tolerant to the site-limiting factors of production – such as soil, moisture, and terrain – rather than through improvement in genetic yield potential *per se*. However, genotype × environment (G × E) interaction has not been generally considered to be a serious factor in oil palm breeding (Rosenquist, 1982; Cochard et al., 1993) where its presence or absence depends on the specific genotypes and specific environments tested (Corley et al., 1993). This misperception that G × E is not important could be attributed to the use of genetically variable commercial sources of planting material (Rajanaidu et al., 1986); the use of related progeny in different environments (Cochard et al., 1993); or the use of related progeny in similar environments (Rosenquist, 1982). Some authors have reported a positive G × E effect (Corley et al., 1993). Meanwhile, results of experiments and analyses (Soh et al., 2003; Corley et al., 1993) have detected the presence of G×E effects using genetically diverse progeny, although their contribution (3–4%) to total experimental variance for yield was small (Rajanaidu et al., 1993; Rafii et al., 2002).

Breeding for Stress Tolerance

Being a profitable and easily grown crop, oil palms have been planted in suboptimal and marginal areas (dry sandy areas, podsols, peat, and highlands). In these areas biotic (pests and diseases) and abiotic (water, temperature, and nutrients) stresses, which delay or reduce production, are likely to be encountered. Among the various challenges faced by farmers of this crop are diseases like *Ganoderma* (common in Asia), bud rot (Latin America), and vascular wilt (Africa) that affects growth and reduce oil yields (Ntsomboh-Ntsefong et al., 2012).

Basal stem rot caused by the basidiomycete fungus *Ganoderma boninense* is the main disease needing resistance breeding in South East Asia (Idris et al., 2004; Susanto, 2009; Virdiana et al., 2011). It used to be a disease of older palms planted on former coconut or oil palm areas with high water tables, but reports of its occurrence on younger plantings in inland areas have become more frequent (Ariffin et al., 2000). *Ganoderma* disease control is the subject of an international cooperative effort coordinated by CABI (formerly CAB International and before that Commonwealth Agricultural Bureaux) in which resistance control is an important objective. There exist possible sources of genetic resistance to *Ganoderma* in oil palm (Durand-Gasselin et al., 2005). Parental, progeny, and clonal differences in resistance/susceptibility were demonstrated in palms planted on *Ganoderma*-infested soil. Significant differences between families for *Ganoderma* incidence have been observed in Indonesia (de Franqueville et al., 2001), though most claims for genetic resistance to *Ganoderma* have been unconvincing (Corley and Tinker, 2003).

Crown disease, a physiological affliction of young 2–3-year-old oil palms in the Far East, causes bending and twisting of young fronds, which can result in loss in early yields when severe and prolonged. Severity is strongly dependent on the environment, and susceptibility is controlled by a single recessive gene (De Berchoux and Gascon, 1963), the expression of which is masked by an epistatic gene conferring incomplete penetrance. Deli Dura material appears to be more susceptible. The disease can be bred out by discarding families with any susceptible progeny.

Bud rot is the main oil palm disease in Latin America. It is lethal and one of the limiting factors to oil palm cultivation. The pathogenic cause of the disease is still in contention (Ariffin et al., 2000), though it is thought to be caused by the oomycete *Phytophthora palmivora*. Knowledge of the fact that the disease behaves differently among regions and planting material from different origins and that poor agronomic management might contribute to oil palm susceptibility to the disease (Edwin et al., 2011) are avenues for breeders to follow to find sustainable solutions to this threat.

Vascular wilt is caused by *Fusarium oxysporum* f. ssp. *elaeidis* and results in serious damage on oil palm culture in Africa. Various methods of control have been tried against the disease: selection of tolerant plant material, soil renewal, chemical control, stimulation of defense reactions, and the use of cover crops and potassium (K) fertilizers. The most feasible and sustainable control method against the disease is research for resistant plant material through artificial inoculation of plantlets at the prenursery stage with pathogenic fungi (Dzido et al., 1978; Ntsomboh-Ntsefong et al., 2012). Because of the availability of a reasonably efficient nursery screening technique, tolerance to *Fusarium* wilt forms an integral part of breeding programs in West Africa (de Franqueville and Renard, 1990).

As far as abiotic stress is concerned, environmental factors – such as rainfall, solar radiation, soil-available nutrients, and other edaphic factors – exert major effects on crop production (Ataga, 1993). Differences in environmental factors do occur from location to location and from year to year. Variation in these factors usually causes genotypes to respond differently from one environment to another giving rise to G × E interactions (Ataga, 1993). Among the several procedures available for characterizing the behavior of genotypes under varying environmental conditions, the commonly used method is conventional analysis of variance followed by joint regression analysis. There are significant genotype and environment effects and G × E interactions for total weight, mean bunch weight, and bunch number in oil palms (Ataga, 1993). Oil palms in West Africa experience reasonably long periods of drought, and hence West African commercial planting materials tend to be more drought tolerant because they have been bred under such conditions (Okwuagwu, 1996). The semiwild collections obtained by MPOB from the drier regions of Nigeria should include drought-tolerant genotypes. Palms introduced from the Bamenda Highlands of Cameroon (Rosenquist, 1986) are presumed to possess genes for cold tolerance, despite their likely precocious flowering behavior under lowland conditions being reported (Chapman et al., 2003). Meanwhile, manifestations of Mg deficiency occur in oil palms planted in sandy areas in some parts of Papua New Guinea and Indonesia. Tolerance to Mg deficiency is an important objective of the breeding program in Papua New Guinea.

SOUTH EAST ASIAN EXPERIENCE OF OIL PALM BREEDING FOR SUSTAINABILITY

Early work on oil palm improvement began in both South East Asia (SEA) and Africa at the beginning of the twentieth century. In Asia, research work was undertaken by plantation companies in Indonesia and Malaysia and by AVROS (Ngando-Ebongue et al., 2012). The oil palm was introduced to SEA (mainly Indonesia, Malaysia, Thailand, and Papua New Guinea) through the four Bogor palms of 1848. The big agroindustrial plantations that are spread all over SEA today owe their existence in part to material obtained from these four palms (Hunger, 1917). Oil palm seeds from these palms were planted at Deli in North Sumatra Province of Indonesia in 1875 where large-scale plantations were established with Deli seeds in 1911 (Ascenso, 1966) in Indonesia. The oil palm plantation Tennamaram Estate was the first to be established in British Malaya in 1917 with seeds obtained from Deli Sumatra.

The rapid expansion of oil palms in SEA began in the early twentieth century when the market share of palm oil increased, especially with exports to Europe. European food manufacturers began to introduce palm oil in snacks, replacing fats from beef. The rubber industry had been the main source of export earnings, but global demand fell sharply after the First World War. The oil palm was thus a useful means of diversification. By 1938, more than 90,000 ha had been planted in Sumatra (Indonesia) and nearly 30,000 ha in Malaya (Lim, 1967; Hartley, 1988). This expansion and increasing demand for oil pushed investors and colonial governments at the time to develop high-yielding seeds in sufficient quantity to cover the high rate of expansion. The first research institution on plantation agriculture was established in Sumatra in 1916. The Algemeene Proefstation der AVROS (APA) was established to carry out research on rubber. The growing importance of the oil palm led to reorientation of its research objectives, and a major part of its goals included the development of breeding material for oil palm plantations.

The independent government of the Republic of Indonesia established the Lembaga Penelitian Marihat in 1963 with the aim of developing high-yielding seeds for the expanding oil palm industry. The APA changed its name to Research Institute of the Sumatra Planters Association (RISPA) in 1957 and later merged with the Marihat Research Institute in 1992 to become the Indonesian Oil Palm Research Institute (IOPRI). Today, all the major oil palm plantations in Indonesia own private research centers working on agronomy, breeding, and pests and diseases. The major research centers are owned by SOCFIN, PT SMART Tbk,[1] PT Astra Agro Lestari Tbk, PT Sampoerna Agro, and many others who have been breeding and producing high-yielding seed varieties for both internal and export markets.

MARDI included the development of oil palm as its main focus of research in 1969. Before this period, oil palm research was carried out separately by individual private plantations – such as United Plantations, Guthrie, Golden Hope, Harrison, and Crosfield – and some universities. Creation of PORIM in 1979 by Act 218 was aimed at centralizing the efforts of individual plantations in the field of research and development. PORIM existed alongside another government agency, the Palm Oil Registration and Licensing Agency (PORLA). In order to be more efficient, the government of Malaysia created a central body merging PORIM and PORLA known as MPOB, which carries out research, regulation, and promotion of the oil palm industry in Malaysia.

Modern breeding of oil palm in SEA is being supported by the use of MAS and genomics (Purba, 2010). Seven companies from Indonesia have formed a 15-member consortium known as the Oil Palm Genome Project (OPGP). The project aims to make use of information from the oil palm genome to assist in the development of conventional breeding. This will go

[1]PT stands for *Perseroan Terbatas* which means "limited company", while Tbk stands for *Terbuka*, the name given to companies listed on the Jakarta Stock Exchange.

a long way toward reducing the amount of time and money spent on varietal development and the amount of land needed for experimental plots.

Each era in the production of oil palms faced particular challenges, and it was a natural progression for the palm oil authorities to respond to those challenges at that particular time. Responding to challenges was done by the creation of research institutions, reorienting research objectives, or combining efforts to face the challenges. We shall present the specific challenges or constraints in the sections that follow and provide sustainable responses that were developed by authorities to those specific challenges. Our major focus is to explain how breeding has been used to respond to challenges without extreme alterations that would pose a threat to both the environment and human society.

Breeding to Improve Oil Yield

Yield improvement was the major aim of all breeding programs in the early days of oil palm expansion in SEA. Yield improvement was aimed at meeting the increasing demand for oil palm for both human consumption and chemical industries. The principal parameters measured in the breeding programs were yield and yield-related components such as average bunch weight (ABW), bunch number (BNO), mesocarp to fruit (MTF), oil to bunch (OTB), shell to fruit (STF), and kernel to fruit (KTF) ratios (Rafii et al., 2002). RRS had been adopted for oil palm varietal development. The first cycle of RRS was carried out between 1970 and 1980 and comprised the introduction of material from exploration and the consolidation of advanced breeding populations in Malaysia, Indonesia, and Papua New Guinea.

In Malaysia, the first cycle of breeding at Chemara and Banting was intensive, while in Indonesia, Marihat and SOCFIN were actively involved in the CIRAD breeding network. The second cycle of oil palm selection was carried out between 1981–1995. It was based on phenotypic evaluation of crosses from superior lines obtained in the first cycle as well as some material introduced from previous exploration. The third cycle was carried out between 1996 and 2003. Third-cycle material included second filial (F_2) generation Dura obtained from successive selfings with the aim of fixing and stabilizing Dura lines for use as female parents in seed production. These materials were planted in Malaysia and Indonesia (Breure, 2002). The fourth cycle ran between 2004 and 2009. It consisted of phenotypic evaluation of the Deli Dura families based on the results of progeny tests planted in Indonesia.

The four Dura palms planted in the Bogor Botanic Gardens in 1848 constituted the base of planting material that circulated in SEA plantations. The genetic base of this material was narrow, and this was worsened by subsequent selfpollination during the RRS breeding scheme employed for oil palms. Parent material used in the breeding programs could not produce the required heterosis in their progeny because of the narrow genetic distance between the parents. Approximately five generations of selection and breeding have been completed in oil palms since the establishment of the first plantations in the 1920s and 1930s (Hartley, 1988). Intensive breeding led to a fourfold increase in Malaysian palm oil production in the last 50 years, half of this being attributable to genetic improvement of material (Corley and Lee, 1992). Selection for this improvement has reduced the coefficient of variation for fruit bunch yield and its components (Thomas et al., 1969b; Rajanaidu and Rao, 1988; Lee et al., 1990).

In recent years, concern has grown that intensive breeding from such a narrow genetic base, particularly in the SEA industry, is leading to limited variation within elite material, potentially reducing the rate of future breeding progress (Rosenquist, 1986). It was noted that the limited genetic base of the oil palm breeding population in Malaysia led to low additive variances among parental material (Hardon, 1970; Hardon et al., 1976). An objective measure of genetic variation would allow selective use of wild and semiwild genotypes to widen the genetic base in a precise manner. As mentioned earlier, SEA breeders decided to undertake exploration in order to introduce new genes into their populations. This would permit them to develop high yielding planting material using the RRS breeding scheme. It is believed that geographical distance could be synonymous with genetic distance because variation and evolution are generally correlated with climatic factors, soil, and geographic range (Hamrick and Godt, 1989). Thus, SEA breeders set out to explore the hinterlands of Africa.

Several expeditions were therefore dispatched by SEA research institutions in their desire to broaden the genetic base of planting material. Cameroonian wild material is believed to contain some interesting genotypes given its agroclimatic diversity and long-standing history as wild genotype provider to many gene banks in the world. Blaak collected material from the Widikum and Ekona regions of Cameroon in 1967, which have been widely used to develop the majority of oil palm planting seeds today. In another collection expedition in Cameroon, Blaak and Sterling (1996) found that some oil palms were able to produce satisfactorily between 1000–2000 m above sea level. After 20 years of breeding using Dami Deli crossed with Cameroonian and Tanzanian selections of African oil palm, precocious-bearing, cold-tolerant oil palms were developed (Chapman et al., 2003). It is in this light that the IPOB or Dewan Minyak Sawit Indonesia (DMSI) explored Cameroon in 2008 looking for wild oil palm genotypes (Ajambang et al., 2012). The results from these expeditions were the breeding of high yielding planting material for SEA plantations.

Having obtained diverse genetic material to be used as parents (especially the Pisifera variety obtained in West and Central Africa), researchers in SEA adopted the North Caroline Mating Design (NCMD) to cross, test, and evaluate the performance of their material (Comstock and Robinson, 1949). They achieved an increase in yield for FFB of 183.63 kg palm^{-1} and an OTB of between 23.77% (Rafii et al., 2002) and 31.2%. FFB yields were between 22 and 30 t^{-1} ha^{-1} yr^{-1} while the OTB ratio ranged between 23.77–31.2% at the end of the second cycle of selection using Deli Dura × AVROS Pisifera progeny (Rajanaidu et al., 2000). Crosses showing parent material brought into SEA from Africa and Latin America are shown in Table 10.1. A list of the varieties developed from these expeditions is presented in Table 10.2.

Breeding to Limit Land Use and Deforestation

Demand for palm oil has been rising because of the increase in world consumption and the conversion of CPO into biofuel. Environmental concerns have also been on the rise against the expansion of oil palm plantations into forest areas. Breeders opted to develop high planting density varieties that would occupy smaller surface areas. They also developed abiotic stress–tolerant varieties that could give reasonable yields on suboptimum land. Compact oil palm varieties with high-density plantings of between 170 trees ha^{-1}and 200 trees ha^{-1} were originally developed in Costa Rica in South America. The second backcross (BC_2) of compact varieties produced an OTB of 29.8%. Although compact varieties were developed in South

TABLE 10.1 Crosses Showing Parent Material Brought into South East Asia (SEA) from Africa and Latin America

Type of cross/variety	Origin of ♀ parent	Origin of ♂ Parent
Dura Dumpy × Pisifera SP 540	Deli	Zaire
Dura Deli × Pisifera SP 540	Deli	Zaire
Dura Deli × Pisifera SP 540 TMA	Deli	Zaire
Dura Deli × Pisifera L2T, L7T, L9T, L14T	Deli	La Me
Dura Deli × Pisifera L239T	Deli	Yangambi
Dura × Pisifera Socfindo LM	Nigeria	La Me
Dura × Pisifera Socfindo YA	Deli	Yangambi
Dura × Pisifera BL	Deli	AVROS
Dura × Pisifera BL 2	Deli	AVROS
Dura × Pisifera BL 3	Deli	La Me
Dura × Pisifera BL 4	Deli	Ekona
Dura × Pisifera AA Topaz 1	Deli	Nigeria
Dura × Pisifera AA Topaz 2	Deli	Ghana
Dura × Pisifera AA Topaz 3	Deli	Ekona
Dura × Pisifera AA Topaz 4	Deli	Yangambi
Dura × Pisifera Sriwijaya 1	Deli	Nigeria
Dura × Pisifera Sriwijaya 2	Deli	Ghana
Dura × Pisifera Sriwijaya 3	Deli	Ekona
Dura × Pisifera Sriwijaya 4	Deli	AVROS
Dura × Pisifera Sriwijaya 5	Deli	Dami
Dura × Pisifera Sriwijaya 6	Deli	Yangambi
Dura × Pisifera TS-1	Deli	AVROS
Dura × Pisifera TS-2	Deli	Ekona
Dura × Pisifera TS-3	Deli	Ghana
Dura × Pisifera TN 1	Deli	AVROS
Evolution ASD	Deli	Composite
Dura × Pisifera Compact	Deli	S. America
Dura Compact × Pisifera	Deli/S. America	Nigeria
Standard ASD	Deli	La Me
Premium ASD	Deli	La Me
Special ASD	Tanzania	Ekona
ASD	Bamenda	Ekona

TABLE 10.2 Oil Palm Varieties Developed in SEA with Material Obtained from Wild Species in Cameroon

Country	Cross
Indonesia	Dura Deli × Pisifera **Ekona BL**
Indonesia	Dura Deli × Pisifera **Ekona 21**
Indonesia	Dura Deli × Pisifera **Ekona BSM**
Indonesia	Deli Dura × Pisifera **Ekona TS**
Indonesia	**Pure Lobe**
Indonesia	**Bamenda** × **Lobe**
Indonesia	**(Bamenda** × **Lobe)** × (Deli × Binga)
Indonesia	**Bamenda** × AVROS) × (Deli × AVROS)
Indonesia	**Lobe** × Deli
Indonesia	**Bamenda** × AVROS
Indonesia	**Bamenda** × **Ekona**
Indonesia	Tanzania × **Ekona**

Words in bold indicate parent material originating from Cameroon.

America, there has been material exchange between SEA and Latin America in recent years (Richardson, 1995; Escobar et al., 1996; Sterling and Alvarado, 2002). The IOPRI has also developed a high-density Yangambi variety that was planted at 181 trees ha^{-1}.

Environmental-based NGOs have been mounting pressure on governments to stop the sale of forests to oil palm investors. In response, breeders have been developing oil palm varieties that can grow in suboptimal environments and on degraded land. Cold-tolerant oil palm varieties developed in Latin America have been brought into Myanmar (Chapman et al., 2003).

Breeding to Improve Human Health

Health issues have been a major concern in modern palm oil consumption. Regulations in most countries are tough regarding the quality of oil supplied into their markets. Some companies in Indonesia have been breeding to improve the iodine value (IV) in oil palm varieties. The IV is an estimation of the unsaturated fatty acid content in vegetable oils. Different oil palm parental material has been identified with IV content between 53.8 and 56.1 (Jacquemard et al., 2010). The MPOB's PS4 and PS8 breeding populations contain high-carotene values greater than 3000 ppm and high vitamin–E values between 1300 ppm and 2500 ppm, respectively (Kushairi and Rajanaidu, 2009).

Breeding to Reduce Palm Oil Production Cost

The rise in palm oil production costs has led to the breeding of oil palm varieties designed to reduce such costs. Research carried out by Sumatra Bioscience in Indonesia has led to the identification of individuals possessing long bunch stalks, which can facilitate mechanical harvesting

(Priwiratama et al., 2010). The MPOB's PS10 population produces bunches with long stalks measuring between 20 cm and 30 cm (Kushairi and Rajanaidu, 2009). Research carried out in Indonesia and Malaysia has identified individual varieties with a greenish fruit color and late abscission fruit traits (Priwiratama et al., 2010). Fruit color (Sunilkumar and Sparjan Babu, 2013; Singh et al., 2014), low lipase activity (Morcillo et al., 2013), and late abscission will reduce fruit loss, which usually occurs between commercial maturity and bunch harvesting.

Sustainability Through Integrated Crop Protection

Sustainable crop production and organic food production are now worldwide concerns (Durand-Gasselin et al., 2011), and the oil palm sector is no exception. The aim is to produce crops in such a way that few or no artificial compounds such as fertilizers and pesticides are used. Many other rules aimed at maintaining a healthy soil and environment (social and physical) have to be followed (Tinker, 2000b, 2001; Corley and Tinker, 2003). The chance of achieving this goal became realistic with the creation of the RSPO, given its multistakeholder approach and detailed orientations spelled out in its principles and criteria to facilitate sustainable oil palm production. An earlier opportunity to breed oil palms for sustainable production revolved around the management of pests and diseases. Integrated pest management is a concept for sustainable production based on selective use of chemical products in the protection system, including the role of natural enemies found in the ecosystem. This concept coordinates the biology of the pest or pathogen, environmental information, and available technology to limit unacceptable levels of damage due to pests, by the most economic means, while presenting the least possible risk to people, goods, resources, and the environment (FAO, 1967; Kogan, 1998; Bajwa and Kogan, 2002).

Before the general use of organosynthetic chemical products in agriculture around the time of the Second World War, problems with pests had generally been managed using crop cultural practices and information on the life cycles and biology of pests. Phytopathology undertakes control against plant pests in association (or while integrating) various traditional control strategies such as crop rotation, the use of seeds free from pathogenic organisms, and the genetic resistance of hosts (Maho-Yalen et al., 2013). Many tropical crops like the oil palm have contributed to development of the current understanding of the principles of integrated pest management (IPM). For instance, it was recognized that insecticides were responsible for pest attacks in oil palms because of upsetting the ecological balance between pests and their natural enemies. In "Pests of oil palms in Malaysia and their control," the principles of what was then known as integrated pest control were enunciated, explaining the reasons for pest outbreaks, methods of monitoring pest populations, and ways of controlling pests without disrupting the natural balance in the agroecosystem (Corley and Tinker, 2003).

Besides pests, oil palms are susceptible to major diseases such as *Fusarium* wilt in several parts of Africa, dry basal rot (*Ceratocystis*) in Nigeria, *Ganoderma* in old and replanted areas in Asia, and fatal yellowing and sudden wither on new plantations in Latin America (Corley and Tinker, 2003). Sustainable management of the damage caused by diseases and pests associated with oil palms can be effected through integrated pest and disease management (IPDM), which usually includes good cultural practices, conservation, and exploitation of natural enemies in biological control (Genty et al., 1983), the use of pesticides when necessary, and plant quarantine measures (Kang and Zam, 1982; Corley and Tinker, 2003).

Breeders are currently working on developing *Ganoderma*-tolerant varieties for oil palm plantations in SEA by eliminating susceptibility and selection. There have been successful examples of breeding and selecting for resistant lines. Resistance breeding has been undertaken with selection based primarily on performance in disease-infested fields, backed up by nursery tests using parents found to consistently transmit resistance or susceptibility to their offspring in field trials (Porter, 1989; Corley and Tinker, 2003; Diabate et al., 2010).

In the case of crown disease, there is a need for Pisifera to be shown to be homozygous for absence of the disease (De Berchoux and Gascon, 1963). It is possible to detect a palm with a susceptible genotype by test-crossing with a palm that is known not to have the inhibiting gene (Blaak, 1970b; Corley and Tinker, 2003). Selection of resistant lines has also been widely applied to control *Fusarium* wilt using a resistance-breeding program involving nursery and rachis tests (Prendergast, 1963; Corley and Tinker, 2003; Diabate et al., 2010). The standard nursery test requires large numbers of plants and, hence, large amounts of nursery space. A method has been proposed that uses clonal plants, which reduces the numbers of plants required (12 per clone) by using inocula with a known, constant concentration of fungal spores, and classifies in great detail the severity of symptoms in each plant (Mepsted et al., 1988; Flood et al., 1989; Prendergast, 1963; Renard et al., 1972; Corley and Tinker, 2003; de Franqueville, 1984). The standard nursery test takes several months. Alternatively, it can be replaced by the rachis test, which involves inoculating 2.5 cm sections from near the tip of the rachis and immersing them in a suspension of *Fusarium* spores under mild vacuum. The results are available within 8 days (Mepsted et al., 1995c; Corley and Tinker, 2003). Contrarily to the nursery test, which is based on the average performance of a group of palms, the rachis test can be applied to individual palms, but only works well on palms that are in good health. In addition, studies have been undertaken on natural enemies, sex ratio, and spatiotemporal distribution of *Coelaenomenodera minuta* Uhmann (Coleoptera: Chrysomelidae), a serious insect pest of oil palms in Cameroon (Mondjeli et al., 2013), with the view of developing an IPM program against the pest.

In sum, the production of oil palms has to adopt accurate IPDM strategies for sustainable crop protection and improvement. IPDM aims to reduce reliance on a single control method; it does so by applying methods that are the least harmful to the environment and most specific to particular pests and diseases. Host plant resistance is the preferred means of minimizing the damage caused by pests or diseases of oil palms. Thus, IPDM can be an economically sound approach that produces quality produce at minimum cost in a sustainable way. Reevaluation of the strategy at the end of every season is necessary using an associate monitoring system for accurate and sustainable oil palm pest and disease control.

CAMEROON'S EXPERIENCE OF OIL PALM BREEDING FOR DISEASE TOLERANCE

In Cameroon where palm oil provides about 90% of edible oil needs, the government has set up a development plan aimed at increasing current palm oil production through massive plantings. However, these efforts may seriously be compromised by *Fusarium* wilt disease, which is the main threat to the development of palm plantations in Africa (Ntsomboh-Ntsefong et al., 2012; Ngando-Ebongue et al., 2013). *Fusarium* wilt incidences as high as 40% have been recorded in field-planted palms at the Ndian Estate of PAMOL Plantations Plc, Cameroon (Hornus et al., 1987). The industrial plots of the Cameroon Development Corporation (CDC) have also suffered severe attacks.

Significant efforts have been made toward developing *Fusarium* wilt–tolerant elite progeny in the breeding programs of PAMOL Plantations Plc and IRAD/Centre Spécialisé de Recherche sur le Palmier à Huile (CEREPAH) at La Dibamba, the two seed producers in Cameroon. La Dibamba is part of the French IRHO/CIRAD network that works alongside other research stations in Africa such as La Me in Ivory Coast or Pobè in Benin. The strategy adopted in the fight against oil palm vascular wilt has been to breed for the multiple defense genes involved in partial resistance rather than breeding for total resistance, as the latter can also be broken down or bypassed easily by the pathogen (Ngando-Ebongue et al., 2013).

Vascular wilt disease was first observed in 1940 at the Cowan Estate in Nigeria (Wardlaw, 1946a, b). When it appeared at PAMOL's Ndian Estate in Cameroon, it then became imperative to undertake breeding and selection program designed to eventually produce oil palm varieties resistant to vascular wilt. With the need for large quantities of D × P seeds for new plantings at that time coupled with very limited outside sources of supply, it became necessary to produce improved wilt-resistant seeds at source. A research program was thus undertaken by PAMOL in 1948 aimed at selecting and breeding for improved yield and testing for *Fusarium* wilt resistance *via* FIPS. The work started in 1949 with the development of foundation stock used to produce the first breeding cycle in 1954–1955. Selections from F_1 plantings were used to create F_2 plantings from 1966 to 1971 along with some introductions from diverse origins (Rosenquist et al., 1990). Most of the progeny were also tested at the nursery for resistance to *Fusarium* vascular wilt (Ngando-Ebongue et al., 2013). The breeding of oil palms is now in the third generation at PAMOL. Phase 1 of the third-generation breeding cycle included a combined breeding program for wilt resistance which began in 1985.

Meanwhile, seed production launched at La Dibamba in the mid-1980s involved reproduction of the best first and second cycle crosses up to early 2000. Since 2005, only crosses from the second selection cycle have been reproduced for commercial seed production purposes. The seeds are made up exclusively from crosses between Deli × La Me BPROs. At CEREPAH, more than 80% of the *Fusarium* wilt–tolerant seeds are produced by combining progeny from (DA 115 D self) × (LM 2T self), as both parents are renowned for their high tolerance to *Fusarium* wilt. Further testing of *Fusarium* wilt tolerance is currently ongoing using improved material from diverse origins (Widikum) and material from recent prospecting missions with the objective of being included in third selection cycle planting material expected to exhibit improved yield and wilt tolerance by 2020. On the basis of number of parent palms available per field, the current production capacity of CEREPAH is estimated at 9 million seeds year^{-1}, two thirds of which are tolerant to *Fusarium* wilt (Ngando-Ebongue et al., 2013).

SMALLHOLDERS AND SUSTAINABLE OIL PALM PRODUCTION

Given that sustainability entails economic viability, social welfare, and environmental friendliness, the involvement and wellbeing of stakeholders in the oil palm industry must be taken into account in breeding programs designed for sustainable production. The oil palm industry can be classified into three main subsectors: the agroindustrial sector, the organized smallholder sector supported by agroindustries, and the heterogeneous informal smallholder sector. According to the RSPO, smallholders are deemed family-based enterprises producing palm oil from fewer than 50 ha of land. Smallholders, in particular, see oil palm as worth the risk, as a means of social mobility into the middle class, as reflected in their increased consumer capacity (Beggs and Moore, 2013). They are not minor players in terms of sustainability and credibility in the oil palm industry. In Indonesia and Malaysia, two countries responsible for over 80% of world oil palm production, smallholders account for 35–40% of total area of planted oil palm and as much as 33% of the output. In West African countries, which produce mainly for domestic and regional markets, smallholders produce up to 90% of the annual harvest (Vermeulen and Goad, 2006).

A major opportunity to meet the rising demand for palm oil in a sustainable manner is through expansion and improvement of smallholder production. Breeders can work out solutions which will facilitate the social insertion of smallholders into the supply chain (Durand-Gasselin et al., 2011). Oil palm production by smallholders, despite the constraints they face, has the potential to secure mutually beneficial outcomes with both large and small producers and processors. They can enhance social and environmental sustainability at the landscape scale, ease land disputes between smallholders and large plantations, and promote credibility among consumers. The following new trends offer greater opportunities for smallholders to secure their place in modern markets with sustainably produced oil (Maria and Helena, 2007):

- *RSPO stewardship*. The RSPO includes a dedicated Smallholder Task Force to work toward inclusion of smallholders in RSPO principles and criteria.
- *Biofuel opportunity*. Accelerating demand for biofuel crops is expected to keep palm oil prices buoyant and perhaps offer new technologies and standard-based niche markets specifically for smallholders.
- *Technology*. Smallholders with access to the latest technology have already demonstrated their ability to produce sustainably and as efficiently as large-scale plantations.
- *Organized representation*. Local and national representative bodies provide a voice for smallholders in business negotiations and wider policy dialog.

CONCLUSIONS AND FUTURE CHALLENGES

Even though the oil palm sector continues to grow, it is viewed as a driver of economic activity that may or may not be sustainable in the long term especially regarding climate change and high demands for CO_2 reduction (Rajanaidu et al., 2011a). In addition to decreasing physical burdens and increasing economic gains, some growers feel that palm plantations are environmentally sustainable (Beggs and Moore, 2013). Collaboration between governmental agencies, grower coops and associations, and local NGOs could generate norms and guidelines for best practices in sustainable oil palm production that take into account the strengths, constraints, and opportunities of all stakeholders. The RSPO initiative is an outstanding opportunity that could be optimized by having a more stringent certification process to incentivize plantation management for conservation of natural resources and biodiversity maintenance, as well as high yield. This process could link growers to more lucrative markets for sustainable foods and cosmetics (Beggs and Moore, 2013).

Oil palm breeders have effectively adopted traditional and modern techniques during the past 50 years to increase oil yield and quality in order to meet growing demand. Many challenges have emerged along the way, which have received attention from research scientists and authorities. These challenges include complaints related to health issues, climate change and instability, environmental protection, and limitation of cultivation area. Although a lot has been done, there still remain some constraints for which breeders will need to use all the science and art they possess to come up with suitable varieties with enhanced yield, specific ecological adaptation, low growth rate, long peduncle, high unsaturated fatty acid content, high beta carotene and iodine value, and resistance to disease (Hartley, 1988; Simmonds, 1993). Breeders have yet to develop varieties tolerant to BSR caused by *Ganoderma boninense*. They need to develop varieties that are adapted to suboptimum environments – such as those tolerant to drought and salinity – if they intend to continue oil palm expansion in today's world. However, the oil palm industry will continue to grow for the following reasons:

- growing world population and still low per capita consumption of oil;
- high productivity of the crop and low cost of production (Corley, 2009);
- versatility of palm oil and palm kernel oil in oleochemical and industrial uses;
- advantage over mineral oil in an increasingly environmentally conscious world as a result of being a renewable resource.

Major stakes made in the crop by many emerging economies demand continuing funding for research and development. Taking all considerations into account, the oil palm sector could be improved by reinforcing strategic breeding approaches to enhance growth and sustainable production.

References

Abdullah, N., Sulaiman, F., 2013. The oil palm wastes in Malaysia (Chapter 3). In: Biomass Now – Sustainable Growth, Use. p. 26. http://dx.doi.org/10.5772/55302

Aberlenc-Bertossi, F., Noirot, M., Duval, Y., 1999. BA enhances the germination of oil palm somatic embryos derived from embryogenic suspension cultures. Plant Cell Tiss. Org. Cult. 56, 53–57.

Aderungboye, F.O., 1982. Significance of vascular wilt in oil palm plantations in Nigeria. In: Pushparajah, Chew, P.S. (Eds.), The Oil Palm and Agriculture in the Eighties. Incorp. Soc Planters, Kuala Lumpur, pp. 475–484, (2).

Adin, A., Weber, J.C., Montes, C.S., Vidaurre, H., Vosman, B., Smulders, M.J.M., 2004. Genetic differentiation and trade among populations of peach palm (*Bactris gasipaes* Kunth) in the Peruvian Amazon – implications for genetic resource management. Theor. Appl. Genet. 108, 1564–1573.

Adlin, L., Kiswito, 1975. New perspectives in oil palm breeding in Indonesia. Symposium on South East Asian Plant Genetic Resources. Bogor, Indonesia, Mar. 20–22, 1975.

Ahmad, T.H., Zamzuri, I., Ooi, S.E., Samsul, K.R., Chan, P.L., Rohani, O., Meilina, O.A., 2011. Forging ahead with clones. Mohd, B.W., Choo, Y.M., Chan, K.W. (Eds.), Further Advances in Oil Palm Research (2000–2010), vol. 1, Malaysian Palm Oil Board, Kuala Lumpur, Malaysia, pp. 102–140.

Ajambang, W., Sudarsono, Asmono, D., Toruan, N.M., 2012. Microsatellite Markers reveal Cameroon's wild oil palm population as a possible solution to broaden the genetic base in the Indonesia-Malaysia oil palm breeding programmes. Afr. J. Biotech. 11 (69), 13244–13249.

Allard, R.W., 1960. Principles of Plant Breeding. John Wiley Sons Inc, New York, p. 485.

Al-Shanfari, A.B., Abdullah, S.N.A., Saud, H.M., Omidvar, V., Napis, S., 2011. Differential gene expression identified by suppression subtractive hybridization during late ripening of fruit in oil palm (*Elaeis guineensis* Jacq.). Plant Mol. Biol. Rep. 30, 768–779.

Arasu, N.T., 1970. A note on the germination of *pisifera* (shellless) oil palm seeds. Malay. Agric. J. 47, 524–527.

Ariffin, D., Idris, A.S., Singh, G., 2000. Status of *Ganoderma* in oil palm. In: Flood, J., Bridge, P.D., Holderness, M. (Eds.), Diseases of Perennial Crops, CAB., International, 2000., pp. 49–68.

Ascenso, J.C., 1966. Outlines of the oil palm breeding programme in Portuguese Guinea. Euphytica 15, 268–277.

Ataga, C.D., 1993. Genotype-environment interaction and stability analysis for bunch yield in the oil palm (*Elaeis guineensis* Jacq.). Oléagineux 48 (2), 59–64.

Ataga, C.D., Fatokun, C.A., 1989. Multivariate studies of the Nigerian Oil Palm (*Elaeis guineensis* Jacq.) germplasm collection. Paper presented at International Conference Palms and Palm Products, November 21–25. Nigerian Institute for Oil Palm Research, Benin City.

Bajwa, W.I., Kogan, M., 2002. Compendium of IPM Definitions (CID). What is IPM and how is it defined in the Worldwide Literature? 19 p.

Bakoumé, C., Aziah, M.Y., Praveena, T., Jangi, M.S., Mohaimi, M., Khairudin, H., Harikrishna, K., Suzaini, Y., 2009a. Identification of different oil palm clones using microsatellite markers. In: Proccedings of PIPOC 2009 International Palm Oil Congress Palm Oil – Balancing Ecologics1 with Economics, vol. 2, 650–660.

Bakoumé, C., Wickneswari, R., Rajanaidu, N., Kushairi, A., Billotte, N., 2009b. Screening Natural Oil Palm (*Elaeis guineensis* Jacq.) Populations Using SSR Markers. Paper Six Presented at the International Seminar on Oil Palm Genomics and its Applications to Oil Palm Breeding. Organized by The International Society for Oil Palm Breeders. Kuala Lumpur Convention Centre, Kuala Lumpur, November 4–5, 2009, 10 p.

Barcelos, E., Miranda Santos, M.d., Elizabeth, M., Varoncellos, C., 1986. Phenotypic variation in natural populations of Caiaue, Elaeis oleifera, H.B.K. Cortes, in the Brazilian Amazon. In: Proceedings of International Workshop Oil palm germplasm and utilization, pp. 102–116, Palm Oil Res. Inst. Malaysia, Kuala Lumpur.

Basri, M.W., 2003. Prioritizing future products for the oil palm industry. Brainstorming session for strategic plan to hasten breeding for 10 priority traits, March 31, 2003. Renaissance Palm Grden Hotel, IOI Resort, Putrajaya.

Beggs, E., Moore, E., 2013. The Social Landscape of African Oil Palm Production in the Osa and Golfito Region, Costa Rica. INOGO, Stanford woods institute for the environment, Stanford University, San José, Costa Rica, 32 p.

Beirnaert, A., Vanderweyen, R., 1941. Contribution a l'étude génétique et biométrique des variétés d'*Elaeis guineensis* Jacquin. Publ. Inst. Nat. Etude Agron. Congo Belge Ser. Sci. 27, 1–101.

Beyegue, H.D., Kushairi, A., Rajanaidu, N., Jalani, B.S., 2011. Assessment of bunch yield and its components for Cameroon-based *Dura* × *Pisifera* oil palm (*Elaeis guineensis* Jacq.) populations. International seminar on breeding for sustainablility in oil palm. November 18, 2011, KLCC Kuala Lumpur, Malaysia. pp. 144–154.

Billotte, N., Risterucci, A.M., Barcelos, E., Noyer, J.L., Amblard, P., Baurens, F.C., 2001. Development, characterization, and across-taxa utility of oil palm (*Elaeis guineensis* Jacq.) microsatellite markers. Genome 44, 413–425.

Billotte, N., Jourjon, M.F., Marseillac, N., Berger, A., Flori, A., Asmady, H., Adon, B., Singh, R., Nouy, B., Potier, F., Cheah, S.C., Rohde, W., Ritter, E., Courtois, B., Charrier, A., Mangin, B., 2010. QTL detection by multi-parent linkage mapping in oil palm (*Elaeis guineensis* Jacq.). Theor. Appl. Genet. 120, 1673–1687.

Biodiversity International, 2010. Sustainable agriculture and the sustainable use of agricultural biodiversity: concepts, trends and challenges. An information note submitted by Biodiversity International for the Fourteenth Meeting of the Subsidiary Body on Scientific, Technical and Technological Advice of the Convention on Biological Diversity, May 10–21, 2010, Nairobi, Kenya.

Blaak, G., 1970b. Epistasis for crown disease in the oil palm (*Elaeis guineensis* Jacq.). Euphytica 19, 22–24.

Blaak, G., Sterling, F., 1996. The prospects of extending oil palm cultivation to higher elevations through using cold-tolerant planting material72The Planter, Kuala Lumpur, pp. 645–652.

Blaak, G., Sparnaaij, L.D., Menendez, T., 1963. Breeding and inheritance in oil palm, Part II. Methods of bunch analysis. J. W. Afr. Int. Oil Palm Res. 4, 146.

Black, A.C., 2006. Dictionary of Agriculture, third ed. A&C Black, London, 289 p.

Bourgis, F., Kilaru, A., Cao, X., Ngando-Ebongue, G.-F., Drira, N., Ohlrogge, J.B., Arondel, V., 2011. Comparative transcriptome and metabolite analysis of oil palm and date palm mesocarp that differ dramatically in carbon partitioning. Proc. Natl. Acad. Sci. USA 108, 12527–12532.

Brett, W., 2009. Dietary fatty acids. Am. Fam. Physician 80 (4), 345–350.

Breure, C.J., 1985. Relevant factors associated with crown expansion in oil palm (*Elaeis guineensis* Jacq.). Euphytica 34, 161–175.

Breure, C.J., 1986. Parent selection for yield and bunch index in the oil palm in West New Britain. Euphytica 35, 65–72.

Breure, C.J., 2002. The search for yield of oil palm: basic principles. In: Fairhurst, T.H., Hardter, R. (Eds.), The Oil Palm: Management for Large, Sustainable, Yields. Potash & Phosphate Institute of Canada (ESEAP), Singapore, 382 p.

Breure, C.J., Corley, R.H.V., 1983. Selection of oil palms for high density planting. Euphytica 32, 177–186.

Brondani, R.P.V., Brondani, C., Tarchini, R., Grattapaglia, D., 1998. Development, characterization and mapping of microsatellite markers in *Eucalyptus grandis* and *E. urophylla*. Theor. Appl. Genet. 97, 816–827.

Budidarsono, S., Dewi, S., Sofiyuddin, M., Rahmanulloh, A., 2012. Socioeconomic Impact Assessment of Palm Oil Production. Technical Brief No. 27: palm oil series. Bogor, Indonesia. World Agroforestry Centre – (ICRAF), SEA Regional Office. 4 p.

Bürün, B., Çoban, P.E., 2002. Embryo culture in Barley (*Hordeum vulgare* L.). Turk. J. Biol. 26, 175–180.

Castilho, A., Vershinin, A., Heslop-Harrison, J.S., 2000. Repetitive DNA and the chromosomes in the genome of oil palm (*Elaeis guineensis*). Ann. Bot. 85, 837–844.

Chandusingh, P.R., Singh, N.K., Prabhu, K.V., Vinod, K.K., Singh, A.K., 2013. Molecular mapping of quantitative trait loci for grain chalkiness in rice (*Oryza sativa* L.). Indian J. Genet. 73 (3), 244–251.

Chang, S.H., Yang, J.C., 1996. Enhancement of plant formation from embryo cultures of *Taxus mairei* using suitable culture medium and PVP. Bot. Bull. Acad. Sin. 37, 35–40.

Chapman, K.R., Escobar, R., Griffee, P., 2003. Cold tolerant or altitude adapted oil palm hybrid development initiatives in the asia/pacific region. AU. J. T. 6 (3), 134–138.

Cheah, S.C., Wooi, K.C., 1995. Application of molecular marker techniques in oil palm tissue culture. In: Rao, V., Henson, I.E., Rajanaidu, N. (Eds.), Recent Developments in Oil Palm Tissue Culture and Biotechnology. Palm Oil Research Institute, Malaysia, Kuala Lumpur, pp. 163–170.

Chin, T.L., Zaharah, A.R., Cheuk, W.C., Mohamed, M.H., Mohd, S.N., Choon, C.T., Mum, K.W., 2011. Variation of leaf nutrient concentrations in oil palm genotypes and their implication on oil yield. Souvenir programme – International seminar on breeding for sustainability in oil palm. November 18, 2011, KLCC Kuala Lumpur, Malaysia, ISOPB/MPOB. pp. 62–75.

Chinchilla, C.M., Richardson, D.L., 1990. Polinizacion en palma aceitera (*Elaeis guineensis* Jacq.) en Centroamerica 1. Poblacion de insectos y conformacion de racimos. Turrialba 40, 452–460.

Choo, Y.M., Cheah, K.Y., 2000. Biofuel. Basiron, Y., Jalani, B.S., Chan, K.W. (Eds.), Advances in Oil Palm Research, Vol. 2, Malaysian Palm Oil Board, Kuala Lumpur, pp. 1293–1345.

Chua, K.L., Singh, R., Cheah, S.C., 2001. Construction of oil palm (*Elaeis guineensis* Jacq.) linkage maps using AFLP markers. In: Proceedingsof 2001 International Palm Oil Congress – Agriculture. Malaysian Palm Oil Board, Kuala Lumpur. pp. 461–465.

Cleber da Silva, P., Lemes da Costa, R.M., Bueno de Moraes, C., Cristiane de Pieri, Evandro, V.T., Furtado, E.L., Mori, E.S., 2014. Genetic variability in progenies of *Eucalyptus dunnii* Maiden for resistance to *Puccinia psidii*. Crop Breed. Appl. Biotechnol. 14, 187–193.

Cochard, B., Noiret, J.M., Baudouin, L., Flori, A., Amblard, P., 1993. Second cycle reciprocal selection in oil palm, results of Deli x La Me hybrid tests. Oleaginaux 48, 441–452.

Cochard, B., Durand-Gasselin, T., Amblard, P., Konan, E., Gogor, S., 1999. Performance of adult oil palm clones. Preprints, 1999 PORIM International Palm Oil Conference. Palm Oil Research Institute, Malaysia, Kuala Lumpur. pp. 12–22.

Cochard, B., Durand-Gasselin, T., Adon, B., 2000. Oil palm genetic resources in the Côte d'Ivoire – composition, assessment and use. Paper presented at International Symposium on Oil Palm Genetic Resources and Utilization. June 8–10, Malaysian Palm Oil Board, Kuala Lumpur.

Cochard, B., Amblard, P., Durand-Gasselin, T., 2005. Oil palm genetic improvement and sustainable development. OCL Oléagineux corps gras lipids 12 (2), 141–147.

Codex Alimentarius/FAO/OMS, 2005. Normes alimentaires pour huiles et graisses. Commission du Codex Alimentarius/FAO/OMS, CODEX-STAN 210, FAO/OMS.

Comstock, R.E., Robinson, H.F., 1949. The components of genetic variance in populations of biparental progenies and their use in estimating the average degree of dominance. Biometrics 4, 254–266.

Comstock, R.E., Robinson, H.F., Harvey, P.H., 1949. A breeding procedure designed to make maximum use of both general and specific combining ability. Agron. J. 41, 360–367.

Corley, R.H.V., 2006. Progress with non-shedding oil palms. Planter 82, 799–807.

Corley, R.H.V., 2009. How much palm oil do we need? Environ. Sci. Policy 12, 134–139.

Corley, R.H.V., Lee, C.H., 1992. The physiological basis for genetic improvement of oil palm in Malaysia. Euphytica 60, 179–184.

Corley, R.H.V., Tinker, P.B., 2003. The Oil Palm, fourth ed. Blackwell publishing, Oxford, Berlin, Germany, 562 p.

Corley, R.H.V., Lee, C.H., Wong, C.Y., 1986. Abnormal flower development in oil palm clones. Planters 62, 233–240.

Corley, R.H.V., Tan, Y.P., Timti, I.N., de Greef, W., 1993. Yield of oil palm progenies in Zaire, Cameroun and Malaysia. In: Proceedings of the 1991 International Society of Oil Palm Breeders Workshop Genotype-Environment Interaction Studies in Perennial Tree Crops. Palm Oil Research Institute Malaysia, Kuala Lumpur. pp. 46–54.

Corley, R.H.V., Boonrak, T., Donough, C.R., Nelson, S., Dumortier, F., Soebagiyo, F.X., Vallejo, G., 1995b. Yield of oil palm clones in different environments. In: Rao, V., Henson, I.E., Rajanaidu, N. (Eds.), Recent Developments in Oil Palm Tissue Culture and Biotechnology. Palm Oil Research Institute, Malaysia, Kuala Lumpur, pp. 145–157.

Das, S., Jha, T.B., Jha, S., 1999. Factors affecting *in vitro* development of embryonic axes of cashewnut. Sci. Hortic. 18 (1-2), 135–144.

Davidson, L., 1993. Management for efficient, cost-effective and productive oil palm plantations. In: Y. Basiron et al., (Ed.). Proceedings of 1991 PORIM International Palm Oil Conference – Agriculture, pp. 153–167, Palm Oil Research Institute, Malaysia, Kuala Lumpur.

De Berchoux, C., Gascon, J.P., 1963. L'arcure défoliée du palmier a huile – éléments pour l'obtention de lignées résistantes. Oléagineux 18, 713–715.

de Blank, S., 1952. A reconnaissance of the American oil palm. Trop. Agric. Trinidad 29, 1–14.

de Franqueville, H., 1984. Vascular wilt of the oil palm: relationship between nursery and field resistance. Oléagineux 39, 513–518.

de Franqueville, H., Diabate, S., 2004. Status on the oil palm vascular wilt. In: MPOB Proceedings of the International Conference on Pests, Diseases of Importance to the Oil Palm, Industry, Kuala Lumpur.

de Franqueville, H., Renard, J.L., 1990. Improvement of oil palm vascular wilt tolerance. Results and the development of the disease at R. Michaux plantation. Oleagineux 45, 399–405.

de Franqueville, H., Asmady, H., Jacquemard, J.C., Hayun Zaelani, H., Durand, G.T., 2001. Indications on sources of oil palm genetic resistance and susceptibility to Ganoderma sp., the cause of basal stem rot. In: Proceedings of the 2001 PIPOC. MPOB, Kuala Lumpur. pp. 420–431.

de Franqueville, H., Durand, G.T., Cochard, B., Nouy, B., Breton, F., Dossa, J., 2011. Integrated approach for management of Fusarium disease on the oil palm. In: Proceedings of the Third MPOB-IOPRI International Seminar on Integrated Oil Palm Pest and Diseases Management Ed. MPOB. Kuala Lumpur. pp. 52–62.

de Touchet, B., Duval, Y., Pannetier, C., 1991. Plant regeneration from embryogenic suspension cultures of oil palm (*Elaeis guineensis* Jacq). Plant Cell Rep. 10, 529–532.

Debergh, P., Maene, L.J., 1981. Ascheme for commercial propagation of ornamental plants by tissue culture. Sci. Hortic. 14, 335–345.

Demol, J., 2002. Amélioration des plantes. Application aux principales espèces cultivées en régions tropicales. Les presses agronomiques de Gembloux, Gembloux (Belgique).

Diabate, S., Ledeme, P., de Franqueville, H., Kouame, B., 1992. Effet d'une souche avirulente de Fusarium oxysporum sur l'expression de la fusariose et sur les propriétés fungitoxiques des extraits racinaires de palmier a huile. In: Proceedings of Conference, 17–22 February, Dakar, Senegal, Interactions plantes-microorganismes, pp. 307–320.

Diabate, S., Traoré, A., Boaké, K., 2010. Evaluation of the performance of tolerant crosses of oil palm selected in prenursery and and replanted on wilt disease area. Agric. Biol. J. N. Am. 1 (6), 1273–1277.

Diana, A., Carmenza, M., Leonardo, R., Hernán, R., 2012. Genetic similarity among commercial oil palm materials based on microsatellite markers. Agronomía Colombiana 30 (2), 188–195.

Donough, C.R., Lee, C.H., 1995. Longer term results from clone trials at PAMOL plantations and golden hope plantations. In: Rao, V., Henson, I.E., Rajanaidu, N. (Eds.), Recent Developments in Oil Palm Tissue Culture and Biotechnology. Palm Oil Research Institute, Malaysia, Kuala Lumpur, pp. 116–133.

Dumet, D., Engelmann, F., Chabrillange, N., Richaud, F., Beule, T., Durand-Gasselin, T., Duval, Y., 1993a. Development of cryopreservation for oil palm somatic embryos using an improved process. Oléagineux 48, 274.

Dumet, D., Engelmann, F., Chabrillange, N., Duval, Y., 1994a. Effect of desiccation and storage temperature on the conservation of cultures of oil palm somatic embryos. Cryo-Letters 15, 85–90.

Durand-Gasselin, T., Guen, V.L., Konan, E., Duval, Y., 1990. Oil palm (*Elaeis guineensis* Jacq.) plantations in Côte d'Ivoire obtained through in vitro culture: first results. Oleagineux 45, 1–11.

Durand-Gasselin, T., Duval, Y., Baudouin, L., Maheran, A. B., Konan K., Noiret, J. M., 1993. Description and degree of the mantled flowering abnormality in oil palm (*Elaeis guineensis* Jacq.) clones produced using the Orstorm-CIRAD procedures. In: Rao, V., Henson, I.E., Rajanaidu, N. (Eds.), Recent developments in oil palm tissue culture, biotechnology. ISOPB Workshop. Kuala Lumpur, Malaysia: Palm Oil Research Institute of Malaysia, 48–61.

Durand-Gasselin, T., Duval, Y., Baudouin, L., Maheran, A.B., Konan, K., Noiret, J.M., 1995. Description and degree of the mantled flowering abnormality in oil palm (*Elaeis guineensis* Jacq.) clones produced using the Orstom-CIRAD procedure. In: Rao, V., Henson, I.E., Rajanaidu, N. (Eds.), Recent Developments in Oil Palm Tissue Culture and Biotechnology. Palm Oil Research Institute, Malaysia, Kuala Lumpur, pp. 48–63.

Durand-Gasselin, T., Cochard, B., Amblard, P., de Franqueville, H., 2002. Un regard sur quarante ans d'amélioration génétique du palmier à huile (*Elaeis guineensis*) et son impact sur la filière. Le sélectionneur Français, A. S. F. 1962–2002, 40 ans de sélection végétale, ISSN 0374-1621, 2002 pp. 133–148.

Durand-Gasselin, T., Asmady, H., Flori, A., Jacquemard, J.C., Hayun, Z., Breton, F., de Franqueville, H., 2005. Possible sources of genetic resistance in oil palm (*Elaeis guineensis* Jacq.) to basal stem rot caused by Ganoderma boninense – prospects for future breeding. Mycopathologia 159, 93–100.

Durand-Gasselin, T., Blangy, L., Picasso, C., de Franqueville, H., Breton, F., Amblard, P., Cochard, B., Louise, C., Nouy, B., 2010. Sélection du palmier à huile pour une huile de palme durable et responsabilité sociale. Oléagineux, Corps Gras, Lipides 17 (6), 385–392.

Durand-Gasselin, T., de Franqueville, H., Breton, F., Amblard, P., Jacquemard, J.C., Syaputra, I., Cochard, B., Louise, C., Nouy, B., 2011. Breeding for sustainable palm oil. International Seminar on Breeding for Sustainablility in Oil Palm. November 18, 2011, KLCC Kuala Lumpur, Malaysia. pp. 155–170.

Dussert, S., Guerin, C., Andersson, M., Joët, T., Tranbarger, T.J., Pizot, M., Gautier Sarah, Omore, A., Durand-Gasselin, T., Morcillo, F., 2013. Comparative transcriptome analysis of three oil palm fruit and seed tissues that differ in oil content and fatty acid composition. Plant Physiol. 162, 1337–1358.

Duval, Y., Amblard, P., Rival, A., Konan, E., Gogor, S., Durand-Gasselin, T., 1997. Progress in oil palm tissue culture and clonal performance in Indonesia and the Cote d'Ivoire. In: Pushparajah, E. (Ed.), Plantation Management for the 21st Century. Incorporated Society of Planters, Kuala Lumpur, pp. 291–307.

Dzido, J.L., Genty, P., Ollagnier, M., 1978. Les principales maladies du palmier à huile en Equateur. Oléagineux 33, 55–63.

Edwin, A., Navia, R., Hernan, M.R., Edwin, F.R., 2011. Response of six Malaysian oil palm planting materials to bud rot disease in the Eastern region of Colombia. Souvenir programme-International seminar on breeding for sustainability in oil palm. November 18, 2011, KLCC Kuala Lumpur, Malaysia, ISOPB/MPOB. p. 76, Summary.

Ekins, P., Dresener, S., Dahlström, K., 2008. The four capital methods of sustainability development evaluation. Policy Studies Institute, London, UK.

Engelmann, F., 1990. Utilisation d'atmosphères a teneur en oxygène réduite pour la conservation de cultures d'embryons somatiques de palmier à huile (*Elaeis guineensis* Jacq.). C. R. Acad. Sci. Ser. III 310, 679–684.

Engelmann, F., Duval, Y., 1986. Cryopreservation of oil palm somatic embryos (*Elaeis guineensis* Jacq.): results and application prospects. Oléagineux 41, 169–174.

Enongene-Ndobe, D., 2014. Sustainable biofuels in Cameroon. The influence of the European Union Renewable Energy Directive (EU-RED) on the production and use of biofuels in Cameroon. Master Thesis, Department of Human Geography, Lund University, Sweden, 68 p.

Escobar, R., 1982. Preliminary results of the collection and evaluation of the American oil palm *E. oleifera* (HBK, Cortes) in Costa Rica. Pushparajah, E., Chew, P.S. (Eds.), The Oil Palm in Agriculture in the Eighties, vol. 1, Incorporated Society of Planters, Kuala Lumpur, pp. 79–93.

Escobar, R., Sterling, F., Peralta, F., 1996. Oil palm planting materials by ASD de Costa Rica. ASD Oil Palm Papers 14, 1–12.

Eujayl, I., Sorrels, M.E., Baum, M., Wolters, P., Powell, W., 2001. Assessment of genotypic variation among cultivated durum wheat based on EST-SSRs and genomic SSRs. Euphytica 119, 39–43.

Fadila, A.M., Norziha, A., Mohd Din, A., Rajanaidu, N., 2011. Evaluation of bunch index in the MPOB germplasm collections. Souvenir programme-International seminar on breeding for sustainability in oil palm. November 18, 2011, KLCC Kuala Lumpur, Malaysia, ISOPB/MPOB. pp. 52–61.

Fairhurst, T., Hardter, R., 2003. From area expansion to yield intensification. In: Fairhurst, T.H., Hardter, R. (Eds.), The Oil Palm: Management for Large and Sustainable Yields. Potash & Phosphate Institute of Canada (ESEAP), Singapore, p. 382.

Fairhurst, T.H., Rankine, I., Gfroerer, A.K., 2003. A conceptual framework for precision agriculture in oil palm plantations. In: Fairhurst, T.H., Hardter, R. (Eds.), The oil palm: management for large and sustainable yields. Potash & Phosphate Institute of Canada (ESEAP), Singapore, 382 p.

FAO, 1967. Report of the first session of FAO Panel of Experts on Integrated Pest Control, Rome (Italy), September 18–22, 19 p.

Feintrenie, L., 2012. Transfer of the Asian model of oil palm development: from Indonesia to Cameroon. World Bank conference on land and poverty. April 23–25, Washington DC.

Fernanda, A.S., Freire da Silva, M., Freitas Guimarães, J.K., Flores da Silva Ferreira, M., Dias Pereira, W., Piovesan, N.D., Gonçalves de Barros, E., 2014. Marker-assisted selection strategies for developing resistant soybean plants to cyst nematode. Crop Breed. Appl. Biotechnol. 14, 180–186.

Fisher, P.J., Gardner, R.C., Richardson, T.E., 1996. Single locus microsatellites isolated using 5′ anchored PCR. Nucleic Acids Res. 24, 4369–4371.

Flood, J., Cooper, R.M., Lees, P.E., 1989. An investigation of the pathogenecity of four isolates of *Fusarium oxysporum* from South America, Africa and Malaysia to clonal oil palm. J. Phytopathol. 124, 80–88.

Frankel, O.H., Hawkes, J.G., 1975. Crop Genetic Resources for Today and Tomorrow. Cambridge University Press, Cambridge.

Gascon, J.P., De Berchoux, Ch., 1964. Caractéristiques de la production d'*Elaeis guineensis* (Jacq.) de diverses origines et leurs croissements. Applicatio à la sélection du palmier à huile. Oléagineux 19, 75–84.

Gascon, J.P., Wuidart, W., 1975. Amélioration de la production et de la qualité de l'huile d'*Elaeis guineensis* Jacq. Oléagineux 30 (1), 1–4.

Gascon, J.P., Le Guen, V., Nouy, B., Asmady, Kamga, F., 1987. Results of second cycle recurrent reciprocal selection trials on oil palm. In: Halim Hassan, A., Chew, P.S., Wood, B.J., Pushparajah, E. (Eds.), Proceedings of 1987 International Oil Palm Conference. Palm Oil Research Institute of Malaysia, Kuala Lumpur, pp. 112–117.

Gascon, J.P., Le Guen, V., Nouy, B., Asmady, Kamga, F., 1988. Résultats d'essais de second cycle de sélection récurrente réciproque chez le palmier à huile *Elaeis guineensis* Jacq. Oléagineux 43, 1–7.

Genty, P., Garzon, M.A., Garcia, R., 1983. Dégâts et contrôle du complexe Leptopharsa-Pestalotiopsis chez le palmier à huile. Oleagineux 38, 291–299.

George, E.F., 1993. Plant propagation by Tissue Culture. The Technology Exegetics, England, pp. 41–44.

Ghafoor, A., Sharif, A., Ahmad, Z., Zahid, M.A., Rabbani, M.A., 2001. Genetic diversity in blackgram (*Vigna mungo* L. Hepper). Field Crop Res. 69, 183–190.

Gill, K.S., 1984. Research Imperatives beyond the Green Revolution in the Third World. In: Phelt, D. (Ed.), Human Fertility, Health and Food Impact on Molecular Biology and Technology. United Nations Fund for Population Activities, New York, pp. 195–231.

Godfray, H.C.J., Beddington, J.R., Crute, I.R., Haddad, L., Lawrence, D., Muir, J.F., Pretty, J., Robinson, S., Thomas, S.M., Toulmin, C., 2010. Food security: the challenge of feeding 9 billion people. Science 327, 812–818.

Gorret, N., Bin, R.Q.K., Oppenheim, S.F., Willis, L.B., Lessard, P.A., Rha, C.K., Sinskey, A.J., 2004. Bioreactor culture of oil palm (*Elaeis guineensis*) and effects of nitrogen source, inoculum size, and conditioned medium on biomass production. J. Biotechnol. 108, 253–263.

Grout, B.W.W., Shelton, K., Pritchard, H.W., 1983. Orthodox behaviour of oil palm seed and cryopreservation of the excised embryo for genetic conservation. Ann. Bot. 52, 381–384.

Guldentops, R.E., 1962. Contribution à l'étude de la trachéomycose du palmier à huile. Parasitica 18, 244–263.

Hamrick, J.L., Godt, M.J.W., 1989. Allozyme diversity in plants. In: Plant Population Genetics, Sinderland, MA, USA.

Hardon, J.J., 1969. Interspecific hybrids in the genus *Elaeis*. II. Vegetative growth and yield of F1 hybrids *E. guineensis* × *E. oleifera*. Euphytica 18, 380–388.

Hardon, J.J., 1970. Inbreeding in populations of the oil palm (*Elaeis guineensis* Jacq.) and its effects on selection. Oléagineux 25, 449–456.

Hardon, J.J., 1974. The Oil Palm. In: Leon, J. (Ed.), Handbook of Plant Introduction in Tropical Crops. F.A.O. Agricultural Studies. No. 93. Rome.

Hardon, J.J., Tan, G.Y., 1969. Interspecific hybrids in the genus *Elaeis*. I. Crossability, cytogenetics and fertility of F1 hybrids of *E. guineensis* x *E. oleifera*. Euphytica 18, 372–379.

Hardon, J.J., Turner, P.D., 1967. Observations on natural pollination in commercial plantings of oil palm (*Elaeis guineensis*) in Malaya. Expl. Agric. 3, 105–116.

Hardon, J.J., Gascon, J.P., Noiret, J.M., Meunier, J., Tan, G.Y., Tam, T.K., 1976. Major oil palm breeding programmes. In: Corley, R.H.V., Hardon, J.J., Wood, B.J. (Eds.), Oil Palm Research. Elsevier, Amsterdam, pp. 109–125.

Hardon, J.J., Corley, R.H.V., Lee, C.H., 1987. Breeding and selecting the oil palm. In: Abbott, A.J., Atkin, R.K. (Eds.), Improving Vegetatively Propagated Crops. Academic Press, London, pp. 63–81.

Hartley, C.W.S., 1977. The Oil Palm (second ed.) Longman, 265 pp.

Hartley, C.W.S., 1988. The Oil Palm (Tropical agriculture series), third ed. Longman Scientific & Technical, Harlow, London, 761 pp.

Hawkes, J.G., 1981. Germplasm Collection, Preservation and use. In: Frey, K.J. (Ed.), Plant Breeding II. Iowa State University Press, Ames, pp. 57–83.

Hayati, A., Wickneswari, R., Maizura, I., Rajanaidu, N., 2004. Genetic diversity of Oil palm (*Elaeis guineensis* Jacq.) germplasm collections from Africa: Implications for improvement and conservation of genetic resources. Theor. Appl. Genet. 108 (7), 1274–1284.

Helene, A., Myriam, C., Frederique, R., Thierry, B., David, C., Alphonese, O., Leifi, N., Bruno, N., James, W.T., 2011. Environmental regulation of sex determination in oil palm: current knowledge and insights from other species. Ann. Bot. 108 (8), 1529–1537.

Heller, J., 1993. Physic nut Jatropha curcas L. Promoting the conservation and use of underutilized and neglected crops. Institute of Plant Genetic and Crop Plant Research, Gatersleben/International Plant Genetic Resource Institute, Rome, 1993.

Heri, A.S., Hernawan, Y.R., NanangSupena, Y.Y. dan, Razak, P., 2011. Effects of inbreeding in oil palm. Souvenir programme-International seminar on breeding for sustainability in oil palm. November 18, 2011, KLCC Kuala Lumpur, Malaysia, ISOPB/MPOB. p. 132.

Ho, Y.W., Varghese, G., 1986. Pathogenic potential of soil *Fusaria* from Malaysian oil palm habitats. J. Phytopathol. 115, 325–328.

Ho, Y.W., Varghese, G., Taylor, G.S., 1985. Pathogenicity of *Fusarium oxysporum* isolates from Malaysia and *F. oxysporum* f. sp. *elaeidis* from Africa to seedlings of oil palm (*Elaeis guineensis*). Phytopath. Z. 114, 193–202.

Holden, J.H.W., Williams, J.T., 1984. Crop Genetic Resources: Conservation and Evaluation. George Alien and Unwin, London.

Hornus, P., Kamga, F., Chaillard, C., 1987. La recherche agronomique et le développement de la culture du palmier à huile au Cameroun. Oléagineux 42, 307–315.

Hunger, F.W.T., 1917. De Oliepalm (*Elaeis guineensis* Jacq.). Historich onderzoek over den oliepalm in Nerderlandsch-Indie. 206 pp. E.J. Bill Leiden.

Idris, A.S., Kushairi, A., Ismail, S., Ariffin, D., 2004. Selection for partial resistance in oil palm progenies to ganoderma basal stem rot. J. Oil Palm Res. 16 (2), 12–18.

Isa, Z.A., Kushairi, A., Mohd Din, A., Suboh, O., Junaidah, J., Noh, A., Choo Kien, W., Musa, B., 2011. A critical re-examination of the method of bunch analysis in oil palm breeding – an update. Souvenir programme-International seminar on breeding for sustainability in oil palm. November 18, 2011, KLCC Kuala Lumpur, Malaysia, ISOPB/MPOB. pp. 19–42.

Jacquemard, J.C., 1995. Le palmier à huile. Collection le technicien d'agriculture tropicale. Maisonneuve & Larose, Paris.

Jacquemard, J. C., 2011. Le palmier à huile. Versailles ; Wageningen (Pays-Bas) ; Gembloux (Belgique): Editions Quæ ; CTA ; Presses Agronomiques de Gembloux.

Jacquemard, J.C, Baudouin, L, Noiret, J.M., 1997. In: L'amélioration des plantes, tropicale. Charrier, A., Jacquot, M., Hamon, S., et Nicolas, D. (Eds.), pp. 507–531, CIRAD et ORSTOM, Monpellier, France.

Jacquemard, J.C., Suryana, E., Cochard, B., De Franqueville, H., Breton, F., Syaputra, I., Dermawan, E., Permadi, P., 2010. Intensification of oil palm (*Elaeis guineensis* Jacq.) plantation efficiency through planting material: new results and developments. International Oil palm Conference Yogyakarta, Indonesia, pp 1–34.

Jacquin, N.J., 1763. Selectarum stirpium. Americanarum historia.

Jalani, B.S., Cheah, S.C., Rajanaidu, N., 1998. Genetic breakthroughs and new frontiers in oil palm breeding. Third National Congress on Genetics, November 18–19, 1998. Generics Society of Malaysia, Palm Oil Research Institute of Malaysia (PORIM), Bangi, pp. 1–22.

Jones, L.H., 1974. Propagation of clonal oil palms by tissue culture. Planter, Kuala Lumpur, 50, 374–381.

Jones, L.H., Hanke, D.E., Eeuwens, C.J., 1995. An evaluation of the role of cytokinins in the development of abnormal inflorescences in oil palms (*Elaeis guineensis* Jacq.) regenerated from tissue culture. J. Pl. Growth Reg. 14, 135–142.

Junaidah, J., Kushairi, A., Jones, B., Kho, L.E., Isa, Z.A., Rusmin, J., 2011. Innovation for oil extraction method using NMR in bunch analysis. Souvenir programme-International seminar on breeding for sustainability in oil palm. November 18, 2011, KLCC Kuala Lumpur, Malaysia, ISOPB/MPOB. pp. 1–18.

Kang, S.M., 1999. The *Elaeidobius kamerunicus story*75Planter, Kuala Lumpur, pp. 143–150.

Kang, S.M., Zam, A.K., 1982. Quarantine aspects of the introduction into Malaysia of an oil palm insect pollinator. Paper presented at International Conference on Plant protection in the tropics. March 1–4, Kuala Lumpur.

Khosla, P., Sundram, K., 1996. Effects of dietary fatty acid composition on plasma cholesterol. Prog. Lipid Res. 35, 93–132.

Khush, G.S., 2002. Molecular genetics-plant breeder's perspective. In: Jain, S.M., Brar, D.S., Ahloowalia, B.S. (Eds.), Molecular Techniques in Crop Improvement. Kluwer Academic Publishers, Dordrecht, The Netherlands, pp. 1–8.

Kogan, M., 1998. Integrated pest management: historical perspectives and contemporary developments. Annu. Rev. Entomol. 43, 243–270.

Korzun, V., Roder, M.S., Vendekake, K., Pasqualone, A., Lotti, C., Ganal, M.W., Blanco, 1999. Integration of dinucleotide microsatellites from hexaploid bread wheat into a genetic linkage map of durum wheat. Theor. Appl. Genet. 98, 1202–1207.

Kushairi, A., Rajanaidu, N., 2009. Current status of oil palm planting materials in the world and future challenges. Sixteenth International Oil Palm Conference and Expopalma. Cartagena de Indias, Colombia. pp. 1–62.

Kushairi, A., Rajanaidu, N., Jalani, B.S., Zakri, A.H., 1999. Agronomic performance and genetic variability of *Dura* x *Pisifera* progenies. J. Oil Palm Res. 11 (2), 24–48.

Kushairi, A., Tarmizi, A. H., Zamzuri, I., Ong-Abdullah, M., Samsul-Kamal, R., Ooi, S. E., Rajanaidu, N., 2010. Production, Performance and Advances in Oil Palm Tissue Culture. Paper presented at the International Seminar on Advances in Oil Palm Tissue Culture, held on May 29, 2010 in Yogyakarta, Indonesia. Organized by the International Society for Oil Palm Breeders (ISOPB), 23 pp.

Kushairi, A., Mohd Din, A., Rajanaidu, N., 2011. Oil palm breeding and seed production. Mohd, B.W., Choo, Y.M., Chan, K.W. (Eds.), Further Advances in Oil Palm Research (2000–2010), vol. 1, Malaysian Palm Oil Board, Kuala Lumpur, Malaysia, pp. 57–101.

Lee, C.H., Donough, C.R., 1993. Genotype-environment interaction in oil palm clones. In: Proceedings of 1991 International Society of Oil Palm Breeders Workshop Genotype–Environment Interaction Studies in Perennial Tree Crops. Palm Oil Research Institute, Malaysia, Kuala Lumpur. pp. 33–45.

Lee, C.H., Toh, P.Y., 1992. Yield performance of the Golden Hope OPRS DxP planting materials. In: Proceedings of 1990 ISOPB International Workshop on Yield Potential in the Oil Palm. Phuket, Thailand, PORIM, Kuala Lumpur. pp. 24–29.

Lee, C.H., Yong, Y.Y., Donough, C., Chiu, S.B., 1990. Selection progress in the Deli dura population. In: Proceedings of International Workshop on Oil Palm Breeding Populations. Palm Oil Institute of Malaysia, Kuala Lumpur, Malaysia. pp. 81–89.

Lim, H.K., 1967. Animal feeding stuffs. Part 3. Compositional data of feeds and concentrates. Malay. Agric. J. 46 (1), 63–79.

Low, L.E., Alias, H., Boon, S., Shariff, M.E., Tan, A.C, Ooi, L.C. L., Cheah, S., Raha, A., Wank, Singh, R., 2008. Oil palm (*Elaeis guineensis* Jacq.) tissue culture ESTs: Identifying genes associated with callogenesis and embryogenesis.

Lubis, A.A., 1988. History and origin of oil palm breeding material used in Marihat research station. ISOPB Workshop on dura and tenera populations, Marihat, 1988.

Lubis, A.U., Lubis, R.A., Akiyat, B.M., Nouy, B., Noiret, J.M., 1990. Preliminary result of the oil palm improvement program using reciprocal recurrent selection in Marihat research station Indonesia, In: Proceedings of the Workshop on Progress of Oil Palm Breeding, Populations, A.C., Soh, N., Rajanaidu, H.B., Mohr Nasir, K., Pamin, C., Muluk, (Eds.), Permatabg Siantar, Sumatra, Indonesia, November 25, 1988. Oil palm Research Institute of Malaysia, pp. 5–25.

Maho-Yalen, J.E., Ntsomboh-Ntsefong, G., Ambang, Z., 2013. Development of integrated pest management system in agricultural production in Cameroon and the Central African sub region. World J. Agric. Res. 1 (6), 133–142.

Maizura, I., Cheah, S.C, Rajanaidu, N., 2001. Genetic diversity of oil palm germplasm collection using RFLPs. Proceedings of the 2001 PIPOC International Palm Oil Congress (Agriculture), pp. 526–534.

Maizura, I., Rajanaidu, N., Zakri, A.H., Cheah, S.C., 2006. Assesment of genetic diversity in oil palm (*Elaeis guineensis* Jacq.) using restriction fragment length polymorphism (RFLP). Genet. Res. Crop Evol. 53, 187–195.

Manju, P.R., Sreelathakumary, I., 2002. Genetic variability, heritability and genetic advance in hot Chilli (*Capsicum Chinense* Jacq.). J. Trop. Agric. 40, 4–6.

Marhalil, M., Rajanaidu, N., 2011. Oil palm (Elaeis guineensis) germplasm in Angola: second prospection. Souvenir programme-International seminar on breeding for sustainability in oil palm. November 18, 2011, KLCC Kuala Lumpur, Malaysia, ISOPB/MPOB. pp. 133–142.

Maria, A.A., Helena, C., 2007. Biofuels trade and sustainable development: The case of Ecuador's palm oil biodiesel. Working document, second ed., 65 p.

Mepsted, R., Flood, J., Cooper, R.M., 1988. Biocontrol of *Fusarium* wilt of oil palm by an avirulent isolate. Paper presented at Conference of British Society of Plant Pathology. Novel and unusual methods of disease control, 13–15 December, Reading.

Mepsted, R., Flood, J., Paul, T., Airede, C., Cooper, R.M., 1995c. A model system for rapid selection for resistance and investigation of resistance mechanisms in *Fusarium* wilt of oil palm. Pl. Pathol. 44, 749–755.

Meunier, J., 1969. Etude des populations naturelles d'*Elaeis guineensis* en Cote d'Ivoire. Oléagineux 24, 195–201.

Meunier, J., 1975. Le 'palmier à huile' americain *Elaeis melanococca*. Oléagineux 30, 51–62.

Meunier, J., Gascon, J.P., 1972. Le schéma général d'amélioration du palmier à huile à l'IRHO. Oléagineux 27, 1–12.

Meunier, J., Jacquemard, J.C., 1988. Improvement of the basic oil palm populations at IRHO. In: Proceedings of workshop on progress of oil palm breeding populations. Permatang Siantar, Sumatra, Indonesia, November 25, 1988. Palm Oil Research Institute of Malaysia. pp. 1–4.

Meunier, J., Potier, F., Amblard, P., Tailliez, B., 1989. Relationship between oil production and the number of bunches in oil palm (*Elaeis guineensis* Jacq.). Consequences for pollination in young plantations. Oléagineux 44, 269–279.

Miranda Santos, M.d., Barcelos, E., Nascimento, J.C., 1986. Genetic resources of *Elaeis oleifera* (H.B.K.) Cortes in the Brazilian Amazon. In: Proceedings of International Workshop on Oil Palm Germplasm and Utilisation. Palm Oil Research Institute, Malaysia, Kuala Lumpur. pp. 95–101.

Mohaimi, M., 2011. Updates on the evaluation of semi-clonal DxP progenies in Sime Darby. Souvenir programme – International seminar on breeding for sustainability in oil palm. November 18, 2011, KLCC Kuala Lumpur, Malaysia, ISOPB/MPOB. pp. 101–111.

Mohd Din, A., Kushairi, A., Maizura, I., Isa, Z.A., Noh, A., Rajanaidu, N., 2005b. MPOB strategic plan for fast track breeding programme. In: Proceedings of the 2005 National Seminar on Breeding and Clonal Technologies. Seri Kembangan, Selangor, Malaysia, MPOB, Bangi. pp. 43–53.

Mohd, B.W., Choo, Y.M., Chan, K.W., 2011. Strategic directions in oil palm research. Mohd, B.W., Choo, Y.M., Chan, K.W. (Eds.), Further Advances in Oil Palm Research (2000–2010), vol. 1, Malaysian Palm Oil Board, Kuala Lumpur, Malaysia, pp. 1–27.

Mondjeli, C., Ntsomboh-Ntsefong, G., Ngando-Ebongue, G.F., Ajambang-Nchu, W., Amah, P.I., Dibog, L., 2013. Spatio-temporal distribution of *Coelaenomenodera minuta* Uhmann (Coleoptera: Chrysomelidae), a serious insect pest of oil palm (*Elaeis guineensis* Jacq.) in the South-west region of Cameroon. Albanian J. Agric. Sci. 12 (3), 479–483.

Montoya, C., Lopes, R., Flori, A., Cros, D., Cuellar, T., Summo, M., Espeout, S., Rivallan, R., Risterucci, A.M., Bittencourt, D., Zambrano, R.J., Alarcon, G.W.H., Villeneuve, P., Pina, M., Nouy, B., Amblard, P., Ritter, E., Leroy, T., Billotte, N., 2013. Quantitative trait loci (QTLs) analysis of palm oil fatty acid composition in an interspecific pseudo-backcross from Elaeis oleifera (H.B.K) and oil palm (*Elaeis guineensis* Jacq.). Tree Genet Genomes.

Montoya, C., Cochard, B., Flori, A., Cros, D., Lopes, R., Cuellar, T., Espeout, S., Syaputra, I., Villeneuve, P., Pina, M., Ritter, E., Leroy, T., Billotte, N., 2014. Genetic architecture of palm oil fatty acid composition in cultivated oil palm (*Elaeis guineensis* Jacq.) compared to its wild relative *E. oleifera* (H.B.K) Cortés. PLoS One 9 (5), e95412.

Morcillo, F., Gagneur, C., Adam, H., Richaud, F., Singh, R., Cheah, S.-C., Rival, A., Duval, Y., Tregear, J.W., 2006. Somaclonal variation in micropropagated oil palm. Characterization of two novel genes with enhanced expression in epigenetically abnormal cell lines and in response to auxin. Tree Physiol. 26, 585–594.

Morcillo, F., Cros, D., Billotte, N., Ngando-Ebongue, G.-F., Domonhédo, H., Pizot, M., Cuéllar, T., Espéout, S., Dhouib, R., Bourgis, F., Claverol, S., Tranbarger, T.J., Nouy, B., Arondel, V., 2013. Improving palm oil quality through identification and mapping of the lipase gene causing oil deterioration. Nat. Commun. 4 (2160), 1–8.

Moretzsohn, M.C., Nunes, C.D.M., Ferreira, M.E., Grattapaglia, D., 2000. RAPD linkage mapping of the shell thickness locus in oil palm (*Elaeis guineensis* Jacq). Theor. Appl. Genet. 100, 63–70.

Musa, B., Kandha, S., Xaviar, A., 2011. Performance of bi-clonal DxP planting material at United Plantations Berhad. Souvenir programme-International seminar on breeding for sustainability in oil palm. November 18, 2011, KLCC Kuala Lumpur, Malaysia, ISOPB/MPOB. pp. 78–100.

Muthamia, Z.K., Morag, F.E., Nyende, A.B., Mamati, E.G., Wanjala, B.W., 2013. Estimation of genetic diversity of the Kenyan yam (*Dioscorea* spp.) using microsatellite markers. Afr. J. Biotechnol. 12 (40), 5845–5851.

Ngalle-Bille, H., Bell, J.M., Ngando-Ebongue, G.F., Nyobe, L., Ngangnou, F.C., Ntsomboh-Ntsefong, G., 2013. Morphogenesis of oil palm fruit (*Elaeis guineensis* Jacq.) in mesocarp and endocarp development. J. Life Sci. 7 (2), 153–158.

Ngando-Ebongue, G.F., 2009. Etude de la lipase endogène du mésocarpe du fruit du palmier à huile (*Elaeis guineensis* Jacq.): application à la sélection de lignées à faible acidité de l'huile. Thèse de doctorat, Université Victor Segalen Bordeaux 2.

Ngando-Ebongue, G.F., Dhouib, R., Carriere, F., Amvam-Zollo, P.H., Arondel, V., 2006. Assaying lipase activity in oil palm fruit (*Elaeis guineensis* Jacq.) mesocarp. Plant Physiol. Biochem. 44, 611–617.

Ngando-Ebongue, G.F., Koona, P., Nouy, B., Carriere, F., Amvam-Zollo, P.H., Arondel, V., 2008. Identification of oil palm breeding lines producing oils with low acid value. Eur. J. Lipid Sci. Technol. 110, 505–509.

Ngando-Ebongue, G.F., Ajambang, W.N., Koona, P., Lalu-Firman, B., Arondel, V., 2012. Oil Palm. In: Technological innovations in major world oil crops (Ed. Gupta), vol. 1 Breeding, pp. 165–200.

Ngando-Ebongue, G.F., Etta, C.E., Ntsomboh-Ntsefong, G., Oben, T.T., 2013. Breeding oil palm (*Elaeis guineensis* Jacq.) for fusarium wilt tolerance: an overview of research programmes and seed production potentialities in Cameroon. Int. J. Agric. Sci. 3 (5), 513–520.

Noorhariza, M.Z., Rajinder, S., Rozana, R., Ismanizan, I., 2012. *E. oleifera* Genomic-SSR Markers: exploitation in oil palm germplasm diversity and cross-amplification in *Arecaceae*. Int. J. Mol. Sci. 13, 4069–4088.

Norziha, A., Marhalil, M., Maizura, 2011. Development of *in vitro* technique to preserve oil, palm genetic materials. International seminar on breeding for sustainablility in oil palm. November 18, 2011, KLCC Kuala Lumpur, Malaysia. p. 51.

Nouy, B., Lubis, R.A., Kusnadi, T.T., Akiyat, Samaritaan, G., 1991. Potentiel de production chez le palmier à huile *Elaeis guineensis*. Résultats des hybrides Déli x La Mé à Nord Sumatra. Oléagineux 46, 91–97.

Ntsomboh-Ntsefong, G., 2007. Fruit status of the oil palm (*Elaeis guineensis* Jacq. var. pisifera) in controlled pollination. Master's Thesis (DEA), Plant Biotechnology, University of Yaounde I, 45 p.

Ntsomboh-Ntsefong, G., Ngando-Ebongue, G.F., Koona, P., Bell, J.M., Youmbi, E., Ngalle-Bille, H., Bilong, E.G., Madi, G., Anaba, B., 2012. Control approaches against vascular wilt disease of *Elaeis guineensis* Jacq. caused by *Fusarium oxysporum* f. sp. *elaeidis*. J. Bio. Life Sci. 3 (1), 160–173.

Oil World, May 2008.

Okoye, M.N., Okwuagwu, C.O., Uguru, M.I., 2009. Population improvement for fresh fruit bunch yield and yield components in oil palm (*Elaeis guineensis* Jacq.). Am. Eur. J. Sci. Res. 4 (2), 59–63.

Okwuagwu, C.O., 1985. The genetic base of the NIFOR Oil Palm breeding programme. In: Proceedings of International Workshop on Oil Palm Germplasm and Utilization. Kualar Lumpur. PORIM 10, pp. 228–237.

Okwuagwu, C.O., 1996. The genetic improvement of the Deli *Dura* breeding population of the oil palm (*Elaeis guineensis* Jacq.). Elaeis J. 8, 55–63.

Okwuagwu, C.O., Ataga, C.D., Okoye, M.N., Okolo, E.C., 2011. Germplasm collection of highland palms of Afikpo in eastern Nigeria. Bayero J. Pure Appl. Sci. 4 (1), 112–114.

Ooi, S.C., Da Silva, E.B., Muller, A.A., Nascimento, J.C., 1981. Oil palm genetic resources – native *E. oleifera* populations in Brazil offer promising sources. Pesq. Agropec. Brasil. 16, 385–395.

Osborne, D.J., Henderson, J., Corley, R.H.V., 1992. Controlling fruit-shedding in the oil palm. Endeavour 16, 173–177.

Paranjothy, K., Rohani, O., Tarmizi, A.H., Tan, C.S., Tan, C.C., 1990. Current status and strategies of oil palm tissue culture research. In: Jalani, B.S. et al., (Ed.), Proceedings 1989 International Palm Oil Development Conference – Agriculture. Palm Oil Research Institute, Malaysia, Kuala Lumpur, pp. 109–121.

Paranjothy, K., Rohani, O., Tan, C.C., Wong, G., Soh, A.C., 1995. Incidence of abnormalities in relation to in vitro protocols. In: Rao, V., Henson, I.E., Rajanaidu, N. (Eds.), Recent developments in oil palm tissue culture and biotechnology. Palm Oil Res. Inst, Malaysia, Kuala Lumpur, pp. 77–85.

Peakall, R., Gilmore, S., Keys, W., Morgante, M., Rafalski, A., 1998. Cross-species amplification of soybean (*Glycine max*) simple sequence repeats (SSRs) within the genus and other legume genera: implications for the transferability of SSRs in plants. Mol. Bio. Evol. 15, 1275–1287.

Perera, L., Russell, J.R., Provan, J., Powell, W., 2000. Use of microsatellite DNA markers to investigate the level of genetic diversity and population genetic structure of coconut (*Cocos nucifera* L.). Genome 43, 15–21.

Perera, L., Russell, J.R., Provan, J., Powell, W., 2001. Level and distribution of genetic diversity of coconut (*Cocos nucifera* L., var. Typica form typica) from Sri Lanka assessed by microsatellite markers. Euphytica 122, 381–389.

Petronella, G.A., Mohd, K.Y., Nik, M.M., Goh, K.J., Gan, H.H., 2009. Effect of N and K fertilizers on nutrient leaching and groundwater quality under mature oil palm in Sabah during the monsoon period. Am. J. Appl. Sci. 6 (10), 1788–1799.

Ponnemma, K.N., Asha, V., Sajeebkhan, A., 2004. Impact of the release of oil palm pollinating weevil, *Elaeidobius kamerunicus* (Coleoptra: Curculionidae) and the statue of weevil population and fruit set in various oil palm plantations in India. J. Plant. Crops 32, 379–384.

Porter, R., 1989. Development of high yielding oil palm planting material resistant to vascular wilt caused by *Fusarium oxysporum* f. sp. *elaeidis*. Paper Presented at International Conference Palms and Palm Products. November 21–25, Nigerian Institute of Oil Palm Research, Benin City.

Potier, F., Nouy, B., Flori, A., Jacquemard, J.C., Edyana, H.S., Durand-Gasselin, T., 2006. Yield potential of oil palm (*Elaeis guineensis* Jacq.) clones: preliminary results observed in the Aek Loba Timur genetic block in Indonesia. International Seminar on Yield Potential in Oil Palm. 27–28 November 2006, Banthai Beach Resort & Spa Phuket, Thailand. The International Soceity for Oil Palm Breeders (ISOPB) (Ed.).

Powell, W., Morgante, M., Andre, C., Hanafey, M., Vogel, J., Tingey, S., Rafalski, A., 1996. The comparison of RFLP, RAPD, AFLP, SSR (microsatellite) markers for germplasm analysis. Mol. Breed. 2, 225–238.

Prendergast, A.G., 1957. Observations on the epidemiology of vascular wilt disease of the oil palm. J. W. Afr. Inst. Oil Palm. Res. 2, 148–175.

Prendergast, A.G., 1963. A method of testing oil palm progenies at the nursery stage for resistance to vascular wilt disease caused by *Fusarium oxysporum*. J. W. Afr. Inst. Oil Palm Res. 4, 156–175.

Priwiratama, H., Djuhjana, J., Stephen, P.C.N., Caligari, P.D.S., 2010. Progress of oil palm breeding for novel traits: late abscission, virescence and long bunch stalk. In: Proceeding of the International Oil Palm Conference. Yogyakarta, June 1–3, 2010. p. 11.

Provan, J., Powell, W., Waugh, R., 1996. Microsatellite analysis of relationships within cultivated potato (*Solanum tuberosum*). Theor. Genet. 92, 1078–1084.

Purba, A.R., 2010. Oil Palm Breeding: Indonesian experience. In: Proceedings of the International Oil Palm Conference. Yogjakarta, Indonesia.

Raberchault, H., Ahee, J., Guenin, G., 1970. Colonies cellulaires et formes èmbryos *in vitro* apatir de cultures d'embryons de palmier a huile (*Elaeis guineensis* Jacq.) à l'aide de cultures de tissue foliaires. Comptes Rendus de l'Académie des Sciences Paris Série 283, 1735–1737.

Rafii, M.Y., Rajanaidu, N., Jalani, B.S., Kushairi, A., 2002. Performance and heritability estimations on oil palm progenies tested in different environments. J. Oil Palm Res. 14 (1), 15–24.

Rajanaidu, N., 1986. The oil palm (*Elaeis guineensis*) collections in Africa. In: Proceedings of Int. Workshop on Oil palm germplasm and utilization. Palm Oil Research Institute, Malaysia, Kuala Lumpur. pp. 59–83.

Rajanaidu, N., 1987. Collection of oil palm (*Elaeis guineensis*) genetic material in Tanzania and Madagascar. Palm Oil Research Institute, Malaysia, Bull. 15, 1–6.

Rajanaidu, N., 1994. PORIM Oil palm genebank, PORIM, Bangi, 19 pp.

Rajanaidu, N., Jalani, B.S., 1994. Oil palm genetic resources – collection, evaluation, utilization and conservation. Paper Presented at Colloquium Oil palm Genetic Resources. September, 13, Palm Oil Research Institute, Malaysia, Bangi, Malaysia.

Rajanaidu, N., Jalani, B.S., 1999. World-wide production and performance of $D \times P$ planting materials. Proceedings of the International Seminar on Sourcing of Oil Palm Planting Materials for Local and Overseas Joint-Venture. Rajanaidu, N., Jalani, B.S. (Eds.). PORIM, Bangi 28–70.

Rajanaidu, N., Rao, V., 1988. Oil palm genetic collections: their performance and use to the industry. In: Proceedings of the 1987 International Oil Palm Conference – Agriculture, Abdul Halim Hassan, Chew, P.S., Wood, B.J., Pushparajah, (Eds.), pp. 59–85.

Rajanaidu, N., Tan, Y.P., Ong, E.C., Lee, C.H., 1986. The performance of inter-origin commercial D x P planting material. In: Proceedings of International Workshop on Oil Palm Germplasm and Utilization. Palm Oil Research Institute, Malaysia, Kuala Lumpur. pp. 155–161.

Rajanaidu, N., Jalani, B.S., Domingos, M., 1991. Collection of oil palm (*Elaeis guineensis*) germplasm in Angola. Int. Soc. Oil Palm Breeders Newsl. 8 (2), 2–3.

Rajanaidu, N., Jalani, B. S., Rao, V., Kushairi, A., 1993. Genotype × environment interaction (G × E) studies in oil palm (*Elaeis guineensis*) progenies. In: Proceedings of 1991 International Society of Oil Palm Breeders Workshop Genotype × environment interaction studies in perennial tree crops, Palm Oil Research Institute, Malaysia, Kuala Lumpur, pp. 12–32.

Rajanaidu, N., Jalani, B.S., Kushairi, A., Rao, V., 1999. Oil palm genetic resources-collection, evaluation, utilization and conservation. In: Proceedings of the Symposium on the Science of Oil Palm, Breeding, Rajanaidu, N., Jalani, B.S. (Eds.), PORIM, Bangi, Malaysia, pp. 219–255.

Rajanaidu, N., Kushairi, A., Rafii, M.Y., Din, A.M., Maizura, I., Jalani B. S. 2000. Oil palm breeding and genetic resources. Malaysian Palm Oil Board, Bangi Selangor, Malaysia. In: Advances in oil palm Research, 1, 171–237.

Rajanaidu, N., Kushairi, A., Rafii, M.Y., Moh'd Din, A., Maizura, I., Jalani, B.S., 2001. Oil Palm Breeding and Genetic Resources. In: Basiron, Y., Jalani, B.S., Chan, K.W.(Eds.) Malaysia Palm Oil Board, Kualar Lumpur, pp. 171–237.

Rajanaidu, N., Kushairi, A., Chan, K.W., 2011. Oil palm biomass: its genetic variability, certification and sustainability. In: Mohd, B.W., Choo, Y.M., Chan, K.W. (Eds.), Further advances in oil palm research (2000–2010), vol. 1, Malaysian Palm Oil Board, pp. 29–46.

Rajanaidu, N., Kushairi, A., Mohd Din, A., Marhalil, M., Fadila, A.M., Mohd, I.Z.A., 2011b. Selection criteria (15-15-15) to develop sustainable oil palm planting materials. Souvenir programme-International seminar on breeding for sustainability in oil palm. November 18, 2011, KLCC Kuala Lumpur, Malaysia, ISOPB/MPOB. p. 112.

Ramsay, L., Macaulay, M., Ivannissevich, D., Maclean, K., Cardle, L., Fuller, J., Edwards, K.J., Tuvesson, S., Morgante, M., Massari, A., Maestri, E., Marmiroli, N., Sjakste, T., Ganal, M.W., Powell, W., Waugh, R., 2000. A simple sequence repeat based linkage map of barley. Genetics 156, 1997–2005.

Rao, V., Chang, K.C., 1982. Anthesis and fruit set in *E. oleifera* (HBK) Cortes. Palm Oil Res. Inst. Malaysia Bull. 4, 27–34.

Rao, V., Donough, C.R., 1990. Preliminary evidence of a genetic cause for the floral abnormalities in some oil palm ramets. Elaeis 2, 199–207.

Rao, V., Chia, C.C., Law, I.H., 1998. Progeny response to underplanting. Paper Presented at International Conference on Developments in Oil Palm Industry for the Twentieth Century. September 21, Bali.

Rao, V., Law, I.H., Zuraini, S., Chia, C.C., 1999. Ekona and AVROS – a tale of two pisiferas. Paper Presented at 1999 PORIM International Palm Oil Congress. Kuala Lumpur.

Reddy, V.G., Upadhyaya, H.D., Gowda, C.L.L., 2006. Current status of Sorghum genetic resources at ICRISAT: their sharing and impacts. Int. Sorghum Millets Newsl. 47, 9–13.

Redshaw, M., 2003. Utilization of field residues and mill by-products. In: Fairhurst, T.H., Hardter, R. (Eds.), The Oil Palm: Management for Large and Sustainable Yields. Potash & Phosphate Institute of Canada (ESEAP), Singapore, p. 382.

Rees, A.R., 1963d. Some factors affecting the germination of oil palm seeds under natural conditions. J. W. Afr. Inst. Oil Palm Res. 4, 201–207.

Renard, J.L., Gascon, J.P., Bachy, A., 1972. Recherches sur la fusariose du palmier à huile (Bilingue fr.–angl.). Oléagineux 27 (12), 581–591.

Richardson, D.L., 1995. The history of oil palm breeding in the United Fruit Company. ASD Oil Palm Papers 11, 1–22.

Rival, A., Levang, P., 2013. La palme des controverses: palmier à huile et enjeux de développement. Essais, editions Quae. France, 98 p.

Rival, A., Buele, T., Barre, P., Hamon, S., Duval, Y., Noirot, M., 1997. Comparative flow cytometric estimation of nuclear DNA content in oil palm (*Elaeis guineensis* Jacq.) tissue cultures and seed-derived plants. Plant Cell Rep. 16, 884–887.

Rival, A., Bertrand, L., Beul, Combes, T.M.C., Trouslot, P., Lashermes, P., 1998. Suitability of RAPD analysis for the detection of somaclonal variants in oil palm (*Elaeis guineensis* Jacq.). Plant Breed. 117 (1), 73–76.

Rival, A., Tregear, J., Jaligot, E., Morcillo, F., Aberlenc, F., Billotte, N., Richaud, F., Beule, T., Borgel, A., Duval, Y., 2001. Oil palm biotechnology: progress and prospects. OCL-Ol. Corps Gras Lipids 8, 295–306.

Rosenquist, E.A., 1982. Performance of identical oil palm progenies in different environments. In: Pushparajah, P., Chez, P.S. (Eds.), The Oil Palm in Agriculture in the Eighties. The Incorporated Society of Planters, Kuala Lumpur, pp. 131–143.

Rosenquist, E.A., 1986. The genetic base of oil palm breeding populations. In: Proceedings of International Workshop on Oil palm germplasm and utilization. Palm Oil Research Institute of Malaysia 10, 27–56.

Rosenquist, E.A., 1989. An overview of breeding technology and selection in *Elaeis guineensis*. Doc Harrisons Fleming Advisory Services Ltd.

Rosenquist, E.A., 1990. An overview of breeding technology and selection in *Elaeis guineensis*. In: Jalani, B.S. et al., (Ed.), In: Proceedings of 1989 International Palm Oil Development Conference – Agriculture. Palm Oil Research Institute, Malaysia, Kuala Lumpur, pp. 5–25.

Rosenquist, E.A., Corley, R.H.V., de Greef, W., 1990. Improvement of tenera populations using germplasm from breeding programmes in Cameroon and Zaire. In: Proceedings of Workshop on Progress of Oil Palm Breeding Populations. Palm Oil Research Institute, Malaysia, Kuala Lumpur. pp. 37–69.

RSPO, 2008. RSPO fact sheets: The growth and use of sustainable palm oil. 8 p.

RSPO, 2013. Principles and Criteria for the Production of Sustainable Palm Oil. Endorsed by the RSPO Executive Board and accepted at the Extraordinary General Assembly by RSPO Members on April 25, 2013 (Including Major Indicators for Endorsement by RSPO Executive Board), 68 pp.

Sambanthamurthi, R., Chong, C.L., Khark, C., Yeu, K.H., Premavathy, R., 1991. Chilling-induced lipid hydrolysis in the oil palm (*Elaeis guineensis*) mesocarp. J. Exp. Bot. 42, 1199–1205.

Sambanthamurthi, R., Sundram, K., Tan, Y., 2000. Chemistry and biochemistry of palm oil. Prog. Lipid Res. 39, 507–558.

Santos, M.M.d., Barcelos, E., Nascimento, J.C., 1986. Genetic resources of *E. oleifera* (H.B.K.) Cortes in the Brazilian Amazon. In: Proceedings of the International Workshop on Oil Palm Germplasm and Utilization. Palm Oil Research Institute, Malaysia, Kuala Lumpur. pp. 95–101.

Sapey, E., Adusei-Fosu, K., Agyei-Dwarko, D., Okyere-Boateng, G., 2012. Collection of oil palm (*Elaeis guineensis* Jacq.) germplasm in the northern regions of Ghana. Asian J. Agric. Sci. 4 (5), 325–328.

Schlegel, R.H.J., 2010. Dictionary of plant breeding, second ed. 586 p.

Seng, S.Y., Donough, C.R., Rajanaidu, N., 2011. Performance of MPOB Nigeria prospection materials in Sabah. Souvenir programme-International seminar on breeding for sustainability in oil palm. November 18, 2011, KLCC Kuala Lumpur, Malaysia, ISOPB/MPOB.

Shama, M., 2000. Exploitation of *E. oleifera* germplasm in improving the oil quality of oil palm. In: MPO.B., Rajanaidu, N., Ariffin, D. (Eds.), Proceedings of International Oil Palm Symposium on Oil Palm Genetic Resources, Utilization, Kuala Lumpur, p. 1.

Sharma, M., Gurmit, S., Toh, T.S., Singh, T., 1999. Optimizing on FFB, oil, OER and oil quality. Paper Presented at 1999 PORIM International Palm Oil Congress. Feb. 1–6, Kuala Lumpur.

Simmonds, N.W., 1993. Introgression and incorporation: strategies for the use of crop genetic resources. Biol. Rev. 68, 539–562.

Singh, A.K., Smart, J., Simpson, C.E., Raina, S.N., 1998. Genetic variation vis-a-vis molecular polymorphism in groundnut, *Arachis hypogaea* L. Genet. Resour. Crop. Evol. 45, 119–126.

Singh, R., Jayanthi, N., Soon-Guan, T., Jothi, M.P., Cheah, S.-C., 2007. Development of simple sequence repeat (SSR) markers for oil palm and their application in genetic mapping and fingerprinting of tissue culture clones. Asia Pacific J. Mol. Bio. Biotech. 15 (3), 121–132.

Singh, R., Noorhariza, M.Z., Ting, N.C., Rozana, R., Tan, S.G., Low, E.T.L., 2008. Exploiting an oil palm EST database for the development of gene-derived and their exploitation for assessment of genetic diversity. Biologia 63, 227–235.

Singh, R., Tan, S.G., Panandam, J.M., Abdul Rahman, R., Ooi, L., Eng-Ti, C.L., Sharma, L.L., Jansen, M., Suan-Choo Cheah, J., 2009. Mapping quantitative trait loci (QTLs) for fatty acid composition in an interspecific cross of oil palm. BMC Plant Bio. 9 (114), 1–19.

Singh, R., Ong-Abdullah, M., Eng-Ti, L.L., Abdul Manaf, Mohamad Arif, Rosli, Rozana, Rajanaidu, N., Cheng-Li Ooi, L., Ooi, Siew–Eng, Chan, Kuang-Lim, Halim, Mohd Amin, Azizi, Norazah, Nagappan, Jayanthi, Bacher, Blaire, Lakey, Nathan, Smith, Steven W., He, Dong, Hogan, Michael, Budiman, Muhammad A., Lee, Ernest K., DeSalle, Rob, Kudrna, David, Goicoechea, Jose Luis, Wing, Rod A., Wilson, Richard K., Fulton, Robert S., Ordway, Jared M., Martienssen, Robert A., Sambanthamurthi, R., 2013a. Oil palm genome sequence reveals divergence of interfertile species in old and new worlds. Nature 500, 335–339.

Singh, et al., 2013b. The oil palm SHELL gene controls oil yield and encodes a homologue of SEEDSTICK. Nature 500, 340–344.

Singh, R., Eng-Ti, L.L., Cheng-Li, O.L., Ong-Abdullah, M., Rajanaidu, N., Ngoot-Chin Ting, Marjuni, M., Pek-Lan, C., Maizura, I., Mohd Arif, A.M., Jayanthi, N., Kuang-Lim, C., Rozana, R., Mohd Amin, H., Norazah, A., Muhammad, A.B., Lakey, N., Bacher, B., Van Brunt, A., Chunyan, W., Hogan, M., Dong He, Jill, D., MacDonald, Smith, S.W., Jared, M.O., Martienssen, R.A., Sambanthamurthi, R., 2014. The oil palm VIRESCENS gene controls fruit colour and encodes a R2R3-MYB. Nat. Comm. 5, 1–8, 4106.

Siti Nor Akmar, A., Zubaidah, R., Manaf, M.A., Nurniwalis, W., 2003. Targeting gene expression in mesocarp and kernel of the oil palm fruits. In: Murata, N., Yamada, M., Nishida, I., Okuyama, H., Sekiya, J., Hajime, W. (Eds.), Advanced Research on Plant Lipids. Kluwer Academic Publishers, Dordrecht, The Netherlands, pp. 411–414.

Skurtis, T., Aïnaché, G., Simon, D., 2010. Le financement du secteur de l'huile de palme : pourquoi les institutions financières de développement doivent continuer à investir en Afrique. Oléagineux, Corps Gras. Lipides 17 (6), 400–403.

Smith, B.G., 1993a. Oil palm breeding – the potential for physiological selection criteria. In: Basiron, Y., et al. (Ed.), Proceedings of the 1991 PORIM International Palm Oil Conference – Agriculture. Palm Oil Research Institute, Malaysia, Kuala Lumpur, pp. 57–62.

Smith, B.G., 1993b. Correlations between vegetative and yield characteristics and photosynthetic rate and stomatal conductance in the oil palm (*Elaeis guineensis* Jacq.). Elaeis 5, 12–26.

Smith, W.K., Thomas, J.A., 1973. The isolation and in vitro cultivation of cells of *Elaeis guineensis*. Oléagineux 28, 123–127.

Soh, A.C., 1986. Expected yield increase with selected oil palm clones from current $D \times P$ seedling materials and its implications on clonal propagation, breeding and ortet selection. Oléagineux 41 (2), 51–56.

Soh, A.C., 1987. Current issues in oil palm breeding. Malay. Appl. Biol. 16, 101–108.

Soh, A.C., 1998. Review of ortet selection in oil palm. Planter, Kuala Lumpur, 74, 217–226.

Soh, A.C., 1999. Breeding plans and selection methods in oil palm. In: Rajanaidu, N., Jalani, B.S. (Eds.), Proceedings of the Seminar on Science of Oil Palm Breeding (Montpellier, 1992), Palm Oil Research Institute, Malaysia, Kuala Lumpur, pp. 65–95.

Soh, A.C., 2010. Review of strategies in breeding for oil palm clonal propagation. In: Proceedings on Advances in Oil Palm Tissue Culture. The International Soceity for Oil Palm Breeders (ISOPB). May 29, 2010, Hotel Santika, Yogyakarta Indonesia.

Soh, A.C., Lee, C.H., Yong, Y.Y., Chin, C.W., Tan, Y.P., Rajanaidu, N., Phuah, P.K., 1990. The precision of oil palm breeding trials in Malaysia. In: Soh, A.C., Rajanaidu, N., Hasan Basri, M.N. (Eds.), International Symposium on Application of Statistics to Perennial Tree Crops, Research, pp. 41–50, Palm Oil Research Institute, Malaysia, Kuala Lumpur.

Soh, A.C., Rajanaidu, N., Cheah, S.C., 1994b. Oil palm breeding: the previous and the next 50 years. In: Chee, K.H. (Ed.), Management for Enhanced Profitability in Plantations. Incorporated Society of Planters, Kuala Lumpur, pp. 35–52.

Soh, A.C., Yong, Y.Y., Ho, Y.W., Rajanaidu, N., 1995. Commercial potential of oil palm clones: early results of their performance in several locations. In: Rao, V., Henson, I.E., Rajanaidu, N. (Eds.), Recent Developments in Oil Palm Tissue Culture, Biotechnology. Palm Oil Research Institute, Malaysia, Kuala Lumpur, pp. 134–144.

Soh, A.C., Wong, G., Tan, C.C., Chew, P.S., Hor, P.Y., Chong, S.P., Gopal, K., 2001. Recent advances towards commercial production of elite oil palm clones. In: 2001 PIPOC International P.O. Congress – Agriculture, Malaysian Palm Oil Board (MPOB), Kuala Lumpur, pp. 33–44.

Soh, A.C., Wong, G., Hor, T.Y., Tan, C.C., Chew, P.S., 2003. Oil palm genetic improvement. Janick, J. (Ed.), Plant Breeding Reviews, vol. 22, John Wiley & Sons, New Jersey, pp. 165–219, Online at http://www.aarsb.com.my/oil-palm-plant-breeding (20/8/2014).

Staritsky, G., 1970. Tissue culture of oil palm (*Elaeis guineensis*) as a tool for its vegetative propagation. Ephytica 19, 288–292.

Sterling, F., Alvarado, A., 2002. Historical account of ASD's oil palm germplasm collections. ASD Oil Palm Papers 24, 1–16.

Sujadi, Nani, H., Agung, K., Abdul, R.P., 2011. Evaluation and selection of β-carotene from 18 F1 genotypes of oil palm (*Elaeis guineensis* Jacq.) in North Sumatera. International seminar on breeding for sustainablility in oil palm. November 18, 2011, KLCC Kuala Lumpur, Malaysia. p. 143.

Sunilkumar, K., Sparjan Babu, D.S., 2013. Surface color based prediction of oil content in oil palm (*Elaeis guineensis* Jacq.) fresh fruit bunch. Afr. J. Agric. Res. 8 (6), 564–569.

Susanto, A., 2009. Basal stem rot in Indonesia. Biology, economic importance, epidemiology, detection and control. International Workshop on Awareness, Detection and Control on Oil Palm, pp. 1–25.

Syed, R.A., 1979. Studies on oil palm pollination by insects. Bull. Ent. Res. 69, 213–224.

Syed, R.A., 1982. Insect pollination of oil palm: feasibility of introducing *Elaeidobius* spp. in Malaysia, In: The oil palm in agriculture in the 80s, vol. 1., Pushparajah, E., Chew, P.S., (Eds.), Incorporated Society of Planters, Kuala Lumpur, pp. 263–290.

Syed, R.A., 1984. Los insectos polinizadores de la palma africana. Palmas (Bogota) 5 (3), 19–87.

Syed, R.A., Law, I.H., Corley, R.H.V., 1982. Insect pollination of oil palm: introduction, establishment and pollinating efficiency of *Elaeidobius kamerunicus* in Malaysia. Planter, Kuala Lumpur, 58, pp. 547–561.

Tan, Y.P., 1985. Weevil pollination in the *E. oleifera* x *Elaeis guineensis* hybrid. In: Proceedings of the Symposium on Impact of the Pollinating Weevil on the Malaysian Oil Palm Industry. Palm Oil Research Institute, Malaysia, Kuala Lumpur. pp. 34–40.

Tan, Y.P., Basri-Wahid, M., 1985. Another pollinating insect of oil palm – Pyroderces. In: Proceedings of the Symposium on Impact of the Pollinating Weevil on the Malaysian Oil Palm Industry. Palm Oil Research Institute, Malaysia, Kuala Lumpur. pp. 50–57.

Tan, K.S., Rao, A.N., 1979. Certain aspects of developmental morphology and anatomy of oil palm. In: Histochemistry, developmental and structural anatomy of Angiosperms: a symposium, pp. 267–285.

Tandon, R., Manohara, T.N., Nijalingappa, B.H.M., Shivanna, K.R., 2001. Pollination and pollen – pistil interaction in oil palm, *Elaeis guineensis*. Ann. Bot. 87, 831–838.

Tarmizi, A.H., Marziah, M., 1995. The influence of low temperature treatment on growth and proline accumulation in polyembryogenic cultures of oil palm (*Elaeis guineensis* Jacq.). Elaeis 7, 107–117.

Taufiq, C.H., Retno, D.S., Ernayunita, A.N., Simamora, E.N.F., Iman, Y.H., 2011. Current condition of oil palm tissue culture clones. Souvenir programme – International seminar on breeding for sustainability in oil palm. November 18, 2011, KLCC Kuala Lumpur, Malaysia, ISOPB/MPOB. p. 77.

Teixeira, J.B., Sondahl, M.R., Nakamura, T., Kirby, E.G., 1995. Establishment of oil palm cell suspensions and plant regeneration. Plant Cell Tiss. Org. Cult. 40, 105–111.

Tester, M., Langridge, P., 2010. Breeding technologies to increase crop production in a changing world. Science 327, 818–822.

Thomas, R.L., Watson, I., Hardon, J.J., 1969b. Inheritance of some components of yield in the 'Deli dura variety' of oil palm. Euphytica 18, 92–100.

Ting, N.-C., Noorhariza, M.Z., Rozana, R., Low, E.-T., Ithnin, M., Cheah, S.-C., Tan, S.-G., Singh, R., 2010. SSR mining in oil palm EST database: application in oil palm germplasm diversity studies. J. Genet. 89, 135–145.

Tinker, P.B., 1984. Site-specific yield potentials in relation to fertilizer use in nutrient balances and fertilizer needs in temperate agriculture. In: Von Peter, A. (Ed.), Proceedings of Eighteenth Colloquium, International Potash Institute, Bern, 1984, pp 193–208.

Tinker, P.B., 2000b. Shades of Green – A Review of UK Farming Systems. Royal Agricultural Society of England, Stoneleigh Park, U.K.

Tinker, P.B., 2001. Organic farming – nutrient management and productivity. Paper presented at International Fertilizer Society Symposium. Lisbon, March 5, March.

Tranbarger, T.J., Dussert, S., Joet, T., Argout, X., Summo, M., Champion, A., Cros, D., Omore, A., Nouy, B., Morcillo, F., 2011. Regulatory mechanisms underlying oil palm fruit mesocarp maturation, ripening, and functional specialization in lipid and carotenoid metabolism. Plant Physiol. 156, 564–584.

Turner, P.D., 1981. Oil Palm Diseases and Disorders. Oxford University Press, Kuala Lumpur.

Ulrich, G.M., Wolfenbarger, L.L., 1999. AFLP genotyping and fingerprinting. Tree 14 (10), 389–394.

UNDP, 2007. United Nations Development Programme. Malaysia Generating Renewable Energy from Oil Palm Wastes. UNDP, Kuala Lumpur.

Vallejo, G., Cassalett, C., 1975. Perspectivas del cultivo de los hibridos interspecificos de noli (*E. oleifera* (H.B.K.) Cortes) × palma africana de aceite (*Elaeis guineensis* Jacq.) en Colombia. Revista Inst. Colomb. Agropec. 10, 19–35.

Varshney, R.K., Nayak, S.N., May, G.D., Jackson, S.A., 2009c. Next generation sequencing technologies and their application for crop genetics and breeding. Trends Biotechnol. 27, 522–530.

Vermeulen, S., Goad, N., 2006. Towards Better Practice in Smallholder Palm Oil Production. Natural Resource Issues Series No. 5. International Institute for Environment and Development, London, UK, 57 p.

Virdiana, I., Flood, J., Sitepu, B., Hasan, Y., Aditya, R., Nelson, S., 2011. Integrated disease management to reduce future Ganoderma infection during oil palm replanting. Simposium nasional dan Lokakarya Ganoderma. Bogor Indonesia, pp. 1–9.

Wahid, M.B., Abdullah, S.N.A., Henson, I.E., 2005. Oil palm-achievements and potential. Plant Prod. Sci. 8, 288–297.

Wardlaw, C.W., 1946a. A wilt disease of the oil palm. Nature 158, 56.

Wardlaw, C.W., 1946b. Fusarium oxysporum on the oil palm. Nature 158, 712.

Williams, C.N., Thomas, R.L., 1970. Observation on sex differentiation in the oil palm, *Elaeis guineensis* L. Ann. Bot. 34, 957–963.

Wong, C.K., Mariam, M.S., 2011. Hexane free oil extraction for estimation of iodine value and carotene content. Souvenir programme-International seminar on breeding for sustainability in oil palm. November 18, 2011, KLCC Kuala Lumpur, Malaysia, ISOPB/MPOB. pp. 43–50.

Wong, G., Tan, C.C., Soh, A.C., 1997. Large scale propagation of oil palm clones – experiences to date. Acta Hort. (ISHS) 447, 649–658.

WWF Report, 2012. Profitability and Sustainability in Palm Oil Production. Analysis of Incremental Financial Costs and Benefits of RSPO Compliance. 60 pp.

Xu, M., Xiangqian, L., Schuyler, S.K., 2000. AFLP-Based Detection of DNA Methylation. Plant Mol. Bio. Rep. 18, 361–368.

Xu, L., Yuan, Y., Zhang, L., Wan, L., Zheng, Y., Zhou, P., Li, D., 2011. Identification and characterization of differential gene expression in the mesocarp and kernel of oil palm nuts using suppression subtractive hybridization. Tree Genet. Genomes 7, 999–1010.

Yalamoussa, T., Hervé, K.K., Nklo, H., 2011. Biology of *Elaeidobius kamerunicus* and *Elaeidobius plagiatus* (Coleoptera: Curculionidae), main pollinators of oil palm in West Africa. Eur. J. Sci. Res. 49 (3), 426–432.

Yusoff, M.S.A., 2000. Refining and modifications of Palm and Palm kernel oil. In: MPO.B., Basiron, Y., Jalani, B.S., Chan, K.W. (Eds.), Advances in oil palm research, Kuala Lumpur, pp. 783–805.

Zamzuri, I., 2011. MPOB clonal propagation programme. International seminar on breeding for sustainablility in oil palm. November 18, 2011, KLCC Kuala Lumpur, Malaysia. pp. 171–184.

Zehdi, S., Sakka, H., Rhouma, A., Ould Mohamed Salem, A., Marrakchi, M., Trifi, M., 2004. Analysis of Tunisian date palm germplasm using simple sequence repeat primers. Afr. J. Biotech. 3, 215–219.

Zehdi, S., Trifi, M., Billotte, N., Merrakchi, M., Pintaud, J.C., 2005. Genetic diversity of Tunisian date palms (*Phoenix dactylifera* L.) revealed by nuclear microsatellite polymorphism. Hereditas 141, 278–287.

Zeven, A.C., 1964. On the origin of the oil palm. Grana palynol. 5, 121–123.

Zeven, A.C., 1965. Oil palm groves in southern Nigeria, Part I. Types of groves in existence. J. Nigerian Inst. Oil Palm Res. 4, 226–249.

CHAPTER

11

Olives

Aurora Díaz

Unidad de Hortofruticultura, Instituto Agroalimentario de Aragón (IA2) (CITA-Universidad de Zaragoza), Av. Montañana, Zaragoza, Spain

INTRODUCTION

The restriction of olive tree (*Olea europaea* L.) cultivation to the Mediterranean Basin could explain the delay in research advances and technology transfer to rural communities when compared with other fruit trees.

Olives have been traditionally grown in the Mediterranean area in monocultivar orchards of various sizes (Barranco, 1997). Many varieties grown locally (i.e., around their hypothetical areas of origin) and with very limited diffusion can still be found. However, this situation is starting to change. In the last two decades, the consumption of olive oil has experienced an increase of 116% (FAOSTAT, 2014). To satisfy this boom in demand, olive trees have been planted in countries with no tradition in olive-growing like China, where production reached 2000 tons in 2013 (FAOSTAT, 2014), which is expected to be much higher when trees mature to full production stage. Additionally, the traditional olive-growing system is being substituted by an intensive growing system based on monovarietal orchards. The success of this strategy in regions with a Mediterranean climate, where rainfall at critical moments of olive production is required, is highly dependent on efficient irrigation design.

There is today a transition toward more sustainable agriculture in many crops. New technologies and practices aimed at improving food safety and productivity while minimizing environmental damage need to be implemented for this to happen. Olive tree cultivation is no exception; such a tendency is likely in the near future. The transformation of traditional and/or ecologically managed olive orchards seems feasible and will probably not demand major economic investment. However, intensive olive plantations as they are conceived nowadays are very far from being sustainable as they rely heavily on a high supply of different inputs (water, fertilizers, herbicides, etc.). Furthermore, some authors state that the hedgerow system, in its current form and management, is less productive and less sustainable than orchards planted at lower densities such as 408 trees ha^{-1} (Pastor et al. 2007). Gamma radiation has been proposed as a means of inducing dwarfism in olives, which could suit modern intensive olive tree plantations and at the same time reduce harvesting costs (Lavee, 2013).

Breeding Oilseed Crops for Sustainable Production. http://dx.doi.org/10.1016/B978-0-12-801309-0.00011-2

In fact, these benefits are clearly demonstrated in the dwarf table olive cultivar Kadeshon, which was obtained using gamma radiation within the Olive-Breeding Program of Israel (Lavee, 2013).

This paints a scenario in which different olive-growing systems coexist with a possible intersection model based on sustainable intensive agriculture. In this context, genetic breeding has great potential to help generate a more sustainable olive-growing model.

CHALLENGES

Optimizing the Use of Water Resources and Developing Drought Tolerance

The total volume of fresh water used to produce a certain good or service is known as the water footprint and is considered to be a consumption-based indicator of water use (Chapagain and Hoekstra, 2008). The water footprint consists of three components: the blue water footprint, which is defined as the volume of surface or groundwater consumed to obtain a product; the green water footprint, which refers to the volume of rainwater used for the same purpose; and the gray water footprint, which is the volume of water necessary to assimilate pollution generated during production or growth. Traditionally, the blue trace has been employed to try to optimize the use of water, although increasing attention has been paid in recent years to the green water footprint to assess efficient uses of hydrological resources. Values of the green–blue water footprint for the most important oil crops are shown in Figure 11.1. No distinction has been made regarding the final product (e.g., olives or oil) in the case of olive tree cultivation. However, the operational water footprint needed to obtain the final oil product has been shown to be negligible (Ercin et al., 2009; Salmoral et al., 2011) compared with the growing and oil extraction stages, which are the main contributors to the final water footprint.

Despite olive cultivation and manufacturing having an average green–blue water footprint (2969 m^3 ton^{-1}) that is below the mean value for the oil crops shown in Figure 11.1 (3648 m^3 ton^{-1}), there are still opportunities for further improvements in water usage. In this sense, the greater the area of land irrigated, the larger the crop footprint becomes. Moreover, irrigated olive orchards raise ecological questions. This happened in the main olive oil production region (Andalusia, Spain), where within a period of 11 years 70% of new land dedicated to olive-growing was irrigated, which led to an increase of 286% in groundwater consumption and added significant pressure on the upstream Guadalquivir Basin (Salmoral et al., 2011). Another environmental disadvantage is that nitrogen input in irrigated olive tree plantations is nearly three times higher than that in rainfed olive orchards (IDAE, 2007). However, this is not such a simple scenario because, when the water footprint per unit olive fruit produced is calculated, the one in rainfed orchards (green footprint) is higher than that in irrigated plantations (green plus blue footprint) (Salmoral et al., 2011). That is, the efficiency of water consumption in relation to crop production is higher in land under irrigation due to greater yield. The intermediate strategy of regulated deficit irrigation (RDI) applied from fruit set to harvest seems to improve water use efficiency and certain olive oil quality parameters – such as oleic acid, total phenol, and tocopherol content; monounsaturated fatty acid/polyunsaturated fatty acid (MUFA/PUFA) ratios; and oil stability indices (OSI) – but with the disadvantage of decreasing oil yield (Serman et al., 2011). In areas where precipitation is scarce during

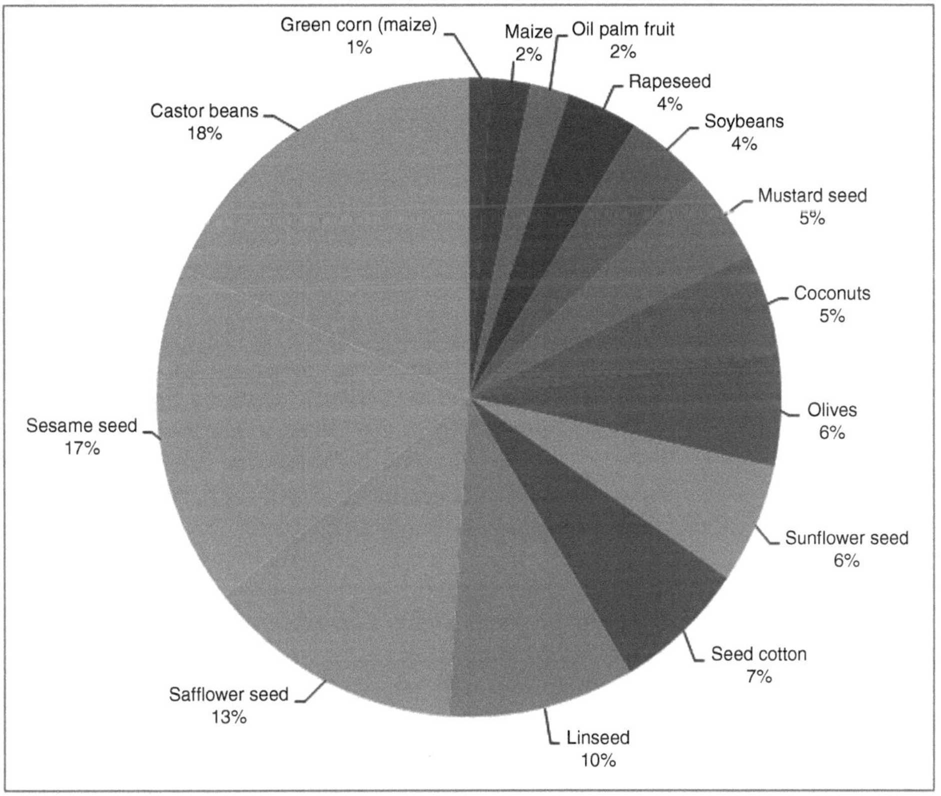

FIGURE 11.1 Green–blue water footprint percentages for the most important oil crops. *Adapted from Mekonnen and Hoekstra, 2013.*

winter and spring months – such as the arid and semiarid Argentinian olive-growing areas, which are quite unlike the Mediterranean Basin – high to full (17–100% estimated crop evapotranspiration or ETc) irrigation rates seem to be needed 2 months before anthesis to avoid the detrimental effects of water stress on fruit yield (Pierantozzi et al., 2013).

Olive trees have adapted to areas in which prolonged droughts, high temperatures, and vegetative growing periods coincide by developing physiological mechanisms to tolerate hydric stress. Olive trees have been so successful in doing this that they are the only crop capable of growing in certain arid or semiarid regions. However, the price that has to be paid is significant decrease in fruit set and yield under severe water scarcity. This is an important drawback to consider in the context of climate change since, being a perennial species, olives are expected to adapt slowly. In this sense, olives harbor enormous genetic variability, which translates into wide diversity in many different traits, drought tolerance and crop productivity included. Therefore, selection for cultivars highly tolerant to dehydration and maintaining reasonably high fruit production becomes an attractive breeding objective. For this to be achieved, it will be necessary to evaluate olive cultivars to determine the degree to which they tolerate dehydration in much the same way as carried out by Faraloni et al. (2011). The value of Faraloni et al.'s work lies not only in classifying the six cultivars studied as putatively drought tolerant (Maurino and Ascolana Tenera), intermediate (Leccino and Nocellara del Belice), and susceptible (Carolea and Biancolilla), but also in the development of an easy and fast assay of *in vitro* screening for drought tolerance, whose results correlate perfectly with those in adult plants. A number of studies comparing the level of tolerance of a few cultivars subject to water deficit have been carried out in the last 15 years (Bacelar et al., 2009; Bosabalidis and Kofidis, 2002; Boughalleb and Hajlaoui, 2011; Boussadia et al., 2008; Chartzoulakis et al., 1999; Ennajeh et al., 2008, 2010; Faraloni et al., 2011; Grisafi et al., 2004; Rezaie et al., 2004; Sofo, 2011). Ranking olive cultivars according to their drought tolerance (Table 11.1) has great value for researchers or farmers interested in starting up an olive-breeding program for tolerance to dehydration stress.

Furthermore, it is essential to understand the molecular and physiological bases underlying the ability of certain cultivars to perform better than others under water scarcity conditions. It would be enlightening to compare olive cultivars native to dry areas with those coming from more temperate regions since the former evolved to acquire drought tolerance, as suggested by Bacelar et al. (2009), and to compare the effects of different treatments (irrigation versus drought) on the same cultivar. Sofo et al. (2008) carried out such an experiment and found that olive plants subject to water deficit at first stop canopy growth but not photosynthetic activity and transpiration, making continuous production of assimilates possible. However, when the situation becomes critical, the nonstomatal component of photosynthesis is inhibited and light-dependent inactivation of photosystem II occurs. Before this stage is reached, it has been demonstrated that photosynthesis depression is genotype dependent (Bacelar et al., 2009), which is a relevant finding from a breeding perspective. Those cultivars capable of maintaining a higher photosynthetic rate with lower values of

TABLE 11.1 Order of Olive Cultivars According to Their Level of Drought Tolerance Within Each Experiment

Rank of drought tolerance	References
Biancolilla > Coratina	Sofo (2011)
Ascolana Tenera = Maurino > Leccino = Nocellara del Belice > Biancolilla = Carolea	Faraloni et al. (2011)
Zalmati > Chemlali	Boughalleb and Hajlaoui (2011)
Chemlali > **Meski**	Ennajeh et al. (2010)
Cobran Double Dagger Osa = **Manzanilla = Negrinha** > **Arbequina = Blanqueta**	Bacelar et al. (2009)
Chemlali > **Meski**	Ennajeh et al. (2008)
Meski > Koroneiki	Boussadia et al. (2008)
Dezfuli > Zard > Conservolia = Fishomi = Manzanilla	Rezaie et al. (2004)
Biancolilla > Cerausola > Nocellara del Belice	Grisafi et al. (2004)
Manzanilla = Negrinha = Cobrancoça > **Arbequina** > **Blanqueta**	Bacelar et al. (2004)
Koroneiki > **Mastoidis**	Bosabalidis and Kofidis (2002)
Koroneiki > **Mastoidis**	Chartzoulakis et al. (1999)

Matching relationships among cultivars in different experiments are shown in bold.

stomatal conductance (reduced transpiration) seem to be able to tolerate water stress better by inducing anatomical changes in leaves and other parts, which have been proposed as selection characters (Ennajeh et al., 2010; Chartzoulakis et al., 1999). All considered, some leaf-morphoanatomical adaptations (upper palisade, spongy parenchyma, upper and lower epidermis thickness, stomatal and trichome density, leaf size) improve the water use efficiency of olive trees (Ennajeh et al., 2010). Another mechanism of leaves as a response to drought is their movement upward to expose their silvery underside, which increases light reflection, although this is dependent on the water status of the plant (Schwabe and Lionakis, 1996). Apart from these transitory changes, some cultivars have more morphological and structural leaf adaptations to protect themselves against water loss than others (Bacelar et al., 2004). However, leaf photosynthetic capacity is greatly affected by other factors like leaf nitrogen content. The slope of this positive linear regression has been demonstrated to vary with irrigation intensity (Diaz-Espejo et al., 2006). Hence, for a system to be sustainable, it is necessary not only to have proper water management to avoid soil salinization and natural source overexploitation, but also to take into account the use of efficient and nonenvironmentally damaging fertilization methods, which will be addressed in the section "Increase in Productivity and Oil Quality without Increasing the Use of Chemicals" of this chapter.

The expression of antioxidant enzymes results in higher tolerance to oxidative damage induced by a number of stresses such as dehydration. This has been verified in leaves of the olive cultivar Coratina, in which SOD (superoxide dismutase), CAT (catalase), APX (ascorbate peroxidase), POD (guaiacol peroxidase), and IAAox (indoleacetate oxidase) were upregulated, whereas PPO (polyphenol oxidase) was downregulated (Sofo et al., 2005, 2008). As peroxidases are involved in lignin biosynthesis, any increase in their activity could reflect changes in cell wall properties, which are required to tolerate drought. Any decrease in PPO activity reinforces the idea of phenols themselves having an antioxidant effect. Furthermore, the higher activity of lipoxygenase (LOX) and malondialdehyde (MDA) (Sofo et al., 2008), two markers of oxidative stress, during a time of drought is associated with the oxidation of membrane lipids and, therefore, with damage to photosynthetic complexes. It seems that different genotypes show dissimilar levels of some of these enzymes under different treatments (Chatzistathis et al., 2010; Ennajeh et al., 2009; Guerfel et al., 2009), which supplies the variability necessary to be considered as a selectable trait in a potential breeding program. Interestingly, the use of arbuscular mycorrhizal fungi (AMF) not only noticeably improved the growth of olive trees under water deficit conditions when compared with plants without AMF colonization, but also strengthened drought tolerance by enhancing the activity of antioxidant enzymes that diminished the damage caused by oxidative stress usually associated with dehydration (Fouad et al., 2014; Bompadre et al., 2013). Nowadays, olive-growing systems do not use AMF to alleviate the effects of drought but, in light of these results, it is worth introducing fungi that naturally colonize many tree species in olive orchards, perhaps coupled with noncontaminating alleviating products such as heat-irradiating clay particles. Moreover, due to the complexity of plant response to oxidative stress caused by drought, a multitrait selection strategy should be adopted. The accumulation of certain soluble protectors in olive leaves has been reported not only as a defense mechanism but also as a reserve to be used during poststress recovery. Once again, differences in the accumulation of some of these solutes (i.e., proline, starch, etc.) during times of water deficit have been observed in some cultivars under study (Bacelar et al., 2009).

Enhancing Disease Tolerance Without Increasing the Use of Chemicals

Drastic decrease in the use of pesticides is essential to bringing about sustainable olive cultivation, not only because their persistent residues are environmental pollutants but also because of the rapidly growing problem of pesticide resistance developed by many causal agents. Genetic breeding offers a means of obtaining new cultivars with enhanced resistance to disease. The first step toward achieving this is to have reliable diagnosis tools available. Real-time quantitative polymerase chain reaction (qPCR) and real-time quantitative reverse-transcription PCR (qRT-PCR) have revealed themselves to be very efficient at detecting the most important viruses, bacteria, and fungi affecting olive trees due to their high specificity and sensitivity (reviewed in Díaz, 2012b).

Using different inoculation, detection, and evaluation techniques, olive cultivars have been classified according to their resistance level to diseases causing the greatest losses in this crop: *Verticillium* wilt (*Verticillium dahlia* Kleb.), leaf spot (*Spilocaea oleaginea* (Cast.) Hughes), and olive knot (*Pseudomonas savastanoi*) (Table 11.2). The use of varieties that show resistance to these diseases seems to be an inexpensive, efficient, and environmental friendly approach. They are in fact currently included in many olive-breeding programs. Furthermore, susceptible cultivars valuable for their yield or quality traits could be grafted onto resistant rootstocks and planted in infected and/or infested soils. Another fact to take into account is that some cultivars show the capacity to recover naturally from some diseases, as in the case of Barnea (but not Souri) trees infected with *Verticillium* (Levin et al., 2003).

In an olive tree–growing scenario that is sustainable, the biological control of diseases using microorganisms naturally associated with the species reveals itself as a promising strategy with no ecological impact. Mercado-Blanco et al. (2004) demonstrated that the treatment of olive trees with certain *Pseudomonas* sp. isolated from olive roots drastically reduced *Verticillium* wilt incidence and severity. In the case of olive knot, it has been observed that another *Pseudomonas* sp. (i.e., *P. fluorescens*), also present in olive roots, effectively suppresses *Verticillium* wilt. It is able to decrease the population size of the pathogen and the number of tumors, as well as to confine the causal agent inside the knots (Maldonado-Gonzalez et al., 2013), supporting its antagonistic role against *P. savastonoi* previously described by Krid et al., (2010). Other species of microorganisms – such as *Lawsonia inermis* (Trigui et al., 2013), *Bacillus subtilis* (Krid et al., 2010, 2012), *Pantonea agglomerans* (Marchi et al., 2006), *Rhizobium* spp. (Mourad et al., 2009), and even olive mill waste (OMW) (Krid et al., 2011) – showed bactericidal activity against *P. savastanoi*. The use of OMW has revealed itself as not only successful at controlling *Verticillium* (Papasotiriou et al., 2013), but also having a positive side effect, as it is a useful way of recycling these residues. These examples of biocontrol are not restricted to diseases caused by microorganisms, as it has been verified that AMF reduced significantly the severity of root galling caused by nematodes (Castillo et al., 2006). Some of these bacteria produced salicylic acid (SA), whose key role in biotic stress responses is well known. Interestingly, SA was the most effective product tested by Obanor et al. (2013) to induce resistance to leaf spot. Actually, it reduced disease severity by a similar percentage to that obtained with traditional copper sulfate treatments. SA endogenous production stimulation and/or exogenous applications could make olive trees acquire resistance by getting round problems such as pesticide resistance and copper soil contamination.

TABLE 11.2 Classification of Cultivars, Rootstocks, and Wild Accessions for Three of the Most Important Diseases Affecting Olive Trees, Including a Ranking When Available

Diseases	Susceptible	Resistant[1]	References
Verticillium wilt	Bical, Corralones de Andújar, Manzanilla de Abla, Manzanilla del Centro, Manzanillo de Antiesteban Pto, Negrillo de Iznalloz, Perillo de Jaén, Picual, Racimal de Jaén	Escarabajillo, Frantoio, Menya, Sevillana de Abla	Garcia-Ruiz et al. (2014)
	Sikitita	Frantoio > Arbosana, Koroneiki	Trapero et al. (2014)
	Bodoquera, Cornicabra, Hojiblanca, Manzanilla de Sevilla, Picual	Arbequina, Koroneiki, Sevillenca < Changlot Real, Empeltre, Frantoio	Trapero et al. (2013)
	Leccino > Coratina	Frantoio	Bubici and Cirulli (2012)
	Ak Zeytin, Ascolana, Belluti, Butko, Chemlali, D9, Gemlik, Leccino, Otur, Samanli, Saurani, Sinop No. 2, Yag Celebi, Yag Zeytini	Arbequina, Dilmit, D36, Egriburun Nizip, Egriburun Tatayn, Erdek Yaglik, Erkence, Frantoio, Girit Zeytini, Hurma Karaca, Kan Celebi, Marantelli, Melkabazi, Sam, Sari Habesi, Sinop No. 1, Siyah Salamuralik, Yaglik Celebi, Yun Celebi, Zoncuk	Erten and Yildiz (2011)
	Manzanilla > Ayvalyk	Gemlik	Dervis et al. (2010)
	Tonda Iblea	—	Lo Giudice et al. (2010)
	Domat, Edremit, Gemlik 5, Manzanilla 1, Manzanilla 3, Memecik, Uslu	Gemlik 2[2]	Sesli et al. (2010)
	Amphissis	Kalamon, Koroneiki	Markakis et al. (2009)
	Amphissis	Ambrakia[3], Kalamon, Koroneiki, Lianolia of Corfu, Nicopolis[3]	Antoniou et al. (2008)
	Leccino	Coratina[4], Frantoio	Cirulli et al. (2008)
	Cima di Mola, wild accessions: Paveg 1–11, 13, 14, 16, 17 > Paveg 12, 15, 18–24	Wild accessions: Paveg 53–57 > 47, 51–52	Colella et al. (2008)
	Arbequina, Arbosana, Ascolana Tenera, Ayvalik, Azapa, Borriolenca, Bouteillan, Callosina, Carrasqueño de Lucena, Chemlal de Kabilye, Dokkar, Dulzal de Carmona, Fulla de Salce, Gordal de Hellín, Imperial de Jaén, Kalogerida, Manzanilla, Manzanilla Cacereña, Megaritiki, Morona, Nabli, Negro del Carpio, Nevadillo Blanco de Jaén, Pendolino, Picual, Zarza	Changlot Real, Cipresino[5], Empeltre, Frantoio, Koroneiki[5], Oblonga[5], Sevillenca[5]	Martos-Moreno et al. (2006)

(Continued)

TABLE 11.2 Classification of Cultivars, Rootstocks, and Wild Accessions for Three of the Most Important Diseases Affecting Olive Trees, Including a Ranking When Available *(cont.)*

Diseases	Susceptible	Resistant[1]	References
	Arbequina, Cornicabra, Hendeño de Hinojosa del Duque, Hojiblanca, Leccino, Lechín de Granada, Manzanilla del Piquito, Meski, Negral, Nevadillo Negro de Jaén, Ocal, Pajarero, Picual, Picudo, Valanolia, Villalonga	Cobrancoça[6], Empeltre[7], Frantoio[7], Manzanilla de Sevilla[6], Morisca[6], Oblonga[7], Verdial de Alcaudete[6]	Blanco-López and López-Escudero (2005)
	Arbequina, Cornicabra, Hendeño, Hojiblanca, Manzanilla Piquito, Leccino, Lechín de Granada, Meski, Negral, Nevadillo Negro, Ocal, Pajarero, Picudo, Picual, Valanolia, Villalonga	Cobrancosa[8], Empeltre[9], Frantoio[9], Manzanilla de Sevilla[8], Morisca[8], Oblonga[9], Verdial Alcaudete[8]	Lopez-Escudero et al. (2004)
	Ayvalik, Memecik, Uslu	Gemlik	Erten (2004)
	Cornicabra	—	Soriano et al. (2002)
	Sevillano	Oblonga	Hartmann et al. (1971)
	Bella di Cerignola	—	Ruggieri (1946)
Leaf spot	Picholine Marocaine	Picholine du Languedoc	Rahioui et al. (2013), El Aabidine et al. (2010)
	Alameño de Cabra, Aloreña de Atarfe-1, Blanqueta, Callosina, Carrasqueño de la Sierra, Changlot Real, Cornezuelo, Cornezuelo de Jaén, Cornicabra, Gordal de Granada, Hojiblanca, Imperial de Jaén, Lechín de Granada, Loaime, Lucio, Manzanilla de Sevilla, Marteño de la Carlota, Meski, Morisca, Morona, Negrillo de Estepa, Nevadillo Negro de Jaén, Ocal, Pajarero, Palomar, Picholine Marroquí, Pico Limón de Grazalema, Picual, Picudo, Verdial de Huévar, Verdial de Vélez Málaga, Verdiell	Arbosana, Borriolenca, Caninese, Chetoui, Escarabajillo, Frantoio, Galega Vulgar, Gordal de Archidona, Koroneiki, Leccino, Lechín de Sevilla, Lentisca, Maelia, Manzanilla de Montefrío, Nevado Rizado, Oblonga, Pequeña de Casas Ibáñez, Racimal, Rapasayo, Razzola, Temprano, Vallesa, Zarza	Trapero and López-Doncel (2005)
	Attica, Barouni, Mission, Novo, Picual	Corregiola, Koroneiki, Leccino	Mekuria et al. (2002)[10]
	Muhasan, Souri	Maalot > Barnea > Manzanillo	Lavee et al. (1999), Mekuria et al. (2001)
	Carolea, Cassanese > Leccino, Noceralla del Belice	Nostrale del Rigali[11], Roggianella	Ciccarese et al. (2002)

TABLE 11.2 Classification of Cultivars, Rootstocks, and Wild Accessions for Three of the Most Important Diseases Affecting Olive Trees, Including a Ranking When Available *(cont.)*

Diseases	Susceptible	Resistant[1]	References
Olive knot	Arbequina, Arróniz, Nevadillo Banco de Jaén, Pajarero, Picudo, Vallesa > Ascolana Tenera, Changlot Real, Gordal Sevillana, Koroneiki, Lechín de Sevilla, Morisca, Picual, Royal de Cazorla > Azapa, Cerezuela, Chemlali, Dulzal de Carmona, Frantoio, FS-17, Gordal de Archidona, Gordal de Hellín, Lechín de Granada, Manzanilla Cacereña, Manzanilla de Sevilla, Mollar de Cieza, Nevadillo Negro de Jaén, Villalonga	—	Peñalver et al. (2005)
	Manzanillo, Picholine, S.A. Verdale > Carolea, Koroneiki, Pendolino	—	Young (2004)
	Barnea, Manzanillo, Picholine, Picual, Verdale > Carolea, Koroneiki, Leccino, Pendolino	—	Young et al. (2004)
	Cellini di Nardo, Frantoio, FS-17, Morcana, Nociara, Ogliarola, Pendolino > Bella di Spagna, Carolea, Cerasela, Cima di Melfi, Coratina, Corniola, Dolce Agogia, Leucocarpa, Maiatica di Ferrandina, Nolca and San Felice, N3	—	Sisto et al. (2001)
	Cordovil de Serpa, Galga Vulgar > Branquita, Santulhana	Cobrancosa, Rendonil	Marcelo et al. (1999)
	Nocellara del Belice > Leccino	Coratina	Varvaro and Martella (1993)
	Meslala > Picholine Marocaine	Gordale	Benjama et al. (1987)
	Ajrasi, Ajrasi bazri, Bashiky	Hawega-2	Osman et al. (1980)

[1] *No olive cultivars are known that are completely resistant to olive knot.*
[2] *Moderately susceptible.*
[3] *Rootstock sucker.*
[4] *Partially resistant to nondefoliating pathotypes, but susceptible to defoliating pathotypes.*
[5] *Moderately resistant.*
[6] *Resistant to nondefoliating pathotypes, but extremely susceptible to defoliating pathotypes.*
[7] *Resistant to nondefoliating pathotypes, but moderately susceptible to defoliating pathotypes.*
[8] *Extremely susceptible to defoliating pathotypes, but resistant to nondefoliating pathotypes.*
[9] *Moderately susceptible to defoliating pathotypes, but resistant to nondefoliating pathotypes.*
[10] *Information from FAO (1996).*
[11] *Only in the greenhouse.*

Olive tree populations isolated by geographical barriers – such as those in East China, East Australia, South Africa (Green and Wickans, 1989; Breton et al., 2008), and the Central Sahara Mountains (Besnard et al., 2007) – are nowadays considered *Olea europaea* subspecies and could be an important source of gene variants carrying interesting characteristics like disease resistance.

Genetic transformation has great potential for sustainable agriculture. Transgenic olive plants with osmotin genes have been obtained (Rugini et al., 1999). This led Rugini et al. to think that such plants could show increased resistance to fungal diseases, as the genes coding for this protein are induced in response to pathogenesis processes. However, only cold protection induced by osmotin has been verified to date in transgenic olive plants (D'Angeli and Altamura 2007). Interestingly, cryoprotection involves programmed cell death (PCD) and cytoskeleton organization as happens with an osmotin-like protein in a potato cell's defense against the fungus that causes late blight of potato (Takemoto et al., 1997). A major advance here will be the availability of the olive genome, which is currently being sequenced by several research groups, including those within the OLEAGEN project.

Integrated and sustainable management of these and other diseases should combine the above proposed strategies with organic soil amendments and farming techniques.

Increase in Productivity and Oil Quality Without Increasing the Use of Chemicals

The productivity of olive trees is determined by a combination of factors, mainly weather conditions, water availability, nutrient status, and crop management techniques like pruning. This is the reason hedgerow models for continuing productivity are based on irrigation, fertilization, pruning to maintain bush structure, and pest and disease control. Superhigh-density orchards are to date mainly planted with four varieties: Arbequina, Arbosana, Koroneiki, and the recently released Sikitita. However, there is also a strong genotype effect. Much variability was observed here in assays carried out at two different germplasm banks – the Centro de Investigación y Formación Agraria (CIFA) World Germplasm Bank called "Alameda del Obispo" in Córdoba (Del Río et al., 2005a) and the Catalonia Germplasm Bank (Tous et al., 2005b) – where the mean productivity of different cultivars was measured. The rank of cultivars studied in both locations was reasonably similar with expected differences attributable to environmental conditions. In both cases, the cultivar reaching the highest mean production was Blanqueta. Similar studies were carried out at the same germplasm banks. These aimed at measuring oil content (Del Rio et al., 2005b; Tous and Romero, 2005) and analyzing oil composition (Tous et al., 2005a; Uceda et al., 2005). Again, both sets of data were mainly in agreement when it came to cultivars with the highest and lowest oil content. From the oil quality point of view, these results are very useful at identifying outstanding cultivars, such as Picual, which showed the highest values for the MUFA/PUFA ratio, stability, and total polyphenol content. The latter greatly influences the sensorial properties and nutritional value of olive oil.

Recent studies show that the genotype effect is also significant in oil content–related traits and is the main factor contributing to variation of fatty acids in olive oil, in addition to other fruit traits like weight (De la Rosa et al., 2013; Leon et al., 2011). The values of almost all fatty acids obtained in advanced selection from the breeding program in Córdoba exceeded the variation range observed in the genitors (Arbequina and Picual), making them good targets for genetic improvement. In fact, some of these genotypes have been claimed to be promising for producing fruits of a good size and oil content (UC-I-42-48); with a high quantity of tocopherols (UC-I 7-8), polyphenols (UC-I 5-44), and C18:1 fatty acids (UC-I 2-68); and with increased oleic acid content (UC-I-2-35), which makes the oil less susceptible to oxidative changes. The latter genotype could also be a good candidate for planting in low-latitude regions where the oleic acid content in oil from adapted cultivars tends to be too low.

Cross breeding is giving results, as is the case with the new variety Askal (Lavee et al., 2003), which rendered higher values of oil per fruit weight, fruit yield per hectare, and oil yield per hectare than cultivars like Barnea and Manzanillo. Much as happens when searching for disease resistance, the use of wild olive trees as genitors in breeding programs could contribute to improving certain characteristics of modern cultivars. In line with this, the progeny obtained by open pollination of the wild accession Alga05 showed more abundant flowering and higher intensity of bearing than that from open-pollinated Picual trees (Klepo et al., 2013), which obviously have profitable consequences on production. Furthermore, some isolated wild populations have survived by inbreeding, which is expected to have increased their self-fertility. As fruit set is also determined by cultivar selfcompatibility when there are no other means of pollination in the surroundings (Díaz et al., 2006), an increased ability for self-fertilization could also cause improvements in production. A direct consequence of enhanced self-fertility is greater degree of homozygosis, which could be exploited to handle characters that exhibit superior values by heterosis, as commonly happens for yield.

Much as happens with disease resistance, trees with AMF have higher fruit and oil yields than nonmycorrhitic controls (Kapulnik et al., 2010) or exhibit higher N, P, and K concentrations (Porras-Soriano et al., 2006), all of which ultimately affects crop yield in a positive way (Ben Rouina et al., 2002). These findings should be taken into account when evaluating the results of genetic breeding.

Combining two objectives, productivity and oil quality, in a single breeding strategy seems feasible, as in the case of the breeding program undertaken in Tunisia in 1989 in which the parents were chosen because of their good performance in one or a few trait components (Trigui et al., 2006). Therefore, dominant cultivars in Tunisia, such as Chemlali, and others of some importance, such as Chemchali, Chetoui, and Meski – characterized by high productivity and providing an oil of tolerable quality according to international referees (due to high levels of palmitic acid and low levels of oleic acid) – were crossed with foreign cultivars that could counteract negative features. As a result, crossing varieties with improved oil characteristics have been obtained – such as 6H and 9D (Baccouri et al., 2007), Hd 034 and Hd 039 (Manai et al., 2007), Hd 031, Hd 034, Hd 038, Hd 039, Hd 044, and Hd 045 (Manai et al., 2008), and CHMC 1 and SCHL 1 (Rjiba et al., 2010) – two of which came from a cross between two Tunisian cultivars.

Many characteristics pursued by olive-breeding programs are related to yield (fruit and oil) and oil quality; therefore, they are polygenic. Hence, the identification and mapping of quantitative trait loci (QTL) turn out to be a feasible strategy, as is currently being applied to several oil crops (reviewed in Díaz et al., 2012a). QTL for important agronomic traits in olives have recently been identified (Atienza et al., 2014), one of which explains around 12.7–15.2% of variability found for fruit weight.

Isolation of a number of genes that encode for key enzymes in fatty acids, antioxidant biosynthesis and modification, triacylglycerol storage, and respective expression assays (Hatzopoulos et al., 2002) could speed up the breeding process for oil content and quality through genetic transformation.

Biofortification

One of the goals of sustainable agriculture is to fight hunger. Biofortification, the breeding of crops to increase their nutritional value, is growing in relevance here. Biofortification can have preventive purposes in such areas as cardiac pathologies. For instance, there is strong evidence that the health benefits of olive oil play a part in preventing heart disease (Visioli and Galli, 1998; Covas et al., 2006) (e.g., by lowering blood pressure; Gilani et al., 2005). The beneficial effect of omega-3 fatty acids on patients with cardiovascular disease as well as on healthy individual is well known. The Olive-Breeding Program undertaken at the Horticultural Research Institute in Yalova (Turkey) led to genotypes from crosses of six different cultivars, which succeeded in producing oils that were rich in omega-3 (Ozdemir et al., 2013). The high linolenic content of some of these oils could satisfy the daily need for omega-3 intake in the human diet. Furthermore, some of these genotypes rendered oils with outstanding characteristics. In addition to high omega-3 content, they had low values for the omega-6/omega-3 ratio. As Ozdemir et al. claim, producing omega-3–enriched oil by cross breeding is one of the cheapest, simplest, and most natural ways of obtaining omega-3 and delivering it to the consumer.

Olive oil is unique in having an oleic acid content that can be as high as 72%. The role played by oleic acid in cancer prevention is still not clear as it is not known whether it is due to its protective properties against oxidative stress or whether it has a direct anticancer effect (reviewed in Waterman and Lockwood, 2007).

Moreover, foods are fortified with health-promoting additives like vitamins. Oil from the recently released cultivar Sikitita, the culmination of a genetic breeding program, has higher β-carotene content than its genitors, Arbequina and Picual (Roca et al., 2011). β-carotene is the best-known provitamin A carotenoid, has high antioxidant capacity, inhibits low-density lipoprotein (LDL) oxidation (much like some polyphenols) (Visioli et al., 2002), and might have an important role to play in the prevention of atherosclerosis.

Polyphenolic compounds are potent antioxidants and have radical scavenging capacity, which confers protection against damage brought about by oxidative stress (Waterman and Lockwood, 2007). They give the oil a characteristic flavor and aroma and contribute to its stability (Ayton et al., 2007; Gutiérrez et al., 2001) and, hence, to its shelf life. Interestingly, some polyphenols present in olive oil, like hydroxytyrosol and oleuropein, have been reported to be able to prevent DNA damage. In doing so, they might be able to prevent mutagenesis and, hence, carcinogenesis (Visioli et al., 2002). These two compounds plus tyrosol are polyphenols that exhibit antimicrobial activity, providing protection against intestinal and respiratory infections (Tuck and Hayball, 2002). In line with the comments it is not just the polyphenolic content that is genotype dependent, the polyphenol profiles are too. This is the reason, these compounds are usually measured in oils from breeding progeny aimed at obtaining varieties rendering high-quality oils.

CONSTRAINTS

Sustainable olive cultivation faces the same general limitations that burden other crops, as well as some that are specific to it. Soil erosion and impoverishment is a critical problem in agricultural and environmental sustainability. Avoiding erosion in olive orchards is possible by tillage management or even designing plantations with a hedgerow structure. However, species rotation usually suggested for maintaining the quality of organic soil matter is obviously not possible in a tree crop.

An olive cultivation model based on limited use of pesticides, herbicides, and fertilizers (even with the genetic development of resistant, highly productive varieties) may have to accept reduced yields. So, the question that needs to be asked before proposing a new olive cultivation model is whether we are prepared to accept some losses.

Being a perennial tree with a long juvenile period, the turnover rate of olive trees is extremely low. This is another reason large-scale changes introduced in olive-growing need to be planned thoughtfully. Moreover, technological designs and field evaluations are considerably slower, more difficult, and costly due to the longer time horizons needed. So, administration involvement (e.g., incentives that encourage farmers to adopt new technologies and models that do not give profitable returns in the near future) is essential. Much of the land dedicated to olive-growing remains in the hands of small farmers. This is the reason it will have to be viable for them, as they will be playing an important role if sustainable olive cultivation is the objective. Another actor in this challenging situation is the research community. Again, a sustained compromise from governments is required to support this model and for it to be able to react in time to adverse situations, such us cultivar resistance being overcome by pathogens.

Abbreviations

AMF	Arbuscular mycorrhizal fungi
APX	Ascorbate peroxidase
CAT	Catalase
ETc	Estimated crop evapotranspiration
IAAox	Indoleacetate oxidase
LDL	Low density lipoproteins
LOX	Lipoxygenase
MDA	Malondialdehyde
MUFA/PUFA	Monounsaturated Fatty Acid/Polyunsaturated Fatty Acid
OMW	Olive mill waste
PCD	PROGRAMMED Cell Death
POD	Guaiacol peroxidase
PPO	Polyphenol oxidase
qPCR	Real-time quantitative PCR
qRT-PCR	Real-time quantitative reverse-transcription PCR
QTL	Quantitative Trait Loci
SOD	Superoxide dismutase

References

Antoniou, P.P., Markakis, E.A., Tjamos, S.E., Paplomatas, E.J., Tjamos, E.C., 2008. Novel methodologies in screening and selecting olive varieties and root-stocks for resistance to *Verticillium dahlia*. Eur. J. Plant Pathol. 122, 549–560.

Atienza, S.G., De la Rosa, R., Leon, L., Martin, A., Belaj, A., 2014. Identification of QTL for agronomic traits of importance for olive breeding. Mol. Breed. 34, 725–737.

Ayton, J., Mailer, R.J., Haigh, A., Tronson, D., Conlan, D., 2007. Quality and oxidative stability of Australian olive oil according to harvest date and irrigation. J. Food Lipids 14, 138–156.

Baccouri, B., Ben Temime, S., Taamalli, W., Daoud, D., M'Sallem, M., Zarrouk, M., 2007. Analytical characteristics of virgin olive oils from two new varieties obtained by controlled crossing on Meski variety. J. Food Lipids 14, 19–34.

Bacelar, E.A., Correia, C.M., Moutinho-Pereira, J.M., Goncalves, B.C., Lopes, J.I., Torres-Pereira, J.M., 2004. Sclerophylly and leaf anatomical traits of five field-grown olive cultivars growing under drought conditions. Tree Physiol. 24, 233–239.

Bacelar, E.A., Moutinho-Pereira, J.M., Goncalves, B.C., Lopes, J.I., Correia, C.M., 2009. Physiological responses of different olive genotypes to drought conditions. Acta Physiol. Plant. 31, 611–621.

Barranco, D., 1997. Variedades y patrones. In: Barranco, D., Fernández-Escobar, R., Rallo, L. (Eds.), El cultivo del olivo. Mundi-Prensa y Junta de Andalucía, Madrid, pp. 59–80.

Ben Rouina, B., Trigui, A., Boukhris, M., 2002. Effect of tree growth and nutrients status of "Chemlali de Sfax" olive trees and their productivity. Acta Hortic. 586, 349–352.

Benjama, A., Walali, L., Jananti, L., Moukhli, A., 1987. Field reaction of different varieties of olive (*Olea europaea* L.) to olive knot disease caused by *Pseudomonas syringae* pv. savastanoi (E.F. Smith) Stevens. Third International Working Group on *Pseudomonas syringae* pathovars. Lisbon, Portugal, 1–4 September 1987. pp. 41–52.

Besnard, G., Christin, P.A., Baali-Cherif, D., Bouguedoura, N., Anthelme, F., 2007. Spatial genetic structure in the laperrine's olive (*Olea europaea* subsp. *laperrinei*) along-living tree from the central Saharan mountains. Heredity 99, 649–657.

Blanco-López, M.A., López-Escudero, F.J., 2005. In: Rallo, L., Barranco, D., Caballero, J.M., Del Río, C., Martín, A., Tous, J., Trujillo, I. (Eds.), Variedades de olivo en España (Book II: Variabilidad y selección). Junta de Andalucía, MAPA y ediciones Mundi-Prensa, Madrid, pp. 329–338.

Bompadre, M.J., Rios De Molina, M.C., Colombo, R.P., Fernandez Bidondo, L., Silvani, V.A., Pardo, A.G., Ocampo, J.A., Godeas, A.M., 2013. Differential efficiency of two strains of the arbuscular mycorrhizal fungus *Rhizophagus irregularis* on olive (*Olea europaea*) plants under two water regimes. Symbiosis 61, 105–112.

Bosabalidis, A.M., Kofidis, G., 2002. Comparative effects of drought stress on leaf anatomy of two olive cultivars. Plant Sci. 163, 375–379.

Boughalleb, F., Hajlaoui, H., 2011. Physiological and anatomical changes induced by drought in two olive cultivars (cv Zalmati and Chemlali). Acta Physiol. Plant. 33, 53–65.

Boussadia, O., Ben Mariem, F., Mechri, B., Boussetta, W., Braham, M., El Hadj, S.B., 2008. Response to drought of two olive tree cultivars (cv Koroneki and Meski). Sci. Hortic. 116, 388–393.

Breton, C., Guerin, J., Ducatillion, C., Medail, F., Kull, A.C., Berville, A., 2008. Taming the wild and wilding the tame: tree breeding and dispersal in Australia and the Mediterranean. Plant Sci. 175, 197–205.

Bubici, G., Cirulli, M., 2012. Control of Verticillium wilt of olive by resistant rootstocks. Plant Soil 352, 363–376.

Castillo, P., Nico, A.I., Azcon-Aguilar, C., Rincon, C., Del Rio, Calvet, C., Jimenez-Diaz, R.M., 2006. Protection of olive planting stocks against parasitism of root-knot nematodes by arbuscular mycorrhizal fungi. Plant Pathol. 55, 705–713.

Chapagain, A.K., Hoekstra, A.Y., 2008. The global component of freshwater demand and supply: an assessment of virtual water flows between nations as a result of trade in agricultural and industrial product. Water Int. 33, 19–32.

Chartzoulakis, K., Patakas, A., Bosabalidis, A.M., 1999. Changes in water relations, photosynthesis and leaf anatomy induced by intermittent drought in two olive cultivars. Environ. Exp. Bot. 42, 113–120.

Chatzistathis, T., Therios, I., Diamantidis, G., 2010. Polyphenol oxidase activity of two olive (*Olea europaea* L.) cultivars as an early criterion of Mn toxicity. J. Sci. Food Agric. 90, 676–682.

Ciccarese, F., Ambrico, A., Longo, O., Schiavone, D., 2002. Search for resistance to *Verticillium*-wilt and leaf spot in olive. Acta Hortic. 586, 717–720.

Cirulli, M., Colella, C., D'Arnico, M., Amenduni, M., Bubici, G., 2008. Comparison of screening methods for the evaluation of olive resistance to *Verticillium dahliae* Kleb. J. Plant Pathol. 90, 7–14.

Colella, C., Miacola, C., Amenduni, M., D'Amico, M., Bubici, G., Cirulli, M., 2008. Sources of *Verticillium* wilt resistance in wild olive germplasm from the Mediterranean region. Plant Pathol. 57, 533–539.

Covas, M.I., Nyyssönen, K., Poulsen, H.E., Kaikkonen, J., Zunft, H.J., Kiesewetter, H., Gaddi, A., de la Torre, R., Mursu, J., Bäumler, H., Nascetti, S., Salonen, J.T., Fitó, M., Virtanen, J., Marrugat, J., EUROLIVE Study Group, 2006. The effect of polyphenols in olive oil on heart disease risk factors: a randomized trial. Ann. Intern. Med. 145, 333–341.

D'Angeli, S., Altamura, M.M., 2007. Osmotin induces cold protection in olive trees by affecting programmed cell death and cytoskeleton organization. Planta 225, 1147–1163.

De la Rosa, R., Talhaoui, N., Rouis, H., Velasco, L., Leon, L., 2013. Fruit characteristics and fatty acid composition in advanced olive breeding selections along the ripening period. Food Res. Int. 54, 1890–1896.

Del Rio, C., Caballero, J.M., Garcia-Fernandez, M.D., 2005b. Rendimiento graso de la aceituna (Banco de Germoplasma de Córdoba). In: Rallo, L., Barranco, D., Caballero, J.M., Del Rio, C., Martin, A., Tous, J., Trujillo, I. (Eds.), Variedades de olivo en España (Book II: Variabilidad y selección). Junta de Andalucía, MAPA and Ediciones Mundi-Prensa, Madrid, pp. 347–356.

Del Río, C., Caballero, J.M., García-Fernández, M.D., 2005a. Producción. In: Rallo, L., Barranco, D., Caballero, J.M., Del Río, C., Martín, A., Tous, J., Trujillo, I. (Eds.), Variedades de olivo en España (Book II: Variabilidad y selección). Junta de Andalucía, MAPA y ediciones Mundi-Prensa, Madrid, pp. 257–274.

Dervis, S., Mercado-Blanco, J., Erten, L., Valverde-Corredor, A., Perez-Artes, E., 2010. Verticillium wilt of olive in Turkey: a survey on disease importance, pathogen diversity and susceptibility of relevant olive cultivars. Eur. J. Plant Pathol. 127, 287–301.

Díaz, A., 2012a. Future perspectives. Gupta, S.K. (Ed.), Technological Innovations in Major World Oil Crops, Breeding, vol. 1, Springer, Nueva York, pp. 305–313.

Díaz, A., 2012b. Olive (*Olea europaea* L.). Gupta, S.K. (Ed.), Technological Innovations in Major World Oil Crops, Breeding, vol. 1, Springer, Nueva York, pp. 267–292.

Díaz, A., Martin, A., Rallo, P., Barranco, D., De la Rosa, R., 2006. Self-incompatibility of 'Arbequina' and 'Picual' olive assessed by SSR markers. J. Am. Soc. Hortic. Sci. 131, 250–255.

Diaz-Espejo, A., Walcroft, A.S., Fernandez, J.E., Hafidi, B., Palomo, M.J., Giron, I.F., 2006. Modeling photosynthesis in olive leaves under drought conditions. Tree Physiol. 26, 1445–1456.

El Aabidine, A.Z., Baissac, Y., Moukhli, A., Jay-Allemand, C., Khadar, B., El Modafar, C., 2010. Resistance of olive tree to *Spilocaea oleagina* is mediated by the synthesis of phenolic compounds. Int. J. Agric. Biol. 12, 61–67.

Ennajeh, M., Tounekti, T., Vadel, A.M., Khemira, H., Cochard, H., 2008. Water relations and drought-induced embolism in olive (*Olea europaea*) varieties 'Meski' and 'Chemlali' during severe drought. Tree Physiol. 28, 971–976.

Ennajeh, M., Vadel, A.M., Khemira, H., 2009. Osmoregulation and osmoprotection in the leaf cells of two olive cultivars subjected to severe water deficit. Acta Physiol. Plant. 31, 711–721.

Ennajeh, M., Vadel, A.M., Cochard, H., Khemira, H., 2010. Comparative impacts of water stress on the leaf anatomy of a drought-resistant and a drought-sensitive olive cultivar. J. Hortic. Sci. Biotechnol. 85, 289–294.

Ercin, A.E, Aldaya, M.M., Hoekstra, A.Y., 2009. A pilot in corporate water footprint accounting and impact assessment, Value of Water Research Report Series No: 39, UNESCO-IHE, Delft, The Netherlands.

Erten, L. 2004. Determination of susceptibility of some olive and rootstocks against Verticillium wilt (*Verticillium dahliae* Kleb.). PhD Thesis, Aegean University, Turkey.

Erten, L., Yildiz, M., 2011. Screening for resistance of Turkish olive cultivars and clonal rootstocks to Verticillium wilt. Phytoparasitica 39, 83–92.

Food and Agricultural Organization (FAO), 1996. Olive germplasm: cultivars and world-wide collections. Plant prod and Protection Div.[http://apps3.fao.org/wiews/olive/oliv.jsp].

Food and Agricultural Organization of the United Nations: FAOSTAT, 2014. [http://faostat3.fao.org/faostat-gateway/go/to/home/E].

Faraloni, C., Cutino, I., Petruccelli, R., Leva, A.R., Lazzeri, S., Torzillo, G., 2011. Chlorophyll fluorescence technique as a rapid tool for in vitro screening of olive cultivars (*Olea europaea* L.) tolerant to drought stress. Environ. Exp. Bot. 73, 49–56.

Fouad, M.O., Essahibi, A., Benhiba, L., Qaddoury, A., 2014. Effectiveness of arbuscular mycorrhizal fungi in the protection of olive plants against oxidative stress induced by drought. Spanish J. Agric. Res. 12, 763–771.

Garcia-Ruiz, G.M., Trapero, C., Del Rio, C., Lopez-Escudero, F.J., 2014. Evaluation of resistance of Spanish olive cultivars to *Verticillium dahliae* in inoculations conducted in greenhouse. Phytoparasitica 42, 205–212.

Gilani, A.H., Khan, A.U., Shah, A.J., Connor, J., Jabeen, Q., 2005. Blood pressure lowering effect of olive is mediated through calcium channel blockade. Int. J. Food Sci. Nutr. 56, 613–620.

Green, P.S., Wickans, G.E., 1989. The Olea europaea complex. In: Tan, K. (Ed.), The Davis and Hadge Festschrift. Edinburgh University Press, Edinburgh, pp. 287–299.

Grisafi, F., Bonafede, E., Vecchia, F.D., Rascio, N., 2004. Some morphological, anatomical, physiological responses of different olive cultivars to high temperatures and drought stress. Acta Botanica Gallica 151, 241–253.

Guerfel, M., Ouni, Y., Boujnah, D., Zarrouk, M., 2009. Photosynthesis parameters and activities of enzymes of oxidative stress in two young 'Chemlali' and 'Chetoui' olive trees under water deficit. Photosynthetica 47, 340–346.

Gutiérrez, F., Arnaud, T., Garrido, A., 2001. Contribution of polyphenols to the oxidative stability of virgin olive oil. J. Sci. Food Agric. 81, 1463–1470.

Hartmann, H., Schnatho, W.C., Whisler, J., 1971. Oblonga …a clonal olive rootstock resistant to verticillium wilt. Calif. Agric. 25, 12–15.

Hatzopoulos, P., Banilas, G., Giannoulia, K., Gazis, F., Nikoloudakis, N., Milloni, D., Haralampidis, K., 2002. Breeding, molecular markers and molecular biology of the olive tree. Eur. J. Lipid Sci. Technol. 104, 574–586.

IDAE, 2007. Ahorro, eficiencia energética y fertilización nitrogenada. Instituto para la Diversificación y Ahorro de la Energía, Madrid.

Kapulnik, Y., Tsror, L., Zipori, I., Hazanovsky, M., Wininger, S., Dag, A., 2010. Effect of AMF application on growth, productivity and susceptibility to Verticillium wilt of olives grown under desert conditions. Symbiosis 52, 103–111.

Klepo, T., De la Rosa, R., Satovic, Z., Leon, L., Belaj, A., 2013. Utility of wild germplasm in olive breeding. Sci. Hortic. 152, 92–101.

Krid, S., Rhouma, A., Mogou, I., Quesada, J.M., Nesme, X., Gargouri, A., 2010. *Pseudomonas* savastanoi endophytic bacteria in olive tree knots and antagonistic potential of strains of *Pseudomonas fluorescens* and *Bacillus subtilis*. J. Plant Pathol. 92, 335–341.

Krid, S., Bouaziz, M., Triki, M.A., Gargouri, A., Rhouma, A., 2011. Inhibition of olive knot disease by polyphenols extracted from olive mill waste water. J. Plant Pathol. 93, 561–568.

Krid, S., Triki, M.A., Gargouri, A., Rhouma, A., 2012. Biocontrol of olive knot disease by *Bacillus subtilis* isolated from olive leaves. Ann. Microbiol. 62, 149–154.

Lavee, S., 2013. Evaluation of the need and present potential of olive breeding indicating the nature of the available genetic resources involved. Sci. Hortic. 161, 333–339.

Lavee, S., Harshemesh, H., Haskal, A., Meni, V., Wodner, M., Ogrodovich, A., Avidan, B., Wiesman, Z., Avidan, N., Trapero Casas, A., 1999. 'Maalot' a new orchard-resistant cultivar to peacock eye leaf spot (*Spilocaea oleagina* Cast.). Olivae 78, 51–59.

Lavee, S., Avidan, B., Meni, Y., 2003. 'Askal' a new high performing oil variety for intensive and supper-intensive olive orchards. Olivae 97, 53–59.

Leon, L., Beltran, G., Aguilera, M.P., Rallo, L., Barranco, D., De la Rosa, R., 2011. Oil composition of advanced selections from an olive breeding program. Eur. J. Lipid Sci. Technol. 113, 870–875.

Levin, A.G., Lavee, S., Tsror, L., 2003. Epidemiology and effects of Verticillium wilt on yield of olive trees (cvs. Barnea and Souri) irrigated with saline water in Israel. Phytoparasitica 31, 333–343.

Lo Giudice, V., Raudino, F., di San Lio, R., Magnano, Cacciola, S.O., Faedda, R., Pane, A., 2010. First report of a decline and wilt of young olive trees caused by simultaneous infections of *Verticillium dahliae* and *Phytophthora palmivora* in Sicily. Plant Dis. 94, 1372–1373.

Lopez-Escudero, F.J., del Rio, C., Caballero, J.M., Blanco-Lopez, M.A., 2004. Evaluation of olive cultivars for resistance to *Verticillium dahlia*. Eur. J. Plant Pathol. 110, 79–85.

Maldonado-Gonzalez, M.M., Prieto, P., Ramos, C., Mercado-Blanco, J., 2013. From the root to the stem: interaction between the biocontrol root endophyte *Pseudomonas fluorescens* PICF7 and the pathogen *Pseudomonas savastanoi* NCPPB 3335 in olive knots. Microbial Biotech. 6, 275–287.

Manai, H., Haddada, F.M., Trigui, A., Daoud, D., Zarrouk, M., 2007. Compositional quality of virgin olive oil from two new Tunisian cultivars obtained through controlled crossings. J. Sci. Food and Agric. 87, 600–606.

Manai, H., Mahjoub-Haddada, F., Oueslati, I., Daoud, D., Zarrouk, M., 2008. Characterization of monovarietal virgin olive oils from six crossing varieties. Sci. Hortic. 115, 252–260.

Marcelo, A., Fernandes, M., Potes, M.F., Serrano, J.F., 1999. Reactions of some cultivars of *Olea europaea* L. to experimental inoculations with *Pseudomonas syringae* pv. *savastanoi*. Acta Hortic. 474, 581–584.

Marchi, G., Sisto, A., Cimmino, A., Andolfi, A., Cipriani, M.G., Evidente, A., Surico, G., 2006. Interaction between *Pseudomonas savastanoi* pv. *savastanoi* and *Pantoea agglomerans* in olive knots. Plant Pathol. 55, 614–624.

Markakis, E.A., Tjamos, S.E., Antoniou, P.P., Paplomatas, E.J., Tjamos, E.C., 2009. Symptom development, pathogen isolation and Real-Time QPCR quantification as factors for evaluating the resistance of olive cultivars to *Verticillium* pathotypes. Eur. J. Plant Pathol. 124, 603–611.

Martos-Moreno, C., Lopez-Escudero, F.J., Blanco-Lopez, M.A., 2006. Resistance of olive cultivars to the defoliating pathotype of *Verticillium dahlia*. Hortscience 41, 1313–1316.

Mekonnen, M.M., Hoekstra, A.Y., 2013. Water footprint benchmarks for crop production, Value of Water. Research Report Series No. 50, UNESCO-IHE, Delft, The Netherlands.

Mekuria, G.T., Collins, G., Sedgley, M., Lavee, S., 2001. Identification of genetic markers in olive linked to olive leaf spot resistance and susceptibility. J. Am. Soc. Hortic. Sci. 126, 305–308.

Mekuria, G.T., Sedgley, M., Collins, G., Lavee, S., 2002. Development of a sequence-tagged site for the RAPD marker linked to leaf spot resistance in olive. J. Am. Soc. Hortic. Sci. 127, 673–676.

Mercado-Blanco, J., Rodriguez-Jurado, D., Hervas, A., Jimenez-Diaz, R.M., 2004. Suppression of Verticillium wilt in olive planting stocks by root-associated fluorescent *Pseudomonas* spp. Biol. Control 30, 474–486.

Mourad, K., Fadhila, K., Chahinez, M., Meriem, R., de Lajudie, P., Abdelkader, B., 2009. Antimicrobial activities of *Rhizobium* sp strains against *Pseudomonas savastanoi*, the agent responsible for the olive knot disease in Algeria. Grasas y Aceites 60, 139–146.

Obanor, F.O., Walter, M., Jones, E.E., Jaspers, M.V., 2013. Efficacy of systemic acquired resistance inducers in olive leaf spot management. Australas. Plant Pathol. 42, 163–168.

Osman, W.A., Tarabeih, A.M., Michail, S.H., 1980. Studies on olive knot disease in Iraq with reference to response of different cultivars. Mesopotamia J. Agric. 15, 245–261.

Ozdemir, Y., Tangu, N.A., Akcay, M.E., 2013. Generating omega-3 rich olive oil by cross breeding. Eur. J. Lipid Sci. Technol. 115, 977–981.

Papasotiriou, F.G., Varypatakis, K.G., Christofi, N., Tjamos, S.E., Paplomatas, E.J., 2013. Olive mill wastes: a source of resistance for plants against *Verticillium dahliae* and a reservoir of biocontrol agents. Biol. Control 67, 51–60.

Pastor, M., Garcia-Vila, M., Soriano, M.A., Vega, V., Fereres, E., 2007. Productivity of olive orchards in response to tree density. J. Hortic. Sci. Biotechnol. 82, 555–562.

Peñalver, R., García, A., Pérez-Panadés, J., Del Río, C., Caballero, J.M., Pinochet, J., Piquer, J., Carbonell, E.A., López, M.M., 2005. In: Rallo, L., Barranco, D., Caballero, J.M., Del Río, C., Martín, A., Tous, J., Trujillo, I. (Eds.), Variedades de olivo en España (Book II: Variabilidad y selección). Junta de Andalucía, MAPA y ediciones Mundi-Prensa, Madrid, pp. 339–346.

Pierantozzi, P., Torres, M., Bodoira, R., Maestri, D., 2013. Water relations, biochemical-physiological and yield responses of olive trees (*Olea europaea* L. cvs. Arbequina and Manzanilla) under drought stress during the pre-flowering and flowering period. Agric. Water Manag. 125, 13–25.

Porras-Soriano, A., Marcilla-Goldaracena, I., Soriano-Martin, M.L., Porras-Piedra, A., 2006. Development and resistance to *Verticillium dahliae* of olive plantlets inoculated with mycorrhizal fungi during the nursery period. J. Agric. Sci. 144, 151–157.

Rahioui, B., Aissam, S., Messaouri, H., Moukhli, A., Khadari, B., El Modafar, C., 2013. Role of phenolic metabolism in the defense of the olive-tree against leaf-spot disease caused by *Spilocaea oleaginae*. Int. J. Agric. Bio. 15, 273–278.

Rezaie, M., Lessani, H., Soleimani, A., 2004. The linked effect of saline and drought stress on olive (*Olea europaea* L.) plants. Acta Hortic. 791, 493–496.

Rjiba, I., Dabbou, S., Gazzah, N., Hammami, M., 2010. Effect of crossbreeding on the chemical composition and biological characteristics of Tunisian new olive progenies. Chem. Biodivers. 7, 649–655.

Roca, M., Leon, L., De la Rosa, R., 2011. Pigment metabolism of 'Sikitita' olive (*Olea europaea* L.): a new cultivar obtained by cross-breeding. J. Agric. Food Chem. 59, 2049–2055.

Ruggieri, G., 1946. Una nuova malatia dell'olivo. L'Italia Agricola 83, 369–372.

Rugini, E., Gutierrez-Pesce, P., Spampinato, P.L., Metzidakis, I., 1999. New perspective for biotechnologies in olive breeding: morphogenesis, *in vitro* selection and gene transformation. Acta Hortic. 474, 107–110.

Salmoral, G., Aldaya, M.M., Chico, D., Garrido, A., Llamas, M.R., 2011. The water footprint of olives and olive oil in Spain. Spanish J. Agric. Res. 9, 1089–1104.

Schwabe, W.W., Lionakis, S.M., 1996. Leaf attitude in olive in relation to drought resistance. J. Hortic. Sci. 71, 157–166.

Serman, F.V., Pacheco, D., Pringles, A.O., Bueno, L., Carelli, A., Capraro, F., 2011. Effect of regulated deficit irrigation strategies on productivity, quality and water use efficiency in a high-density 'Arbequina' olive orchard located in an arid region of Argentina. Acta Hortic. 888, 81–88.

Sesli, M., Onan, E., Oden, S., Yener, H., Yegenoglu, E.D., 2010. Resistance of olive cultivars to *Verticillium dahlia*. Sci. Res. Essays 5, 1561–1565.

Sisto, A., Cantore, P., Iacobellis, N. 2001. Preliminary results on the response of olive cultivars to artificial inoculation with *Pseudomonas syringae* subsp. *savastanoi*. In Proceedings of the Eleventh Congress of the Mediterranean Phytopathological Union and Third Congress of the Sociedade Portuguesa de Fitopatologia. Evora, 17–20 September, 240–242.

Sofo, A., 2011. Drought stress tolerance and photoprotection in two varieties of olive tree. Acta Agric. Scandinavica 61, 711–720.

Sofo, A., Dichio, B., Xiloyannis, C., Masia, A., 2005. Antioxidant defences in olive trees during drought stress: changes in activity of some antioxidant enzymes. Func. Plant Bio. 32, 45–53.

Sofo, A., Manfreda, S., Fiorentino, M., Dichio, B., Xiloyannis, C., 2008. The olive tree: a paradigm for drought tolerance in Mediterranean climates. Hydrol. Earth Syst. Sci. 12, 293–301.

Soriano, M.L., Porras, A., Canizares, C., Gonzalez, M.M., Linares, A., Vitagliano, C., Martelli, G.P., 2002. Incidence and severity of wilt in artificially inoculated cv. Cornicabra young olive trees with defolianting and non-defolianting *Verticillium dahliae* pathotypes. Acta Hortic. 586, 791–795.

Takemoto, D., Furuse, K., Doke, N., Kawakita, K., 1997. Identification of chitinase and osmotin-like protein as actin-binding proteins in suspension-cultured potato cells. Plant Cell Physiol. 38, 441–448.

Tous, J., Romero, A., 2005. Rendimiento graso de la aceituna (Banco de Germoplasma de Córdoba). In: Rallo, L., Barranco, D., Caballero, J.M., Del Rio, C., Martin, A., Tous, J., Trujillo, I. (Eds.), Variedades de olivo en España (Book II: Variabilidad y selección). Junta de Andalucía, MAPA and Ediciones Mundi-Prensa, Madrid, pp. 347–356.

Tous, J., Romero, A., Diaz, I., 2005a. Composición del aceite (Banco de Germoplasma de Cataluña). In: Rallo, L., Barranco, D., Caballero, J.M., Del Rio, C., Martin, A., Tous, J., Trujillo, I. (Eds.), Variedades de olivo en España (Book II: Variabilidad y selección). Junta de Andalucía, MAPA and Ediciones Mundi-Prensa, Madrid, pp. 357–372.

Tous, J., Romero, A., Plana, J., 2005b. Producción. In: Rallo, L., Barranco, D., Caballero, J.M., Del Río, C., Martín, A., Tous, J., Trujillo, I. (Eds.), Variedades de olivo en España (Book II: Variabilidad y selección). Junta de Andalucía, MAPA y ediciones Mundi-Prensa, Madrid, pp. 257–274.

Trapero, A., López-Doncel, L.M., 2005. Resistencia y susceptibilidad al repilo. In: Rallo, L., Barranco, D., Caballer, J.M., Del Río, C., Martín, A., Tous, J., Trujillo, I. (Eds.), Variedades de olivo en España (Book II: Variabilidad y selección). Junta de Andalucía, MAPA y ediciones Mundi-Prensa, Madrid, pp. 321–328.

Trapero, C., Serrano, N., Arquero, O., Del Rio, C., Trapero, A., Lopez-Escudero, F.J., 2013. Field resistance to Verticillium wilt in selected olive cultivars grown in two naturally infested soils. Plant Dis. 97, 668–674.

Trapero, C., Díez, C.M., Rallo, L., López-Escudero, F.J., Barranco, D., 2014. Selecting Verticillium wilt-resistant olive genotypes by classical breeding. Orlando, Florida. Proceedings of the ASHS Annual Conference. , 28–31 July.

Trigui, A., Yengui, A., Belguith, H., 2006. Olive germplasm in Tunisia. Olea 25, 19–23.

Trigui, M., Hsouna, A.B., Hammami, I., Culioli, G., Ksantini, M., Tounsi, S., Jaoua, S., 2013. Efficacy of *Lawsonia inermis* leaves extract and its phenolic compounds against olive knot and crown gall diseases. Crop Prot. 45, 83–88.

Tuck, K.L., Hayball, P.J., 2002. Major phenolic compounds in olive oil: metabolism and health effects. J. Nutr. Biochem. 13, 636–644.

Uceda, M., Beltran, G., Jimenez, A., 2005. Composición del aceite (Banco de Germoplasma de Córdoba). In: Rallo, L., Barranco, D., Caballero, J.M., Del Rio, C., Martin, A., Tous, J., Trujillo, I. (Eds.), Variedades de olivo en España (Book II: Variabilidad y selección). Junta de Andalucía, MAPA and Ediciones Mundi-Prensa, Madrid, pp. 357–372.

Varvaro, L., Martella, L., 1993. Virulent and avirulent isolates of *Pseudomonas syringae* subsp. *savastanoi* as colonizers of olive leaves: evaluation of possible biological control of the olive knot pathogen. EPPO Bull. 23, 423–442.

Visioli, F., Galli, C., 1998. Olive oil phenols and their potential effects on human health. J. Agric. Food Chem. 46, 4292–4296.

Visioli, F., Poli, A., Galli, C., 2002. Antioxidant and other biological activities of phenols from olives and olive oil. Med. Res. Rev. 22, 65–75.

Waterman, E., Lockwood, B., 2007. Active components and clinical applications of olive oil. Alt. Med. Rev. Vol. 12, 331–342.

Young, J.M., 2004. Olive knot and its pathogens. Aust. Plant Pathol. 33, 33–39.

Young, J.M., Wilkie, J.P., Fletcher, M.J., Park, D.C., Pennycook, S.R., Triggs, C.M., Watson, D.R.W., 2004. Relative tolerance of nine olive cultivars to *Pseudomonas savastanoi* causing bacterial knot disease. Phytopathol. Mediterr. 43, 395–402.

CHAPTER

12

Soybean

Aditya Pratap, Surinder Kumar Gupta**, Jitendra Kumar*, Suhel Mehandi*, Vankat R. Pandey**

*Crop Improvement Division, Indian Institute of Pulses Research, Kanpur, India

**Division of Plant Breeding & Genetics, Faculty of Agriculture, Sher-e-Kashmir University of Agricultural Sciences & Technology Chatha, Jammu (J&K), India

INTRODUCTION

Soybean (*Glycine max* (L.) Merrill) belongs to the family Leguminosae and subfamily Papilionaceae. It is probably one of the most cultivated grains in the world. As a legume crop, soybean is capable of utilizing atmospheric nitrogen through biological nitrogen fixation and is therefore less dependent on synthetic nitrogen fertilizers. Although its first commercial use was for its oil, now it is considered an important protein crop also. In addition to having high protein content (25–45%), soybean seeds contain 18–23% oil and thus add to the importance of the species as an edible, oil-yielding crop. Soybean has also been used to prepare a variety of fresh, fermented, and dried foods such as tofu, soy milk, soy sauce, and miso, both in Asia and other parts of the world for many centuries (Probst and Judd, 1973). Assorted health food snacks, energy foods, and cereals are also produced from soybeans while its other uses include bean sprouts and soy nuts. After oil is extracted the remaining soy meal is used as a nutritious animal feed while soy oil finds several uses in industries related to the production of pharmaceuticals, plastics, papers, inks, paints, varnishes, pesticides, and cosmetics (Pratap et al., 2012). Use of soy oil as a biodiesel has opened up another possibility of renewable sources of energy for industrial uses.

Soybean cultivation is highly concentrated geographically in only four countries – the United States, Brazil, Argentina, and China. Production in these countries alone accounts for almost 90% of the world's output. Countries in Asia excluding China and Africa together account for only 5% of the total production of soybean. Among the oilseed crops, soybean alone has the maximum global production share (53%) followed by rapeseed mustard (15%), cottonseed (10%), and peanut (9%) (Pratap et al., 2012). Soybean is highly sensitive to environmental fluctuations. Among its abiotic stresses, water is the biggest limiting factor having great impact on its productivity. Simultaneously, temperature and photoperiod are also

Breeding Oilseed Crops for Sustainable Production. http://dx.doi.org/10.1016/B978-0-12-801309-0.00012-4

important factors, which determine the cultivation of soybean over space and time. Nevertheless, tremendous strides have been made in the production and productivity of soybean over the past several years thanks to the developments in soybean transgenics, genomics, and breeding. In this chapter we discuss the wonder crop of soybean with emphasis placed on its breeding and development.

PRODUCTION AND PRODUCTIVITY TRENDS

Soybean has been grown as a commercial crop mainly in temperate ecologies for thousands of years. Initially it was cultivated in northern Asia and in more recent decades in North America and in countries of the Southern Cone of Latin America. During the past decades, concerted efforts have been made by crop scientists and soybean growers leading to an increase in the world production of soybean from 155.1 million metric tons in 1999 to 284 million metric tons in 2013 (www.soystats.com). Currently, the share of soybean in global oilseed production is around 55%, and over the last 10 years its production has expanded at a rate of over 5% per year on average. Among all oilseed crops, soybean alone has the maximum global production share (53%) while other crops such as rapeseed, cotton, and peanut contribute 15, 10, and 9%, respectively. The United States grows soybean over the largest area and holds a share of about 32% of the world's soybean production, followed by Brazil (31%), Argentina (19%), China (6%), and India (4%) (www.soystats.com, Table 12.1).

HISTORY, ORIGIN, AND EVOLUTION

Several workers have discussed soybean history in the past since this crop has tremendous importance worldwide (Hymowitz, 1970; Guo, 1993; Singh and Hymowitz, 1999; Guo et al., 2010). Evidence suggests that soybean has been domesticated since the Shang Dynasty in the eastern half of northern China during ca. 1700–1100 BC (Singh and Hymowitz, 1999). It is one of the oldest cultivated food legumes and has been known in India for over

TABLE 12.1 World Soybean Production

Country	Production (million metric tons)
United States	89.5
Brazil	87.5
Argentina	54.0
China	12.2
India	11.0
Paraguay	8.1
Canada	5.2
Other	16.5
Total	284.0

5000 years. Therefore, this place has been proposed as a candidate place for its domestication (Hymowitz, 1970). However, based on molecular diversity studies, it has been suggested that domestication happened in South China (Ding et al., 2008). However, bearing in mind the evidence and the variability available, both North and South China regions are considered to be involved in domestication of soybean since ancient times. The oldest records of soybean cultivation appear in bronze inscriptions, which are not dated earlier than 1100 BC as mentioned in early writings. Further trade of soybean migrated to South China, Korea, Japan, and South East Asia with the expansion of the Shang Dynasty. The writings of European visitors to China and Japan have mentioned several references to native soy foods in the sixteenth and seventeenth centuries. Soybeans were brought to the United States by a seaman, Samuel Bowen, in 1765. Mr. Bowen used the soybean to produce soy sauce and a soybean noodle for export to England (http://www.soymeal.org/pdf/HistorySoybeanUse.pdf). The late-type soybean identified in south China showed closeness with wild types and hence wild soybean is expected as a common ancestor for both cultivated types from south and north China. These early cultivated types probably originated during the process of dissemination to north China (Gai et al., 2000). Recent studies also support the idea that soybean originated in south China because the soybean population of south China has been observed to be more diverse compared with the north China population (Ding et al., 2008).

CROP BIOLOGY AND BREEDING BEHAVIOR

Soybean belongs to the family Leguminosae and subfamily Papilionaceae. The cultivated soybean is known as *G. max* (L.) Merrill (Gazzoni, 1994). The genus *Glycine* consists of two subgenera: namely, *Glycine* (perennials) and *Soja* (annuals). The perennials consist of 22 recognized species and the annuals 2 species, *G. max* L. Merrill. (cultigen) and *Glycine soja* Sieb. & Zucc. (wild species and progenitor of *G. max*) (Hymowitz, 2004).

Soybean is a hairy annual having an extensive tap root system. The root system may grow as deep as 2 m and adventitious roots grow from the hypocotyls (Chaturvedi et al., 2011). Cultivated soybeans are mostly erect in growth habit, bushy, 20–180 cm tall, and usually with a few primary branches and no secondary branches. The leaves are trifoliate and alternate with long petioles and small stipules and stiples; the leaflets are ovate to lanceolate with mucronate tips. Flowering and maturity are highly influenced by photoperiod and flowering times fluctuate with genotypes (Sekizuka and Yoshiyama, 1960; Fukui and Kaizuma, 1971). The flowers are papilionaceous, white or pale purple, with a tubular calyx of five unequal sepal lobes and a five-member corolla that consists of a posterior standard petal, two lateral wing petals, and two anterior keel petals (Guard, 1931). The androecium is diadelphous having a 9 + 1 arrangement. The pistil is unicarpellate and has 1–4 campylotropous ovules (Palmer et al., 2001). Each flower is subtended by two bracteoles and has a hairy calyx of five pointed sepals united for about half of their length. The pods are short stalked and occur in groups of 3–15, are 3–7 cm long and hairy, light brown at maturity, and slightly constricted between the seeds. The seeds vary greatly in shape, size, and color though these are most often round and yellowish, brown, or black with epigeal germination.

Soybeans are mostly self-pollinated although, natural cross-pollination has also been observed with a rate of 0.03–1.14% in cultivated species (Culter, 1934; Caviness, 1966;

Chiang and Kiang, 1987), and 2.4–3.0% in wild soybeans (Kiang et al., 1992). The wild annual species *G. soja* is predominantly self-pollinated and flowers are self-fertilized cleistogamous. However, it has been observed that in the perennial wild relative, *G. argyrea* (Ting.), and its closely related species, *G. clandestine* (Wendl.), cleistogamous as well as chasmogamous flowers occur on the same plants (Brown et al., 1986; Schoen and Brown, 1991; Palmer et al., 2001). Protogyny is found in self-pollinating soybean flowers having 3–4 ovules and thus these mature before anthesis (Stelly and Palmer, 1985). Flowers open and normally self-pollinate at anthesis.

PLOIDY STATUS

The genus *Glycine* has basically originated as an ancient polyploid. Haploid genome studies suggested that soybean is a diploidized ancient tetraploid (Safari and Schlueter, 2011). Studies suggested that two major rounds of duplication events occurred in its genome during speciation (Van et al., 2008). Probably these duplication events occurred at approximately 14.56 and 45 MYA (Schlueter et al., 2004; Blanc and Wolfe, 2004). However, it has been observed that an even more ancient round of polyploidization occurred at some point in the ancestry of the genus because the genetic map of soybean revealed multiple nested duplications (Shoemaker et al., 2006).

The soybean genome is considered to be allo- and autopolyploid in origin. The initial cytogenetic and molecular studies (Singh and Hymowitz, 1985; Shoemaker et al., 1996) hypothesized an allopolyploid soybean genome. However, the phylogenetic analysis of nuclear genes suggested its autopolyploid in nature (Doyle et al., 2003; Straub et al., 2006). Udall and Wendel (2006) concluded that soybean originated from two diploid progenitors. Based on molecular studies, a close phylogenetic relationship has been reported between cultivated species *Glycine max* and wild species *G. soja* (a progenitor of this species). The north Asian subgenus *soja* has been suggested to be the probable wild progenitor of the cultigen *G. max* (L.) Merrill (Doyle et al., 2003). The occurrence of a number of perennial diploid relatives of *Glycine* throughout Australia and Papua New Guinea probably originated some allopolyploid taxa due to intercrossing among the diploid species (Doyle et al., 2004).

Molecular cytogenetic tools have also been used to suggest polyploidy in *Glycine max* and identify two soybean centromere-specific satellite repeat classes in its genome although small chromosome size (1–2 μm), large chromosome number ($2n = 40$), and less morphological diversity make the cytological study of metaphase chromosomes difficult (Sen and Vidyabhusan, 1960; Clarindo et al., 2007). In soybean, all chromosomes are metacentric or submetacentric except a single acrocentric pair. This makes them difficult to distinguish through cytological investigations. However, it was suggested that soybean consists of two subgenomes (Gill et al., 2009). The ancestor of soybean, and the remainder of the genus *Glycine*, has been hypothesized as being formed via a polyploidy event within the last 15 million years (Shoemaker et al., 2006). However, it remains unclear whether the event was allo or autopolyploidy (Kumar and Hymowitz, 1989; Straub et al., 2006). Analysis of chromosome number in soybean suggested that the soybean genome has gone through several rounds of polyploidization and segmental duplication (Lackey, 1980). This has also subsequently been confirmed on the basis of multiple hybridizing restriction fragment length polymorphism

(RFLP) fragments (Shoemaker et al., 2006) and on the basis of implicated expressed sequence tags (ESTs) (Blanc and Wolfe, 2004; Schlueter et al., 2004).

GENETIC IMPROVEMENT

Most of the breeding efforts in soybean have been concentrated on improvement of morphology, physiology/biochemistry, seed composition, quality traits, and resistance to biotic and abiotic stresses. Seed yield is one of the most important economic traits for farmers. The scientific community has taken up the challenge of increasing the seed yield of soybean and their efforts have resulted in an improvement of seed yields in soybean at a rate of 23 kg $ha^{-1}year^{-1}$ due to improved genetic gain, improved cultural practices, and a rise in atmospheric CO_2 concentrations. Most gains in soybean yield improvement have been achieved as a result of conventional breeding techniques such a recurrent selection (Burton et al., 1990) and development of glyphosate-resistant cultivars. Tremendous strides have also been made toward improvement of other yield-related traits, plant morphology, nutrient assimilation, water and radiation use efficiency, nodulation and nitrogen fixation, etc.

Like most of the exotic crops, soybean improvement started with domestication that probably occurred initially in the eastern half of China during ca. 1700–1100 BC (Singh and Hymowitz, 1999). It was subsequently brought into Europe and was likely introduced into North America in 1765. In the United States, the introduced soybeans were grown in all agricultural research stations during 1860–1900, with the aim of making improvements through research (Probst and Judd, 1973). Consequently, a number of introduced lines have made significant genetic improvement in yield potential (Pathan and Sleper, 2008). Introduced lines or cultivars from the United States were also important sources of adapted materials for growing soybean in Brazil (Ferraz de Toledo et al., 1994).

Since soybean is a self-pollinated crop, breeding for development of improved cultivars has mainly been focused on hybridization, selection, and testing yield potential of improved lines. Modification in bulk method through early generation testing has also been shown to be very successful improving those characteristics, which have significant additive and additive × additive genetic components of variance in soybean (Cooper, 1990). This could help to reduce population load by discarding inferior lines in early generations. However, use of early generation selection in F_2, F_3, and even F_4 families depends upon the target trait and environmental conditions.

Various recurrent selection methods have also been used for improving agronomically important traits. For example, S1 family selection was used for yield (Kenworthy and Brim, 1979; Rose et al., 1992) and protein (Brim and Burton, 1979). Mass selection was used for oil (Burton and Brim, 1981) and seed weight (Tinius et al., 1991). Half-sib family selection was used for seed yield (Burton and Carver, 1993) and oil quality (Carver et al., 1986). Recurrent selection in soybean could be successfully applied due to the availability of male-sterile lines. Its use has been employed for making genetic improvements to yield (Tinius et al., 1991), oil and protein content (Burton and Brim, 1981), and fatty acid (Carver et al., 1986). In the United States, superior cultivars are selected and released in soybean following a process of cyclic recurrent selection. Protein and oil content in seeds of soybean are unarguably the most important traits. However, oil content has a negative correlation with protein content (Brim and

Burton, 1979: Hartwig and Kilen, 1991; Helms and Orf, 1998; Cober and Voldeng, 2000). Therefore, the recurrent selection method has been suggested as the most effective way for increasing oil content (Burton and Brim, 1981).

High-phytic acid (PA) in soybean seeds causes mineral malnutrition in human beings. An inheritance study of this trait showed that dominant recessive epistasis controls the total phosphorus (P) and phytate P (PhyP). This knowledge of genetics can help us to develop low phytate–containing varieties (Sompong et al., 2010). In soybean, the ratio of polysaturated fatty acid/saturated fatty acid, monounsaturated fatty acid, and essential fatty acids, such as linoleic/linolenic, determines the quality of soybean oil. Therefore, these traits have also been studied systematically. High-linolenic acid in soybean oil has poor oxidative stability (Patil et al., 2004). Isoflavon in soybean oil, an important target for improving oil quality, showed epistatic interactions except for malonyldaidzin (MDZ). Therefore, additive genetic variances in superior lines or the cytoplasmic effect and the epistatic interactions between cytoplasmic and nuclear genes can be exploited to obtain the largest selection gains for this trait (Chiari et al., 2006). Soybean seeds also contain a major carotenoid, lutein, that is beneficial to eye health. Studies showed that this has a positive correlation with oleic acid and a negative correlation with linoleic and linolenic acid content (Lee et al., 2009).

Lodging resistance and erect growth habit are other important traits that require improvement in soybean cultivars because these help to reduce mechanical harvest loss and to increase maximum light penetration through the plant canopy. Soybean breeders also focused on other traits such as narrow leaflets, brachytic stem (short internode), stem termination change to alter height, and more fibrous rooting (Wells et al., 1993). Genetic manipulation in reproductive period has also been a focus for developing improved cultivars adapted to specific environments. However, in practice, these efforts toward changing the length of the pod-filling period and/or changing the rate of dry matter accumulation in pods have resulted in minor improvements to yield, although a positive correlation has been observed between reproductive period and yield traits (Smith and Nelson, 1986).

Distant Hybridization and Alien Gene Transfer

The concept of gene pools, as proposed by Harlan and de Wet, 1971 can be applied well to the genus *Glycine*. GP-1, or the primary gene pool, is defined as biological species that can easily be crossed within the gene pool producing F_1 hybrids that are vigorous, exhibit normal meiotic chromosome pairing, and possess total seed fertility, such that segregation is normal and gene exchange is basically easy. The primary gene pool (GP-1) for soybean includes cultivars, land races, and *G. soja* genotypes. GP-2, or the secondary gene pool, as defined by Harlan and de Wet, 1971 consists of species that can be crossed with GP-1 to produce F_1 hybrids that have some fertility. However, there are no currently described *Glycine* species in GP-2. GP-3, or the tertiary gene pool, is the extreme limit of potential genetic resources. Using the tertiary gene pool, gene transfer is almost impossible, or requires rescue techniques that result in sterile (or lethal) hybrids using species from GP-3. In *Glycine* the 23 wild perennial species have been considered to be in GP-3. These species are geographically isolated from *G. max* and *G. soja* (Singh and Hymowitz, 1999) and are extremely genetically diverse. Therefore, they have the ability to grow under very diverse conditions and have a very wide geographical distribution (Kollipara et al., 1997; Singh and Hymowitz, 1999).

Among different gene pools, GP-1 has been used widely for the development of most of the present-day cultivars. Although the wild species/relatives are useful reservoirs of genes for various quality traits including resistance genes to many biotic and abiotic stresses, transfer of these alien genes into the elite cultivars through conventional breeding has been limited, due to pre and postfertilization barriers. Sometimes introgression of these carries the associated transfer of undesired alleles (linkage drag). Therefore, efforts have been made for introgression of favorable alleles from alien genes of distantly related wild species to cultivated background, more efficiently by using embryo rescue techniques and the advanced backcross quantitative trait locus (AB-QTL) approach.

Studies have shown that it is possible to transfer alien gene(s) from wild perennial species to soybean. This has resulted in intersubgeneric hybridization, which has led to the production of fertile, modified diploid lines. Nevertheless, among a number of attempts made to hybridize wild perennial *Glycine* spp. with soybean, only a few sterile intersubgeneric F_1 hybrids could be obtained. Initially, use of *G. soja* for broadening the genetic base of *G. max* was reported by a number of workers (Hartwig, 1973; Ertl and Fehr, 1985; Carpenter and Fehr, 1986). However, subsequently limited efforts were made for making interspecific crosses between *G. max* and *G. soja* (Palmer et al., 1987; Singh and Hymowitz, 1988). Alien genes for soybean cyst nematode resistance have been introgressed from *G. tomentella* into modified derived diploid soybean lines. Using recent molecular marker tools, several QTLs for seed yield, seed protein, oil content, late maturity, and tallness have been identified in soybean among 296 $BC_2F_{4:6}$ backcross introgression lines (BILs) generated from the cross *G. max* (Dunbar) × *G. soja* (PI 326582A) using the AB-QTL approach.

Hybrid Development

Soybean is a highly self-pollinated crop and therefore success of hybrid development has been achieved to a very limited extent. Absence of stable male sterility–female fertility systems, lack of efficient pollen transfer mechanisms, low seed set, and poor natural crossing have also limited the use of hybrid breeding in soybean (Singh and Hymowitz, 1999; Palmer et al., 2001). Despite this, extensive efforts have been made to identify male sterility systems and a number of genic male sterile lines (*ms1, ms2, ms3, ms4, ms5,* and *ms6*) have been identified (Skorupska and Palmer, 1989; Palmer and Skorupska, 1990; Palmer et al., 2001). The first male-sterile, female-fertile mutant was described by Brim and Young (1971) who initiated efforts toward the development of commercial hybrids in soybean.

Efforts were made to identify cytoplasmic male sterility (CMS) systems by several workers (Zhang et al. 1999a,b; Sun et al., 1994, 1997; Bai and Gai, 2003; Zhao and Gai, 2006; Li et al., 1995; Xu et al., 1999). In China, identification of stable CMS, restorer, and maintainer lines (Wang et al., 2009) led to the development of the world's first commercial hybrid in soybean, which was released in 2003 for cultivation. Intraspecific (*G. max* × *G. max*) and interspecific (*G. max* × *G. soja*) hybridizations have also been used to identify cytoplasmic–nuclear, male-sterile lines along with their maintainers and restorers in soybean (Davis, 1987; Sun et al., 1994; Zhang and Dai, 1997; Ding et al., 1998; Zhao et al., 1998; Bai and Gai, 2003; Zhao and Gai, 2006). Pollinator insects such as honeybees (*Apis mellifera)* and alfalfa leaf cutter bees (*Megachile rotundata*) are used to facilitate entomophilous cross-pollination of male-sterile soybean plants for the production of hybrid seed (Nelson and Bermard, 1984;

Ortiz-Perez et al., 2007). In this connection, some wild native bees, primarily belonging to families Megachilidae, Halictidae, Anthophoridae, and Andrenidae, have been observed as efficient pollinators (Ortiz-Perez et al., 2007).

BIOTECHNOLOGY

Among the major yield constraints, high influence of genotype $\times$ environment (G $\times$ E) interactions with the expression of important quantitative traits and susceptibility of soybean genotypes to biotic and abiotic stresses affect the genetic improvement and yield stability of soybean. Recent biotechnological tools have complemented traditional plant breeding producing an accelerated improvement to soybean. Various biotechnological tools, such as plant tissue cultures, genetic transformation, molecular breeding, and marker-assisted selection, have played a major role in developing superior cultivars (for a review see Pratap et al., 2012).

Micropropagation and Somaclonal Variation

Successful micropropagation greatly helps with the generation of additional variability through isolation of somaclonal/gametoclonal variants as well as genetic transformants. There are reports on morphological variants in soybean through cell and tissue cultures (Graybosch and Palmer, 1987; Bailey et al., 1993). Cotyledonary nodes from mature seeds have been the most responsive for the induction of multiple shoots via organogenesis in soybean (Barwale et al., 1986). Initially, Barwale et al. (1986) succeeded in obtaining fertile plants in 54 soybean genotypes using callus cultures derived from immature embryos on basal MS medium (Murashige and Skoog, 1962) with B5 vitamins (Gamborg et al., 1968), supplemented either by 8 mg L^{-1} naphthaleneacetic acid (NAA) or 3 mg L^{-1} benzylanimopurine (BAP) and 0.037 mg L^{-1} NAA. These embryos were successfully regenerated into plants on the medium supplemented with 0.38 mg L^{-1} BAP and 0.04 mg L^{-1} indol-3-butyric acid (IBA). This protocol was also applied for soybean transformation (Trick and Finer, 1998; Santarem and Finer, 1999) and *in vitro* mutagenesis (Van et al., 2008). Bailey et al. (1993) made further improvements to the protocol, testing additional growth regulators, sources of carbohydrates, and other medium additives. Several other modifications were made to the media and culture conditions to improve plantlet recovery frequency from cultured explants (Walker and Parrott, 2001; Schmidt et al., 2005). A genotype has been reported to influence protocol efficiency (Barwale et al., 1986; Parrott et al., 1989; Tomlin et al., 2002; Van et al., 2008).

Several attempts were also made to develop anther and microspore culture systems for soybean (Cardoso et al., 2004; Ye et al., 1994). Many initial studies reported the induction of callus from anthers (Ivers et al., 1974; Liu and Zhao, 1986), shoot organogenesis (Yin et al., 1982; Jian et al., 1986), and embryo-like structures (ELS) from anther-derived callus (Hu et al., 1996). In a few cases, a low number of plants were regenerated, though the haploid origin of the plants was uncertain (Yin et al., 1982; Jian et al., 1986; Hu et al., 1996; de Moraes et al., 2004; Tiwari et al., 2004). In most of the double-haploid (DH) production protocols in soybean, anthers were collected from the field (de Moraes et al., 2004; Cardoso et al., 2004) in contrast with most other species where donor plants are grown under controlled conditions. Furthermore, in general, B5 medium with 16 organic compounds and with Yeung's amino acids (Yeung and Sussex, 1979) was found appropriate for anther culture. de Moraes et al. (2004) obtained one confirmed haploid plant ($n = 20$) following induction of embryogenic calli from anthers on this basal medium supplemented with 2.0 mg L^{-1} 2,4-D, 0.5 mg L^{-1} BAP, 9% sucrose, and 0.25% phytagel. This result further confirms the finding of Hu et al. (1996) that 2,4-D is essential for soybean microspore callus induction. Furthermore, several studies were conducted regarding the stage of anther to be cultured (Yin et al., 1982; Ye et al., 1994, Cardoso et al., 2004), pretreatments (Liu and Zhao, 1986; Rodrigues et al., 2005), etc. However, any major breakthrough is yet to be achieved in haploid breeding in soybean.

Molecular Breeding

Soybean is the most successful food legume in which molecular markers in breeding programs have been used routinely. As a consequence, a number of improved lines/varieties for resistance to different soybean cyst nematode (SCN) races (Arelli and Young, 2009), *Phytophthora* root rot, brown stem rot, insect resistance (Walker et al., 2002; Warrington et al., 2008), low linolenic-acid content (Sauer et al., 2008), yield (Concibido et al., 2003), mosaic virus resistance (Saghai Maroof et al., 2008; Shi et al., 2009) have been developed. Further, a number of varieties (JTN-5503, JTN5303, DS-880, JTN-5109) have been released in soybean for resistance to diseases and SCN resistance, most of them in the United States (Arelli et al., 2006, 2007; Arelli and Young, 2009; Smith, 2010). Further, identification of molecular markers associated with traits of interest to breeders has witnessed tremendous progress in soybeans. Linkage mapping–based approaches have been extensively used for mapping genes for various biotic stresses such as *Sclerotinia* stem rot (Guo et al., 2008), brown stem rot (Patzoldt et al., 2005), *Phytophthora* stem rot (Han et al., 2008; Wang et al., 2010), Asian soybean rust (Hyten et al., 2009; Chakraborty et al., 2009), soybean mosaic virus (Shi et al., 2008), sudden death syndrome (Kazi et al., 2008), and cyst nematodes (Wu et al., 2009; Vuong et al., 2010). Similarly for biotic stresses, QTL mapping has been successful for waterlogging (Githiri et al., 2006), salt stress (Lee et al., 2009; Tuyen et al., 2010), manganese toxicity (Kassem et al., 2004), and aluminum tolerance (Qi et al., 2008).

Genetic Transformation

Transgenic soybeans have been one of the biggest commercial successes of transgenic plants globally. Since the first reports on their genetic transformation (Hinchee et al., 1988; McCabe et al., 1988), there have been continuous efforts toward development of herbicide-tolerant soybeans. In 2014, the global area under soybean cultivation was 111 million ha, of which 82% was under transgenic soybean cultivation. Argentina was growing 100% transgenic soybeans at the end of 2014 (Table 12.2). This shows the success of transgenic soybeans and the willingness of farmers to adopt it. For soybean transformation, particle bombardment is one of the most popular technologies because it is less genotype dependent than *Agrobacterium*-mediated transformation. However, a disadvantage of this technique is that it sometimes results in complex transgene integration patterns, thus enhancing the likelihood of transgene silencing (Travella et al., 2005; Yang et al., 2005). An example of this phenomenon is a study concerning transformation using isoflavone-biosynthetic genes in soybean (Zernova et al., 2009). The transgenic lines carried multiple transgene inserts and although the lines

TABLE 12.2 Countrywise Presence of Transgenic Soybeans in 2013

	Cultivation since	Cultivation area 2013 (in million ha)	GMO ratio 2013 (%)
Worldwide	1997	84.5	79
United States	1997	29.3	93
Argentina	1997	20.8	100
Brazil	1999	26.9	92
Canada	1997	1.6	90
Mexico	2000	<0.1	n.a.
Paraguay	2004	3.0	95
South Africa	2001	0.5	n.a.
Uruguay	2002	1.3	100
Bolivia	2008	1.0	91

Source: www.soystats.com.

were transformed with sense constructs aimed at overexpression of isoflavone-biosynthetic enzymes, the transgenic lines actually contained lower levels of isoflavones, suggesting co-suppression of homologous soybean genes (Zernova et al., 2009). Over the last decade, a number of transgenic lines have been produced in soybeans for different traits (Table 12.3).

Kinney (1997) produced a high oleic-acid (84%) soybean line. Padgette et al. (1995) reported that a stable glyphosate-tolerant soybean line had been developed using *Agrobacterium*-mediated gene transfer. Using genetic transformation, fatty acid metabolism has been manipulated in soybeans with a reduction in saturated and polyunsaturated fatty acids and a concomitant strong increase in oleic acid (Buhr et al., 2002), and high amounts of long-chain polyunsaturated fatty acids (Chen et al., 2006; Eckert et al., 2006). Soybeans have also been used during the last decade as a platform to produce biopharmaceuticals such as vaccines and antibodies (Ma et al., 2005; Kaiser, 2008) resulting in the production of edible vaccines (Moravec et al., 2007). Plastid transformation in soybeans was first reported by Dufourmantel et al. (2004) and has subsequently been used to obtain high-level expression of Cry1Ab protein and 4-hydroxyphenylpyruvate dioxygenase, conferring strong insecticidal activity tolerance and herbicide tolerance, respectively (Dufourmantel et al., 2005; Dufourmantel et al., 2007).

OIL CONTENT AND PROTEIN QUALITY

Soybeans comprise approximately 40% protein, 20% lipid, 17% cellulose and hemicellulose, 7% sugar, 5% crude fiber, and about 6% ash. Soybean oil and soya meal account for about 25 and 65% of global consumption of oil/fat and meal/cake. Although soybeans are produced in a limited number of countries they are widely traded and the respective oil and meal are available in almost every country in the world. As a result, world soybean oil and meal consumption has grown steadily at 5–6% per year on average during the last 15–20 years.

TABLE 12.3 Some Recent Examples of Soybean Lines Improved Through Genetic Engineering

Introduced gene(s)	Trait	References
Ribozyme-terminated fatty acid desaturase and thioesterase	Modified seed oil composition	Buhr et al. (2002)
Gly m Bd 30 K	Allergen elimination	Herman et al. (2003)
Borago officinalis fatty acid Δ^6 desaturase	Modified seed oil composition	Sato et al. (2004)
cry1Ab	Insect resistance	Dufourmantel et al. (2005)
Coat protein of soybean mosaic virus	Virus resistance	Furutani et al. (2006)
Inverted repeat of coat protein of soybean dwarf virus	Virus resistance	Tougou et al. (2006)
Fatty acid Δ^6 desaturase and Δ^{15} desaturase	Modified seed oil composition	Eckert et al. (2006)
RNAi construct targeting cyst nematode MSP gene	Nematode resistance	Steeves et al. (2006)
Fatty acid Δ^6 desaturase, fatty acid elongase, and fatty acid Δ^5 desaturase	Modified seed oil composition	Chen et al. (2006)
4-hydroxyphenylpyruvate dioxygenase	Weed control	Dufourmantel et al. (2007)
Dicamba monooxygenase	Weed control	Behrens et al. (2007)
Heat-labile toxin (LT) B subunit	Oral vaccine	Moravec et al. (2007)
Modified β-conglycinin α subunit gene containing bioactive peptides	Accumulation of bioactive peptides	Nishizawa et al. (2008)
Modified β-conglycinin α subunit gene	Accumulation of bioactive peptides	Yamada et al. (2008)
ω-3 Fatty acid desaturase gene (GmFAD3)	Reduction of α-linolenic acid (18:3)	Flores et al. (2008)
Δ^6 Desaturase gene, Δ^6 elongase gene, Δ^5 desaturase gene	Production C20-LCPUFAs	Kajikawa et al. (2008)
Diacylglycerol acyltransferase 2A gene (*UrDGAT2A*)	Increase of oil content	Lardizabal et al. (2008)
Δ^{12} Fatty acid desaturase gene (*GmFAD2-1*)	Increase of oleic acid	Wang and Xu (2008)
SLC1	Increased oil content	Rao and Hildebrand (2009)
Molecular chaperone BiP (binding protein) gene	Tolerance to drought stress	Valente et al., 2009
Phosphinothricin (PPT) *N*-acetyltransferase	PPT tolerance	Kita et al. (2009)
Ac transposase gene	Induction of transposition of Ds delineated element	Mathieu et al. (2009)
Oxalate decarboxylase	Fungal resistance	Cunha et al. (2010)
Mutated anthranilate synthase	Nutritional quality improvement	Ishimoto et al. (2010)
GmTFL1b (terminal flower 1b) for Dt1	Complementation of the stem growth habit in determinate	Liu et al., 2010

(Continued)

TABLE 12.3 Some Recent Examples of Soybean Lines Improved Through Genetic Engineering *(cont.)*

Introduced gene(s)	Trait	References
β-Conglycinin α subunit gene	Reduction of β-conglycinin	Nishizawa et al. (2010)
Fatty acid ω-6 desaturase 2 gene, acyl–acyl carrier protein thioesterase 2 gene, diacyglycerol acyltranferase gene, dihydrodipicolinate synthetase gene, high-lysine protein gene, truncated cysteine synthase gene	Improvement of oil content and composition	Li.Z. et al. (2010)
Epoxygenase gene, diacylglycerol acyltranferase gene	Increase of epoxy fatty acid	Li R. et al. (2010)
Mutated anthranilate synthase gene (*OASAID*)	Increase of free tryptophan	Kita et al. (2010)
Oxalate decarboxylase gene (oxdc)		Cunha et al. (2010)
Mutated aspartate kinase gene	Increase of threonine	Qi et al. (2011)
Homogentisate geranylgeranyltransferase gene (OsHGGT)	Accumulation of tocotrienol	Kim et al. (2011)
β-Amyrin synthase gene (GmBASI)	Reduction of seed saponin content	Takagi et al. (2011)
Dicer-like gene (DCL4a and DCL4b)	Targeted mutagenesis	Curtin et al. (2011)
Human growth hormone gene (hgh)	Accumulation of mature form of HGH	Cunha et al. (2011)
Cry1Ac	Controlling Lepidopteran pests	Yu et al. (2013)
GmFad2-1b	Increased oleic acid and reduced palmitic acid	Zhang et al. (2014)

Their demand has risen primarily in developing countries where economic growth and an expanding population have boosted the consumption of soy products.

Soybean seed oil is favored by many seeking a healthier life. The reasons for this are the quality and functionality of soy oil – being the precursor of Omega-3, Omega-6, and vitamin E – and its comparatively low price (Mounts et al., 1986). Globally, annual per capita soya oil consumption is estimated at around 3 kg. Nevertheless, consumption levels vary widely in different countries, with average consumption levels in industrialized countries almost three times those observed in developing nations. Moreover, the share of soybean oil in total vegetable oil consumption or total oil/fat intake varies considerably between regions and countries, depending on numerous factors such as the availability of locally produced oils, consumer habits and preferences, local and international market prices, and national trade policies. Soybean oil has high linoleic-acid content (an essential polyunsaturated fatty acid) and high linolenic-acid content (Table 12.4). Attempts are being made to breed the linolenic acid out directly by genetic transformations and indirectly by breeding for high-oleic lines. Hymowitz et al. (1972), Hammond and Fehr (1983a,b) have reported variability in linolenic acid content of up to 3%. Reduced palmitic acid has been reported by Fehr et al., 1991 and Wilcox and Cavins (1990). In addition to the desired high concentration of polyunsaturated fatty acids, soybean oil has several minor constituents that are valuable commercial products.

TABLE 12.4 Fatty Acid Composition of Soybean Oil

	Fatty acid composition (wt. %)	
Component fatty acid	**Range**	**Average**
Saturated		
Lauric	—	0.1
Myristic	<0.5	0.2
Palmitic	6–12	10.7
Stearic	2.0–5	3.9
Arachidic	<1.0	0.2
Bohenic	<0.5	—
Total	10–19	15.0
Unsaturated		
Palmitoleic	<0.5	0.3
Oleic	20–50	22.8
Linoleic	38–60	50.8
Linolenic	2–13	6.8
Eicosenioc	<0.1	—
Total	—	80.7

These minor constituents, which include lecithin, phytosterols, and tocopherols, are made available as a result of the high volume of soybean oil processed.

The protein content in soybeans varies from 35% to 50%. Soy protein is rich in lysine, which is markedly deficient in cereals and low in sulfur-containing amino acids such as methionine and cystine. Howell et al. (1972) have attempted to increase the methionine content in soybean protein. Mutation breeding has been found useful and has been applied to improve the oil and protein content in soybeans. Oil content and protein content in soybeans are negatively correlated traits, therefore increasing one usually results in decreasing the other (Hanson et al., 1961; Thorne and Fehr, 1970; Shannon et al., 1972; Brim and Burton, 1979; Sebern and Lambert, 1984; Hartwig and Kilen, 1991; Helms and Orf, 1998; Cober and Voldeng, 2000). Bhatnagar et al. (1992) were able to break this negative association through mutagenesis and obtained stable high protein and high oil-containing genotypes. Index selection has also been used to simultaneously increase them both. Genetic factors are major contributors to protein and oil content with some influence of environment in which the variety is grown (Fehr et al., 1992). Genotypes having high oil content invariably contain relatively higher oil content under different growing environments. In low-temperature environments, soybean seeds contain relatively high oil percentages.

The oil content in soybeans has been reported to be controlled by maternal influence (Brim et al., 1968; Singh and Hadley, 1968) while additive gene action has been reported to be important for this trait (Singh and Hadley, 1968; McKendry et al., 1985; Raut et al., 2000a). Percentage of oil also greatly depends on the extraction method used. The solvent extraction method is reported to extract more oil. Owing to breeding efforts, oil content has increased by as much as 21% in modern soybean varieties. In India, elite lines with higher oil percentages (approximately 24%) have been developed at the Agharkar Research Institute (ARI) in Pune through the conventional breeding method using early-generation testing (Raut et al., 2000b).

Soya Meal

Soya meal is widely consumed around the world and has penetrated markets in most regions, mainly due to progressive expansion of intensive livestock production methods in developed countries as well as in parts of the developing world. Livestock production markets require high-protein feedstock, and soy meal is ideal in these situations. Although, traditionally, over 50% of global utilization occurs in developed countries, annual consumption growth in developing countries far exceeds the expansion recorded in developed countries, mainly reflecting changes in consumer habits triggered by income growth.

OIL EXTRACTION

Hydraulic presses, screw presses, and solvent extraction are the commonly used means of extracting oil from soybean seeds (Pratap et al., 2012). Although, in earlier times hydraulic or screw presses were more commonly used, the modern oil extraction industries prefer to use the solvent extraction process, which removes more oil (hexane being the most common solvent used) (Carrao Panizzi and Gontijo Mandarino, 1994).

The extraction process is completed in several steps. First, the seeds are cleaned to remove foreign objects, stones, dirt, etc. and dried to a moisture level of 9.5%. After this, the seeds are passed through corrugated rolls in roller mills for cracking and dehulling the cracks. This follows sheating the cracked soybean meal to about 165°F to soften it before flaking. The heated and cracked meal is then passed through a roller mill to produce soy flakes, which are subsequently placed into a vapor-sealed percolation extractor, where solvent is percolated through a bed of soy flakes, dissolving the oil (Pratap et al. 2012). The mixture of oil and solvent (micella) leaves the bottom of the bed through a perforated plate. After this, hexane is removed leading to production of crude soybean oil, which is further subjected to a refining process that includes degumming, neutralization, bleaching, deodorization, and hydrogenation. The refining process removes impurities such as waxes, free fatty acids, sterols, pigments, and minerals from the crude oil. Byproduct soy flakes with the oil removed are conveyed to a desolventizer toaster to remove undrained hexane. This process removes the hexane and destroys antinutritional factors such as trypsin inhibitors, ureases, and hemagglutinins. Then, the meal is dried to about 13–14% moisture in a dry cooler and finally screened and ground to produce a uniform size prior to shipment to the end user.

Soya Oil as a Source of Biofuel

In recent years, soybean oil has received increased attention globally from governments and energy companies either through direct burning as a fuel or as a raw material for biodiesel production, besides using it in various intermediary forms. This was especially the case following the rise in mineral oil prices. Currently, global biofuel consumption is dominated by bioethanol, which is derived primarily from starch crops such as sugarcane and corn. Although vegetable oil is also used as a feedstock for producing biodiesel, its use is secondary only. Today, the most widely used feedstock for biodiesel production is rapeseed oil, followed by sunflower oil. The United States is the only country where significant amounts of soybean oil are regularly used for biodiesel production at the commercial scale.

Since the use of different types of plant oils as feedstocks for biodiesel production largely depend upon the policies of individual countries, cost of vegetable oil in the international market, and availability, etc., these directly affect the cultivation pattern of soybean.

In addition to biodiesel production, soybeans can also be used for other important purposes like ethanol production. Soybean hulls contain significant amounts of carbohydrates. Moreover, producers prefer to use soybean hulls for animal feeding owing to the rich protein content of soybean oil (Mielenz et al., 2009). Soybean oil is converted to biodiesel using the process of transesterification, which is usually facilitated by large-scale reaction of the oil to alcohol and catalysts.

SOYBEAN OIL FOR INDUSTRIAL USES

There is considerable potential for increased use of soybean oil as a renewable source of industrial oil, provided its physical and chemical properties are modified. Soybean oil is a complex mixture of five fatty acids (palmitic, stearic, oleic, linoleic, and linolenic), all of which have different melting points, oxidative stabilities, and chemical functionalities (Cahoon, 2003). The traditional solution to oil instability has been to partially hydrogenate the oil by adding hydrogen atoms across the double bonds in unsaturated fatty acids. However, keeping in mind the huge demand for soybean oil and the cost involved in its hydrogenation, changing the oil composition through breeding efforts offers a long-term solution. Breeders have worked successfully to increase the proportion of oleic acid versus linoleic and linolenic acids in soybean (Kinney, 1997). High-oleic soybean oil has about 80% oleic acid compared with 25% in conventional soybean oil, which is the highest among all oilseed crops. For this, the expression of *FAD2* genes – which code the enzyme that converts monounsaturated oleic acid to polyunsaturated linoleic acid – was downregulated. This prevents the addition of a second double bond to oleic acid to form linoleic acid, resulting in greatly increased levels of oleic acid. High-oleic soybean oil has a lower saturated fatty acid content than conventional soybean oil, making it more useful for consumer purposes. More recently, soybeans with even higher oleic acid content (85%) have also been generated by downregulating the *FAD2* gene along with the *FATB* genes that control the production of palmitic acid (Buhr et al., 2002). High-oleic soy oils and their blends are very good alternatives to partially hydrogenated oils and have found very good applications in a number of industries: snack food preparation, feedstock, pharmaceuticals, cosmetics, and machine lubricants.

Genetic engineering can also be used to produce high linolenic-acid soybeans (high-linolenic acid is a polyunsaturated fatty acid with low oxidative stability). By increasing the expression of the *FAD3* gene, soybean seeds with linolenic acid content more than 50% of total oil have been generated (Cahoon et al., 2001). These soybean lines have linolenic acid comparable with linseed oil (Cahoon, 2003) and are suitable for coating applications such as paints, varnishes, polishes, and inks.

References

Arelli, P.R., Young, L.D., 2009. JTN-5109 soybean germplasm resistant to nematode population infecting cv. Hartwig. Agron. Soc. Am., The Abstracts 268–218, p.133, (*http://a-c-s.confex.com/crops/2009am/*).

Arelli, P.R., Young, L.D., Mengistu, A., 2006. Registration of high yielding and multiple disease resistant soybean germplasm JTN-5503. Crop Sci. 46, 2723–2724.

Arelli, P.R., Pantalone, V.R., Allen, F.L., Mengistu, A., 2007. Registration of soybean germplasm JTN-5303. J. Plant Regist. 1, 69–70.

Bai, Y.N., Gai, J.Y., 2003. Development of soybean cytoplasmic-nuclear male sterile line NJCMS2A and restorability of its male fertility. Sci. Agric. Sinica 36, 740–745.

Bailey, M.A., Boerma, H.R., Parrott, W.A., 1993. Genotype effects on proliferative embryogenesis and plant regeneration of soybean. *In Vitro* Cell. Dev. Biol. Plant 29, 102–108.

Barwale, U.B., Kerns, H.R., Widholm, J.M., 1986. Plant regeneration from callus cultures of several soybean genotypes via embryogenesis and organogenesis. Planta 167, 473–481.

Behrens, M.R., Mutlu, N., Chakraborty, S., Dumitru, R., Jiang, W.Z., LaVallee, B.J., Herman, P.L., Clemente, T.E., Weeks, D.P., 2007. Dicamba resistance: enlarging and preserving biotechnology-based weed management strategies. Science 316, 1185–1188.

Bhatnagar, P.S., Tiwari, S.P., Singh, C., 1992. Disrupting the negative association between oil and protein content in soybean seeds through mutagenesis. Mut. Breed. Newsl. 39, 7.

Blanc, G., Wolfe, K.H., 2004. Widespread paleopolyploidy in model plant species inferred from age distributions of duplicate genes. Plant Cell 16, 1667–1678.

Brim, C.A., Burton, J.W., 1979. Recurrent selection in soybeans II. Selection for increased percent protein in seeds. Crop Sci. 19, 494–498.

Brim, C.A., Young, M.F., 1971. Inheritance of a male-sterile character in soybeans. Crop Sci. 11, 564–566.

Brim, C.A., Schutz, W.M., Collins, F.I., 1968. Maternal effect on fatty acid composition and oil content of soybean, *Glycine max* (L.) Merrill. Crop Sci. 8, 517–518.

Brown, A.H.D., Grant, J.E., Rullen, R., 1986. Outcrossing and paternity in *Glycine argyrea* by paired fruit analysis. Biol. J. Linn. Soc. 29, 283–294.

Buhr, T., Sato, S., Ebrahim, F., Xing, A., Zhou, Y., Mathiesen, M., Schweiger, B., Kinney, A., Staswick, P., Clemente, T., 2002. Ribozyme termination of RNA transcripts down-regulate seed fatty acid genes in transgenic soybean. Plant J. 30, 155–163.

Burton, J.W., 1987. Soybean [(*Glycine max* (L.) Merr.)]. Field Crops Res. 53, 171–186.

Burton, J.M., Brim, C.A., 1981. Recurrent selection in soybeans III. Selection for increased oil in seeds. Crop Sci. 21, 31–34.

Burton, J.W., Carver, B.F., 1993. Selection amongst families vs. selfed half-sib or full-sib families in autogamous crops. Crop Sci. 33, 21–28.

Burton, J.W., Koinange, E.M.K., Brim, C.A., 1990. Recurrent selfed progency selection for yield in soybean using male sterility. Crop Sci. 30, 1222–1226.

Cahoon, E.B., 2003. Genetic enhancement of soybean oil for industrial uses: prospects and challenges. AgBioForum 6, 11–13.

Cahoon, E.B., Ripp, K.G., Hall, S.E., Kinney, A.J., 2001. Formation of conjugated delta 8, delta10 double bonds by delta12-oleic acid desaturase related enzymes: Biosynthetic origin of calendic acid. J. Biol. Chem. 276, 2637–2643.

Cardoso, M.B., Kaltchuk-Santos, E., de Mundstock, E.C., Bodanese-Zanettini, M.H., 2004. Initial segmentation patterns of microspores and pollen viability in soybean cultured anthers: Indication of chromosome doubling. Braz. Arch. Biol. Tech. 47, 703–712.

Carpenter, J.A., Fehr, W.R., 1986. Genetic variability for desirable agronomic traits in populations containing *Glycine soja* germplasm. Crop Sci. 26, 681–686.

Carrao Panizzi, M.C., Gontijo Mandarino, J.M., 1994. Soybean for human consumption: nutritional quality, processing and utilization. Tropical soybean: Improvement and Production. Food and Agricultural Organization of the United Nations, Rome, pp. 241–254.

Carver, B.F., Burton, J.W., Willson, R.F., Carter, Jr., J.E., 1986. Cumulative response to various recurrent selections schemes in soybean: oil quality and correlated agronomic traits. Crop Sci. 26, 853–858.

Caviness, C.E., 1966. Estimates of nature cross-pollination in Jackson soybeans in Arkanasss. Crop Sci. 6, 211–212.

Chakraborty, N., Curley, J., Frederick, R.D., Hyten, D.L., Nelson, R.L., Hartman, G.L., Diers, B.W., 2009. Mapping and confirmation of a new allele at Rpp1 from soybean PI 594538A conferring RB lesion type resistance to soybean rust. Crop Sci. 49, 783–790.

Chaturvedi, S.K., Gupta D.S., Jain, R., 2011. Biology of Food Legumes. In: Pratap, A., Kumar, J. (eds.). Biology and Breeding of Food Legumes, CABI, Oxfordshire, pp. 35–48.

Chen, R., Matsui, K., Ogawa, M., Oe, M., Ochiai, M., Kawashima, H., Sakuradani, E., Shimizu, S., Ishimoto, M., Hayashi, M., Murooka, Y., Tanaka, Y., 2006. Expression of Δ6, Δ5 desaturase and GLELO elongase genes from *Mortierella alpina* for production of arachidonic acid in soybean [*Glycine max* (L.) Merrill] seeds. Plant Sci. 170, 399–406.

Chiang, Y.C., Kiang, Y.L., 1987. Geometric position of genotypes, honeybee foraging pattern and outcrossing in soybean. Bot. Bull. Acad. Sinica 28, 1–11.

Chiari, L., Naoe, L.K., Piovesan, N.D., José, I.C., Cruz, C.D., Moreira, M.A., and Gonçalves de Barros, E., 2006. Inheritance of isoflavone contents in soybean seeds. Euphytica 150, 141–147.

Clarindo, W.R., Carvalho, C.R., De Alves, B.M.G., 2007. Mitotic evidence for the tetraploid nature of *Glycine max* provided by high quality karyograms. Plant Syst. Evol. 26, 101–107.

Cober, E.R., Voldeng, H.D., 2000. Developing high-protein, high-yield soybean populations and lines. Crop Sci. 40, 39–42.

Concibido, V.C., La Vallee, B., Mclaird, P., Pineda, N., Meyer, J., Hummel, L., Yang, J., Wu, K., Delannay, X., 2003. Introgression of a quantitative trait locus for yield from *Glycine soja* into commercial soybean cultivars. Theor. Appl. Genet. 106, 575–582.

Cooper, R.L., 1990. Modified early generation testing procedure for yield selection in soybean. Crop Sci. 30, 417–419.

Culter, G.H., 1934. A simple method for making soybean hybrids. J. Am. Soc. Agron. 26, 252–253.

Cunha, W.G., Tinoco, M.L.P., Pancotti, H.L., Ribeiro, R.E., Aragão, F.J.L., 2010. High resistance to *Sclerotinia sclerotiorum* in transgenic soybean plants transformed to express an oxalate decarboxylase gene. Plant Pathol. 59, 654–660.

Cunha, N.B., Murad, A.M., Cipriano, T.M., Araujo, A.C.G., Aragao, F.J.L., Leite, A., Vianna, G.R., McPhee, T.R., Souza, G.H.M.F., Water, M.J., et al.,2011. Expression of functional recombinant human growth hormone in transgenic soybean seeds. Transgenic Res. 20, 811–826.

Curtin, S.J., Zahang, F., Sander, J.D., Haun, W.L., Starker, C., Baltes, N.J., Revon, D., Dahlborg, E.J., Goodwin, M.J., Coffman, A.P., et al.,2011. Targeted mutagenesis of duplicated gene in soybean with zinc-finger nuclease. Plant Physiol. 156, 466–473.

Davis, W.H., 1987. Process for forming seeds capable of growing hybrid soybean plants. United States Patent, 4, pp. 648–204.

de Moraes, A.P., Bonadese-Zanettini, M.H., Callegari-Jacques, S.M., Kaltchuk-Santos, E., 2004. Effect of temperature shock on soybean microspore embryogenesis. Braz. Arch. Biol. Tech. 47, 537–544.

Ding, D., Cui, Z., Gai, J., 1998. Development of cytological features of the cytoplasmic-nuclear male-sterile soybean line NJCMS1A. Soybean Genet. Newsl. 25, 34–35.

Ding, Y.L., Zhao, T.J., Gai, J.Y., 2008. Genetic diversity and ecological differentiation of Chinese annual wild soybean (*Glycine soja*). Biodiv. Sci. 16, 133–142.

Doyle, J.J., Doyle, J.L., Harbison, C., 2003. Chloroplast-expressed glutamine synthetase in Glycine and related Leguminosae: phylogeny, gene duplication, and ancient polyploidy. Syst. Bot. 28, 567–577.

Doyle, J.J., Doyle, J.L., Rauscher, J.T., Borwn, A.H.D., 2004. Evolution of the perennial soybean polyploidy complex (*Glycine* subgenus *Glycine*): a study of contrasts. Biol. J. Linn. Soc. 82, 583–597.

Dufourmantel, N., Pelissier, B., Garcon, F., Peltier, G., Ferullo, J.M., Tissot, G., 2004. Generation of fertile transplastomic soybean. Plant Mol. Biol. 55, 479–489.

Dufourmantel, N., Tissot, G., Goutorbe, F., Garçon, F., Muhr, C., Jansens, S., Pelissier, B., Peltier, G., Dubald, M., 2005. Generation and analysis of soybean plastid transformants expressing *Bacillus thuringiensis* Cry1Ab protoxin. Plant Mol. Bio. 58, 659–668.

Dufourmantel, N., Dubald, M., Matringe, M., Canard, H., Garçon, F., Job, C., Kay, E., Wisniewski, J.P., Ferullo, J.M., Pelissier, B., Sailland, A., Tissot, G., 2007. Generation and characterization of soybean and marker-free tobacco plastid transformants over-expressing a bacterial 4-hydroxyphenylpyruvate dioxygenase which provides strong herbicide tolerance. Plant Biotech. J. 5, 118–133.

Eckert, H., La Vallee, B., Schweiger, B.J., Kinney, A.J., Cahoon, E.B., Clemente, T., 2006. Co-expression of the borage Δ^6 desaturase and the Arabidopsis Δ^{15} desaturase results in high accumulation of stearidonic acid in the seeds of transgenic soybean. Planta 224, 1050–1057.

Ertl, D.S., Fehr, W.R., 1985. Agronomic performance of soybean genotypes from *Glycine max* × *Glycine soja* crosses. Crop Sci. 25, 589–592.

Fehr, W.R., Welke, G.K., Hammond, E.G., Duvick, D.N., Cianzio, S.R., 1991. Inheritance of reduced Palmitic acid content in seed oil of soybean. Crop Sci. 31, 88–89.

Fehr, W.R., Welke, G.A., Hammond, E.G., Duvick, D.N., Cinzio, S.R., 1992. Inheritance of reduced linolenic acid content in soybean genotypes A-16 and A-17. Crop Sci. 32, 903–906.

Ferraz de Toledo, J.F., Alves de Almeida, L., de Souza Kiihl, R.A., Carrao Panizzi, M.C., Miranda, L.C., Menosso, O.G., 1994. Genetics and Breeding. Tropical Soybean: Improvement and Production. Food and Agricultural Organization of the United Nations, Rome, pp. 19–36.

Flores, T., Karpova, O., Su, X., Zeng, P., Bilyeu, K., Sleper, D.A., Nguven, H.T., Zahang, Z.J., 2008. Silencing of GmFAD3 gene by siRNA leads to low α-linolenic acid (18:3) of fad3-mutant phenotype in soybean. Trangenic Res. 17, 839–850.

Fukui, J., Kaizuma, N., 1971. Morphological and ecological differentiation of various characters of the Japanese wild soybean *Glycine soja*. J. Fac. Agr. Iwate Univ. 10, 195–208, in Japanese with English abstract.

Furutani, N., Hidaka, S., Kosaka, Y., Shizukawa, Y., Kanematsu, S., 2006. Coat protein gene-mediated resistance to soybean mosaic virus in transgenic soybean. Breed. Sci. 56, 119–124.

Gai, J.Y., Xu, D.H., Gao, Z., Abe, Y.S.J., Fukushi, H., Kitajima, S., 2000. Studies on the evolutionary relationship among eco-types of *G. max* and *G. soja* in China. Acta Agron. Sinica 26, 513–520.

Gamborg, O.L., Miller, R.A., Ojima, K., 1968. Nutrient requirements of suspension cultures of soybean cells. Exp. Cell Res. 50, 151–158.

Gazzoni, D.L., 1994. Botany. Tropical Soybean: Improvement and Production. Food and Agricultural Organization of the United Nations, Rome, pp. 1–12.

Gill, N., Findley, S., Walling, J.G., Hans, C., Ma, J., Doyle, J., Stacey, G., Jackson, S.A., 2009. Molecular and chromosomal evidence for allopolyploidy in soybean. Plant Physiol. 151, 1167–1174.

Githiri, S.M., Watanabe, S., Harada, K., Takahashi, R., 2006. QTL analysis of flooding tolerance in soybean at an early vegetative growth stage. Plant Breed. 125, 613–618.

Graybosch, R.A., Palmer, R.G., 1987. Analysis of a male-sterile character in soybeans. J. Hered. 78, 66–70.

Guard, A.T., 1931. Development of floral organs of the soybean. Bot. Gaz. 91, 97–102.

Guo, W.T., 1993. The History of Soybean Cultivation. Hehai University Press, Nanjing, China.

Guo, X., Wang, D., Gordon, S.G., Helliwell, E., Smith, T., Berry, S.A., Martin, St., S.K., Dorrance, A.E., 2008. Genetic mapping of QTLs underlying partial resistance to *Sclerotinia sclerotiorum* in soybean PI 391589A and PI 391589B. Crop Sci. 48, 1129–1139.

Guo, J., Wang, Y., Song, C., Zhou, J., Qiu, L., Huang, H., Wang, Y., 2010. A single origin and moderate bottleneck during domestication of soybean (*Glycine max*): implications from microsatellites and nucleotide sequences. Ann. Bot. 106 (3), 505–514.

Hammond, E.G., Fehr, W.R., 1983a. Registration of A_5 germplasm line of soybean. Crop Sci. 23, 192.

Hammond, E.G., Fehr, W.R., 1983b. Registration of A_6 germplasm line of soybean. Crop Sci. 23, 192–193.

Han, Y., Teng, W., Yu, K., Poysa, V., Anderson, T., Qiu, L., Lightfoot, D.A., Li, W., 2008. Mapping QTL tolerance to Phytophthora root rot in soybean using microsatellite and RAPD/SCAR derived markers. Euphytica 162, 231–239.

Hanson, W.D., Leffel, R.C., Howell, R.W., 1961. Genetic analysis of energy production in the soybean. Crop Sci. 1, 121–126.

Harlan, J.R., de Wet, J.M.J., 1971. Towards a rational classification of cultivated plants. Taxon 20, 509–517.

Hartwig, E.E., 1973. Varcietal development. In: Caldwell, B.E. (Ed.), Soybeans: improvement, production, and uses. American Society of Agronomy Publ. No. 16, Madison, WI, pp. 187–210.

Hartwig, E.E., Kilen, T.C., 1991. Yield and composition of soybean seed from parents with different protein, similar yield. Crop Sci. 31, 290–292.

Helms, T.C., Orf, J.H., 1998. Protein, oil, and yield of soybean lines selected for increased protein. Crop Sci. 38, 707–711.

Herman, E.M., Helm, R.M., Jung, R., Kinney, A.J., 2003. Genetic modification removes an immunodominant allergen from soybean. Plant Physiol. 132, 36–43.

Hinchee, M.A.W., Connor-Ward, D.V., Newell, C.A., McDonnell, R.E., Sato, S.J., Gasser, C.S., Fischhoff, D.A., Re, D.B., Fraley, R.T., Horsch, R.B., 1988. Production of transgenic soybean plants using *Agrobacterium*-mediated DNA transfer. Biotechnology 6, 915–921.

Howell, R.W., Brim, C.A., Rinne, R.W., 1972. The plant geneticist's contribution toward changing lipid and amino acid composition of soybeans. J. Am. Oil Chemists Soc. 49, 30–32.

Hu, C.Y., Yin, G.C., Bodanese Zanettini, M.H., 1996. Haploid of soybean. In: Jain, S.M., Sopory, S.K., Veilleux, R.E. (Eds.), *In Vitro* Haploid Production in Higher Plants. Kluwer Academic Publisher, Dordrecht, Netherlands, pp. 377–395.

Hymowitz, T., 1970. On the domestication of the soybean. Eco. Bot. 24, 408–421.

Hymowitz, T., 2004. Speciation and cytogenetic. In: Boerma, H.R., Specht, J.E. (Eds.), Soybeans: Improvement Production, and Uses. Agronomy Monographs. third ed. ASA-CSSA-SSSA, Madison, WI, pp. 97–136, No.16.

Hymowitz, T., Palmer, R.G., Hadley, H.H., 1972. Seed weight, protein, oil and fatty acid relationship within genus *Glycine*. Trop. Agric. 49, 245–250.

Hyten, D.L., Smith, J.R., Frederick, R.D., Tucker, M.L., Song, Q., Cregan, P.B., 2009. Bulked segregant analysis using the GoldenGate assay to locate the Rpp3 locus that confers resistance to soybean rust in soybean. Crop Sci. 49, 265–271.

Ishimoto, M., Rahman, S.M., Hanafy, M.S., Khalafalla, M.M., El-Shemy, H.A., Nakamoto, Y., Kita, Y., Takanashi, K., Matsuda, F., Murano, Y., Funabashi, T., Miyagawa, H., Wakasa, K., 2010. Evaluation of amino acid content and nutritional quality of transgenic soybean seeds with high-level tryptophan accumulation. Mol. Breed. 25, 313–326.

Ivers, D.R., Palmer, R.G., Fehr, W.R., 1974. Anther culture in soybeans. Crop Sci. 14, 891–893.

Jian, Y.Y., Liu, D.P., Luo, X.M., Zhao, G.L., 1986. Studies on induction of pollen plants in *Glycine max* (L.) Merr. J. Agric. Sci. 2, 26–30.

Kaiser, J., 2008. Is the drought over for pharming? Science 320, 473–475.

Kassem, M.A., Meksem, K., Kang, C.H., Njiti, V.N., Kilo, V., Wood, A.J., Lightfoot, D.A., 2004. Loci underlying resistance to manganese toxicity mapped in a soybean recombinant inbred line population of 'Essex' × 'Forest'. Plant Soil 260, 197–204.

Kazi, S., Shultz, J., Afzal, J., Johnson, J., Njiti, V.N., Lightfoot, D.A., 2008. Separate loci underlie resistance to root infection and leaf scorch during soybean sudden death syndrome. Theor. Appl. Genet. 116, 967–977.

Kenworthy, W.J., Brim, C.A., 1979. Recurrent selection in soybeans I. Seed yield. Crop Sci. 19, 315–318.

Kiang, Y.T., Chiang, Y.C., Kaizuma, N., 1992. Genetic diversity in natural populations of wild soybean in Iwate prefecture, Japan. J. Hered. 83, 325–329.

Kim, Y.H., Lee, Y.Y., Kim, Y.H., Choi, M.S., Jeong, K.H., Lee, S.K., Seo, M.J., Yun, H.T., Lee, C.K., Kim, W.H., et al.,2011. Antioxydant activity and inhibition of lipid peroxydation in germinating seeds of transgenic soybean expressing OsHGGT. J. Agric. Food Chem. 59, 584–591.

Kinney, A.J., 1997. Genetic engineering of oilseeds for desired traits. Setlow, J.K. (Ed.), Genetic Engineering, 19, Plenum Press, New York, pp. 149–166.

Kita, Y., Hanafy, M.S., Deguchi, M., Hasegawa, H., Terakawa, T., Kitamura, K., Ishimoto, M., 2009. Generation and characterization of herbicide-resistant soybean plants expressing novel phosphinothricin N-acetyltranferase gene. Breed Sci. 59, 245–251.

Kita, Y., Nakamoto, Y., Takahashi, M., Kitamura, K., Wakasa, K., Ishimoto, M., 2010. Manipulation of amino acid composition in soybean seed by combination of deregulated tryptophan biosynthesis and storage protein deficiency. Plant Cell Rep. 29, 87–95.

Kollipara, K.P., Singh, R.J., Hymowitz, T., 1997. Phylogenetic and genomic relationship in the genus *Glycine willd.* Based on sequences from the ITS region of nuclear rDNA. Genome 40, 57–68.

Kumar, P.S., Hymowitz, T., 1989. Where are the diploid ($2n = 2x = 20$) genome donors of *Glycine willd.* (Leguminosae, Papilionoideae)? Euphytica 40, 221–226.

Lackey, J.A., 1980. Chromosome numbers in the Phaseoleae (Fabaceae,Faboideae) and their relationship to taxonomy. Am. J. Bot. 67, 595–602.

Lardizabal, K., Effertz, R., Levering, C., Mai, J., Pedroso, M.C., Jury, T., Aasen, E., Gruys, K., Bennett, K., 2008. Expression of Umbelopsis ramanniana DGAT2A in seed increases oil in soybean. Plant Physiol. 148, 86–89.

Lee, J.D., Shannon, J.G., So, Y.S., Sleper, D.A., Nelson, R.L., Lee, J.H., Choung, M.G., 2009. Environmnent effects on lutein content and relationship of lutein and other compounds in soybean. Plant Breed. 128, 97–100.

Li, L., Yang, Q., Hu, Y., Zhu, L., Ge, H., 1995. Discovery of parent interaction sterile material of soybean cultivars and its genetic inference. J. Anhui Agric. Sci. 23, 304–306.

Liu, D.P., Zhao, G.L., 1986. Callus formation from pollen and culture *in vitro* of soybean. Soybean Sci. 5, 17–20.

Liu, B., Watanabe, S., Uchiyama, T., Kong, F., Kanazawa, A., Xia, Z., Nagamastsu, A., Arai, M., Yamada, T., Kitamura, K., et al.,2010. The soybean stem growth habit gene Dtl is an orthologe of Arabidopsis terminal floweri. Plant Physiol. 153, 198–210.

Ma, J.K.C., Chikwamba, R., Sparrow, P., Fischer, R., Mahoney, R., Twyman, R.M., 2005. Plant-derived pharmaceuticals – the way forward. Trends Plant Sci. 10, 580–585.

Mathieu, M., Winters, E.K., Kong, F., Wan, J., Wang, S., Eckert, H., Luth, D., Paz, M., Donovan, C., Zhang, Z., et al.,2009. Establishment of a soybean transposon-based mutagenesis repository. Planta 229, 279–289.

McCabe, D.E., Swain, W.F., Martinell, B.J., Christou, P., 1988. Stable transformation of soybean (*Glycine max*) by particle acceleration. Biotechnology 6, 923–926.

McKendry, A.L., McVetty, P.B.E., Voldeng, H.D., 1985. Inheritance of seed protein and seed oil content in early maturing soybean. Can. J. Genet. Cytol. 37, 603–607.

Mielenz, J.R., Bardsley, J.S., Wyman, C.E., 2009. Fermentation of soybean hulls to ethanol while preserving protein value. Bioresource Technol. 100, 3532–3539.

Moravec, T., Schmidt, M.A., Eliot, M., Herman, E.M., Woodford-Thomas, T., 2007. Production of *Escherichia coli* heat labile toxin (LT) B subunit in soybean seed and analysis of its immunogenicity as an oral vaccine. Vaccine 25, 1647–1657.

Mounts, T.L., Warner, K., List, G.R., Kleiman, R., Fehr, W.R., Hammond, E.G., Wilcox, J.R., 1986. Effect of fatty acid composition on soybean oil stability. J. Am. Chemist's Soc. 65, 624–628.

Murashige, T., Skoog, F., 1962. A revised medium for rapid growth and bioassays with tobacco tissue cultures. Physiol. Plant. 15, 473–497.

Nelson, R.L., Bermard, R.L., 1984. Production and performance of hybrid soybean. Crop Sci. 23, 549–553.

Nishizawa, K., Kita, A., Doi, C., Yamada, Y., Ohinata, K., Toshikawa, M., Ishimoto, M., 2008. Accumulation of the bioactive peptides, novokinin, LPYPR and Rubiscolin, in seeds of genetically modified soybean. Biosci. Biotechnol. Biochem. 72, 3301–3305.

Nishizawa, K., Takagi, A., Teraishi, M., Kita, A., Ishimoto, N., 2010. Application somatic embryos to rapid and reliable analysis of soybean seed components by RNA interference-mediated gene silencing. Plant Biotech. 27, 409–420.

Ortiz-Perez, E., Cianzio, S.R., Wiley, H., Horner, H.T., Davis, W.H., Palmer, R.G., 2007. Insect-mediated cross-pollination in soybean [*Glycine max* (L.) Merr.]: I. Agronomic performance. Field Crops Res. 101, 259–268.

Padgette, S.R., Kolacz, K.H., Delannay, X., Re, D.B., LaVallee, B.J., Tinius, C.N., Rhodes, W.K., Otero, Y.I., Barry, G.F., Eichholtz, D.A., Peschke, V.M., Nida, D.L., Taylor, N.B., Kishore, G.M., 1995. Development, identification, and characterization of a glyphosate-tolerant soybean line. Crop Sci. 35, 1451–1464.

Palmer, R.G., Skorupska, H., 1990. Registration of a male-sterile genetic stock (T295H) of soybean. Crop Sci. 30, 241.

Palmer, R.G., Gai, J., Sun, H., Burton, J.W., 2001. Production and evaluation of hybrid soybean. Plant Breed. Rev. 21, 263–307.

Parrott, W.A., Williams, E.G., Hildebrand, D.F., Collins, G.B., 1989. Effect of genotype on somatic embryogenesis from immature cotyledons of soybean. Plant Cell Tiss. Org. Cult. 16, 15–21.

Pathan, M.S., Sleper, D.A., 2008. Advances in soybean breeding. In: Stacey, G. (Ed.), Genetics and Genomic of Soybean. Springer Science, New York, pp. 113–133.

Patil, A., Taware, S.P., Raut, V.M., 2004. Quality of Indian soybean (*Glycine max* (L.) Merrill) varieties in relation to fatty acids composition. Indian J. Genet. Plant Breed. 64, 245–246.

Patzoldt, M.E., Grau, R., Stephens, P.A., Kurtzweil, N.C., Carlson, S.R., Diers, B.W., 2005. Localization of a quantitative trait locus providing brown stem rot resistance in the soybean cultivar Bell. Crop Sci. 45, 1241–1248.

Pratap, A., Gupta, S.K., Jitendra Kumar, R.K., Solanki, 2012. Soybean. In: Gupta, S.K. (Ed.), Technological innovations in major world oil crops, Vol. I Breeding. Springer Science, New York, pp. 293–321.

Probst, A.H., Judd, R.W., 1973. Origin, US History and Development, and World Distribution. In: Caldwell, B.E. (Ed.), Soybean: Improvement, Production, and Uses. Agron Mongr. 16. ASA, CSSA, and SSSA, Madison, WI, pp. 1–15.

Qi, B., Korir, P., Zhao, T., Yu, D., Chen, S., Gai, J., 2008. Mapping quantitative traits loci associated with aluminium toxin tolerance in NJRIKY Recombinant Inbred Line Population of Soybean (Glycine max). J. Integrative Plant Biol. 50, 1089–1095.

Qi, Q., Huang, J., Crowlev, J., Ruschke, L., Goldman, B.S., Wen, L., Rapp, W.D., 2011. Metabolically engineered soybean seed with inhanced threonine levels : biochemical characterization and seed specific expression of lysine, insensitive variants of asparted kinases from the interic bacterium Xenorhabdus bovienii. Plant Biotech. J. 9, 193–204.

Rao, S.S., Hildebrand, D., 2009. Changes in oil content of transgenic soybeans expressing the yeast *SLC1* gene. Lipids 44, 945–951.

Raut, V.M., Taware, S.P., Halvankar, G.B., 2000a. Gene effects for some quantitative characters in soybean (*Glycine max*) crosses. Indian J. Agric. Sci. 70, 334–335.

Raut, V.M., Taware, S.P., Halvankar, G.B., Varghese P., 2000b. Development of high yielding variety MACS-450 by using Kalitur mutant MACS-111. DAE-BRNS. Symposium. BARC-Bombay, pp. 103–110.

Rodrigues, L.R., Oliveira, J.M.S., Mariath, J.E.A., Iranco, L.B., Bodanese-Zanettini, M.H., 2005. Anther culture and cold treatment of floral buds increased symmetrical and extra-nuclei frequencies in soybean pollen grains. Plant Cell Tiss. Org. Cult. 81, 101–104.

Rose, J.L., Butler, D.G., Ryley, M.J., 1992. Yield improvement in soybeans using recurrent selection. Aust. J. Agric. Res. 4, 135–144.

Safari, A., Schlueter, J.A., 2011. Polyploidy. In: Pratap, A., Kumar, J. (eds.) Biology and Breeding of Food Legumes. CABI, Oxfordshire, pp. 111–119.

Saghai Maroof, M.A., Jeong, S.C., Gunduz, I., Tucker, D.M., Buss, G.R., Tolin, S.A., 2008. Pyramiding of soybean mosaic virus resistance genes by marker assisted selection. Crop Sci. 48, 517–526.

Santarem, E.R., Finer, J.J., 1999. Transformation of soybean (*Glycine max* (L.) Merrill) using proliferative embryogenic tissue maintained on semi-solid medium. In Vitro Cell. Dev. Bio. Plant 35, 451–455.

Sato, S., Xing, A., Ye, X., Schweiger, B., Kinney, A., Graef, G., Clemente, T., 2004. Production of γ-Linolenic Acid and Stearidonic Acid in Seeds of Marker-Free Transgenic Soybean. Crop Sci. 44, 646–652.

Sauer, M.L.A., Scott, R.A., Cheesbrough, T.M., 2008. Marker assisted selection for low linolenic acid content in soybean. J. Crop Improvement 21, 139–155.

Schlueter, J.A., Dixon, P., Granger, C., Grant, D., Clark, L., Doyle, J.J., Shoemaker, R.C., 2004. Mining EST databases to resolve evolutionary events in major crop species. Genome 47, 868–876.

Schmidt, M.A., Tucker, D.M., Cahoon, E.B., Parrott, W.A., 2005. Towards normalization of soybean somatic embryo maturation. Plant Cell Rep. 24, 383–391.

Sebern, N.A., Lambert, J.W., 1984. Effect of stratification for percent protein in two soybean populations. Crop Sci. 24, 225–228.

Sekizuka, S., Yoshiyama, T., 1960. Studies on the native wild grasses for fodder (IV). Crop-scientific studies on wild species of *Glycine soja* in Japan. J. Kanto-Tosan Agr. Exp. Stn. 15, 57–73, in Japanese with English abstract.

Sen, N.K., Vidyabhusan, R.V., 1960. Tetraploid soybeans. Euphytica 9, 317–322.

Shannon, J.G., Wilcox, J.R., Probst, A.H., 1972. Estimated gains from selection for protein and yield in the F_4 generation of six soybean populations. Crop Sci. 12, 824–826.

Shi, A., Chen, P., Li, D.X., Zheng, C., Hou, A., Zhang, B., 2008. Genetic confirmation of 2 independent genes for resistance to soybean mosaic virus in J05 soybean using SSR markers. J. Hered. 99, 598–603.

Shi, A., Chen, P., Li, D., Zheng, C., Zhang, B., Hou, A., 2009. Pyramiding multiple genes for resistance to soybean mosaic virus in soybean using molecular markers. Mol. Breed. 23, 113–124.

Shoemaker, R., Polzin, K., Labate, J., Specht, J., Brummer, E.C., Olson, T., Young, N., Concibido, V., Wilcox, J., Tamulonis, J., Kochert, G., Boerma, H.R., 1996. Genome duplication in soybean (*Glycine* subgenus *soja*). Genetics 144, 329–338.

Shoemaker, R.C., Schlueter, J., Doyle, J.J., 2006. Paleopolyploidy and gene duplication in soybean and other legumes. Curr. Opin. Plant Biol. 9, 104–109.

Schoen, D.J., Brown, A.H.D., 1991. Whole and Part- Flower self–pollination in Glycine clandestine and G. argyrea and Evolution of Autogamy. Evolution 45, 1651–1664.

Shultz, J.L., others, 2006. The soybean genome database (SoyGD): a browser for display a duplicated, polyploidy, *regions* and sequence tagged sites on integrated physical and genetic maps of *Glycine max*. Nucleic Acid Res. 34, D758–D765.

Singh, B.B., Hadley, H.H., 1968. Maternal control of oil synthesis in soybeans, (*Glycine max* (L.) Merr.). Crop Sci. 8, 622–625.

Singh, R.J., Hymowitz, T., 1985. The genomic relationships among six wild perennial species of the genus *Glycine* subgenus *Glycine willd*. Theor. Appl. Genet. 71, 221–230.

Singh, R.J., Hymowitz, T., 1988. The genomic relationship between *Glycine max* (L.) Merr. and *G. soja* Sieb. and Zucc. as revealed by pachytene chromosome analysis. Theor. Appl. Genet. 76, 705–711.

Singh, R.J., Hymowitz, T., 1999. Soybean genetic resources and crop improvement. Genome 42, 605–616.

Skorupska, H., Palmer, R.G., 1989. Genetics and cytology of the *ms6* male-sterile soybean. J. Hered. 80, 304–310.

Skrzypek,E., Czyczyło-Mysza, I., Węgzony, M., 2011. Micropropagation. In: Pratap, A., Kumar, J., (eds.) Biology and Breeding of Food Legumes, CABI, Oxfordshire, pp. 147–158.

Smith, J., 2010. USDA, ARS, National Genetic Resources Program. Germplasm Resources Information Network – (GRIN). National Germplasm Resources Laboratory, Beltsville, Maryland. (http://www.ars-grin.gov/cgi-bin/npgs/acc/search.pl?accid=PI+659348).

Smith, J.R., Nelson, R.L., 1986. Relationship between seed filling period and yield among soybean breeding lines. Crop Sci. 26, 469–472.

Sompong, U., Kaewprasit, C., Nakasathien, S., Srinives, P., 2010. Inheritance of seed phytate in mungbean (*Vigna radiata*). Euphytica 171, 389–396.

Steeves, R.M., Todd, T.C., Essig, J.S., Trick, H.N., 2006. Transgenic soybeans expressing siRNAs specific to a major sperm protein gene suppress *Heterodera glycines* reproduction. Funct. Plant Biol. 33, 991–999.

Stelly, D.M., Palmer, R.G., 1980. A partially male-sterile mutant line of soybeans, *Glycine max* (L.) Merr.: inheritance. Euphytica 29, 295–303.

Stelly, D.M., Palmer, R.G., 1985. Relative development of basal, medial, and apical ovules in soybean. Crop Sci. 25, 877–879.

Straub, S.C.K., Pfeil, B.E., Doyle, J.J., 2006. Testing the polyploid past of soybean using a low-copy nuclear gene – is Glycine (Fabaceae: Papilionoideae) and auto- or allopolyploid? Mol. Phylogenet. Evol. 39, 580–584.

Sun, H., Zhao, L., Huang, M., 1994. Studies on cytoplasmic-nuclear male sterile soybean. Chin. Sci. Bull. 39, 175–176.

Sun, H., Zhao, L., Huang, M., 1997. Cytoplasmic-nuclear male-sterile soybean line from interspecific crosses between *G. max* and *G. soja*. World Soybean Research Conference VKasetsart University Press, Bangkok, Thailand, pp. 99–102.

Takagi, K., Nishizawa, K., Hirose, A., Kita, A., Ishimoto, M., 2011. Manipulation of saponin biosynthesis by RNA interference mediated silencing of β-amyrin synthase gene expression in soybean. Plant Cell Rep. 30, 1835–1846.

Thorne, J.C., Fehr, W.R., 1970. Incorporation of high-protein, exotic germplasm into soybean populations by 2- and 3-way crosses. Crop Sci. 10, 652–655.

Tinius, C.N., Burton, J.W., Carter, Jr., T.E., 1991. Recurrent selection for seed size in soybean, I. Response to selection in replicate populations. Crop Sci. 31, 1137–1141.

Tiwari, S., Shanker, P., Tripathi, M., 2004. Effects of genotype and culture medium on *in vitro* androgenesis in soybean (*Glycine max* Merr.). Indian J. Biotech. 3, 441–444.

Tomlin, E.S., Branch, S.R., Chamberlain, D., Gabe, H., Wright, M.S., Stewart, Jr., C.N., 2002. Screening of soybean, *Glycine max* (L.) Merrill, lines for somatic embryo induction and maturation capability from immature cotyledons. *In Vitro* Cell. Dev. Biol. Plant 38, 543–548.

Travella, S., Ross, S.M., Harden, J., Everett, C., Snape, J.W., Harwood, W.A., 2005. A comparison of transgenic barley lines produced by particle bombardment and *Agrobacterium*-mediated techniques. Plant Cell Rep. 23, 780–789.

Trick, H.N., Finer, J.J., 1998. Sonication-assisted *Agrobacterium*-mediated transformation of soybean [*Glycine max* (L.) Merrill] embryogenic suspension culture tissue. Plant Cell Rep. 17, 482–488.

Tuyen, D.D., Lal, S.K., Xu, D.H., 2010. Identification of a major QTL allele from wild soybean (Glycine soja Sieb. & Zucc.) for increasing alkaline salt tolerance in soybean. Theor. Appl. Genet 121 (2), 229–236.

Udall, J.A., Wendel, J.F., 2006. Polyploidy and crop improvement. Crop Sci. 46 (S1), S3–S14.

Valente, M.A.S., Faria, J.A.Q.A., Soares, J.R.L., Ramos, P.A.B., Reis, G., Pinheiro, L., Piovesan, N.D., Morais, A.T., Menezes, C.C., Cano, M.A.O., Fietto, L.G., et al.,2009. The ER luminal binding protein (B_1P) mediates an increase in drought tolerance in soybean and delays drought indouced leaf senescence in soybean and tobacco. J. Exp. Bot. 60, 533–546.

Van, K., Jang, H., Young-Eun, J., Lee Suk-Ha, J., 2008. Regeneration of plants from EMS-treated immature embryo cultures in Soybean [*Glycine max* (L.) Merr.]. Crop Sci. Biotech. 11, 119–126.

Vuong, T.D., Sleper, D.A., Shannon, J.G., Nguyen, H.T., 2010. Novel quantitative trait loci for broad-based resistance to soybean cyst nematode (*Heterodera glycines* Ichinohe) in soybean PI 567516C. Theor. Appl. Genet. 121 (7), 1253–1266.

Walker, D.R., Parrott, W.A., 2001. Effect of polyethylene glycol and sugar alcohols on soybean somatic embryo germination and conversion. Plant Cell Tiss. Org. Cult. 64, 55–62.

Walker, D.R., Boerma, H.R., All, J.N., Parrott, W.A., 2002. Combining cry1Ac with QTL alleles from PI 229358 to improve soybean resistance to Lepidopteran pests. Mol. Breed. 9, 43–51.

Wang, G., Xu, Y., 2008. Hypocotyl-based Agrobacterium-mediated transformation of soybean and application RNA interference. Plant Cell Rep. 27, 1177–1184.

Wang, S., Sun, H., Zhao, L., Wang, Y., Peng, B., Fan, X., Zhang, B., 2009. Progress and problem analysis on soybean male sterility and heterosis exploitation in China. Soybean Sci. 28, 1089–1096.

Wang, H., Waller, L., Tripathy, S., Martin, St., S.K., Zhou, L., Krampis, K., Tucker, D.M., Mao, Y., Hoeschele, I., Saghai Maroof, M.A., Tyler, B.M., Dorrance, A.E., 2010. Analysis of genes underlying soybean quantitative trait loci conferring partial resistance to *Phytophthora sojae*. Plant Genome 3, 23–40.

Warrington, C.V., Zhu, S., Parrot, W.A., All, J.N., Boerma, H.R., 2008. Seed yield of near-isogenic soybean lines with introgressed quantitative trait loci conditioning resistance to corn earworm (Lepidoptera: Noctuidae) and soybean looper Lepidoptera: Noctuidae) from PI 229358. J. Eco. Entomol. 101, 1471–1477.

Wells, R., Burton, J.W., Kilen, T.C., 1993. Soybean growth and light interception: response to differing leaf and stem morphology. Crop Sci. 33, 520–524.

Wilcox, J.R., Cavins, J.F., 1990. Registration of C1726 and C1727 Soybean germplasm with altered levels of palmitic acid. Crop Sci. 30, 240.

Wu, X., Blake, S., Sleper, D.A., Shannon, J.G., Cregan, P.B., Nguyen, H.T., 2009. QTL, additive, and epistatic effects for SCN resistance in PI 437654. Theor. Appl. Genet. 118, 1093–1105.

Xu, Z., Li, L., Qiu, L., Chang, R., Wang, M., Li, Z., Guo, P., 1999. Selection of three lines and localization of the restorer genes in soybean using SSR markers. Sci. Agric. Sinica 32, 32–38.

Yamada, Y., Nishizawa, K., Yokoo, N., Zhao, H., Onishi, K., Teraishi, M., Utsumi, S., Ishimoto, M., Yoshikawa, M., 2008. Antihypertensive activity genetically modified soybean seeds accumulating novokinin. Peptides 29, 331–337.

Yang, G., Lee, Y.H., Jiang, Y., Kumpatla, S.P., Hall, T.C., 2005. Organization, not duplication, triggers silencing in a complex transgene locus in rice. Plant Mol. Bio. 58, 351–366.

Ye, X.G., Fu, Y.Q., Wang, L.Z., 1994. Study on several problems of soybean anther culture. Soybean Sci. 13, 193–199.

Yeung, E.C., Sussex, I.M., 1979. Embryogeny of *Phaseolus coccineus*: the suspensor and the growth of the embryo-proper *in vitro*. Zeitschrift für Pflanzenphysiologie 91, 423–433.

Yin, G.C., Zhu, Z.Y., Xu, Z., Chen, L., Li, X.Z., Bi, F.Y., 1982. Studies on induction of pollen plant and their androgenesis in *Glycine max* (L.). Soybean Sci. 1, 69–76.

Zernova, O.V., Lygin, A.V., Widholm, J.M., Lozovaya, V.V., 2009. Modification of isoflavones in soybean seeds via expression of multiple phenolic biosynthetic genes. Plant Physiol. Biochem. 47, 769–777.

Zhang, L., Dai, O., 1997. Selection of cytoplasm-nuclear male-sterile soybean line W931A. Sci. Agric. Sinica 30, 90–91.

Zhang, L., Dai, O., Zhang, L., 1999a. Breeding of soybean male-sterile line of nucleo-cytoplasmic interaction. Soybean Sci. 18, 327–330.

Zhang, L., Dai, O., Huang, Z., Li, J., 1999b. Selection of soybean male-sterile line of nucleo-cytoplasmic interaction and its fertility. Sci. Agric. Sinica 32, 34–38.

Zhao, L., Sun, H., Huang, M., 1998. The development of preliminary studies on cytoplasmic male sterile soybean line ZA. Soybean Sci. 17, 268–270.

Zhao, T.J., Gai, J.Y., 2006. Discovery of new male-sterile cytoplasm sources and development of a new cytoplasmic nuclear male-sterile line NJCMS 3A in soybean. Euphytica 152, 387–396.

CHAPTER

13

Omics – A New Approach to Sustainable Production

Sajad Majeed Zargar, Nancy Gupta**, Muslima Nazir*[†], *Rakeeb Ahmad Mir*[‡], *Surinder Kumar Gupta*[§], *Ganesh Kumar Agrawal*[¶], *Randeep Rakwal*[¶,††]

*Centre for Plant Biotechnology, Division of Biotechnology, Sher-e-Kashmir University of Agricultural Sciences & Technology of Kashmir, Shalimar, Srinagar, J&K, India

**School of Biotechnology, Sher-e-Kashmir University of Agricultural Sciences & Technology of Jammu, Chatha, Jammu, J&K, India

[†]Department of Botany, Faculty of Science, Jamia Hamdard University, Jamia Nagar, New Delhi, India

[‡]School of Biosciences and Biotechnology, BGSB University, Rajouri, India

[§]Division of Plant Breeding & Genetics, Faculty of Agriculture, Sher-e-Kashmir University of Agricultural Sciences & Technology Chatha, Jammu (J&K), India

[¶]Research Laboratory for Biotechnology and Biochemistry (RLABB), Kathmandu, Nepal

[††]Organization for Educational Initiatives, University of Tsukuba, Tennoudai, Tsukuba, Ibaraki, Japan

INTRODUCTION

Oilseed crops are a diverse group of crop species which are utilized for the production of oil that can be used for human consumption. About 40 oilseeds, belonging to different families and genera, are known to be consumable (Lennerts, 1983), however, only a few are significant in total world trade. They are grown world over under varied agroclimatic situations and are vital commodities in world trade and commerce. Oil seed crops fit into three distinct categories: the annual or biennial group which includes soybeans, sunflower, groundnuts, and rapeseeds; the perennial tree crops which include coconuts and oil palms; and the third group in which an embryo is a byproduct, including cotton and corn germ. Other oilseeds are of minor importance in world trade, however, they do play a significant role in local markets or serve as raw materials for special products. These oils include castor, safflower, and linseed. Dietary fat, a concentrated source of energy, supplies about half of our calorific dietary

Breeding Oilseed Crops for Sustainable Production. http://dx.doi.org/10.1016/B978-0-12-801309-0.00013-6

requirement, carrying fat-soluble vitamins. Besides the consumption of oils used as ingredients in human food, many of the oilseeds are used unprocessed in animal feed as well as in processed forms such as cakes or meals which are an important source of dietary proteins for animals. The production of oilseeds is a major constraint mainly in developing countries. Moreover, the increasing world population, which is expected to rise to 9.1 billion by 2050 (UNPF, 2005), will further make the situation worse. In order to feed this larger population, food and oil production must increase by 70%. In addition, the per capita consumption of oil and fat, as recommended by the World Health Organization (WHO) and the Food and Agriculture Organization (FAO), for maintaining the nutritional requirements of the human body, is 21 kg a year. However, the present world average for per capita consumption of oil and fat stands at 16 kg. Science-based crop improvement efforts will endeavor to enhance yields in order to feed this increasing population. Moreover, the impacts of climate change, such as frequent droughts, floods, and temperature variations, also need to be considered in order to improve oilseed crops.

Conventional plant breeding has been used to develop new varieties of crops for hundreds of years. However, conventional plant breeding can no longer sustain the global oilseed production demands from this increasing population, nor the decline in agricultural resources such as land and water. There is an apparent plateauing of the yield curve of oilseed crops. The world production of major oilseed crops has remained static over the past decade at around 455.7 million tons. Keeping all this in mind, the production and quality of oilseeds needs to be improved at an accelerated rate in order to meet requirements of the present and future. Improvement in oilseed production, productivity, and quality are dependent on the rational utilization of new technologies. In this connection, advances made in the omics approach have opened an avenue to tackle many of the unresolved problems pertaining to oilseed crop sustainability. In this chapter, we discuss various high-throughput techniques, which can be put to use for attaining sustainability in oilseed production. We also emphasize the need for precision phenotyping in order to have suitable outcomes from omics-based approaches.

GENOMIC APPROACH

Before the 1980s, conventional methods of plant breeding made a very significant contribution to oilseed crop improvement, but these methods were slow at improving complex traits like crop yield, oil yield, some quality traits, and biotic and abiotic stress resistance. Moreover, it was not possible to detect the position of the genes controlling the phenotypic expression of these complex traits. In order to introgress the desirable trait, it is important to have information about the location of a gene(s) as well as linked markers. Hence plant breeding assisted by genomics-based techniques is an important step for achieving this goal. Genomics-assisted crop improvement (GACI) offers unprecedented opportunities to identify major loci influencing target traits and to select for plants with desirable combinations of alleles via marker-assisted selection (MAS), marker-assisted backcrossing (MABC), or MARS (Varshney and Tuberosa, 2013). Good progress has been made in the genomics of major oilseed crops such as soybean, brassica, groundnut, and sunflower but other oilseed crops need further biotechnological interventions for improvement (Yadava et al., 2012). These advances include the availability of a large repertoire of molecular markers, genotyping platforms, advances in genetic mapping procedures, study of marker–trait associations leading to the discovery of genes/quantitative trait loci (QTL), sequencing technologies leading to the reduction in sequencing costs, and available genome sequences of major crop genomes, etc. Advances in genomic tools and technologies have made gene discovery possible in oilseed crop plants and has helped lead to a better understanding/genetic dissection of gene networks/complex quantitative traits that contribute to the development of a superior plant (Yadava et al., 2012). Therefore, these techniques have increased precision in conventional breeding programs and in several cases have led to the development of superior cultivars with enhanced resistance/tolerance to biotic/abiotic stresses through genomics-assisted breeding (Walker et al., 2002; Saghai-Maroof et al., 2008; Yue et al., 2007). Figure 13.1 gives the advantages of genomics approaches over conventional breeding. In this chapter, we will go on to discuss some major requirements for genomics assisted crop improvement and give examples.

Genomic Approaches to the Genetic Dissection of Complex Traits

Most of the economically important traits of oilseed crops, such as economic yield, oil yield, various quality traits, and biotic and abiotic stress tolerance, are genetically complex and governed by polygenes distributed throughout the genome. These polygenes were impossible to map through Mendelian genetics or classical genetics because of their minor or

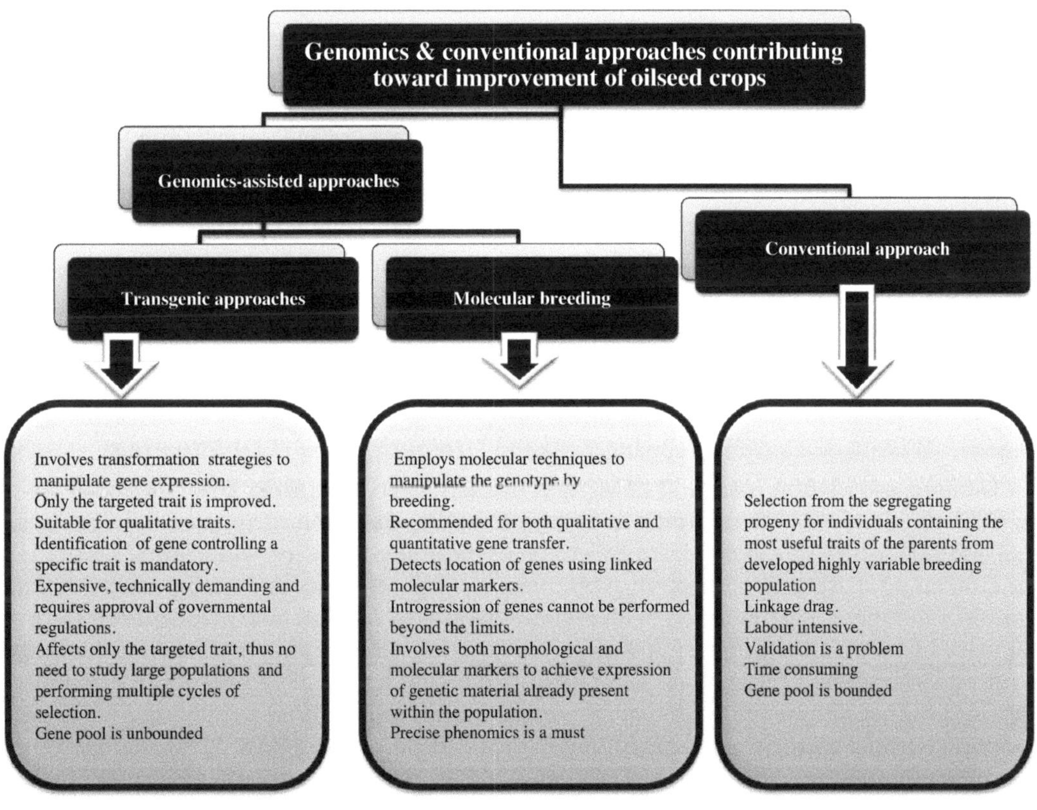

FIGURE 13.1 **Genomics & conventional approaches for improvement of oilseed crops.**

small effect on phenotypic expression. Little was known about such genes including their location on the chromosomes, the individual effect of the genes/QTLs, and how they controlled traits (Kearsey and Farquhar, 1998). With the evolution of new markers and the continuous advancement that occurs in genomics, it is now possible for us to locate/map the position of genes/QTLs on chromosomes and study their individual effect. Mostly, two genomic approaches have been used for identifying marker–trait associations in oilseed crops, which include QTL or linkage mapping and association or linkage disequilibrium (LD) mapping. Each approach has its merits and demerits (Neale and Savolainen, 2004).

Linkage Mapping/QTL Mapping

For linkage mapping you need to have genetic resources in the form of a biparental mapping population and genomic resources in the form of molecular markers and genome maps. A number of molecular markers and genome maps have been developed for most of the major oilseed crops such as soybean (Lark et al., 1993; Xia et al., 2007; Cregan et al., 1999; Song et al., 2004), groundnut (Garcia et al., 1995; Jayashree et al., 2005; Wang et al., 2007; Varshney et al., 2009), *Brassica napus* (Landry et al., 1991; Foisset et al., 1996; Lydiate and Sharpe 2003; Sun et al., 2007), sunflower (Berry et al., 1995; Jan et al., 1998, Langer et al., 2003; Yu et al., 2003; Tang et al., 2002; Kolkman et al., 2007; Fusari et al., 2008), sesamum (Dixit et al., 2005; Laurentin and Karlovasky, 2006, 2007; Sharma et al., 2009), *B. juncea* (Sharma et al., 1994, 2002; Kalita et al., 2007), Linseed (Oh et al., 2000; Fu et al., 2002; Roose et al., 2006), and safflower (Amiri et al., 2001; Johnson et al., 2007; Sehgal and Raina 2005; Yang et al., 2007).

For mapping of QTLs/genes of complex traits the QTL mapping approach has been adopted in various oilseed crops. QTL mapping has been done for seed and oil yield (Diers et al., 1992; Lark and Mansur, 1994; Hu et al., 1995; Ecke et al., 1995; Zhao et al., 2006; Delourme et al., 2006; Shi et al., 2009; Qui et al., 2006; Rahman et al., 2008; Nath and Goswami 2009; Sharma et al., 2002; Perez et al., 2004), abiotic stress tolerance (Lee et al., 2004; Varshney et al., 2009; Herve et al., 2001; Kiani et al., 2007; Jamaux et al., 1997; Githiri et al., 2006; Tuyen et al., 2010; Kassem et al., 2004; Li et al., 2000; Qi et al., 2008; Li et al., 2005), biotic stress tolerance (Diers et al., 1991; Li et al., 1998; Guo et al., 2005; Leflon et al., 2007; Zhao and Meng 2003; Happstadius et al., 2003; Ferreira et al., 1995; Panjabi et al., 2010; Varma et al., 2005; Mondal et al., 2007; Lenardon et al., 2005; Chen et al., 2001), and oil quality (Bachlava et al., 2008; Li et al., 2008; Hasan et al., 2008; Gupta et al., 2004; Bisht et al., 2009).

Genome-Wide Association Mapping/Linkage Disequilibrium (LD) Mapping

LD-based association mapping (AM) was initially used in human genetics and is an alternative approach to QTL mapping in crop species. Based on the correlation between genotype and phenotype in a germplasm collection or natural population, associating mapping is undertaken. QTL detection, through AM, offers several advantages over biparental linkage mapping, namely: (i) exploitation of all recombination events that take place during the evolutionary history of a crop species, resulting in much higher mapping resolution; (ii) less time required in mapping QTL as there is no need to develop a specialized mapping population; and (iii) a higher number of alleles can be sampled compared to linkage mapping where usually only two alleles are detected or studied (Neale and Savolainen, 2004; Breseghello and Sorrells, 2006; Remington et al., 2001). However, AM has its own disadvantages, which include occurrence of false positive or spurious associations between a marker and a trait, which can be caused by the existence of a population structure causing a seeming LD between loci that are not linked or only show loose linkage. For dealing with such a hidden population structure, statistical options have been developed (Falush et al., 2003). Such problems mainly arise due to the wrong selection (natural or artificial), genetic drift, and mating system (Flint-Garcia et al., 2003). Some of the oilseed crops where genome-wide association mapping has been used for the discovery of genes/QTLs for different economically important traits are given in Table 13.1. These QTLs/genes need to be first confirmed and validated before being used to improve oilseed crops.

Marker-assisted Introgression of Genes

In the case of oilseed crops, few successful attempts have been made in which genes linked with molecular markers are introgressed through molecular breeding, inspite of the availability of a large number of QTLs/genes being in the public domain (Walker et al., 2002; Saghai-Maroof et al., 2008; Yue et al., 2007). This may be attributed to several problems such as a low level of polymorphism between parents with contrasting traits, recombination between the marker and the target gene, QTL ×environment interaction, epistatic interactions, an epigenetic effect, low resolution, and also the QTLs may be population specific and the genetic

TABLE 13.1 Details of Linked Molecular Markers Associated with Stress Tolerance and Quality Traits in Oilseed Crops

S. no.	Crop	Trait	Linked markers	References
1	Soybean	Salinity tolerance	84	Zhang et al. (2014)
2	Soybean	Alkalinity tolerance	86	Zhang et al. (2014)
3	Soybean	Iron deficiency chlorosis	130	Mamidi et al. (2011)
4	Soybean	Cutworm resistance	7	Wang et al. (2011)
5	Soybean	Quality traits	79	Hou et al. (2011)
6	Soybean	Oil content, protein content, drought tolerance, SCN resistance, SMV resistance	21	Ying-Hui et al. (2011)
7	Groundnut	Agronomic traits	89	Jiang et al. (2014)
8	Groundnut	Morphological traits	12	Mandoulakani et al. (2010)
9	Rapeseed	Yield related traits	43	Cai et al. (2014)
10	Rapeseed	Seedling development traits	42	Korber et al. (2013)
11	Rapeseed	Quality traits	34	Honsdorf et al. (2010)
12	Rapeseed	Seed protein, oil content	65	Hwang et al. (2014)
13	Sesame	Quality traits	35	Wei et al. (2013)
14	Sesame	Macrophomina phaseolina resistance	43	Yan-Xin et al. (2012)

variation found in the specific biparental population might not be shared in other genetic populations. There is a need to validate and fine map these QTLs to identify tightly linked markers. Detection of QTL and their validation has to be carried out using large populations (>200 individuals) across several locations. In addition, their expression needs to be confirmed in different genetic backgrounds. The epigenetic effects are confirmed by using more than one population in parallel, with each having different genetic backgrounds. To identify the desirable QTL combinations to be used in different situations it is essential to know the kind, and extent, of epistatic interactions. All this can be achieved by the combination of a number of factors like high-throughput precise phenotyping, high-throughput genotyping, greater research efforts, proper experimental strategies, and liberal funding.

Soybean mosaic virus (SMV) causes a disease of soybean that is prevalent in all the soybean growing areas of the world. Through MAS various resistance genes *Rsv* were pyramided conferring resistance to SMV in the background of the susceptible cultivar Essex (Saghai-Maroof et al., 2008). The two and three *Rsv* genes, i.e., *RsvRsv3, Rsv1Rsv4,* and *Rsv1Rsv3Rsv4,* acted in a complementary manner, conferring resistance against all strains of SMV. It was also observed that two- and three-gene pyramided lines show higher levels of resistance to SMV than single-gene isolines. Similarly, MAS was used in soybean to pyramid QTL conditioning corn earworm resistance (Walker et al., 2002). This work demonstrated that combining transgene and QTL-mediated resistance to lepidopteran pests gives more resistance than the transgene alone. MAS was also used in sunflower to identify maintainer lines carrying the *rf1/rf1* genotype and demonstrated that this technique can be successfully used to identify individuals from a large segregating population (Yue et al., 2007).

Transgenic Breeding for Crop Improvement

In conventional plant breeding gene transfer occurs through crossing between the same or closely related species, thereby limiting gene transfer for crop improvement. However, transgenic techniques allow gene transfer between completely unrelated species, even across the kingdom. As such, traits that are not normally available within a species can be incorporated. Transgenic organisms exhibit properties that would be impossible to obtain by conventional breeding techniques. Transgenic techniques make plant-breeding programs more effective in two important ways. First, it allows transfer of specific genes, into a new variety without linkage drag. Second, it makes the process of trait transfer faster, more exact, cheaper, and less likely to fail than traditional cross-breeding methods. Additionally, it gives breeders the freedom to incorporate genes from unrelated species into the target plant, a possibility that is unprecedented in plant breeding. Although many countries are still against the cultivation of genetically modified crops the fact is that biotech crop areal coverage (hectares) has increased by an unprecedented 100-fold from 1.7 million hectares in 1996, to over 170 million hectares in 2012 (http://isaaa.org/). In addition, the global areal extent of biotech canola reached an estimated 5.5 million hectares in 2007 (James, 2007), with the majority of it being under herbicide-resistant canola. The basic strategies behind transgenic development are the introduction of modified endogenous genes or foreign genes, and overexpression or suppression (antisense or cosuppression) of endogenous genes. It is now possible to develop genetically engineered oilseed crop plants. This is cumbersome, and the rate of recovery of transgenic lines is still limited, however, a successful genetic transformation system has been developed for many economically important oilseed crops (Moloney et al., 1989; Barfield and Pua 1991; Narasimhula et al., 1992; Snow et al., 2003; Radonic et al., 2008; Malathi et al., 2006).

Transgenic breeding has been used for the improvement of a number of characteristics of oilseed crops that include resistance to herbicides (Manjunatha et al., 2008; *en.wikipedia.org/wiki/Genetically_modified_food*), tolerance to biotic/aboitic stress (Misra and Gedamu 1989; Sakhno et al., 2012; Sang et al., 2013; Mondal et al., 2007; Zhu et al., 1999; Keshavareddy et al., 2013; Bag et al., 2007; Snow et al., 2003; Radonic et al., 2008; Malathi et al., 2006), yield increase (Elhiti et al., 2012; Ma et al., 2013), and oil quality improvement (Weier et al., 1997; Yunus et al., 2008; Bhore and Shah, 2012). Hence, transgenics and transgenic breeding is also a genomics based option for improvement of oilseed crops (Table 13.2).

TRANSCRIPTOMIC APPROACH

Studying the transcriptome of a particular tissue, organ, or organism under certain specific conditions is termed as transcriptomics. It actually represents a global or whole transcriptome in which all expressed genes are studied in a single go, under a particular developmental stage or physiological condition. The transcriptome is more complex than a genome that encodes it because a gene can produce more than one type of mRNA molecule that encodes different types of proteins through alternate splicing. Through transcriptomics one can determine when and where the expression of genes are turned on or off with in cells of an organism. The difference in the patterns of gene expression in different cells and tissues are the reasons for physical, biochemical, and developmental differences among them. Comparative transcriptome analysis of different types of cells/tissue and at different developmental stages can be used to understand what makes the cell different from other cell types, the normal functioning of that cell type, and how changes in the normal level of gene activity may reflect or contribute to a disease. The data generated through transcriptomic-based experiments can also be used to infer the function of the gene by thoroughly searching the transcriptomic-based databases.

The older methods of gene expression analysis such as northern blotting and (Q) RT-PCR, which can analyze one gene at a time, have recently been replaced by high-throughput technologies like hybridization-based and sequencing-based approaches. These high-throughput technologies are used to study the gene expression of hundreds of thousands of genes simultaneously at the whole genome level. Among the hybridization-based methods, microarray technology is most extensively used for large-scale gene expression studies in oilseed crops such as soybean (Le et al., 2012a,b; Schaarschmidt et al., 2013; Khan et al., 2004; Alvord et al., 2007; Maguire et al., 2002; O'Rourke et al., 2009; Mazarei et al., 2011; Duressa et al., 2011), rapeseed (Gaeta et al., 2009, Yin et al., 2006; Niu et al., 2009), groundnut (Guo et al., 2006, 2011; Payton et al., 2009; Luo et al., 2005; Bi et al., 2010; Xiao-Ping et al., 2011), and sunflower (Lai et al., 2006; Fernandez et al., 2012).

The sequence-based methods for gene expression analysis including expressed sequence tags (ESTs), serial analysis of gene expression (SAGE), massively parallel signature sequencing (MPSS), open reading frame EST (ORESTES) and digital expression analysis (RNA-seq) using next generation sequencers, are a better alternative to quantifying gene expression (Hoen et al., 2008; Marioni et al., 2008). Initially, the sequencing of ESTs was used to study

TABLE 13.2 Details of Transgenes for Specific Traits Used in the Development of Transgenic Oilseed Crops

S. no.	Oilseed crop	Transgene	Trait	References
1	Soybean	Glyphosate-resistance gene	Herbicide resistance	*en.wikipedia.org/wiki/Genetically_modified_food*
2	Soybean	Cry1Ac	Insect resistance	*en.wikipedia.org/wiki/Genetically_modified_food*
3	Soybean	Delta 9 and delta 12 desaturases	Oil quality	*en.wikipedia.org/wiki/Genetically_modified_food*
4	Rapeseed	Human metallothio-nein-II	Cadmium resistance	Misra and Gedamu (1989)
5	Rapeseed	LPAAT	Oil quality	Weier et al. (1997)
6	Rapeseed	Human alpha 2b inter-feron gene	Viral resistance	Sakhno et al. (2012)
7	Rapeseed	Shoot meristem less	Oil yield and quality	Elhiti et al. (2012)
8	Rapeseed	miR156b	Seed carotenoid	Wei et al. (2010)
9	Rapeseed	*MSI-99m*	Disease resistance	Sang et al. (2013)
10	Indian mustard	Glucanase	Disease resistance	Mondal et al. (2007)
12	Indian mustard	*gshII*	Cadmium resistance	Zhu et al. (1999)
13	Groundnut	cry1AcF	Insect resistance	Keshavareddy et al. (2013)
14	Groundnut	cry1X	Insect resistance	Entoori et al. (2008)
15	Rapeseed	DRR206	Disease resistance	Wang and Fristensky (2001)
16	Rapeseed	katE	Disease resistance	El-Awady et al. (2008)
17	Rapeseed	cry1A	Insect resistance	Stewart et al. (1997)
18	Groundnut	EPSPS	Herbicide resistance	Manjunatha et al. (2008)
19	Groundnut	CP	Viral resistance	Bag et al. (2007)
20	Groundnut	*cpo-p*	Disease resistance	Niu et al. (2009)
21	Groundnut	mtlD	Disease resistance	Bhauso et al. (2014)
22	Groundnut	Mustard defensin	Disease resistance	Anuradha et al. (2008)
23	Groundnut	Synthetic cry1EC gene	Insect resistance	Yang et al. (2011)
24	Soybean	cry1Ac	Insect resistance	Yu et al. (2013)
25	Sunflower	crylAc	Insect resistance	Snow et al. (2003)
26	Sunflower	Glucanase and chitinase	Disease resistance	Radonic et al. (2008)
27	Oil palm	PAT, KASII, SAD	Oil quality	Yunus et al. (2008)
28	Oil palm	CryIA(b)	Insect resistance	Lee et al. (2006)
29	Oil palm	*At*WRI1	Oil yield	Ma et al. (2013)
30	Oil palm	Antisense PATE	Oil quality	Bhore and Shah (2012)
31	Safflower	elta-6-desaturase	Oil quality	www.australianoilseeds.com/__data/.../final_Swarnalatha_oral_paper.pdf

TABLE 13.2 Details of Transgenes for Specific Traits Used in the Development of Transgenic Oilseed Crops (*cont.*)

S. no.	Oilseed crop	Transgene	Trait	References
32	Cotton	ASAL	Insect resistance	Vajhala et al. (2013)
33	Cotton	Cry genes	Insect resistance	*www.cicr.org.in/pdf/transgenic_bt_cotton.pdf*
34	Cotton	antisense *AV2*	Viral resistance	Sanjaya et al. (2005)
35	Castor	cryIAb	Insect resistance	Malathi et al. (2006)

gene expression and identify differentially expressed transcripts. However, this method is time consuming, highly expensive, and has a low-throughput, hence it has been replaced by SAGE and MPSS for transcriptomic analysis. Although, the SAGE and MPSS methods are high-throughput they have not been widely implemented as a microarray for gene expression analysis in oilseed crops, being used only in few studies (Obermeier et al., 2008; Obermeier et al., 2009; Yang et al., 2011). The advances in sequencing technology, especially the emergence of next generation sequencing (NGS), has made whole transcriptome sequencing (RNA-seq) more convenient and enabled the quick study of gene expression on the entire genome level, even in a nonmodel organism without a reference genome or for organisms having limited genomic resources available (Strickler et al., 2012; Collins et al., 2008). Hence, RNA-seq technology is equally applicable to those experimental systems and species where specific genetic information may be limited or almost lacking, something which was impossible using previous technologies (e.g., microarrays) (Ward et al., 2012; Collins et al., 2008). RNA-seq technology has already been used for gene expression analysis in many oilseed crops such as soybean (Severin et al., 2010; Kim et al., 2011), rapeseed (Jiang et al., 2013), groundnut (Chen et al., 2013; Xia et al., 2013), and oil palm (Shearman et al., 2013; Alagna et al., 2009). Therefore, it is clear that the microarray technique has been more widely used in transcriptomic analysis than RNA-seq in oilseed crops. This is due to the ease of data analyses using programs and modules that quickly turn raw microarray data into spreadsheets of gene expression values and significant differentially expressed genes. However, improvements in RNA-seq data analysis, which is still in its infancy, will eventually make microarrays obsolete for gene expression analyses (Ramskold et al., 2012).

PROTEOMICS APPROACH

Proteomics is a large-scale comprehensive study of a specific proteome, including information on protein abundances, their variations, and modifications, along with their interacting partners and networks, used in order to understand cellular processes. It refers to the entire compliment of proteins, including modifications made to a particular set of proteins, produced by an organism or a cellular system. This varies with time and the distinct requirements, or stresses, that a cell or organism undergoes, and represents fundamental components of the physiological and metabolic pathways that are vital for growth, development, and the interactions of a living organism with its environment. Proteomics is more complicated than genomics because an organism's genome is more or less constant, whereas the proteome differs from cell to cell and from time to time. Distinct genes are expressed in different cell types, which means that even the basic set of proteins that are produced in a cell need to be identified. It is now known that mRNA is not always translated into protein, and the amount of protein produced for a given amount of mRNA depends on the gene it is transcribed from and on the current physiological state of the cell. Proteomics confirms the presence of the protein and provides a direct measure of the quantity present and hence provides a different, and often more precise, level of understanding than genomics. In general, proteomic approaches can be used (i) for proteome profiling to understand how proteins execute biological processes; (ii) for comparative expression analysis of two or more protein samples, where proteins may fold into specific 3D structures that determine function; (iii) for understanding posttranslational modifications which are regulated after synthesis and are major regulatory mechanisms controlling many basic cellular processes; (iv) for the study of protein–protein interactions; and (v) for identifying new biomarkers to detect and monitor specific stress manifestations (Chandramouli and Qian, 2009). Proteome coverage has steadily increased over the past decade due to great effort, especially at the technical level, in the plant proteomics field (readers are referred to further studies in Agrawal et al., 2011; Agrawal and Rakwal, 2008).

A large number of proteomic studies have been carried out for major oilseed crops such as soybean, rapeseed, groundnut, sunflower, and oil palm for various traits/different types of treatment (Table 13.3). Associating the physiologically significant candidate proteins for various biotic and abiotic stress tolerance mechanisms in oilseed crops is the best strategy for their improvement. Both gel-based and gel-free proteomics approaches have been used for proteomic identification in oilseed crops (see Chandramouli and Qian, 2009). Gel-based approaches that have been used include one-dimensional gel electrophoresis (SDS-PAGE) (Han et al., 2013; Pereira et al., 2013; Giavalisco et al., 2006; Kakaei and Kahrizi, 2011; Messaitfa et al., 2014) and 2D-polyacrylamide gel electrophoresis (2-DGE) that have been frequently used for proteomic separation and identification in almost all major oilseed crops (Mesquita et al., 2012; Wan et al., 2005; Alam et al., 2010; Kim et al., 2013; Cheng et al., 2013; Jiang et al., 2012; Katam et al., 2010; Wang et al., 2012; Ghaffari et al., 2013; Syahanim et al., 2013). Due to the availability of genomic sequence data of more and more organisms and EST information, MS techniques such as MALDI TOF and ESI proteomics-based studies have become more accurate and precise. However, there are certain technical challenges such as limited dynamic resolution when thousands of proteins are analyzed, employing procedures to obtain a pure proteome, and the separation and identification of hydrophobic proteins. Employment of affinity chromatography, use of fluorescent dyes, certain fractionation techniques, and reverse phase high-performance liquid chromatography (HPLC) can eliminate these drawbacks. Characterization of the complete proteome is possible only by employing high-throughput technology (i.e., multidimensional chromatography, spectrometers, and robotics). Proteomics analyses have been performed to evaluate the functions of proteins in various plant organs of rice (Tanaka et al., 2004; Fukuda et al., 2003; Komatsu et al., 1993; Islam et al., 2004; Zhao et al., 2005; Komatsu et al., 1999; Imin et al., 2001; Kerim et al., 2003; Shen et al., 2002; Tanaka et al., 2004; Nozu et al., 2006), the seed protein of soyabean (Herman et al., 2003; Mooney et al., 2004; Natarajan et al., 2006), etc. D'Hooghe and coworkers performed an experiment on oilseed rape in which the proteome analysis of the leaf, combined with a physiological study, was carried out to analyse the effect of sulfur depletion on the metabolic processes of

TABLE 13.3 Various Proteomics-Based Approaches Employed in Different Oilseed Crops for Identification of Differentially Expressed Proteins

Crop	Approach	Organ/ tissue	Trait/treatment	No. of proteins identified / differentially expressed	References
Soybean	2-DE	Leaf	Physiological traits	122	Mesquita et al. (2012)
Soybean	2-DE	Root hair	*Bradyrhizobium japonicum*	96	Wan et al. (2005)
Soybean	2-DE	Root	Waterlogging stress	24	Alam et al. (2010)
Soybean	2-DE	Seed coat	Seed coat colour	36	Kim et al. (2013)
Soybean	2-DE	embryonic axis	Embryonic development	401	Natarajan et al. (2013)
Soybean	2-DE	Roots, nodules	*Bradyrhizobium japonicum*	1300, 1800	Lim et al., (2010)
Soybean	SDS-PAGE	Leaf	*Phakopsora pachyrhizi*	22	Pereira et al. (2013)
Rapeseed	2-DE	Leaf, stem, root	Comparative analysis	93, 385, 266	Albertin et al. (2009)
Rapeseed	2-DE	Anther	Male saterility	141	Cheng et al. (2013)
Rapeseed	2-DE	Seed	Seed color difference	27	Jiang et al. (2012)
Rapeseed	2-DE	Leaf	Salt stress	12	Fang et al. (2011)
Rapeseed	2-DE	Anther	Male sterility	167	Sheoran and Sawhney (2010)
Rapeseed	2-DE	Leaf	*Sclerotinia sclerotiorum*	32	Liang et al. (2008)
Rapeseed	2-DE	Leaf	Male sterility	16	Fang et al. (2012)
Rapeseed	2-DE	Xylem sap and leaf apoplast	*Verticillium longisporum*	12	Floer et al. (2008)
Rapeseed	2-DE	Seed	Seed filling	794	Hajduch et al. (2007)
Groundnut	2-DE	Leaf	Physiological traits	133	Katam et al. (2010)
Sunflower	2-DE	Leaf	Cold stress	168	Balbuena et al. (2011)
Sunflower	2-DE	Leaf	Drought stress	350	Castillejo et al. (2008)
Sunflower	2-DE	Seed	N^+ ion implantation	369	Guijun et al. (2006)
Sunflower	2-DE	Leaf	Drought stress	46	Fulda et al. (2011)
Sunflower	2-DE	Seed	Oil traits	77	Hajduch et al. (2007)
Oil palm	2-DE	Fruit	Fruit development	121	Bianco et al. (2013)
Oil palm	2-DE	Root	*Ganoderma boninense*	2411	Al-Obaidi et al. (2014)
Oil palm	2-DE	Root	*Ganoderma boninense*	25	Syahanim et al. (2013)
Soybean	SDS-PAGE	Seed	Physiological traits	764	Han et al. (2013)

(Continued)

TABLE 13.3 Various Proteomics-Based Approaches Employed in Different Oilseed Crops for Identification of Differentially Expressed Proteins (*cont.*)

Crop	Approach	Organ/ tissue	Trait/treatment	No. of proteins identified / differentially expressed	References
Rapeseed	SDS-PAGE and 2-DE	Phloem	Physiological traits	140	Giavalisco et al. (2006)
Rapeseed	SDS-PAGE	Seed	Physiological traits	17	Kakaei and Kahrizi (2011)
Groundnut	SDS-PAGE and 2-DE	Leaf	Drought stress	49	Kottapalli et al. (2009)
Sunflower	SDS-PAGE	Leaf	Salt stress	40	Messaitfa et al. (2014)
Soybean	2-D-DIGE	Root	Aluminum stress	49	Duressa et al. (2011)
Soybean	Label-free quantitative proteomic technique	Root tip	Flooding stress	94	Oh et al. (2014)
Soybean	2-DGE and Sec-Mud-PIT	Seed	Seed filling	478	Agrawal et al. (2008)
Soybean	iTRAQ	Leaf	Sex determination	1269	Qin et al. (2013)
Rapeseed	iTRAQ	Guard cell	Methyl jasmonate	84	Zhu et al. (2012)

plants (D'Hooghe et al., 2013). Sulfur reduction in soil leads to the inhibition of photosynthesis and thus the growth of plants is affected to a great extent. Chlorophyll content estimation in stressed and controlled plants showed no variation. 2D electrophoresis revealed 36 differential expression proteins when comparison was made between plants under controlled conditions and the plants given low-sulphur treatment. Furthermore, increase in anthocyanins, synthesis of H_2O_2, increased production of enzymes such as 12-oxophitadienoate reductase (involved in jasmonate biosynthesis), and ACC synthase (involved in ethylene biosynthesis) determined by proteomic analysis suggested that S reduction leads to oxidative stress. However, enhanced synthesis of ATP synthase in chloroplasts indicates that energy production is maintained. This study revealed that despite the induction of oxidative stress, leaves carried out photosynthesis due to accumulation of ATP synthase F1 complex. However, reactive oxygen species (ROS) production and detoxification are not very effective in assisting the plant to survive for long periods of sulphur reduction.

Recently, high-throughput, gel-free technologies for proteomic separation have been introduced that will gradually replace the low-throughput approaches and are able to identify a greater number of proteins. These gel-free approaches, used for proteomic separation in oilseed crops, include multidimensional protein identification technology (MudPit) (Agrawal and Rakwal 2008), isotope-coded affinity tagging (ICAT) (Oh et al., 2014), stable isotope labeling by amino acids in cell culture (SILAC), and isobaric tagging for relative and absolute quantitation (iTRAQ) (Qin et al., 2013; Zhu et al., 2012; Loei et al., 2013; Zargar et al., 2013).

For the identification of low-abundance proteins, something which is difficult through conventional proteomics approaches, new methods have been introduced that include iTRAQ OFFGEL fractionation, 2D-LC, and the long-column method. These three different methods of peptide separation use 5–10 μg proteins from protoplasts and identify large numbers of proteins compared with conventional approaches (Fukao et al., 2013; Zargar et al., 2013). To date these methods have not been used for proteomic identification in oilseed crops but with the reduction of costs required by these methods, in addition to their communalization, will definitely replace older methods of protein separation. Advanced automated peptide purification systems with high precision and greater reproducibility is the major challenge in plant proteomics (Zargar et al., 2013).

Proteomics Applications in Plant Breeding

In crop improvement programs, proteomics can help with the identification of molecular switches that may assist in the development of stress-tolerant genotypes. The study of the protein expression of different lines under both stressed and controlled conditions can expose various candidate proteins that are synthesized under stressed conditions. This differential expression, studied by various separation and detection techniques, can elucidate the possible pathways activated under defense mechanism and stress conditions. We can study the plant proteome response by analyzing the effects of stress on various crops. Different lines showing differential expression under varied conditions can be selected to understand the molecular regulation for stress tolerance. The stress-induced differential proteins can be identified by comparing plants under stress and controlled conditions. The abundance of stress-induced responsive genes, and the phenotypic performance of a plant, can be compared to verify the results. Some proteins may show a direct correlation with the degree of stress. This correlation can be further verified by analyzing an advanced mapping population like recombinant inbred lines (RILs), near isogenic lines (NILs), and double haploid lines. Other relevant information including the expression pattern at mRNA and metabolomics may help to further verify the correlation of these candidate proteins with desirable traits. Finally, introgression of these putative genes can be done in marker-assisted breeding or gene transformation programs to improve crop tolerance to stress.

METABOLOMICS APPROACH

The simultaneous and systematic determination of the complete set of metabolites and their alteration in response to external stimuli is refered to as metabolomics. The study of gene functions necessitates an understanding of metabolite production under different developmental stages and other plant environment interactions. A metabolomics approach unveils complete biochemical reaction networks, production of metabolites, and enzymes involved in the pathway. This approach depicts various plant metabolites that are involved in stress response, various molecular interactions, and altered phenotype. This metabolite synthesis indicates precursors involved in numerous biological pathways. Transcriptomics and/or proteomics techniques can predict the phenotype and possible metabolic pathways to some extent but analyzing the metabolome can elucidate biological mechanisms more effectively.

Metabolomic approaches employ various separation techniques such as gas chromatography (GC), capillary electrophoresis (CE), HPLC, and ultra-performance liquid chromatography (UPLC) as well as detection techniques such as nuclear magnetic resonance (NMR) and mass spectroscopy (MS). A lot of work has been reported regarding stress responses, nutrient uptake, and plant–pathogen interactions by metabolomics. However, there is more to be done to understand the molecular processes in a more effective manner. Plant metabolism is a dynamic system and each metabolite might be associated with a higher or lower expression of other metabolites – additionally, a pathway may produce sometimes more than one metabolite (Fischbach and Clardy, 2007). Therefore, in metabolic studies it is very important to look at all the metabolites at the same time.

Muhammed Ali Hossain reported the genomic regions/QTL responsible for disease resistance in *B. napus* against *Verticillium longisporum* disease and identified phenolic metabolites associated with susceptibility (Hossain, 2014). The composition of soluble phenolics, cell wall phenolics, and lignin fractions from hypocotyls of the mapping populations were determined by reversed phase (RP)-HPLC and GC/MS. Comparative quantitative trait mapping for resistance and phenolic metabolites was performed. Liquid chromatography coupled with mass spectrometry analysis was carried out and finally phenolic metabolites involved in resistance mechanisms and pathways were detected. The study provided the evidence for alteration in synthesis of phenolic compounds on the onset of infection. Some compounds were positively correlated whereas others were negatively correlated with the resistance mechanism of the plant. Identification of phenolics involved in the resistance mechanism paves the way to understand the metabolic pathways effectively involved during host–pathogen interaction. Based on the pattern of metabolite expression putative candidate genes involved in key interactions of the resistance reaction were identified.

Recently, metabolic studies of *B. napus* for drought resistance were reported (see Hatzig et al., 2014). They analyzed biomass traits, hormone metabolites, and osmolytes in two genetically unrelated, drought-tolerant genotypes and two unrelated, drought-sensitive genotypes and studied the impact of osmotic stress on growth, physiology, and the metabolism of winter oilseed rape (*B. napus* L.) drought leads to cellular dehydration and damage to cellular structures due to the production of ROS, which triggers various adaptive metabolic pathways. The metabolite profiling on the application of drought indicated alteration in growth properties, hormone signaling, and metabolite patterns. The plants under stressed conditions showed enhanced proline synthesis but reduced ornithine biosynthesis. ABA concentration and its metabolites were assessed in both the conditions, as ABA is known to be responsible for osmolyte accumulation. However, ABA synthesis was upregulated in only drought-susceptible varieties and not in drought-tolerant varieties, which clearly indicates the involvement of other pathways for increased proline and carbohydrate accumulation in drought resistant varieties. It was also noticed that accumulation of glucose, fructose, sucrose, and myo-inositol only occurred in the susceptible oilseed rape genotypes, but not to the same extent in resistant varieties. Metabolite and hormone profiling, along with additional traits such as leaf temperature, provided the evidence that stress sensing is different between the resistant and sensitive genotypes.

Thus, measuring the level of proteome, transcriptome, along with metabolic profiling will give a better insight to the regulatory pathways occurring inside the organism and as such can help in the improvement of crop traits.

IONOMICS APPROACH

Ionomics is referred as the quantitative measurement of the complete set of ions in a living organism and deals with the changes in ion production under varied external stimuli. Biochemical pathways elucidate the production of ions that play a crucial role in maintaining cell integrity, transportation of minerals, and act as cofactors for enzymes catalyzing important regulatory pathways. Thus, alteration in the production of ions in a process can lead to large changes in metabolic pathways. It is very important to analyze the ions simultaneously as the production of some ions influence the production of other ions. The vital role of these ions in carrying out life processes necessitates studying the plant ionome. Thus, ionomic-based studies will help in understanding gene regulations in a better and effective manner.

The ionomics approach has been reported to differentiate mutants (having variation in a single element as well as multielements) and natural alleles (Lahner et al., 2003; Chen et al., 2009). Identification and cloning of the genomic region/loci responsible for the two multielement phenotypes made it possible to detect genes affecting the composition of elements in soybean seeds (Lahner et al., 2003). Two major genes (ESB1 and TSC10a) were explored which control suberin deposition in the root endodermis (Baxter et al., 2009; Chao et al., 2011). These genes were observed to alter ionome or elements present in the leaf. An altered expression of these genes was shown to affect the transport of water and ions in the root, resulting in an altered multielement profile. Elemental profiling has been carried out in soybean varieties to detect the mutants containing varied amounts of elements in their seed (Ziegler et al., 2013). Two major factors, that is, genotypic constitution and plant–environment interaction, contribute toward seed composition since most of the elements except carbon and oxygen are obtained from the soil. Soil–plant interaction was demonstrated by growing the plants with different elemental composition in the field. Ion coupled plasma mass spectrometry analysis was performed among 947 lines for 20 elements. Soil variation across the field was reduced to some extent by spatial correction and mutants were identified. This study revealed the potential of high throughput approach to detect elemental composition of seeds. Such high-throughput, omics-based technologies will help to reveal various hidden regulatory pathways and expressions of genomes more accurately. Genomics approaches can be used to predict the products of genes but due to certain posttranscriptional and posttranslational modifications, the proteins, metabolites, and ions formed may be totally different. Hence it is necessary to utilize proteomics, transcriptomics, metabolomics, and ionomics approaches to depict the gene function and biochemical pathways more efficiently. An understanding of all the expressed genes, proteins, metabolites, and ions will provide a better insight into metabolic networks. As such, omics-based approaches can be a great boon, improve the nutritional status, disease resistance, and stress tolerance mechanisms of the plant.

PRECISE PHENOMICS – A MUST FOR ALL OMICS-BASED APPROACHES

Tremendous progress has been made in the area of genomics in recent years to unravel and understand crop genomes. The advances made in genomics include the development of different types of molecular markers, genotyping platforms, genetic mapping procedures, the study of marker–trait associations leading to the discovery of genes/QTLs, advances in sequencing technologies leading to reduction in sequencing costs and making available genome sequences of major crop genomes, etc. The advancements, which occurred in sequencing technologies made the sequencing of plant and crop genomes routine (Jackson et al., 2011). The identification of allelic variation through resequencing of crop genomes is also expected to become routine in the near future (Furbank and Tester, 2011). Additionally, the cost of genotyping plant genomes will continuously decrease due to the evolution and emergence of high-throughput markers and genotypic platforms. Moreover, the results from transcriptomics, proteomics, and other omics approaches will be precise only if proper phenotyping is done. Therefore, to take advantages of omics-based information for crop improvement and harness its full benefit, one needs to link and integrate this information with plant phenotype in an open-field environment (Furbank and Tester, 2011). However, our capacity to collect reliable phenotypic data is still lagging behind compared with our present-day capacity to generate high-throughput molecular genotyping data. This is considered a major hindrance facing plant breeders (Edwards et al., 2013; Stamp and Visser, 2012). Some of the most important phenotyping bottlenecks include: (i) phenotyping of large mapping populations for QTL/gene discovery, followed by the cloning of major QTLs; (ii) phenotyping in replicated trials in several environments over a number of years; (iii) phenotyping using destructive tools at fixed times/growth stages; (iv) slow and costly phenotyping; and (v) less precise estimates of phenotypic data for testing allelic variations of a candidate gene in a germplasm set (Furbank and Tester, 2011).This creates a gap between genotype and phenotype (GP gap). Therefore, there is a need for a phenomics revolution that can take care of this problematic bottleneck. Efforts are being made throughout the world to alleviate this problem by developing plant phenomics facilities that can scan and record data for thousands of plants in a day in a sophisticated manner (http://www.plantphenomics.org.au/). These high-throughput phenomics facilities make use of sophisticated noninvasive imaging, spectroscopy, image analysis, robotics, and high-performance computing facilities thus saving time, labor, and cost.

CONCLUSIONS

Exploitation of natural genetic resources using established omics-based technologies is a must to attain sustainability in the production of oilseed crops. Although the first Green Revolution, which was based on breeding, significantly enhanced production as well as the productivity of crops, in order to attain sustainability in crop production, especially oilseed crops, the adoption of new technologies is must. We cannot ignore any technology, which does not have a negative impact on sustainability. As such, using omics-based technologies, a desired agronomic trait can be transferred to an oilseed crop in order for its improvement, something that may lead to what we call the next-generation crop. However, it is essential to understand the basic impact of next-generation crops. Finally, creating next-generation oilseed crops is crucial, where integration of markers with breeding, and other omics-based approaches, will play a major role (Figure 13.2). Moreover, precise phenotyping is critical and there is a need for a phenomics revolution, which would serves as the backbone for other omics-based approaches.

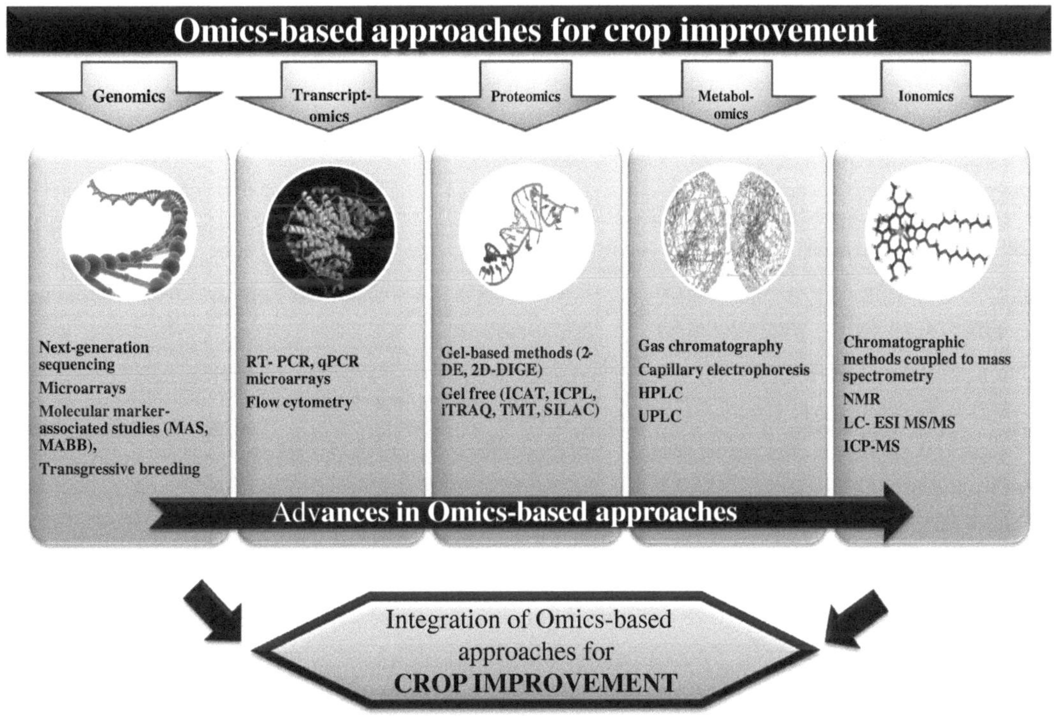

FIGURE 13.2 **Representation of various omics based approaches that can be used for improvement of oilseed crops.**

References

Alvord, W.G., Roayaei, J.A., Quiñones, O.A., Schneider, K.T., 2007. A microarray analysis for differential gene expression in the soybean genome using bioconductor. Brief. Bioinform. 8 (6), 415–431.

Agrawal, G.K., Bourguignon, J., Rolland, N., 2011. Plant organelle proteomics: collaborating for optimal cell function. Mass Spec. Rev. 30, 772–853.

Agrawal, G.K., Hajduch, M., Graham, K., Thelen, J.J., 2008. In-depth investigation of the soybean seed-filling proteome and comparison with a parallel study of rapeseed. Plant Physiol. 148, 504–518.

Agrawal, G.K., Rakwal, R., 2008. Plant Proteomics: Technologies, Strategies, and Applications. John Wiley & Sons, Hoboken, NJ.

Alagna, F., D'Agostino, N., Torchia, L., Servili, M., Rao, R., Pietrella, M., Giuliano, G., Chiusano, M.L., Baldoni, L., Perrotta, G., 2009. Comparative 454 pyrosequencing of transcripts from two olive genotypes during fruit development. BMC Genomics 10, 399.

Alam, I., Lee, D.G., Kim, K.H., Park, C.H., Sharmin, S.A., Lee, H., Oh, K.W., Yun, B.W., Lee, B.H., 2010. Proteome analysis of soybean roots under water logging stress at an early vegetative stage. J. Biosci. 35, 49–62.

Albertin, W., Langella, O., Joets, J., Négroni, L., Zivy, M., Damerval, C., Thiellement, H., 2009. Comparative proteomics of leaf, stem, and root tissues of synthetic *Brassica napus*. Proteomics 9, 793–799.

Al-Obaidi, J.R., Mohd-Yusuf, Y., Razali, N., Jayapalan, J.J., Tey, C.C., Md-Noh, N., Junit, S.M., Othman, R.Y., Hashim, O.H., 2014. Identification of proteins of altered abundance in oil palm infected with *Ganoderma boninense*. Int. J. Mol. Sci. 15, 5175–5192.

Amiri, R.M., Azdi, S.B., Ghanadha, M.R., Abd, M.C., 2001. Detection of DNA polymorphism in landrace populations of safflower in Iran using RAPD-PCR technique. Iran J. Agri. Sci. 32, 737–745.

Anuradha, S.T., Divya, K., Jami, S.K., Kirti, P.B., 2008. Transgenic tobacco and peanut plants expressing a mustard defensin show resistance to fungal pathogens. Plant Cell Rep. 27, 1777–1786.

Bachlava, E., Dewey, R.E., Auclair, J., Wang, S., Burton, J.W., Cardinal, A.J., 2008. Mapping genes encoding microsomal ω-6 desaturase enzymes and their cosegregation with QTL affecting oleate content in soybean. Crop Sci. 48, 640–650.

Bag, S., Singh, R.S., Jain, R.K., 2007. Agrobacterium-mediated transformation of groundnut with coat protein gene of tobacco streak virus. Indian J. Virol. 18 (2), 65–69.

Balbuena, T.S., Salas, J.J., Martínez-Force, E., Garcés, R., Thelen, J.J., 2011. Proteome analysis of cold acclimation in sunflower. J. Proteome Res. 10, 2330–2346.

Balbuena, T.S., Salas, J.J., Martínez-Force, E., Garcés, R., Thelen, J.J., 2011. Proteome analysis of cold acclimation in sunflower. J. Proteome Res. 10 (5), 2330–2346.

Barfield, D.G., Pua, E.C., 1991. Gene transfer in plants of *Brassica juncea* using *Agrobacterium tumefaciens* mediated transformation. Plant Cell Rep. 10, 308–314.

Baxter, I., Hosmani, P.S., Rus, A., Lahner, B., Borevitz, J.O., Muthukumar, B., Mickelbart, M.V., Schreiber, L., Franke, R.B., Salt, D.E., 2009. Root suberin forms an extracellular barrier that affects water relations and mineral nutrition in Arabidopsis. PLoS Genet. 5 (5), e1000492, doi: 10.1371/journal.pgen.1000492.

Berry, S.T., Leon, A.J., Hanfrey, C.C., Challis, P., Burkholz, A., Barnes, S.R., Rufener, G.K., Lee, M., Caligari, P.D.S., 1995. Molecular markers analysis of *Helianthus annuus* L.2. Construction of an RFLP map for cultivated sunflower. Theor. Appl. Genet. 91, 195–199.

Bhauso, T.D., Thankappan, R., Kumar, A., Mishra, G.P., Dobaria, J.R., Rajam, M.V., 2014. Over-expression of bacterial mtlD gene confers enhanced tolerance to salt-stress and water-deficit stress in transgenic peanut (*Arachis hypogaea*) through accumulation of mannitol. AJCS 8 (3), 413–421.

Bhore, S.J., Shah, F.H., 2012. Genetic Transformation of the American Oil Palm (*Elaeis oleifera*) Immature Zygotic Embryos with Antisense Palmitoyl-Acyl Carrier Protein Thioesterase (PATE) Gene. World Appl. Sci. J. 16 (3), 362–369.

Bi, Yu-Ping, Liu, Wei, Xia, Han, Lei, Su, Zhao, Chuan-Zhi, Wan, Shu-Bo, Wang, Xing-Jun, 2010. EST sequencing and gene expression profiling of cultivated peanut (*Arachis hypogaea* L.). Genome 53 (10), 832–839.

Bianco, L., Alagna, F., Baldoni, L., Finnie, C., Svensson, B., Perrotta, G., 2013. Proteome regulation during Olea europaea fruit development. PLoS One 8 (1), e53563, doi: 10.1371/journal.pone.0053563.

Bisht, N.C., Gupta, V., Ramchiary, N., Sodhi, Y.S., Mukhopadhyay, A., Arumugam, N., Pental, D., Pradhan, A.K., 2009. Fine mapping of loci involved with glucoinolate biosynthesis in oilseed mustard (*Brassica juncea*) using genomic information from allied species. Theor. Appl. Genet. 118, 413–421.

Breseghello, F., Sorrells, M.E., 2006. Association mapping of kernel size and milling quality in wheat (*Triticum aestivum* L.) cultivars. Genetics 172 (2), 1165–1177.

Breseghello, F., Sorrells, M.E., 2006. Association analysis as a strategy for improvement of quantitative traits in plants. Crop Sci. 46, 1323–1330.

Cai, D., Xiao, Y., Yang, W., Ye, W., Wang, B., Younas, M., Wu, J., Liu, K., 2014. Association mapping of six yield–related traits in rapeseed (*Brassica napus* L.). Theor. Appl. Genet. 127, 85–96.

Castillejo, M.A., Maldonado, Ogueta, S., Jorrin, J.V., 2008. Proteomic analysis of responses to drought stress in sunflower (*Helianthus annuus*) leaves by 2DE gel electrophoresis and mass spectrometry. Open Proteomics J. 1, 59–71.

Chandramouli, K., Qian, P.Y., 2009. Proteomics: challenges, techniques and possibilities to overcome biological sample complexity. Hum. Genom. Proteomics 8, 23920.

Chao, D.Y., Gable, K., Chen, M., Baxter, I., Dietrich, C.R., Cahoon, E.B., Guerinot, M.L., Lahner, B., Lu, S., Markham, J.E., Morrissey, J., Han, G., Gupta, S.D., Harmon, J.M., Jaworski, J.G., Dunn, T.M., Salt, D.E., 2011. Sphingolipids in the root play an important role in regulating the leaf ionome in *Arabidopsis thaliana*. Plant Cell 23, 1061–1081.

Chen, X., Zhu, W., Azam, S., Li, H., Zhu, F., Li, H., Hong, Y., Liu, H., Zhang, E., Wu, H., Yu, S., Zhou, G., Li, S., Zhong, N., Wen, S., Li, X., Knapp, S.J., Ozias-Akins, P., Varshney, R.K., Liang, X., 2013. Deep sequencing analysis of the transcriptomes of peanut aerial and subterranean young pods identifies candidate genes related to early embryo abortion. Plant Biotechnol. J. 11 (1), 115–127.

Chen, Y., Kenaschuk, E., Dribnenki, B., 2001. Inheritance of rust resistance genes and molecular markers in microspore derived populations of flax. Plant Breed. 120, 82–84.

Chen, Z., Shinano, T., Ezawa, T., Wasaki, J., Kimura, K., Osaki, M., Zhu, Y., 2009. Elemental interconnections in Lotus japonicus: A systematic study of the affects of elements additions on different natural variants. Soil Sci. Plant Nutr. 55, 91–101, doi:10.1111/j.1747-0765.2008.00311.x.

Cheng, Y., Wang, Q., Li, Z., Cui, J., Hu, S., Zhao mail, H., Chen, M., 2013. Cytological and comparative proteomic analyses on male sterility in *Brassica napus* L. induced by the chemical hybridization agent monosulphuron ester sodium. PLoS One 8 (11), e80191, doi:10.1371/journal.pone.0080191.

Collins, L.J., Biggs, P.J., Voelckel, C., Joly, S., 2008. An approach to transcriptome analysis of non-model organisms using short-read sequences. Genome Inform. 21, 3–14.

Cregan, P.B., Jarvik, T., Bush, A.L., Shoemaker, R.C., Lark, K.G., Kahler, A.L., Kaya, N., 1999. An integrated genetic linkage map of the soybean genome. Crop Sci. 39 (5), 1464–1490.

Delourme, R., Falentin, C., Hateau, V., Clouet, V., Horvais, R., Gandon, B., Specel, S., Hanneton, L., Dheu, J.E., Deschamps, M., Margale, E., Vincourt, P., Renard, M., 2006. Genetic control of oil content in oilseed rape *Brassica napus* L. Theor. Appl. Genet. 113 (7), 1331–1345.

D'Hooghe, P., Escamez, S., Trouverie, J., Avice, J.C., 2013. Sulphur limitation provokes physiological and leaf proteome changes in oilseed rape that lead to perturbation of sulphur, carbon and oxidative metabolisms. BMC Plant Biol. 13, 23.

Diers, B.W., Mansur, L., Imsande, J., Shoemaker, R., 1991. Mapping *Phytophthora* resistance loci in soybean with restriction fragment length polymorphism markers. Crop Sci. 32, 377–383.

Diers, B.W., Fehr, W., Keim, P., Shoemaker, R.C., 1992. RFLP analysis of soybean seed protein and oil content. Theor. Appl. Genet. 83, 608–612.

Dixit, A., Jin, M.H., Chung, J.W., Yu, J.W., Chung, H.K., Ma, K.H., Park, Y.J., Cho, E.G., 2005. Development of polymorphic microsatellite markers in sesame (*Sesamum indicum* L.). Mol. Ecol. Notes 5 (4), 736–738.

Duressa, D., Soliman, K.M., Taylor, R.W., Chen, D., 2011. Gene expression profiling in soybean under aluminum stress: genes differentially expressed between Al-tolerant and Al-sensitive genotypes. Am. J. Mol. Biol. 1, 156–173.

Ecke, W., Uzunova, M., Weissleder, K., 1995. Mapping the genome of rapeseed (*Brassica napus* L.) II. Localization of genes controlling erucic acid synthesis and oil content. Localization of genes controlling erucic acid synthesis and oil content. Theor. Appl. Genet. 91, 972–977.

Edwards, D., Batley, J., Snowdon, R.J., 2013. Accessing complex crop genomes with next-generation sequencing. Theor. Appl. Genet. 126 (1), 1–11, doi: 10.1007/s00122-012-1964-x. Epub 2012.

El-Awady, M., Moghaieb, R.E.A., Haggag, W., Youssef, S.S., El-Sharkawy, A.M., 2008. Transgenic canola plants overexpressing bacterial catalase exhibit enhanced resistance to *Peronospora parasitica* and *Erysiphe polygoni*. Arab. J. Biotech. 11 (1), 71–84.

Elhiti, M., Yang, C., Chan, A., Durnin, D.C., Belmonte, M.F., Ayele, B.T., Tahir, M., Stasolla, C., 2012. Altered seed oil and glucosinolate levels in transgenic plants overexpressing the *Brassica napus* shoot meristem less gene. J. Exp. Bot. 63 (12), 4447–4461.

Entoori, K., Sreevathsa, R., Arthikala, M.K., Kumar, P.A., Kumar, A.R.V., Madhusudhan, B., Makarla, U., 2008. A chimeric cry1X gene imparts resistance to *Spodoptera litura* and *Helicoverpa armigera* in the transgenic groundnut. Eur. Asian J. BioSci. 2, 53–65.

Falush, D., Stephens, M., Pritchard, J.K., 2003. Inference of population structure using multilocus genotype data: linked loci and correlated allele frequencies. Genetics 164, 1567–1587.

Fang, K., ShanJing, M., Kun, D., Mian, W., XiaoYan, Z., ChengCai, C., YouPing, W., 2011. Comparative proteomics analysis of *OsNAS1* transgenic *Brassica napus* under salt stress. Chin. Sci. Bull. 56, 2343–2350.

Fang, K., Zhenglian, X., Chaoying, Y., 2012. Comparative proteomic analysis of bar-transgenic *Brassica napus* L. and cultivated rapeseed. Biotechnol. Bull. 10, 75–82.

Fernandez, P., Soria, M., Blesa, D., DiRienzo, J., Moschen, S., Rivarola, M., Clavijo, B.J., Gonzalez, S., Peluffo, L., Príncipi, D., Dosio, G., Aguirrezabal, L., García-García, F., Conesa, A., Hopp, E., Dopazo, J., Heinz, R.A., Paniego, N., 2012. Development, characterization and experimental validation of a cultivated sunflower (*Helianthus annuus* L.) gene expression oligonucleotide microarray. Plos One 7 (10), e45899, doi: 10.1371/journal.pone.0045899.

Ferreira, M.E., Rimmer, S.R., Williams, P.H., Osborn, T.C., 1995. Mapping loci controlling *Brassica napus* resistance to Leptosphaeria maculans under different screening conditions. Genetics 85, 213–217.

Fischbach, M.A., Clardy, J., 2007. One pathway, many products. Nat. Chem. Biol. 3, 353–355.

Flint-Garcia, S.A., Thornsberry, J.M., Buckler, E.S., 2003. Structure of linkage disequilibrium in plants. Annu. Rev. Plant Biol. 54, 357–374.

Floer, S., Druebert, C., Majcherczyk, A., Karlovsky, P., Kües, U., Polle, A., 2008. Defencereactions in the apoplastic proteome of oilseed rape (*Brassica napus* var. *napus*) attenuate *Verticillium longisporum* growth but not disease symptoms. BMC Plant Biol. 8, 129.

Foisset, N., Delourme, R., Barret, P., Hubert, N., Landry, B.S., Renard, M., 1996. Molecular-mapping analysis in *Brassica napus* using isozyme, RAPD and RFLP markers on a doubled-haploid progeny. Theor. Appl. Genet. 93 (7), 1017–1025.

Fu, Y.B., Diederichsen, A., Richards, K.W., Peterson, G., 2002. Genetic diversity within a range of cultivars and landraces of flax (*Linum usitatissimum* L.) as revealed by RAPDs. Genet. Resour. Crop Evol. 49, 167–175.

Fukao, Y., Yoshida, M., Kurata, R., Kobayashi, M., Nakanishi, M., Fujwara, M., Nakajima, K., Ferjani, A., 2013. Peptide separation methodologies for in-depth proteomics in Arabidopsis. Plant Cell Physiol. 54 (5), 808–815, doi:10.1093/PCP/PCT033.

Fukuda, M., Islam, N., Woo, S.H., Yamagishi, A., Takaoka, M., Hirano, H., 2003. Assessing matrix assisted laser desorption/ionization-time of flight-mass spectrometry as a means of rapid embryo protein identification in rice. Electrophoresis 24, 1319–1329.

Fulda, S., Mikkat, S., Stegmann, H., Horn, R., 2011. Physiology and proteomics of drought stress acclimation in sunflower (*Helianthus annuus* L.). Plant Biol. 13 (4), 632–642.

Furbank, R.T., Tester, M., 2011. Phenomics-technologies to relieve the phenotyping bottleneck. Trends Plant Sci. 16, 635–644.

Fusari, C.M., Veronica, V.L., Hopp, H.E., Heinz, R.A., Paniego, N.B., 2008. Identification of single nucleotide polymorphisms and analysis of linkage disequilibrium in sunflower elite inbred lines using the candidate gene approach. BMC Plant Biol. 8, 1–14.

Gaeta, R.T., Yoo, S.Y., Pires, J.C., Doerge, R.W., Chen, Z.J., Osborn, T.C., 2009. Analysis of gene expression in resynthesized *Brassica napus* allopolyploids using *Arabidopsis* 70mer oligo microarrays. Plos One 4 (3), 4760.

Garcia, G.M., Stalker, H.T., Kochert, G., 1995. Introgression analysis of an interspecific hybrid population in peanut (*Arachis hypogaea* L.) using RFLP and RAPD markers. Genome 38, 166–176.

Ghaffari, M., Toorchi, M., Valizadeh, M., Komatsu, S., 2013. Differential response of root proteome to drought stress in drought sensitive and tolerant sunflower inbred lines. Funct. Plant Biol. 40 (6), 609–617.

Giavalisco, P., Kapitza, K., Kolasa, A., Buhtz, A., Kehr, J., 2006. Towards the proteome of *Brassica napus* phloem sap. Proteomics 6 (3), 896–909.

Githiri, S.M., Watanabe, S., Harada, K., Takahashi, R., 2006. QTL analysis of flooding tolerance in soybean at an early vegetative growth stage. Plant Breed. 125 (6), 613–618.

Guijun, D., Weidong, P., Gongshe, L., 2006. The analysis of proteome changes in sunflower seeds induced by N^+ implantation. J. Biosci. 31 (2), 247–253.

Guo, B., Fedorova, N.D., Chen, X., Wan, C.H., Wang, W., Nierman, W.C., Bhatnagar, D., Yu, J., 2011. Gene expression profiling and identification of resistance genes to *Aspergillus flavus* infection in peanut through EST and microarray strategies. Toxins (Basel) 3 (7), 737–753.

Guo, B., Luo, M., Liang, X., Holbrook, C., 2006. Microarray gene expression analysis of peanut responding to drought stress and *Aspergillus* infection. In: Proceedings of the International Conference on Groundnut Aflatoxin Management and Genomics, Guangzhou, China. p. 41.

Guo, B., Sleper, D.A., Arelli, P.R., Shannon, J.G., Nguyen, H.T., 2005. Identification of QTLs associated with resistance to soybean cyst nematode races 2, 3 and 5 in soybean PI 90763. Theor. Appl. Genet. 111, 965–971.

Gupta, V., Mukhopadhaya, A., Arumugam, N., Sodhi, Y.S., Pental, D., Pradhan, A.K., 2004. Molecular tagging of erucic acid trait in oilseed mustard (*Brassica juncea*) by QTL mapping and single nucleotide polymorphism in *FAE1* gene. Theor. Appl. Genet. 108, 743–749.

Hajduch, M., Casteel, J.E., Tang, S., Hearne, L.B., Knapp, S., Thelen, J.J., 2007. Proteomic analysis of near-isogenic sunflower varieties differing in seed oil traits. J. Proteome Res. 6 (8), 3232–3241.

Han, C., Yin, X., He, D., Yang, P., 2013. Analysis of proteome profile in germinating soybean seed, and its comparison with rice showing the styles of reserves mobilization in different crops. PloS One 8 (2), e56947, doi: 10.1371/journal.pone.0056947.

Happstadius, I., Ljungberg, A., Kristiansson, B., Dixelius, C., 2003. Identification of *Brassica oleracea* germplasm with improved resistance to *Verticillium* wilt. Plant Breed. 122, 30–34.

Hasan, M., Friedt, W., Kuhnemann, P.J., Freitag, N.M., Link, K., Snowdon, R.J., 2008. Association of gene-linked SSR markers to seed glucosinolate content in oilseed rape (*Brassica napus* ssp. *napus*). Theor. Appl. Genet. 116 (8), 1035–1049.

Hatzig, S., Zaharia, L.I., Abrams, S., Hohmann, M., Legoahec, L., Bouchereau, A., Nesi, N., Snowdon, R.J., 2014. Early osmotic adjustment responses in drought-resistant and drought-sensitive oilseed rape. J. Integr. Biol. 56 (8), 797–809.

Herman, E.M., Helm, R.M., Jung, R., Kinney, A.J., 2003. Genetic modification removes an immunodominant allergen from soybean. Plant Physiol. 132, 36–43.

Herve, D., Fabre, F., Berrios, E.F., Leroux, N., Chaarani, G.A., Planchon, C., Sarrafi, A., Gentzbittel, L., 2001. QTL analysis of photosynthesis and water status traits in sunflower (*Helianthus annuus* L.) under greenhouse conditions. J. Exp. Bot. 52 (362), 1857–1864.

Hoen, P.A., Ariyurek, Y., Thygesen, H.H., Vreugdenhil, E., Vossen, R.H., de Menezes, R.X., Boer, J.M., van Ommen, G.J., den Dunnen, J.T., 2008. Deep sequencing-based expression analysis shows major advances in robustness, resolution and inter-lab portability over five microarray platforms. Nucleic Acids Res. 36, 141.

Honsdorf, N., Becker, H.C., Ecke, W., 2010. Association mapping for phenological, morphological, and quality traits in canola quality winter rapeseed (*Brassica napus* L.). Genome 53 (11), 899–907.

Hossain, M.A., 2014. Genetic and Metabolomic Analysis of phenolic compounds associated with *Verticillium longisporum* resistance in oilseed rape (*Brassica napus* L.), Thesis, University of Giessen, ISBN/ISSN: 978-3-8359-6152-4.

Hou, J., Wang, C., Hong, X., Zhao, J., Xue, C., Guo, N., Gai, J., Xing, H., 2011. Association analysis of vegetable soybean quality traits with SSR markers. Plant Breed. 130 (4), 444–449.

Hu, J., Quiros, C., Arus, P., Struss, D., Robbelen, G., 1995. Mapping of a gene determining limolenic acid concentration in rapeseed with DNA-based markers. Theor. Appl. Genet. 90, 258–262.

Hwang, E.Y., Song, Q., Jia, G., Specht, J.E., Hyten, D.L., Costa, J., Cregan, P.B., 2014. A genome-wide association study of seed protein and oil content in soybean. BMC Genomics 15, 1, doi:10.1186/1471-2164-15-1.

Imin, N., Kerim, T., Weinman, J.J., Rolfe, B.G., 2001. Characterisation of rice anther proteins expressed at the young microspore stage. Proteomics 1 (9), 1149–1161.

Islam, N., Lonsdale, M., Upadhyaya, N.M., Higgins, T.J., Hirano, H., Akhurst, R., 2004. Protein extraction from mature rice leaves for two dimensional gel electrophoresis and its application in proteome analysis. Proteomics 4, 1903–1908.

Jackson, S.A., Iwata, A., Lee, S.-H., Schmutz, J., Shoemaker, R., 2011. Sequencing crop genomes: approaches and applications. New Phytol. 191 (4), 915–925, doi: 10.1111/j.1469-8137.2011.03804.x.

Jamaux, I., Steinmetz, A., Belhassan, E., Farrar, J., 1997. Looking for molecular and physiological markers of osmotic adjustment in sunflower. Putting plant physiology on the map: genetic analysis of development and adaptive traits. In: Proceedings of the second new phytologist symposium, Bangor, pp. 117–127.

James, C., 2007. Global status of commercialized biotech/GM crops: 2007, ISAAA Brief No. 37, Executive Summary, International Service for the Acquisition Applications (ISAAA), New York.

Jan, C.C., Vick, B.A., Miller, J.F., Kahler, A.L., Butler, E.T., 1998. Construction of an RFLP linkage map for cultivated sunflower. Theor. Appl. Genet. 96, 15–22.

Jayashree, B., Morag, F., Dan, I., Doyle, J., Crouch, J.H., 2005. Analysis of genomic sequences from peanut (*Arachis hypogaea*). J. Biotechnol. 8 (3), 226–237.

Jestin, C., Lodé, M., Vallée, P., Domin, C., Falentin, C., Horvais, R., Coedel, S., Manzanares-Dauleux, M.J., 2011. Association mapping of quantitative resistance for Leptosphaeria maculans in oilseed rape (*Brassica napus* L.). Mol. Breed. 27 (3), 271–287, doi: 10.1007/s11032-010-9429-x.

Jiang, H., Huang, L., Ren, X., Chen, Y., Zhou, X., Xia, Y., Huang, J., Lei, Y., Yan, L., Wan, L., Liao, B., 2014. Diversity characterization and association analysis of agronomic traits in a Chinese peanut (*Arachis hypogaea* L.) mini-core collection. J. Integr. Plant Biol. 56 (2), 159–169.

Jiang, J., Wang, J., Li, A., Zhang, Y., Sokolov, V., Wang, Y., 2012. Proteomic differences in seed filling between yellow-seeded progeny of *Brassica napus-Sinapis alba* (Brassicaceae) and black-seeded parent *B. napus*. Russ. J. Genet. 48 (4), 396–403.

Jiang, J., Shao, Y., Du, K., Ran, L., Fang, X., Wang, Y., 2013. Use of digital gene expression to discriminate gene expression differences in early generations of resynthesized Brassica napus and its diploid progenitors. BMC Genomics 1 14, 72.

Johnson, R.C., Kisha, T.J., Evans, M.A., 2007. Characterizing safflower germplasm with AFLP molecular markers. Crop Sci. 47 (4), 1728–1736.

Kakaei, M., Kahrizi, D., 2011. Study of seed proteins pattern of *Brassica napus* varieties via sodium dodecyl sulfate polyacrylamid gel electrophoresis. Int. Res. J. Biotechnol. 2 (1), 026–028.

Kalita, M.C., Mohapatra, T., Dhandapani, A., Yadava, D.K., Srinivasan, K., Mukherjee, A.K., Sharma, R.P., 2007. Comparative evaluation of RAPD, ISSR and anchored-SSR markers in the assessment of genetic diversity and fingerprinting of oilseed Brassica genotypes. J. Plant Biochem. Biotechnol. 16 (1), 41–48.

Kassem, My A., Meksem, K., Kang, C.H., Njiti, V.N., Kilo, V., Wood, A.J., Lightfoot, D.A., 2004. Loci underlying resistance to manganese toxicity mapped in a soybean recombinant inbred line population of 'Essex2019; x 'Forrest'. Plant Soil 260 (1–2), 197–204.

Katam, R., Basha, S.M., Suravajhala, P., Pechan, T., 2010. Analysis of peanut leaf proteome. J. Proteome Res. 9 (5), 2236–2254.

Kerim, T., Imin, N., Weinman, J.J., Rolfe, B.G., 2003. Proteome analysis of male gametophyte development in rice anthers. Proteomics 3, 738–751.

Kearsey, M.J., Farquhar, A.G.L., 1998. QTL analysis in plants. Heredity 80, 137–142.

Keshavareddy, G., Rohini, S., Ramu, S.V., Sundaresha, S., Kumar, A.R., Kumar, P.A., Udayakumar, M., 2013. Transgenics in groundnut (*Arachis hypogaea* L.) expressing cry1AcF gene for resistance to *Spodoptera litura* (F.). Physiol. Mol. Biol. Plants 19 (3), 343–352.

Khan, R., Alkharouf, N., Beard, H., MacDonald, M., Chouikha, I., Meyer, S., Grefenstette, J., Knap, H., Matthews, B., 2004. Microarray Analysis of gene expression in soybean roots susceptible to the soybean Cyst nematode two days post invasion. J. Nematol. 36 (3), 241–248.

Kiani, S.P., Grieu, P., Maury, P., Hewezi, T., Gentzbittel, L., Sarrafi, A., 2007. Genetic variability for physiological traits under drought conditions and differential expression of water-associated genes in sunflower (*Helianthus annuus* L.). Theor. Appl. Genet. 114 (2), 193–207.

Kim, K.H., Kang, Y.J., Kim, D.H., Yoon, M.Y., Moon, J.K., Kim, M.Y., Van, K., Lee, S.H., 2011. RNA-Seq analysis of a soybean near-isogenic line carrying bacterial leaf pustule-resistant and susceptible alleles. DNA Res. 18 (6), 483–497.

Kim, Y.J., Lee, S., Lee, H.M., Lee, B.W., Ha, T.J., Bae, D.W., Son, B.Y., Kim, Y.H., Baek, S.B., Kim, Y.C., Kim, S.G., Kim, S.T., 2013. Comparative proteomics analysis of seed coat from two black colored soybean cultivars during seed development. Plant Omics J. 6 (6), 456–463.

Kolkman, J.M., Berry, S.T., Leon, A.J., Slabaugh, M.B., Tang, S., Gao, W., Shintani, D.K., Burke, J.M., Knapp, S.J., 2007. Single nucleotide polymorphisms and linkage disequilibrium in sunflower. Genetics 177, 457–468.

Komatsu, S., Kajiwara, H., Hirano, H., 1993. A rice protein library: a data file of rice proteins separated by two-dimentional electrophoresis. Theor. Appl. Genet. 86, 935–942.

Komatsu, S., Rakwal, R., Li, Z., 1999. Separation and characterization of proteins in rice (*Oryza sativa* L.) suspention cultured cells. Plant Cell Tiss. Org. Cult. 55, 183–192.

Korber, N., Wittkopp, B., Bus, A., Snowdon, R., Friedt, W., Stich, B., 2013. Seedling development in a Brassica napus diversity set and its relation to agronomic performance. Theor. Appl. Genet. 125, 1275–1287.

Kottapalli, K.R., Rakwal, R., Shibato, J., Burow, G., Tissue, D., Burke, J., Puppala, N., Burow, M., Payton, P., 2009. Physiology andproteomics of the water-deficit stress response in three contrasting peanut genotypes. Plant Cell Environ. 32, 380–407.

Lahner, B., Gong, J., Mahmoudian, M., Smith, E.L., Abid, K.B., Rogers, E.E., Guerinot, M.L., Harper, J.F., Ward, J.M., McIntyre, L., Schroeder, J.I., Salt, D.E., 2003. Genomic scale profiling of nutrient and trace elements in *Arabidopsis thaliana*. Nature Biotechnol. 21, 1215–1221.

Lai, Z., Gross, B.L., Yizou, Andrews, J., Rieseberg, L.H., 2006. Microarray analysis reveals differential gene expression in hybrid sunflower species. Mol. Ecol. 15 (5), 1213–1227.

Landry, B.S., Hubert, N., Etoh, T., Harada, J.J., Lincoln, S.E., 1991. A genetic map for *Brassica napus* based on restriction fragment lemgth polymorphism detected with expressed DNA sequences. Genome 34, 543–552.

Langer, K., Lorieux, M., Desmarais, E., Griveau, Y., Gentzbittel, L., Berville, A., 2003. Combined mapping of DALP and AFLP markers in cultivated sunflower using F_9 recombinant inbred lines. Theor. Appl. Genet. 106, 1068–1074.

Lark, K.G., Weisemann, J.M., Matthews, B.F., Palmer, R., Chase, K., Macalma, T., 1993. A genetic map of soybean (*Glycine max* L.) using an intraspecific cross of two cultivars: 'Minosy'and 'Noir 1'. Theor. Appl. Genet. 86 (8), 901–906.

Lark, K.G., Mansur, L.M., 1994. Epistatic expression of quantitative trait loci (QTL) in soybean [Glycine max (L) Merr.] determine by QTL association with RFLP alleles. Theor. Appl. Genet. 88, 486–489.

Laurentin, H.E., Karlovasky, P., 2006. Genetic relationship and diversity in a sesame (*Sesamum indicum* L.) germplasm collection using amplified fragment length polymorphism. BMC Genet. 7, 10.

Laurentin, H.E., Karlovasky, P., 2007. AFLP fingerprinting of sesame (*Sesamum indicum* L.) cultivars: identification, genetic relationship and comparison of AFLP informativeness parameters. Genet. Resour. Crop Evol. 54 (7), 1437–1446.

Le, D.T., Nishiyama, R., Watanabe, Y., Tanaka, M., Seki, M., Ham, L.H., Shinozaki, K.Y., Shinozaki, K., Tran, L.S.P., 2012a. Differential gene expression in soybean leaf tissues at late developmental stages under drought stress revealed by genome-wide transcriptome analysis. PLoS One 7 (11), e49522, doi: 10.1371/journal.pone.0049522.

Le, D.T., Nishiyama, R., Watanabe, Y., Vankova, R., Tanaka, M., Seki, M., Ham, L.H., Shinozaki, K.Y., Shinozaki, K., Tran, L.S.P., 2012b. Identification and expression analysis of cytokinin metabolic genes in soybean under normal and drought conditions in relation to cytokinin levels. PLoS One 7 (8), e42411, doi:10.1371/journal.pone.0042411.

Lee, G.J., Boerma, H.R., Villagarcia, M.R., Zhou, X., Carter, T.E., Li, Z., Gibbs, M.O., 2004. A major QTL conditioning salt tolerance in S-100 soybean and descendent cultivars. Theor. Appl. Genet. 109 (8), 1610–1619.

Lee, M.P., Yeun, L.H., Abdullah, R., 2006. Expression of Bacillus thuringiensis insecticidal protein gene in transgenic oil palm. Electron. J. Biotechnol. 9 (2), 117–126.

Leflon, M., Brun, H., Eber, F., Delourme, R., Lucas, M.O., Vallee, P., Ermel, M., Balesdent, M.H., Chevre, A.M., 2007. Detection, introgression and localization of genes conferring specific resistance to Leptosphaeria maculans from *Brassica rapa* into *B. napus*. Theor. Appl. Genet. 115 (7), 897–906.

Lenardon, S.L., Bazzalo, M., Abratti, G., Cimino, C., Galella, M., Grondona, M., Giolitti, F., Leon, A.J., 2005. Screening sunflower for resistance to sunflower chlorotic mottle virus and mapping Rmco-1 resistance gene. Crop Sci. 45, 735–739.

Lennerts, L., 1983. Oelschrote, Oelkuchen, Pflanzliche, Oele und Fette, Herkunft, Gewinning, Verwendung, Bonn Verlag Alfred, Strothe, Hannover.

Li, M.T., Li, Z.Y., Zhang, C.Y., Qian, W., Meng, J.L., 2005. Reproduction and cytogenetic characterization of interspecific hybrids derived from crosses between *Brassica carinata* and *B. rapa*. Theor. Appl. Genet. 110, 1284–1289.

Li, Y., Guan, R., Liu, Z., Ma, Y., Wang, L., Li, L., Lin, F., Luan, W., Chen, P., Yan, Z., Guan, Y., Zhu, L., Ning, X., Smulders, M.J.M., Li, W., Piao, R., Cui, Y., Yu, Z., Guan, M., Rector, B.G., All, J.N., Parrott, W.A., Boerma, H.R., 1998. Identification of molecular markers linked to quantitative trait loci for soybean resistance to corn earworm. Theor. Appl. Genet. 96, 786–790.

Li, Y., Guan, R., Liu, Z., Ma, Y., Wang, L., Li, L., Lin, F., Luan, W., Chen, P., Yan, Z., Guan, Y., Zhu, L., Ning, X., Smulders, M.J., Li, W., Piao, R., Cui, Y., Yu, Z., Guan, M., Chang, R., Hou, A., Shi, A., Zhang, B., Zhu, S., Qui, L., 2008. Genetic structure and diversity of cultivated soybean Glucine *max* (L.) Merr landraces in China. Theor. Appl. Genet. 117 (6), 857–871.

Liang, Y., Srivastava, S., Rahman, M.H., Strelkov, S.E., Kav, N.N., 2008. Proteome changes in leaves of *Brassica napus* L. as a result of Sclerotinia sclerotiorum challenge. J. Agric. Food Chem. 56 (6), 1963–1976, doi: 10.1021/jf073012d.

Li, X., XiTing, Z., MingXing, Z., 2000. Changes in endogenous plant hormones in the apical buds of soybean cultivar Zao 12 during inflorescence initiation and formation. Chin. J. Oil Crop Sci. 22 (3), 48–50.

Lim, C.W., Park, J.Y., Lee, S.H., Cheol, H.H., 2010. Comparative proteomic analysis of soybean nodulation using a supernodulation mutant SS2-2. Biosci. Biotechnol. Biochem. 74 (12), 2396–2404.

Loei, H., Lim, J., Tan, M., Lim, T.K., Lin, Q.S., Chew, F.T., Kulaveerasingam, H., Chung, M.C., 2013. Proteomic analysis of the oil palm fruit mesocarp reveals elevated oxidative phosphorylation activity is critical for increased storage oil production. J. Proteome Res. 12 (11), 5096–5109.

Luo, M., Liang, X.Q., Dang, P., Holbrook, C.C., Bausher, M.G., Lee, R.D., Guo, B.Z., 2005. Microarray-based screening of differentially expressed genes in peanut in response to *Aspergillus parasiticus* infection and drought stress. Plant Sci. 169, 695–703.

Lydiate, D., Sharpe, A., 2003. Aligning Genetic Maps of *Brassica napus* Using Microsatellite Markers. Plant and Animal Genome XI, San Diego, USA.

Maguire, T.L., Grimmond, S., Forrest, A., Iturbe-Ormaetxe, I., Meksem, K., Gresshoff, P., 2002. Tissue-specific gene expression in soybean (*Glycine max*) detected by cDNA microarray analysis. J. Plant Physiol. 159 (12), 1361–1374.

Malathi, B., Ramesh, S., Rao, K.V., Reddy, V.D., 2006. *Agrobacterium*-mediated genetic transformation and production of semilooper resistant transgenic castor (*Ricinus communis* L.). Euphytica 147 (3), 441–449.

Ma, W., Kong, Q., Arondel, V., Kilaru, A., Bates, P.D., Thrower, N.A., Benning, C., Ohlrogge, J.B., 2013. *WRINKLED1*, A ubiquitous regulator in oil accumulating tissues from Arabidopsis embryos to oil palm mesocarp. PLoS One 8 (7), e68887, doi:10.1371/journal.pone.0068887.

Mamidi, S., Chikara, A., Goos, R.J., Hyten, D.L., Annam, D., Moghaddam, S.M., Lee, R.K., Cregan, P.B., McClean, P.E., 2011. Genome-wide association analysis identifies candidate genes associated with iron deficiency chlorosis in soybean. Plant Genome 4 (3), 154–164.

Mandoulakani, B.A., Alami, A., Esfahani, M., 2010. Association analysis for morphological traits in peanut (*Arachis hypogea* L.) using microsatellite markers. Iran. J. Crop Sci. 12 (4), 510–519.

Manjunatha, S.B., Suma, T.C., Sreevathsa, R., Devendra, R., Udaya Kumar, M., Prasad, T.G., 2008. Evaluation of advanced generation transgenic groundnut lines resistant to herbicide-glyphosate. Indian J. Weed Sci. 40 (3&4), 162–165.

Marioni, J.C., Mason, C.E., Mane, S.M., Stephens, M., Gilad, Y., 2008. RNA-seq: an assessment of technical reproducibility and comparison with gene expression arrays. Genome Res. 18, 1509–1517.

Mazarei, M., Li, W., Al-Ahmad, H., Arelli, P.R., Pantalone, V.R., Stewart, Jr., C.N., 2011. Gene expression proWling of resistant and susceptible soybean lines infected with soybean cyst nematode. Theor. Appl. Genet. 123, 1193–1206.

Mesquita, R.O., de Almeida Soares, E., de Barros, E.G., Loureiro, M.E., 2012. Method optimization for proteomic analysis of soybean leaf: improvements in identification of new and low-abundance proteins. Genet. Mol. Biol. 35 (1), 353–361.

Messaitfa, Z.H., Shehata, A.I., El Quraini, F., Al Hazzani, A.A., Rizwana, H., El wahabi, M.S., 2014. Proteomics analysis of salt stressed Sunflower (*Helianthus annuus*). Int. J. Pure App. Biosci. 2 (1), 6–17.

Misra, S., Gedamu, L., 1989. Heavy metal tolerant transgenic *Brassica napus* L. and *Nicotiana tabacum* L. plants. Theor. Appl. Genet. 78 (2), 161–168.

Moloney, M.M., Walker, J.M., Sharma, K.K., 1989. High efficiency transformation of *Brassica napus* using Agrobacterium vectors. Plant Cell Rep. 8, 238–242.

Mondal, K.K., Bhattacharya, R.C., Koundal, K.R., Chatterjee, S.C., 2007. Transgenic Indian mustard (*Brassica juncea*) expressing tomato glucanase leads to arrested growth of *Alternaria brassicae*. Plant Cell Rep. 26 (2), 247–252.

Mondal, S., Badigannavar, A.M., Murty, G.S.S., 2007. RAPD markers linked to a rust resistance gene in cultivated groundnut (*Arachis hypogaea* L.). Euphytica 159, 233–239.

Mooney, B.P., Krishnan, H.B., Thelen, J.J., 2004. High-throughput peptide mass fingerprinting of soybean seed proteins: automated workflow and utility of UniGene expressed sequence tag databases for protein identification. Phytochemistry 65 (12), 1733–1744.

Natarajan, S.S., Xu, C., Bae, H., Caperna, T.J., Garrett, W.M., 2006. Characterization of storage proteins in wild (*Glycine soja*) and cultivated (*Glycine max*) soybean seeds using proteomic analysis. J. Agric. Food Chem. 54, 3114–3120.

Nath, U.K., Goswami, G., 2009. Inheritance and variation of erucic acid content in a transgenic rapeseed (*Brassica napus* L.) doubled haploid population. Mol. Breed. 23, 125–138.

Neale, D.B., Savolainen, O., 2004. Association genetics of complex traits in conifersi. Trends Plant Sci. 9, 325–330.

Niu, C., Akasaka-Kennedy, Y., Faustinelli, P., Joshi, M., Rajasekaran, K., Yang, H., Chu, Y., Cary, J., Ozias-Akins, P., 2009. Antifungal activity in transgenic peanut (*Arachis hypogaea* L.) conferred by a nonheme chloroperoxidase gene. Peanut Sci. 36, 126–132.

Nozu, Y., Tsugita, A., Kamijo, K., 2006. Proteomic analysis of rice leaf, stem and root tissues during growth course. Proteomics 6, 3665–3670.

Obermeier, C., Hosseini, B., Friedt, W., Snowdon, R., 2008.Serial analysis of gene expression (SAGE) during *Brassica napus* seed development. ISHS Acta Horticulturae 867V International Symposium on Brassicas and XVI International Crucifer Genetics Workshop, Brassica.

Obermeier, C., Hosseini, B., Friedt, W., Snowdon, R., 2009. Gene expression profiling via Long SAGE in a non-model plant species: a case study in seeds of *Brassica napus*. BMC Genomics 10, 295.

Oh, M.W., Nanjo, Y., Komatsu, S., 2014. Gel free proteomic analysis of soyabean root proteins affected by calcium under flooding stress. Front. Plant Sci. 5, 559, doi: 10.3389/fpls.2014.00559.

Oh, T., Gorman, M., Cullis, C., 2000. RFLP and RAPD mapping in flax (*Linum usitatissimum*). Theor. Appl. Genet. 101, 590–593.

O'Rourke, J.A., Nelson, R.T., Grant, D., Schmutz, J., Grimwood, J., Cannon, S., Vance, C.P., Graham, M.A., Shoemaker, R.C., 2009. Integrating microarray analysis and the soybean genome to understand the soybeans iron deficiency response. BMC Genomics 10, 376.

Panjabi, P., Yadava, S.K., Sharma, P., Kaur, A., Kumar, A., Arumugam, N., Sodhi, Y.S., Mukhopadhyay, A., Gupta, V., Pradhan, A.K., Pental, D., 2010. Molecular mapping reveals two independent loci conferring resistance to *Albugo candida* in the east European germplasm of oilseed mustard *Brassica juncea*. Theor. Appl. Genet. 121, 137–145.

Payton, P., Kottapalli, K.R., Rowland, D., Faircloth, W., Guo, B., Burow, M., Puppala, N., Gallo, M., 2009. Gene expression profiling in peanut using high density oligonucleotide microarrays. BMC Genomics 10, 265.

Pereira, M.R., Gouvêa, B.C., Marcelino-Guimarães, F.C., de Oliveira Ramos, H.J., Moreira, M.A., de Barros, E.G., 2013. Proteomic analysis of soybean leaves in a compatible and an incompatible interaction with *Phakopsora pachyrhizi*. Organelles Proteomics 1 (1), 16–27.

Perez, V.B., Knapp, S.J., Leon, A.J., Fernandez, M.J.M., Berry, S.T., 2004. Mapping minor QTL for increased stearic acid content in sunflower seed oil. Mol. Breed. 13, 313–322.

Qi, Z., Wang, W., Min-jian, C.A.O., 2008. The morphological and physiological diversities in roots of different potassium efficient soybean lines. J. Shenyang Agric. Univ. 2, 001.

Qin, J., Gu, F., Liu, D., Yin, C., Zhao, S., Chen, H., Zhang, J., Yang, C., Zhan, X., Zhang, M., 2013. Proteomic analysis of elite soybean Jidou17 and its parents using iTRAQ based quantitative approaches. Proteome Sci. 11, 12.

Qui, D., Morgan, C., Shi, J., Long, Y., Liu, J., Li, R., Zhuang, X., Wang, Y., Tan, X., Dietrich, E., Weihmann, T., Everett, C., Vanstraelen, S., Beckett, P., Fraser, F., Trick, M., Barnes, S., Wilmer, J., Schmidt, R., Li, J., Li, D., Meng, J., Bancroft, I., 2006. A comparative linkage map of oilseed rape and its use for QTL analysis of seed oil and erucic acid content. Theor. Appl. Genet. 114 (1), 67–80.

Radonic, L.M., Zimmermann, J.M., Zavallo, D., López, N., Bilbao, M.L., 2008. Introduction of antifungal genes in sunflower via *Agrobacterium*. Electron. J. Biotechnol. 11 (5).

Rahman, M., Sun, Z., McVetty, P.B.E., Li, G., 2008. High throughput genome-specific and gene-specific molecular markers for erucic acid genes in *Brassica napus* (L.) for marker-assisted selection in plant breeding. Theor. Appl. Genet. 117 (6), 895–904.

Ramskold, D., Luo, S., Wang, Y.C., Li, R., Deng, Q., Faridani, O.R., Daniels, G.A., Khrebtukova, I., Loring, J.F., Laurent, L.C., Schroth, G.P., Sandberg, R., 2012. Full-length mRNA-Seq from single-cell levels of RNA and individual circulating tumor cells. Nat. Biotechnol. 30, 777–782.

Remington, D.L., Thornsberry, J.M., Matsuoka, Y., Wilson, L.M., Whitt, S.R., Doebley, J., Kresovich, S., Goodman, M.M., Buckler, E.S., 2001. Structure of linkage disequilibrium and phenotypic associations in the maize genome. Proc. Natl. Acad. Sci. USA 98, 11479–11484.

Roose, A.C., Cariou, P.E., Vautrin, D., Tavernier, R., Solignac, M., 2006. Polymorphic microsatellite loci in *Linum usitatissimum*. Mol. Ecol. Notes 6, 796–799.

Saghai-Maroof, M.A., Jeong, S.C., Gunduz, I., Tucker, D.M., Buss, G.R., Tolin, S.A., 2008. Pyramiding of soybean mosaic virus resistance genes by marker-assisted selection. Crop Sci. 48, 517–526.

Sakhno, L.O., Kvasko, O.Y., Olevinska, Z.M., Spivak, M.Y., Kuchuk, M.V., 2012. Creation of transgenic *Brassica napus* L. plants expressing human alpha 2b interferon gene. Cytol. Genet. 46 (6), 342–346.

Sang, X., Jue, D., Yang, L., Bai, X., Chen, M., Yang, Q., 2013. Genetic transformation of *Brassica napus* with msi-99m gene increases resistance in transgenic plants to *sclerotinia sclerotiorum*. Mol. Plant Breed. 4 (30), 247–253.

Sanjaya, Satyavathi, V.V., Prasad, V., Kirthi, N., Maiya, S.P., Savithri, H.S., Lakshmi Sita, G., 2005. Development of cotton transgenics with antisense *AV2* gene for resistance against cotton leaf curl virus (CLCuD) via *Agrobacterium tumefaciens*. Plant Cell Tiss. Organ Cult. 81 (1), 55–63.

Schaarschmidt, S., Gresshoff, P.M., Hause, B., 2013. Analyzing the soybean transcriptome during autoregulation of mycorrhization identifies the transcription factors GmNF-YA1a/b as positive regulators of arbuscular mycorrhization. Genome Biol. 14, 62.

Sehgal, D., Raina, S.N., 2005. Genotyping safflower (*Carthamus tinctorius*) cultivars by DNA fingerprints. Euphytica 146 (1/2), 67–76.

Severin, A.J., Woody, J.L., Bolon, Y.T., Joseph, B., Diers, B.W., Farmer, A.D., Muehlbauer, G.J., Nelson, R.T., Grant, D., Specht, J.E., Graham, M.A., Cannon, S.B., May, G.D., Vance, C.P., Shoemaker, R.C., 2010. RNA-seq atlas of glycine max: a guide to the soybean transcriptome. BMC Plant Biol. 10, 160, doi: 10.1186/1471-2229-10-160.

Sharma, A., Mohapatra, T., Sharma, R.P., 1994. Molecular mapping and character tagging in *Brassica juncea*–I. Degree, nature and linkage relationship of RFLPs and their association with quantitative traits. J. Plant Biochem. Biotechnol. 3 (2), 85–89.

Sharma, R., Aggarwal, R.A.K., Kumar, R., Mohapatra, T., Sharma, R.P., 2002. Construction of an RAPD linkage map and localization of QTLs for oleic acid level using recombinant inbreds in mustard (*Brassica juncea*). Genome 45 (3), 467–472.

Sharma, S.N., Kumar, V., Mathur, S., 2009. Comparative analysis of RAPD and ISSR markers for characterization of sesame (*Sesamum indicum* L.) genotypes. J. Plant Biochem. Biotechnol. 18 (1), 37–43.

Shearman, J.R., Jantasuriyarat, C., Sangsrakru, D., Yoocha, T., Vannavichit, A., Tragoonrung, S., Tangphatsornruang, S., 2013. Transcriptome analysis of normal and mantled developing oil palm flower and fruit. Genomics 101 (5), 306–312.

Shen, S., Matsubae, M., Takao, T., Tanaka, N., Komatsu, S.J., 2002. A proteomic analysis of leaf sheaths from rice. J. Biochem. 132, 613–620, (Tokyo).

Sheoran, I.S., Sawhney, V.K., 2010. Proteome analysis of the normal and *Ogura* (*ogu*) CMS anthers of *Brassica napus* to identify proteins associated with male sterility. Botany 88 (3), 217–230.

Shi, J., Li, R., Qui, D., Jiang, C., Long, Y., Morgan, C., Bancroft, I., Zhao, J., Meng, J., 2009. Unraveling the complex trait of crop yield with quantitative trait loci mapping in *B. napus*. Genetics 182, 851–861.

Snow, A.A., Pilson, D., Rieseberg, L.H., Paulsen, M.J., Pleskac, N., Reagon, M.R., Wolf, D.E., Selbo, S.M., 2003. A Bt transgene reduces herbivory and enhances fecundity in wild sunflowers. Ecol. Appl. 13 (2), 279–286.

Song, Q.J., Marek, L.F., Shoemaker, R.C., Lark, K.G., Concibido, V.C., Delannay, X., Specht, J.E., Cregan, P.B., 2004. A new integrated genetic linkage map of the soybean. Theor. Appl. Genet. 109 (1), 122–128.

Stamp, P., Visser, R., 2012. The twenty-first century, the century of plant breeding. Euphytica 186, 585–591.

Stewart, C.N., All, J.N., Raymer, P.L., Ramachandran, S., 1997. Increased fitness of transgenic insecticidal rapeseed under insect selection pressure. Mol. Ecol. 6, 773–779.

Strickler, S.R., Bombarely, A., Mueller, L.A., 2012. Designing a transcriptome next-generation sequencing project for a nonmodel plant species. Am. J. Bot. 99 (2), 257–266.

Sun, Z., Wang, Z., Tu, J., Zhang, J.Z., Yu, F., McVetty, P.B.E., Li, G., 2007. An ultradense genetic recombination map for *Brassica napus*, consisting of 13551 SRAP markers. Theor. Appl. Genet. 114 (8), 1305–1317.

Syahanim, S., Abrizah, O., Mohamad Arif, A.M., Idris, A.S., Mohd Din, A., 2013. Identification of differentially expressed proteins in oil palm seedlings artificially infected with *Ganoderma*: A proteomics approach. J. Oil Palm Res. 25, 298–304.

Tanaka, N., Konishi, H., Khan, M.M., Komatsu, S., 2004. Proteome analysis of rice tissues by two-dimentional electrophoresis: an approach to the investigation of gibberellin regulated proteins. Mol. Genet. Genomics 270, 485–496.

Tang, S., Yu, J.K., Slabaugh, M.B., Shintani, D.K., Knapp, S.J., 2002. Simple sequence repeat map of the sunflower genome. Theor. Appl. Genet. 105, 1124–1136.

Tuyen, D.D., Lal, S.K., Xu, D.H., 2010. Identification of a major QTL allele from wild soybean (Glycine soja Sieb. & Zucc.) for increasing alkaline salt tolerance in soybean. Theor. Appl. Genet. 121 (2), 229–236.

Vajhala, C.S.K., Sadumpati, V.K., Nunna, H.R., Puligundla, S.K., Vudem, D.R., Khareedu, V.R., 2013. Development of transgenic cotton lines expressing *Allium sativum* agglutinin (ASAL) for enhanced resistance against major sap-sucking pests. Plos One 8 (9), e72542, doi: 10.1371/journal.pone.0072542.

Varma, T.S.N., Dwivedi, S.L., Pande, S., Gowda, M.V.C., 2005. SSR markers associated with resistance to rust (Puccinia arachidis Speg.) in groundnut (*Arachis hypogaea* L.). SABRAO J. Breed. Genet. 37 (2), 107–119.

Varshney, R.K., Bertioli, D.J., Moretzsohn, M.C., Vadez, V., Krishnamurthy, L., Aruna, R., Nigam, S.N., Moss, B.J., Seetha, K., Ravi, K., He, G., Knapp, S.J., Hoisington, D.A., 2009. The first SSR-based genetic linkage map for cultivated groundnut (*Arachis hypogaea* L.). Theor. Appl. Genet. 118, 729–739.

Varshney, R.K., Tuberosa, R., 2013. In: Varshney, R.K., Tuberosa, R. (Eds.), Translational Genomics for Crop Breeding: Abiotic Stress Tolerance, Yield, and Quality, an Introduction, in Translational Genomics for Crop Breeding. Abiotic Stress, Yield and Quality 2. John Wiley & Sons Ltd, Chichester, UK, doi: 10.1002/9781118728482.ch1.

Walker, D., Boerma, H.R., All, J., Parrott, W., 2002. Combining cry1Ac with QTL alleles from PI229358 to improve soybean resistance to Lepidoptera pests. Mol. Breed. 9, 43–51.

Wan, J., Torres, M., Ganapathy, A., Thelen, J., DaGue, B.B., Mooney, B., Xu, D., Stacey, G., 2005. Proteomic analysis of soybean root hairs after infection by *Bradyrhizobium japonicum*. Mol. Plant Microbe. Interact. 18 (5), 458–467.

Wang, C.T., Yang, X.D., Chen, D.X., Yu, S.L., Liu, G.Z., Yue, Y.T., Jain, Z.X., 2007. Isolation of simple sequence repeats from groundnut. J. Biotechnol. 10 (3), 473–480.

Wang, H., Gao, Z., Zhang, D., Cheng, H., Yu, D., 2011. Identification of genes with soybean resistance to common cutworm by association analysis. Chin. Bull. Bot. 46 (5), 514–524.

Wang, M.L., Sukumaran, S., Barkley, N.A., Chen, Z., Chen, C.Y., Guo, B., Pittman, R.N., Stalker, H.T., Holbrook, C.C., Pederson, G.A., Yu, J., 2011. Population structure and marker-trait association analysis of the US peanut (Arachis hypogaea L.) mini-core collection. Theor. Appl. Genet. 123 (8), 1307–1317.

Wang, Y., Fristensky, B., 2001. Transgenic canola lines expressing pea defense gene DRR206 have resistance to aggressive blackleg isolates and to *Rhizoctonia solani*. Mol. Breed. 8, 263–271.

Wang, Z., Yan, S., Liu, C., Chen, F., Wang, T., 2012. Proteomic analysis reveals an aflatoxin-triggered immune response in cotyledons of *Arachis hypogaea* infected with *Aspergillus flavus*. J. Proteome Res. 11 (5), 2739–2753.

Ward, J.A., Ponnala, L., Weber, C.A., 2012. Strategies for transcriptome analysis in nonmodel plants. Am. J. Bot. 99 (2), 267–276.

Wei, S., Yu, B., Gruber, M.Y., Khachatourians, G.G., Hegedus, D.D., Hannoufa, A., 2010. Enhanced seed carotenoid levels and branching in transgenic *Brassica napus* expressing the *ArabidopsismiR156b* gene. J. Agric. Food Chem. 58 (17), 9572–9578.

Wei, W., Zhang, Y., Lü, H., Li, D., Wang, L., Zhang, X., 2013. Association analysis for quality traits in a diverse panel of Chinese sesame (*Sesamum indicum* L.) germplasm. J. Integr. Plant Biol. 55 (8), 745–758.

Weier, D., Hanke, C., Eickelkamp, A., Lühs, W., Dettendorfer, J., Schaffert, E., Mollers, C., Friedt, W., Wolter, F.P., Frentzen, M., 1997. Trierucoylglycerol biosynthesis in transgenic plants of rapeseed (*Brassica napus* L.). Eur. J. Lipid Sci. Technol. 99 (5), 160–165.

Xia, H., Zhao, C., Hou, L., Li, A., Zhao, S., Bi, Y., An, J., Zhao, Y., Wan, S., Wang, X., 2013. Transcriptome profiling of peanut gynophores revealed global reprogramming of gene expression during early pod development in darkness. BMC Genomics 14, 517.

Xia, Z., Tsubokura, Y., Hoshi, M., Hanawa, M., Yano, C., Okamura, K., Harada, K., 2007. An integrated high-density linkage map of soybean with RFLP, SSR, STS, and AFLP markers using a single F2 population. DNA Res. 14 (6), 257–269.

Xiao-Ping, Z., Fang, He., Yan-Bin, H., Hai-Yan, L., Er-Hua, Z., Gui-Yuan, Z., Shao-Xiong, L., Ni, Z., Shi-Jie, W., Xing-Yu, L., Xuan-Qiang, L., 2011. Analysis of gene expression profiles in pod and leaf of two major peanut cultivars in Southern China. Acta Agron. Sin. 37 (8), 1378–1388.

Yadava, D.K., Vasudev, S., Singh, N., Mohapatra, T., Prabhu, K.V., 2012. Breeding major oil crops: present status and future research needs. Technological Innovations in Major World Oil Crops, 1, Springer, New York, pp. 17–51.

Yang, L.X., Xue, W., ShaoPeng, Z., DaWei, L., YuXi, D., Wei, D., 2011. Comparative profiling of the transcriptional response to soybean cyst nematode infection of soybean roots by deep sequencing. Chin. Sci. Bull. 56, 1904–1911.

Yang, Y., Xia, W.W., Zheng, Y.L., Chen, L., Liu, R.J., Huang, C.Y., 2007. Genetic diversity and relationships among safflower (*Carthamus tinctorius* L.) analysed by inter-simple sequence repeats (ISSRs). Genet. Resour. Crop Evol. 54 (5), 1043–1051.

Yan-Xin, Z., Lin-Hai, W., Dong-Hua, L., Wen-Liang, W., Yuan, G., Xiu-Rong, Z., 2012. Association Mapping of Sesame (*Sesamum indicum* L.) Resistance to macrophomina phaseolina and identification of resistant accessions. China Agric. Sci. 45 (13), 2580–2591.

Yin, H., Li, S., Zhao, X., Du, Y., Ma, X., 2006. cDNA microarray analysis of gene expression in *Brassica napus* treated with oligochitosan elicitor. Plant Physiol. Biochem. 44 (11-12), 910–916.

Ying-Hui, L., Marinus, S., Ru-Zhen, C., Li-Juan, Q., 2011. Genetic diversity and association mapping in a collection of selected Chinese soybean accessions based on SSR marker analysis. Conservation Genet. 12 (5), 1145.

Yu, H., Li, Y., Li, X., Romeis, J., Wu, K., 2013. Expression of Cry1Ac in transgenic Bt soybean lines and their efficiency in controlling lepidopteran pests. Pest Manag. Sci. 69 (12), 1326–1333.

Yu, J.K., Tang, S., Slabaugh, M.B., Heesacker, A., Cole, G., Herring, M., Soper, J., Han, F., Chu, W.C., Webb, D.M., Thompson, L., Edwards, K.J., Berry, S., Leon, A.J., Grondona, M., Olungu, C., Maes, N., Knapp, S.J., 2003. Towards a saturated molecular genetic linkage map for cultivated sunflower. Crop Sci. 43, 367–387.

Yue, B., Miller, J.F., Hu, J., 2007. Experimenting with marker-assisted selection in confection sunflower germplasm enhancement.www.sunflowernsa.com/research/research-workshop/documents/Yue.

Yue, B., Miller, J.F., Hu, J., 2007. Experimenting with marker-assisted selection in confection sunflower germplasm enhancement. Twenty-Ninth Sunflower Research Workshop, January 10–11, 2007, Fargo, ND. Available: http://www.sunflowernsa.com/research/research-workshop/documents/Yue_Experiment_Marker_07.pdf.

Yunus, A.M.M., Parveez, A., Kadir, G., 2008. Development of transformation vectors for the production of potentially high oleate transgenic oil palm. Electron. J. Biotechnol. 11 (3), ISSN: 0717-3458.

Zargar, S.M., Kurata, R., Rakwal, R., Fukao, Y., 2013. Unraveling the iron deficiency responsive proteome in Arabidopsis shoot by iTRAQ-OFFGEL approach. Plant Signal Behav. 8 (10), e26892.

Zhang, W.J., Yuan, N., Bu, S.H., Li, M., Jian-Ying, F., Jin, Z., Sheng-Xian, Y., Odinga, M.M., Shi-Ping, W., Xiao-Feng, L., Yuan-Ming, Z., 2014. Epistatic association mapping for alkaline and salinity tolerance traits in the soybean germination stage. Plos One 9 (1), e84750, doi: 10.1371/journal.pone.0084750.

Zhao, C., Wang, J., Cao, M., Zhao, K., 2005. Proteomic changes in rice leaves during development of field-grown rice plants. Proteomics 5, 961–972.

Zhao, J., Meng, J., 2003. Genetic analysis of loci associated with partial resistance to Sclerotinia sclerotiorum in rapeseed (*Brassica napus* L.). Theor. Appl. Genet. 06, 759–764.

Zhao, J., Becker, H.C., Zhang, D., Zhang, Y., Ecke, W., 2006. Conditional QTL mapping of oil content in rapeseed with respect to protein content and traits related to plant development and grain yield. Theor. Appl. Genet. 113 (1), 33–38.

Zhu, M., Dai, S., Zhu, N., Booy, A., Simons, B., Yi, S., Chen, S., 2012. Methyl jasmonate responsive proteins in *Brassica napus* guard cells revealed by iTRAQ-based quantitative proteomics. J. Proteome Res. 11 (7), 3728–3742.

Zhu, Y.L., Pilon-Smits, E.A.H., Jouanin, L., Terry, N., 1999. Overexpression of glutathione synthetase in Indian mustard enhances cadmium accumulation and tolerance. Plant Physiol. 119 (1), 73–80.

Ziegler, G., Terauchi, A., Becker, A., Armstrong, P., Hudson, K., Baxter, I., 2013. Ionomic screening of field-grown soyabean identifies mutants with altered seed elemental composition. Plant Genome 6 (2), doi:10.3835/plantgenome2012.07.0012.

CHAPTER

14

Forecasting Diseases and Insect Pests for a Value-Added Agroadvisory System

Amrender Kumar, Chirantan Chattopadhyay**, Bimal Kumar Bhattacharya*[†]*, Vinod Kumar*[‡]*, Amrendra Kumar Mishra**

***Indian Agricultural Research Institute, Indian Council of Agricultural Research (ICAR), New Delhi, India**

****National Centre for Integrated Pest Management, Indian Council of Agricultural Research (ICAR), New Delhi, India**

[†]Crop Inventory and Agro-ecosystem Division, Space Applications Centre, Indian Space Research Organization (ISRO), Ahmedabad, India

[‡]Directorate of Rapeseed-Mustard Research, Indian Council of Agricultural Research (ICAR), Sewar, Bharatpur, Rajasthan, India

Humankind has always been eager to predict the future to plan suitably for the same. As knowledge has built up, concepts such as the disease triangle and disease tetrahedron have been considered. They involve the interaction between host and pathogen in a given environment over time. Plant disease forecasting is a mechanism to predict the occurrence or change in severity of plant diseases. At the field scale, these systems are used by growers to make economic decisions about disease management. The environment is usually the factor that controls whether disease develops or not, the vulnerability of the host, and the presence of pathogens in a particular season through its effects on processes such as overseasoning or the ability of the pathogen to cause disease. In these cases a disease-forecasting system attempts to define when the environment will be conducive to disease development.

Breeding Oilseed Crops for Sustainable Production. http://dx.doi.org/10.1016/B978-0-12-801309-0.00014-8

WHY STUDY EPIDEMIOLOGY/EPIZOOLOGY AND FORECASTING OF CROP PESTS?

A wide gap exists between potential yield and that realized in a farmer's field. Among the factors contributing toward this yield gap are the biotic stresses that affect crops. The severity of infestation of insect pests and diseases differs among seasons, regions, and individual crops within a region. In the absence of stable, desirable, and diverse sources of resistance to biotic menaces, pesticides remain the only effective means of managing them. Knowledge about the timing of start of infestation of these pests and their gradual progress in advance could enable planning of a necessary pesticide schedule for the season, region, or the particular crop against the specific menace expected. This could be enabled by development of region, crop, and pest-specific prediction models to forewarn about these menaces. Since these biotic menaces are weather dependent, weather-based prediction models could be developed to enable management of these pests.

WHERE TO USE FORECAST MODELS?

Development of a forecast model consumes a lot of resources such as manpower and time; hence, such resources are restricted to important crops. The candidate disease or insect pest should be sporadic, and its occurrence, severity, and progression should be influenced by weather factors and, hence, should vary accordingly and cause economically significant yield losses over large areas. The availability of timely and quality forecasts could enable mitigation of the risks due to the occurrence of the disease by application of effective economic prophylactic measures. Forecast models provide an alternative to calendar spray schedules to bring need-based precision (e.g., instead of spraying at 7–14 day intervals, spraying precisely when and where the disease or insect pest is likely to appear or has just initiated to enable input costs to be cut). Thus, precision pest management may reduce the amount of spraying and so provide economic and environmental benefits. Finally, the forecast system should enable taking economically acceptable actions as an integral part of an integrated pest management package, while growers should be capable and flexible enough to take due advantage of forewarning about a disease or insect pest.

A disease or insect pest-forecasting model is a set of formulae, rules, or algorithms patterned after the biology of the specific pathogen or insect pest and at the same time keeping host and standard crop management practices in mind. The stages of modeling start with development [assumptions and selection of key environmental and host variables such as resistance to disease and the gene-for-gene theory of Flor (1942) in addition to considering a farmer's socioeconomic situation such as constraints and perceptions], devising equations based on laboratory and field studies to predict the risk of occurrence of disease and/or development of inoculum, validation (testing assumptions over agroclimatic zones and years, and revising and refining them based on the results or needs of different locations), and implementation (such as transferring grower trials and issuing of advisories in the public interest to the private domain – namely, consultants, industry, etc. – for improved crop management). Thus, to devise a forecast model, thorough study of temporal and spatial changes in the development of epidemics is important, as is a reliable sampling procedure, wherein assessment of initial inoculum (by spore traps and use of serological, molecular tools) could make the job easy. Forecasting plant disease epidemics is intellectually stimulating. For the most part, having as much information as possible about diseases or insect pests is useful before venturing to predict its development. Sometimes, one or two factors affecting disease development predominate to such a degree that knowledge about them is sufficient to formulate a reasonably accurate forecast. Thus, forecasts could be based on initial inoculum, meteorological parameters, and their combinations. Thus, the criteria for making predictions should be based on sound investigation in the laboratory and field. If information on favorable weather conditions is known, subjective weights based on this information can be used to construct weather indices. In the absence of such information, correlation coefficients between targeted forecasts and respective weather variables/product of weather variables can be used (Agrawal et al., 2004).

REGIONAL FORECASTING FOR CROP PROTECTION ADVISORIES

The crux of crop protection advisory services for farmers depends on real-time availability of quality regional short to medium–range weather forecasts and tested epidemiological/epizoological models. To enable forecasting for likely occurrence of any disease or insect pest to be effective, risk-related information must be made available to the targeted farmer at least 7–10 days ahead of actual occurrence of the menace so that the farmer is able to put in place the requisite economic action (i.e., pesticide spraying, etc.) to mitigate the impact of the pest. Thus, farmers need to be advised of mitigation strategies for likely impacts of weather parameters on insect pest or disease infestation. Earlier, most of the work focused on the effects of a single meteorological variable on the host, pathogen, or the interaction of the two under controlled conditions. Since the advent of Dutch rules, thumb rules or even empirical models have been used to forecast such plant diseases as potato late blight; such models are presently in vogue (http://web.pau.edu/potato/). There are several disease-forecasting networks across the globe: maize (EPICORN, for southern corn leaf blight), tomatoes and potatoes (EPIDEM, TOMCAST, BLITECAST, for early and late blights), and apple (Maryblight, EPIVEN, for fireblight and scab). Then came advanced weather-based location-specific forewarning models based on multiple or stepwise regression, discriminant function analysis, artificial neural networks (ANNs) developed for major diseases (*Alternaria* blight, white rust, powdery mildew), and aphid of oilseed *Brassica* in India (Desai et al., 2004; Chattopadhyay et al., 2005a,b; 2011; Laxmi and Kumar, 2011; Jain et al., 2009; Rao et al., 2012, 2013, 2014). These were successfully validated between 2005–2008 at Bharatpur (Rajasthan) for occurrence of *Alternaria* blight, white rust, and aphid on Indian mustard crop; agroadvisories were even issued for public use.

Fuzzy regression analysis can be applied to many real-life problems in which the strict assumptions of classical regression analysis cannot be satisfied. In regression analysis, the errors between a regression model and observed data are generally assumed as an error that is a random variable having a normal distribution, constant variance, and a zero mean. In fuzzy regression analysis, the same unfitted errors are viewed as fuzziness. Fuzzy regression can be quite useful in estimating the relationship among variables in which data are imprecise and fuzzy. It gives a fuzzy functional relationship between dependent and independent variables where vagueness is present in some form. There are three situations in which fuzzy analysis can be viewed: (i) crisp parameters and fuzzy data, (ii) fuzzy parameters and crisp data, (iii) fuzzy parameters and fuzzy data. The fuzzy regression method is based on minimizing fuzziness as an optimal criterion, which can be achieved by linear programming procedures (Kumar et al., 2014a).

Machine-learning techniques offer many methodologies such as decision tree (DT) induction algorithms, genetic algorithms, neural networks, rough sets, and fuzzy sets as well as many hybridized strategies for classification and prediction (Komorowski et al., 1999; Witten and Frank, 1999; Han and Kamber, 2000; Pujari, 2001; Jain et al., 2009). Decision tree induction represents a simple and powerful method of classification that generates a tree and a set of rules, representing a model of different classes from a given dataset. A DT is a flowchart-like tree structure, in which each internal node denotes a test on an attribute, each branch represents an outcome of the test, and each leaf node represents the class. The topmost node in a tree is the root node. The DT ID3 algorithm and its successor, the C4.5 algorithm (Quinlan, 1993), are widely used. One of the strengths of DTs over other methods of induction is the ease with which they can be used for both numeric and nonnumeric domains. Another advantage of a DT is that it can easily be mapped to rules. ANNs are attractive tools (machine-learning techniques) for forecasting and classification purposes. ANNs are data-driven self-adaptive methods in that there are few *a priori* assumptions about the models designed for the problems under study. These learn from examples and capture subtle functional relationships from the data even if the underlying relationships are unknown or hard to describe. After learning from the available data, ANNs can often correctly infer the unseen part of a population even if the data contain noisy information. As forecasting is performed *via* prediction of future behavior (unseen part) based on the examples of past behavior (seen part), it is an ideal application area for ANNs, at least in principle (De wolf and Francl, 2000; Laxmi and Kumar, 2011a; Laxmi and Kumar, 2011b). However, the technique requires a large database.

For qualitative data or mere availability of epidemic status, logistic regression models could be developed to forewarn about the occurrence of diseases by categorizing the data into two or three categories – occurrence and nonoccurrence as 1/0 (two categories) or low, medium, and high severity as 0, 1, 2 (three categories), etc. – of diseases and insect pests (Agrawal et al., 2014).

SHORT-RANGE WEATHER FORECASTING FROM AGROMET STATION OBSERVATIONS USING A GENETIC ALGORITHM – A CASE STUDY

A study has been conducted to develop a nonlinear data-adaptive technique based on a genetic algorithm (GA) for the prediction of routine weather parameters (Kumar et al., 2014b) using observations from satellite-linked agromet stations (AMSs) from a network set up by the Indian Space Research Organization (ISRO). The algorithm produces equations that best describe the temporal evolutions of daily minimum, maximum near-surface air temperature, relative humidity, and daily averaged wind speed at selected AMS locations. These enable forecasts of these weather parameters, which could have possible use in location-specific insect pest or disease forecast models. The forecast equations developed in the present study use only past observations of these parameters. This approach, unlike other prediction methods, provides an explicit analytical forecast equation for each parameter. Predictions up to 3 days in advance have been validated using independent datasets, unknown to the training algorithm, with impressive results. The power of the algorithm has also been demonstrated by its superiority over persistent forecasts used as benchmarks.

FORECASTING PODFLY IN LATE PIGEONPEA – A CASE STUDY

Qualitative and quantitative models have been developed for damage due to podfly on late-maturing pigeonpea in Kanpur. Historical data from 1987–1988 to 2009–2010 on percent pod damage and weekly weather variables were considered for model fitting. Weather-based indices were generated and used as explanatory variables. Models were validated on subsequent (2010–2011 and 2011–2012) data and found to be satisfactory for both qualitative (epidemic/nonepidemic year) and quantitative (extent of damage) forewarning of damage due to podfly in late pigeonpea at Kanpur. The pest data comprised percent pod damage due to podfly from 1987–1988 to 2011–2012 for Kanpur. The data pertaining to weather variables (maximum and minimum temperature, morning and afternoon relative humidity, and rainfall: X_1 to X_5) for the period 1987–1988 to 2011–2012 were considered as explanatory variables in the study.

MODEL FOR QUALITATIVE DATA – LOGISTIC MODEL

Pest population/damage below the threshold value can be taken as the nonepizootic status of a pest, whereas that above the threshold value can be taken as the epizootic status of a pest. Accordingly, quantitative data on percent pod damage due to podfly (Y) has been converted into two categories (i.e., a dichotomous form $Y' = 0$ for the nonepizootic status and $Y' = 1$ for the epizootic status). To forecast the epizootic status of crops, a logistic regression model was developed. Pest epizootic status (Y') was considered the dependent variable and weather indices as explanatory variables. For each weather variable two indices were developed, one as a simple total of values of weather variables in different weeks and the other as a weighted total, weights being correlation coefficients between percent pod damage and weather variables in respective weeks. The first index represents the total number of weather variables received by the crop during the period under consideration while the other takes care of the distribution of weather variables with special reference to their importance in different weeks in relation to the extent of damage. On similar lines, indices were computed using the products of weather variables (taken two at a time) for joint effects. The model used was

$$\mathrm{P}(Y' = 1) = \frac{1}{1 + \mathrm{e}^{-\mathrm{L}}} + \varepsilon$$

$$L = a_0 + \sum_{i=1}^{p} \sum_{j=0}^{1} a_{ij} Z_{ij} + \sum_{i \neq i'}^{p} \sum_{j=0}^{1} b_{ii'j} Z_{ii'j} + \varepsilon$$

where L is a linear function of the weather indices defined as follows:

$$Z_{ij} = \sum_{w=n_1}^{n_2} r_{iw}^{j} X_{iw}$$

$$Z_{ii'j} = \sum_{w=n_1}^{n_2} r_{ii'w}^{j} X_{iw} X_{i'w}$$

where X_{iw} is the value of the ith weather variable in the wth week; r_{iw} is the correlation coefficient between Y and X_{iw}; $r_{ii'w}$ is the correlation coefficient between Y and the product of X_{iw} and $X_{i'w}$; p is the number of weather variables considered; n_1 is the initial week for which weather data were included in the model; n_2 is the final week for which weather data were included in the model; and ε is the error term.

Forecast/Prediction was obtained using the following rule:

$P > 0.5$ possibility of a pest epizootic
$P \leq 0.5$ possibility of nonoccurrence of a pest epizootic

MODELS FOR QUANTITATIVE DATA

A quantitative model has been developed that takes percent pod damage as the dependent variable and weather indices (as defined in Section 3.1) as explanatory variables. The form of the model is

$$Y = a_0 + \sum_{i=1}^{p}\sum_{j=0}^{1} a_{ij} Z_{ij} + \sum_{i \neq i'}^{p}\sum_{j=0}^{1} b_{ii'j} Z_{ii'j} + \varepsilon$$

The performance evaluation measure considered is mean absolute percentage error (MAPE)

$$\text{MAPE} = \frac{1}{m}\sum_{i=1}^{m}\left|\frac{Y_i - F_i}{Y_i}\right| \times 100$$

where Y_i and F_i are observed and forecast values, respectively; and m is the number of observations for which forecasts were worked out.

QUALITATIVE MODEL RESULTS

Data in quantitative form were classified into two categories by taking epizootic status as 1 for pod damage greater than 15% and 0 otherwise. A logistic regression model was used to develop a model forewarning about epizootic status. With a view to forecasting epizootic status well in advance, a forecast was attempted by models based on different groups of weeks starting with the 45th standard meteorological week (SMW) and taking weather data for the period 1987–1988 to 2009–2010. Weather indices were considered independent variables, while qualitative data – comprising two classes (*viz.* 1 epizootic status and 0 nonepizootic status) for podfly in pigeonpea – were used as the dependent variable. Forecasts along with probabilities have been obtained for subsequent years, which was not included in model development. Linear functions (to be used in logistic models) along with forewarning at different weeks of forecast are presented in Table 14.1.

Perusal of the table indicates that the models predicted the status correctly for both groups of years (2010–2011 and 2011–2012) at all weeks of forecasts (1st SMW to 4th SMW).

TABLE 14.1 Models Forewarning About Epizootic Status

Data used (SMW)	Linear function	Epizootic status for 2010–2011			Epizootic status for 2011–2012		
		Forewarning		Observed status	Forewarning		Observed status
		Probability	Status		Probability	Status	
45–1	$L = -2.475 - 0.016\ Z_{141}$	0.905	1	1	0.931	1	1
45–2	$L = -1.802 - 0.014\ Z_{141}$	0.803	1	1	0.998	1	1
45–3	$L = 1.368 - 0.008\ Z_{141}$	0.819	1	1	0.912	1	1
45–4	$L = 3.128 + 0.007\ Z_{141}$	0.883	1	1	0.972	1	1

Therefore, reliable predicting of epizootic status can be provided using data from the 45th to the 1st SMW. A weighted index of the interaction between maximum temperature and evening relative humidity was found to be important. Using this model, predictions for years 1987–1988 to 2009–2010 and forecasts for years 2010–2011 and 2011–2012 are presented in Table 14.2. The results indicate that for 86% of years the approach has provided correct epizootic status.

QUANTITATIVE MODEL RESULTS

Weather variables for the period from 1987–1988 to 2009–2010 were considered for developing models to forecast percent pod damage due to podfly. Weather index-based regression models were developed using weather indices as independent variables, while percent pod damage was used as the dependent variable. Forecasts have been obtained for subsequent years, something that was not included in model development. Models that can be used to forecast percent pod damage due to podfly at different weeks of forecast are presented in Table 14.3.

The coefficient of determination (R^2) was 0.52 at the 1st SMW, which increased to 0.76 at the 4th SMW. Therefore, the model at the 4th SMW was used to forecast percent pod damage. In this model, both unweighted and weighted indices of the interaction between maximum temperature and evening relative humidity and a weighted index of the interaction between maximum and minimum temperature were found important. Forecasts for subsequent years (not included in model development) using data up to the 4th SMW were 38.18 and 35.1 against observed values of 31.7 and 32.5 for years 2010–2011 and 2011–2012, respectively. For studying the stability, models were fitted by deleting each observation one by one and forecasts for deleted observations were obtained. Predictions (using models based on complete data) and forecasts (based on models using data excluding the year of forecast) are given in Table 14.4.

The results indicate that percent deviation of forecasts from observed values was low (below 20%) for most years. The MAPE of forecasts was obtained as 17.4, which was quite reasonable for quantitative forewarning about the pest. Predictions and forecasts were in close agreement, thus indicating the stability of the model.

TABLE 14.2 Predictions and Forecasts – Qualitative Model

			Predictions/forecasts	
Year	**Percent pod damage**	**Observed status**	**Probability**	**Status**
1987	22.1	1	0.95	1
1988	21.2	1	0.97	1
1989	20.0	1	0.94	1
1990	20.2	1	0.85	1
1991	24.6	1	0.94	1
1992	21.6	1	0.98	1
1993	10.0	0	0.93	1
1994	42.5	1	0.95	1
1995	32.6	1	0.89	1
1996	35.0	1	0.48	0
1997	26.9	1	0.99	1
1998	22.3	1	0.81	1
1999	10.3	0	0.32	0
2001	14.0	0	0.11	0
2002	20.0	1	0.38	0
2003	18.0	1	0.50	1
2007	24.7	1	0.97	1
2008	34.7	1	0.98	1
2009	34.3	1	0.99	1
2010	31.7	1	0.99	1
2011	32.5	1	0.96	1

TABLE 14.3 Models for Forecasting Percent Pod Damage at Different Weeks of Forecast

Data used (SMW)	Model	R^2
45–1	$Y = 20.40 + 0.078\ Z_{121} + 0.0197\ Z_{141}$	0.52
45–2	$Y = 21.27023 - 0.70425\ Z_{20} + 0.23411\ Z_{121}$	0.56
45–3	$Y = 5.278 + 0.0013\ Z_{140} + 0.0708\ Z_{121} + 0.0316\ Z_{141}$	0.70
45–4	$Y = -2.755 + 0.00097\ Z_{140} + 0.0829\ Z_{121} + 0.028\ Z_{141}$	0.76

TABLE 14.4 Predictions and Forecasts – Quantitative Model

Year	Pod damage (observed, %)	Prediction	Forecast	Deviation from observed (%)
1987	22.1	24.64	25.01	13.2
1988	21.2	21.50	23.68	11.7
1989	20.0	19.01	19.11	4.5
1990	20.2	21.91	21.66	7.2
1991	24.6	20.77	21.31	13.4
1992	21.6	26.48	26.53	22.8
1993	10.0	20.72	19.90	99.0
1994	42.5	38.22	40.22	5.4
1995	32.6	27.38	30.34	6.9
1996	35.0	27.86	26.96	23.0
1997	26.9	25.05	24.20	10.0
1998	22.3	24.71	23.31	4.5
1999	10.3	11.75	8.70	15.5
2001	14.0	14.41	12.39	11.5
2002	20.0	26.32	27.66	38.3
2003	18.0	17.79	17.61	2.2
2007	24.7	24.03	21.68	12.2
2008	34.7	22.79	22.51	35.1
2009	34.3	29.55	32.96	3.9
2010	31.7	37.74	37.82	19.3
2011	32.5	36.47	34.27	5.4

MAPE = 17.4.

WHY USE A COMPUTER-BASED DECISION SUPPORT SYSTEM?

Forecast information needs to be simple and user friendly. Interpretation and use of prediction models is difficult for anyone lacking proper statistical knowledge. Further, keeping in mind the need to help plant researchers, extension personnel, and farmers in the forecasting of disease or insect pests and timely application of management interventions, computer and web-based systems have been developed. The software uses statistical models to predict disease or insect pest occurrence well in advance of the time of actual arrival on the crop. The software architecture can be multilayered: *viz.* Client Side Interface Layer (CSIL, using HTML and JavaScript with forms for accepting information from the user and validating those forms), Server Side Application Layer (SSAL), and Database Layer (DBL, by using the MS-SQL Server database for storing user information such as login name and login password).

Online decision support systems are used worldwide to forecast different diseases: *viz.* tan spot, *Septoria* leaf blotch, leaf rust, and *Fusarium* head blight/scab diseases of wheat (http://www.ag.ndsu.edu/ndsuag/features/small-grain-disease-forecasting), canola light leaf spot (*Pyrenopeziza brassicae*) (http://www.rothamsted.bbsrc.ac.uk/Research/Centres/Content.php?Section=Leafspot&Page=llsforecast), and phoma stem canker (*Leptosphaeria maculans*, *Leptosphaeria biglobosa*) (http://www.rothamsted.bbsrc.ac.uk/Research/Centres/Content.php?Section=Leafspot&Page=phomaforecast). With support from ICAR, we too have designed and implemented web-based forecast software (Kumar et al., 2012; www.drmr.res.in/aphid-forecast/index.php) for prediction of mustard aphid (*Lipaphis erysimi*) on oilseed brassicas for different locations (Bharatpur, Morena, Hisar, Ludhiana, Pantnagar, Berhampur) in India. This software, developed by deploying ubiquitous unbeatable open-source LAMP technology, uses weather parameters as independent variables to predict crop age at time of first appearance of aphid on crop, peak number of aphid, and crop age at peak population as dependent variables, well ahead of actual occurrence of the event. Online evaluation of the system is in process and initial user response has been very positive due to effective forecasts and a user-friendly interface. Parallel to this, we have also developed a computerized image-based Rapeseed Mustard Disease Identification and Management (RMDI&M) expert software to identify and manage oilseed *Brassica* diseases, which provides necessary support for the mustard grower. Similarly, GrapeNet is another forecasting system success story in which the National Research Centre (NRC) for Grapes (Indian Council of Agricultural Research, ICAR) issues (i) a crop protection advisory to farmers associated with the Grape Growers Association and the Agricultural and Processed Food Products Export Development Agency (APEDA, http://www.apeda.gov.in/apedawebsite/Grapenet/GrapeNet_new.htm) and (ii) monitors chemical pesticide residue irrespective of the need-based application made. These systems can be used on any machine that has access to the internet.

WHY USE REMOTE SENSING IN THE FORECASTING OF CROP PESTS?

We have discussed the success of regional or location-specific networks or models in vogue to forecast the occurrence of different plant diseases and insect pests. These are all based on observations recorded at surface level for disease and meteorological factors. Data recorded at surface meteorological observatories remain valid for a maximum radius of 75 km. In order to cover all agroecological zones of India (3,287,240 km^2), weather data recording from at least 1200 observatories would be required (presently ~1000 meteorological observatories are functional in India with some linked through satellite communication, but they are not uniformly distributed) on top of multiyear observations on disease epidemics/insect pest epizootics in those locations. In India, 70% of farmers holding land have plots that average 0.39 ha (some as little as 20 × 20 m) and only 1% of crop growers hold >10 ha (mean: 17.3 ha). The patchiness of disease incidence could pose problems for proper assessment. Thus, such an exercise could be highly time consuming and labor intensive for the seventh largest country in the world as a consequence of its difficult terrain, 66% gross crop area under food crops, shortage of skilled manpower, and shrinking resources. Remote sensing overcomes such limitations with its ability to access all parts of the country and can often achieve high spatial resolution (5 × 5 m by multispectral Linear Imaging Self-Scanner, LISS IV, at 25-day intervals), thus leading to accurate estimation of the area affected. Further, in case of forecast failures, accurate assessment of damage as a result of pests is possible so that adequate compensation can be provided to farmers. Of course, RS does involve standardization of techniques based on groundtruthing and, hence, could be crosschecked with the actual ground situation.

The first multispectral airborne study for the identification of plant diseases using RS in India was conducted jointly by ICAR and the ISRO. The study identified coconut wilt using aerial false color photography (Dakshinamurti et al., 1971). At present, RS (ISRO) data are also being used to generate weather forecasts, providing crop estimates in terms of net sown area and yield, issued in operational mode for the last few years with reasonable accuracy for such crops as rice, wheat, mustard, potato, etc. thanks to collaboration between ICAR and ISRO at the Space Applications Centre (SAC) under the aegis of the Department of Agriculture and Cooperation (DAC), Ministry of Agriculture (MoA)-funded project FASAL (forecasting agricultural output using space, agrometeorological, and land-based observations). Mustard production has been forecast at the national level under FASAL using multidate temporal AWiFS (Advanced Wide Field Sensor on IRS-P6; 56 × 56 m) and radar data. Two forecasts were made during the growing season at different crop growth stages. Encouraged by these successes, the India Meteorological Department (GoI) envisages implementation of FASAL initially at 46 centers, which is likely to be extended to 130 in due course (IMD, 2014).

The success of RS at forecasting plant diseases depends on high-resolution (1–5 m) multispectral/hyperspectral or microwave observations from a satellite platform. Research efforts have been made to apply or refine ground-based models using satellite-based spatial weather and high-resolution RS observations for mustard aphid infestation (Bhattacharya et al., 2007b; Dutta et al., 2008). Use of RS and a geographic information system (GIS) could be explored to analyze satellite-based agromet data products, map the geographical distribution of pests, and delineate hotspot zones. Superimposition of causative abiotic and biotic factors on visual pest maps can be useful for disease and insect pest forecasting. Since diseased plants increase reflectance, particularly in chlorophyll absorption bands (0.5–0.7 μm) and water absorption bands (1.45–1.95 μm), forecasting plant disease is possible by RS. Though information on this aspect is scant, disease severity assessment and yield loss estimation using the changed reflectance pattern of diseased plants can be attempted. RS (greenness vegetation index derived from Landsat MSS digital data, four bands) has been successfully used to distinguish healthy wheat in India from diseased wheat in Pakistan. Favorable weather from January to April and a sudden rise in temperature at mid-April are the main causes for yellow rust disease. Routine surface monitoring of the weather and by remote sensing could help predict epidemics/epizootics long before first appearance of the disease in the crop, advantageous to making accurate decisions related to pest (including disease) management. Similarly, preparation of mustard crop masks, mapping the spatial distribution (population density) of aphid-growing zones and growth prediction, and provision of dates of severe pest infestation (peak population) at each grid level in the Bharatpur region of Rajasthan state (Bhattacharya et al., 2007b; Dutta et al., 2008) has been possible. It has also been possible to detect *Sclerotinia* rot–affected mustard using RS technology (Dutta et al., 2006; Bhattacharya et al., 2007a; Bhattacharya and Chattopadhyay, 2013). Such positive experiences augur well for any future endeavor.

However, the potential benefits of short to medium–range weather forecasts from numerical weather prediction (NWP) models or future climate projections have been least harnessed in India for regional crop protection services. The recent momentum for assimilating more satellite-based spatiotemporal atmospheric and land surface products (Bhattacharya et al., 2013; Kumar et al., 2013a; Bhattacharya et al., 2014) from Indian geostationary satellites (Kalpana-1, Insat 3A) for high-resolution (5–15 km) weather forecasts from advanced NWP models such as that of the WRF (Weather Research and Forecasters) is encouraging. Such regular high-resolution forecast products are available to registered users (http://www.mosdac.gov.in). The Mahalanobis National Crop Forecasting Centre (MNCFC) has been set up within the Indian Agricultural Research Institute (IARI) campus in New Delhi and has started functioning under the behest of MoA (DAC). Therefore, an integrated decision support system (IDSS) for crop protection services (Fig. 14.1) can be envisaged. It would likely evolve in a phased manner and comprise the following three components: (i) operationally, it would involve periodic production of alarm zones encompassing 127 agroclimatic zones by well-tested models, weather forecasts, high-resolution RS data, and operational crop maps within the GIS framework. (ii) Research priorities such as (a) development of forecast models for major diseases and insect pests with large-scale applicability, validation in farmers' fields, and model refinements; (b) evaluation and improvement of the quality of well-validated satellite-based products, improved data assimilation approaches; (c) field to satellite-based RS with high-resolution observations to differentiate crops, phenological stages within the crop growth period, biotic stresses from abiotic stresses (moisture and nutrients), normal health; and (d) development of models for minor insect pests and diseases in view of climate change. (iii) Human resource renewal such as (a) creation of experts who can handle spatial data, who are brave enough to think differently, who are bold enough to believe that as a team they could bring positive change in the present practices of integrated pest management, and who are talented enough

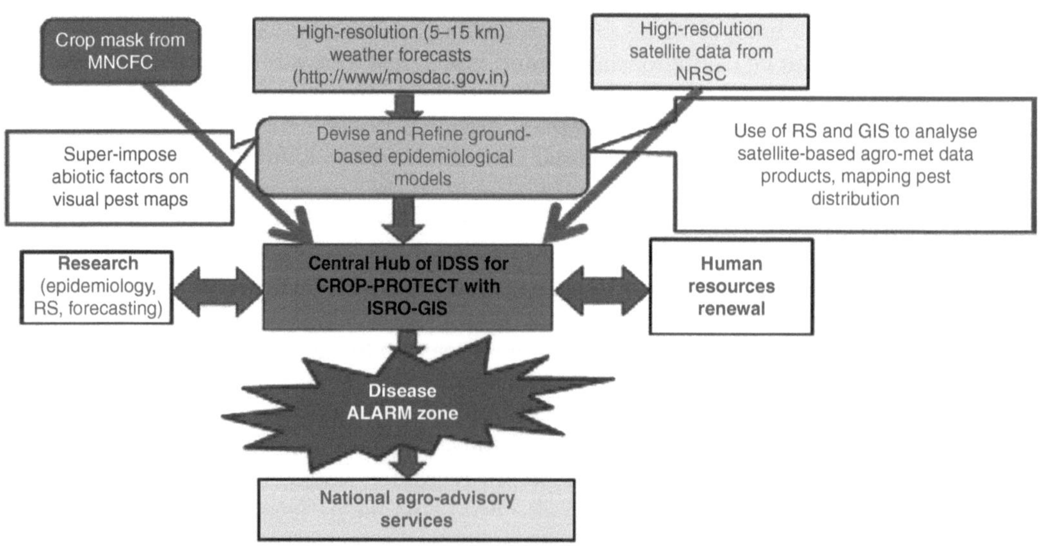

FIGURE 14.1 **Proposed architecture of IDSS for crop protection services.**

to do it; (b) familiarization of policy makers with more digital products for interpretation; and (c) providing interactive services for regular alert dissemination and getting feedback from farmers through the Village Resources Centre (VRC) network using satellite communication. From the VRC, alerts can also be disseminated to farmers through the India Meteorological Department (IMD)'s growing network of AMFUs (agromet field units), through mobile users, and through such geoportals as Bhuvan. To facilitate this, ISRO collaborated with IMD and ICAR partners such as the National Centre for Integrated Pest Management (NCIPM) and the All India Coordinated Research Project on Agrometeorology (AICRPAM) to form three working groups tasked with using satellite data to (i) improve weather forecasting, (ii) develop agromet products, and (iii) develop value-added products for dissemination.

COPING WITH CLIMATE CHANGE AND SUSTAINING ACCURATE FORECASTS

As a result of climate change, current models and those still to be born need to be oriented to dynamic mode. The models already developed are based on some meteorological, disease, or insect pest observations recorded in the past and on previous correlations between weather and insect pests or disease. However, as a result of climate change, the pest–weather relationship is bound to change, as will the behavior of hosts, newer varieties, cropping practices, etc. (Chakraborty et al., 2008). Dynamic models incorporate the recorded data of each crop season for a particular disease or insect pest to suitably revise itself and thus remains stable, relevant enough to continue providing accurate forecast (Kumar et al., 2013b, 2014d).

The decisions made by farmers are of vital importance to good crop yields. Forecast weather products and area-wide weather networks are becoming more prevalent (Ghosh et al., 2012; Chattopadhyay et al., 2012). The challenge now is to bring about continuous improvement in the productivity, profitability, stability, and sustainability of major farming systems, wherein scientific management of plant diseases holds a pivotal role (Swaminathan, 1995). Crop loss models, representing the dynamic interaction between pathogen or insect pest and host, are essential to forecasting losses due to plant diseases and insect pests. Accurate information concerning possible yield losses due to occurrence of a disease or insect pest is needed by growers or plant protection specialists so that the most cost-effective control measures can be put in place. With ever increasing concern for a cleaner environment and discouragement of chemical pesticide use, there is a need to approach disease and insect pest management by accumulating as much knowledge on the dynamics of plant pests as possible (Zadoks, 1985). Thus, future research and education in crop protection need to include this aspect of pest management; funding for this would certainly be less than many ambitious studies.

References

Agrawal, R., Mehta, S.C., Bhar, L.M., Kumar, A. 2004. Development of weather-based forewarning system for crop pests and diseases. Report, MM-III-17, NATP, IASRI, ICAR.

Agrawal, R., Kumar, Amrender, Singh, S.K., 2014. Forecasting podfly in late pigeonpea. Indian J. Agr. Sci. 84, 214–217.

Bhattacharya, B.K., Chattopadhyay, C., 2013. A multi-stage tracking for mustard rot disease combining surface meteorology and satellite remote sensing. Comput. Electron. Agr. 90, 35–44.

Bhattacharya, B.K., Chattopadhyay, C., Dutta, S., Meena, R.L., Roy, S., Parihar, J.S., Patel, N.K., 2007b. Detection of Sclerotinia rot incidence in Indian mustard from polar orbiting satellite platform. In: *Proceedings, 12th International Rapeseed Congress, Vol. IV* (Plant Protection), Wuhan, China, Mar 26–30, 2007 (Eds. Fu Tingdong, Guan Chunyun), Science Press USA Inc., pp. 79–83.

Bhattacharya, B.K., Dutta, S., Dadhwal, V.K., Parihara, J.S., Chattopadhyay, C., Agrawal, R., Kumar, V., Khan, S.A., Roy, S., Shekhar, C., 2007b. Predicting aphid (*Lipaphis erysimi*) growth in oilseed *Brassica* using near surface meteorological data from NOAA TOVS – a case study. Int. J. Remote Sens. 28, 3759–3773.

Bhattacharya, B.K., Nigam, R., Kumar, P., Gairola, R.M., Pandya, M.R., Tripathy, R., Dutta, S., Vyas, S., 2014. Sat-met products and their demonstrated agro-meteorological applications over India. SAC/EPSA/ABHG/CAD/SR/02/2014.

Bourke, P.M.A., 1970. Use of weather information in the prediction of plant disease epiphytotics. Ann. Rev. Phytopathol. 8, 345–369.

Chakraborty, S., Luck, J., Hollaway, G., Freeman, A., Norton, R., Garrett, K.A., Percy, K., Hopkins, A., Davis, C., Karnosky, D.F., 2008. Impacts of global change on diseases of agricultural crops and forest trees. CAB Rev. 3, 1–15.

Chattopadhyay, C., Agrawal, R., Kumar, A., Bhar, L.M., Meena, P.D., Meena, R.L., Khan, S.A., Chattopadhyay, A.K., Awasthi, R.P., Singh, S.N., Chakravarthy, N.V.K., Kumar, A., Singh, R.B., Bhunia, C.K., 2005a. Epidemiology and forecasting of Alternaria blight of oilseed *Brassica* in India – a case study. Zeitschrift für Pflanzenkrankheiten und Pflanzenschutz 112, 351–365.

Chattopadhyay, C., Agrawal, R., Kumar, A., Singh, Y.P., Roy, S.K., Khan, S.A., Bhar, L.M., Chakravarthy, N.V.K., Srivastava, A., Patel, B.S., Srivastava, B., Singh, C.P., Mehta, S.C., 2005b. Forecasting of *Lipaphis erysimi* on oilseed Brassicas in India – a case study. Crop Prot. 24, 1042–1053.

Chattopadhyay, C., Agrawal, R., Kumar, A., Meena, R.L., Faujdar, K., Chakravarthy, N.V.K., Kumar, A., Goyal, P., Meena, P.D., Shekhar, C., 2011. Epidemiology and development of forecasting models for white rust of *Brassica juncea* in India. Arch. Phytopathology Plant Protect. 44, 751–763.

Chattopadhyay, C., Bhattacharya, B.K., Kumar, V., Kumar, A., Meena, P.D., 2012. Impact of climate change on pests and diseases of oilseeds *Brassica* – the scenario unfolding in India. J. Oilseed Brassica 2, 48–55.

Dakshinamurti, C., Krishnamurthy, B., Summanwar, A.S., Shanta, P., Pisharoty, P.R., 1971. Remote sensing for coconut wilt. In: Proc. 6th Intl. Symposium on Remote Sensing of Environment. Environmental Res. Inst., Michigan, Ann Arbor, USA. pp. 25–29.

De Wolf, E.D., Francl, L.J., 2000. Neural network classification of tan spot and Stagonospora blotch infection periods in a wheat field environment. Phytopathology 90, 108–113.

Desai, A.G., Chattopadhyay, C., Agrawal, R., Kumar, A., Meena, R.L., Meena, P.D., Sharma, K.C., Rao, M.S., Prasad, Y.G., Ramakrishna, Y.S., 2004. *Brassica juncea* powdery mildew epidemiology and weather-based forecasting models for India – a case study. Zeitschrift für Pflanzenkrankheiten und Pflanzenschutz 111, 429–438.

Dutta, S., Bhattacharya, B.K., Rajak, D.R., Chattopadhyay, C., Patel, N.K., Parihar, J.S., 2006. Disease detection in mustard crop using EO-1 hyperion satellite data. J. Indian Soc. Remote Sensing (Photonirvachak) 34, 325–329.

Dutta, S., Bhattacharya, B.K., Rajak, D.R., Chattopadhyay, C., Dadhwal, V.K., Patel, N.K., Parihar, J.S., Verma, R.S., 2008. Modelling regional level spatial distribution of aphid (*Lipaphis erysimi*) growth in Indian mustard using satellite-based remote sensing data. Int. J. Pest. Manage. 54, 51–62.

Flor, H.H., 1942. Inheritance of pathogenicity in *Melampsora lini*. Phytopathology 32, 653–669.

Ghosh, A., Das, A., Bhattacharya, B.K., Kumar, V., Kumar, A., Meena, P.D., Chattopadhyay, C., 2012. Impact of Climate Change on Pests and Diseases – the Scenario Unfolding in Indian Agriculture. SATSA Mukhopatra Annual Technical Issue 16, 15–29.

Gregory, P.H., 1968. Interpreting plant disease dispersal gradients. Ann. Rev. Phytopathol. 6, 189–212.

Han, J., Kamber, M., 2000. Data Mining: Concepts and Techniques. Morgan Kaufmann Publisher, California, USA.

Hirst, J.M., 1952. An automatic volumetric spore trap. Annals Appl. Biol. 39, 257–265.

IMD (India Meteorological Department), 2014. Future projections in AAS. http://www.imd.gov.in/services/agri_new/Activities.htm [accessed 15.10.2014].

Jain, R., Minz, S., Ramasubramanian, V., 2009. Machine learning for forewarning crop diseases. J. Indian Soc. Agril. Stat. 63, 97–107.

Komorowski, J., Pawlak, Z., Polkowki, L., Skowron, A., 1999. Rough sets: a tutorial. In: Pal, S.K., Skowron, A. (Eds.), Rough Fuzzy Hybridization. Springer-Verlag, Singapore, pp. 3–99.

Kumar, V., Kumar, A., Chattopadhyay, C., 2012. Design and implementation of web-based aphid (*Lipaphis erysimi*) forecast system for oilseed Brassicas. Indian J. Agr. Sci. 82, 608–614.

Kumar, P., Bhattacharya, B.K., Pal, P.K., 2013a. Impact of vegetation fraction from Indian Geostationary Satellite on hort-range weather forecast. Agric. For. Meteorol. 168, 82–92.

Kumar, A., Kumar, V., Bhattacharya, B.K., Singh, N., Chattopadhyay, C., 2013b. Integrated disease management: need for climate-resilient technologies. J. Mycol. Plant Pathol. 43, 28–36.

Kumar, A., Srinivas, P.S., Mishra, A.K., Chandrasekhran, H., 2014a. Fuzzy regression interval models for forewarning onion thrips. IEEE – International Conference on Computing for Sustainable Global Development (INDIACom), pp. 197–201.

Kumar, P., Bhattacharya, B.K., Nigam, R., Kishtawal, C.M., Pal, P.K., 2014b. Impact of Kalpana-1 derived land surface albedo on short-range weather forecasting over the Indian subcontinent. J. Geophys. Res. Atmos. 119, 2764–2780.

Kumar, A., Chattopadhyay, C., Singh, K.N., Vennila, S., Rao, V.U.M., 2014c. Trend analysis of climatic variables in pigeonpea growing regions in India. Mausam 65, 161–170.

Laxmi, R.R., Kumar, A., 2011a. Forecasting of powdery mildew in mustard (*Brassica juncea*) crop using artificial neural networks approach. Indian J. Agric. Sci. 81, 855–860.

Laxmi, R.R., Kumar, A., 2011b. Weather based forecasting model for crops yield using neural network approaches. Stat. Appl. 9 (1&2 New Series), 55–69.

Mehta, K.C., 1933. Rusts of wheat and barley in India – a study of their annual recurrence, life- histories and physiologic forms. Indian J. Agric. Sci. 3, 939–962.

Nagarajan, S., Singh, D.V., 1990. Long-distance dispersion of rust pathogens. Ann. Rev. Phytopathol. 28, 139–153.

Pujari, A.K., 2001. Data Mining Techniques. Universities Press, Hyderabad, India.

Quinlan, J.R., 1993. C4.5: Programs for Machine Learning. Morgan Kauffman, San Mateo, CA, USA.

Rao, B.P., Ramaraj, A.P., Chattopadhyay, C., Prasad, Y.G., Rao, V.U.M., 2012. Predictive model for mustard aphid infestation for eastern plains of Rajasthan. J. Agrometeorology 14, 60–62.

Rao, B.B., Rao, V.U.M., Nair, L., Prasad, Y.G., Ramaraj, A.P., Chattopadhyay, C., 2013. Assessing aphid infestation in Indian mustard (*Brassica juncea* L.) under present and future climate scenarios. Bangladesh J. Agril. Res. 38, 373–387.

Rao, B.B., Rao, V.U.M., Nair, L., Prasad, Y.G., Chattopadhyay, C., 2014. Mustard aphid infestation in India: development of forewarning models. J. Environ. Biol. 35, 683–688.

Schröeder, H., 1960. Dispersal by air and water – the flight and landing. In: Horsfall, J.G., Dimond, A.E. (Eds), *Plant Pathology – An Advanced Treatise, vol. 3*, pp. 169–227.

Swaminathan, M.S., 1995. Foreword. In: Nagarajan, S., Muralidharan, K. (Eds.), Dynamics of Plant Diseases. Allied Publ. Ltd., New Delhi, p. v.

van Everdingen, E., 1926. Het verband de wears gesterheid ende aardappelzeiekte (*Phytophthora infestans*). Tijdschr. Plantennziekten. 32, 129–140.

Vanderplank, J.E., 1963. Plant Disease – Epidemics and Control. Academic Press, New York, pp. 349.

Waggoner, P.E., Horsfall, J.G., 1969. Epidem – a simulator of plant disease written for a computer. In: *Bull. Conn. Agric. Expt. Stn. No. 698*, Connecticut, USA, pp. 80.

Witten, I.H., Frank, E., 1999. Data Mining: Practical Machine Learning Tools and Techniques with Java Implementations. Morgan Kaufmann Publishers, California, USA.

Zadoks, J.C., 1985. On the conceptual basis of crop loss assessment: the threshold theory. Ann. Rev. Phytopathol. 23, 455–473.

CHAPTER

15

Designer Oil Crops

Mukhlesur Rahman, Monika Michalak de Jiménez

Department of Plant Sciences, North Dakota State University, Fargo, ND, USA

INTRODUCTION

Oilseed crops, such as soybean (*Glycine max* L.), rapeseed/canola (*Brassica napus* L.), sunflower (*Helianthus annuus* L.), peanut (*Arachis hypogaea* L.), safflower (*Carthamus tinctorius* L.), flax (*Linum usitatissimum* L.), cotton (*Gossypium hirsutum* L.), castor (*Ricinus communis* L.), sesame (*Sesamum indicum* L.) etc., are predominantly grown for the oil contained in their seeds. The major oilseed producing countries in the world are the United States, Brazil, China, Argentina, and India. Oil crops are primarily used as a source of edible oil and as protein-rich meal for livestock feed; however, they are also utilized to produce pharmaceuticals, surfactants, plasticizers, emulsifiers, detergents, lubricants, adhesives, cosmetics, oleochemicals, fuels, etc. (Metzger and Bornscheuer, 2006, FAOSTAT, 2014). Some of the vegetable oils, such as sunflower, rapeseed, soybean, and palm oil, have been validated as potential oils for biodiesel (Moser, 2008; Yin et al., 2008; Pleanjai and Gheewala, 2009; Qiu et al., 2011). Due to their application in biodiesel production, oilseed crops have received more attention in recent decades, which has resulted in an increased cultivation area as well as an increased production due to improved yields per unit area (Lu et al., 2011).

Between 1983 and 2013 there was an 82% increase in oilseed crop cultivation area (161 million hectares to 294 million hectares) (Fig. 15.1). Similarly, average yield (t/ha) increased about 48% (1.6–2.4 t/ha). A threefold increase in the total world oilseed production was also observed – from 53 million tons in 1983 to 197 million tons in 2013. Figure 15.1 depicts the increasing trend of world oilseed production through its expansion in cultivated area, and increased yield, which is mostly a result of intensive breeding efforts. By comparison, without a significant change in cultivation area, the total world production of cereal crops as well as average seed yields increased 71 and 57%, respectively (FAOSTAT, 2014).

Although overall oilseed production increased about threefold in the last 30 years, the production growth trends are variable for individual oilseed crops. Table 15.1 illustrates the changes in production, cultivation area, and average seed or fruit yield of the major oilseed crops. Oil palm has the highest production increase followed by rapeseed, soybean,

Breeding Oilseed Crops for Sustainable Production. http://dx.doi.org/10.1016/B978-0-12-801309-0.00015-X

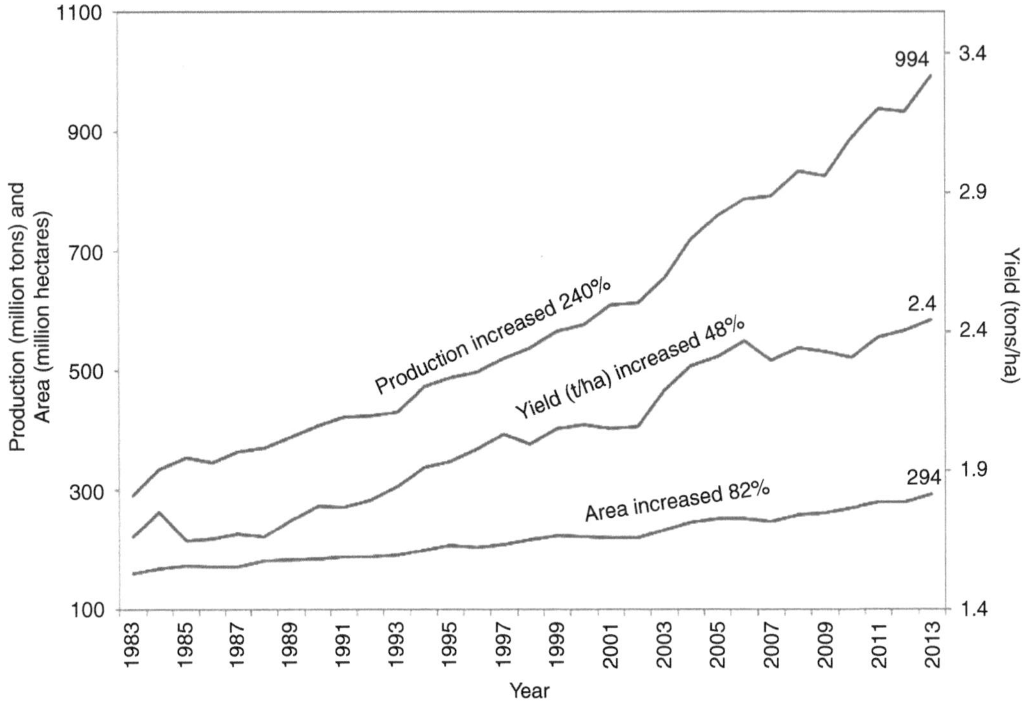

FIGURE 15.1 World trends in oil crop production, yield, and growing area over the last 30 years. *Source: FAOSTAT, 2014*

and sunflower. On the other hand the production of linseed and safflower declined. Sharp increases in cultivation area from the 5-year average between 1979 and 1983 to the 5-year average between 2009 and 2013 were 297, 177, 106, and 96% for oil palm, rapeseed, soybean, and sunflower, respectively (Table 15.1). Again, cultivation area declined for linseed and safflower.

Although, both production and cultivation area decreased for linseed, it had the highest increase in seed yield per unit area in comparison with other oilseed crops. A high change of seed yield per unit area was noticeable for castor oil seed, oil palm, and rapeseed, whereas moderate progress was identified in sesame seed, groundnut, seed cotton, mustard, and soybean.

Oilseed crop consumption has increased drastically (Table 15.2) in the last decade due to an increased demand for food applications as well as a rising demand for biodiesel and other nonfood industrial products (Lu et al., 2011). For example, between 2000 and 2010, there was a drastic increase (from 1 to 7 million metric tons) in nonfood utilization of rapeseed, mainly for biodiesel production (Lu et al., 2011).

The biggest challenge for the vegetable oil industry is to meet the growing demand for oil by producing affordable and sufficient amounts of plant-derived oils (Lu et al., 2011). Sustainable oilseed production, a production that satisfies the requirement of human beings, has become a very pressing subject. Sustainable production can be defined as "industrial processes

TABLE 15.1 Changes in Major World Oilseed Production, Cultivation Area, and Average Seed or Fruit Yield over the Last 30 Years

	Production (million tons)			Area (million hectares)			Yield (kg/ha)		
Crops	**1979–1983**	**2009–2013**	**Change (%)**	**1979–1983**	**2009–2013**	**Change (%)**	**1979–1983**	**2009–2013**	**Change (%)**
Soybean	86.0	253.5	195	50.7	104.4	106	1,696	2427	43
Oil palm fruit	31.4	243.1	673	4.2	16.6	297	7514	14,630	95
Seed cotton	43.3	71.7	66	33.6	33.8	0	1290	2123	65
Rapeseed	12.7	64.5	410	12.2	33.7	177	1036	1915	85
Coconut	32.7	60.9	86	8.9	11.9	34	3691	5114	39
Groundnut	18.6	41.2	122	18.6	24.8	33	995	1658	67
Sunflower seed	15.1	37.5	149	12.6	24.8	96	1192	1509	27
Olive	9.6	19.2	100	5.4	10.0	84	1792	1918	7
Sesame seed	1.9	4.5	131	6.3	8.4	35	309	529	71
Linseed	2.5	2.1	-17	5.3	2.2	-59	475	973	105
Castor oil seed	0.9	2.0	121	1.5	1.6	8	590	1197	103
Safflower seed	0.9	0.7	-24	1.3	0.8	-38	693	843	22
Mustard seed	0.3	0.6	123	0.5	0.7	44	525	806	54

Source: FAOSTAT (2014).

TABLE 15.2 World Consumption of Vegetable Oils (Million Metric Tons) from 1995 to 2013, by Oil Type

	Year									
Vegetable oil	**1996**	**2006**	**2007**	**2008**	**2009**	**2010**	**2011**	**2012**	**2013**	**2014**
Palm oil	15.8	35.3	37.7	39.4	42.1	44.5	47.9	50.9	54.3	56.3
Soybean oil	19.7	33.6	35.6	37.5	36.0	38.3	40.7	41.3	42.5	44.4
Canola oil	11.2	16.9	17.6	18.3	20.1	22.4	23.5	23.8	23.6	24.5
Sunflower oil	8.6	9.8	10.3	9.4	10.6	11.4	11.8	13.1	13.6	14.5
Other	16.4	19.8	20.0	20.5	20.9	21.2	22.5	23.7	23.9	24.2

Source: Statista (2014; http://www.statista.com/statistics/263937/vegetable-oils-global-consumption/).

that transform natural resources into products that society needs in ways that minimize resources and energy used, the wastes produced, and the effects of work practices and wastes on communities" (as described by the Australian Government Department of the Environment and Heritage; http://www.environment.gov.au/system/files/resources/1b93d012-6dfb-4ceb-a37f-209a27dca0e0/files/sustainable-future.pdf). Therefore, a goal of sustainable production in agriculture is to meet human needs, both nutritional as well as nonfood, with an emphasis on environmental protection. In order to reach and maintain sustainable oil production, classic breeding efforts need to be coupled with biotechnological approaches. The expansion of oilseed growing areas could be another approach to meeting increased demand. Current acreage could be expanded by incorporating the use of less fertile lands receiving lower rainfall and by modifying existing crops to successfully grow in less optimal conditions. Alternatively, new oilseed crops, adapted to marginal land could be utilized (Lu et al., 2011). Furthermore, current knowledge of oilseed metabolism, as well as the availability of genetic or metabolic engineering technology, can help in the creation of designer oilseed crops containing more desirable fatty acid profiles with a superior nutritional value as well as providing new applications for the nonfood industry (Lu et al., 2011).

The following sections discuss different goals for designer oil crops as well as giving multiple examples of engineered plants in an effort to meet an always increasing demand for oil and oilseed crops.

BIOTECHNOLOGY AND METABOLIC ENGINEERING OF DESIGNER OIL CROPS

Genetic engineering gave us various tools to introduce a wide choice of genes into a desired plant in a precise genomic region without or with minimal negative effects. Therefore, in the future the application of biotechnology in agriculture will be a key driver for sustainable food production (Chilton, 2014). In recent years, biotechnology and metabolic engineering has developed significantly and allowed for the development of new, improved crops. For example, technology has created an opportunity to modify genes in the metabolic pathways of fatty acid biosynthesis and to produce desired fatty acids with C8–C24 chain lengths within a specific oil crop, which can then widely be used in both food and nonfood industries. A similar genetic engineering approach has been applied in multiple oilseed crops, such as, soybean, rapeseed, sunflower, and cotton, to modify the fatty acid composition, to improve the oil quality, and to increase oil content in seeds in order to fulfil the world demand for oil.

Vegetable Oil and Its Fatty Acid Composition

Vegetable oils are composed almost entirely of triacylglycerides (TAG), formed through the esterification of different fatty acids in three different positions of the glycerol backbone (Fig. 15.2). Therefore, the quality of vegetable oil completely depends on the composition of fatty acids in TAG, also referred to as the fatty acid profile. The major fatty acid constituents of vegetable oils are palmitic acid (C16:0), stearic acid (C18:0), oleic acid (C18:1), linoleic acid (C18:2), linolenic acid (C18:3), arachidic acid (C20:0), eicosenoic acid (C20:1), and erucic acid (C22:1). Fatty acids are classified into two types – saturated and unsaturated fatty acids. Chemically, the saturated fatty acids do not have double bonds between carbon atoms, whereas, unsaturated fatty acids contain at least one double bond with hydrogen. Depending on the number of such double bonds the fatty acids are referred to as monounsaturated or polyunsaturated.

Table 15.3 depicts the dietary fat composition of different vegetable oils, including saturated (e.g., palmitic acid, stearic acid), monounsaturated (e.g., oleic acid, ecosenoic acid, erucic acid), and polyunsaturated fatty acids (PUFA, e.g., linoleic acid, linolenic acid).

In the human diet, high levels of saturated fats have been positively correlated with cardiovascular disease (Siri-Tarino et al., 2010). Oleic acid is a monounsaturated fatty acid. High oleic acid content in seed oil increases the thermostability of the oil, making it more suitable as a cooking oil and therefore a desirable target for quality improvement (Tanhuanpää et al., 1998). Oleic acid also plays an important role in the biosynthesis of ecosenoic, erucic, linoleic, and linolenic acid (Rakow and McGregor, 1973). Erucic acid is one of the major fatty acids in rapeseed oil. Low erucic acid rapeseed oil is a very digestible and healthy vegetable oil (Beare et al., 1963). In contrast, a high level of high erucic acid rapeseed (HEAR) oil intake in food is associated with fibrotic myocardium and increased blood cholesterol levels (Gopalan et al., 1974). Nevertheless, HEAR oil has several applications in the oleochemical industry for the production of high-temperature lubricants, nylons, plastics, slip and coating agents, soaps, painting inks, as well as surfactants (Princen and Rothfus, 1984; Topfer et al., 1995). Linolenic acid is a PUFA, and is one of the essential fatty acids that reduces plasma cholesterol levels in the human body (Eskin et al., 1996), controls blood clotting, brain cell membrane

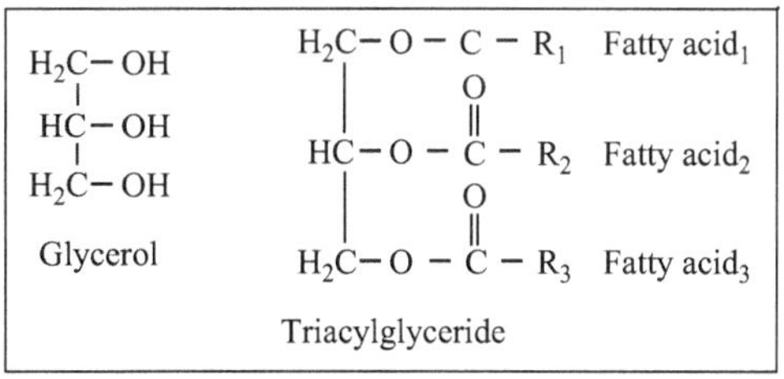

FIGURE 15.2 **Molecular structures of glycerol and TAG.**

formation, reduces inflammation, and helps lowering the risk of heart disease, cancer, arthritis, etc. (Fares et al., 2014). Deficiency of omega-3 fatty acids may produce fatigue, poor memory, dry skin, mood swings, or depression, etc. (http://umm.edu/health/medical/altmed/supplement/omega3-fatty-acids). Because linolenic acid contains three double bonds it can be easily oxidized, causing rancidity and flavors which taste off, as well as shortening the shelf life and frying life of the oil (Eskin et al., 1989; Prevot et al., 1990; Przybylski et al., 1993). Partial hydrogenation of the double bond in the *cis* configuration of unsaturated fatty acids (e.g., linolenic acid) generates trans fatty acids, raises serum low-density lipoprotein cholesterol (bad cholesterol) and lowers high-density lipoprotein cholesterol (good cholesterol) in humans and is directly associated with the increased risk of coronary heart disease (Zock et al., 1998). Linoleic acid (omega-6) and linolenic acid (omega-3) are essential in our daily diet since the human body cannot synthesize them. Linoleic acid is involved in the biosynthesis of signaling molecules in the body (Neitzel, 2010). It was found that a deficiency of linoleic acid in the diet causes mild skin scaling, hair loss (Cunnane and Anderson, 1997), and poor wound healing in rats (Ruthig and Meckling-Gill, 1999). Western diets contain high linoleic to linolenic acid ratios (~ 20:1), whereas a desired ratio should be around 2:1. Oil crops are a good source of both linoleic and linolenic acid. However, only canola oil contains roughly a 2:1 ratio of linoleic to linolenic acid (Table 15.3).

Increasing Seed Oil Content Through Breeding and Genetic Engineering

Multiple oil crop species (candlenut, sesame, oiticica and ucuhuba) contain up to 60–76% oil in their seeds (Murphy, 1996). Therefore, there is a potential to increase oil content in other oil crops. Some predictions state that the oil content of rapeseed, which is currently 45–48%

TABLE 15.3 Comparison of Dietary Fats in Different Vegetable Oils

Dietary fat	Saturated fat	Linoleic acid	α-Linolenic acid	Monounsaturated fat
Canola oil	7	21	11	61
Safflower oil	10	76	Trace	14
Sunflower oil	12	71	1	16
Corn oil	13	57	1	29
Olive oil	15	9	1	75
Soybean oil	15	54	8	23
Peanut oil	19	33	Trace	48
Cottonseed oil	27	54	Trace	19
Lard	43	9	1	47
Beef tallow	48	2	1	49
Palm oil	51	10	Trace	39
Butterfat	68	3	1	28
Coconut oil	91	2	0	7

Source: POS Pilot Plant Corporation Saskatoon, Saskatchewan, Canada, June, 1994.

in Canada and around 42% in China and Australia, might even reach 65% (Shen et al., 2007; Seberry et al., 2008; Wang, 2010). Recently Hu et al. (2013) obtained the ultrahigh oil content rapeseed line YN171 with a 64.8% oil content. The study of the YN171 line of *B. napus*, focusing on the structural analysis of its seeds, indicated a high positive correlation between the oil body organelle to seed ratio and oil content of the seed. Hu et al. (2013) estimated that rapeseed oil content could even reach 75%. Unfortunately, some studies reported an inverse relationship between oil and protein accumulation in the seeds of some species, such as rapeseed and soybean (Chung et al., 2003; Cober and Voldeng, 2000; Hu et al., 2013). Additionally, Vollman and Rajcan (2009) noted other traits also correlated with oil content, such as time to flowering, seed weight, and fatty acid concentrations, which complicate the process of oil crop breeding.

Genetic engineering studies, with goals to increase seed oil content, have focused on acyltransferases in the TAG synthesis pathway (Lardizabal et al., 2008; Maisonneuve et al., 2009; Weselake et al., 2009). The role of acyltransferases in TAG synthesis has been previously established when *Arabidopsis* diacylglycerolacyltransferase 1 (DGAT1) mutants showed lowered seed oil content (Katavic et al., 1995; Zou et al., 1999), and later on the overexpression of DGAT1 caused an increased oil content in *Arabidopsis* and maize (Zheng et al., 2008).

Lardizabal et al. (2008) reported transgenic soybean with increased oil without a negative impact on protein content or yield. Overexpression of a "codon-optimized version of a DGAT2A" originating from soil fungus resulted in a 1.5% increase in total seed oil. Similarly, expression of yeast sphingolipid compensation genes in soybean showed an average 1.5% increase in TAG and a 3.2% increase of total seed oil content in seeds (Rao and Hildebrand, 2009).

Kelly et al. (2013) reported a novel method of suppression of lipolysis for enhancing oil yield in *B. napus* which can be also utilized in other oilseed crops. The inhibition of lipolysis was facilitated by RNAi silencing of the SUGAR-DEPENDENT1 triacylglycerol lipase gene family during seed development that resulted in an 8% increase in oil yield with minimal adverse effects on seed vigor. Recently, van Erp et al. (2014) conveyed a study in *Arabidopsis* which involved multigene engineering of triacylglycerol metabolism. In the study seed-specific overexpression of WRINKLED1 and DGAT1 together with suppression of the triacylglycerol lipase SUGAR-DEPENDENT1 yielded higher seed oil content than changing the three genes separately.

Different studies also explored the option of increasing plant oil yield for the purpose of biodiesel production by enhancing accumulation of TAG in leaves and other vegetative tissues (Durrett et al., 2008). Some increased oil level in leaves was achieved using different genetic engineering methods (Andrianov et al., 2010; Vanhercke et al., 2013; Winichayakul et al., 2013; Fan et al., 2013), however, using integrated metabolic approaches it was possible to achieve over 15% increase in oil content of dry weight vegetative tissue (Kelly et al., 2013; Vanhercke et al., 2013).

Genetic Engineering of Fatty Acid Profiles

Since fatty acid composition of oil is a major determinant of seed quality (Ohlrogge and Browse, 1995), oilseed research has mainly focused on increasing oil content as well as designing its fatty acid composition. Several genes in the fatty acid biosynthesis pathway, such as thioesterases, transacylases, desaturases, and desaturase-related enzymes, were cloned and transformed into oilseed crops leading to the production of designer oil crops (Lessire, 2005).

Vegetable oil can be improved with the development of high oleic, low linolenic–acid lines. Scarth and McVetty (1999) discussed that canola oil containing 75% oleic acid and 4% linolenic acid improved the quality of edible oil while at the same time allowing its use in the oleochemical industries. High oleic acid lines of safflower were reported in the past to have 64–83% (of total fatty acids) oleic acid content (Knowles and Hill, 1964). A safflower germplasm with even higher oleic acid (86–91% of total fatty acids) was later on identified by Fernandez-Martınez et al. (1993). Hamdan et al. (2009) found that the recessive ol gene, identified as oleoylphosphatidylcholinedesaturase FAD2-1, together with modifying genes was responsible for the very high level of the fatty acid. Interestingly, the modifying these genes could also reduce oleic acid below expected levels in lines with the ol alleles (Hamdan et al., 2009). A negative effect of modifying genes on the level of oleic acid has also been discovered in sunflower (Urie, 1985; Fernandez-Martınez et al., 1989)

The soybean fatty acid profile, with 24% monounsaturated fats, is less desirable than the fatty acid composition of canola and olive oil, comprising of 61% and 75% monounsaturated fats, respectively (Teres et al., 2008; White, 2007). Haun et al. (2014) generated a high oleic acid soybean line using a targeted mutagenesis approach. The new line produced 80% of the oleic acid which is four times higher than that produced by the wild type. The mutant line was created by introducing small deletions of coding sequence in two fatty acid desaturase 2 genes (FAD2-1A and FAD2-1B), facilitating conversion of the monounsaturated fat, oleic acid, to a polyunsaturated fat, linoleic acid. The mutated high oleic acid soybean lines are not transgenic because they do not contain any foreign DNA, therefore, they can be introduced into any breeding program, even in countries with stringent genetically modified organism (GMO) policies. High oleic, low linoleic–acid lines have also been generated in canola and some *Brassica* species using a similar approach of suppressing the activity of the FAD2 enzyme (Velasco et al., 2003; Sivaraman et al., 2004; Peng et al., 2010). Peng et al. (2010) reported that a single transformation with a silencing transcript generated a canola line with 85% (of total fats) oleic acid, lowered PUFAs content (10%), and with undetectable levels of erucic acid. Hybrid F_1 seeds generated from the reciprocal crossing of the transgenic line and parents with diverse genetic backgrounds had 80% oleic acid, around 10% PUFAs, as well as very low or undetectable levels of erucic acid (Peng et al., 2010).

A higher content of stearic acid in vegetable oil is beneficial for solid fat applications due to its cooking, as well as health, properties. Several soybean lines were generated with high stearic acid (9–29% stearic acid) mostly via mutagenesis (Pantalone et al., 2002). Recently, Ruddle et al. (2013) reported a novel mutation in a 9-stearoyl-ACP-desaturase in soybean as a new source of high stearic acid, determined its relationship with another high stearic locus, and provided molecular markers that can be used in breeding programs. Typically, soybean oil contains 2–5 % stearic acid (ILSI, 2010; USDA-ARS, 2012), whereas it is 14.6% in the mutant line (Ruddle et al., 2013). Unfortunately, most of the new high stearic acid lines have been associated with agronomic problems, such as lower seed yield, poor germination, and reduced seedling growth rate (Lundeen et al., 1987; Rahman et al., 1997; Wang et al., 2001). In cotton, Liu et al. (2002) utilized hairpin RNA-mediated, posttranscriptional gene silencing to modify the stearic and oleic acid content. The genetically engineered cotton plants increased stearic acid from their normal level of 3% to as high as 40%, and oleic acid from 15% to 77%.

Calgene, a biotechnology company, has developed the first engineered high lauric acid rapeseed. Lauric oil is primarily utilized in the production of soaps and detergents (Kumar, 1998).

This fatty acid is normally not present in this crop. The high level of lauric acid was achieved by introducing a gene encoding specific C12 thioesterase (Kumar, 1998). Since, lauric acid originates from coconut and palm tree, both grown in tropical regions, the genetically engineered high lauric rapeseed provided an opportunity to produce crops with fatty acids in temperate regions as well.

Other Modifications

The goal of creating designer oil crops is not only limited to the production of superior lines with high oil contents having desirable fatty acid compositions. Other modifications are related to the usage of oilseeds as a protein meal or for nonfood products. These changes target meal quality traits such as higher protein content, protein composition, as well as a decreased content of antinutritional compounds. The quality of oil crops for meal depends on many components such as protein content, fiber content, and the level of antinutritional components in the seed.

Oilseed Crop Meals

Oil meals, also referred to as oil cakes, are byproducts of the oil extraction process from oilseed crops (Ramachandran et al., 2007). Since oilseed meals are rich in protein they are highly nutritious and are used as animal feed for ruminants, poultry, and fish (Ramachandran et al., 2007). Additionally, the meals are rich in macronutrients (nitrogen, phosphorus, potassium, calcium, magnesium, and sulfur), as well as micronutrients (zinc, iron, manganese, and boron). Protein meals are also highly decomposable and their properties improve plant nitrogen intake from the soil, making them very useful as fertilizers, especially for use in organic agriculture (Moore, 2001).

INCREASING PROTEIN AND DESIRABLE AMINO ACID CONTENT

Seed protein content is one of the most important characteristic of meal, therefore, higher protein contents increase the value of oil crops. However, the major challenge in breeding for higher protein content in a seed is that often the trait is negatively correlated with seed yield (Vollman and Rajcan, 2009) as well as with oil accumulation in seeds (Chung et al., 2003; Cober and Voldeng, 2000; Hu et al., 2013). Soybean is a great source of protein that varies from 35% to 50% depending on the genotype and growing conditions (Krishnan, 2005). However, the nutritional quality of soybean proteins is not optimal due to a low level of sulfur amino acids, such as methionine and cysteine, which cannot be normally synthesized by monogastric animals (Krishnan, 2005). Even though traditional breeding managed to increase total seed protein content, it was unsuccessful at increasing the level of the essential sulfur amino acids (Jez and Krishnan, 2009). The application of genetic engineering approaches could increase sulfur-rich proteins in soybean by a small amount, using the expression of heterologous proteins, such as methionine-rich zeins and Brazil Nut Albumin (BNA). Unfortunately, with the increase in the heterologous sulfur amino acids in transgenic soybean there was a decrease in the endogenous sulphur-rich proteins (Streit et al., 2001), suggesting that the introduction of sulfur-rich proteins is not a viable solution to enhance the nutritional value of soybean. Other studies reported the assimilation of sulfur can also take place in developing soybean seeds (Sexton and Shibles, 1999;Chronis and Krishnan, 2004) with two key enzymes, serine acetyl transferase (SATase) and OASTL, involved in cysteine synthesis (Leustek et al., 2000; Noji and Saito, 2002). Previous studies focused on generating transgenic soybean plants expressing high levels of lysine and tryptophan (Falco et al., 1995; Rapp et al., 2003). However, to increase the cysteine level, or any other amino acids level, the key enzymes involved in their biosynthesis pathway need to be engineered to be expressed in developing seeds (Krishnan, 2005).

DECREASING THE LEVEL OF ANTINUTRITIONAL COMPOUNDS IN MEAL

Factors which negatively affect the quality of oilseed meals are referred to as antinutritional compounds. Glucosinolates, fiber, sinapine, present in *B. napus*, are some examples of such compounds (Hannoufa et al., 2014).

Glucosinolates belong to a family consisting of over 100 aliphatic and aromatic forms in Brassicaceae. Glucosinolates adversely affect the nutritional value, taste, and smell of food and animal feed (Hansen et al., 1997). However, it has been shown that higher intake of glucosinolates may also have positive effects on humans, such as an anticancer effect (Cartea et al., 2008; Jeffery and Araya, 2009) as well as acting as an antioxidant (Jeffery and Araya, 2009; Traka and Mithen, 2009). Low and ultralow glucosinolate cultivars of rapeseed (10–30 umol/g) are currently available (Uppström, 1995; Khajali and Slominski, 2012). However, the low glucosinolate level varieties have only a reduced level of aliphatic glucosilates, therefore, there is still a need to reduce aromatic forms. Few metabolic engineering studies targeted lowering indole-glocosinolates (Chavadej et al., 1994; Mikkelsen et al., 2002, 2003), however, Chavadej et al. (1994) managed to successfully transform *B. napus* with tryptophan decarboxylase, decreasing the level of aromatic indole-glucosinolate.

Sinapine is a phenolic compound found in *B. napus* seeds (Clauss et al., 2011) associated with a bitter flavor in feed, leading to a poor feed taste for livestock. It is also known to accumulate in the milk and meat of animals fed on *B. napus* (Ismail et al., 1981). Large amounts of this compound may also cause growth and reproductive problems (Pearson et al., 1980). Sinapine in meal can be reduced using chemical treatments, however, due to the associated costs of meal processing creation of a low-sinapine *B. napus* germplasm would be a more economical approach (Hüsken et al., 2005). A study by Nair et al. (2000), utilizing silencing of the ferulic acid 5-hydroxylase(FAH) in *B. napus* yielded a 40 % reduction in sinapine content in seeds. An even higher reduction was obtained by the suppression of UDP-glucose: sinapate glucosyltransferase (Hüsken et al., 2005). Bhinu et al. (2009a) achieved a 90% reduction of sinapine by simultaneous down-regulation FAH and sinapoyl glucose:choline sinapoyl transferase (SCT) in *B. napus* seeds.

Dietary fiber content is variable in different oil crops. A significant portion of dietary fiber is in the form of indigestible lignin (Bell, 1995). Therefore, lower fiber content in seed meal is related to its improved digestibility and nutritional value (Slominski et al., 1994). *B. napus* contains around 8% (of oil-free meal) lignins, which is relatively high compared with other oilseed crops (Hannoufa et al., 2014). Different breeding efforts yielded the production of the yellow-seeded *B. napus* line with lower lignin content due to a thinner seed hull (Relf-Eckstein et al., 2007; Akhov et al., 2009). Additionally, biotechnological approaches, utilizing RNAi silencing, allowed Bhinu et al. (2009b) to generate transgenic *B. napus* lines with decreased lignin contents up to 40% relative to wild types.

Designer Vegetable Oil–derived Biodiesel

Rapeseed and soybean oil showed the potential to be successfully used for biodiesel production (Demirbas, 2002; Yin et al., 2008; Pleanjai and Gheewala, 2009). In order to improve vegetable oil for biodiesel production there is a need to enhance the oxidative stability of the oil as well as its lubricity (Kinney and Clemente, 2005). Soybean is an attractive crop for biodiesel production because it produces more usable energy and less greenhouse gases than corn-based ethanol (Hill et al., 2006). However, soybean oil has comparatively high oxidative reactivity, due to its PUFA content, which negatively impacts the storage life of the soybean-derived biodiesel, leading to the formation of compounds which can clog fuel filters (Canakci et al., 1999; Mittelbach and Gangl, 2001). It was shown that decreasing linoleic and linolenic acid content in the oil would improve stability of the vegetable oil (Kinney and Clemente, 2005). It has been also established, through standard oxidation measurements, that the high oleic and low palmitic oil content of the oil makes it oxidatively more stable than the traditional soybean oil (Kinney and Knowlton, 1997). Severe reduction in PUFAs paralleled with an increase in oleic acid content was achieved in soybean where the Fad2-1 gene was down-regulated blocking the flux from oleic acid into PUFA increasing at the same time the content of monounsaturated fatty acids (Kinney, 1998). High oleic acid (84–88%) transgenic soybean was also generated using sense Fad2-1 by means of microprojectile bombardment (Kinney and Knowlton, 1997). Importantly, the newly modified line of soybean did not show any negative impact on agronomic traits in field trials (Kinney and Knowlton, 1997).

It is also worth mentioning that both the content and composition of tocopherols in vegetable oil also contributes to its stability. These lipid-soluble antioxidants add to the nutritional value as well as to the oxidative stability of the oil (Hunter and Cahoon, 2007). Transgenic soybean, which expressed four different transgenes, yielded more than a 10-fold increase in vitamin E-type molecules, referred to as tocotrienols (Karunanandaa et al., 2005). Additionally, transgenic expression of homogentisate geranylgeranyl transferase (HGGT) in soybean caused tocotrienol accumulation as well as a sixfold to tenfold increase in the total tocopherol and tocotrienol content of soybean seeds (Cahoon et al., 2006; Meyer, 2007). Another transgenic approach is to focus on introducing tocopherol cyclases from *Arabidopsis* and *Zea mays* to *B. napus*. The overexpression of the cyclases in seeds of *B. napus* led to a 28% increase in total tocopherols and the overexpression of the *Arabidopsis* c-tocopherol methyl transferase in *Brassica juncea* yielded a sixfold increase in α-tocopherol (Kumar et al., 2005; Yusuf and Sarin, 2007).

The lubricity of the vegetable oil is the second component which affects the quality of biodiesel fuel (Kinney and Clemente, 2005). It has been shown that a higher content of ricinoleic acid facilitates a better lubricity of the oil. Production of soybean oil with 10–20% ricinoleic acid and 70–80% oleic acid would be a realistic target which would enhance the performance of biodiesel fuel. Since ricinoleic acid is made from oleic acid using an enzyme called oleate 12-hydroxylase (Bafor et al., 1991), this enzyme was targeted and a transgenic soybean-carrying castor-bean oleate 12-hydroxylase gene has been generated (Kinney and Clemente, 2005).

CONCLUSIONS

Oilseed crops are primarily grown for edible oil. Recently, oilseeds attracted more attention due to an increasing demand for their healthy vegetable oils, livestock feeds, pharmaceuticals, biofuels, and other oleochemical industrial uses. The increased interest resulted in an 82% expansion of oilseed crop cultivation areas and about a 240% increase in total world production over the last 30 years. Therefore, to satisfy the increasing world demand, sustainable oil production, through classic breeding efforts needs to be coupled with biotechnological approaches in order to expand oil yield per unit area. The expansion of oilseed growing areas can be another approach utilized to meet this increased demand. Genetic engineering of oilseeds will allow not only the sustainable production of oilseed crops but also enhanced nutritional value as well as enhanced quality for industrial purposes. TAGs, composed of various fatty acids, are the main component of vegetable oil. Many genes in TAG biosynthesis pathways have been identified and studied well. New biotechnology methods allow insertion or modification of genes involved in the biosynthesis of a desired fatty acid, in order to accumulate a higher level of fatty acid or even to produce a novel fatty acid. Genetic engineering started a new era for designer oil crops and has created opportunities for sustainable oilseed crop production around the world.

References

Andrianov, V., Borisjuk, N., Pogrebnyak, N., Brinker, A., Dixon, J., Spitsin, S., Flynn, J., Matyszczuk, P., Andryszak, K., Laurelli, M., Golovkin, M., Koprowski, H., 2010. Tobacco as a production platform for biofuel: overexpression of *Arabidopsis* DGAT and LEC2genes increases accumulation and shifts the composition of lipids in green biomass. Plant Biotechnol. J. 8, 277–287.

Akhov, L., Ashe, P., Tan, Y., Datla, R., Selvaraj, G., 2009. Proanthocyanidin biosynthesis in the seed coat of yellow seeded, canola quality *Brassica napus* YN01-429 is constrained at the committed step catalyzed by dihydroflavonol 4-reductase. Botany 87, 616–625.

Bafor, M., Smith, M.A., Jonson, L., Stobart, A.K., Stymne, S., 1991. Ricinoleic acid biosynthesis and triacylglycerol assembly in microsomal preparations from developing castor-bean (*R. communis*) endosperm. Biochem. J. 280, 507–514.

Bell, J.M., 1995. Meal and by-product utilization in animal nutrition. In: Kimber, D., McGregor, D.I. (Eds.), Brassica oilseeds, production and utilization. CAB International, Oxon, UK, pp. 301–337.

Beare, J.L., Campbell, J.A., Youngs, C.G., 1963. Effects of saturated fat in rats fed rapeseed oil. Can. J. Biochem. Physiol. 41, 605–612.

Bhinu, V.S., Schafer, U.A., Li, R., Huang, J., Hannoufa, A., 2009a. Targeted modulation of sinapine biosynthesis pathway for seed quality improvement in *Brassica napus*. Transgenic Res. 18, 31–44.

Bhinu, V.S., Li, R., Huang, J., Kaminskyj, S., Sharpe, A., Hannoufa, A., 2009b. Perturbation of lignin biosynthesis in *Brassica napus* (canola) plants using RNAi. Can. J. Plant. Sci. 89, 441–453.

Cahoon, E.B., Coughlan, S.J., Cahoon, R.E., Butler, K.H., 2006. Compositions and methods for altering tocotrienol content. United States Patent No. 7154029.

Canakci, M., Monyem, A., Van Gerpen, J., 1999. Accelerated oxidation processes in biodiesel. Trans. ASAE 42, 1565–1572.

Cartea, M.E., Velasco, P., Obregón, S., Padilla, G., de Haro, A., 2008. Seasonal variation in glucosinolate content in *Brassica oleracea* crops grown in northwestern Spain. Phytochemistry 69, 403–410.

Chavadej, S., Brisson, N., McNeil, J.N., De Luca, V., 1994. Redirection of tryptophan leads to production of low indole glucosinolate canola. Proc. Natl. Acad. Sci. USA 91, 2166–2170.

Chilton, M.D., 2014. Genetic engineering will drive food security. Crops, Soils, Agronomy News, 18.

Chronis, D., Krishnan, H.B., 2004. Sulfur assimilation in soybean (*Glycine max* [L.] Merr.):molecular cloning and characterization of a cytosolic isoform of serine acetyltransferase. Planta 218, 417–426.

Chung, J., Babka, H.L., Graef, G.L., Staswick, P.E., Lee, D.J., Cregan, P.B., Shoemaker, R.C., Specht, J.E., 2003. The seed protein, oil, and yield QTL on soybean linkage group I. Crop Sci. 43, 1053–1067.

Clauß, K., von Roepenack-Lahaye, E., Böttcher, C., Roth, M.R., Welti, R., Erban, A., Kopka, J., Scheel, D., Milkowski, C., Strack, D., 2011. Overexpression of sinapine esterase BnSCE3 in oilseed rape seeds triggers global changes in seed metabolism. Plant Physiol. 155, 1127–1145.

Cober, E.R., Voldeng, H.D., 2000. Developing high-protein, high yield soybean populations and lines. Crop Sci. 40, 39–42.

Cunnane, S., Anderson, M., 1997. Pure linoleate deficiency in the rat: influence on growth, accumulation of n-6 polyunsaturates, and (1-14C) linoleate oxidation. J. Lipid Res. 38, 805–812.

Demirbas, A., 2002. Biodiesel from vegetable oils via transesterification in supercritical methanol. Energ. Convers. Manage. 43, 2349–2356.

Durrett, T.P., Benning, C., Ohlrogge, J., 2008. Plant triacylglycerols as feedstocks for the production of biofuels. Plant J. 54, 593–607.

van Erp, H., Kelly, A.A., Menard, G., Eastmond, P.J., 2014. Multigene engineering of triacylglycerol metabolism boosts seed oil content in *Arabidopsis*. Plant Physiol. 165, 30–36.

Eskin, N.A.M., Vaisey-Genser, M., Durance-Todd, S., Przybylski, R., 1989. Stability of low linolenic acid canola oil to frying temperatures. J. Am. Oil Chemists Soc. 66, 1081–1084.

Eskin, N.A.M., Macdonald, B.E., Przybylski, R., Malcomson, L.J., Scarth, R., Mag, T., Ward, K., Adolphe, D., 1996. Canola Oil. Edible oil and fat products: oils and oilseeds, vol. 2, fifth ed. John Wiley & Sons, Inc., New York, pp. 1–95.

Falco, S.C., Guida, T., Locke, M., Mauvais, J., Sanders, C., Ward, R.T., Webber, P., 1995. Transgenic canola and soybean seeds with increased lysine. Biotechnology 13, 577–582.

Fan, J., Yan, C., Zhang, X., Xu, C., 2013. Dual role for phospholipid: diacylglycerolacyltransferase: enhancing fatty acid synthesis and diverting fatty acids from membrane lipids to triacylglycerol in *Arabidopsis* leaves. Plant Cell 25, 3506–3518.

FAOSTAT, 2014. Food and Agriculture Organization of the United Nations Statistics Division (http://faostat3.fao.org/home/E) (accessed on 20.10.14).

Fares, H., Lavie, C.J., DiNicolantonio, J.J., O'Keefe, J.H., Milani, R.V., 2014. Omega-3 fatty acids: a growing ocean of choices. Curr. Atheroscler. Rep. 16, 389.

Fernández-Martínez, J.M., Jiménez, A., Domínguez, J., Garces, R., Mancha, M., 1989. Genetic analysis of the high oleic content in cultivated sunflower (*Helianthus annuus* L.). Euphytica 41, 39–51.

Fernández-Martínez, J.M., del Rio, M., de Haro, A., 1993. Survey of safflower (*Carthamus tinctorius* L.) germplasm for variants in fatty acid composition and other seed characters. Euphytica 69, 115–122.

Gopalan, C., Krishnamurthi, D., Shenolikar, I.S., 1974. Myocardial changes in monkeys fed mustard oil. Nutr. Metab. 16, 352–365.

Hamdan, Y.A.S., Pérez-Vich, B., Velasco, L., Fernández-Martínez, J.M., 2009. Inheritance of high oleic acid content in safflower. Euphytica 168, 61–69.

Hannoufa, A., Pillai, B.V.S., Chellamma, S., 2014. Genetic enhancement of *Brassica napus* seed quality. Transgenic Res. 23, 39–52.

Hansen, M., Lausten, A.M., Olsen, C.E., Poll, L., Sørensen, H., 1997. Chemical and sensory quality of broccoli (*Brassica oleracea* L. var. *italica*). J. Food Quality 20, 441–459.

Haun, W., Coffman, A., Benjamin, M., Demorest, Z.L., Lowy, A., Ray, E., Retterath, A., Stoddard, T., Juillerat, A., Cedrone, F., Mathis, L., Voytas, D.F., Zhang, F., 2014. Improved soybean oil quality by targeted mutagenesis of the fatty acid desaturase 2 gene family. Plant Biotech. J. 12, 934–940.

Hill, J., Nelson, E., Tilman, D., Polasky, S., Tiffany, D., 2006. Environmental, economic, and energetic costs and benefits of biodiesel and ethanol biofuels. Proc. Natl. Acad. Sci. USA 103, 11206–11210.

Hu, Z., Hua, W., Zhang, L., Deng, L., Wang, X., Liu, G., Hao, W., 2013. Seed structure characteristics to form ultrahigh oil content in rapeseed. PLOS ONE 8 (4), e62099. doi: 10.1371/journal. pone.006209.

Hunter, S.C., Cahoon, E.B., 2007. Enhancing vitamin E in oilseeds: unraveling tocopherol and tocotrienol biosynthesis. Lipids 42, 97–108.

Hüsken, A., Baumert, A., Strack, D., Becker, H.C., Möllers, C., Milkowski, C., 2005. Reduction of sinapate ester content in transgenic oilseed rape (*Brassica napus*) by dsRNAi-based suppression of BnSGT1 gene expression. Molec. Breeding 16, 127–138.

ILSI, 2010. International life sciences institute crop composition database, Version 4. 2. International Life Sciences Institute, Washington, DC, http://www.cropcomposition.org.

Ismail, F., Vaisey-Genser, M., Fyfe, B., 1981. Bitterness and astringency of sinapine and its components. J. Food Sci. 46, 1241–1244.

Jeffery, E.H., Araya, M., 2009. Physiological effects of broccoli consumption. Phytochem. Rev. 8, 283–298.

Jez, J.M., Krishnan, H.B., 2009. Sulfur assimilation and cysteine biosynthesis in soybean seeds: towards engineering sulfur amino acid content. In: Krishnan, H.B., ed. Modification of seed composition to promote health and nutrition. Agronomy Monograph 51, pp. 249–261.

Karunanandaaa, B., Qia, Q., Hao, M., Baszis, S.R., Jensen, P.K., Wong, Y.H., Jiang, J., Venkatramesh, M., Gruysa, K.J., Moshiria, F., Post-Beittenmillera, D., Weissa, J.D., Valentin, H.D., 2005. Metabolically engineered oilseed crops with enhanced seed tocopherol. Metab Eng. 7, 384–400.

Katavic, V., Reed, D.W., Taylor, D.C., Giblin, E.M., Barton, D.L., Zou, J., MacKenzie, S.L., Covello, P.S., Kunst, L., 1995. Alteration of seed fatty acid composition by an ethyl methanesulfonate-induced mutation in *Arabidopsis thaliana* affecting diacylglycerolacyltransferase activity. Plant Physiol. 108, 399–409.

Kelly, A.A., Shaw, E., Powers, S.J., 2013. Suppression of the SUGAR-DEPENDENT1 triacylglycerol lipase family during seed development enhances oil yield in oilseed rape (*Brassica napus* L.). Plant Biotechnology J. 11, 355–361.

Khajali, F., Slominski, B.A., 2012. Factors that affect the nutritive value of canola meal for poultry. Poultry Sci. 91, 2564–3257.

Kinney, A.J., 1998. Plants as industrial chemical factories – new oils from genetically engineered soybeans. Lipid/Fett 100, 173–179.

Kinney, A.J., Clemente, T.E., 2005. Modifying soybean oil for enhanced performance in biodiesel blends. Fuel Process. Technol. 86, 1137–1147.

Kinney, A.J., Knowlton, S., 1997. Designer oils for food applications. In: Harlander, S., Roller, S. (Eds.), Genetic Engineering for Food Industry: A Strategy for Food Quality Improvement. Blackie Academic, London, pp. 193–213.

Knowles, P.F., Hill, A.B., 1964. Inheritance of fatty acid content in the seed oil of a safflower introduction from Iran. Crop Sci. 4, 406–409.

Krishnan, H.B., 2005. Engineering soybean for enhanced sulfur amino acid content. Crop Sci. 45, 454–461.

Kumar, H.D., 1998. Modern concepts of biotechnology. VIKAS Publishing House Pvt. Ltd., New Delhi.

Kumar, R., Raclaru, M., Schüsseler, T., Gruber, J., Sadre, R., Lühs, W., Zarhloul, K.M., Friedt, W., Enders, D., Frentzen, M., Weier, D., 2005. Characterisation of plant tocopherolcyclases and their overexpression in transgenic *Brassica napus* seeds. FEBS Lett. 579, 1357–1364.

Lardizabal, K., Effertz, R., Levering, C., 2008. Expression of *Umbelopsisramanniana* DGAT2A in seed increases oil in soybean. Plant Physiol. 148, 89–96.

Lessire, R., 2005. Genomics applied to oilseed crops, any benefit? European J. Lipid Sci. Tech. 107, 211–212.

Leustek, T., Martin, M.N., Bick, J.-A., Davies, J.P., 2000. Pathways and regulation of sulfur metabolism revealed through molecular and genetic studies. Ann. Rev. Plant Physiol. Plant Molec. Biol. 51, 141–165.

Liu, Q., Singh, S.P., Green, A.G., 2002. High-stearic and high-oleic cottonseed oils produced by hairpin RNA-mediated post-transcriptional gene silencing. Plant Physiol. 129, 1732–1743.

Lu, C., Napier, J.A., Clemente, T.E., Cahoon, E.B., 2011. New frontiers in oilseed biotechnology: meeting the global demand for vegetable oils for food, feed, biofuel, and industrial applications. Curr. Opin. Biotechnol. 22, 252–259.

Lundeen, P.O., Fehr, W.R., Hammond, E.G., Cianzio, S.R., 1987. Association of alleles for high stearic acid with agronomic characters of soybean. Crop Sci. 27, 1102–1105.

Maisonneuve, S., Bessoule, J.-J., Lessireet, R., Delseny, M., Roscoe, T.J., 2009. Expression of rapeseed microsomal lysophosphatidic acid acyltransferaseisozymes enhances seed oil content in *Arabidopsis*. Plant Physiol. 152, 670–684.

Metzger, J.O., Bornscheuer, U., 2006. Lipids as renewable resources: Current state of chemical and biotechnological conversion and diversification. Appl. Microbiol. Biotechnol. 71, 13–22.

Meyer, K., 2007. Compositions and methods for altering alpha- and beta-tocotrienol content. World Patent Application No. WO 2007/059077.

Mikkelsen, M.D., Petersen, B.L., Olsen, C.E., et al.,2002. Biosynthesis and metabolic engineering of glucosinolates. Amino Acids 22, 279–295.

Mikkelsen, M.D., Peterse, B.L., Glawischnig, E., Jensen, A.B., Andreasson, E., Halkier, B.A., 2003. Modulation of CYP79 genes and glucosinolate profiles in *Arabidopsis* by defense signaling pathways. Plant Physiol. 131, 298–308.

Mittelbach, M., Gangl, S., 2001. Long storage stability of biodiesel made from rapeseed and used frying oil. J. Am. Oil Chem. Soc. 78, 573–577.

Moore, A., 2011. Fertilizer potential of biofuel byproducts. In: Dos Santos Bernardes, M.A. (Ed.), Biofuel production – recent developments and prospects. InTech Rijeka, Croatia.

Moser, B.R., 2008. Influence of blending canola, palm, soybean, and sunflower oil methyl esters on fuel properties of biodiesel. Energy Fuels 22, 4301–4306.

Murphy, D.J., 1996. Engineering oil production in rapeseed and other oil crops. Curr. Trends Biotechnol. 14, 206–213.

Nair, R.B., Joy, R.W., Kurylo, E., Shi, X., Schnaider, J., Datla, R.S., Keller, W.A., Selvaraj, G., 2000. Identification of a CYP84 family of cytochrome P450-dependent mono-oxygenase genes in *Brassica napus* and perturbation of their expression for engineering sinapine reduction in the seeds. Plant Physiol. 123, 1623–1634.

Neitzel, J.J., 2010. Fatty acid molecules: fundamentals and role in signaling. Nature Education 3, 57.

Noji, M., Saito, K., 2002. Molecular and biochemical analysis of serine acetyltransferaseandcysteine synthase towards sulfur metabolic engineering in plants. Amino Acids 22, 231–243.

Ohlrogge, J., Browse, J., 1995. Lipid synthesis. Plant Cell 7, 957–970.

Pantalone, V.R., Wilson, R.F., Novitzky, W.P., Burton, J.W., 2002. Genetic regulation of elevated stearic acid concentration in soybean oil. J. Am. Oil Chem. Soc. 79, 549–553.

Pearson, A.W., Butler, E.J., Fenwick, R.G., 1980. Rapeseed meal and egg taint: The role of sinapine. J. Sci. Food Agric. 31, 898–904.

Peng, Q., Hu, Y., Wei, R., Zhang, Y., Guan, C., Ruan, Y., Liu, C., 2010. Simultaneous silencing of FAD2 and FAE1 genes affects both oleic acid and erucic acid contents in *Brassica napus* seeds. Plant Cell Rep. 29, 317–325.

Pleanjai, S., Gheewala, S.H., 2009. Full chain energy analysis of biodiesel production from palm oil in Thailand. Appl. Energy 86, S209–S214.

Prevot, A., Perrin, J.L., Laclaverie, G., Auge, Ph., Coustille, J.L., 1990. A new variety of low linolenic rapeseed oil; characteristics and room-odor tests. J. Am. Oil Chem. Soc. 67, 161–164.

Princen, L.H., Rothfus, J.A., 1984. Development of new crops for industrial raw materials. J. Am. Oil Chem. Soc. 61, 281–289.

Przybylski, R., Malcolmson, L.J., Eskin, N.A.M., Durance-Tod, S., Mickle, J., Carr, R., 1993. Stability of low linolenic acid canola oil to accelerated storage at 60°C. Lebensmittel-Wissenschaft und -Technologie. Food Sci. Tech. 26, 205–220.

Qiu, F., Li, Y., Yang, D., Lib, X., Sun, P., 2011. Biodiesel production from mixed soybean oil and rapeseed oil. Appl. Energy 88, 2050–2055.

Rahman, S.M., Takagi, Y., Kinoshita, T., 1997. Genetic control of high stearic acid content in seed oil of two soybean mutants. Theor. Appl. Genet. 95, 772–776.

Rakow, G., McGregor, D.I., 1973. Opportunities and problems in modification of levels of rapeseed C18 unsaturated fatty acids. J. Am. Oil Chem. Soc. 50, 400–403.

Ramachandran, S., Singh, K.S., Larroche, C., Soccol, C.R., Pandey, A., 2007. Oil cakes and their biotechnological applications – A review. Bioresource Technol. 98, 2000–2009.

Rao, S.S., Hildebrand, D., 2009. Changes in oil content of transgenic soybeans expressing the yeast SLC1 gene. Lipids 44, 945–951.

Rapp, B.W., Weaver, L., Wang, Q., Crow, L., Ream, J., Hill, B., Brown, W., Oulmassov, T., Gruys, K., 2003. Enhancing the nutritional value of soybean. Plant Biol 2003, Annual meeting of the American Society of Plant Biologists, abstract 38001.

Relf-Eckstein, J., Raney, J. P., Rakow, G., 2007. Meal quality improvement in *Brassica napus* canola through the development of yellow-seeded germplasm. In: Proceedings of 12th International Rapeseed Congress, Wuhan, China, March 26–30, 2007. Poster presentation, vol 5, pp. 289–291.

Ruddle, II, P., Whetten, R., Cardinal, A., Upchurch, R.G., Miranda, L., 2013. Effect of a novel mutation in a Delta9-stearoyl-ACP-desaturase on soybean seed oil composition. Theor. Appl. Genet. 126, 241–249.

Ruthig, D.J., Meckling-Gill, K.A., 1999. Both (n-3) and (n-6) fatty acids stimulate wound healing in the rat intestinal epithelial cell line, IEC-6. J. Nutrition 129, 1791–1798.

Scarth, R., McVetty, P.B.E., 1999. Designer oil canola-A review of new food-grade Brassicaoils with focus on high oleic, low linolenic types. In: Proceeding of the Tenth International Conference of Rapeseed Congress. Canberra, Australia.

Seberry, D.E., Mailer, R.J., Parker, P.A., 2008. The quality of Australian canola. Australian Oilseeds Federation, NSW Dept of Primary Industries, Australia.

Sexton, P.J., Shibles, R.M., 1999. Activity of ATP sulfurylase in reproductive soybean. Crop Sci. 39, 131–135.

Siri-Tarino, P.W., Sun, Q., Hu, F.B., Krauss, R.M., 2010. Saturated fat, carbohydrate, and cardiovascular disease. Am. J. Clin. Nutr. 91, 502–509.

Sivaraman, I., Arumugam, N., Sodhi, Y.S., Gupta, V., Mukhopadhyay, A., Pradhan, A.K., Burma, P.K., Pental, D., 2004. Development of high oleic and low linoleic acid transgenics in a zero erucic acid *Brassica juncea* L. (Indian mustard) line by antisense suppression of the fad2 gene. Mol. Breeding 13, 365–375.

Shen, J.X., Fu, T.D., Tu, J.X., Ma, C., 2007. Potential in production and genetic improvement of rapeseed and prospect for rape oil-based biodiesel in China. J. Huazhong Agric. Univ. 26, 894–899.

Slominski, B.A., Campbell, L.D., Guenter, W., 1994. Carbohydrates and dietary fiber components of yellow- and brown-seeded canola. J. Agric. Food Chem. 42, 704–707.

Streit, L.G., Beach, L.R., Register, III, J.C., Jung, R., Fehr, W.R., 2001. Association of the Brazil nut protein gene and Kunitz trypsin inhibitor alleles with soybean protease inhibitor activity and agronomic traits. Crop Sci. 41, 1757–1760.

Tanhuanpää, P., Vilkki, J., Vihinen, M., 1998. Mapping and cloning of FAD2 gene to develop allele-specific PCR for oleic acid in spring turnip rape (*Brassica rapa* ssp. *oleifera*). Molec. Breeding 4, 543–550.

Teres, S., Barcelo-Coblijn, G., Benet, M., Álvarez, R., Bressani, R., Halver, J.E., Escribá, P.V., 2008. Oleic acid content is responsible for the reduction in blood pressure induced by olive oil. Proc. Natl. Acad. Sci. USA 105, 13811–13816.

Topfer, R., Nartini, N., Schell, J., 1995. Modification of plant lipid synthesis. Science 268, 681–686.

Traka, M., Mithen, R., 2009. Glucosinolates, isothiocyanates and human health. Phytochem Rev. 8, 269–282.

Uppström, B., 1995. Seed Chemistry. In: Kimber, D.S., McGregor, D.I. (Eds.), Brassica oilseeds production and utilization. Cab International, Wallingford, UK, p. 223.

Urie, A.L., 1985. Inheritance of high oleic acid in sunflower. Crop Sci. 25, 986–989.

USDA-ARS, 2012. National genetic resources program. Germplasm resources information network – GRIN. National germplasm resources laboratory, Beltsville, Maryland. http://www.ars-grin.

Vanhercke, T., El Tahchy, A., Shrestha, P., Zhou, X.R., Singh, S.P., Petrie, J.R., 2013. Synergistic effect of WRI1 and DGAT1 coexpression on triacylglycerol biosynthesis in plants. FEBS Lett. 587, 364–369.

Velasco, L., Fernández-Martínez, J.M., De Haro, A., 2003. Inheritance of increased oleic acid concentration in high-erucic acid ethiopian mustard. Crop Sci. 43, 106–109.

Vollmann, J., Rajcan, I., 2009. Oil Crop Breeding and Genetics. In: Vollmann, J., Rajcan, I. (Eds.), Oil crops, handbook of Plant Breeding 4. Springer Science+ Business Media, LLC, p. 1.

Wang, T., Harp, T., Hammond, E.G., Burrisaa, J.S., Fehr, W.R., 2001. Seed physiological performance of soybeans with altered saturated fatty acid contents. Seed Sci. Res. 11, 93–971.

Wang, H.Z., 2010. Review and future development of rapeseed industry in China. Chinese J Oil Crop Sci. 32, 300–302.

Weselake, R.J., Taylor, D.C., Rahman, M.H., Shah, S., Laroche, A., McVetty, P.B.E., Harwood, J.L., 2009. Increasing the flow of carbon into seed oil. Biotechnol. Adv. 6, 866–878.

White, P.J., 2007. Fatty acid in oilseeds (vegetable oils). In: Chow, C.K. (Ed.), Fatty Acids in Foods and Their Health Implications. CRC Press, Marcel Dekker, Inc., New York, pp. 210–263.

Winichayakul, S., Scott, R.W., Roldan, M., Hatier, J.H., Livingston, S., Cookson, R., Curran, A.C., Roberts, N.J., 2013. In vivo packaging of triacylglycerols enhances *Arabidopsis* leaf biomass and energy density. Plant Physiol. 162, 626–639.

Yin, J.Z., Xiao, M., Wang, A.Q., Xiu, Z.-L., 2008. Synthesis of biodiesel from soybean oil by coupling catalysis with subcritical methanol. Energy Conver. Manage. 49, 3512–3516.

Yusuf, M.A., Sarin, N.B., 2007. Antioxidant value addition in human diets: genetic transformation of *Brassica junceawith* gamma-TMT gene for increased alpha-tocopherol content. Transgenic Res. 16, 109–113.

Zheng, P., Allen, W.B., Roesler, K., Williams, M.E., Zhang, S., Li, J., Glassman, K., Ranch, J., Nubel, D., Solawetz, W., Bhattramakki, D., Llaca, V., Deschamps, S., Zhong, G.-Y., Tarczynski, M.C., Shen, B., 2008. A phenylalanine in DGAT is a key determinant of oil content and composition in maize. Nat. Genet. 40, 367–372.

Zock, P.L., Urgert, R., Hulshof, P.J., Katan, M.B., 1998. Dietary trans-fatty acids: a risk factor for coronary disease. Ned. Tijdschr. Geneeskd. 142, 1701–1704.

Zou, J., Wei, Y., Jako, C., Kumar, A., Selvaraj, G., Taylor, D.C., 1999. The *Arabidopsis thaliana* TAG1 mutant has a mutation in adiacylglycerolacyltransferase gene. Plant J. 19, 645–653.

CHAPTER

16

Genetic Improvement of Rapeseed Mustard through Induced Mutations

Vinod Choudhary, Sanjay J. Jambhulkar

Nuclear Agriculture & Biotechnology Division, Bhabha Atomic Research Centre, Mumbai, India

INTRODUCTION

Extensive plasticity of morphological traits in *Brassica* crops is responsible for its adaptation in to a wide range of environments. However, yield potential is still the limiting factor. Possible reasons for this could be the unexplored genetic potential, changing environmental factors imposing drought and heat conditions, and stress due to insect pests and diseases. Germplasm strengthening and the development of high-yielding varieties able to tolerate biotic and abiotic stresses and fulfil nutritional requirements are the major objectives of the rapeseed–mustard improvement program. Induced mutagenesis is one of the approaches used to widen and utilize the beneficial alleles for effective crop improvement. It has been successfully employed to enhance the production and productivity of crop plants (Chopra, 2005; Maluszynski et al., 1995). Mutagenesis has also been successfully employed in the improvement of qualitative and quantitative traits in oleiferous *Brassica* species (Robbelen, 1990; Bhatia et al., 1999). An overview of the modifications in morphological, biochemical, and yield attributes through mutagenesis, their direct and indirect use to develop high-yielding varieties, and the molecular mechanism for the selected mutations is given in this chapter, which is a revision of my previous review of the topic (Jambhulkar, 2007) in light of the mutation breeding work carried out to 2014.

MUTATIONS FOR MORPHOLOGICAL TRAITS

Morphological characteristics are phenotypic markers and could be used for purity maintenance of genotype and linkage mapping. *Brassica* crops consist of a large spectrum of variation for morphological characteristics. However, desirable mutations could be induced

Breeding Oilseed Crops for Sustainable Production. http://dx.doi.org/10.1016/B978-0-12-801309-0.00016-1

for a specific characteristic without affecting the rest of the genome and could be used in genetic, biochemical, and molecular analysis as well as for the development of high-yielding varieties.

Chlorophyll Mutations

Chlorophyll mutations are described as having variable chlorophyll content to the parent. This could be utilized to develop plants for efficient photosynthesis where more energy is wasted in higher chlorophyll production but with lower photosynthetic capability. Stringam (1973) isolated seven chlorophyll deficient mutations in *Brassica campestris*. Ethyl methanesulfonate induced a large spectrum of chlorophyll mutations isolated by Fowler and Stefansson (1975) in *B. napus*. Zhao et al. (2000) obtained chlorophyll-reduced (Cr) mutations using fast neutrons and diethyl sulfonate (DES) and used them in the hybrid development of *B. napus*. Eleven different types of chlorophyll mutants, namely, albina, xantha, viridis, tigrina, maculata, variegated, albo-xantha, xantho-alba, xantho-viridis, virido-xantha, and virido-alba were isolated in *Brassica juncea* with different doses of EMS (Gupta et al., 2012).

Dwarf Mutations

Dwarfing genes in rice and wheat are responsible for the Green Revolution (Khush, 2001). Similarly, dwarfing genes could be exploited in *Brassica* crops to increase seed yield by reducing lodging and increasing harvest index. Dwarf mutants compared to their parents have been isolated in *B. campestris* (Hawk and Crowder, 1978; Tyagi et al., 1983; Chauhan and Kumar, 1986; Rai and Singh, 1993; Javed et al., 2003), *B. juncea* (Das and Rahman, 1988; Khatri et al., 2005), and *B. napus* (Shah et al., 1999; Zanewich et al., 1991; Thaganal et al., 2013) using physical and chemical mutagens, whereas Devlin et al. (1997) isolated elongated internode (ein) mutations in *B. rapa*. However, it is not reported whether these mutations have been exploited for development of high-yielding varieties or not. A 70 cm dwarf mutant, NDF-1, was isolated using chemical inducers and bombardment of fast neutrons from a 200 cm tall parent DH line 3529 in *B. napus* (Wang et al., 2004). Recently, we have isolated dwarf, early, yellow seed coat, and reduced erucic acid mutations in *B. juncea* variety "varuna" using gamma rays. The height of this mutation has been reduced to half, i.e., 90 cm, whereas yield potential remains equal to the parent. Characterization for harvest index is being undertaken.

Flower Mutations

Rapeseed–mustard in general possesses yellow colored flowers. An X-ray induced white flower mutation was isolated by Rai and Jacob (1956). A large spectrum of flower color mutations were isolated in *B. napus* using EMS (Fowler and Stefansson, 1975). A novel yellow–white flower mutation was isolated from the male sterile progenies derived from the commercial *B. napus* hybrid CO22 (Yu et al., 2004). A male sterile mutant in *Brassica juncea* was obtained through the use of chemical mutagens, such as EMS, ENU, Ebr (Bhat et al., 2001), and gamma rays and EMS combined (Chauhan and Singh, 1998). The apetalous *Brassica* plant has better physiological characteristics and disease avoidance mechanisms. The apetalous trait is a kind of floral abnormality that was firstly reported in India. Ramanujam (1940) reported a stray plant of *B. campestris* in a lucerne field devoid of bright yellow petals. Mutations for increased or decreased number of petals (apetalous) in the flower were isolated in *B. rapa* (Buzza, 1983; Cours and Williams, 1977; Singh, 1961a, b), *B. napus* (Lu and Fu, 1990; Fray et al., 1997; Jiang, 2001), and *Brassica carinata* (Rana, 1985).

Siliqua Characteristics

Mutations for siliqua characteristics are mainly for angle of siliqua and number of loculs. Reduced siliqua angle to the raceme branch could avoid aphid infestation in rapeseed–mustard. X-ray induced appressed pod mutants were isolated in *B. juncea* varieties Rai5 (Rai, 1958) and RL9 (Nayar and George, 1969). Using gamma rays, Kamala and Rao (1984) obtained a three valved pod mutation in yellow "sarson." Lavania (1979) obtained a four-siliqua mutant from T-25 using EMS. Bhat et al. (2001) isolated tri- and tetra-locular siliqua and non-shattering mutations from variety pusa "jai kisan" using various chemical mutagen, such as EMS, ENU, and Ebr. Bunching and appressed pod mutants were isolated using combined treatment of gamma rays and EMS from variety RH30 by Singh and Sareen (2004).

Seed Coat Color

In general, seeds of the *Brassica* species are brown/black in color. A thinner seed coat, higher oil content, high protein content, and lower fiber content are the specific characteristics of yellow seeded mutations in rapeseed–mustard over brown seeded parents and therefore are more desirable than brown/black seeded varieties (Shirzadegan and Robbelen, 1985; Stringam et al., 1974; Woods, 1980; Xiao, 1982). This also improved the nutritive value of meal after oil extraction (Simbaya et al., 1995; Slominski et al., 1999). Yellow seeded genotypes are available for *B. rapa*, *B. juncea*, and *B. carinata*. No natural or induced mutations have been reported in *B. napus* and therefore interspecific hybridization was extensively used to develop yellow seed coat genotypes (Rahman, 2001)

Up to the late 1960s, all *B. juncea* genotypes available in the germplasm collection had brown or black seed coats. Induced mutation to isolate yellow seed coat was initiated at Bhabha Atomic Research Centre (BARC), Mumbai, India by Nayar (1968) after isolating a yellow seed coat mutant from a brown seed coat variety Rai5 using ^{35}S radioisotope. Another yellow seeded mutant was isolated from the same variety Rai5 using ^{32}P radioisotope (Nair, 1968) and named Trombay Mustard 1 (TM1). Using this mutant in a cross-breeding program, improved high-yielding genotypes were developed (Nayar, 1976, 1979). The yellow seed coat mutants and their derivatives were extensively used in the cross-breeding program throughout India and a large number of high-yielding, bold-seeded genotypes were developed (Abraham and Bhatia, 1986). In recent years, we have also isolated a yellow seed coat mutant from the most popular varieties "varuna," pusa bold, and "kranti" using gamma rays (data unpublished). These mutants are being used extensively in crossbreeding to develop high-yielding, yellow seed coat varieties (Jambhulkar et al., 2005; Jambhulkar and Shitre, 2007; Jambhulkar, 2012).

Root Morphology

Improved root morphology may help develop drought tolerant genotypes. Basak and Prasad (2004) isolated mutants for high root biomass upon treatment of the "rose red" variety of turnip (*B. rapa*, L.) using gamma rays and EMS. Gamma ray induced mutations in the parent variety "varuna" showed preliminary tolerance to drought because the plant possesses a greater root length.

Insect Pest Resistant Mutation

Resistance to insect pests is one of the major constraints in realizing the yield potential of rapeseed–mustard. Among the various insect pests, aphids are important. Systematic breeding approaches are required to develop resistant insect pest varieties (Singh and Rana, 2002). Mutation breeding stands a better chance because limited or no variability was observed in the germplasm. However, information in this aspect is scant. Waxy leaves were found to be susceptible to aphid attack. Using gamma rays, ethylene imine, and hydrazine, a non-waxy mutant was isolated in the varieties "pusa sweti" and PTW Globe of *B. rapa,* which had shown resistance to aphids (Srinivasachar and Malik, 1972).

Mutations for Seed Yield and Its Related Traits

Various yield attributes contribute directly or indirectly to seed and oil yields. Mutations have been successfully isolated for desirable economic traits such as plant height, number of siliquae per plant, number of seeds per siliqua, seed weight, seed yield, and oil content (Chauhan and Kumar, 1986; Javed et al., 2003; Mahla et al., 1990, 1991; Rehman et al., 1987; Rehman, 1996; Robbelen, 1990; Shah et al., 1990, 1998, 1999). The most important yield contributing characteristics in rapeseed–mustard are numbers of primary and secondary branches, siliquae per plant, siliquae on the main fruiting axis, number of seeds per siliqua, and seed weight. Mutants with more branches and siliquae per plant have been reported (Chauhan and Kumar, 1986; Javed et al., 2003; Naz and Islam, 1979; Shah et al., 1990) in rapeseed–mustard. Seventeen mutations for early, dwarf, and high grain yield were developed using gamma ray and EMS in *B. juncea* cv.S-9, three of them were significantly superior to the others (Khatri et al., 2005). High-yielding mutations were isolated in *B. napus* (Hao-jie et al., 2005; Raza et al., 2009; Malek et al., 2012; Uddin et al., 2012; Ali and Shah, 2013) and *B. juncea* (Barve et al., 2009; Malek et al., 2012). Variability for reduced height and pods per plant was generated using gamma rays in *B. napus* (Thaganal et al., 2013).

EARLY-FLOWERING MUTATIONS

Early maturing varieties are suitable to use in specific cropping patterns although it is at the expense of yield potential. However, a mutation for early flowering in agronomically superior lines could be useful across various cropping patterns. Early-flowering mutants were obtained in *B. napus* (Thurling and Depittayanan, 1992; Shah et al., 1999) and *B. juncea* (Nayar and George, 1969).

Oil Content

Increasing oil content in a high-yielding variety is the right approach to take to increase oil yield. Khatri et al. (2005) treated seeds of cv. S-9 of *B. juncea* L. with gamma rays and EMS and isolated mutants with 3% more oil content. Gamma ray induced mutants had 1–5% higher oil content than those of their parent variety "Pant Rai 5" (Verma and Rai, 1980).

Fatty Acid Mutations

Edible oil provides energy and thus is an important component of the human diet. In vegetable oils, oleic acid (C18:1) is the major part of the fatty acid followed by linoleic acid (C18:2) and α-linolenic acid (C18:3), along with other fatty acids, such as palmitic and stearic acids. Traditional rapeseed–mustard oil contains a high proportion (~50%) of erucic acid (C22:1), which is considered as an antinutritional factor and thus it is an exception to other vegetable oils. Mutation breeding has been successful in tailoring oil crops for desirable fatty acid compositions (Robbelen, 1990) because oil crop plants tolerate a wide range of fluctuations in fatty acid composition without losing viability and a single mutation can result in a desirable oil composition. Liho, the first zero erucic acid mutant, was reported by Stefanson et al. (1961) in *B. napus*. This started an era of mutant assisted, quality improvement in oilseed crops. Downey (1964) suggested that erucic acid–free oil should also occur in *B. campestris* as it is one of the parents of *B. napus* and reported zero erucic acid natural mutant/variability in *B. campestris*. Two zero erucic acid natural mutants were found in the Chinese accession of *B. juncea* (Kirk and Oram, 1981) and were termed zem1 and zem2. Olsson (1984) also reported natural mutation/variability for LEA in *B. juncea*. A high level of erucic acid also has industrial applications, however, a mutation for high erucic acid has not yet been reported.

Linolenic acid is easily oxidized and its oil cannot be stored for long period. However, no variety or species of cruciferae was found free from linolenic acid. In view of this Robbelen and Rakow (1970) in Germany started a mutation-breeding program and isolated a reduced linolenic acid mutant in *B. napus* (Rakow, 1973; Robbelen and Nitsch, 1975). High oleic acid in oil is considered nutritionally desirable for human health. Auld et al. (1992) and Rucker and Robbelen (1997) isolated a high oleic acid mutant in *B. napus*. This initial success of mutant isolation laid the foundation stone for the improvement of oil quality in *Brassica* crops. Mutations for various fatty acids, such as palmitic acid (Rucker and Robbelen, 1997; Schnurbusch et al., 2000), oleic acid (Auld et al., 1992; Rucker and Robbelen, 1997; Spasibionek, 2006; Schnurbusch et al., 2000; Raza et al., 2009; Velasco et al., 1997; Barro et al., 2001), linoleic acid (Robbelen and Nitsch, 1975; Spasibionek, 2006; Auld et al., 1992; Velasco et al., 1997), linolenic acid (Auld et al., 1992; Rucker and Robbelen, 1997; Robbelen and Nitsch, 1975; Rakow, 1973; Spasibionek, 2006; Schnurbusch et al., 2000; Raza et al., 2009; Robbelen and Nitsch, 1975; Auld et al., 1992; Velasco et al., 1997), and oleic acid (Velasco et al., 1995; Velasco et al., 1998; Sheikh et al., 2009; Barro et al., 2001; Stefanson et al., 1961; Downey, 1964; Kirk and Oram, 1981) have been reported. These mutations have been used extensively to develop high-yielding cultivars with desirable fatty acid compositions. Development of yellow mustard (*Sinapis alba* L.) with superior quality traits (low erucic and linolenic acid content and low glucosinolate content) can make this species as a potential oilseed crop. Tian et al. (2014) have recently isolated three inbred lines Y1127, Y514, and Y1035 with low (3.8%), medium (12.3%), and high (20.8%) linolenic acid (C18:3) content, respectively, in this species. Inheritance studies detected two fatty acid desaturase3 (*FAD3*) gene loci controlling the variation of C18:3 content.

Double Haploids and Mutagenesis

DH techniques have many attractive features – rapid creation of mutations and immediate fixation of them in the first generation thereby achieving the goal of homozygosity at the earliest time. The development of efficient *in vitro* culture methods, have facilitated the use of mutation techniques for crop improvement (Maluszynski et al., 1995). The haploid system can successfully be used for the induction of genetic variation through mutagenesis. It has advantages like immediate fixation of mutated genotypes in the first generation and *in vitro* selection and increased selection efficiency. Mutagenic treatment can be given to spikes, buds, anthers, microspores, and haplod calli, embroys, or protoplasts. The dose of mutagen needs to be adjusted for the sensitivity of material used. Such studies for chemical mutagens, such as EMS (Barro et al., 2001; Beversdorf and Kott, 1987), ENU (Swanson et al., 1988, 1989), NaN_3 (Polsoni et al., 1988), MNH (MNU) (Jedrzejaszek et al., 1997), as well as physical mutagens, such as gamma rays (Beversdorf and Kott, 1987; McDonald et al., 1991; Swanson et al., 1988), X-rays (McDonald et al., 1991), and UV rays (McDonald et al., 1991; Jedrzejaszek et al., 1997) were undertaken in various *Brassica* species to isolate mutations. It was opined that microspore mutagenesis has been most successful in rapeseed at isolating desirable stable mutants. A dwarf mutant, NDF-1, was derived from a DH line, 3529, of *B. napus* L. after treating the seeds with chemical inducers and subjecting them to fast-neutron bombardment (Wang et al., 2004). Prem et al. (2012) isolated reduced height, appressed pod, altered fatty acid composition, higher protein proportion, and lower glucosinolate content mutants from *B. juncea* parents after treating the microspore with EMS.

Combining mutagenic treatment with *in vitro* selection at haploid level can improve recovery of desired mutants. Direct application of herbicide as the selection agent to the culture medium enforces immediate expression of tolerance in the tissue. Selection for herbicide resistance is likely to be the simplest test for this system, since a single-point mutation at any one of the number of genes could effectively interfere with the uptake, assimilation, or translocation of the herbicide, resulting in a resistant/tolerant plant. *In vitro* mutagenesis and selection have resulted in the development of mutants for herbicide resistance (Ahmad et al., 1991; Beversdorf and Kott, 1987; Kott, 1995, 1998; McDonald et al., 1991; Palmer et al., 1996; Polsoni et al., 1988; Swanson et al., 1988, 1989), disease resistance (Ahmad et al., 1991; Bansal et al., 1998; Liu et al., 2005; Newsholme et al., 1989; McDonald and Ingram, 1986; Sacriston, 1982), long pod and short plant (Shi et al., 1995) in *B. napus*.

Improvement of oil and meal quality through microspore mutagenesis in rapeseed has been successfully demonstrated by Kott et al. (1996). Mutations for increased level of oleic acid and reduction of linolenic acid isolated via the DH technique have been reported by Auld et al. (1992) and Wong and Swanson (1991). Similarly, Turner and Facciotti (1990) and Huang (1992) isolated mutants for high oleic acid and decreased level of saturated fatty acids from a mutagenized microspore culture. Mutations for high and low levels of erucic acid isolated from microspore culture, have been reported by Barro et al. (2001). Beaith et al. (2005) treated microspore with ultraviolet light and isolated mutants for reduced palmitic and stearic acid content in *B. napus*.

Molecular Basis of Mutations

Phenotypical expression is governed by biochemical processes, which are controlled by genes. Spontaneous as well as induced mutations are due to deletion, duplication, inversion, translocation, and substitution of a chromosome or base pairs in DNA (deoxyribonucleic acid) and also due to transposable elements, which are responsible for alteration in the biochemical products that result in the mutant phenotype. Characterization of mutations at protein and DNA levels strengthen the basic understanding of mutations.

Fatty Acids

LEA is characterized by a near absence of very long chain fatty acids (VLCFA) in the seed oil, which has been correlated with a lack of acyl-CoA elongation activity (Roscoe et al., 2001). Using site directed mutagenesis, Katavic et al. (2002) validated the speculation of the presence of serine at position 282 in all functional proteins of HEA, which is replaced by phenylalanine in LEA, *B. napus* are under the control of *FAE1* enzyme. The phenylalanine 282 residue was substituted with a serine residue in the *FAE1* polypeptide from LEA *B. napus* cv.Westar. This mutated gene was expressed in yeast and GC analysis revealed the presence of very long chain monounsaturated fatty acids (VLCMFAs), indicating that elongase activity was restored in the LEA *FAE1* enzyme by the single amino acid serine. Thus, the LEA trait in *B. napus* can be attributed to phenylalanine, which prevents the biosynthesis of eicosenoic and erucic acids.

Synthesis of oleic acid is under the influence of *fad2* gene that codes an enzyme, endoplasmic delta-12 oleate desaturase, responsible for the desaturation of oleic acid (C18:1) into linoleic acid (C18:2). Tanhuanpaa et al. (1998) reported single nucleotide mutation in the *fad2* gene of *B. rapa,* which caused an increase in the C18:1 content. This mutation substituted leucine with proline. Hu et al. (2006) also identified single nucleotide mutation from C to T in the gene *fad2* of *B. napus*. This mutation generated the stop codon (TAG) leading to premature termination of the peptide chain during translation. As a result, only 185 amino acids were incorporated into the polypeptide instead of all 384 amino acids representing the full-length polypeptide. Thus, this truncated polypeptide may not function as an active desaturase for the desaturation of C18:1 to C18:2 and, therefore, will result in the accumulation of C18:1 in the seeds of the mutant line.

Dwarfs

Mutations in the biosynthetic pathways of gibberellin (GA) can cause the dwarfing phenotype in plants (Hedden, 2003; Peng et al., 1999; Sun and Gubler, 2004). GA promotes stem growth by causing degradation of the DELLA protein in the ubiquitin-proteasome pathway. The most widely utilized semidwarf wheat cultivars have the Rht 131b or Rht D1b allele that encodes a mutant form of a DELLA protein, which is a GA signaling repressor (Peng et al., 1999; Hedden, 2003). Five DELLA protein genes (GA1, RGA, RGL1, RGL2, and RGL3) are responsible for GA regulated plant growth and development in *Arabidopsis*. Muangprom and Osborn (2004) found that the *dwf2* mutation in *B. rapa* was insensitive to exogenous GA_3 for both plant height and flowering time and inferred that it is not a mutation in the GA biosynthetic pathway.

Muangprom et al. (2005) reported that a mutant of *B. rapa* (Brrga1-d) is caused by a single nucleotide substitution of conserved amino acid in the C-terminal domain of a DELLA protein. Brrga1-d retains its repressor function in the presence of GA and does not interact with a protein component required for degradation, suggesting that the mutated amino acid causes dwarfism by preventing an interaction needed for degradation. The ein mutation of *B. rapa* leads to a deficiency in immunochemically detectable phytochrome B (Devlin et al., 1997). Molecular analysis of the *PHYB* gene from ein indicates deletion in the flanking DNA 5′ of the ATG start codon, which could interfere either with *PHYB* transcription or processing of the *PHYB* transcript. Restriction fragment length polymorphism and inverse polymerase chain reaction (PCR) fragments generated from the *PHYB* gene of the wild type and ein seedlings demonstrated the deletion to be 500 bp in length. Other than the DELLA protein, Li et al. (2011) identified a mutation in the *cis* acting element in the promoter region of the *GID1* gene in the NDF-1 dwarf mutant of *B. napus* where they found that a three-base mutation in the pyrimidine box of the promoter region suppresses the transcription level of the *BnGID1* gene. The GID1 protein is a soluble GA receptor. It binds with high affinity to only bioactive GAs and DELLA proteins that induce DELLA protein degradation via ubiquitin E3ligase complex SCF in 26S proteasome (Sun, 2011).

Development of High-Yielding Varieties

The induced mutagenesis technique has resulted in the development of more than 3218 varieties in various crops and horticultural plants all over the world. Among these, 163 varieties belong to oilseed crops. The greatest number of varieties have been released for soybean (58), followed by groundnut (44), seasame (16), linseed (15), castor (4), and sunflower (1). Mutation breeding in rapeseed–mustard has also resulted in the development of a total of 31 high-yielding varieties comprising 12 in *B. juncea*, 14 in *B. napus*, 2 in *B. rapa*, and 3 in white mustard. Sixteen varieties have been developed using gamma rays, four by X-rays, two by colchicines, one by ^{32}P, and one each by EMS, DMS, and MNH. In *B. juncea*, eight varieties have been developed in India, five each in Bangladesh, Sweden, and China, and two each in Canada, Japan, the former USSR, and Pakistan. Labana (1976) obtained a gamma ray induced mutant, RLM-198, with 25% higher seed yield, moderately resistant to aphids and leaf miners, a higher oil content, and a 5–6 day earlier maturity. Labana (1981) evolved another mutant, RLM-214, with 22% more seed yield and with a high oil content, while being shattering resistant from parent RL-18 using X-rays. Mutations for increased seed yield, resulting in the development of high-yielding varieties, have also been reported by Rehman (1996) and Shah et al. (1999). Backcrossing of M-11, a low linolenic acid mutant with variety Regent resulted in the development of a high-yielding, low linolenic acid variety Stellar (Scrath et al., 1988) and Apollo (Scrath et al., 1995).

Brassica TILLING

Targeting-induced local lesions in genomes (TILLING) is a reverse genetic strategy, which helps to locate an allelic series of induced point mutations in genes of interest. It allows the rapid and inexpensive detection of induced point mutations in populations of physically/chemically mutagenized individuals. In addition to allowing efficient detection of mutations by the TILLING approach, EcoTILLING technology is also ideal for examining natural variation. Both TILLING and EcoTILLING are attractive strategies for a wide range of applications from the basic functional genomic study to practical crop breeding. This technique was first utilized in the *Arabidopsis* TILLING Project (ATP) during 2001. The

ATP project has detected, sequenced, and delivered over 1000 mutations in more than 100 genes (Till et al., 2003). First results on TILLING in rapeseed were presented by Wang et al. (2008). Screening of 1344 M_2 plants resulted in 19 mutants of the BnFAE1 gene family, which were phenotypically verified by M_3 analysis. Among them, three were functionally conceded with reduced seed erucic acid content. Using a TILLING technique, Stephenson et al. (2010) developed EMS treated a *B. rapa* TILLING population and analyzed six genes. A high level of mutation, with a density of about one per 60 kb, was isolated. An average of 68 mutations were isolated by screening a 1 kb amplicon in just one third (3072 M_2 plants) of the population with a probability of 97% for obtaining a stop codon mutation resulting in a truncated protein. Harloff et al. (2012) identified mutations in two major genes, BnaX.SGT and BnaX.REF1, which were responsible for the sinapine-biosynthetic pathway. Screening of M_2 populations resulted in the isolation of 229 and 341 mutations within the BnaX.SGT sequences (135 missense and 13 nonsense mutations) and the BnaX.REF1 sequences (162 missense, 3 nonsense, 8 splice site mutations), respectively. These mutants provide a new resource for breeding low-sinapine oilseed rape. The frequencies of missense and nonsense mutations corresponded to the frequencies of the target codons. Mutation frequencies ranged from 1/12 kb to 1/22 kb for the Express 617 population and from 1/27 kb to 1/60 kb for the YN01-429 population. Gilchrist et al. (2013) identified 432 unique mutations in 26 different genes in *B. napus* (canola cv. DH12075). This reflects a mutation density ranging from 1/56 kb to 1/308 kb (depending on the locus) with an average of 1/109 kb. Guo et al. (2014) screened 3488 EMS treated M_2 progeny of *B. napus* by the TILLING technique and identified 55, 14, and 34 single nucleotide mutations in the *BnC6FTb, BnC6FTa,* and *BnTFL1-2* genes, respectively. A total of 43 mutations are located in introns, 19 are silent mutations, and 3 are located within the untranslated regions (UTRs). Mutation rates ranged between 1/72 kb and 1/24 kb per 1000 plants. Most of the mutations were due to substitution and change in nucleotide positions. Nonsense and missense mutations were also isolated.

References

Abraham, V., Bhatia, C.R., 1986. Development of strains with yellow seed coat in Indian mustard (*Brassica juncea* Czern. & Coss.). Plant Breed. 97, 86–88.

Ahmad, I., Day, J.P., McDonald, M.V., Ingram, D.S., 1991. Haploid culture and UV mutagenesis in rapid-cycling *Brassica napus* for the generation of resistance to chlorsulfuron and *Alternaria rassicicola*. Ann. Bot. 67, 521–525.

Ali, H.M.A., Shah, S.A., 2013. Evaluation and selection of rapeseed (*Brassica napus* L.) mutant lines for yield performance using augmented design. J. Anim. Plant Sci. 23, 1125–1130.

Auld, D.L., Heikkinen, M.K., Erickson, D.A., Sernyk, J.L., Romero, J.E., 1992. Rapeseed mutants with reduced level of polyunsaturated fatty acids and increased levels of oleic acid. Crop Sci. 32, 657–662.

Bansal, V.K., Thiagarajah, M.R., Stringam, G.R., Hardin, R.T., 1998. Haploid plantlet screening in the development of blackleg resistant DH lines of *Brassica napus*. Plant Breed. 117, 103–106.

Barro, F., Fernandez-Escobar, J., De La Vega, M., Martin, A., 2001. Double haploid lines of *Brassica carinata* with modified erucic acid content through mutagenesis by EMS treatment of isolated microspores. Plant Breed. 120, 262–264.

Barve, Y.Y., Gupta, R.K., Bhadauria, S.S., Thakre, R.P., Pawar, S.E., 2009. Induced mutations for development of *B.* Juncea canola quality varieties suitable for Indian agro-climatic conditions. In: Shu, Q.Y. (Ed.), Induced Plant Mutations in the Genomics Era. Food and Agriculture Organization of the United Nations, Rome, pp. 373–375.

Basak, S., Prasad, C., 2004. Screening for root mutant in turnip (Brassica rapa L.). Cruciferae Newsl. 25, 9–10.

Beaith, M.E., Fletcher, R.S., Kott, L.S., 2005. Reduction of saturated fats by mutagenesis and heat selection in *Brassica napus* L. Euphytica 144, 1–9.

Beversdorf, W.D., Kott, L.S., 1987. An *in vitro* mutagenesis/selection system for *Brassica napus*. Iowa State J. Res. 64, 435–443.

Bhat, S.R., Haque, A., Chopra, V.L., 2001. Induction of mutations for cytoplasmic male sterility and some rare phenotypes in Indian mustard (*Brassica juncea* L.). Indian J. Genet. 61, 335–340.

Bhatia, C.R., Nichterlein,K., Maluszynski, M., 1999. Oilseed cultivars development from induced mutations and mutation altering fatty acid composition. Mutation Breeding Review No.11, pp. 1–36.

Buzza, G.C., 1983. The inheritance of an apetalous characters in canola (B. napus L.). Cruciferae Newsl. 8, 11–12.

Chauhan, Y.S., Kumar, K., 1986. Gamma ray induced chocolate B seeded mutant in *Brassica campestris*. Curr. Sci. 55, 410.

Chauhan, S.V.S., Singh, N.K., 1998. Gamma ray and EMS induced male sterility mutants in *Brassica juncea*. J. Cytol. Genet. 33, 29–34.

Chopra, V.L., 2005. Mutagenesis: Investigating the process and processing the outcome for crop improvement. Curr. Sci. 89, 353–359.

Cours, B.J., Williams, P.H., 1977. Genetic studies in *Brassica campestris* L. Plant Breed Abstr. 51, 1533.

Das, M.L., Rahman, A., 1988. Induced mutagenesis for the development of high yielding varieties in mustard. J. Nucl. Agric. Biol. 17, 1–4.

Devlin, P.F., Somers, D.E., Quail, P.H., Whitelam, G.C., 1997. The *Brassica rapa* elongated internode (*EIN*) gene encodes phytochrome B. Plant Mol. Biol. 34, 537–547.

Downey, R.K., 1964. Selection of *Brassica campestris* L. containing no erucic acid in its seed oil. Can J. Plant Sci. 44, 295.

Fowler, D.B., Stefansson, B.R., 1975. Ethyl-methane sulfonate induced mutations in rape (*Brassica napus*). Can. J. Plant Sci. 55, 817–821.

Fray, M.J., Puangsomlee, P., Goodrich, J., Coupland, G., Evans, E.J., Arthur, A.E., Lydiate, D.J., 1997. The genetics of stamenoid petal production in oilseed rape (*Brassica napus*) and equivalent variation in *Arabidopsis thaliana*. Theor. Appl. Genet. 94, 731–736.

Gilchrist, E.J., Christine, H.D., Sidebottom, Koh, C.S., MacInnes, T., Sharpe, A.G., Haughn, G.W., 2013. A mutant *Brassica napus* (canola) population for the identification of new genetic diversity via TILLING and next generation sequencing. Plos One 8, e8430.

Guo, Y., Harloff, H., Christian, J., Carlos, M., 2014. Mutations in single FT- and TFL1-paralogs of rapeseed (*Brassica napus* L.) and their impact on flowering time and yield components. Front. Plant Sci. 5 (282), 1–12.

Gupta, S., Tyagi, D.V.S., Rawat, R.S., 2012. Induced Chlorophyll Mutations and Mutagenic Effectiveness in Indian Mustard (*Brassica juncea* L.Czern & Coss). Progress. Agr. 12, 128–131.

Hao-jie, L.I., Xiao-bin, P.U., Jin-fang, Z., Qi-xing, Z., Liang-cai, J., 2005. Primary study on radiation induced mutation in *Brassica napus*. Southwest China J. Agr. Sci. 18, 542–546.

Harloff, H.J., Lemcke, S., Mittasch, J., Frolov, A., Guo, J.W., Dreyer, F., Leckband, G., Jung, C., 2012. A mutation screening platform for rapeseed (*Brassica napus* L.) and the detection of sinapine biosynthesis mutants. Theor. Appl. Genet. 124, 957–969.

Hawk, J.A., Crowder, L.V., 1978. The inheritance of four mutants in early – flowering *Brassica campestris* L. J. Hered. 69, 125–127.

Hedden, P., 2003. The genes of the green revolution. Trends Genet. 19, 5–9.

Hu, X., Sullivan-Gilbert, M., Gupta, M., Thompson, S.A., 2006. Mapping of loci controlling oleic and linolenic acid contents and development of *fad2* and *fad3* allele-specific marker in canola. Theor Appl. Genet. 113, 497–501.

Huang, B., 1992. Genetic manipulation of microspores and microspore derived embryos. *In Vitro* Cell Dev. Biol. 28, 53–58.

Jambhulkar, S.J., 2007. Mutagenesis: generation and evaluation of induced mutations. Gupta, P.C. (Ed.), Advances in Botanical Research (Rapeseed Breeding), vol. 45, Elsevier Publication Ltd, USA, pp. 418–428.

Jambhulkar, S.J., 2012. Induced mutagenesis for genetic improvement of Indian mustard. First National Brassica Conference on Production Barriers Technological options in Oilseed Brassica, March 2–3, CCSHAU. Hisar, Haryana, India. pp. 22–23.

Jambhulkar, S.J., Shitre, A.S., 2007. Genetic improvement of rapeseed mustard through induced mutations. IANCAS Bull. 6, 327–329.

Jambhulkar, S.J., Shitre, A.S., Pathak, B., 2005. Heterosis in Indian mustard using induced mutations. Conference on Resource Development and Marketing issues in rapeseed mustard, March 28–29. Jaipur, India. pp. 27–28.

Javed, M.A., Siddiqui, M.A., Khan, M.K.R., Khatri, A., Khan, I.A., Dahar, N.A., Khanzada, M.H., Khan, R., 2003. Development of high yielding mutants of *Brassica campestris* L. cv. Toria selection through gamma rays irradiation. Asian J. Plant Sci. 2, 192–195.

Jedrzejaszek, K., Kruczkowska, H., Pawlowska, H., Skucinska, B., 1997. Stimulating effect of mutagens on *in vitro* plant regeneration. MBNL 43, 10–11.

Jiang, L., 2001. A mutant with apetalous flowers in oilseed rape (*Brassica napus*): mode of inheritance and influence on crop physiology and *Sclerotinia* infection. PhD Thesis.

Kamala, T., Rao, R.N., 1984. Gamma ray induced 3-valved mutant in yellow sarson. J. Nuclear Agric. Biol. 13, 28.

Katavic, V., Mietkiewska, E., Borton, D.L., Giblin, E.M., Reed, D.W., Taylor, D.C., 2002. Restoring enzyme activity in nonfunctional low erucic acid *Brassica napus* fatty acid elongase 1 by single amino acid substitution. Eur. J. Biochem. 269, 2631–2652.

Khatri, A., Khan, I.A., Siddiqui, M.A., Raza, S., Nizamani, G.S., 2005. Evaluation of high yielding mutants of *Brassica juncea* Cv. S-9 developed through gamma rays and EMS. Pak. J. Bot. 37, 279–284.

Khush, G.S., 2001. Green revolution: the way forward. Nat. Rev. Genet. 2, 815–822.

Kirk, J.T.O., Oram, R.N., 1981. Isolation of erucic acid free lines of *Brassica juncea*: Indian mustard now a potential oilseed crop in Australia. J. Austral. Inst. Agric. Sci. 47, 51–52.

Kott, L., 1995. Production of mutants using the rapeseed double haploid system. Induced mutations and Molecular Techniques for Crop Improvement. IAEA, Vienna, pp. 505–515.

Kott, L., 1998. Application of double haploid technology in breeding of oilseed *Brassica napus*. AgBiotech News Inf. 10, 69N–74N.

Kott, L., Wong, R., Swanson, E., Chen, J., 1996. Mutation and selection for improved oil and meal quality in *Brassica napus* utilizing microspore culture. Jain, S.M., Sopory, S.K., Veilleux, R.E. (Eds.), *In Vitro* Haploid Production in Higher Plants, vol. 2, Springer, Berlin, Heidelberg, New York, pp. 151–167.

Labana, K.S., 1976. Release of mutant variety of raya (*Brassica juncea*). MBNL 7, 11.

Labana, K.S., 1981. Release of mutant variety in raya (*Brassica juncea*). MBNL 17, 13.

Lavania, U.C., 1979. Interphase sensitivity and specificity of siliqua mutations induced by EMS in Indian mustard. Curr. Sci. 48, 101.

Li, H., Wang, Y., Li, X., Gao, Y., Wang, Z., Zhao, Y., Wang, M., 2011. A GA-insensitive dwarf mutant of *Brassica napus* L. correlated with mutation in pyrimidine box in the promoter of GID1. Mol. Biol. Rep. 38, 191–197.

Liu, S., Wang, H., Zhang, J., Fitt, B.D.L., Xu, Z., Evans, N., Liu, Y., Yang, W., Guo, X., 2005. *In vitro* mutation and selection of double haploid *Brassica napus* lines with improved resistance to *Sclerotinia sclerotiorum*. Plant Cell Rep. 24, 133–144.

Lu, Z.J., Fu, S.Z., 1990. Inheritance of apetalous character in rape(*Brassica napus* L.) and its implication in breeding. Jiangsu J. Agric. Sci. 6, 30–36.

Mahla, S.V.S., Mor, B.R., Yadava, J.S., 1990. Effect of mutagen on yield and its component characters in mustard. Haryana Agric. Univ. J. Res. 20, 259–264.

Mahla, S.V.S., Mor, B.R., Yadava, J.S., 1991. Mutagen induced polygenic variability in some mustard (*Brassica juncea* L.) varieties and their hybrids. J. Oilseed Res. 8, 173–177.

Malek, M.A., Ismail, M.R., Monshi, F.I., Mondal, M.M.A., Alam, M.N., 2012. Selection of promising rapeseed mutants through multi-location trials. Bangladesh J. Bot. 41, 111–114.

Maluszynski, M., Ahloowalia, B.S., Sigurbjornsson, B., 1995. Application of *in vivo* and *in vitro* mutation techniques for crop improvement. Euphytica 85, 303 315.

McDonald, M.V., Ingram, D.S., 1986. Towards the selection *in vitro* for resistance to *Alternaria brassicicola* (Schw) Welts., in *Brassica napus* ssp. Oleifera (Metzg.) Sinsk., winter oilseed rape. New Phytol. 104, 621–629.

McDonald, M.V., Ahmad, I., Menten, J.O.M., Ingram, D.S., 1991. Haploid culture and *in vitro* mutageneis (UV light, X-rays and gamma rays) of rapid cycling *Brassica napus* for improved resistance to diseases. Plant Mutations Breeding for Crop Improvement, vol. 2, IAEA, Vienna, pp. 129–138.

Muangprom, A., Osborn, T.C., 2004. Characterisation of dwarf gene in *Brassica rapa*, including the identification of a candidate gene. Theor. Appl. Genet. 108, 1378–1384.

Muangprom, A., Thomas, S.G., Sun, T.P., Osborn, T.C., 2005. A novel dwarfing mutation in a green revolution gene from *Brassica rapa*. Plant Physiol. 137, 931–938.

Nayar, G.G., 1968. Seed colour mutation in *Brassica juncea* Hook. F. and Thomas induced by radioactive phosphorus ^{32}P. Sci. Cult. 34, 421–422.

Nayar, G.G., 1976. Studies on radiation induced mutants and their recombinant types in mustard. Oilseed J., 46–47.

Nayar, G.G., 1979. Breeding strategy for the improvement of mustard (*Brassica juncea*) Coss. In: Proc. Symp. The role of induced mutation in crop improvement, Osmania University September 10–13. Hyderabad, India. pp. 258–270.

Nayar, G.G., George, K.P., 1969. X-ray induced early flowering, appressed pod mutant in *Brassica juncea* Coss. In: Proceedings of the Symposium on Radiation and Radiomimetic Substances in Mutation Breeding, September 26–29. Bombay, India. pp. 409–413.

Naz, R.M.A., Islam, R.Z., 1979. Effect of irradiation of *Brassica trilocularis* (yellow sarson). Pak. J. Agril. Res. 17, 87–93.

Newsholme, D.M., MacDonald, M.V., Ingram, D.S., 1989. Studies of selection *in vitro* for novel resistance to phytotoxic products of *Leptospheria maculans* (Desm.) Ces. & De Not in secondary embryogenic lines of *Brassica napus* ssp. Oleifera (Metzg.) Sinsk., winter oilseed rape. New Phytol. 113, 117–126.

Olsson, G., 1984. Selection for low erucic acid in *Brassica juncea*. Sveriges Utsadesf. Tidskr. 94, 187.

Palmer, C.E., Keller, W.A., Arnison, P.G., 1996. Utilization of *Brassica* haploids. In: Jain, S.M., Sopory, S.K., Veilleux, R.E. (Eds.), *In Vitro* Haploid Production in Higher Plants. Kluwer Academic Publisher, Dordrecht, Netherlands, pp. 173–192.

Peng, J., Richards, D.E., Hartley, N.M., Murphy, G.P., Devos, K.M., Flintham, J.E., Beales, J., Fish, L.J., Worland, A.J., Pelica, F., 1999. 'Green revolution' genes encode mutant gibberellin response modulators. Nature 400, 256–261.

Polsoni, L., Kott, L.S., Beversdorf, W.D., 1988. Large scale microspore culture techniques for mutation/selection studies in *Brassica napus*. Can. J. Bot. 66, 1681–1685.

Prem, D., Gupta, K., Agnihotri, A., 2012. Harnessing mutant donor plants for microspore culture in Indian mustard [*Brassica juncea* (L.) Czern and Coss]. Euphytica 184, 207–222.

Rahman, M.H., 2001. Production of yellow-seeded *Brassica napus* through interspecific crosses. Plant Breed. 120, 463–472.

Rai, U.K., 1958. X-ray induced appressed pod mutant in *Brassica juncea*. Sci. Cult. 24, 46–47.

Rai, U.K., Jacob, K.T., 1956. Induced mutation studies in sesamum and mustard. Sci. Cult. 22, 344–346.

Rai, B., Singh, D., 1993. A note on potential sources of dwarfing genes in Indian rapeseed (*Brassica campestris* L. Prain). Ind. J. Genet. 53, 153–156.

Rakow, G., 1973. Selektion auf linol- und linolensauregehalt in rapssamen nach mutagener behandlung. Z. Pflanzenzuchtg 69, 205–209.

Ramanujam, S., 1940. An apetalous mutation in turnip (*Brassica campestris* l.). Nature 145, 552–553.

Rana, M.A., 1985. Developmental morphology of flower and inheritance of an apetalous mutant in *Brassica carinata*. Dessertation, University of California Davis, Abstr. Int. B. (Sci. and Eng.) 45, 3425B.

Raza, G., Siddique, A., Khan, I.A., Ashraf, M.Y., Khatri, A., 2009. Determination of essential fatty acid composition among mutant lines of canola (*Brassica napus*), through high pressure liquid chromatography. J. Integr. Plant Biol. 51, 1080–1085.

Rehman, A., 1996. New mutant cultivar. MBNL 42, 27.

Rehman, A., Das, M.L., Hawlidar, M.A.R., Mansure, M.A., 1987. Promising mutants in *Brassica campestris*. MBNL 29, 14–15.

Robbelen, G., 1990. Mutation breeding for quality improvement a case study for oilseed crops. Mutation Breeding Review No.6, pp. 1–43.

Robbelen, G., Nitsch, A., 1975. Genetical and physiological investigations on mutants for polyeonic fatty acid in rapeseed, *Brassica napus* L. I. Selection and description of new mutants. Z. Pflanzenzuchtg. 75, 93–105.

Robbelen, G., Rakow, G., 1970. Selection for fatty acids in rapeseed. In: Proceedings of the International Conference on Rapeseed. Sainte-Adele, Canada. pp. 476–490.

Roscoe, T.J., Lessire, R., Puyaubert, J., Renard, M., Delseny, M., 2001. Mutations in the fatty acid elongation 1 gene are associated with a loss of beta-ketoacyl-CoA synthase activity in low erucic acid rapeseed. FEBS Lett. 492, 107–111.

Rucker, B., Robbelen, G., 1997. Mutants of *Brassica napus* with altered seed lipid fatty acid composition. In: Proceedings of the Twelfth International Symposium. Plant Lipids, Kluwer Academic Publisher. Dordrecht, The Netherlands. pp. 316–318.

Sacriston, M.D., 1982. Resistance response to *Phoma lingam* of plants regenerated from selected cell and embryogenic cultures of haploid *Brassica napus*. Theor. Appl. Genet. 61, 193–200.

Scarth, R., McVetty, R.B.E., Rimmer, S.R., Stefansson, B.R., 1988. Stellar low linolenic-high linoleic acid summer rape. Can. J. Plant Sci. 68, 509–511.

Scarth, R., Rimmer, S.R., McVetty, P.B.E., 1995. Apollo low linolenic summer rape. Can. J. Plant Sci. 75, 203–204.

Schnurbusch, T., Möllers, C., Becker, H.C., 2000. A mutant of *Brassica napus* with increased palmitic acid content. Plant Breed. 119, 141.

Shah, S.A., Ali, I., Rahman, K., 1990. Induction and selection of superior genetic variables of oilseed rape, *Brassica napus* L. Pak. J. Bot. 7, 37–40.

Shah, S.A., Ali, I., Rahman, K., 1998. Use of induced genetic variability for the improvement of oilseed rape (*Brassica napus* L.). In: Proceedings of the Second International Symposium. New Genetical Approaches to Crop improvement-II, Nuclear Institute of Agriculture. Tandojam, Pakistan. pp. 229–237.

Shah, S.A., Ali, I., Iqbal, M.M., Khattak, S.U., Rahman, K., 1999. Evolution of high yielding and early flowering variety of rapeseed (*Brassica napus* L.) through *in vivo* mutagenesis. In: Proceedings of the Third International Symposium. New Genetical Approaches to Crop improvement–III, Nuclear Institute of Agriculture. Tandojam, Pakistan. pp. 47–53.

Sheikh, F.A., Lone, B., Najeeb, S., Shikari, A.B., Parray, G.A., Rather, A.G., Khudwani, R.R., Khudwani, R.S., 2009. Induced mutagenesis for seed quality traits in Ethiopian mustard (*Brassica carinata* a. braun). J. Agri. Biol. Sci. 4, 42–46.

Shi, S., Wu, J., Liu, H., 1995. *In vitro* mutation for long pod and short plant. Nucleic Agric. 9, 252–253.

Shirzadegan, M., Robbelen, G., 1985. Influence of seed colour and hull proportions of seeds in *Brassica napus* L. Fette. Seifen Anstrichmittel 87, 235–237.

Simbaya, J., Slominski, B.A., Rakow, G., Campbell, L.D., Downey, R.K., Bell, J.M., 1995. Quality characteristics of yellow seeded *Brassica* seed meals: protein, carbohydrates and dietary fibre components. J. Agric. Food Chem. 43, 2062–2066.

Singh, D., 1961a. Heredity changes in number of petals in Brown Sarson. Indian Oilseed J. 5, 190–193.

Singh, D., 1961b. An apetalous mutant in toria brown sarson (*B. campestris* var. brown sarson) and its inheritance. Curr. Sci. 30, 62–63.

Singh, D., Rana, J.S., 2002. Insect-pest resistance breeding in oilseeds *Brassica*. Brassica 4, 52–57.

Singh, K.P., Sareen, P.K., 2004. Induced podding mutants of Indian mustard (Brassica juncea L. Czern & Coss.). Cruciferae Newsl. 25, 17.

Slominski, B.A., Simbaya, J., Campbell, L.D., Rakow, G., Guenter, W., 1999. Nutritive value for broilers of meals derived from newly developed varieties of yellow- seeded canola. Anim. Feed. Sci. Technol. 78, 249–262.

Spasibionek, S., 2006. New mutants of winter rapeseed (*Brassica napus* L.) with changed fatty acid composition. Plant Breed. 125, 259–267.

Srinivasachar, D., Malik, R.S., 1972. An induced aphid reistant non-waxy mutant in turnip, *Brassica rapa*. Curr. Sci. 41, 820–821.

Stefanson, B.R., Hougen, F.W., Downey, R.K., 1961. Note on the isolation of rape plants with seed oil free from erucic acid. Can. J. Plant. Sci. 41, 218–219.

Stephenson, P., Baker, D., Girin, T., Perez, A., Amoah, S., King, G.J., Østergaard, L., 2010. A rich TILLING resource for studying gene function in *Brassica rapa*. BMC Plant Biol. 10, 62–71.

Stringam, G.R., 1973. Inheritance and allelic relationships of seven chlorophyll-deficient mutants in *Brassica campestris* L. Can. J. Genet. Cytol. 15, 335–339.

Stringam, G.R., McGregor, D.I., Pawlowski, S.H., 1974. Chemical and morphological characteristics associated with seed coat colour in rapeseed. In: Proceedings of the Fourth International Rapeseed Congress. Giessen, West Germany. pp. 99–108.

Sun, T.P., 2011. The molecular mechanism and evolution of GA GID1 DELLA signaling module in plants. Current Biology 21, R338–R345.

Sun, T.P., Gubler, F., 2004. Molecular mechanism of gibberellin signaling in plants. Annu. Rev. Plant Biol. 55, 197–223.

Swanson, E.S., Coumans, M.P., Brown, G.L., Patel, J.D., Beversdorf, W.D., 1988. The characterization of herbicide tolerant plants in *Brassica napus* L. after *in vitro* selection of microspores and protoplasts. Plant Cell Rep. 7, 83–87.

Swanson, E.S., Herrgesell, M.J., Arnoldo, M., Sippell, D., Wong, R.S.C., 1989. Microspore mutagenesis and selection: canola plants with field tolerance to the imidazolinones. Theor. Appl. Genet. 78, 525–530.

Tanhuanpaa, P., Vikki, J., Vihinen, M., 1998. Mapping and cloning of *FAD2* gene to develop allele-specific PCR for oleic acid in spring turnip rape (*Brassica rapa* ssp.oleifera). Mol. Breed. 4, 543–550.

Thaganal, W.M., Ndirangu, C.M., Omolo, E.O., Riungu, T.C., 2013. Variability in M_2 generations and characteristics of advanced mutant lines of rapeseed (*Brassica napus* L.). J. Agri. Sci. 4 (2), 83–86.

Thurling, N., Depittayanan, V., 1992. EMS Induction of early flowering mutants in spring rape (*Brassica napus*). Plant Breed. 108, 177–184.

Tian, E., Zeng, F., MacKay, K., Roslinsky, V., Cheng, B., 2014. Detection and molecular characterization of Two *FAD3* genes controlling linolenic acid content and development of allele-specific markers in yellow mustard (*Sinapis alba*). PLoS One 9 (5), e97430.

Till, B.J., Colbert, T., Tompa, R., Enns, L.C., Codomo, C.A., Johnson, J.E., Reynolds, S.H., Henikoff, J.G., Greene, E.A., Steine, M.N., Comai, L, Henikoff, S., 2003. High-throughput TILLING for functional genomics. In: Erich Grotewold (Ed.), Plant Functional Genomics, Springer, Germany, Humana Publisher press, 236, 205–220.

Turner, J., Facciotti, D., 1990. High oleic acid *Brassica napus* from mutagenized microspores. In: McFerson, J.R., Kresovich, S., Dwyer, S.G. (Eds.), Proceedings of the Sixth Crucifer Genetics Workshop, Geneva, NY, p. 24.

Tyagi, D.V.S., Rai, B., Verma, R.B., 1983. A note on the bunching dwarf mutant in toria. Ind. J. Genet. 43, 374–377.

Uddin, M.A., Ullah, M.A., Sultana, F., Rahman, K.M., Rahman, M.Z., 2012. Evaluation of some rapeseed mutants based on morpho-physiological, biochemical and yield attributes. J. Environ. Sci. Nat. Resour. 5, 281–285.

Velasco, L., Fernandez-Martinez, J.M., De Haro, A., 1995. Isolation of induced mutants in Ehtiopean mustard (*Brassica carinata* Braun). Plant Breed. 116, 396–397.

Velasco, L., Fernandez-Martinez, J.M., De Haro, A., 1997. Selection for reduced linolenic acid content in Ethiopian mustard (*Brassica carinata* Braun) with low levels of erucic acid. Plant Breed. 114, 454–456.

Velasco, L., Fernandez-Martinez, J.M., De Haro, A., 1998. Increasing erucic acid content in Ethiopian mustard through mutation breeding. Plant Breed. 117, 85–87.

Verma, V.D., Rai, B., 1980. Mutation in seed-coat colour in Indian mustard. Ind. J. Agri. Sci. 50, 545–548.

Wang, M.L., Zhao, Y., Chen, F., Yin, X.C., 2004. Inheritance and potentials of a mutated dwarfing gene ndf1 in *Brassica napus*. Plant Breed. 123, 449–453.

Wang, N., Wang, Y., Tian, F., King, G.J., Zhang, C., Long, Y., Shi, L., Meng, J., 2008. A functional genomics resource for *Brassica napus*: development of an EMS mutagenised population and discovery of *FAE1* point mutations by TILLING. New Phytol. 180, 751–765.

Wong, R.S.C., Swanson, E., 1991. Genetic modification of canola oil: high oleic acid canola. In: Haberstroh, C., Morris, C.E. (Eds.), Fat and Cholesterol Reduced Food. Gulf, Houston, Texas, pp. 154–164.

Woods, D.L., 1980. The association of yellow seed coat with other characters in mustard Brassica juncea. Cruciferae Newl. 5, 23–24.

Xiao, D., 1982. Analysis of the correlation between seed coat colour and oil contents of *Brassica napus* L. Acta Agron. Sin. 8, 24–27.

Yu, C.Y., Hu, S.W., Zhang, C.H., Yu, Y.J., 2004. The discovery of novel flower color mutation in male sterile rapeseed (*Brassica napus* L.). Yi Chuan 26, 330–332.

Zanewich, K.P., Rood, S.B., Southworth, C.E., Williams, P.H., 1991. Dwarf mutant of *Brassica*: responses to applied gibberellins and gibberellin content. Plant Growth Regul. 10, 121–127.

Zhao, Y., Wang, M.L., Zhang, Y.Z., Du, L.F., Pan, T., 2000. A chlorophyll reduced seedling mutant in oilseed rape, *Brassica napus*, for utilization in F_1 hybrid production. Plant Breed. 119, 131–135.

CHAPTER

17

Pollination Interventions

Uma Shankar, Dharam P. Abrol

Faculty of Agriculture, Division of Entomology, Sher-e-Kashmir University of Agricultural Sciences & Technology Chatha, Jammu, Shalimar, J&K, India

INTRODUCTION

Edible vegetable oils are a high-value agricultural commodity and demand for high-quality seed oils continues to grow as the world's population increases (Wittkop et al., 2009). With insect pollinators enhancing yields in almost 70% of the crops worldwide their decline poses a genuine threat to global food security (Klein et al., 2007). Insect pollination enhances yield in oilseed rape by approximately 25% and selecting varieties with higher nectar secretion is likely to have positive implications for both pollinator populations and agricultural production. Furthermore, oilseed rape is economically important for commercial beekeepers as it is the main source of nectar in spring (Bommarco et al., 2012). Within intensively managed agricultural landscapes, natural or seminatural components provide important nesting and foraging sites for wild pollinators and proximity to such habitats increases pollinator species richness, crop visitation rates, and thus pollination success (Garibaldi et al., 2014). Maintaining pollinator diversity can ensure resilience of pollination as an ecosystem service due to species' showing a differential response to environmental change (i.e., response diversity). Although, pollination improves the yield of most crop species and contributes to one third of the global crop production, the decline of pollinators is alarming in the scenario of agricultural intensification due to loss of floral resources and diminishing pollination services (Klatt et al., 2014; Vanbergen et al., 2013; Wratten et al., 2012).

When the natural process of pollination is mediated through the use of bees in a planned manner such an intervention can significantly influence the quality and quantity of crop yield. This practice has become an essential component of our modern cultivation system in order to enhance the productivity of crops to maintain the quality and quantity of food commodities.

Oilseed crops have a special significance in India due to their contribution to the Indian economy. India occupies a prominent position, both in regard to acreage and production of oilseeds.

Breeding Oilseed Crops for Sustainable Production. http://dx.doi.org/10.1016/B978-0-12-801309-0.00017-3

Cultivation of oilseeds constitutes the second largest agricultural commodity after cereals in India, sharing 14% of the country's gross cropped area. These crops are cultivated on about 16.5 million hectares, with a total production of 10 million tons (Abrol and Shankar, 2012).

The important oilseed crops grown in India are rapeseed mustard, sesame, linseed, safflower, sunflower, niger, and linseed (Guidry, 1964). Most of the oilseed crops are either fully dependent upon cross-pollinating agents for seed production or benefit greatly by insect pollination. Inadequate pollination results not only in reduced yields but also in delayed yield and a high percentage of culls or inferior fruits. Although self-pollination (SP) can set fruits, insect pollination has a number of advantages as far as quality or quantity of seed/fruit production is concerned. Insect pollination increases either: (i) the proportion of fruits set or (ii) the quality of fruits set, because fruit size depends on the number of seeds set or the size of seeds. Seed size is generally greater after cross pollination by insects (Charlesworth and Charlesworth, 1987; Richards, 1997). A number of oilseed crops such as oilseed rape (Free and Ferguson, 1983; Williams et al., 1987), flax, linseed, and sunflower (Free and Williams, 1976) belong to a group sharing this pollination system.

The level of productivity of oilseeds in India is far behind the average productivity in the rest of the world. One of the factors responsible for the low productivity of oilseed crops is the failure of proper seed setting due to a lack of pollination (Rao et al., 1980; Free, 1993; Abrol, 2007, 2008, 2009). Insect pollinators not only enhance the yield of the crop but also contribute to uniform and early setting. Therefore, cross pollination (CP) of entomophilous crops by honey bees is considered as one of the most effective and econoomic methods for triggering good crop yields. Applied pollination, pollinator management, and managed pollination or interventions are among the common efforts most recently being practiced to optimize/maximize production in cross-pollinated crops and to bring the pollinators to target crops (Figs. 17.1 and 17.2).

Among the various pollinating agents, insect pollinators play a predominant role in increasing the yield of oilseed crops (Mishra et al., 1988). Important oil crops that need pollination interventions for enhancing productivity are rapeseed mustard, sunflower, safflower, sesame,

FIGURE 17.1 ***Apis mellifera* apiary nearby a mustard field.**

FIGURE 17.2 **Pollination intervention in mustard.**

niger, taramira, and linseed. The pollination requirements, type of pollinators, and their behavior and efficiency in the pollination and enhanced production of different crops is briefly described further.

RAPESEED MUSTARD AND CANOLA (*BRASSICA* SPP.)

The rapeseed mustard group is mostly cross pollinated, although some varieties are self-fruitful. Selfing has been reported to reduce seed yield, seed size, and yield in subsequent generations (Delaplane and Mayer, 2000). Self-incompatible plants require pollen transfer from plant to plant (Wallace et al., 2002). Pollination interventions in the form of planned pollination (Figs. 17.3 and 17.4) has been found to greatly benefit the quality and quantity of seed production in crucifers. For instance, in male sterile oilseed rape (*Brassica napus*), yields

FIGURE 17.3 **Planned pollination in mustard.**

FIGURE 17.4 **Conservation of non-*Apis* bees for oilseed pollination.**

of the species were increased by honey bee pollination (BP) (Westcott and Nelson, 2001). Similarly, in sarson (*Brassica campestris*), pollination by insects increased the seed yield, caused formation of well-shaped, larger grain, and produced more viable seed (Abrol, 2009). The impact of insect pollination on production of different oil crops is presented in Table 17.1.

Many species of *Brassica*, such as rape (*Brassica napus*), sarson (*Brassica campestris* var. Sarson), toria (*Brassica campestris* var. Toria), Indian mustard or rai (*Brassica juncea*), white mustard (*Brassica alba*), and black mustard (*Brassica nigra*) are cultivated widely as oilseed crops throughout the world. Most of these crops bloom during February–March for over a month. The bright yellow, fragrant flowers are produced in long terminal racemes which are highly attractive to honey bees and other natural insect pollinators. Mohammed (1935)

TABLE 17.1 Impact of Animal Pollination on the Production of Oil Crops

Crop species	Commodity	World production (*t*)	Animal pollination	References
Brassica napus	Rapeseed, oilseed rape	46,770,903	Increase	Free (1993); Adegas and Nogueira Couto (1992); Abel and Wilson (1999); Bürger (2004); Manning and Boland (2000); Abel et al. (2003); Morandin and Winston (2005)
Helianthus annuus	Sunflower	26,460,824	Increase	Bichee and Sharma (1988); Crane (1991); Free (1993); De Grandi Hoffman and Martin (1993); Moreti et al. (1996); Heard (1999); De Grandi-Hoffman and Watkins (2000); Dag et al. (2002); Greenleaf and Kremen (2006)

in West Punjab at Lyallpur recorded 117 visiting species of insects belonging to 7 orders; *Andrena ilerda*, *Apis florea*, and *Halictus* spp. comprised 82% of these insects and were the most important in this order. He also observed that the stigma of the crops remained receptive for 2 days after opening and in 5 days their fertility was lost, whereas the pollen remained viable for 7 days. It was observed in Australia that among *A. mellifera* foragers 72% collected nectar only, 25% collected both nectar and pollen, and 3% collected pollen only (Langridge and Goodman, 1975). The pollen was collected between 10.00–14.00 hours.

Mohammed (1935) reported that when enclosing individual plants in muslin bags, seed set was 12.3% in toria, 20.3% in brown sarson, and 91.0% in yellow seeded sarson. He further reported that in the case of hand pollinated toria flowers, there was 100% pod formation. In a similar study, Sihag (1986) studied seed set in toria plots that were caged and not caged (left open) for natural CP; there developed 25 and 1060 pods per plant, 6.3 and 18.2 seeds per pod, and 4 and 385 seeds per plant. Similar observations for sarson gave 59% and 95% pods containing 3.5 and 12.7 seeds per pod (Mishra et al., 1988). Observations made in field plots of both sarson and toria showed that when the crop was caged, to exclude insects, yield was 68 g; when caged with *Apis cerena* having access the yield was 219 g; and in an open field near honey bee colonies where other insect pollinators were also present the yield was 244 g (Latif et al., 1960). In general, there is quite a high degree of self-incompatibility in these crops and good yields are obtained by CP with the aid of natural fauna in most places where good weather prevails. Yield can most likely be further increased with the help of honey bee colonies to the extent of 10–30% (Rahman, 1940; Singh, 1954) allowing for inclement weather. Singh and Singh (1992) observed that bee pollinated plants, in field cages, compared with self-pollinated plants, in individual bags, produced 3 times heavier pods, 4 times more seed per pod, 50 times more seeds, and 84 times more seeds yielded per plant.

Singh and Singh (1992) studied the impact of BP on seed yield, carbohydrate composition, and lipid composition of mustard seed *Brassica campestris* L. and found that bee pollinated plants produced 3 times heavier pods, 4 times more seeds per pod, 50 times more seeds per plant, 11 times more pods per plant, and 84 times more seed yield per plant than self-pollinated plants. Carbohydrate content was inversely proportional to lipid content. In the seeds of self-pollinated plants the total carbohydrate content was about twice that of seeds from bee pollinated plants. Triglycerides constituted the majority of the neutral lipids. In the seeds of beep ollinated plants triglycerides constituted about 74% of the total nonpolar lipids, which was about 20 times more than in self-pollinated plants. Sterol was the least abundant of all the lipids and phosphatidylcholine was absent from all seeds. High lipid content was directly related to high seed yield, and concentration of total lipid increased according to type of pollination in the order SP < OP < HP < BP.

Deodikar and Suryanarayana (1972) found enhanced yield in oilseed crops when pollinated by bees (Table 17.2). Langridge and Goodman (1975) also found significant differences in production of rapeseed in open and caged plots. Plots receiving unrestrained visits had significantly higher plant density, seed yield, seeds per plant, 1000-seed weight, weight of seeds per plant, germination, and percentage of oil content (Table 17.3) compared with those having insect visits prevented.

From the seed yield data (Table 17.4) it is revealed that percentage pod setting was significantly higher in plants which had access to insect pollination (95.21%) than those which were net caged (70.27%) or muslin bagged (58.81%). Similarly, number of seeds per pod was also

TABLE 17.2 Yield in Self-Pollinated and Bee-Pollinated Crops

Crop botanical name	Reported range of seed increase from bee-pollinated over self-pollinated crops	
	Percentage increase	Times more
Brassica napus L. (rape)	12.8–139.3	1.128–2.39
Brassica campestris L. var. Toria (toria)	66.0–220.9	1.66–3.20
Brassica campestris L. var. Dichotoma (sarson)	222	3.22
Brassica juncea Czern & Coss. (rai, Indian mustard)	18.4	1.184
Brassica alba Boiss (white mustard)	128.1–151.8	2.28–2.51

Source: Deodikar and Suryanarayana (1972).

TABLE 17.3 Rapeseed Crop With and Without Pollinators

Attribute	Open plots	Closed plots	Significance of difference
Plant density	459	406	—
Seed yield	725	453	$P < 0.01$
Seeds per plant	1.61 g	1.10 g	$P < 0.01$
1000-seed weight	1.78 g	1.88 g	$P < 0.01$
No. of seeds per plant	927	588	$P < 0.01$
Percent germination	95.3	97.1	NS
% oil content	37.9	36.1	NS

TABLE 17.4 Yield and Oil Potential of *Brassica campestris* var. Sarson Under Three Conditions of Pollination

Parameter	OP (X+SE)	Net caged (X+SE)	Muslin bagged (X+SE)	Critical difference	Transformation
Percentage pod setting	95.21+2.46	70.27+4.38	58.81+6.57	(10.09)	Angular
Seeds per pod	12.72+0.82	5.55+1.18	3.95+0.85	(0.73)	$\sqrt{n+1}$
Percentage healthy seeds	84.78+2.76	41.45+8.25	48.37+12.28	(18.41)	Angular
Seed weight in mg per100 seeds	478.50+21.83	459.40+23.41	542.20+11.00	(1.31)	$\sqrt{n}$
Percentage oil content	39.02+0.29	40.76+0.90	41.64+0.27	1.81	—
Oil yield in mg per pod	19.03	3.02+	1.95		
Increase in oil yield over muslin bagged	9.76	1.55	1.0		

Source: Mishra et al. (1988).

significantly higher in open-pollinated flowers (912.62 seeds per pod) in comparison with net-caged flowers (5.55 seeds per pod) and muslin-bagged flowers (3.05 seeds per pod). Pods obtained from open-pollinated flowers had significantly more healthy seeds (84.78%) than those from net-caged (41.45%) or muslin-bagged (48.37%) flowers; the latter two, however, did not differ from each other. The average weight of 100 healthy, dry seeds from muslin-bagged flowers was significantly higher (542.20 mg) than the weight of seeds from open-pollinated (478.50 mg), and net-caged flowers (459.40 mg). The percentage oil content of healthy seeds from muslin-bagged (41.64%) flowers was significantly higher than that from open-pollinated (39.02%) flowers. However, the oil content of seeds from muslin-bagged flowers was equal to that from net-caged flowers. When all the yield parameters were taken into consideration for calculating oil content (mg pod^{-1}), it was found that the increase in total oil yield from open-pollinated and net-caged plants, was 9.76 and 1.55 times more, respectively, than muslin-bagged plants.

Singh et al. (2000) found that there was significant improvement in qualitative and quantitative parameters of rape when pollined by *Apis cerana himalaya* (Table 17.5).

Atmowidi et al. (2007) analyzed the diversity of pollinator insects and its effect on the seed set of mustard (*Brassica rapa*) planted in agricultural ecosystem in West Java. At least 19 species of insects pollinated the mustard, and 3 species, that is, *Apis cerana*, *Ceratina* sp., and *Apis dorsata* showed a high abundance. The highest abundance and species richness of pollinators occurred at 08.30–10.30 A.M. and the diversity was related to the number of flowering plants. Insect pollinations increased the number of pods, seeds per pod, seed weight per plant, and seed germination.

Chhuneja et al. (2007) studied the role of *Apis mellifera* in the seed production of *Brassica campestris* var. Toria and found that the mean number of siliqua per plant, number of seeds per 20 siliquae, 1000-seed weight, seed yield per plant, and oil content was significantly higher at 50 m and 75 m distances from bee colonies. It was further reported that one colony of *Apis mellifera* with a strength of 10 frame bees was sufficient for the effective pollination of 1 ha of crops. Araneda et al. (2010) evaluated the yield of *Brassica napus* cv. Artus pollinated by *A. mellifera* in Chile and found that the parameter least affected by bee intervention was the

TABLE 17.5 Quantitative and Qualitative Effects of *Apis cerena Himalaya* Pollination on Rapeseed

Yield parameters	Control (pollination without insects, PWI)	OP	BP	Critical Difference (CD) at 5%	Percentage increase over (PWI)	Percentage increase over (OP)
Siliqua set (%)	20.67 (26.39)	68.42 (56.01)	70.23 (57.24)	(5.00)	239.77	2.65
Siliqua length (cm)	3.94 (2.11)	4.16 (2.12)	4.52 (2.24)	(0.07)	14.72	8.65
No. of seeds per Siliqua	9.12 (3.14)	11.65 (3.52)	13.73 (3.80)	(0.24)	50.55	17.85
1000-seed weight (g)	2.01 (1.66)	2.31 (1.75)	2.36 (1.76)	NS	17.41	2.16
Seed yield (q/ha)	1.96	12.37	13.45	2.90	586.22	8.73
Seed germination (%)	55.00 (47.93)	81.00 (65.98)	83.00 (66.02)	(5.21)	5.90	2.50
Oil content (%)	33.25 (35.18)	38.75 (38.48)	40.33 (39.41)	(0.81)	21.30	4.10

Note: Figures in parentheses are transformed values.
Source: Singh et al. (2000).

grains per silique variable. In contrast, siliqua per plant and 1000-grain weight parameters presented significant differences, contributing to a yield greater than 5 t ha^{-1}; which represented a figure 50.34% higher than in the treatment without bees. He further concluded that the inclusion of bees in crops is fully justified as a production tool.

Sharma and Abrol (2014) found that toria crops benefit greatly from cross pollination. They found that the number of siliqua per plant in toria under open pollination (OP) was twice (279.33 siliqua/plant) that formed under caged conditions (128.99 siliqua/plant). Seed weight was lower under caged conditions than under OP, and a greater number of seeds, and better quality seeds, were obtained from OP. The combined effect was that OP yielded 1.80 times more seed than plants having no access to insect pollinators (Table 17.6). Sabir et al. (2000) found that honey bees maximized seed yield, 1000-grain seed yield, and germination percentage in *Brassica campestris*. Earlier, Sabbahi et al. (2005) found a significant improvement in the seed yield when honey bees were present. Munawar et al. (2009) reported significant increases in a range of plant parameters when caged with bees as compared with plants without bees. Abrol (2009) studied the pollination requirement of different oilseed crops and also the expected increase in yield due to BP. This is summarized in Table 17.7.

For rapeseed, although the wind is the principal vector in terms of the distances over which pollen is transported (Hoyle et al., 2007), bees are the principal pollinators, being the most abundant insects in the cultivars and varieties of rapeseed used for seed production (Westcott and Nelson, 2001; Pordel et al., 2007). Thus, the introduction of hives

TABLE 17.6 Seed Yield and Yield Parameters in Toria as Influenced by Mode of Pollination

Yield attribute	BP	OP	CP
Siliqua per plant	258.33 + 0.25	276.33 + 0.16	122.67 + 0.32
Seeds per siliqua	22.10 + 0.11	29.50 + 0.22	15.67 + 0.17
Pod length (cm)	6.23 + 0.006	7.37 + 0.18	5.26 + 0.01
1000-seed weight (g)	5.65 + 0.01	5.97 + 0.04	2.15 + 0.01
Seed yield (q/ha)	12.77 + 0.02	13.73 + 0.01	7.50 + 0.02

Source: Sharma and Abrol (2014).

TABLE 17.7 Expected Increase in Crop Yield due to BP

Crop	Pollination requirement	Expected percentage increase in yield due to cross pollination
Mustard	Often cross pollinated	13.00–222.00
Safflower	Often cross pollinated	4.00–114.00
Sunflower	Often cross pollinated	21.00–3400
Sesame	Often cross pollinated	24.00–40.00
Niger	Often cross pollinated	17.00–45.00
Linseed	Mostly cross pollinated	2.0–49.00

Source: Abrol (2009).

of *Apis mellifera* in controlled pollination helps to increase the production of crops such as rapeseed (*Brassica napus*) (Sabbahi et al., 2006), since this plant is characterized by the production of abundant pollen and good quality nectar at relatively high concentrations of sugar, in flowers with a color and structure which are attractive to insects, particularly bees (Smith, 2002; Sabbahi et al., 2006).

The studies clearly establish that most of the oilseed crops are cross pollinated and adequate pollination is vital for any significant seed production. An increase in seed yield as a result of insect pollination has been reported in mustard (Mohammed, 1935; Latif et al., 1960). Latif et al. (1960) found that *Apis cerana* colonies near sarson and toria fields increased the seed yield by 60%. Similar increases in fields have been reported for oilseed rape (Langridge and Goodman, 1975; Kisselhagen, 1977) and *Brassica campestris* var. Jambuck (Kubisova et al., 1980). Bisht et al. (1980) found that flowers of rapeseed visited by *Apis* species had higher pod set, a greater number of seeds per pod, with the weight of seed also higher than those deprived of pollinator visits. Mishra et al. (1988) found that in *Brassica campestris* the percentage pod setting, number of pods per plant, and proportion of healthy seeds was significantly higher in open pollinated flowers than in net-caged and muslin-bagged flowers. Similarly, average weight of seeds and oil content was higher in open pollinated flowers. *Apis cerana* was the most common pollinating species (Fig. 17.5). The other pollinators observed included *Apis mellifera, Apis dorsata*, bumblebee, solitary bee, syrphid flies (Figs. 17.6–17.10). In a similar study, Prasad et al. (1989) found that pollination *of Brassica juncea* by *A. cerana* resulted in more siliqua setting, increased length of siliqua, seed weight, increased yield and had a pronounced effect on oil content and germination (Table 17.8).

Mishra and Kaushik (1992) reported that yield and oil content was higher in honey bee pollinated crops compared with self-pollinated crops. The results showed that a higher percentage of pod setting was obtained in open pollinated crops of *Brassica nigra, Brassica campestris, Brassica carinata, Brassica napus*, and *Eruca sativa* having 80.00%, 84.42%, 71.90%, and 70.30%, respectively. Average 1000-seed weight (of healthy seeds) varied between 3.0–5.1 g under OP, whereas it was 1.8–3.7 g under SP. Langridge and Goodman (2003), in a trial on oilseed rape (*Brassica napus*) cv. Midas in northern Victoria, found that no increase in yield of

FIGURE 17.5 ***Apis cerana* foraging on mustard flowers.**

FIGURE 17.6 *Apis mellifera* foraging on mustard.

FIGURE 17.7 *Apis dorsata* on mustard.

FIGURE 17.8 *Bombus haemorrhoidalis* collecting nectar from crucifer.

FIGURE 17.9 **Solitary bee foraging on mustard.**

FIGURE 17.10 **Syrphid fly collecting nectar from mustard.**

TABLE 17.8 Effect of Different Pollination Treatments on Oil and Protein Content and Germination of Seeds in rai cv. Pusa Bold in Pusa, Bihar (India)

Treatment	Oil content±SE (%)*	Protein content±SE (%)**	Germination±SE (%)**
Caged with bees	31.80 (34.03±0.11)	17.80 (4.29±0.00)	93.30 (9.64±0.02)
Uncaged	32.10 (34.51±0.11)	18.44 (4.29±0.01)	98.00 (9.64±0.02)
Caged	26.30 (30.85±0.13)	6.39 (4.05±0.01)	73.00 (8.54±0.03)
Coefficient of variance (CV) (%)	(0.9)	(0.6)	(0.1)
Standard error (mean)	(3.09)	(0.01)	(0.0)
Latin square design (LSD) (5%)	(0.2)	(0.02)	(0.01)

* *Figures in parentheses are values transformed into angles.*
** *Figures in parentheses are values transformed in square roots.*
Source: Prasad et al. (1989).

FIGURE 17.11 *Apis mellifera* **with collected nectar and pollen taken to a beehive.**

seed, oil content, or percentage germination was obtained from plots where bees and larger insects had access compared with plots enclosed in cages of 2.5 mm mesh to exclude these insects. The 1000-seed weight was slightly greater in the enclosed plots. Bees and a syrphid, *Melangyna viridiceps*, were the predominant insects visiting the flowers. The bees stored some surplus honey and pollen and built up the colony population (Fig. 17.11). SP and possibly wind pollination appear to be the norm for this cultivar.

Evidently, pollination is one of the most important natural factors enhancing crop production. Among the various pollinating agents, insect pollinators play a predominant role in increasing the yield of oilseed crops. Almost one third of the total cropped area under oilseeds has been reported to be entomophilous (Mishra et al., 1988). Rapeseed yield can be doubled through pollination by insects. Pollinators not only enhance the yield of the crop but also contribute to uniform and early pod setting. Being a cross-pollinated crop rapeseed attracts a large number of insect pollinators.

SUNFLOWER (*HELIANTHUS ANNUUS* L.; FAMILY COMPOSITAE)

In the case of sunflower (*Helianthus annuus* L.) there is strong evidence that insufficient pollination can significantly minimize its yields (Free, 1999). Sunflower benefits from insects that visit its flowers for pollen or nectar. Honey bee (*Apis mellifera* L.) pollination increases the seed yield by 30% and oil content by more than 6% in hybrid varieties (Furgala et al., 1979; Jyoti and Brewer, 1999). Honey bees are known as the main pollinators of sunflower in most parts of the world (Figs. 17.12 and 17.13). For example, in Viamão (Brazil), 96% of the sunflower insect visitors were *A. mellifera* L. (Hoffmann, 1994). Other bees, especially the non-*Apis* bees are also known to frequent sunflower but are largely perceived as unreliable and ineffective pollinators mainly due to their low activity (Radford et al., 1979). However, De Grandi-Hoffman and Watkins (2000) noted a possible indirect role of non-*Apis* bees, shown by their ability to enhance pollination by *Apis mellifera* (Figs. 17.14–17.16).

FIGURE 17.12 Pollination intervention in sunflower.

FIGURE 17.13 *Apis mellifera* pollinating sunflower.

FIGURE 17.14 *Megachile* bee foraging on sunflower.

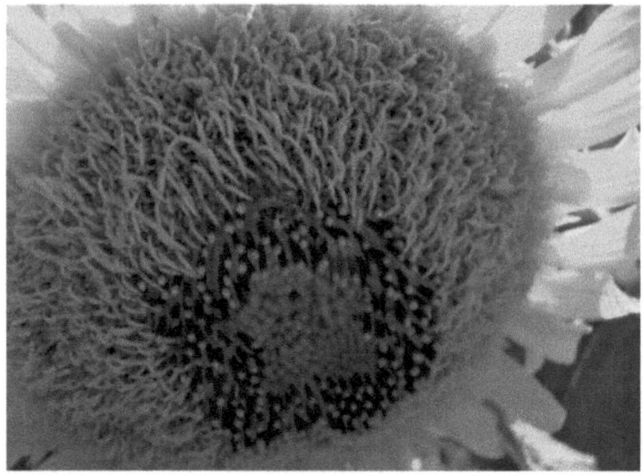

FIGURE 17.15 *Pithetes smargdula* **working on sunflower.**

FIGURE 17.16 **View of nesting materials for non-*Apis* bees.**

Recently, Greenleaf and Kremen (2006) showed clearly that non-*Apis* bees are beneficial in sunflower pollination in their behavioral interactions with honey bees. Honey bees are forced to move to many flowers when several non-*Apis* bees forage at the same time and in the same floral heads. This improves the efficiency with which honey bees pollinate sunflower. Other than bees, different insect species also visit the flowers of sunflower.

In an attempt to determine the effect of honey bee visits on sunflower seed production in Manitoba, Furgala (1954b) placed 2.5 colonies per hectare along the eastern border of sunflower fields. The seed production per hectare in these two fields was as much as that recorded in four fields located 4.8 km away from the honey bee colonies. Furthermore, he noted that seed yield in the farmers field decreased progressively as the distance increased from east to west.

Insect pollination of sunflower has also been studied in Pakistan (Manzoor-ul-haq and Fiaz, 1980). Wakhle et al. (1978), on the basis of chemical analysis of sunflower seeds obtained from different pollination treatments, showed that there was a significant increase in oil (6.5%), as well as oil and protein content together (7%), in seeds resulting from BP when compared with self-seeds (Table 17.9).

Evidently BP is beneficial giving quantitative and also qualitative improvement in sunflower seeds. In a similar study Singh et al. (2000) reported that seeds in BP and CP treatments have higher oil percentages and higher oil and protein contents compared with those in SP or even OP treatments. Seeds under OP, BP, and CP treatments showed nearly 6%, 7%, and 8.5% increases in oil and protein contents over SP treatments, respectively (Table 17.10). Oz et al. (2009) determined the efficiency of pollination with honey bees (*Apis mellifera*) on sunflower hybrid seed production under (1) cages with honeybees, (2) hand pollination (HP) (in cages), and (3) cages without honey bees. They found that seed set ratios were 98–99% for pollination in cages with honey bees or artificially by hand, whereas this ratio was reduced to a level of 4–5% pollination in cages without honey bees. Pollination in cages with honey bees and by hand increased seed yield per head by about 206–226%, respectively, compared with pollination in cages without honey bees. They further added that the use of honey bees for sunflower hybrid seed production improved seed set ratio, 1000-seed weight, number of filled seeds per head, and seed yield per head. Similarly, Rajasri et al. (2012) in a 2-year study

TABLE 17.9 Average Value of Moisture, Oil, and Protein Content in Sunflower Seed Variety, EC 68414

Type of treatment	Number of samples	Moisture	Oil*	Protein*	Oil and protein*
SP	6	8.43 (0.142)	32.44 (2.187)	17.83 (2.48)	49.72 (3.94)
OP	7	10.197 (1.66)	33.978 (4.0.34)	25.55 (3.13)	55.47 (4.33)
BP	9	9.75 (2.048)	38.81 (3.41)	17.90 (2.952)	56.71 (4.35)
CP	12	8.533 (0.0681)	38.94 (3.045)	20.33 (4.051)	58.29 (4.62)

Values in parentheses indicate standard deviation.
* *The figures are percentages of oil and protein content on a moisture-free basis.*
Source: Wakhle et al. (1978).

TABLE 17.10 Quantitative and Qualitative Effects of *Apis Cerana Himalaya* Pollination on Sunflower

Yield parameters	Control (pollination without insects, PWI)	OP	BP	CD at 5%	Percentage increase over (PWI)	Percentage increase over (OP)
Seed set (%)	2.93 (9.84)	60.50 (51.13)	71.58 (57.81)	(2.20)	2343.00	18.31
1000-seed weight (g)	50.32	64.29	54.88	3.26	9.06	17.15
No. of seeds per gram	20.87 (4.63)	16.58 (4.13)	17.04 (4.19)	2.77 (0.37)		
Seed yield (q/ha)	2.06	16.60	18.55	1.47	800.49	11.75
Seed germination (%)	69.88 (56.78)	86.50 (68.80)	91.13 (74.55)	(11.94)	30.41	5.35
Oil content (%)	35.96 (36.85)	46.79 (43.16)	43.70 (41.38)	(1.32)	21.52	7.07

Figures in parentheses are transformed values.
Source: Singh et al. (2000).

on sunflower hybrid NDSH1 found that the crop covered with insect proof netting without honey bees recorded significantly lower seed setting (%) in both years studied, with a mean of 6.33% compared with other methods (Table 17.11).

All other pollination methods where honey bees were introduced were found to be significantly superior to OP (16.52%). The highest mean seed setting percentage of 74.13% was recorded with BP combined with HP (Fig. 17.17) followed by BP with 8 frames and 4 frames of bees (50.89% and 33.26% respectively). Honey bees + HP gave significantly higher seed yield (14.4 q ha^{-1}) than OP (6.4 q ha^{-1}) and for crops caged with nets (2.3 q ha^{-1}). Yields were increased about 526% with honey bee + HP and 178% with OP compared with control

TABLE 17.11 Effect of Honey Bees on Seed Setting (%) and Seed Yield for the Hybrid Sunflower, NDSH1

	Seed set (%)			Seed yield (q/ha)			
Treatments	**2006–2007**	**2007–2008**	**Mean**	**2006–2007**	**2007–2008**	**Mean**	**Percentage increase in yield over control**
OP	23.88	10.23	16.52	4.03	8.70	6.4	178
Covered with a net	11.88	1.16	6.33	2.06	2.50	2.3	—
Caged with 4-frame colony	19.87	47.1	33.26	3.34	10.08	6.7	191
Caged with 8-frame colony	51.6	50.18	50.89	8.64	11.15	9.9	330
BP and HP	87.8	60.46	74.13	15.3	13.54	14.4	526
F-test	Sig	Sig		Sig	Sig		
CD	4.90	3.27		0.71	2.29		
CV	6.20	4.6		5.59	17.64		

Source: Rajasri et al. (2012).

FIGURE 17.17 **HP in sunflower.**

TABLE 17.12 Self-Incompatibility Test in Different Experimental Conditions

Types of pollination	Total number of flowers	Number of filled seeds	Number of unfilled seeds	Percentage of seeds setting
Naturally pollinated heads	54.75+7.70	47.08+13.78	7.67+6.07	85.99
Butter paper bagged head	54.76+5.50	19.46+8.88	35.3+3.38	35.54
Muslin cloth bagged head	51.53+5.35	19.66+4.67	31.87+0.68	38.15

Source: Pandey and Kumari (2007).

plots. There were about 330% and 191% increased yields of sunflower due to the introduction of honey bees of 8-frame and 4-frame colonies respectively. The supplemental honey bee pollination + HP significantly increased the percentage of seed setting and seed yield compared with OP and crops in cages without honey bees. The hybrid sunflower seeds obtained from honey bee pollination coupled with HP showed significant superiority with a higher percentage of germination accounting for 99% followed by OP (96%) and BP (95%) compared to control plots without bees (93%).

SAFFLOWER (*CARTHAMUS TINCTORIUS* L.; FAMILY ASTERACEAE)

Seed production in safflower is directly related to the success of pollination because the plants show SP in the absence of pollinators (Knowles, 1969). Classen (1950) reported 0–100% CP. In most of the plants, CP ranged from 5–40%. Pollinators contribute to various degrees of pollination of the flower (Kadam and Patankar, 1942; Levin and Butler, 1966; Butler et al., 1966; Levin et al., 1967). Safflower is usually considered to be a self-pollinated crop. Insects, particularly bees, are the major agents of pollination (Boch, 1961; Eckert, 1962; Rubis et al., 1966). Temperature and humidity affect the seed setting of bagged flowers (Patil and Chavan, 1948). Pandey and Kumari (2007) found that there was 85.99% seed setting in open pollinated heads followed by 38.15% (in muslin cloth) and 35.54% (in butter paper) bagged conditions (Table 17.12).

SESAME (*SESAMUM INDICUM* L.; FAMILY PEDALIACEAE)

Sesame (*Sesamum indicum* L.) is one of the oldest and most important oilseed crops cultivated on about 7.8 million hectares of the world's total crop area (FAOSTAT, 2012). Kamel et al. (2013) recorded 29 insect pollinators associated with sesame in which *Apis mellifera* and *Ceratina tarsata* were the most dominant species. In the case of sesame, the plants exposed to insects gave 25% higher yield compared with those covered with cages to exclude insects. Both OP and BP treatments were effective in increasing the seed yield of sesame from 22–33% compared to pollination without insects (PWI) (Panda et al., 1988) and quantitative and qualitative parameters had also been improved in OP (Mahmoud, 2012) (Figs. 17.18–17.21). *Apis cerena, Apis*

FIGURE 17.18 **A view of sesame cultivation on Jammu hills.**

FIGURE 17.19 ***Apis dorsata* foraging on a sesame flower.**

FIGURE 17.20 ***Nomia* bee pollinating a sesame flower.**

FIGURE 17.21 ***Bombus haemorrhoidalis*** **on a sesame flower.**

dorsata, and *Apis florea* were the most common insects on the crops and they began foraging at 0600, 0700, 0800 hours visiting 6.3, 5.0, and 1.9 flowers per minute in open conditions. Flowers pollinated by them gave 51, 53, and 4 seeds per pod, respectively. Seed weight was also heavier compared with the control group (Rao et al., 1981). While foraging 72% of *A. cerena* collected nectar, 23% pollen only, and 4% both nectar and pollen. At one time or another they touched anthers bringing about CP and thus increased yields (Phadke et al., 1967). *Apis mellifera* is also known to work on flowers very actively. Abrol (1991) reported *A. dorsata* as an important pollinator of sesame and found that field activities of bees were made more intense at higher temperatures, light intensities, and radiation. Extensive studies were made in Punjab on the genetics and pollination requirement of this crop (Sikka and Gupta 1949).

LINSEED/FLAX (*LINUM USITATISSIMUM* L., FAMILY LINACEAE)

Flax is considered to be normally self-pollinated although some crossing (1–6%) occurs (Dillman, 1938; Dillman and Goar,1937; Masuo, 1958), mostly among large-flower types. Rubis (1970) worked with a male-sterile line having disk-form flowers, and stated that he obtained practically no CP of the male sterile lines with lines having tubular flowers; however, good seed set was obtained, indicating heavy CP with other lines that had large, disk-form flowers.

Several tests have shown that BP improves seed yields in fiber flax. Bezdenezhoykh (1956) reported that in Russia honey bees in cages increased seed production of fiber flax by 22.5% over plants in cages without bees. Gubin (1945) also studied the effect of BP of fiber flax in Russia and reported that bees increased seed production by 22.5–38.5%. Luttso (1957), also in Russia, reported that BP increased seed production by 29%, the number of seeds per capsule by 18%, and the weight per seed by 11% in comparison with fields without BP. Likewise, Smirnov (1956) showed a 19% increase in the number of seeds per capsule, a 22% increase in the total weight of seeds, and a 2.2% increase in the weight per seed. He also reported that plants visited by bees set up the crop and ceased blooming earlier than plants from which bees

TABLE 17.13 Seed Yield and Yield Parameters in Different Oilseed Crops as Influenced by BP

		Pollination	
Crops (variety)	**Yield attributes**	**With bees**	**Without bees**
Mustard (var. M-27)	Pod set (%)	71.90	48.60
	Seeds per pod	10.80	5.90
	1000-seed weight (g)	5.00	2.00
	Seed yield (q/ha)	13.90	1.20
	Oil content (%)	36.40	32.60
Niger (var. M-15)	Seed set (%)	45.80	28.40
	Seed yield (q/ha)	3.79	3.11
	Total florets	45.75	33.42
	Oil content (%)	35.20	32.00
	Average seed set (%)	23.80	14.18
	seed weight (g/head)	0.80	0.07
Sesame (var. Kalika)	Average yield (q/ha)	8.47	5.64
	Oil content (%)	36.80	36.40
Sunflower (var. Modern)	Seed yield (g/head)	23.80	11.60
	No. of seeds per head	540.80	418.25
	Filled seed per head (%)	62.40	58.65
	Seed yield (q/ha)	27.54	9.78

were excluded. Kozin (1954) reported a sizeable increase in seeds per boll and seed weight when 40 colonies were placed near a fiber flax field, but he did not indicate the size of the field.

The studies on pollination of crops clearly establishes that CP not only improves quantitatively but improves the quality of oil crops as well. Even self-compatible varieties give enhanced yields and improve qualitatively when cross pollinated (Table 17.13) (Figures 17.22–17.25).

FIGURE 17.22 **Field view of linseed cultivation.**

FIGURE 17.23 **Flowering in linseed crops.**

FIGURE 17.24 ***Apis mellifera* pollinating a linseed flower.**

FIGURE 17.25 ***Apis cerana* foraging on linseed.**

Therefore, pollination in oil crops assumes the utmost significance for boosting production of oil crops in a world deficient of oil/oilseeds.

Clearly most of the oilseed crops are cross pollinated and exclusively depend upon bees and other pollinators for pollination services.

POLLINATION MANAGEMENT

Problems Associated with BP

Pollinator–plant interaction is a very complex phenomenon and is influenced by many overlapping effects. The protection of pollinators, including honey bees is as essential as the protection of crops from insect-pest damage. The use of pesticides for pest control on the one hand and the role of honey bees (*Apis* spp.) for crop pollination on the other have become essential components of modern agriculture. Unfortunately, these two practices are not always compatible, as honey bees are susceptible to many of the commonly used pesticides (Johansen, 1977; Russell et al., 1998; Cuningham et al., 2002; Sundararaju, 2003), used for the control of insect-pests (Poehling, 1989; Stark et al., 1995). The major constraint confronting pollinator–plant interaction is the indiscriminate and excessive use of pesticides for controlling insect-pests (Bisht et al., 1980; Rana and Goyal, 1991; Zhong et al., 2004). The loss of honey bees directly affects beekeeping, through loss of honey production, and indirectly affects crop production due to inadequate pollination. Reduction of the population of these beneficial insects due to insecticides, therefore, incurs significant environmental, ecological, and economic costs (Bai and Reddy, 1977; Pimentel et al., 1980; Crane and Walker, 1983; Parkash and Kumaraswami, 1984; Khan and Dethe, 2004).

Furthermore, modern agricultural practices have resulted in the reduction of wild insect pollinators and disturbed insect–flower relationships by way of the disappearance of wastelands and uncultivated strips of land, the destruction of certain food sources by weed control, and overall changes in the environment. Wild bees are also damaged by pesticides. Poisoning may result from contaminated food as well as from florets, leaves, soil, or other material used by the bees in nesting. The toxicity of a specific insecticide to honey bees and wild bees is not always the same, and even among wild bees some materials are more toxic to one species than to another. Pesticide application practices that may reduce bee poisoning should be adopted as given below:

1. Apply pesticides only when needed.
2. Choose the pesticide with the lowest hazard rating for bees, particularly the lowest residual toxic effect, from the list of pesticides available for a particular pest control program.
3. Liquid or granule applications are less hazardous than dusts. Microencapsulated forms of pesticides have a significantly longer residual life than other application forms. The minute capsules can be carried back to the colony in the same manner that pollen is carried, and can kill brood and young adult bees.
4. Ground application is less hazardous than aerial application, particularly when applied in close proximity to apiaries. Fine sprays are less toxic than coarse sprays.

5. Where practicable, apply pesticides when bees are not active on the crop. For pesticides considered a low hazard when they have dried, an early morning application may be suitable. For pesticides with a residual toxic effect of a few hours, apply in the late afternoon or early evening.
6. The time of day when a pesticide is applied should be chosen to minimize the risk of spray drift occurring either over apiaries or over plants being foraged by bees. Treatments can and should be applied only when bees are not foraging for nectar or pollen. If plants which are attractive to bees need to be treated while in bloom, they should be treated at night or in the early morning or late evening when the bees are not flying.
7. Where practical, beekeepers should be given prior notice, preferably a minimum of 48 hours, to a pesticide application to allow apiaries to be moved from the area or have their entrances closed.

NUMBER OF COLONIES REQUIRED FOR POLLINATION

Several investigators have attempted to determine the number of colonies of honey bees required for increased yields in rape – their recommendations vary from place to place and for different crop types (Table 17.14). For instance, Hammer (1963, 1966) recommended 3 colonies per hectare; Radchenko (1964), recommended 2; Downey and Bolton (1961), recommended 1; White (1970), recommended 2; and Vesely (1962) recommended 3–4 colonies per hectare. Although hoverflies appear to play some role in pollination of rape we consider the honey bee the more efficient pollinator. The ideal pollinator population and proper distribution of colonies for most efficient pollination needs to be determined for various oil crops.

TABLE 17.14 Pollination Requirements of Different Crops (Number of Colonies/ha)

Crop	Blooming period of the crop	Number of *Apis mellifera* colonies per hectare	Number of *Apis cerana* colonies per hectare	Time of placement of colonies
Mustard and rape	December–January; February–March	3–5	5–8	10–15% bloom
Niger	August–September	3–5	6–8	5–10% bloom
Safflower	March–April	5	4–6	5–10% bloom
Sunflower	June	5	8–10	5–10% bloom
Coconut	Throughout the year	2–3	4-6	5–10% bloom
Sesame	April–May	2–3	4–6	5–10% bloom
Cotton	December–January	3–5	5–8	10–15% bloom

POLLINATION RECOMMENDATIONS

Honey bees are the most effective agents involved in CP of rapeseed mustard and other oil crops. Indiscriminate use of pesticides/fungicides often kills a large number of pollinators. In certain cases, a single crop over a vast area is cultivated. This also reduces the number of wild honey bee colonies in these areas. Therefore, the importance of managed and migratory beekeeping in the field is being realized as an important input to increase the production of oilseed crops. For effective pollination and increased yield of oilseed crops, the efficiency of a bee colony as a pollinator would depend upon certain factors such as colony strength, number and time of placement of colonies, distribution of colonies in the field/orchards, time and placement of colonies, and weather conditions.

CONCLUSIONS AND FUTURE STRATEGIES

This chapter clearly indicates that pollination interventions in terms of supplementing oilseed fields with honey bee colonies can boost oilseed production. A more comprehensive strategy for management of crop pollination is needed to ensure a reliable source of pollinators, both managed as well as native pollinators. Honey bee pollination is not only responsible for enhancement of yield in self-incompatible varieties but also self-fertile varieties are seen to give a many-fold increase in yield due to the intervention of managed pollinators. This requires an understanding of the biology and ecology of pollinating insects, as well as providing appropriate nesting habitats, and ensuring the availability of alternative sources of "foraging" to sustain populations when target crops are not in bloom. A number of other bee and insect pollinators, such as orchard bees, bumblebees, and squash bees, which are not affected by either mites or Africanized bees, are considered as likely candidates for management and use in commercial agriculture. An additional role can be played by native or wild pollinators, provided that attention is given to curtailing population losses caused by both inadvertent insecticide poisoning and habitat destruction (Kevan and Eisikowitch, 1990). Pollinator management and managed pollination are common efforts recently being practiced to achieve the maximization of production in cross-pollinated crops and to bring pollinators to target crops. There is strong evidence that pollinators are declining as a result of local and global environmental degradation (Kluser and Peduzzi, 2007). Recent declines in both wild and domesticated pollinators, and parallel declines in the plants which rely upon them, represent a serious threat to global food security and sustainable agriculture (Potts et al., 2010). Agriculture has become more pollinator dependent because of a disproportionate increase in the area cultivated with pollinator-dependent crops. If the trend toward favoring cultivation of pollinator-dependent crops continues, the need for the service provided by declining pollinators will greatly increase in the near future.

Kevan and Eisikowitch (1990) highlighted the pivotal role played by managed and native insect pollinators in the sustainability of agricultural practices. However, wild insect pollinators, and the pollination services they provide, have declined in agricultural landscapes in some regions (Biesmeijer et al., 2006; Potts et al., 2010). Several factors associated with increased farming intensity to support the growing human population can limit the suitability of farm environments for insect pollinators, including reduction of natural areas, habitat fragmentation, and scarcity of flowering food and nesting resources (Carvell et al., 2007). Monoculture planting of crops lacks floral diversity and can limit the provision of resources for pollinators throughout the season. Compared with more diverse landscapes, the lack of resources in agricultural landscapes can reduce insect pollinator diversity (O'Toole, 1993) and potentially decrease wild bee contributions to crop pollination (Potts et al., 2010). Enhancement of structurally resource-poor environments through the establishment of habitats containing flowering plants and grasses can support beneficial insects in agricultural landscapes (Long et al., 1998; Kells et al., 2001; Sheffield et al., 2008). These agricultural restoration programs are expected to provide the greatest support for bees in simple landscapes with the greatest floral contrast to the background landscape (Scheper et al. 2013). Wildflower plantings can provide pollen and nectar resources when crops are not in bloom and, depending on the bee species biology, may also provide nesting habitats (Carreck and Williams, 2002; Kremen et al., 2004; Heard et al., 2007). Hoverflies are efficient pollinators of other crops, such as mango (Dag and Gazit, 2001) and oilseed rape (Jauker and Wolters, 2008), and the larvae of *aphidophagous* species are also biological control agents of many soft-bodied arthropods (Bugg et al., 2008; Smith et al., 2008), thus providing an additional ecosystem service to crops.

References

Abel, C.A., Wilson, R.L., 1999. The use of diverse plant species for increasing *Osmia cornifrons* (Hymenoptera: Megachilidae) in field cages. J. Kansas Entomol. Soc. 71, 23–28.

Abel, C.A., Wilson, R.L., Luhmann, R.L., 2003. Pollinating efficacy of *Osmia cornifrons* and *Osmia lignaria* subsp. lignaria (Hymenoptera: Megachilidae) on three *Brassicaceae* species grown under field cages. J. Kans. Entomol. Soc. 38, 545–552.

Abrol, D.P., 1991. Conservation of pollinators for promotion of agricultural production in India. J. Anim. Morphol. Physiol. 38 (1–2), 123–139.

Abrol, D.P., 2007. Foraging behaviour of *Apis mellifera* and *Apis cerana* as determined by the energetics of nectar production in different cultivars of *Brassica campestris* var. toria. J. Api. Res. 51 (2), 5–10.

Abrol, D.P., 2008. Plant insect interaction in crucifers. In: Gupta, S.K. (Ed.), Biology and Breeding of Crucifers. CRC Press, Taylor and Francis, USA, pp. 129–149.

Abrol, D.P., 2009. Bees and Beekeeping in India, second ed. Kalyani Publishers, New Delhi, pp.720.

Abrol, D.P., 2009. Honeybee Diseases and their Management, second ed. Kalyani Publishers, Ludhiana, pp. 750.

Abrol, D.P., Shankar, U., 2012. Pollination in oil crops-recent advances and future strategies. Gupta, S.K. (Ed.), Technological Innovations in Major World Oil Crops, 2, Springer Sciences+Business Media, LLC, pp. 221–267.

Adegas, J.E.B., Nogueira Couto, R.H., 1992. Entomophilous pollination in rape (*Brassica napus* L. var. oleifera) in Brazil. Apidologie 23, 203–209.

Araneda, X.D., Ulloa, R.B., Carrillo, J.A., Contreras, J.L., Bastidas, M.T., 2010. Evaluation of yield component traits of honeybee pollinated (*Apis mellifera* L.) rapeseed canola (*Brassica napus* L.). Chil. J. Agric. Res. 70 (2), 309–314.

Atmowidi, T., Buchori, D., Manuwoto, S., Suryobroto, B., Hidayat, P., 2007. Diversity of pollinator insects in relation to seed set of mustard (*Brassica rapa* L.: Cruciferae). HAYATI J. Biosci. 14 (4), 155–161.

Bai, A.R.K., Reddy, C.C., 1977. Laboratory toxicity of some insecticides to *Apis cerana* indica. J. Apicult. Res. 16 (3), 161–162.

Bezdenezhoykh, S.M., 1956. Pollination of fiber flax and training of bees. In: Krishchunas, I.V., Gubin, A.F. (Eds.), Pollination of Agricultural Plants. Gos Izd-vo. Selkhoz Litry, Moskva, pp. 21–24.

Bichee, S.L., Sharma, M., 1988. Effect of different modes of pollination in sunflower *Helianthus annuus* L. (Compositae) sown on different dates by *Trigona iridipennis* Smith. Apiacta 23, 65–68.

Biesmeijer, J.C., Roberts, S.P.M., Reemer, M., Ohlemuller, R., Edwards, M., Peeters, T., Schaffers, A.P., Potts, S.G., Kleukers, R., Thomas, C.D., Settele, J., Kunin, W.E., 2006. Parallel declines in pollinators and insect-pollinated plants in Britain and the Netherlands. Science 313, 351–354.

Bisht, D.D., Naim, M., Mehrotra, K.N., 1980. Studies on the role of honeybees in rapeseed production. In: Proceedings of Second International Conference on Apiculture in Tropical Climate. Indian Agricultural Research Institute, New Delhi, pp. 491–496.

Boch, R., 1961. Honey bee activity on safflower (*Carthamus tinctorius* L.). Canad. J. Plant. Sci. 41, 559–562.

Bommarco, R., Marini, L., Vaissiere, B.E., 2012. Insect pollination enhances seed yield, quality, and market value in oilseed rape. Oecologia 169, 1025–1035.

Bugg, R.L., Colfer, R., Chaney, W.E., Smith, H.A., Cannon, J., 2008. Flower flies (*Syrphidae*) and other biological control agents for aphids in vegetable crops. Agric. Nat. Resour. Publ. 8285, 1–25.

Bürger, C., 2004. Effects of landscape structure on bee diversity and pollination at different spatial scales. Ph.D. thesis. Institute of Agroecology, University of Göttingen, Germany.

Butler, Jr., G.D., Werner, E.G., Leven, M.D., 1966. Native bees associated with safflower in Southercenteral Arizona. Kans Ent. Soc. J. 39 (3), 434–436.

Carreck, N.L., Williams, I.H., 2002. Food for insect pollinators on farmland: insect visits to flowers of annual seed mixtures. J. Insect Conserv. 6, 13–23.

Carvell, C., Meek, W.R., Pywell, R.F., Goulson, D., Nowakowski, M., 2007. Comparing the efficacy of agri-environment schemes to enhance bumble bee abundance and diversity on arable field margins. J. Appl. Ecol. 44, 29–40.

Charlesworth, D., Charlesworth, B., 1987. Inbreeding depression and its evolutionary consequences. Ann. Rev. Ecol. System. 18, 237–268.

Chhuneja, P.K., Singh, J., Gatoria, G.S., Blossom, 2007. Assessing the role of honey bee (*Apis mellifera* L.) in the seed production of *Brassica campestris* var. toria in the Punjab. Indian J. Crop Sci. 2 (2), 34–40.

Classen, C.E., 1950. Natural and controlled crossing in safflower (*Carthamus tinctorius* L.). Argon. J. 42, 381–384.

Crane, E., 1991. *Apis* species of tropical Asia as pollinators and some rearing methods for them. Acta Hort 288, 29–48.

Crane, E., Walker, P., 1983. Pollination of tropical and subtropical crops by bees. In: Crane, E., Walker, P. (Eds.), The Impact of Pest Management on Bees and Pollination. IBRA, Cardiff, pp. 5–21.

Dag, A., Gazit, S., 2001. Mango pollinators in Israel. J. Appl. Hortic. 2, 39–43.

Dag, A., Lior, E., Afik, O., 2002. Pollination of confection sunflowers (*Helianthus annuus* L.) by honey bees (*Apis mellifera*). Am. Bee J. 142, 443–445.

De Grandi Hoffman, G., Martin, J.H., 1993. The size and distribution of the honey bee (*Apis mellifera* L.) cross-pollinating population on male sterile sunflowers (*Helianthus annuus* L.). J. Apic. Res. 32, 135–142.

De Grandi-Hoffman, G., Watkins, J.C., 2000. The foraging activity of honey bees *Apis mellifera* and non-Apis bees on hybrid sunflowers (*Helianthus annuus*) and its influence on cross-pollination and seed set. J. Apic. Res. 39, 37–45.

Delaplane, K.S., Mayer, D.F., 2000. Crop Pollination by Bees. CABI Publishing, New York, USA.

Deodikar, G.B., Suryanarayana, M.C., 1972. Crop yields and bee pollination. Indian Bee J. 34, 53–64.

Dillman, A.C., 1938. Natural crossing in flax. J. Am. Soc. Agron 30, 279–286.

Dillman, A.C., Goar, L.G., 1937. Flaxseed production in the far western states. U.S. Dept. Agr. Farmers' Bul. 1792, pp. 22.

Downey, R.K., Bolton, J.L., 1961. Production of [Polish and Argentine] rape in Western Canada. Canada Dept. Agr. Res. Br. Pub. 1021, pp. 19.

Eckert, J.E., 1962. The relation of honey bees to Safflower. Amer. Bee J. 102, 349–350.

FAOSTAT, 2012. Food and Agriculture Organization of the United Nations. Roma, Italy.

Free, J.B., 1993. Insect pollination of crops, second ed. Academic Press, Harcourt Brace Jovanovich Publ., London, 684 p.

Free, J.B., 1999. Pollination in the tropics. Beekeeping Dev 51, 6–7.

Free, J.B., Ferguson, A.W., 1983. Foraging behaviour of honeybees on oilseed rape. Bee World 64, 2–24.

Free, J.B., Williams, I.H., 1976. Pollination as a factor limiting the yield of field bean (*Vicia faba* L.). J. Agric. Sci., Camb 87, 395–399.

Furgala, B., 1954b. Honey bees increase seed yields of cultivated sunflowers. Glean. Bee Cult. 82, 532–534.

Furgala, B., Noetzel, D.M., Robinson, R.G., 1979. Observations on the pollination of hybrid sunflowers. Proceedings of the Forth International Symposium on Pollination, Maryland Agricultural Experiment Station Special Miscellaneous Publication 1, pp. 45–48.

Garibaldi, L.A., Carvalheiro, L.G., Leonhardt, S.D., Aizen, M.A., Blaauw, B.R., Isaacs, R., Kuhlmann, M., Kleijn, D., Klein, A.M., Kremen, C., Morandin, L., Scheper, J., Winfree, R., 2014. From research to action: enhancing crop yield through wild pollinators. Front. Ecol. Environ. 12 (8), 439–447.

Greenleaf, S.A., Kremen, C., 2006. Wild bees enhance honey bees' pollination of hybrid sunflower. Proc. Nat. Acad. Sci. USA 103, 13890–13895.

Gubin, A.F., 1945. Cross pollination of fibre flax. Bee World 26, 30–31.

Guidry, N.P., 1964. Graphic summary of world agriculture. United States Department Of Agriculture Miscellaneous Publication, 705, pp. 64.

Hammer, O., 1963. Summer rape as a competitor affecting the pollination of clovers. Dansk froavl no 14, 7.

Hammer, O., 1966. Some problems of competition between summer rape and clover, in relation to pollination. In: Second International Symposim Pollination, London, 1964. Bee World, 47 (Suppl. 1), pp. 99–106

Heard, T.A., 1999. The role of stingless bees in crop pollination. Ann. Rev. Entomol. 44, 183–206.

Heard, M.S., Carvell, C., Carreck, N.L., Rothery, P., Osborne, J.L., Bourke, A.F.G., 2007. Landscape context not patch size determines bumble-bee density on flower mixtures sown for agrienvironment schemes. Biol. Lett. 3, 638–641.

Hoffmann, M., 1994. Observações sobre a polinização entomófila de *Helianthus annuus* L. em Viamão, Rio Grande do Sul. An Soc Entomol Bras. Londrina 23 (3), 391–397, [In Portuguese].

Hoyle, M., Hayter, K., Cresswell, J.E., 2007. Effect of pollinator abundance on self-fertilization and gene flow: application to GM Canola. Ecol. Appl. 17, 2123–2135.

Jauker, F., Wolters, V., 2008. Hover flies are efficient pollinators of oilseed rape. Oecologia 156, 819–823.

Johansen, C.A., 1977. Pesticides and pollinators. Ann. Rev. Ent. 22, 177–192.

Jyoti, J., Brewer, G.J., 1999. Effect of honeybee (Hymenoptera: Apidae) pollination on sunflower hybrids. Proc. 21st Sunflower Research Workshop. Nat Sunflower Assoc, 103–107, 14–15.

Kadam, B.S., Patankar, V.K., 1942. Natural cross-pollination in safflower. Indian J. Genet. 2, 69–70.

Kamel, S.M., Blal, A.H., Mahfouz, H.M., Said, M., 2013. The most common insect pollinator species on sesame crop (Sesamum indicum L.) in Ismailia Governorate. Egypt. Arthropods 2 (2), 66–74.

Kells, A., Holland, J., Goulson, D., 2001. The value of uncropped field margins for foraging bumblebees. J. Insect Conserv. 5, 283–291.

Kevan, P.G., Eisikowitch, D., 1990. The effects of insect pollination on canola (*Brassica napus* L. cv. O. A. C. Triton) seed germination. Euphytica 45, 39–41.

Khan, R.B., Dethe, M.D., 2004. Median lethal time of new pesticides to foragers of honeybee. Pestology 28 (1), 28–29.

Kisselhagen, S., 1977. Seed yield is increased if there are bees on the rape. Tidsskrift for Biavl 111, 66–67.

Klatt, B.K., Holzschuh, A., Westphal, C., Clough, Y., Smit, I., Pawelzik, E., Tscharntke, T., 2014. Bee pollination improves crop quality, shelf life and commercial value. Proc. R. Soc. B 281, 20132440.

Klein, A.M., Vaissière, E.B., Cane, J.H., Steffan-Dewenter, I., Cunningham, S.A., Kremen, C., Tscharntke, T., 2007. Importance of pollinators in changing landscapes for world crops. Proceedings of the Royal Society B 274, 303–313.

Kluser, S., Peduzzi, P., 2007. Global pollinator decline: a literature review (Geneva: UNEP/GRID).

Knowles, P.F., 1969. Centers of plant diversity and conservation of crop germplasm, Safflower. Econ. Bot. 23, 324–329.

Kozin, R.B., 1954. Effect of bees on the seed crop of flax. Pchelowdstvo, Mosk 6, 41–43.

Kremen, C., Williams, N.M., Bugg, R.L., Fay, J.P., Thorp, R.W., 2004. The area requirements of an ecosystem service: crop pollination by native bee communities in California. Ecol. Lett. 7, 1109–1119.

Kubisova, S., Nedbalova, V., Plesnik, R., 1980. Cinnost vcely medonosne (*Apis mellifera* L.) na repce (*Brassica napus* L. var. napus). Pol'nohospoddrstvo 26, 744–754.

Langridge, D.F., Goodman, R.D., 1975. A study on pollination of oilseed rape (*Brassica campestris*). Austral. J. Exptl. Agric. Anim. Husband. 15, 285–288.

Langridge, D.F., Goodman, R.D., 2003. Honeybee pollination of oilseed rape, cultivar Midas. J. Exptl. Agric. Anim. Husband. 22 (115), 124–126.

Latif, A., Qayyum, A., Abbas, M., 1960. The role of Apis indica in the pollination of toria and sarson (*Brassica campestris* var. Toria and dichotoma). Bee World 41, 283–286.

Levin, M.D., Butler, G.D., 1966. Bees associated with Safflower in South Central Arizona. Jour. Eco. Ent. 59, 654–657.

Levin, M.D., Butler, G.D., Rubis, D.D., 1967. Pollination of safflower by insects other than honeybees. Jour. Econ. Ent. 60, 1481–1482.

Long, R.F., Corbett, A., Lamb, C., Reberg-Horton, C., Chandler, J., Stimmann, M., 1998. Beneficial insects move from flowering plants to nearby crops. California Agric. 52, 23–26.

Luttso, V.P., 1957. The pollination of flax by honeybees. Dokl. Tskha 30, 327–331.

Mahmoud, M.F., 2012. Insects associated with sesame (*Sesamun indicum* L.) and the impact of insect pollinators on crop production. Pestic. Phytomed. (Belgrade) 27 (2), 117–129.

Manning, R., Boland, J., 2000. A preliminary investigation into honey bee (*Apis mellifera*) pollination of canola (*Brassica napus* cv. Karoo) in Western Australia. Aust. J. Exp. Agr. 40, 439–442.

Manzoor-ul-haq, Fiaz, M., 1980. Insect pollination of sunflower *(Helianthus annuus)* in Pakistan. J. Apic. Res. 19, 83–88.

Masuo, Y., 1958. On natural crossing in flax. Crop Sci. Soc. Japan Proc 27, 321–323, [In Japanese.] Abstract in Plant Breed. 29(4), 817–818.

Mishra, R.C., Kaushik, H.D., 1992. Effect of cross-pollination on yield and oil content of Brassica campestris L. var. sarson with pollination efficiency of honeybees Apis spp. Ann. Entomol. 10, 33–37.

Mishra, R.C., Kumar, J., Gupta, J.K., 1988. The effect of mode of pollination on yield and oil potential of *Brassica campestris* L. var. *sarson* with observation on insect pollinators. J. Apic. Res. 27, 186–189.

Mohammed, A., 1935. Pollination studies in tara (*Brassica napus* L. var. dichotsma Prain) and sarson (*Brassica campestris* L. var. sarson, Prain). Indian J. Agric. Sci 5, 125–154.

Morandin, L.A., Winston, M.L., 2005. Wild bee abundance and seed production in conventional organic and genetically modified canola. Ecol. Appl. 15, 871–881.

Moreti, A.C., de, C.C., da Silva, R.M.B., da Silva, E.C.A., Alves, M.L.T.M.F., Otsuk, I.P., 1996. Aumento na produção de sementes de girassol (*Helianthus annuus*) Pela ação de insetos polinizadores. Sci. Agric. 53, 280–284.

Munawar, M.S., Raja, S., Siddique, M., Niaz, S., Amjad, M., 2009. The pollination by honeybee (*Apis mellifera* L.) increases yield of canola (*Brassica napus* L.). Pakistan Entomol. 31 (2), 103–106.

O'Toole, C., 1993. In: LaSalle, J., Gauld, I.D. (Eds.), Diversity of native bees and agroecosystems. Hymenoptera and Biodiversity. CAB Int, Wallingford, UK, pp. 169–196.

Oz, M., Karasu, A., Cakmak, I., Goksoy, A.T., Turan, Z.M., 2009. Effects of honeybee (Apis mellifera) pollination on seed set in hybrid sunflower (Helianthus annuus L.). African J. Biotechnol. 8 (6), 1037–1043.

Panda, P., Sontakke, B.K., Sarangi, P.K., 1988. Preliminary studies on the effect of bee (*Apis cerana* indica F.) pollination on yield of sesamum and niger. India Bee J. 50, 63–64.

Pandey, A.K., Kumari, A., 2007. Pollination ecology of safflower (*Carthamus tinctorius* Linn.). In: Seventh International Safflower Conference. Waaga Wagga Australia, 1–10.

Parkash, R., Kumaraswami, T., 1984. Toxicity of some insecticide spraying on hymenopertrans pollinators in new Brunswick. Canadian Entomol. 112, 259–269.

Patil, J.A., Chavan, V.M., 1948. Selfing methods in Safflower. Ind. Oilseeds J. 2, 10–12.

Phadke, K.G., Pant, N.C., Sinha, R.B.P., 1967. The foraging behaviour of *Apis indica* F (Hymenoptera: Apidae) visiting flowers of sesame (*Sesamum orientate* L.). Indian J. Ent. 29, 157–163.

Pimentel, D., Andow, D., Dyson-Hudson, R., Gallahan, D., Jacobson, S., Irish, M., Kroop, S., Moss, A., Schreiner, I., Shepard, M., Thompson, T., Vinzant, B., 1980. Environmental and social costs of pesticides: a preliminary assessment. Oikos 34 (2), 127–140.

Poehling, H.M., 1989. Selective application strategies for insecticides in agricultural crops. In: Jepson, P.C. (ed.) Pesticides and non target invertebrates. Intercept, Wimborne, pp. 151–175.

Pordel, M.R., Hatami, B., Mobli, M., Ebadi, R., 2007. Identification of insect pollinators of three different cultivars of winter canola and their effect on seed yield in Isfahan. J. Sci. Technol. Agric. Natur. Resour. 10 (4B), 413–426.

Potts, S.G., Biesmeijer, J.C., Kremen, C., Neumann, P., Schweiger, O., Kunin, W.E., 2010. Global pollinator declines: trends, impacts and drivers. Trends Ecol. Evolut. 25, 345–353.

Prakash, R., Kumaraswami, J., 1984. Toxicity of some insecticides to the Indian bee. Apis cerana Feb. Indian Bee J. 46, 15–17.

Prasad, D., Hameed, S.F., Singh, R., Yazdani, S.S., Singh, B., 1989. Effect of bee pollination on the quantity and quality of rai crop (*Brassica juncea* Coss). Indian Bee J. 51 (2), 45–47.

Radchenko, T.H., 1964. The influence of pollination on the crop and the quality of seed of winter rape. Bdzhil'nitstvo, 1, 68–74.

Radford, B., Nelson, R., Rhodes, J., 1979. Agents of pollination in sunflower crops in the central Darlingdowns, Queensland. Aust. J. Exp. Agric. Anim. Husb. 19, 565–569.

Rahman, K.A., 1940. Insect pollinators of toria (*Brassica napus* L. var dichotoma) and Sarson (*Brassica campestris* L. var Sarson) at Lyallpur. Indian J. Agric. Sci 10, 422–447.

Rajasri, M., Kanakadurga, K., Durga Rani, V., Choudhary, A., 2012. Honey bees–potential pollinators in hybrid seed production of sunflower. Int. J. Appl. Biol. Pharmaceut. Technol. 3 (2), 216–221.

Rana, B.S., Goyal, N.P., 1991. Field toxicity of methyl demeton and dimethoate to the foragers of honeybee (*Apis cerana* indica F.) visitors to *Brassica chinensis* L. Indian Bee J. 53 (1–4), 73–77.

Rao, G.M., Lazar, G.M., Suryanarayana, M.C., 1981. Foraging behavior of honeybees in sesame (*Sesamum indicum* L.). Indian Bee J. 43, 97–100.
Rao, G.M., Suryanarayana, M.C., Thakar, C.V., 1980. Bees can boost oilseed production. Indian Farming 29, 25–26.
Richards, A.J., 1997. Plant breeding systems, second ed. Chapman and Hall, London.
Rubis, D.D., 1970. Breeding insect pollinated crops. Arkansas Agric. Ext. Service 127, 19–24.
Rubis, D.D., Levin, M.D., Mc Gregor, S.E., 1966. Effects of honey bee activity and cages on attributes of thin hull and normal Safflower lines. Crop Sci. 6, 11–14.
Russell, D., Meyer, D., Bukowski, J., 1998. Potential impact of micro encapsulated pesticides on New Jersey Apiaries. Am. Bee J. 38 (3), 207–210.
Sabbahi, R., De Oliveira, D., Marceau, J., 2005. Influence of honey bee (Hymenoptera: Apidae) density on the production of canola (Crucifera: Brassicacae). J. Econ. Entomol. 98 (2), 367–372.
Sabbahi, R., De Oliveira, D., Marceau, J., Houle, E., 2006. Floraison du canola (Crucifera: Brassicaceae) et le rendement des colonies d'abeilles domestiques (Hymenoptera: Apidae). Avalaible from: http://www.agrireseau.qc.ca/apiculture/documents/Rendem_apicole-Rachid_revise.pdf
Sabir, A.M., Suhail, A., Hussain, A., Saeed, A., 2000. Insecticidal mortality, forgaing behaviour and pollination role of honeybee (*Apis mellifera* L.) on brassica (*Brassica campestris* L.). Pakistan J. Zool. 32 (4), 369–372.
Scheper, J., Holzschuh, A., Kuussaari, M., Potts, S.G., Rundlof, M., Smith, H.G., Kleijn, D., 2013. Environmental factors driving the effectiveness of European agri-environmental measures in mitigating pollinator loss – a meta-analysis. Ecol. Lett. 16, 912–920.
Sharma, D., Abrol, D.P., 2014. Effect of insecticides on foraging behaviour and pollination role of *Apis mellifera* L. (Hymenoptera: Apidae) on toria (*Brassica campestris* var. toria) crop. Egypt. J. Biol. 16, 79–86.
Sheffield, C.S., Westby, S.M., Smith, R.F., Kevan, P.G., 2008. Potential of bigleaf lupine for building and sustaining *Osmia lignaria* populations for pollination of apple. Can. Entomol. 140, 589–599.
Sihag, R.C., 1986. Insect pollination increases seed production in cruciferous and umbelliferous crops. J. Agric. Res. 25, 121–126.
Sikka, S.M., Gupta, N.P., 1949. Pollination studies in *Sesamum orientale* L. Indian J. Genetics 9, 33–191.
Singh, S., 1954. Insect pollination and breeding of fruit varieties. Indian J. Hort. Sci. 11, 123–125.
Singh, R.P., Singh, P.N., 1992. Impact of bee pollination on seed yield, carbohydrate composition and lipid composition of mustard seed. J. Apic. Res. 31, 128–133.
Singh, M.P., Singh, K.I., Devi, C.S., 2000. Role of *Apis cerena* Himalaya pollination on yield and quality of rape seed and sunflower crops. In: Proceedings of 4th Asian Apicultural Association, International Conference, Kathmandu, Nepal, March 23–28, 1998.
Smirnov, V.M., 1956. Cross-pollination of fl ax by bees. Pchelovodstvo Mosk 9, 53–55.
Smith, W., 2002. Honey bees on canola. New South Wales Agriculture, Department of Primary Industries, Orange, New South Wales, Australia.
Smith, H.A., Chaney, W.E., Bensen, T.A., 2008. Role of syrphid larvae and other predators in suppressing aphid infestations in organic lettuce on California's Central Coast. J. Econ. Entomol. 101, 1526–1532.
Stark, D.J.D., Jepson, P.C., Mayer, D.F., 1995. Limitation to use of topical toxicity data for production of pesticide side effects in the field. J. Econ. Entomol. 88 (5), 1081–1088.
Sundararaju, D., 2003. Occurrence of bee fauna and extent of pollination in insecticide sprayed ecosystem of cashew. J. Palynol. 39, 121–125.
Vanbergen, A.J., Baude, M., Biesmeijer, J.C., Britton, N.F., Brown, M.J.F., Bryden, J., Budge, G.E., Bull, J.C., Carvell, C., Challinor, A.J., Connolly, C.N., Evans, D.J., Feil, E.J., Garratt, M.P., Greco, M.K., Heard, M.S., Jansen, V.A.A., Keeling, M.J., Kunin, W.E., Marris, G.C., Memmott, J., Murray, J.T., Nicolson, S.W., Osborne, J.L., Paxton, R.J., Pirk, C.W.W., Polce, C., Potts, S.G., Priest, N.K., Raine, N.E., Roberts, S., Ryabov, E.V., Shafir, S., Shirley, M.D.F., Simpson, S.J., Stevenson, P.C., Stone, G.N., Termansen, M., Wright, G.A., 2013. Threats to an ecosystem service: pressures on pollinators. Front. Ecol. Environ. 11 (5), 251–259.
Vesely, V., 1962. The economic effectiveness of bee pollination on winter rape (*Brassica napus* l., var. Oleifera metz.). Min. Zemedel. Lesn. A vodniho hospodar. Ust. Vedtech. Inform. Zemedel. Ekon 8, 659–673.
Wakhle, D.M., Shakuntala Nair, K., Phadke, R.P., 1978. Effect of bee-pollination on the oil and protein content in the seeds of sunflower (*Helianthus annuus* L.). Indian Bee J. 40, 1–2.
Wallace, H.M., Maynard, G.V., Trueman, S.J., 2002. Insect flower visitors, foraging behaviour and their effectiveness as pollinators of *Persoonia virgata* R. Br. (Proteaceae). Austral J. Entomol. 41, 55–59.
Westcott, L., Nelson, D., 2001. Canola pollination: an update. Bee World 82, 115–129.

White, B., 1970. Pollination of commercial rape seed crops. Australasia. Beekeeper 72, 99–100.

Williams, I.H., Martin, A.P., White, R.P., 1987. The effect of insect pollination on plant development and seed production in winter oil-seed rape (Brassica napus L.). J. Agric. Sci. 109, 135–139.

Wittkop, B., Snowdon, R.J., Friedt, W., 2009. Status and perspectives of breeding for enhanced yield and quality of oilseed crops for Europe. Euphytica 170, 131–140.

Wratten, S.D., Gillespie, M., Decourtye, A., Mader, E., Desneux, N., 2012. Pollinator habitat enhancement: benefits to other ecosystem services. Agric. Ecosyst. Environ. 159, 112–122.

Zhong, H.E., Mark, L., Sheve, P., Cate, B., 2004. Minimizing the import if the mosquito adulterine noted or honeybees, *Apis mellifera* (Hymenoptera: Apidae): Aerial Ultra volume application using a high-pressure nozzle system. J. Econ. Entomol. 97 (1), 1–7.

CHAPTER

18

Breeding Oilseed Crops for Climate Change

Abdullah A. Jaradat

USDA-ARS and Department of Agronomy and Plant Genetics, University of Minnesota, MN, USA

INTRODUCTION

Oil-producing crops, whether oilseed or oil-bearing crops, occupy about 10% of global arable land; they are major providers of calories for humans and livestock, and raw material for a wide range of industrial products (e.g., oleochemicals); however, they are increasingly being used as sources of biofuels, especially biodiesel. Five major oil crops: oil palm (*Elaeis guineensis*), soybean (*Glycine max*), oilseed rape (*Brassica napus*), sunflower (*Helianthus annuus*), and olives (*Olea europaea*), in decreasing order, account for ~80% of world oil production (FAO, 2014). Oil palm is a very productive perennial tree, difficult to genetically manipulate; it grows in a limited area in South East Asia; its oil is rich in C16:0/C18:1. Soybean is primarily a protein-producing crop; however, seed of transgenic varieties have high oleate content used in biodiesel production. Oilseed rape is the major temperate oilseed crop; it produces C18:1-rich oil. Sunflowers are widely grown around the world, they produce C18:2-rich oil. The olive is a Mediterranean, drought-resistant perennial fruit tree; its oil is C18:1-rich, and is difficult to genetically manipulate. In general, most oilseed crops are annuals and grown in temperate regions (e.g., soybean, corn, rapeseed, and sunflower). Major oil-rich fruit crops include tree species such as *E. guineensis*, *Cocos nucifera*, and *O. europaea*. Oil-bearing crops include annual (i.e., oilseeds) and perennial plants whose seeds, fruits (mainly mesocarp), and nuts are prized for the edible or industrial oils that are extracted from them. The oil content ranges from as low as 10–15% of the weight of coconuts to >50% of the weight of *Sesamum indicum* seeds and *E. guineensis* kernels; whereas, protein content is very high in soybeans (~40%), but is much lower (15–25%) in most other oilseeds, and is lower still in some other oil-bearing crops (Nogueira, 2011).

The contribution of nonedible plant oils (e.g., from *Jatropha curcas*; 30–50% oil) and soapnut (*Sapindus mukorossi* and *Sapindus trifoliatus*; 52% oil), as new sources for biodiesel production have the advantage of not competing with edible oils produced from crop plants. Other

Breeding Oilseed Crops for Sustainable Production. http://dx.doi.org/10.1016/B978-0-12-801309-0.00018-5

potential oil crops include *Azadirachta indica, Calophyllum inophyllum, Pongamia pinnata,* among 75 oil plants that contain 30% or more oil in their seed, fruit, or nut. Major plant families with oil-producing plants include Amaryllidaceae, Apocynaceae, Asclepiadaceae, Compositae, Convolvulaceae, Cruciferae, Euphorbiaceae, Flacourtiaceae, Lauraceae, Liguminosae, Malvaceae, Moraceae, Myrcaceae, and Palmae. A small portion (5–6%) of the world production of oil crops is used for seed (i.e., oilseeds) and animal feed, while about 8% is used for food, and the remaining 86% are processed into oil.

Economically, oilseed crops are among the most diverse and most important crops in developed as well as developing countries, whether for food, feed, or industrial uses. The complex nature of agroecosystems where these crops are being grown around the world suggests that the impact of global climate change (GCC) on agroecosystem components will vary broadly across geographical and temporal scales. Industrial applications are based on the properties of particular saturated fatty acid (SFAs) components of these oils, while the protein-rich residue remaining after the oil has been extracted from oilseeds is an important source of nutrients for livestock. Oilseed meals from *Glycine max, Arachis hypogaea, Brassica napus,* and *Linum usitatissimum* are rich in protein; and when mixed with cereal grains, they provide nutritionally balanced livestock feeds (Colombini et al., 2014). Recently, a dramatic rise in market worth has been witnessed along the oilseed value chain, which is characterized by several important stages of transformation (i.e., from seeds to oils, meals, and biodiesel). Market forces recently favored oilseed production, especially in Europe and the United States, and positively influenced their worldwide production especially of biodiesel; thereby leading to a very fast growth of exports of oilseeds and their different products from 5% in 1961 to more than 12% in 2011. More than 50% of vegetable oils (mostly from *Elaeis guineensis, G. max,* and *Brassica* spp.) have been traded recently (FAOSTAT, 2012, http://faostat.fao.org). Efforts to increase oil crop productivity and production will have to address increasing demand in the era of abiotic and biotic stresses associated with GCC.

FUTURE OF OILSEED PRODUCTION: IMPACT OF CLIMATE CHANGE

Edible oil and oilseed markets have limited advantage risks in spite of lower fertilizer input cost, especially in the United States. The global outlook of edible oil and meal for livestock, as well as global supplies of edible oils most consumed by humans (Box 18.1) is expected to have reached a record 196 million tons in 2014; a 4.3% increase from last year (FAO, 2014). Similarly, expected world production of major oilseeds in 2014 is 492 million tons, an increase of about 6% from last year. Nevertheless, a negative impact on the production of oilseed crops is expected as a result of GCC. Increased demand for oilseeds and their products during the last few decades resulted in significant changes in global oilseed markets; however, world trade seems to be less restricted in oilseeds than in their products. Biodiesel, which is a methyl ester or ethyl ester derived from the oil of oilseed crops, is the most important renewable bioenergy source with a wide range of applications. Its importance was stimulated by mandatory alternative fuel use legislation in Europe and more recently in the United States, both of which account for 95% of global biodiesel demand. The top 10 producers of biodiesel, in decreasing order, are: Malaysia, Indonesia, Argentina, United States, Brazil, Netherlands, Germany, Philippines, Belgium, and Spain; these countries represent

BOX 18.1

OILSEED CROPS, OIL-BEARING PLANTS, POTENTIAL OIL-PRODUCING PLANTS, AND EXPERIMENTAL WILD PLANTS REFERENCED IN THIS CHAPTER

Arabidopsis thaliana, *Argania spinosa* (argan tree), *Aleurites cordata*, *Aleurites fordii* (tung nuts), *Arachis hypogaea* (peanuts), *Azadirachta indica*, *Brassica carinata* (Palliser's promise), *Brassica napus* (canola, rapeseed), *Brassica carinata*, *Brassica alba*, *Brassica nigra*, *Brassica hirta* (mustard), *Butyrospermum parkii* (karate nut), *Calophyllum inophyllum*, *Camelina sativa* (camelina), *Cannabis sativa* (hemp), *Carthamus tinctorius* (safflower), *Cochlearia officinalis*, *Cocus nucifera* (coconut), *Crambe abyssinica* (crambe), *Cuphea* spp. (cuphea), *Cynaria cardunculus*, *Diplotaxis tenuifolia*, *Elaeis guineensis* (oil palm), *Glycine max* (soybean), *Gossypium hirsutum* (cotton), *Guizotia abyssinica* (niger), *Helianthus annuus* (sunflower), *Jatropha curcas* (jatropha), *Lesquerella fendleri* (Fendler's bladderpod), *Lepidium campestre*, *Limnanthes alba* (meadowfoam), *Linum usitatissimum* (flax), *Lunaria annua* (honesty), *Manihot esculenta*, *Miscanthus × giganteus*, *Olea europaea* (olive), *Papaver somniferum* (poppy seed), *Parthenium argentatum* (guayule), *Plukenetia volubilis* (sacha tree), *Pongamia pinnata*, *Racinus communis* (castor bean), *Sapindus mukorossi*, *Sapindus trifoliatus*, *Sesamum indicum* (sesame), *Simmondsia chinensis* (jojoba), *Simmondsia chinensis* (Stokes' aster), *Sinapis arvensis*, *Thlaspi arvense* (pennycress), *Vernonia galamensis* (vernonia), and *Zea mays* (corn).

a wide range of production potential which depends mainly on growing conditions, land availability, production inputs, technologies, and feedstock. Higher oilseed prices are expected to continue in the foreseeable future due to implications of biodiesel and ethanol policies for developing countries (FAO, 2014).

Global climate change is expected to alter the availability of resources and the conditions that are crucial to oilseed crop performance. One way these crop plants will respond to GCC is through environmentally induced shifts in phenotypes (i.e., phenotypic plasticity); therefore, understanding plastic responses is crucial for predicting and managing the effects of GCC on current and future oilseed crops (Nicotra et al., 2010). Climate change is predicted to bring about increased temperatures across the world in the range of 1.6°C to as much as 6°C by 2050. Although rainfall is predicted to increase globally, some areas will receive less annual rainfall, while others may receive much more. The timing of rains and crop-growing periods will also change. The frequency and duration of extreme weather events are also predicted to increase, although uncertainty exists about the expected degree of changes. These predicted changes in climate are expected to have fairly widespread impacts on agriculture, with poor countries in the south highlighted as being particularly vulnerable, having already weak economies, and limited institutional capacities to adapt. Changes in climate are likely to place new pressures on conservation of wild relatives and land races of oilseed crop species (IPCC, 2013). Climate change is likely to be an additional threat to agricultural biodiversity, increasing genetic erosion of land races and threatening wild species, including crop wild relatives (CWR); for example, up to 61% of *Arachis* spp. could become extinct within 50 years.

The response of oilseed crops to GCC will be primarily dictated by a complex set of interactions to temperature, CO_2, solar radiation, and precipitation. Most oilseed crops are currently grown at temperatures that match their threshold values; however, temperatures are expected to rise over the twenty-first century as a result of global warming and GCC and shifts may occur in production areas due to temperature changes beyond critical thresholds for growth, flowering, seed set, and oil production. Nevertheless, differences among oilseed crop cultivars to temperature extremes seem to be larger than for other abiotic stresses (Rodríguez et al., 2014), thus highlighting the importance of genetic manipulation and plant breeding as important tools for adapting these crops to future GCC. Shifting annual oilseed crop selection may provide a temporary reprieve from GCC; however, breeding and selection of new cultivars, especially of perennial oilseed crops, requires many years, thus greatly limiting the possibility of adapting to GCC.

Increasing carbon dioxide concentration [CO_2] may have a positive effect on oilseed production; however, this positive effect may be negated by higher temperatures, lower water use efficiency, and subsequent accelerated development resulting in lower seed and oil yields. Altered precipitation patterns, coupled with rising temperatures and increased [CO_2], will affect oilseed crop growth, development, and production, especially where rainfall coincides with the growing season (e.g., Midwestern USA). Biotic stresses, including weed, insect, and disease pressure and competition for natural and external inputs, will be exacerbated by GCC. In addition, simulation models suggest that by 2050, energy crops (e.g., oilseeds such as *Brassica napus* and *Helianthus annuus*) and biofuel crops (such as *Sorghum* spp. and *Miscanthus* × *giganteus*) show a northward expansion in potential cropping areas, but a reduction in their current southern growing regions (García-López et al., 2013), while their biodiversity is expected to decline in response to GCC.

GLOBAL GENETIC RESOURCES AND GENETIC DIVERSITY OF OILSEED CROPS

Although the genetic diversity conserved in crop gene banks will be useful in the long run, *ex situ* conserved germplasm may not harbor the new genes that may be needed as an insurance policy against the vagaries of GCC (Gill et al., 2014). Oilseed crops account for 2–3% of the total *ex situ* plant genetic resources for food and agriculture in the FAO *ex situ* germplasm conservation system, and the genetic integrity of long-term stored germplasm is questionable (FAO, 2014); therefore, a critical assessment of stored germplasm and development of new variability for important traits like oil content and its quality could be critical for planned breeding strategies. The viability of oilseed accessions stored *ex situ* declines over time. For example, the viability of stored seed of *Brassica napus* L. declined from 100% to between 50–25% in less than 10 years; whereas, the viability of *ex situ* stored *Linum usitatissimum* seed remained above 80% during the same period. On the other hand, within winter *B. napus* types, a decay of genetic diversity with reduced levels of erucic acid and glucosinolates has been observed over time (Ali and Farhatullah, 2013).

Several oilseed crops experienced bottlenecks after domestication; the genetic diversity of other crops declined as a result of breeding and selection, whereas low genetic diversity in a few crops (e.g., peanut) hindered their improvement using traditional breeding methods (Sun et al., 2013). The vast genetic potential of wild relatives of these crops remains underutilized unless molecular breeding can be employed for their improvement and for the production of adapted commercial hybrids, especially in developing countries. The wild relatives of oilseed crops and oil-bearing plants have the genetic memory of thousands of years of adaption to GCC and are a key resource of alleles for GCC adaptation, providing researchers with genes and traits for multiple biotic and abiotic resistance (Jaradat, 2015). Climate change calls for increased consolidation of germplasm collections of narrowly adapted and endemic wild species and CWR due to increased likelihood of their extinction. Increased demand for novel, diverse, and resilient germplasm in the face of GCC presents a challenge for gene banks to ensure that important germplasm is adequately conserved, as well as an opportunity for stimulating its greater use by breeders and agronomists through adequate characterization and screening for useful traits.

BREEDING OF OILSEED CROPS FOR ABIOTIC STRESS: LEARNING FROM PAST EXPERIENCE

The twentieth century witnessed great innovations and breakthroughs in genetics and plant breeding, the outcomes of which satisfied, to a large extent, the immediate and long-term plant-based needs of society. Spectacular achievements were made especially whenever there was accessible, characterized, and evaluated genetic diversity within the germplasm pool, and when breeders focused on timely measuring the right traits in the right environments. Genetic diversity of oilseed crop germplasm was enhanced using land races, mutants, and occasionally CWR. Molecular marker techniques assisted in genotyping analysis, determination of phylogenetic relationships, population structure, map-based cloning, quantitative trait loci (QTL) mapping, and marker-assisted selection (MAS). However, these techniques were not generally appropriate for measuring the adaptive genetic diversity of oilseed crops (El-beyrouthy et al., 2014).

Historically, breeding of oilseed crops confronted the challenge of bridging a widening gap between demand and supply of oil and its byproducts for food, feed, and industry, by breeding and releasing genetically uniform cultivars (Keneni et al., 2012). As a result, the wide variation present in traditionally cultivated and maintained land races was replaced with a few genetically uniform cultivars of each oilseed crop. Consequently, depleted genetic diversity and large-scale uniformity created ideal conditions for vulnerability to biotic stress. Therefore, a new paradigm of plant breeding advocated a solution based on breeding for specific adaptation instead of wide adaptation, systematic spatiotemporal gene deployment, integrated horizontal and vertical disease resistance, and even the use of interspecific cultivar mixtures (Podevin et al., 2013). Efficient but moderate selection pressure can be used to replenish genetic diversity and enhance adaptation to GCC in oilseed crops, provided that a high effective population size with migration and mutation can be maintained (Cowling, 2013). Selection strategies, such as best linear unbiased prediction and genomic selection, in addition to short generation intervals, may contribute to sustainable oilseed breeding under GCC.

For the foreseeable future, GCC will help prioritize the objectives of the research agenda of oilseed crops breeding. Other factors include increasing prices for nonrenewable inputs, evolving consumer demands (Schulman, 2012), and a desire for higher, sustainable, and stable production with lower nonrenewable inputs. Therefore, breeding oilseed crops requires more efficient tools with which to unlock and apply existing genetic diversity, to understand the relationship and bridge the phenotype–genotype gap, and appropriately apply biotechnological procedures (Schulman, 2012). Transgenics, MAS, and other technologies in the toolbox of molecular breeding may reduce the amount of diversity required in breeding programs; individual genes rather than collections of traits become the target.

Conventional and molecular breeding approaches will have to complement empirical evaluation with genetic prediction (Cooper et al., 2014). New breeding methodologies will continue to unfold as applied application of the scaling of quantitative biology, and will be founded on the integration of quantitative and population genetics, biometry, genotyping and phenotyping, and aided by modeling and computer simulations. The outcomes and outputs achieved by this approach will be measured in terms of sustainable improvements in oilseed and other desirable products that meet the needs of society. Moreover, joint and strong partnership between private and public breeding sectors and large-scale gene banks are required to avoid potential conflicts involving breeders' rights, germplasm use, and conservation (Dias, 2014).

CAN CARBON IN OILSEED CROPS HELP MITIGATE CLIMATE CHANGE?

Oilseed crops differ in their ability to collect solar energy and in their efficiency to store it as chemical energy in the form of reduced carbon (C) (Table 18.1). Along with bioenergy crops, oilseed crops are the basis for biological systems that are expected to contribute to renewable energy production, help stabilize the rising levels of greenhouse gases (GHG), and mitigate the risk of GCC (Jaradat, 2014). Several biochemical pathways in seed oil biosynthesis determine the fate of C in oilseed crops (Bates et al., 2013). Fatty acids (triacylglycerol; FAs) are the most naturally abundant form of reduced C chains; oils from oil-producing crops represent the main renewable resource of these FAs. Knowledge of the regulation of fatty acid production is widely exploited by genetic engineering for oilseed crop improvement (Nayeri and Tarizade, 2014). Several subcellular organelles are involved in the biochemical pathways leading to oil synthesis in seeds; a process requiring extensive lipid trafficking (Bates et al., 2013). Although much has been elucidated about the enzymes and pathways responsible for acyl group flux, there are still major gaps in our understanding of how flux between alternative pathways is controlled, and about the specialized cell biology leading to biogenesis of oil bodies that store up to 80% of C in seeds (Hyun et al., 2013).

Climate change is predicted to have an increasingly negative impact on oilseed crop growth and yield, particularly when the "worst case" CO_2 emission scenario (A1FI) of the Intergovernmental Panel on Climate Change (IPCC) is considered, and in the absence of efforts to mitigate global CO_2 emissions. In spite of a wide range of adaptation options available to mitigate negative impacts of GCC on crop production, impacts are still mostly negative in wide regions of the world (Olesen et al., 2011). Measures of mitigating GCC by bioenergy

TABLE 18.1 Example of Oil Content, Specific Energy Density, and Biofuel Yield of a Selected Number of Oilseed Crops

Bioenergy carrier	Oil content (%)	Specific energy density (MJ kg^{-1})	Biofuel yield (L ha^{-1})
Aleuritus cordata, A. fordii			4700
Arachis hypogaea	39–48		
Brassica napus	37–46	37.8	1190
Calendula sativa			305
Camelina sativa			538
Cannabis sativa			363
Carthamus tinctorius			779
Cocus nucifera	52–60		
Elaeis guineensis	18–26		5950
Glycine max	17–20	37.8	446
Gossypium hirsutum	16–18		325
Helianthus annuus	39–48	40.0	952
Jatropha curcas	24–26/30–50		1892
Linum usitatissimum			478
Olea europaea		39.8	
Ricinus communis	42–45	39.5	1413
Sesamum indicum	38–40		
Simmondsia chinensis			1818

Source: Compiled from Nogueira (2011), Jaradat (2010) and references therein.

crops, including oilseed crops, have been developed on the basis of a comparative GHG emission analysis for biodiesel production (Blackshaw et al., 2011). For example, data on GHG improvements (over fossil fuels) of biofuels derived from *E. guineensis* had the highest positive (40–70%) values over fossil fuels, whereas the GHG load of *Cannabis sativa* cultivation is intermediate between perennial and most annual energy crops. However, net fuel chain GHG abatement potential of 11 t CO_2eq ha^{-1} yr^{-1} in the midyield estimate is comparable with perennial crops (Finnan and Styles, 2013). On the other hand, *Helianthus annuus* and *Brassica napus* were considered as most suitable for biodiesel production in the Mediterranean Basin; *Glycine max* could also be used under certain conditions. Other results suggested that GHG emissions from 1 MJ of *Camelina sativa*-derived biodiesel ranged from 7.61 g CO_2 to 24.72 g CO_2 equivalent and from 3.06 kg CO_2MJ^{-1} to 31.01 kg CO_2MJ^{-1} equivalent for *Camelina sativa* jet fuel, whereas the nonrenewable energy consumption for its biodiesel ranged from 0.4 MJ^{-1} to 0.67 MJ^{-1}. Relatively high conversion efficiencies (~88%) have been reported for oil from oil crops using oil to fuel conversion technology.

Direct and indirect land use change (LUC) associated with oilseed production has been used as a means of mitigating GCC. The wide range of the combined LUC when converted to fatty acid methyl esters (FAME) of *Brassica napus* was estimated at 117–260 in Europe,

45–84 for *E. guineensis* in Indonesia, and 51–100 (gCO_2eq CMJ^{-1}) for *Glycine max* oil in Brazil. These estimates validate the large variability within and among crops, agroecosystems, and countries, and suggested that GHG emissions from indirect LUC were more important than emissions from direct LUC. However, achieving larger seed yield would significantly reduce environmental impacts associated with seed, oil, and fuel production of oilseed crops. Lower GHG emissions and energy consumption associated with some oilseed crops (e.g., *E. guineensis*, *Jatropha curcas*, *B. napus* and *Camelina sativa*) in comparison with other oilseed-derived fuel and petroleum fuel make derived fuel from these crops environmentally attractive (Li and Mupondwa, 2014).

INTERACTION BETWEEN ABIOTIC AND BIOTIC STRESSES: IMPACT ON OILSEED CROPS

The interaction between biotic and abiotic stresses takes place at different spatiotemporal levels, from genes to the field, and may involve changes at the transcriptome, cellular, and physiological levels (Jaradat, 2014). Several molecular mechanisms that act together in a complex regulatory network are involved in the specificity in multiple stress responses; they include transcription factors, kinase cascades, and reactive oxygen species as key components of this crosstalk, as are heat shock proteins and small RNAs. Abiotic stresses associated with GCC (e.g., high temperature, high humidity, and [CO2]), will influence the occurrence, prevalence, and severity of biotic stresses (e.g., diseases, insects, and weeds). Moreover, GCC will affect how biotic stresses will be managed as to timing, preference, and efficacy of chemical, physical, and biological measures of control within integrated pest management (IPM) programs.

One of the positive effects of GCC is that elevated CO_2 (e[CO_2]) may increase the yield and biomass of some oilseed crops (Bishop et al., 2014). For example, seed number per unit area increased by about 26% in oilseed rape, but seed weight was significantly decreased due to a short seed-filling period. This increase, however, may not hold at a global scale due to interactions between increased [CO_2] and temperature (IPCC, 2013). Prediction of future requirements in managing biotic stresses is of great interest to farmers, extension agents, and industry. However, a comprehensive analysis of potential GCC effects on biotic stress control and management is difficult because of limited and fragmented knowledge. Implications of GCC for biotic stress may differentially affect components of complex biological interactions, leading to outcomes that are difficult to predict (Newton et al., 2011), but will help prioritize breeding objectives of oilseed crops.

Plant responses to biotic and abiotic stresses are highly complex and involve changes at different organizational levels, including transcriptome, cellular, and physiological levels. Depending on the type of environmental conditions encountered, plants respond to multiple stresses differently from how they respond to individual stresses by activating a specific program of gene expression. In many cases, the presence of abiotic stress can reduce or enhance susceptibility to pests and pathogens, rather than being additive (Ahuja et al., 2010). The nonlinear responses of oilseed crops, and even individual genotypes, to stresses are expected to complicate the analysis of genotype × environment × management interactions (Heslot et al., 2014).

Plant hormone-signaling pathways (particularly abscisic acid) mediate the interaction between biotic and abiotic stresses that may induce or antagonize one another. Additionally, specificity in multiple stress responses is also controlled by several molecular mechanisms that act together in a complex regulatory network (Urano et al., 2010). Identifying main regulators that connect biotic and abiotic stress response pathways is essential in providing better chances of breeding oilseed crops with broad-spectrum stress tolerance. For example, regulation of intracellular potassium (K^+) homeostasis is indispensable to facilitate plant adaptive responses to a broad range of abiotic and biotic stresses including drought, salinity, and oxidative stress (Anschutz et al., 2014). Obviously, besides being an essential nutrient supporting optimal plant growth and yield, K^+ is also an important signaling factor mediating a wide range of plant adaptive responses to environmental stresses.

DESIGNING OILSEED CROPS FOR A CHANGING CLIMATE

For the foreseeable future, new generations of oilseed crops adapted to more variable environmental conditions, which can yield more with less external inputs need to be developed and adopted by farmers. In addition, more sustainable agronomic practices are required to help adapt new cultivars to GCC. Major changes in breeding and management programs may be required to achieving these interconnected goals (George et al., 2014) and facilitate the development of improved production practices that can cope with GCC and its extremes (Dias, 2014). A number of field-based platforms and advanced phenotyping systems are already available to build and validate functional–structural plant models (FSPMs) for oilseed crops (Kang et al., 2014). Such models provide an innovative approach of investigating genotypic performances under different environmental conditions. Additionally, these models will assist in breeding new cultivars, and can be used as decision support tools to design crop ideotypes adapted to current and projected GCC conditions.

Climate Change Components

By definition (IPCC, 2013), GCC is a change in the long-term weather patterns that characterize different regions of the world, while weather refers to the short-term (daily to weekly) changes in temperature, wind, and precipitation of a particular location. Drivers of GCC through changes in atmospheric composition can influence growth and development of oilseed crops directly by their impacts on plant physiology. In the long run, GCC could affect oilseed crops in several ways such as gas exchange, yield, and quantity and quality of seed and oil. In regions where crops are growing at or close to their physiological maxima, GCC (especially high temperature, warming, and salinity) will have an immediate impact on their growth and yield. Moreover, GCC impacts agroecosystems through changes over the long term in key variables affecting plant growth (e.g. rising temperatures and increased salinity) and through increasing the variability (frequency and intensity) of weather conditions (rainfall, drought, and elevated temperature). These changes also affect crop productivity and product (seed or oil) quality as well as how biotic stresses impact such crops (Ortiz et al., 2014).

High Temperature, Drought, and Heat

Accelerated crop phenology, greater risk of reproductive failure, and greater risk of crop failure are some of the implications when temperatures exceed the critical threshold of a particular crop. Classical and molecular genetics and genomic methods, along with precise field and laboratory phenotyping, have improved the discovery of genes and metabolic pathways conferring drought tolerance in oilseed crops (Zhang et al., 2014). Valuable genes for drought tolerance have been identified, for example, in *Camelina sativa* which makes it an oilseed crop adapted to dry land conditions (Ghamkar et al., 2010). Transgenic technology (e.g., wax-biosynthetic genes in transgenic *Camelina sativa*) is a valid approach in generating drought-resistant genotypes (Lee et al., 2014).

Potential yield reduction due to accelerated phenological development can be countered by selecting longer season cultivars and by manipulating the plant growth cycle through planting date(s) adjustment. However, breeding for tolerance of extreme temperatures will be necessary to counter the increased frequency of extreme temperatures, while greater emphasis on breeding for increased drought resistance and higher water use efficiency (WUE) will lower the impact of drought or reduced rainfall (Zhang et al., 2014). The emerging environmental conditions will tend to reduce water consumption by oilseed crops, but this effect will be approximately negated by the effect of higher temperature on evaporation rates. Screening and evaluation of oilseed germplasm tolerant to high-temperature stress can be indirectly accomplished by membrane stability and excised leaf water loss techniques (Ram et al., 2014); seed yield per plant was positively associated with membrane stability index.

At the whole plant level, high temperature accelerates plant growth and developmental rate to the point that the canopy architecture is modified. High or low precipitation affects the canopy architecture through flooding or drought stress, respectively. These GCC components will have corresponding impacts on biotic stresses. Due to the vulnerability of pollen and developing grains after fertilization, the flowering and grain-filling stages are the most sensitive to high-temperature stress (Singh et al., 2012). At the reproductive stage, the impact of high temperature is manifested at the C balance of developing fruits or seeds (e.g., siliquae of oilseed rape) by reducing the seed growth rate or reducing the duration of seed filling. Lipid profiles and the content of polyunsaturated fatty acids (PUFAs) produced by oilseed crops are influenced by high temperature during plant growth and seed maturation (Schulte et al., 2013). The production of two PUFAs (linoleic and linolenic) was reduced in *B. napus*, *Glycine max*, and *H. annuus* seed in response to an increase of temperature from 10°C to 40°C; genes and their products, which are involved with plant adaption to environmental stresses, have been discovered through comparative functional genomic analyses of plants exposed to low or high-temperature stress (Kovalchuk and Weselake, 2014). Heat-responsive genes have been identified through transcriptional profiling in siliquae of *B. napus* at the seed-filling growth stage (Yu et al., 2014); yield loss and alterations to seed compositions were documented in heat-susceptible genotypes.

Several indicators of tolerance to drought at the germination, seedling, flowering, and mature growth stages have been developed and used for germplasm screening and selection in oilseed breeding programs (Zhang et al., 2014). These included the evaluation of yield-based drought tolerance indices for screening *Carthamus tinctorius* genotypes (Pearl and Burke, 2014). Osmotic adjustment is a key component of drought response in several oilseed crops such as *Brassica napus* (Hatzig et al., 2014); however, other physiological mechanisms enabling avoidance of drought stress symptoms may contribute to drought tolerance. Marker–trait associations (MTAs) developed and validated for *Arachis hypogaea* may be deployed in improving drought tolerance–related traits and yield components (Pandey et al., 2014). Hydrogen sulfide (H_2S) was identified as an emerging signaling molecule involved in the regulation of physiological processes in plants associated with drought tolerance; moreover, potassium iodide (KI) was evaluated as a senescing agent and screening tool for tolerance to terminal drought in *Glycine max* (Bhatia et al., 2014). Several QTLs associated with drought (and waterlogging) tolerance during the seedling stage in *B. napus* have been identified (Li et al., 2014c), whereas *B. carinata* was identified as a highly drought–tolerant species well adapted to semiarid regions (Séguin-Swartz et al., 2013).

Rising Atmospheric Carbon Dioxide

By midcentury and at an increasing rate of about 2.2 μmol year^{-1}, [CO_2] is expected to reach, if not exceed, 500 μmol mol^{-1} (www.co2now.org) as predicted under the A1FI CO_2 emission scenario (IPCC, 2013); consequently, the climate will be warmer by about 2.8°C (Jaggard et al., 2010). Ongoing CO_2 enrichment is likely to increase yields of most C_3 oilseed crops by about 13%; however, yields of C_4 crops will generally be unchanged. Significant genetic variation in response to e[CO_2], when coupled with on-farm adaptation strategies, could help mitigate the expected negative impacts of GCC and potentially improve future crop yields, as demonstrated in *G. max* field studies (Bishop et al., 2014).

Temporal performances of oilseed genotypes with high or low response to e[CO_2] were consistent, thus suggesting a genetic control of response to e[CO_2]. This genetic variation in yield responsiveness to rising [CO_2] in many oilseed crops would provide an opportunity to breed more responsive crop genotypes to e[CO_2]. Oilseed breeders, most probably, will be able to increase seed and oil yields considerably in the CO_2-enriched environment of the future; however, current and emerging biotic stresses need to be controlled, so long as environmental protection policy changes do not remove too many crop protection chemicals. The impact of e[CO_2] on oilseed crop productivity and oil quality (Pal et al., 2014) is very important for global food and nutritional security under GCC scenarios. Exposure to e[CO_2] significantly influenced the rate of photosynthesis (PS), seed yield, and oil quality traits in a wide range of oilseed germplasm, including hybrids, improved cultivars, and land races. Greater whole plant net PS, dark respiration, daily C gain, and thus growth and WUE were achieved under e[CO_2] (Leonardos et al., 2014). For example, increased yield and biomass of *B. napus* due to e[CO_2]was expressed as 26% increase in seed number, but average seed weight was significantly decreased, probably as a result of shorter duration of seed filling. Higher seed yield and biomass (mainly as carbohydrates) accumulation under e[CO_2] may dilute the concentration of seed proteins and macro and micronutrients. However, an increase in e[CO_2]-mediated PUAFs compared with SFAs in some oilseed crops, such as *Helianthus annuus*, is a desirable outcome of e[CO_2] for nutrition and oil quality purposes (Pal et al., 2014).

A major contributor to daily C gain and the responsiveness of growth to e[CO_2] in oilseed crops, such as the model plant *Arabidopsis thaliana*, is the PS capacity of the inflorescence, especially when the expression of mitochondrial pyruvate dehydrogenase kinase is repressed (Leonardos et al., 2014). Daily C gain on a plant basis was a few hundred fold higher when the inflorescence was fully developed as compared with when the rosette leaves alone sustained the growth of *A. thaliana*. Therefore, improved performance of C–N metabolism could be potentially feasible for CO_2 assimilation and adaptation to e[CO_2] (as well as to high and extreme temperature as predicted by GCC scenarios) in current and potential oilseed crops such as *B. juncea* (Tsuda et al., 2014). However, the technical challenges of selecting for [CO_2]$^-$responsive oilseed germplasm, as well as the adverse effects of other abiotic stresses (e.g., salinity) on PS, chlorophyll, fluorescence, and their components (Singh et al., 2012), need to be taken into consideration. Photosynthesis in "green" seeds, pods, and siliquae plays an important role in the improved efficiency of oil accumulation in *A. thaliana* and oilseed crops (e.g., *B. napus* and *G. max*). However, the molecular mechanism behind the coordinated expression of FA biosynthesis-related and PS-related genes in these tissues remains to be elucidated. Nevertheless, it is known that seed-specific overexpression of *BnWRI1* coordinates both FA biosynthesis and PS pathways during oil accumulation in *B. napus* (Wu et al., 2014).

Salinity

Salt-tolerant oilseed crops, similar to other glycophytes, express several physiochemical mechanisms at different developmental stages in response to salinity stress (Zhang et al., 2014); however, seed germination and seedling growth stages are the most sensitive. These responses allowed for the development of several indicators of tolerance to salinity at the germination, seedling, flowering, and physiological maturity stages, and were used in germplasm screening, selection, and prebreeding of oilseed crops. Seed and oil yield, and oil quality of oilseed crops are impacted directly by salinity (Sun et al., 2013) and indirectly by salinity-induced water deficit (Zhang et al., 2014). Besides its immediate impact on PS, chlorophyll, fluorescence, and their components (Singh et al., 2012), salinity is one of the most important abiotic and environmental stresses that promote leaf senescence and thus affect crop yield (Allu et al., 2013).

A number of common signaling molecules, in both developmental and salt-induced leaf senescence, include hydrogen peroxide (Allu et al., 2013), and intracellular K^+, the regulation of their homeostasis is essential to mediate plant-adaptive responses to salinity, as well as to other abiotic stresses (Anschütz et al., 2014). Therefore, improved tolerance to salinity, through avoidance, or by delaying senescence under salinity stress, will have an important role in maintaining high oilseed productivity (Allu et al., 2013). This highlights the significance of identifying crosstalk regulatory components between tolerance to salinity stress and leaf senescence in oilseed crops. Salinity stress increases biosynthesis of secondary metabolites (e.g., antioxidants) and enzyme activity in some oilseed crops, such as *Carthamus tinctorius* (Pearl and Burke, 2014). Increased levels of metabolites, osmolytes, and antioxidant enzymes may have a significant role in protecting plants from salinity stress, thereby ensuring better plant growth and yield. Key changes in physiological response to salinity (e.g., in *Ricinus communis*, as a potential oilseed crop) include stomatal closure, reduced water loss, and osmotic adjustments by leaves and roots (Rodríguez et al., 2014). Moreover, moderate salinity (3–4 dS m^{-1}) promotes the accumulation of secondary metabolites (e.g., flavonoids), suggesting that biosynthesis of secondary metabolites may be associated with abiotic stress (Bruce, 2014).

Genetic manipulation and QTL mapping of oilseed crops for tolerance to salinity stress did not affect the chemical composition of the seed of *Arachis hypogaea*; however, seed yield of the transgenic crop was increased under stress conditions (Sun et al., 2013). Linkage mapping to identify salinity (and drought) tolerant genotypes identified several target QTLs in *B. napus* and *Arabidopsis thaliana* (Zhang et al., 2014). Developing salt-tolerant oilseed crops will likely sustain their production in countries where salinity (and water shortages) are already problems. Efforts were made to genetically engineer several oilseed crops, including *A. hypogaea*, for higher tolerance to multiple abiotic stresses, including salinity (Sun et al., 2013). Finally, efforts are under way to develop halophytes as new oilseed crops, such as *Diplotaxis tenuifolia* and *Cochlearia officinalis* of the Brassicaceae (Zhang et al., 2014). As halophytes, both species survived the highest salinity treatments (300 mM NaCl for *D. tenuifolia* and 400 mM NaCl for *C. officinalis*) and are considered as potential oilseed crops for saline agriculture.

Genetics and Genomics of Oilseed Crops

The increasing role of genomics (i.e., the direct study of DNA sequences) in breeding oilseed crops is accelerating with the rapid advances in genome technology (Lai et al., 2012). Genetic manipulation of the plant genome and the production of genetically modified plants by means of metabolic engineering and genomics, rather than classical plant breeding, is promoted as the most efficient route to improve oilseed crops in the face of GCC and its multiple abiotic stresses (Maheshwari and Kovalchuk, 2014). Therefore, functional genomics is being extensively used in understanding metabolic pathways and genetic improvement of several oilseed crops (Maghuly and Laimer, 2013). Bioinformatics, advanced relational databases, and sophisticated statistical programs are necessary to translate fast genomics data and generate relevant information to help in the breeding process. Improved genome-wide technologies, when supported by precise plant phenotyping and field monitoring (Nelissen et al., 2014) made it possible to efficiently move from *pot to plot*.

The growing genomics toolbox includes technologies for gene discovery, complex trait analysis by genome-wide association studies (GWAS), global gene expression analyses, genomic selection, and predictive breeding strategies (Snowdon and Luy, 2012). Based on genome-sequencing projects of several oilseed crops (e.g., the *Brassica* genome), the genomic toolbox offers much promise for the discovery of favorable allelic combinations (Rahman, 2013), accelerating the breeding process, and successfully achieving breeding objectives (Schulman, 2012). In addition, new genome-based techniques will accelerate the domestication of new crops such as *Diplotaxis tenuifolia*, *Cochlearia officinalis* (Sedbrook et al., 2014), and *Brassica carinata* (Zou et al., 2014).

Molecular Stress Response of Oilseed Crops to Climate Change

During the past few decades, plant molecular biology has been the key driver to elucidate molecular pathways underlying plant growth, development, and response to multiple abiotic stresses. Extensive research carried out using *Arabidopsis thaliana* as a model oilseed crop for the brassicas, generated large molecular databases and elucidated basic molecular mechanisms; however, the translation of research results to oilseed crop improvement is lagging behind (Nelissen et al., 2014). Oilseed crop plants, among others, evolved a wide range of molecular mechanisms to rapidly sense environmental changes and adapt to stress accordingly. Knowledge of these mechanisms is essential to breed crops and produce transgenic cultivars with enhanced tolerance to multiple GCC stresses. Plant response to these stresses is often accompanied by changes at the transcriptome, proteome, and metabolome levels (Jung et al., 2014).

Characterization of molecular mechanisms regulating hormone synthesis, signaling, and action allowed for hormone biosynthetic pathways to be modified for the generation of transgenic crop plants having enhanced abiotic stress tolerance (Peleg and Blumwald, 2011). Plant growth regulators include the classical five – abscisic acid, ethylene, cytokinin, auxin, gibberellin, and jasmonate – as well as brassinosteroids, salicylic acid, nitric oxide, and strigolactone. Controlling the dose:response ratio of these (and other) hormones remains a challenge so that a balance can be maintained between the positive effects of plant hormones on stress tolerance and the negative effects on growth and yield. Physiological and molecular mechanisms are usually reset when crop plants are subjected to stresses (e.g., extreme temperature, e[CO_2], drought, and salinity) so that homeostasis, survival, and growth may continue. One such mechanism involves transcriptional regulators (i.e., heat shock factors, HSFs) activating expression of heat shock proteins (HSPs). HSPs are important in thermotolerance reactions and act as molecular chaperones to facilitate protein refolding; they prevent denaturation or aggregation of target proteins (Ahuja et al., 2010). Heat-responsive genes have been identified in siliquae of *Brassica napus* at the seed-filling stage through transcriptional profiling. The molecular bases of the heat responses in siliquae during later reproductive stages were elucidated; this provided valuable information and gene resources for the genetic improvement of heat tolerance in *B. napus* breeding (Yu et al., 2014).

Genetic and epigenetic regulation, including DNA methylation, plays important roles in abiotic stress gene networks (Urano et al., 2010). The process of DNA methylation is responsive to various biotic and abiotic stresses, including heat stress. However, it differs between heat-tolerant and heat-sensitive genotypes of *B. napus* in response to heat stress, which further illuminates the molecular mechanisms of the adaptation, and explains different responses to heat stress in *B. napus* genotypes. DNA methylation increased more in the heat-sensitive genotype than in the heat-tolerant genotype under heat treatment; more DNA methylation and demethylation events occurred in the heat-sensitive and heat-tolerant genotype, respectively. A large and diverse set of genes were affected by heat stress via cytosine methylation changes, suggesting an important role of these genes in the response and adaption to heat stress in *B. napus* (Gao et al., 2014).

Concomitant molecular adjustments of physiological responses to GCC stresses (e.g., drought) include extensive changes in gene expression leading to adjustments in the biochemical and proteomic machinery of the plant. A key signaling intermediate that controls the expression of many plant genes is abscisic acid; it controls several abiotic stress–inducible genes, and leads to decreased water loss by regulating the stomatal aperture (Palta, 2014). Components of the salt's overly sensitive pathway – such as salt and abscisic acid–induced accumulation of reactive oxygen species; salt-inducible transcription factors; cytoskeleton, peroxisomes, apoplastic proteins, and glycolytic enzymes – are involved in molecular adaptation to salinity stress. When oilseed crops are subjected to salinity stress, the salt's overly sensitive pathway regulates Na^+/K^+ ion homeostasis. In some salt-tolerant crop plants, this adaptive mechanism operates to maintain low cytoplasmic concentrations of sodium by sequestering Na^+ in vacuoles. Pollen development and viability are sensitive to adverse climatic environments and abiotic stresses (De Storme and Geelen, 2014). High levels of microspore abortion and male sterility may develop in response to high or low temperature, salinity stress, osmotic shock, water deficit, and drought. This response usually involves cytoskeletal alterations, altered sugar utilization, abnormalities in auxin metabolism, and accumulation of reactive oxygen species. Although stress-induced male sterility is used in hybrid production, under field conditions it has a negative impact on crop performance and quality.

Cisgenesis in Oilseed Crop and Climate Change

Cisgenesis is a genetic modification process whereby beneficial alleles of cisgenes are transferred from crossable species into a recipient plant; therefore, the latter is not a transgenic (Hou et al., 2014) and there is no risk of its release in the environment. In cisgenesis, the DNA sequence includes the gene of interest flanked by its own promoter and terminator, whereas, in intragenesis, the gene of interest can be combined with regulatory elements from the species itself or from a cross compatible species (Lusser et al., 2012). Nevertheless, recent debate over the "slippery slope of cisgenesis" (Eriksson et al., 2014) cautioned that "..the expertise and time needed to create case-specific genomic clones that include endogenous promoters and terminators, free from selectable marker genes and vector backbones, should not be underestimated."

Cisgenesis combines traditional breeding techniques with modern biotechnology and speeds up the breeding process, while at the same time it avoids linkage drag (i.e., cotransfer of DNA sequences that are linked to the gene of interest), and enhances the use of existing alleles (Hou et al., 2014). Practically, cisgenic insertion of additional copies of native genes may provide a new approach to modify plant genomes, expand genetic variance in plant architecture available to breeders, and accelerate the transfer of alleles between species that are difficult to cross. Recent progress in plant genome sequencing made it possible to isolate plant genes (i.e., cisgenes) from crossable species. Recent cisgenesis and intragenesis-based products include *Zea mays* and *B. napus* (Lusser et al., 2012); the new crops have already passed the construct optimization, trait development, preregulatory data, and large-scale transformation phases, and are at the trait integration, field testing, and regulatory data generation phase.

Genome and Phenotypic Plasticity under Climate Change

A change in genome structure or organization associated with environmental signals, known as genome plasticity, leads to the evolution of new phenotypes; this change might result from mutational hotspots, genome expansion, transposable elements, or somatic recombination (Nicotra et al., 2010). Phenotypic plasticity, on the other hand, indicates the range of phenotypes a single genotype can express as a function of its environment (Schulman, 2012). For example, plastic changes in seed longevity, phenology, leaf duration, and reaction of metabolic process to temperature, are all well documented in response to e[CO_2] associated with GCC. Genome plasticity and phenotypic plasticity are distinct from each other but can provide a mechanism that generates phenotypically plastic responses. Understanding plastic responses, whether in providing a buffer against, or assisting prompt adaptation to, rapid GCC (Bloomfield et al., 2014) is crucial for predicting and managing the effects of GCC on adaptation and performance of current and future oilseed crops (Nicotra et al., 2010). Epigenetic changes proved to be important in responding to environmental stress; they can happen much faster than changes at the DNA level; therefore, they can be more important in the face of rapid climatic changes (Urano et al., 2010). Consequently, new developments in our understanding of epigenetics and signaling cascades hold promise for predicting how plasticity will influence the responses of oilseed crops to GCC (Bloomfield et al., 2014).

Genetic and epigenetic regulation play important roles in abiotic stress gene networks (Urano et al., 2010). New knowledge about signaling cascades and epigenetics is promising for predicting how plasticity will influence the responses of oilseed crops to GCC. Plant plasticity is facilitated by epigenetic processes that modulate chromatin through dynamic changes in DNA methylation, histone variants, small RNAs, and transposable elements (Bloomfield et al., 2014). Directional selection in contemporary oilseed crop breeding may have led to reduced adaptive plasticity in modern cultivars; breeders did not address increased plasticity *per se*, instead, they targeted plant traits to obtain higher yields in particular environments, or for homeostasis under a range of environmental conditions. Breeding for phenotypic plasticity in traits other than seed or oil yield will potentially provide resilience under increasingly unpredictable environmental conditions of GCC.

A number of functional traits that can be used in investigating adaptive phenotypic plasticity in oilseed crops include (Nicotra et al., 2010): leaf traits (specific leaf area as indicator of relative growth rate, PS capacity, leaf duration, and leaf N content; leaf size, shape, thickness, and pigmentation), flowering (time, flower size, and phenology as indicators of species ability to respond to GCC), seed (size and number per fruit as indicator of fitness), root (specific length as relates to precipitation pattern shift under GCC).The larger the genetic variation in an oilseed crop for a particular trait or group of traits, the better its ability to withstand and adapt to abiotic (and biotic) stresses associated with GCC. A portion of the genetic variation is expected to determine the ability of a crop plant to sense a particular environmental change (e.g., high temperature) and produce a relevant plastic response.

Gene Regulation and Expression under Climate Change

The availability of reliable tools for routine analyses of gene expression profiles in oilseed crops enabled researchers to evaluate important traits, especially those related to oil biosynthesis and resistance to abiotic (and biotic) stresses, such as drought, salt, high and low temperature, and nutrient deficiency or exposure. Generally, it is challenging to assign a change or modification in plant metabolism to a specific stress factor; usually, a variety of abiotic (and biotic) stress factors occur at the same time and can induce antagonistic responses resulting in accumulation, degradation, and consumption of primary and secondary metabolites (Bloem et al., 2014). Genome-wide transcription analysis in response to abiotic stresses is a significant development in providing a basis for effective genetic manipulation in order to improve stress tolerance in oilseed crops (Jung et al., 2014). Genes underlying environmental adaptability tend to be overretained in polyploid oilseed crop species (Li et al., 2014a), and functionally redundant multiple-copy genes related to abiotic stress have been found in several oilseed crops.

The genetic and molecular basis of early osmotic adjustment response in drought-resistant and drought-sensitive *B. napus* is still unclear; however, heterosis may convey an improved general drought stress tolerance in this crop (Hatzig et al., 2014). Heterozygous F_1 hybrids of winter *B. napus* showed phenotypes with improved vigor and a general elasticity against drought stress in comparison with homozygous inbred lines. Drought-resistant genotypes showed no hormonal adjustment or osmolyte accumulation. Other physiological routes may exist through which these genotypes can resist drought. The aquaporin *CsPIP2;1* in *Camelina sativa*, which is involved in response to salt and drought stresses, is expressed in roots and leaves (Jang et al., 2014).

Although *Jatropha curcas* is very drought tolerant, only a few drought-inducible genes have been identified (e.g., ethylene-responsive element binding factor, *JcERF*; betaine aldehyde dehydrogenase 1, *JcBD1*; and plasma membrane intrinsic protein 2, *JcPIP2*), whereas a sizable number (332) of differentially expressed genes were found when *J. curcas* was subjected to drought stress, 585 during the recovery stage from stress, and 374 genes were involved at both stages (Cartagena et al., 2014).

Tropical oil-producing crops, such as *E. guineensis*, are sensitive to low temperatures (Lei et al., 2014); analysis of its RNA-Seq under cold stress revealed a C repeat binding factor–mediated gene expression pattern unlike the one found in *Arabidopsis thaliana*. The C repeat binding factor–mediated signal transduction pathway plays a dominant role in cold tolerance. Expression of genes controlling PUFA biosynthesis and oil deposition in developing seeds of *Plukenetia volubilis*, a potential oilseed perennial tree (Wang and Liu, 2014), showed that low temperatures enhance the biosynthesis and deposition of PUFAs in its seed; the same effect was found in *G. max* and *Linum usitatissimum*.

Genes underlying adaptability to nutrient deficiency (e.g., Zn, Fe) or exposure (e.g., Cd) stress in *B. rapa*, and other polyploids, are overretained at the gene dosage level, and functionally redundant at the transcriptional level (Li et al., 2014b). Most of these multiple-copy genes showed the same direction of expression change under stress conditions; however, very little is known about their differential expression when plants are subjected to these stress conditions. Redundancy of such genes allows polyploids to adapt to abiotic stress; however, it makes the regulatory network more complex in *B. rapa* compared with the diploid *Arabidopsis thaliana*. Such diversified expression patterns between overretained and redundant genes could be attributed to an evolutionary mechanism to meet the enhanced demand of polyploids for a wide spectrum of ontogenetic functionality and differentiated cell types within plant tissues.

Gene Flow under Climate Change

Gene flow is a natural biological process that occurs to some degree in all flowering plants; four steps can be delineated in the process: (i) gene flow through pollen or seed, (ii) spontaneous hybridization, (iii) hybrid behavior, fitness cost due to hybridization and introgression, and (iv) fitness benefits due to transgenes. Each step can potentially be impacted by environmental and climatic conditions and stresses. Pollen and grain development are the most sensitive stages for temperature stress (Singh et al., 2012). Therefore, pollen-mediated gene flow and subsequent hybridization and grain development can be impacted by GCC. In addition, shape and composition of a cropping system can impact spatiotemporal gene flow through seed bank dynamics (Liu et al., 2013). Pollen-mediated gene flow between crops, CWR or weeds, can introduce new alleles into oilseed crops with the probability of introgression depending on the degree to which the new alleles alter fitness and therefore are selected for in nature. A cultivated oilseed crop is more likely to hybridize with closely related wild species having a similar ploidy level, genome, and chromosome-pairing ability (De Jong and Rong, 2013).

Oilseed crop biology will determine seed fate (i.e., dormancy, germination, or mortality) and can influence the success of allele introgression. For example, duration of the flowering period of an oilseed crop, especially crops with indeterminate flowering habits with the possibility of viable flowers being present at harvest, determines potential flowering synchronicity with its wild relatives. Additionally, GCC stress components will determine how successful the introgression will be, and what role hybrids will play in metapopulation dynamics (Warwick and Martin, 2013). Oilseed crops differ as to their genetic base, phylogenetic origin, mode of pollination, outcrossing rate, and the rate of gene flow. Compared with *Linum usitatissimum* L., *Arachis hypogaea*, for example, has a narrow genetic base due to its monophylogenetic origin, is selfpollinated, and virtually lacks gene flow. Gene flow and hybridization of transgenic *L. usitatissimum* L. or *B. napus*, for example, with many wild relatives depend on species sympatric distribution, concurrent flowering, ploidy level, sexual compatibility, and potentially on environmental conditions. However, no gene flow has been verified yet, for example, from transgenic *B. juncea* (for herbicide resistance) into the weedy *Sinapis arvensis* (Warwick and Martin, 2013). Nevertheless, in this and similar cases of gene flow, the transgene may persist, as subsequent generations populations are formed that have selfing capacity, normal fertility, and dormancy that may contribute to persistence of the transgene and the herbicide resistance trait in nature.

Gene flow, in the case of spatiotemporal coexistence of genetically modified and non genetically modified oilseed crops (e.g., *B. napus*), may have far-reaching effects on survival, dormancy, emergence, and seed bank evolution (Haile et al., 2014), and in the metapopulation dynamics role of natural hybrids (De Jong and Rong, 2013). Feral populations of *B. napus*, due to siliqua shattering, may persist for up to 8 years without human intervention (Gepts, 2002) with potentially increasing intensity of gene flow. Gene flow from transgenic plants to wild relatives may cause wild plants to acquire traits that improve or impair their fitness in natural habitats. There is a rising concern over crop-to-crop gene flow in *Linum usitatissimum* under field conditions. Gene flow from high α-linolenic acid cultivars to low α-linolenic acid cultivars under the changing climate of western Canada has economic repercussions. Therefore, a better understanding of crop-to-crop or crop-to-weed gene flow is essential for ecological risk assessment (Gressel, 2014). A key environmental risk of transgenic oilseed crops is potential gene flow into feral populations or to wild relatives of the crop and the emergence of invasive plants or the loss of already endangered wild genetic diversity.

Reproductive Fitness in Oilseed Crops under Climate Change

Biological and agronomic characteristics of reproductive fitness in oilseed crops are critical for genetic manipulation and to increase seed and oil yields. Integrated high-throughput plant-phenotyping (HTPP) and genome-wide association analyses in oilseed crops have the potential of accelerating the rate of trait discovery and improving phenotypic predictions of yield sink size (e.g., seed or fruit) and reproductive resilience under GCC stresses. Especially when high heritability is associated with moderate to high genetic advance for the duration of the reproductive phase, then there will be a reasonable possibility for improvement of traits related to reproductive fitness (Jain, 2014). Several reproductive fitness traits are usually strongly correlated and QTL colocalized across environments (e.g., flowering time, dry rosette weight, seed yield, harvest index, ratio of reproductive to vegetative organs, seed C% and N%, and 1000-seed weight). These positive correlations are consistent with life history theory in that genotypic variation in resource acquisition masks tradeoffs (Chardon et al., 2014). Constitutive overexpression of a newly discovered Gγ protein (AGG3; Choudhury et al., 2014) in *Camelina sativa* resulted in faster vegetative and reproductive growth, along with increased PS efficiency. Several components of reproductive fitness (e.g., seed size, seed mass, and seed number per plant) increased by 15–40%, and higher oil yield per plant was achieved when AGG3 was expressed constitutively or specifically in seed tissue.

The risk of high-temperature impact on oilseed crop production is increasing with increased global warming. Impaired fertility and yield loss due to heat stress are widely reported in oilseed crops (Yu et al., 2014). Adverse climatic conditions and environmental stresses particularly impact male reproductive development in oilseed crops. When exposed to stress, male gametophytic organs often show abnormalities at the morphological, structural, and metabolic levels; such changes usually lead to meiotic defects or premature spore abortion and male reproductive sterility. Pollen viability, flower number, bud number, seed and pod number, as components of the reproductive system, are used to identify genotypes with high-temperature tolerance. The number and length of buds emerged on main stem, and pod number produced under high temperatures, might provide a useful preliminary screening criterion for high-temperature tolerance in oilseed crops such as *B. napus* and identify genetic resources (e.g., *B. rapa*) to provide genes for heat stress tolerance (Annisa et al., 2013).

Depending on the type of stress involved (e.g., heat, cold, or drought) and the duration of exposure, the underlying cellular defects are highly variable and involve a wide range of aberrations (De Storme and Geelen, 2014). Interactions between leaf senescence and resource allocation to seeds, hence reproductive fitness, were revealed by QTL metaanalysis of traits related to senescence, resource allocation, and seed yield in *Arabidopsis thaliana*. Sequential and monocarpic senescence observed at vegetative and reproductive stages, respectively facilitate N remobilization and control the duration of C fixation. However, genetic and environmental factors control N and C resource allocation to seeds (Chardon et al., 2014). Heat-responsive genes, with organ-specific expression, have been identified through transcriptional profiling at the seed-filling stage in siliqua wall and seed of *B. napus* (Yu et al., 2014). The molecular bases of response to heat during late reproductive stages provided valuable information and identified gene resources for the genetic improvement of heat tolerance in *B. napus*. When reproductive tolerance to high temperature was increased by 2°C (Singh et al., 2012), reproductive efficiency was improved, especially after the plants were subjected to drought stress. Large genotypic variation for heat tolerance was found at the reproductive stage in several oilseed accessions (spring-type *B. rapa* accessions and one *B. juncea* accession). Heat tolerance was determined by the ability of accessions to set seed equally at normal (i.e., control) and high-temperature treatments (Annisa et al., 2013).

Carbon Footprint and GHG Emissions of Oilseed Crops

Oilseed crops, as solar energy collectors and thermochemical energy storage systems, are a component of biological systems that are expected to contribute to renewable energy production, help stabilize the rising levels of GHGs, and mitigate the risk of GCC (Jaradat, 2010). The production of oilseed crops is promoted as a means of mitigating global warming by decreasing dependency on fossil energy (Pugesgaard et al., 2014). The fate of atmospheric carbon, including the C footprint and GHG emissions, after being fixed in oilseed crops through PS, is largely determined by a number of (bio)-chemical pathways within the plant–soil–atmosphere continuum; other factors – such as management, external inputs, and land use change – are also involved (Jaradat, 2014). For oilseed crops to be C negative, a reduction in the C footprint and in GHG emissions is needed throughout the life cycle of crops on the farm, and their primary and secondary products need to be utilized (i.e., life cycle assessment) (Uusitalo et al., 2014).

A quantitative measure, which takes into consideration the spatiotemporal changes of the C footprint and GHG emissions of oilseed crops (kg CO_2eq ha^{-1}, or CO_2eq C MJ^{-1}) is a measure used for comparative and assessment purposes (Shrestha et al., 2013). Crops, such as *E. guineensis* and *B. napus*, have been bred and developed to divert and store a large part of PS-fixed C into their seed as oil. Therefore, higher seed and oil yield, and the energy density of an oilseed crop (Gan et al., 2012) would significantly lower the environmental impacts associated with its seed, oil, and biofuel production. The lower GHG emissions and energy consumption associated with an oilseed crop in comparison with other sources (e.g., biomass or petroleum) renders its biofuel environmentally more sustainable (Li and Mupondwa, 2014). For example, the GHG emissions from 1 MJ of biodiesel from *Camelina sativa* ranged from 7.61 g CO_2eq to 24.72 g CO_2eq. Similarly, GHG emission reduction (in relation to diesel) for generating 1 GJ energy from *Jatropha curcas* varied from 40–107% and net energy ratio from 1.4 to 8.0. Although GHG emission reduction of 54 and 40% and net energy ratio of 1.7 and 1.4 for irrigated and rainfed scenarios, respectively, may indicate environmentally friendly sustainability of biodiesel from *J. curcas* (Johnson et al., 2011), these results have been disputed elsewhere (von Maltitz et al., 2014). A smaller C footprint and higher environmental benefit could accrue if there is C flux into triacylglycerol (TAG) in vegetative plant tissue (Weselake et al., 2009).

GHG improvements of biofuels derived from *E. guineensis* were 40–70% above fossil fuels. Specific energy density (MJ kg^{-1}; Table 18.1) reflects major differences between several oilseed crops. In particular, the wide range of the combined direct and indirect LUC, estimated in g CO_2eq C MJ^{-1}, when converted to *B. napus* FAME, was estimated at 117–260 in Europe, 45–84 for *E. guineensis* in Indonesia, and 51–100 for *G. max* in Brazil. On the other hand, total water requirements estimated as evapotranspiration (m3 GJ^{-1}) for *B. napus* (100–175), *E. guineensis* (46–250), and *G. max* (143–500) exhibited large differences. These estimates demonstrate the large variability within and among crops, agroecosystems, and countries, and suggested that GHG emissions from indirect LUC were more important than emissions from direct LUC (Bishop et al., 2014).

The form, timing, and quantity of applied N fertilizer, after being delivered to the farm gate, comprise a sizable portion of the C footprint and GHG emissions, depending on the cropping system as well as edaphic and environmental conditions. The C footprint of oilseed crops can be 25–45% greater in wet sites than in dry sites (Gan et al., 2012). However, in spite of better management practices, reduced GHG emission intensities (40 and 65% on area and grain dry matter basis due to higher yields, respectively), and genetic improvement in resource acquisition and allocation, the GHG emission of some oilseed crops increased over time (e.g., 13% in the Canadian Prairies) due to recent expansion in planted area (Shrestha et al., 2013).

Oil vs. Protein in Oilseed Crops

The basic relationship between oil and protein in oilseed crops can be found in oil bodies, such as oleosin and caleosin, which are common structural proteins derived from oilseed. Oil bodies store oil as C resource; they have been exploited for a variety of industrial applications including biofuels (Hyun et al., 2013). As surrogates of oil and protein, respectively, C and N resource allocation to seeds is generally controlled by genetic and environmental factors (Chardon et al., 2014). Oilseed crops respond to e[CO_2] and temperature environments through changes in C–N metabolism and gaseous exchange. Therefore, improved performance in C–N metabolism may improve CO_2 assimilation and adaptation of oilseed crops (e.g., *B. juncea* L.) to high temperature and e[CO_2] as GCC stressors.

Oil content in oilseed or oil-bearing crops ranges from low (e.g., 10–15% by weight of coconuts) to high (>50% in *Sesamum indicum* seeds and *E. guineensis* kernel). Protein content is relatively high in *G. max* (~40%), much lower (15 – 25%) in most oilseed crops, and lower still (>10% in oil bodies of olive embryos) in some oil-bearing crops. However, seed oil of some wild (or semidomesticated) oil-bearing plants, such as *Plukenetia volubilis*, is rich in both oil (40–60%) and protein (27–30%). Low temperatures tend to decrease oil content in favor of more protein, while the contents of PUFAs and linolenic acid tend to increase (Wang and Liu, 2014). Environmental and genetic effects on oil and fatty acid composition of *Camelina sativa* genotypes have been delineated (Pecchia et al., 2014), and genotypes having high-protein content (31.7%) associated with 29.4–31.9% oil and high PUFAs (C18:3 and C:18:2; 59.4 and 17.1%, respectively) have been identified.The protein-rich residue remaining after the oil has been extracted from oilseeds is an important source of nutrients for farm animals. Oilseed meals from *G. max*, *Arachis hypogaea*, *B. napus*, and *Linum usitatissimum* are rich in protein (~35%); when mixed with cereal grains, they provide nutritionally balanced feeds. The presence of toxic proteins in some oilseed crops or oil-bearing plants (e.g., ricin in *Ricinus communis*; Haililu and Ogunwole 2013) banned its reintroduction for oil production.

A wide range of genetic, environmental, and management factors and their multiway interactions determine, to a large extent, the oil and protein contents, and the tradeoff between them in oilseed crops (Peltonen-Sainio et al., 2011). Some oilseed crops have more potential than others to accumulate both oil and protein in their seed or fruit (e.g., *A. hypogaea*); however, the dynamics and relationships between their content in the seed, with seed weight, seed yield, oil yield, and protein yield are not universally predictable, especially under GCC.

Genotypic differences in the relationship between oil and protein contents were documented in *G. max* and *B. napus* (Peltonen-Sainio et al., 2011). In both species, protein content increased at a small, or no expense of oil content; and reduction in oil content was not proportional to the concomitant increase in protein content, even when the larger energy per unit weight in oil (37 KJ g^{-1}) compared with the much lower energy (17 KJ g^{-1}) in protein was taken into consideration, suggesting that oil and protein production can be optimized under favorable agronomic and climatic conditions (Fig. 18.1). Genes and QTLs that increase oil content without decreasing protein, or with increased seed yield in *G. Max* have been identified along with molecular markers associated with oil-related QTLs. These QTLs have positive effects on other important traits such as seed yield and protein content. A transgenic *Jatropha curcas* plant, where a sugar-dependent1 gene (*JcSDP1*) was silenced, accumulated 13–30% more total oil, which was accompanied by a small (7%) reduction in protein content (Kim et al., 2014). Oilseed crops and oil-bearing plants, depending on gene expression, differ in the timing and rate of oil and protein biosynthesis and accumulation. High temperature and e[CO_2] impact their C–N metabolism; therefore, singly or in combination, these GCC stressors impact oil and protein biosynthesis and accumulation accordingly. Elevated temperature interferes with seed filling in favor of protein accumulation at the expense of oil and possibly carbohydrate accumulation (Wang and Liu, 2014). A metabolic engineering model was suggested to evaluate the tradeoff between high oil and high protein, by enhancing one pathway at the expense of another that competes for substrates (Sweetlove et al., 2014).

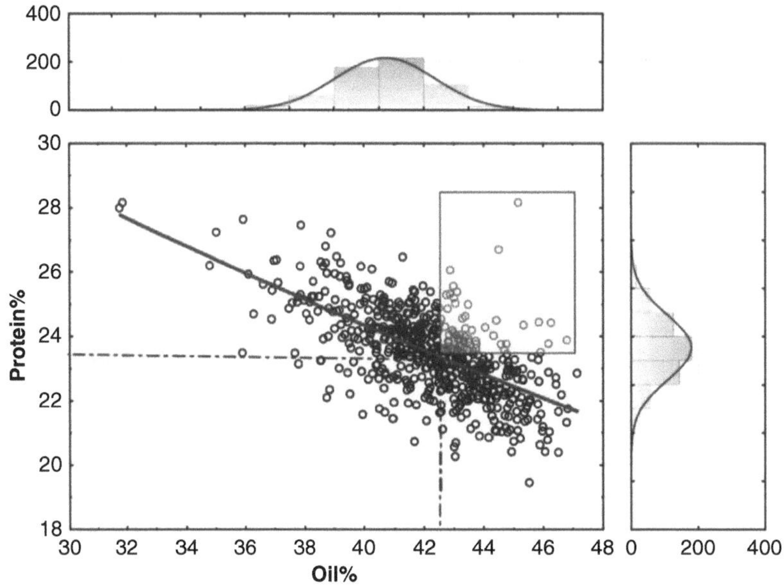

FIGURE 18.1 Relationship between oil and protein contents ($r = -0.64$; $p < 0.001$) in a large phenotype-screening experiment showing accessions with above mean content (shaded area, oil% = 42.2, protein% = 23.5) of both seed storage macromolecules (Jaradat, unpublished data).

Genetic Architecture of Seed Weight and Quality

Oilseed quality is a complex trait that may be influenced by many interacting factors, the most important of which are assimilate partitioning and storage as well as pre and postharvest technological and environmental factors. The seed or the fruit of oilseed and oil-bearing crops is the key product; its shape, size, and architecture affect several aspects of plant ecology and evolution. Pleiotropy, more than linkage, may constitute the genetic basis of seed weight and morphometrics of the bearing fruit (i.e., pod or siliqua) (Li et al., 2014c), while a QTL network accounts for the genetic basis of seed weight (in addition to seed number and siliqua length in *B. napus*). Additive and epistatic genetic effects were implicated in the genetic control of these traits; therefore, MAS may help improve seed and fruit (siliqua or pod) traits in oilseed crops. In hybrid breeding, seed weight is usually positively associated with seed weight heterosis of the better parent (Xing et al., 2014); therefore, it may provide a better selection criterion for higher yield than other yield components under abiotic stress (Zhang et al., 2014). Although seed weight exhibits a relatively low level of variation in some oilseed species (Shinwari et al., 2014), it can be enhanced, along with oil accumulation, through *GmMYB73* overexpression as demonstrated in *A. thaliana* (Liu et al., 2014).

Seed weight, depending on species and genotype, may respond differently to the source:sink ratio; during the active seed-filling period it may experience source:sink colimitations (Takashima et al., 2013). Phenotypic and biochemical seed traits of oilseed crops determine seedling, seed, and product quality, including seed weight, which is largely determined by the genotype. Seed weight was identified as one of the most important phenotypic traits in defining an oilseed crop ideotype, affecting embryo metabolism, and conferring tolerance to heat stress (Ram et al., 2014). As a major determinant of seed yield, seed weight exerts the largest positive indirect effect on economic yield, in decreasing order, through biological yield, harvest index, flower number per plant, capsule number per plant, reproductive phase length, secondary branch number, and plant height at maturity (Jain, 2014). Physiological seed quality, especially under abiotic stress, reflects seed germination speed, emergence, seedling growth, and early vigor, hence the need to merge physiological and genetic methodologies at the seed and at whole plant levels, to improve tolerance to abiotic stress (Palta, 2014). Plant and seed physiology research has already progressed and adopted HTPP as well as quantitative and phenomic technologies on fully sequenced genomes of oilseed populations (Brown et al., 2014).

Genome-wide association studies have dissected the genetic architecture of seed quality (Li et al., 2012) or anti-quality traits, such as glucosinolate content (Sønderby et al., 2010), PUAFs in *Jatropha curcas* seed (Maghuly and Laimer 2013), erucic acid in *B. napus* (Ali and Farhatullah, 2013), suboptimal oil for biofuel (Sedbrook et al., 2014), or meal quality (Colombini, 2014), and detected associations with genomic regions underlying quality traits in several oilseed crops. Oilseed quality can be impacted by GCC directly by abiotic stresses (e.g., drought, salinity, cold, heat, and heavy metal toxicity) and indirectly through biotic stresses. Therefore, tradeoffs between yield, quality traits, and environmental impact need to be considered in breeding oilseed crops for adaptation to GCC. For example, the yellow-seeded *B. napus* has a thinner seed coat, higher oil and protein content, higher feed value, and reduced antinutritional polyphenolics and proanthocyanidins when compared with higher yielding black-seeded phenotypes. Recent molecular advances and quality studies offered new opportunities to "domesticate" and improve the nutritional quality of new crops such as *Guizotia abyssinica* (Geleta and Ortiz, 2013) and *Plukenetia volubilis*, a potential oilseed perennial tree (Wang and Liu, 2014). Finally, lack or shortage of quality seed (e.g., newly domesticated oilseed species, newly developed oilseed crops, or new cultivars) constitute bottlenecks that may delay oilseed crop adoption, production, and productivity (Sheoran et al., 2013).

Oilseed Crop Ideotype(s) for a Changing Climate

Environmental and GCC concerns, especially in the developed world, have already established a favorable context for developing environmentally friendly cropping systems and placed demands on plant breeders to design ideotype(s) (i.e., ideal combination of traits in a specific genotype) that can meet more complex objectives. Due to the complexity of such an endeavor, modeling has a major role in designing annual (Kang et al., 2014) and perennial (Kantar et al., 2014) oilseed crop ideotypes and in predicting the impact of changes to agricultural systems in view of GCC. Simulation models are imbedded in the process of designing an ideotype because the impact of GCC variability on plant growth is closely linked to the design of the ideotype. A functional–structural model for genotypic and phenotypic characterization of *H. annuus*, for example (Kang et al., 2014), provides an opportunity to design and test an oilseed crop ideotype. In so doing, robust trait combinations are needed to fit specific agroecological conditions under future climate(s), while long-term experimental platforms and HTPP and genotyping systems are indispensable components (George et al., 2014). The ideotype concept has been used by breeders when screening plant populations for multiple morphometric and agronomic traits. In ideotype breeding, the relationship between crop yield and its determining factors is indirectly modeled, and the ideotype becomes a virtual, intermediary object, useful to formulate, challenge, and reconcile the different visions of classical breeders, geneticists, agronomists, plant pathologists, ecophysiologists, and others. Historically, however, plant-breeding programs, due to resource and time limitations, tend to breed out of modern oilseed cultivars some valuable traits. Simulation models and structural equation models (SEMs) can be valuable tools in developing *de novo* oilseed crop ideotypes that do not appear in currently available germplasm, and in optimizing their phenotypes through tradeoffs between different plant traits in order to suit future climate(s).

Selection criteria may differ from one group of oilseed crops to another, depending on which morphophysiological and biochemical traits are correlated with life cycle, adaptation, and productivity. Arguably (Redden 2013), the most important traits for phenotyping/ideotyping are those that can integrate the effects of time (i.e., ontogeny) and space (i.e., canopy architecture). High yield, high oil content, low or no toxicity in oil and meal, high seed weight, determinacy, and tolerance to biotic and abiotic stresses constitute a minimum set of traits to develop a commercial oilseed cultivar. Traits comprising a perennial oilseed crop ideotype (e.g., *H. annuus*) include seed and capitulum traits contributing to high seed yield (i.e., high pollen fertility, high seed set and seed weight, large capitulum size); plant architecture traits enabling uniform seed maturity (i.e., no branching and one central capitulum, or synchronous flowering if more than one capitulum); and tuber traits contributing to a manageable perennial habit (i.e., low tuber number, high individual tuber weight). The perennial ideotype is the domesticated phenotype with the addition of perennial traits (Kantar et al., 2014). Therefore, a combination of perennial and high oil content traits, while retaining or improving adaptive traits to GCC, are desirable components of a perennial ideotype. Such an ideotype would fit into cropping systems in marginal areas in order to maximize environmental benefits.

Path analysis and SEMs are used to determine direct and indirect relationships among and within groups of traits (quantitative and qualitative phenotypic markers), including yield components, in ideotype breeding (Zhang et al., 2014). An ideotype may be defined as a combination of genetic markers, whereby methodologies that link dynamic system modeling with biometrical techniques used in molecular breeding are employed in the process. The outcome of the process is twofold: (i) optimization of phenotypes through tradeoffs between the large number of plant traits being used in ideotyping; and (ii) the development of phenotypic ideotype(s) suitable for future GCC scenarios and farming systems that are not currently available in germplasm collections of oilseed breeders. A less efficient methodology of developing an ideotype can be based on measurements of metabolic biochemical processes and reaction to abiotic stresses. Such a "physiological ideotype" may contribute to the understanding of metabolism of oilseed crops.

Regardless of GCC scenarios, an oilseed ideotype is expected to have positive effects on current and future agroecosystems; therefore, traits that may protect or improve the environment need to be integrated into its genetic makeup. Ideotyping for wide vs. specific adaptation to key environmental conditions and stresses can be based on simulation model(s) that calculate the probability distribution of expected yields on the basis of agroecological and environmental characteristics and best management practices of target production regions for a particular oilseed crop. Predicting the genotype × environment × management (G × E × M) interaction is a prerequisite to integrating environmental covariates and oilseed crop modeling into a genomic selection context (Heslot et al., 2014). One objective of this process is to improve the accuracy of predicting genotypic performance under environments for which weather data are available, then provide insights into the genetic architecture of G × E × M interactions to predict ideotype performance using past and future weather data and GCC scenarios. Nevertheless, different responses to GCC may occur both between populations throughout a species range, and between cooccurring individuals within a population. Therefore, plasticity could facilitate the expression of a relatively well-adapted oilseed crop ideotype under future environments, and phenotypic plasticity may play a bigger role than genetic diversity in its environmental adaptation.

BREEDING OBJECTIVES OF OILSEED CROPS UNDER A CHANGING CLIMATE

To help identify and prioritize breeding objectives of oilseed crops under a changing climate, it is prudent to define and quantify how differentially GCC stressors, directly and indirectly, may affect the complex biological components of seed yield, oil yield, and their interactions. Practically, these effects are difficult to dissect and use in simulation model(s) due to incomplete knowledge of their mechanistic bases (Newton et al., 2011). In addition, oilseed crop breeders need to identify germplasm with broader and novel genetic diversity, whether for currently desirable or new traits. Such diverse germplasm will be needed to strengthen tolerance, if not resistance, to biotic and abiotic stresses, and to adapt oilseed crops to future GCC (Dempewolf et al., 2014). The potential to revolutionize trait discovery and improve phenotypic prediction will dramatically increase when HTPP and genome-wide association studies are integrated. New traits may be available in traditional land races, old cultivars, elite cultivars, breeding populations, *ex situ* or *in situ* conserved crop wild relatives, or can be produced *de novo* using appropriate biotechnologies. Inevitably, advanced simulation models that enable accurate prediction of the effects of GCC on oilseed productivity and quality at different spatiotemporal scales are required to accelerate and optimize breeding of oilseed crops under multiple GCC stresses.

Phenology and Plasticity under Climate Change

The ability of an oilseed plant to adjust to seasonal variations during development is an indicator of its phenological plasticity (Bloomfield et al., 2014), which is a desirable trait in the context of breeding crops resilient to GCC. Developmental plasticity (i.e., the capacity of a genotype to express phenotypic variation without DNA sequence mutation) and functional plasticity, a complex system of plasticity found in polyploid oilseed crops (i.e., the ability of a genotype to relocate vital biological functions among tissues as a result of paralogous genes; Bloomfield et al., 2014), may become more important adaptation strategies to seasonal environmental (and edaphic) variations under GCC. Oilseed crop phenology will accelerate in response to higher temperatures and will be accompanied by reduced yield as a result of increased risk of reproductive failure (De Storme and Geelen, 2014). Coupling downscaled weather scenarios with phenology model(s) to study GCC stress during critical phenological stages of oilseed crops emphasizes the importance of earliness for heat tolerance in these crops.

However, higher temperatures will most likely be associated with e[CO_2]. Plastic changes in phenology due to e[CO_2] will be manifested in fluctuations in flowering time, seed longevity, leaf duration, and a multitude of metabolic processes. Plasticity could facilitate the expression of relatively well-adapted phenotypes under new environmental conditions brought about by GCC (Merilä and Hendry, 2013). Phenotypic plasticity, rather than genetic diversity, is likely to have a more significant role for oilseed crops in tolerating stress conditions caused by rapid GCC in their current agroecosystems (Redden, 2013). Plasticity can be facilitated by epigenetic processes that modulate chromatin through dynamic changes in DNA methylation, histone variants, small RNAs, and transposable elements (Bloomfield et al., 2014).

Although plants may respond to GCC through phenotypic plasticity (Nicotra et al., 2010), new phenotypes may however evolve through change(s) in genome structure or genome reorganization (i.e., genome plasticity) in response to environmental signals. Both plasticity types have underlying molecular and genetic mechanisms relevant to GCC. However, no conclusive evidence has been presented yet implicating GCC as the definite and only underlying cause, or to determining whether phenotypic changes are adaptive, genetically based, or the outcome of phenotypic plasticity (Merilä and Hendry, 2013). Thus, characterization of these mechanisms will help facilitate the production of genotypes with enhanced capacity to adapt to GCC (Peleg and Blumwald, 2011). Phenology indicators of oilseed crops, when included in SEMs to determine relationships between and within groups of plant traits, help determine how yield components and canopy traits relate to seed and oil yield in phenotypically diverse germplasm (Zhang et al., 2014). Hierarchical or positive correlations in the plasticity of agronomic, yield, and quality traits may exist in current oilseed germplasm; therefore, plasticity associated with different eras of breeding for yield potential and agronomic traits under abiotic stresses may exist. Future changes in ecogeographical regions suitable for current and future oilseed cultivars and crops, as well as their potential productivity under GCC, can be predicted by linking phenotypic indicators and information based on GPS–GIS technology.

Fitness in oilseed crops (e.g., oil content, seed size, seed germination, seed number per siliqua, seed color, seed mass, days to bolting, days to flowering; Figure 18.2) and final seed or oil yield are determined by complex interactions among traits expressed throughout plant ontogeny (Dechaine et al., 2014). Environment-specific effects on fitness were confirmed using QTL mapping. For example, days to bolting, siliqua length, and seed color were highly correlated and displayed QTL colocalization across environments. Dense genetic linkage maps for flowering traits in oilseed (Li et al., 2014c) and potential oilseed crops (e.g., *B. carinata*) (Zou et al., 2014) are valuable tools in studying the phenology of flowering phenotypes. Out of 24 QTLs for flowering and budding time identified in the double-haploid population of *B. carinata*, one (*qFT.B4-2*) was identified under a wide range of environmental conditions in *Arabidopsis thaliana*. Flowering and grain-filling stages are the most sensitive to GCC stresses when pollen is most vulnerable to high temperature and desiccation (Singh et al., 2012). Late flowering and delayed maturity are two drawbacks when winter-type high-yielding traits are introduced into spring-type *B. napus* (Rahman and Kebede, 2012). Days to first flower from first bud and average daily flower production in some oilseed crops (e.g., *Linum* spp.) showed moderate heritability and low genetic advance (Jain, 2014), whereas days to first flower varied widely (29–73 days) in resynthesized *B. napus* genotypes (Karim et al., 2014). Polyploidy in oilseed crops, and therefore the large number of flowering-related genes and the complexity of the gene networks controlling flowering, may be behind the morphological

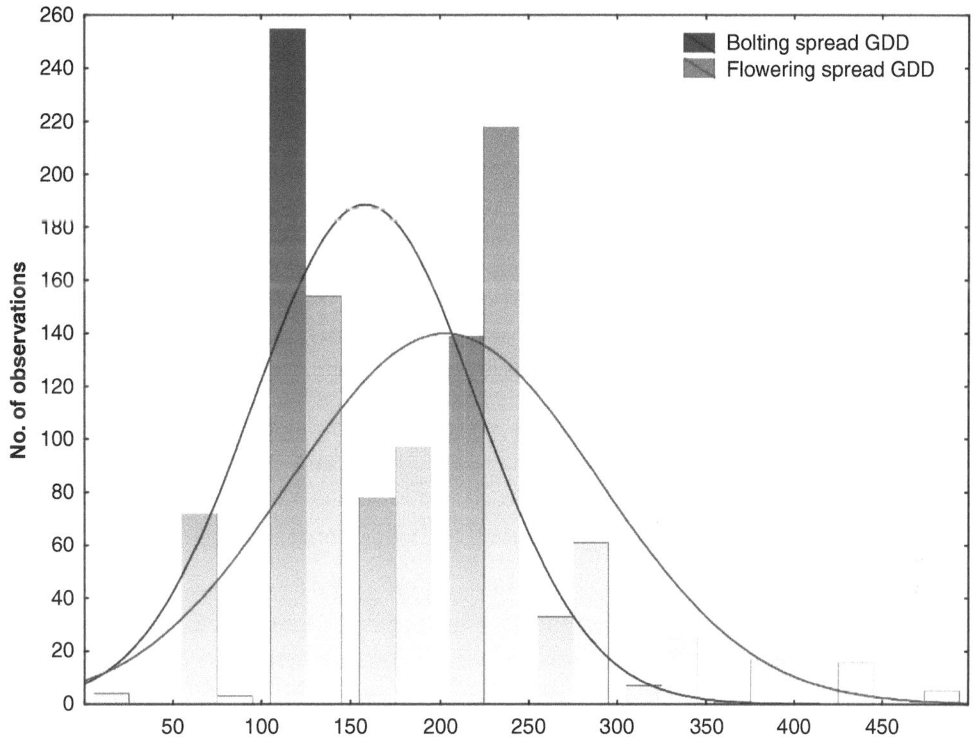

FIGURE 18.2 Relationship ($r = 0.29$; $p < 0.05$) between bolting and flowering spread (in growing degree days, GDD, showing large variation and overlapping between both traits) in a large *B. napus* phenotyping experiment (Jaradat, unpublished data).

differentiation for flowering-related traits, particularly in allopolyploids, such as *B. napus* (Schiessl et al., 2014).

The genetic architecture of flowering phenology as an indicator of plant fitness, as well as reproductive output, are more environmentally sensitive than other measured fitness traits; therefore, identification of proximate environmental suitability that may signal flowering and its pattern will help in selecting oilseed crops to fit future climates. The flowering phenology of perennial oil-bearing plants, such as *Olea europaea*, will be more affected by late-spring temperature than winter or early-spring temperature; therefore, it may be less affected by global warming than other early-spring flowering species (Diez et al., 2014). Rising moisture deficit, high temperatures, and frequency of extreme weather events in the Mediterranean Basin will lead to changes in olive phenological cycles and a significant decrease in oil yield.

Plant Architecture under Climate Change

The adaptive potential of oilseed crops to GCC needs to be explored for selecting appropriate genotypes and breeding productive cultivars. The capacity and efficiency of plant organs to capture and use environmental resources depend, to a large extent, on canopy architecture

(Kang et al., 2014). The ability of the plant to compensate for reduced resources acquisition by the production of new organs or by remobilization of reserves from other parts of the plant may mitigate GCC stress effects. Biotic and abiotic stresses influence oilseed crops through their impact on the initiation and development of plant traits that can integrate time (ontogeny) and space (architecture); on the other hand, GCC influences on biotic stressors are mainly plant mediated. Rising temperature, e[CO_2], altered precipitation, and their interactions modify plant growth and development with simultaneous changes in canopy architecture, size, density, microclimate, and quantity of tissue susceptible to GCC stressors. Restricted expansion of plant architecture due to GCC stress will reduce its ability to compete for natural and external resources and inputs, thereby reducing its seed and oil yield potential (Zhang et al., 2014). Plant architecture traits that enable synchronized flowering and uniform seed maturity may be more advantageous than yield itself in selecting and breeding for adaptation to GCC.

Functional–structural plant models (Kang et al., 2014) have been used to design, characterize, and provide new insights into how oilseed genotypes and crops may respond to environmental stimuli or GCC stressors. The capacity for FSPM as a decision-support tool can be enhanced by utilizing plant trait's captured under different environmental or stress conditions throughout all growth and development phases. However, in spite of the current utility of FSPMs as sophisticated tools for predicting crop yield and its relationship to the structure and function of the plant canopy, the significant impact that plant genetics can have on yield needs to be integrated into such models. Therefore, the next phase of model development and testing is to integrate HTPP and GWAS in order to maximize the rate of trait discovery and improve phenotypic prediction under GCC stress (Brown et al., 2014). Moreover, SEMs and path analyses can be valuable tools in determining how plant architectural and quantitative traits are related to seed and oil yield in phenotypically diverse oilseed genotypes and cultivars (Zhang et al., 2014). Canopy architecture, in addition to seed or fruit yield, are among the most important determinants of oil quality in oil-bearing plants, such as *Jatropha curcas* (Kim et al., 2014), *Plukenetia volubilis* L. (Wang and Liu, 2014), and *Olea europaea* (Diez et al., 2014).

Seed Yield, Oil Yield, and Fatty Acid Profiles under Climate Change

Crop yield and quality are influenced by GCC directly through abiotic stresses and indirectly through biotic stresses, especially pathogens that they themselves will change as climate changes, but may increase in virulence (Newton et al., 2011). Seed and oil yield of annual and perennial oilseed crops and oil-bearing plants are interrelated and complex traits that result from multivariate relationships and compensation between many yield components and crop characteristics, as well as interaction with management and environmental factors. Seed yield, stability, and response to external inputs have been and remain key traits in many oilseed-breeding programs. Achieving higher seed yield would considerably reduce environmental impacts associated with seed and oil production of oilseed crops (Li and Mupondwa, 2014). The challenge under GCC stress is to simultaneously optimize seed yield and maintain, if not increase, oil content, oil yield, protein yield, and optimize fatty acid profiles whether for food (PUFAs) or industry (SFAs).

Several factors, including genetic effects of the embryo, cytoplasm, and maternal tissues, as well as genotype × environment × management interactions determine to a large extent the oil content in oilseed crops and oil-bearing plants. Oilseed crops having the C_3 metabolic pathway may respond positively to e[CO_2] by producing more biomass and accumulating more carbohydrates, thus diluting seed proteins and macro and micronutrients, and may improve oil quality by raising PUFAs at the expense of SFAs (Pal et al., 2014). However, the favorable effect of e[CO_2] may be negated by the attendant higher temperatures, lower WUE, and the subsequent accelerated plant development that could result in reduced seed and oil yields. Therefore, key targets in oilseed improvement under GCC stress include increasing overall oil yield and stability on a per seed or per fruit basis and maintaining very high PUFA content for premium edible oil and high SFA contents for oleochemicals used in industrial and pharmaceutical applications (Murphy, 2014).

Oil physical properties (e.g., fluidity) and metabolism are affected by its SFA and PUFA composition. Plant species, genotypes, temperature, environmental conditions, and management practices, in decreasing order, influence FA profiles (Schulte et al., 2013). The large variation in oil content and FA profiles in seed or fruit of wild (e.g., *Manihot esculentus* and *Plukenetia volubilis* L), semi-domesticated (e.g., *Jatropha curcas*), and domesticated oilseed crops and oil-bearing plants (e.g., *B. napus* and *Olea europaea*) emphasize the role of genetic diversity within and among species. For example, oil content from *Manihot* spp. seed range from 17.2% in *Manihot caerulescens* to 30.7% in *Manihot flabellifolia* (Alves et al., 2014). Five FAs were found in all *Manihot* spp. with the average content of linoleic (C18:2) 61.5%, oleic (C18:1) 20.0%, palmitic (C16:0) 12.3%, stearic (C18:0) 4.5%, and linolenic (C18:3) 1.7%. FA contents varied significantly between and within oilseed and oil-bearing plant species. For example, the highest content of linoleic acid (average 65%) was found in seeds of *Manihot carthaginensis, M. cecropiaefolia, M. flabellifolia, M. glaziovii, M. peruviana*, and *M. pseudoglaziovii*; and the highest level of oleic acid (average 23%) was found in seed of *M. anomala, M. caerulescens, M. dichotoma, M. esculenta*, and *M. tomentosa*. Comparable levels of variation within and among SFA and PUFA content in other oilseed crops may reflect adaptive strategies for seed survival and seedling establishment under contrasting climate conditions. Higher PUFA content may have evolved as an adaptation to cold climates (Wilkes et al., 2013); however, a wide range of these FAs was observed under warmer climates as well. The FA composition of oilseed crops, such as *G. max, B. napus, Camelina sativa*, and *H. annuus*, synthesized and accumulated more monounsaturated and less PUFAs when growing temperature was experimentally raised from 10°C to 40°C (Schulte et al., 2013).

Differences among plant species, rather than climate, may explain a large part of the relative abundance of common FAs in oilseed crops, while the role of genetics in regulating plastic response to GCC is well documented. In regional studies (e.g., Zhang et al., 2015), the level of PUFAs increased with latitude and decreased with both mean annual temperature and precipitation; temperature's impact was more pronounced than that of precipitation. Oil content (27.5–37.2%) and FA composition, mostly PUFAs, varied among *Camelina sativa* genotypes with year and growing seasons (Pecchia et al., 2014). The contents of two PUFAs (oleic and linoleic) were reduced in response to cold climate and were inversely correlated with the contents of one SFA (erucic; C22:1), which itself increased in cold climates in *B. juncea* (Wilkes et al., 2013). The elevation and geographical location of growing sites and their attendant temperature and solar radiation influenced the contents and ratios of several FAs within PUFAs and SFAs in *B. napus* (Ali and Farhatullah, 2013) and *C. sativa* (Ghamkar et al., 2010).

The relationship found between phylogenetics and climate variables in relation to PUFAs and SFAs may lead to the discovery of potential oil plants with particular FA composition for food and industrial applications. Moreover, it is speculated that the variation of FA content and composition in oilseeds may reflect adaptive strategies for seed survival and seedling establishment under contrasting climate conditions. As a competitive and adaptive advantage under abiotic stress (e.g., chilling or low temperature), membrane fluidity and stability, due to higher PUFAs, may allow oilseeds to germinate earlier and grow faster, although they store less total energy than seeds with a higher proportion of SFAs.

PREBREEDING OF OILSEED CROPS FOR CLIMATE CHANGE

Historically, plant breeding was based on the introduction of new genetic variability based on the old saying "cross the best with the best and hope for the best" (Podevin et al., 2013), provided that the right traits were evaluated and measured under the right climatic conditions and management practices. This strategy succeeded in releasing elite cultivars whenever genetic variation was available to plant breeders within accessible and manageable germplasm. The need for diverse germplasm will increase dramatically with the advent of GCC to sustain and strengthen adaptation, tolerance, or resistance to increasing abiotic and biotic stresses, and to optimize the seed and oil yield of current and future oilseed crops and oil-bearing plants. Hence, the availability of a more diverse germplasm represents both a challenge for gene banks to ensure that important gene pools are properly conserved and an opportunity for stimulating greater use of *ex situ* as well as *in situ* conserved germplasm.

In order to diversify the primary gene pool and conduct genetic enhancement of oilseed crops, prebreeding provides a unique opportunity through the introgression of desirable genes into genetic backgrounds readily used by breeders, with minimum linkage drag (Sharma et al., 2013). Genes may come from a wide range of genetic resources including the primary and secondary gene pool (e.g., land races, old cultivars, breeding lines) and the tertiary gene pool (e.g., wild and weedy relatives and genetic stocks) of the oilseed crop. Linkage drag, usually but not exclusively associated with secondary and tertiary gene pools, is the most important factor responsible for low use of germplasm by breeders in oilseed crop improvement; what is more, it is the major reason behind the need for prebreeding. Nevertheless, recent developments in genetics and genomics have made it possible to obtain genes much easier from novel sources of germplasm. For example, using cisgenesis (Hou et al., 2014), breeders can avoid linkage drag, enhance the use of existing alleles, and combine traditional breeding techniques with modern biotechnology, while efficiently speeding up the breeding process. The presence of molecular markers will greatly assist in reducing linkage drag and increasing the efficiency of introgression (Sharma et al., 2013).

Implementation of prebreeding by gene banks will encourage greater use of genetic diversity and reduce the cost and increase the efficiency of breeding programs. Germplasm characterization and preliminary evaluation are the basic activities in prebreeding to meet the germplasm needs of oilseed crop breeders. Spatial analysis, using GPS–GIS technologies, has been suggested (Redden, 2013) as a means of "predictive characterization" to envisage which germplasm, especially CWR, might have the desired traits. Prebreeding provides a unique opportunity to overcome linkage drag through the introgression of desirable genes from CWR into genetic background readily used by the breeder. However, several functional activities are integral parts of prebreeding; these include in consecutive order: identification of hotspots for specific traits using ecogeographical/GPS–GIS techniques; germplasm acquisition through exchange and collection from centers of origin and centers of diversity; maintenance of existing genetic diversity in germplasm collections; germplasm enhancement to identify useful genes and traits to be transferred into an acceptable genetic background; and evaluation of, and research on, newly transferred genetic diversity.

Linkage and Genetic Mapping

Mapping QTLs, as the foundation for molecular MAS and map-based gene cloning (Zhou et al., 2014), has made it possible for oilseed breeders to develop genotypes and cultivars with high yield potential, improved oil content, and oil quality. Usually, only larger effect QTLs are detected, but are unlikely to account for a large part of genetic variation controlling polygenic traits such as seed or oil yield (Wood, 2014). Genetic mapping of quantitative traits using genetic linkage maps is an efficient method to identify QTLs underlying yield and oil-related agronomic traits. A unique QTL usually reflects a significant positive (e.g., seeds per siliqua and siliqua length in *B. napus*) or negative correlation (e.g., seed weight and seeds per siliqua) between agronomic traits (Li et al., 2014c). Then, the genomic location and effects of the detected QTL are mapped on a particular linkage group; moreover, the portion of phenotypic variance explained by the QTL is estimated. Mapping the genetic architecture of a phenotypic trait was made possible by combined use of HTPP and genotyping technologies; thus, it was possible to predict the expected genetic worth of a genotype based on its genetic fingerprint. However, epistatic or pleiotropic effects may exist and may vary in magnitude according to the trait(s) being mapped (Li et al., 2014a), in which case more emphasis should be placed on selection directly for the trait rather than for the QTL (Wood, 2014).

A large number of studies identified major and minor QTLs in oilseed crops accounting for small, medium, and large portions of genetic variances in several agronomic traits. These traits were directly or indirectly related to seed yield, oil yield, and oil composition, or were involved in adaptation to biotic and abiotic stresses. Dense genetic linkage maps have been constructed to identify QTL(s) responsible for: color determination in *B. rapa* L. (Kebede and Rahman, 2014); flower development in *B. carinata* (Zou et al., 2014); days to flowering in *H. annuus* (Dechaine et al., 2014); high oil content in *G. max* without decreasing protein content; seed oil biosynthesis in *Jatropha curcas* (Javed et al., 2014); salt and drought tolerance in *B. napus* and *A. thaliana* (Zhang et al., 2014); and several agronomic traits in *Linum usitatissimum* L. (Soto-Cerda et al., 2014). A high-density linkage map would be useful not only to reveal genetic variation in a particular plant species using QTL analysis and genome sequencing, but also for MAS and genomic selection for breeding and improvement of oilseed crops.

Recent technologies for linkage and genetic mapping include combined linkage and regional association and GWAS (Brown et al., 2014); the latter is an efficient strategy for direct mapping, validation, and fine mapping of pleiotropic QTLs for complex traits. Genetic diversity, expressed as allelic polymorphism, is related to phenotypic variation using linkage disequilibrium in the breeding population of an oilseed crop. Compared with classical QTL mapping, this procedure captures more variation in the breeding population without the need to develop segregation populations, and the mapping resolution is usually higher.

However, linkage disequilibrium may be present due to other factors such as genetic drift, selection, and the presence of subpopulations within the breeding population.

Phenotypic traits can be captured at different spatiotemporal scales, across environments, and throughout plant ontogeny, using multidimensional phenotyping platforms (Merilä and Hendry, 2014); when integrated with FSPMs and GWAS, the causal genetic basis of complex traits can be identified, eventually allowing for phenotypes to be predicted from genotypes (Brown et al., 2014). GWAS were used to dissect the genetic structure of seed weight, seed quality, and glucosinolate content in *B. napus* (Hannoufa et al., 2013) and, if integrated with HTPP and FSPM, they will advance the rate of trait discovery and improve phenotypic predictions under controlled and field environments. In addition, inexpensive and accessible genotyping by sequencing (Cheng et al., 2014) will accelerate GWAS use in trait mapping of new oilseed crops under abiotic stresses.

Germplasm Enhancement

Sustained oilseed production to meet increasing demand for different uses and applications under current and future GCC and LUC may have to rely increasingly on genetic enhancement of all categories of diverse germplasm including local cultivars, land races, breeding lines, and CWR (Jaradat, 2015). The need to broaden the genetic base of current oilseed crops, advance our understanding of functional germplasm diversity (Cooper et al., 2014), and enhance the genetic diversity of underutilized oilseed crops, is more urgent than ever due to GCC. The genetic base of a typical underutilized crop (e.g., *Carthamus tinctorius*, *Crambe abyssinica*, and *Guizotia abyssinica*) is narrow (Pearl and Burke, 2014); most likely, these crops originated from a single pool of genetic diversity in a secondary center of origin in the Old World. However, their genetic resources harbor largely undocumented genetic diversity for tolerances to single or multiple abiotic stresses such as tolerance to drought and high temperature in *Brassica juncea* (Ram et al., 2014) and *Jatropha curcas* (Cartagena *et al.* 2014), low temperature in *B. rapa* (Jung et al., 2014), and salinity in *A. hypogaea* (Sun et al., 2013), *Ricinus communis* (Rodríguez et al., 2014), *Diplotaxis tenuifolia*, and *Cochlearia officinalis* (Sedbrook et al., 2014). However, the efficiency of genetic enhancement using conventional methods is limited due to difficulties in transferring desirable traits from the secondary or tertiary gene pool of an oilseed crop, linkage drag, and expense associated with the long time required to conventionally develop new cultivars (Hannoufa et al., 2013).

Generally, local oilseed germplasm has better quality traits (e.g., FA profiles and beneficial endogenous secondary metabolites such as carotenoids, choline, and tochopherols), adaptation, and tolerance to stresses, but lower yield potential than exotic germplasm. Some of this germplasm has been moved around the world through germplasm collection and exchange, and it is highly likely that it will be moved more than ever to facilitate the process of breeding new oilseed crops and oil-bearing plants for adaptation to GCC. However, germplasm of CWR remains a relatively low priority for enhancement due to financial and political restrictions on germplasm collecting in, and transfer from, developing countries. A general lack of knowledge about trait genetics in CWR germplasm (Jaradat, 2015) may accentuate the problem.

Adaptation to GCC and tolerance to abiotic stresses require new traits that might not be available in the primary or even the secondary gene pool of oilseed crops; this will further increase the importance of and demand for CWRs. The gene pool of oilseed crops can be selectively broadened with directed genetic introgression from CWRs for improved tolerance to drought, high temperature, salinity, and frost (Redden 2013); the opportunistic weedy traits of some CWR, such as *A. thaliana*, may indicate underlying genetic variation in plasticity-related genes, which may be used for oilseed crop improvement. Although CWRs provide unique alleles not found in the primary gene pool of oilseed crops, exotic germplasm may present problems associated with linkage drag, maladaptation due to photoperiod sensitivity, lack of flowering synchrony, and the need for gene introgression into locally adapted germplasm before its valuable alleles can be recovered (Jain, 2014). However, new biotechnologies, including cisgenesis, can avoid linkage drag and enhance the use of existing alleles in exotic germplasm (Hou et al., 2014).

Modeling Oilseed Crop Response to Abiotic Stress

Research questions that can be addressed using computer modeling within the context of GCC are those concerned with delineating future ecogeographical boundaries for oilseed crop adaptation and performance (Crane et al., 2011): How can yield and quality traits be improved? What conducive, species-specific management practices should be adopted under GCC stresses? How to assess the potential risk of greater climate variability to oilseed crop production? Simulating GCC impact on oilseed crop yield at a regional scale requires intensive data and extensive field experiments. Agroclimatic-zoning models can determine potential growing areas of oilseed crops (e.g., *Camelina sativa*; Ghamkar et al., 2010) using GPS–GIS technologies and databases associated with environmental boundaries for each crop. Simulation of regional field trials to assess and predict the performance of oilseed cultivars under adverse environments by capturing plant traits using HTPP and applying GWAS to identify causal genes and background variation. When combined with the FSPM, crop growth and fitness can also be predicted in target environments with higher probability (Brown et al., 2014); this trait capture strategy can be applied under controlled and field conditions and allows breeders to preselect oilseed genotypes for increased productivity under GCC.

Most studies that address the potential impacts of GCC and potential adaptation strategies to GCC are largely based on computer simulations and biometric modeling (e.g., path analysis and SEM). Crop phenology, canopy traits, yield, and its components are necessary for building simulation models and SEMs to determine how certain traits relate to seed and oil yield in phenotypically diverse germplasm (Zhang et al., 2014). This approach, however, is largely devoid of direct human judgment (Crane et al., 2011); a synergistic relationship between both approaches may meet the challenge of uncertainties associated with GCC more efficiently. Crop simulation models offer research tools for evaluating tradeoffs between C footprints of oilseed crops and their potential adaptation to GCC (Matthews et al., 2013); the process is facilitated by integrating quantitative genetics, biometry, and gene-to-phenotype knowledge of traits embedded within crop simulation models (Cooper et al., 2014). The most important plant traits in this regard can be reliably identified and prioritized when weather scenarios from a climate model are downscaled and combined with an FSPM (Gouache et al., 2012). Such FSPMs have been used to simulate plant growth and development as dictated by physiological processes and their response to local environmental conditions at the organ (e.g., leaf, inflorescence, fruit, or seed) level (Kang et al., 2014). The SUNLAB model (Kang et al., 2014), for example, was adapted and successfully used to model the phenotypic responses of different genotypic variants of *H. annuus* to diverse environmental factors, including temperature stress and water deficit. Moreover, as a support decision tool, the model potentially can help design ideotypes for oilseed crop species. Oil content in *H. annuus* was adequately predicted using a statistical model as a function of variety, management, and environment; an improved version of the model was capable of predicting oil content in a large panel of *H. annuus* genotypes grown in contrasting environments.

A crop simulation model using genetic data to estimate allelic effects under present and future GCC scenarios can theoretically simulate the yield of several genotypes in response to GCC (e.g., temperature and rainfall) and to spatiotemporal management factors (Parent and Tardieu, 2014). Whole plant adaptation to, or protection against, abiotic stresses often occurs at the expense of its performance (i.e., growth, yield, and quality). Consequently, a given trait might have a positive, negative, or no effect on yield or adaptation depending on the stress scenario being simulated (Tardieu and Tuberosa, 2010). By leveraging agronomic knowledge and the large historical datasets generated by breeding programs, a simulation model can capture the nonlinear response of QTL to abiotic stress, and provide insights into the genetic architecture of G × E and could predict genotype performance based on past and future weather data derived from GCC models (Heslot et al., 2014). Moreover, this modeling approach can be extended to target a population of environments where a particular oilseed crop will be sustainably grown. Nevertheless, sophisticated algorithms are still needed to model and statistically evaluate G × E × M; however, the reliability of such models is limited by the large number of correlated predictors, each explaining a small portion of total variance, and by nonlinear genotypic responses to stresses which are expected to further complicate the analysis.

Simulation models that can incorporate genetic information and predict GCC impact on oilseed crop growth and production will become increasingly available because of the vast body of data accumulated over recent decades using HTPP and genotyping (Araus and Cairns, 2014). Likewise, advances in sophisticated modeling and fast computers, coupled with bioinformatics, will make it realistic to integrate metabolomics with whole plant physiology and growth under GCC in future simulation models (Sweetlove et al., 2014). However, combinations of long-term experimental platforms, accurate phenotyping, and modeling have already succeeded in developing the on-farm technologies currently used by farmers. Projections of GCC impact on oilseed crops are more reliable if they integrate, in addition to crop vulnerability assessment in response to spatiotemporal climate variability, farmer income and other indicators of agricultural performance (e.g., ecosystem services) as part of the simulation models used to predict crop growth, development, and yield. Progressive farmers are already using affordable HTPP technologies, built around in-season data collected in their fields, as decision support tools to make appropriate management decisions and crop choices to fit their local prevailing conditions and climate predictions. However, the modeling approach to GCC adaptation has been challenged by the concept of "agriculture as performance" (Crane et al., 2011); the latter is a socially embedded dynamic process that emphasizes that short and long-term adaptive processes and technologies are more than simple technical responses to biophysical conditions.

BREEDING AND SELECTION STRATEGIES UNDER CHANGING CLIMATES

Crop yield, quality, and nutritional value, including those of oilseed crops and oil-bearing plants, have been steadily increasing and improving since the Neolithic (Podevin et al., 2013). Farmers/plant breeders have by and large delivered in feeding an ever-growing population, although significant challenges, including GCC, remain. Most recently, however, the depletion of genetic diversity has been found to be one of the negative outcomes of major improvements in plant breeding and the release of pure genotypes, which are more prone to biotic and abiotic stresses than the genetically diverse land races and old cultivars that they replaced (Keneni et al., 2012). Therefore, the outcomes required and successes accomplished by plant breeding during the current "Anthropocene" will be measured in terms of sustainable improvements in producing vegetable oils and their derivatives that meet the growing needs of society, while increasing the genetic diversity of these crops (Cooper et al., 2014). Oilseed crop–breeding programs may have to undergo major changes in order to develop a new generation of crops that yield more with fewer external inputs and are adapted to changing climate (George et al., 2014).

Careful and critical valuation of currently available germplasm and creation of new genetic diversity in ecophysiological, developmental, and quality traits, followed by acquiring and enhancement of novel germplasm with traits that can be used to develop oilseed crop cultivars adapted to GCC and its extremes. Hence, the urgency to develop crop-specific and ecoregion-specific strategies so that new traits and improved cultivars will be relevant to problems of abiotic and biotic stresses and projected GCC conditions in 10–15 years time. A positive and active public–private partnership in germplasm enhancement and prebreeding will be essential for future success (Dias, 2014); the objectives of this partnership may range from improving a single key trait to prebreeding a new ideotype. Germplasm development, enhancement, and prebreeding, supported by new (bio)technologies (e.g., phenotyping and genotyping as well as spatiotemporal modeling of agroecosystems) can advance on a time scale that enables close coordination between oilseed crop breeding and end user demand (Runck et al., 2014). While oilseed crop breeding can be more efficient and competitive through advanced and precise selection methods, sustainability of the breeding process needs to be guaranteed in order to replenish the genetic diversity it lost due to intense selection, genetic drift, and small population sizes. Large effective population size, in particular, is vital to achieving sustainable plant breeding (Cowling, 2013).

Multiple Approaches to Oilseed Crop Breeding for a Changing Climate

A combination of traditional plant breeding, classical genetic and modern genomic approaches, and precise phenotyping enhanced the discovery of genes and metabolic pathways conferring tolerance to abiotic stresses in several oilseed crops (Zhang et al., 2014). However, future oilseed-breeding approaches need to complement empirical evaluation with genetic prediction (Cooper et al., 2014) by combining the use of traditional breeding, new genomic tools, and farmer-participatory methods. The overall objective is to give breeders new tools in their quest to breed oilseed crops that can cope with GCC. Approaches such as scaling up system biology to the whole plant level using scientific breakthroughs in genetics and ecophysiology (Tardieu and Tuberosa, 2010), integrating functional gene-to-phenotype modeling, and gene-to-phenotype trait relationships (Cooper et al., 2014) would complement empirical evaluation with genetic prediction. Furthermore, approaches that link dynamic system modeling with the statistical and biometric methods used in molecular breeding will provide a reasonably effective strategy to support the development of a new breed of oilseed crops; the development and parameterization of the system model should be integrated with an omics analysis (Urano et al., 2010).

Breeding for phenotypic plasticity, a prerequisite to building resilience in an increasingly variable environment, depends on the possibility of identifying key environmental sensing genes in future oilseed crops. Therefore, accelerated gene discovery using several biotechnologies and associated omics proved to be an efficient applied approach to breed for adaptation to abiotic stresses (Eldakak et al., 2013). Future advances in this regard depend on further improvements in omics databases as a prerequisite to effective utilization of these biotechnologies to produce the next generation of oilseed crops. Developments in the field of omics are expected to support the "reflective plant-breeding paradigm" as a means of diversifying agroecosystems based on a pipeline of new and diverse germplasm (Runck et al., 2014). Inserting additional copies of native genes (i.e., cisgenisis) may provide a new approach to modifying plant genomes, expanding genetic variance for specific traits, and accelerating the transfer of alleles between species that are difficult to cross (Hou et al., 2014). Alternatively, more options for modifying gene expression and trait development are available through intragenesis (Lusser et al., 2012); both options may be socially more acceptable than transgenics when edible oil products and environmental health are concerned.

A systematic approach to improving abiotic stress tolerance by combining physiological, genetic, and morphological methodologies was the basis for the development of crop ideotypes (Palta, 2014). Questions that need answering include: Should oilseed breeders design ideotypes for specific environments or for specific production systems (e.g., conventional vs. organic)? How can HPPT and long-term experiments be optimized to select appropriate physiological, phenotypic, and genetic traits for each option? Breeding for broad adaption, where tolerances to multiple stresses are desired, may be achieved by breeding for one factor at a time, and genes can be stacked into a single genome over time.

Creating New Genetic Diversity

Assuming all plant-breeding techniques have limitations (Podevin et al., 2013) and the need for new genetic diversity continues to rise, then the requirement for broader and new genetic diversity will have to be met using more robust combinations of traditional and novel methods. Traditional plant breeding takes advantage of the gene pool available to a particular crop through intraspecific crosses to create new allelic combinations and broaden genetic diversity. The broader and deeper the gene pool, the higher the probability of making unique allelic combinations through planned crosses. Much genetic variation has been created and utilized in breeding new oilseed cultivars or hybrids through domestication, planned parental crosses, genetic enhancement, mutation-assisted breeding, prebreeding, and molecular breeding. Newly introduced alleles would allow for creating novel allelic combinations for further crop improvement (Rahman, 2013). However, when natural variation is not available (e.g., homozygous mutations in two waxy homeologs), new technologies, such as targeting-induced local lesions in genomes (TILLING) can be used to create and identify genetic variation that can be used in oilseed crop improvement (Himelblau et al., 2009). Nevertheless, the expense, time, and expertise needed to create case-specific genomic clones that include endogenous promoters and terminators, free from selectable marker genes and vector backbones, should not be underestimated (Eriksson et al., 2014). Genetic transformation could provide an avenue to create new genetic diversity (e.g., a plant more similar to a phenotypic or physiological ideotype) for a particular ecoregion or production system, should genetic limitations make it difficult for conventional oilseed breeding to achieve this objective (Kantar et al., 2014). Facilitated by a seamless flow of genetic variation across genomes, genetic diversity can be created by combining two genomes from related allotetraploid species such as the A and B genomes from *Brassica* spp. (Tsuda et al., 2014); the derived amphiploid helped synthesize agronomically *de novo* germplasm (e.g., of *B. juncea*) which showed high heterosis when crossed with the wild type.

Domestication

Since the Neolithic, humans have used genetic adaptation of wild plant species (i.e., domestication) to meet their growing needs for food and other essentials. Typically, the process was accompanied by genetic and phenotypic changes whereby the newly domesticated crop became genetically separate from its wild progenitor and totally dependent on humans for its survival. Plant genes that are associated with the domestication process have been identified using QTL mapping, genome-wide association, and whole genome resequencing studies (Meyer and Purugganan, 2013). The domestication process, as a form of biological evolution, suggested that new competitive interactions between recent gene duplications can contribute to evolutionary innovation (Blackman, 2013). However, following domestication, crop plants experienced a period of reduced genetic diversity due to domestication bottlenecks. It is suggested that new insights into evolutionary principles could provide an improved understanding of domestication (Lenser and Theißen, 2013).

Although the first oilseed crop (*Linum usitatissimum*) was domesticated for its fiber, capsular dehiscence was among the first traits modified by humans after initial domestication; ever since, it served as an oilseed crop and source of a unique and healthy FA profile (Fu, 2011). Olives (*Olea europaea* ssp. *europaea* var. *sativa*) were domesticated as oil-bearing plants (Diez et al., 2014); despite being vegetatively propagated, their genetic diversity might have been shaped by hybridization with wild *Olea* spp. around the Mediterranean basin. Rapid domestication of wild plants as new oilseed crops is a prerequisite to expanding the oil production base, broadening the genetic base, and adapting to GCC. A recent example of a potential new domesticate is *Jatropha curcas*, which is still under domestication. Knowledge of the ecology, reproductive system, and dispersal patterns of *J. curcas* is relatively recent and incomplete. Recent studies on genetic diversity among accessions collected from several countries using molecular markers indicated the presence of low genetic diversity in *J. curcas* (Maghuly and Laimer, 2013). Although a certain level of diversity was expressed at the phenotypic, physiological, and biochemical levels, it was largely an epigenetic response to environmental variation (Kant and Wu, 2011).

Relatively recently, the domestication of new oilseed crops – such as *G. max* and *B. napus* – was a long and costly endeavor; however, new genome-based techniques promise a rapid and less costly crop domestication process (Sedbrook et al., 2014). Currently, HTPP and genomics technologies offer ample opportunities for rapid crop domestication through gene discovery, complex trait analysis by GWAS, global gene expression analyses, genomic selection, and predictive breeding strategies (Snowdon and Luy, 2012). Eventually, any success in domesticating new crops will depend on prebreeding and germplasm enhancement activities in support of suitable methods for germplasm screening and selection (Kant and Wu, 2011). Nevertheless, domestication, breeding, and improvement of new oilseed crops, especially for GCC adaptation, need to maintain, if not improve, phenotypic plasticity throughout the whole breeding cycle as it allows for phenotypic and physiological adjustments to environmental stresses.

Molecular Breeding

Molecular breeding, or MAS, refers to the technique of using DNA markers that are tightly linked to phenotypic traits to assist in a selection scheme for a particular breeding objective. Therefore, the identification and characterization of suitable genetic markers – such as random amplified polymorphic DNA (RAPD), intersimple sequence repeats (ISSRs), and amplified fragment length polymorphism (AFLP) – are the most important components of molecular breeding (Cooper et al., 2014). Notably, two oilseed crops, *G. max* and *B. napus*, are well advanced due to contributions from molecular breeding, while others such as *H. annuus* and *Carthamus tinctorius* are still lagging behind (Pearl and Burke, 2014). However, in spite of the problems encountered in improving these two oilseed crops, the information generated from molecular markers, transcriptome profiling, and genome sequencing could facilitate genetic improvement and lead to full domestication of wild oilseed plants (e.g., *Thlaspi arvense* and *Jatropha curcas*) or to improving underutilized oilseed crops (e.g., *Sesamum indicum* and *Arabidopsis hypogaea*).

Genomic selection, a fast emerging molecular breeding method for crop improvement, assists in the identification of superior genotypes with higher breeding value in segregating breeding populations. It is based on genome-wide marker profile data, a prerequisite to using the method (Schulman, 2012). Due to its efficiency, MAS was suggested as a breeding method for drought and salt-tolerant *Brassica* oilseed crops (Zhang et al., 2014). In addition, MAS can provide valuable contributions to the design and development of oilseed crop ideotype(s). The statistical methods used in MAS, when linked to dynamic system modeling, provide a realistic procedure of defining an ideotype as a combination of genetic markers. Molecular breeding has made spectacular progress in a wide range of applications, such as genetic transformation, genetic diversity assessment, large-scale transcriptome and proteome studies, identification of candidate genes for trait improvement, and whole genome sequencing (Johnson et al., 2011).

Mutation-assisted Breeding

Valuable agronomic (Cheng et al., 2014), phenotypic (Gilchrist et al., 2013), quality (Rahman et al., 2013), and adaptation to abiotic stress (Gepts, 2002) traits have been improved in oilseed crops using forward screens of phenotypes or reverse screens of mutations in target genes of oilseed crops. As an alternative to selection breeding, mutation-assisted breeding contributed to oilseed improvement in which natural genetic variation was limited. Most changes brought about by G $\times$ E involve gain-of-function mutations (Gepts, 2002); however, during domestication, loss-of-function mutations were predominant (i.e., a functional enzyme or structural protein was converted into an inactive one), and were easily observed by humans. Hence, domestication syndrome traits resulted from selection of spontaneous mutations that occurred in wild populations and were selected at various stages of plant growth. Mutational hotspots (Nicotra et al., 2010), in addition to genome expansion, transposable elements, or somatic recombination, may result in genome plasticity which involves a change in genome structure or genome organization in response to environmental signals, and would lead to the development of new phenotypes adapted to GCC.

Gene targeting can be used for precise modification of plant genomes and provides a powerful tool for understanding gene function. Through homologous recombination, gene targeting can induce a variety of mutations, including substitution of several nucleotides as well as insertions and deletions (Cheng et al., 2014). A mutation can be induced and discovered by the reverse genetic approach, which does not rely on transgenic technology; it was used to create useful genetic variability required for oilseed crop breeding and improvement without transgenic modifications using endogenous *B. napus* (Gilchrist et al., 2013) and *B. oleracea* (Himelblau et al., 2009) genes. The resulting populations are valuable sources of distinct phenotypic and genetic variability that can be used to improve many traits of economic and environmental values. Mutation-assisted breeding applied to an underutilized crop, such as *Crambe abyssinica*, resulted in valuable changes in its FA profile appropriate for industrial applications (Cheng et al., 2014). A valuable improvement resulted from an increase of erucic acid (C22:1) and reduction of PUFA content. Downregulating the expression of FA desaturase2 genes (*CaFAD2*), which code for the enzyme that converts C18:1 into C18:2 was suggested as a strategy to increase oleic acid (C18:1), reduce PUFAs, and possibly increase C22:1.

Targeted mutation detection was successfully applied in oilseed improvement; a nonsense mutation in even a single *CaFAD2* gene produced changes in oil composition. Stacking the mutations in different *CaFAD2* may result in additional increase in C18:1 and decrease in PUFAs. A mutation in FA desaturase 3 (*FAD3*), which encodes the desaturase responsible for catalyzing the formation of α-linolenic acid, in a diploid *Brassica* species resulted in producing useful germplasm for creating an amphidiploid that produces low α-linolenic acid content in its oil. The genetic control of PUFAs, which is undesirable for industrial purposes, was identified using a mutagenized population of *Brassica napus* (Wells et al., 2014); several mutant lines were identified with oil having very low PUFA levels (~6%). Mutation-assisted breeding in perennial oil-bearing plants shortens the long generation time and long juvenile phase, and overcomes the negative effects of large heterozygosity in potential oil-bearing plants such as *Jatropha curcas* (Laviola et al., 2010) and *Plukenetia volubilis* L. (Wang and Liu, 2014).

Participatory Breeding

Participatory plant breeding, as it applies to oilseed crops or oil-bearing plants, is a strategy for breeding crops which has a unique set of procedures that apply in situations where the demand for specific traits among farmers, industries, and consumers is not fully understood and challenging to identify with market research methods (Bhargav and Meena, 2014). This breeding method depends on the involvement of farmers and other stakeholder at all stages of the process. Because it is aimed at multiple breeding targets, it may accelerate the development of climate-resilient oilseed crop cultivars, as well as cropping systems where the crops will be produced (Banga and Kang, 2014). When combined with available germplasm, traditional plant-breeding expertise, and genomics tools, participatory breeding approaches succeed in developing high-yielding oilseed crop cultivars with wide acceptance by the farming community and adaptation to GCC stresses (Geleta and Ortiz, 2013).

Farmers participating in oilseed crop breeding (e.g., *A. hypogaea*) confirmed empirical findings – such as synchronous and early-flowering genotypes that may escape drought stress – and phenotypic traits of pods and kernels as indicators of quality (Vindhiyavarman et al., 2010). These and other findings (e.g., Geleta and Ortiz, 2013) indicate that farmer participation in the breeding process, including selection of parents, monitoring and evaluating crops on-farm and on-station, and seed production, has a much more formal character than often expected by scientists. Oilseed crop replacement rates, as a means of modifying crop genotypes when resource-poor farmers cannot afford to alter the production environment, may be determined, directly or indirectly, through farmers participating in crop breeding and selection. The rate of change may be based on performance, quality, or adaptation to changing climates.

INNOVATIVE BREEDING STRATEGIES TO COMBAT CLIMATE CHANGE

Future achievements in breeding oilseed crops, within a GCC context, will benefit from further advances and innovations (Table 18.2) in low-cost sequencing, metabolic engineering, HPPT, genotyping, and spatiotemporal modeling of agroecosystems in order to account for environmental and edaphic variation. Genetic diversity, optimized for local adaptation and introduced at different levels in the form of intravarietal, intervarietal, interparental, or interspecific diversity, will support future advancement in oilseed crop breeding and industry (Dias, 2014). A modified form of systematic spatiotemporal gene deployment, interspecific varietal mixtures, and integrated horizontal and vertical resistance may be of value in combating abiotic stresses as was the case in combating biotic stresses (Keneni et al., 2012).

Agroclimatic-zoning models, aided by GPS–GIS technology, will help determine agroclimatic boundaries and future production regions for single or multiple oilseed crops, especially under marginal environmental and production conditions where current breeding approaches, cultivar release, registration and certification procedures are leading to more genetic uniformity (Keneni et al., 2012). The reflective plant-breeding paradigm, described as a robust system of germplasm development to support strategic diversification of agroecosystems (Runck et al., 2014), was proposed to enable close coordination of breeding and commercialization to coincide with germplasm development and lead to the emergence of a cost-effective production system and supply value chains for end user markets, whether for food, pharmaceuticals, or industry.

Reverse Genetic Tools

The last decade witnessed much attention on genomic approaches to plant breeding (Murphy, 2014) and the deployment of a new generation of technologies, such as MAS, next-generation sequencing, transgenesis, and TILLING (Himelblau et al., 2009). The reverse genetic method for mutation generation and discovery (TILLING) does not rely on transgenic technology and enables the identification of putative loss-of-function mutations at each locus

TABLE 18.2 A Brief Description of Innovative Breeding Techniques with Potential Use in Oilseed Crop and Oil-Bearing Plants

Innovative breeding techniques	Objectives
Zinc finger domain	To create a double-strand break at a specific site in the DNA. The double-strand break stimulates the cell's repair mechanism, the process of homologous recombination, and the insertion of DNA to generate site-specific mutation, point mutation, or to insert a DNA stretch in a specific site in the plant genome.
Oligonucleotide-directed mutagenesis	Induces site-specific mutation at a target sequence within the DNA to introduce a new mutation, reverse an existing mutation, or induce short deletions.
Cisgenesis and intragenesis	To confer a new property to the modified plant. Cisgenics could achieve results also possible by traditional breeding methods. Intragenesis offers more options for modifying gene expression and trait development; it includes the use of gene-silencing techniques, such as RNA interference through the introduction of inverted DNA repeats.
RNA-dependent DNA methylation	To produce plants that do not contain foreign DNA sequences and in which no changes or mutations are made in the nucleotide sequence but in which gene expression is modified epigenetically.
Grafting	To produce a chimeric plant by grafting a nongenetically modified scion on a genetically modified rootstock (e.g., in perennial oil-bearing plants).
Reverse breeding	To produce homozygous parental lines of a selected heterozygous plant.
Agroinfiltration	To screen for plants with valuable phenotypes and use them in breeding programs.

Source: Modified from Cooper et al. (2014).

regardless of phenotype. It was proposed to identify genetic mutations in key oilseed crops (e.g., *B. oleracea*), explore the evolution of gene function, generate breeding lines with new phenotypes for commercial application, and predict the function of paralogous genes in related species, such as *B. napus*. Of the many enzymes that are potential TILLING targets to produce genes encoding more active forms, those involved in carbon flow into seed oil – such as pyruvate dehydrogenase kinase, D-glucose-6-phosphate dehydrogenase, and diacylglycerol transacylase – are the most important (Weselake et al., 2009).

Reverse genetic screens in *Arabidopsis thaliana* generated reliable data to draw strong inferences about the occurrence and randomness of chemically induced mutations and unbiased determination of mutation statistics (Greene et al., 2003). New genetic resources were developed in *Helianthus annuus*, as a model for seed oil biosynthesis, using a mutant population and reverse genetics tool to identify novel traits for crop improvement (Kumar et al., 2013). Reverse genetics provided a unique opportunity to integrate desirable genes from land races and CWR into genetic backgrounds readily used by breeders with minimum linkage drag.

When combined with information generated about gene function (e.g., in *A. thaliana*), it is expected to identify mutations in, and explore the evolution of, gene function and may generate phenotypes with commercial value (Himelblau et al., 2009); preliminary results indicate that mutations of many individual loci in *B. oleracea* lead to the development of distinct phenotypes despite genetic redundancy.

Metabolic Engineering

A strategic objective of metabolic research in oilseed crops is to understand the relationships between genotypes and metabolic outputs in the context of plant growth and chemical composition, particularly FA profiles, and how the latter are influenced by environmental stresses. Arguably (Sweetlove et al., 2014), the nature of supply and demand is a bigger driver of fluxes in central metabolism than are variations in the environment. Several methods have been used to explain the regulation mechanisms of oil accumulation in oilseed crops (Harwood et al., 2013); these include conventional breeding and selection, QTL mapping, transcription factors, biochemistry, genetic manipulation (e.g., gene overexpression or silencing), and metabolic control analysis. Metabolic and physiological differences between drought-resistant and drought-sensitive *B. napus* genotypes, for example, may not be attributed to typical hormonal adjustments and osmolyte accumulation, but to alternative physiological mechanisms used by resistant genotypes (Hatzig et al., 2014). Better understanding of such anomalies may result from integrating metabolic performance with whole plant physiology.

Metabolic control analyses have been applied to help increase yield in several oilseed crops, including *B. napus*; the new knowledge has been used to update genetic manipulation for the improvement of other oilseed crops (Harwood et al., 2013), whereas a metabolic biosynthesis pathway is being engineered into transgenic oilseed crops to produce seed oils containing γ-linolenic acid, omega-3, and very long–chain PUFAs, not naturally available from plant sources, with increased health benefits (Ruiz-López et al., 2012). In metabolically engineered *Crambe abyssinica* lines, for example, the average levels of C18:1, C18:2, and C18:3 were markedly decreased and that of C22:1 (erucic acid, a major feedstock for the oleochemical industry) was increased from 60% in the wild type to 73% in the best transgenic line (Li et al., 2012).

Functional genomics is being extensively applied in understanding metabolic pathways and in genetic improvement and breeding of oilseed crops (Maghuly and Laimer, 2013). Plant genomes are manipulated using metabolic engineering to produce genetically modified oilseed crops whose oil quality, content (Maheshwari and Kovalchuk, 2014), and composition are optimized for use as industrial feedstocks (Blackshaw et al., 2011). As a result, many genes and gene products, which are involved in adapting plants to environmental stresses, have been discovered through comparative functional genomic analyses of crop plants exposed to abiotic stresses, such as low or high temperature (Kovalchuk and Weselake, 2014). Although the biochemical and molecular mechanisms of crop tolerance to certain abiotic stresses are not fully understood (Kovalchuk and Weselake, 2014), different indicators of tolerance to drought or salinity, at different growth stages (germination, seedling, flowering, and mature stages) have been developed and used for germplasm screening and selection in breeding practices (Zhang et al., 2014). Nevertheless, further developments in metabolic engineering are expected to enhance crop tolerance to multiple abiotic stresses and may extend crop adaptation and production into drier regions where temperature extremes and limited moisture are frequently barriers to effective plant growth.

Perennial Oilseed Crops

Perennial crops, including oil-bearing plants, may provide additional ecosystem services besides oil production, such as maintaining healthy soil, controlling erosion, improving water quality, and enhancing wildlife habitat (Kantar et al., 2014). More importantly, they are more efficient C sequesters than annuals, provide high-energy yields, significantly reduce N losses (Pugesgaard et al., 2014), and may combat GCC more efficiently. For example, the yield potential of *Lepidium campestre* (5 t ha^{-1}), a cold-hardy member of the Brassicaceae, is about 30% higher than mean yield of winter *B. napus*. The plant has better architectural traits, and its oil has larger erucic acid content than *B. napus* (Karim et al., 2014). However, new perennial oilseed crops or oil-bearing plants may be slower to adapt to GCC than annuals due to lengthy life cycless and the time needed to develop a new cultivar.

The genetic resources base of CWR and semidomesticated perennial oil-bearing plants is rich in plants with fast growth, tolerance to biotic and abiotic stresses, and low requirements for external inputs (Jaradat, 2010). For example, there are 37 tetraploid and hexaploid perennial species of the 51 *Helianthus* spp.; other potential oil crops include the perennial tree species *Terminalia catappa*, *Couroupita guianensis* (Iha et al., 2014), *Jatropha curcas* (Laviola et al., 2010), progeny of the interspecific cross *H. annuus* × *H. tuberosus* (Kantar et al., 2014), *Cynara cardunculus* (Vasilakoglou and Dhima, 2014), and the biennial *Lepidium campestre*. The contribution of nonedible plant oils from *J. curcas* (~30% oil), *Sapindus mukorossi* and *Sapindus trifoliatus* (~52% oil), as new sources for biodiesel production have the advantage of not competing with edible oils produced from oilseed crop plants. Other perennial oil-bearing plants, among 75 oil plants containing 30% or more oil in their seeds, fruits, or nuts, include *Azadirachta indica*, *Calophyllum inophyllum*, and *Pometia pinnata* (Jaradat, 2010). Domestication and breeding of perennial oilseed crops are under way to understand gene function in relation to adaptation, C footprint, reproductive traits, and seed and oil traits (e.g., *J. curcas*; Maghuly and Laimer, 2013), with the aim of improving the physicochemical oil properties of potential oil-bearing wild trees (Iha et al., 2014).

The use of new genome-based technologies will expand the genetic base of perennial oilseed crops that can mitigate GCC, contribute to GHG emission reduction, accelerate CWR domestication, facilitate wide crosses, and increase the probability of selecting new genetically superior cultivars for oil production. For example, it was possible to select viable progeny from the cross between the annual *H. annuus* and the perennial *H. tuberosus* (Kantar et al., 2014). Recurrent selection for seed yield resulted in immediate improvement of a population destined for use in a perennial *Helianthus* spp. breeding program; tuber-related traits from the perennial parent and seed-related traits from the annual parent were independently inherited, thus contributing to better phenotypic traits in the progeny in a relatively short time.

Breeding Leafy Oil Biofactories

Plant biotechnologists have the primes to produce high levels of common vegetable oils in leaves at levels comparable with those in oilseed crops. Plant leaves are considered better organs to produce industrial quality oil than seeds, which are destined for reproduction. In addition, plants as biofactories will be harvested before the onset of flowering, thus capturing maximum energy in the oleochemicals produced. Leaf-based assays are being used to rapidly screen complex transgenic metabolic pathways, whereby single genes are stacked into multicomponent pathways and assessed for function. It is anticipated (Wood, 2014) that leaf-based assays could be used to build lipids and oleochemicals to be produced in nonseed biomass using leaf biofactories. These new products may be produced using high-energy density biomass from *Miscanthus* × *giganticum* and *Nicotiana tabacum,* which accumulate up to 17.7% TAG in leaves due to the expression of three genes involved in different aspects of TAG production.

Some leafy genotypes in the Brassicaceae have been identified with high oil content and heat tolerance (Annisa et al., 2013), whereas ectopic overexpression of *LEC2* from *Ricinus communis* triggered the expression of genes that encode regulators of seed maturation and oil body proteins in vegetative tissues of *Arabidopsis thaliana* (Kim et al., 2014). Leaf-functional traits, such as specific leaf area, size, shape, thickness, in addition to their role in adaptive phenotypic plasticity (Nicotra et al., 2010), may contribute to more efficient PS, C accumulation, and potentially to higher oil yield. The integrated approach of C flux into FAs, TAG assembly, and turnover needs to be optimized to achieve industrially relevant levels of storage in plant leaves. The process can be optimized if high TAG accumulation can be achieved and extended to other plant organs (e.g., stems and tubers) at the same time, so that the whole biomass can be used for renewable production of storage lipids.

FUTURE OF OILSEED BREEDING FOR CLIMATE CHANGE

Genetic improvement of oilseed crops during the last century contributed ~50% of yield increases; it will continue to contribute substantially to any future yield and quality improvement. However, breeding will rely more on enhanced traits to be developed by current and emerging biotechnologies. Oilseed-breeding methods and technologies will most likely continue to unfold as an application for scaling the quantitative–biology continuum in response to multiple socioeconomic drivers, as well as climate change stresses. Future advances in the quantitative component will necessarily be founded on integrated bases of quantitative genetics, biometry, and gene-to-phenotype knowledge of plant ecophysiological and phenotypic traits, whereas the biological component will depend on integrated experimental and functional gene-to-phenotype modeling approaches that advance the understanding of functional germplasm diversity, and gene-to-phenotype trait relationships for all sources of variation that can be utilized in oilseed crop breeding.

Due to the complexity of breeding for multiple abiotic stresses, and to the large diversity within and among taxa and species of oilseed crops, there will inevitably be an increase in the scale of breeding programs to accommodate wider germplasm diversity, including crop wild relatives. The breeding process will have to be enabled by more complex models and genetic prediction methodologies, starting with prebreeding and germplasm enhancement until new cultivars are released. Genetic prediction methods, in turn, will have to be supported by, and integrated with, high-throughput plant phenotyping, the cost of which is expected to become affordable by most oilseed-breeding programs. This strategy will furnish breeders with large relational databases on multiple traits, including physiological determinants of adaptation to climate change. Simulation models can be more effective in predicting genotypic responses to abiotic stresses if allelic effects are simulated in current and future climate change scenarios and if individual or multiple phenotypic traits are assessed to guide breeding of oilseed crops, especially those with a narrow genetic base due to their monophyletic origin and selfpollination. Moreover, model algorithms that are closer to actual plant mechanisms are required in order to adequately and realistically estimate the effects of major and minor QTL, provided that relevant traits are measurable in phenotyping facilities.

Oilseed crop plants will respond through phenotypic plasticity to environmental conditions and changes in resource availability caused by climate change. Plasticity, besides providing a buffer against rapid climate change, unlike genetic diversity contributes to rapid adaptation. Understanding plastic response is critical for predicting and managing climate change impacts on current and future oilseed crops. Therefore, it will be useful to discover whether past breeding practices led to a reduction in adaptive plasticity of modern cultivars and whether we should breed for plasticity in key traits such as phenology, flowering time, and reproduction. Nevertheless, it is prudent to explore and identify hierarchies or positive correlations in the plasticity of agronomic, yield, and oil quality traits, and exploit those associated with single or multiple abiotic stresses.

Climate change will bring about greater demand for genetic resources, whether from related cultivated, semidomesticated, or wild species. Past and recent genetic bottlenecks reduced the genetic diversity of many oilseed crops; therefore, crop wild relatives, as largely untapped genetic resources with genetic affinity to cultivated oilseed crops, remain a relatively low priority due to genetic barriers; however, their "genetic memory" of adaptation to abiotic stresses can now be more easily transferred and employed in breeding more resilient oilseed crops. The adaptive value conferred by largely undocumented locally rare or widespread alleles of underutilized oilseed crops and their relatives will increase in importance. Advances in molecular genetics will facilitate moving genes from more distant species to improve abiotic stresses in future oilseed crops.

References

Ahuja, I., de Vos, R., Bones, A., Hall, R., 2010. Plant molecular stress responses face climate change. Trends Plant Sci. 15, 664–674.

Ali, I., Farhatullah, 2013. Environmental and induced mutations influences on oil content, oleic and erucic acid rapeseed. Sarhad J. Agric. 29, 187–193.

Allu, A., Soja, A., Wu, A., Szymansli, J., Balazadeh, S., 2014. Salt stress and senescence; identification of cross-talk regulatory components. J. Exp. Bot. 65 (14), 4023–4036.

Alves, A., Manthey, L., Isbell, T., Ellis, D., Jenderek, M., 2014. Diversity in oil content and fatty acid profile in seed of wild cssava germplasm. Ind. Crops and Prod. 60, 310–315.

Annisa, Chen, S., Turner, N., Cowling, W., 2013. Genetic variation for heat tolerance during the reproductive phase in *Brassica rapa*. J. Agron. Crop Sci. 199, 424–435.

Anschütz, U., Berker, D., Shabala, S., 2014. Going beyond nutrition: regulation of potassium homoeostasis as common denominator of plant adaptive responses to environment. J. Plant Physiol. 171, 670–687.

Araus, J., Cairns, J., 2014. Field high-throughput phenotyping: the new crop breeding frontier. Trends Plant Sci. 19, 52–61.

Banga, S., Kang, M., 2014. Developing climate-resilient crops. J. Crop Improvement 28, 57–87.

Bates, P., Stymne, S., Ohlrogge, J., 2013. Biochemical pathways in seed oil synthesis. Curr. Opin. Plant Biol. 16, 358–364.

Bhargav, D., Meena, H., 2014. Participatory plant breeding: farmers as breeders. Popular Kheti 2, 7–14.

Bhatia, V., Jumrani, K., Pandey, G., 2014. Evaluation of the usefulness of senescing agent potassium iodide (KI) as a screening tool for tolerance to terminal drought in soybean. Plant Know. J. 3, 23–30.

Bishop, K., Betzelberger, A., Long, S., Ainsworth, E., 2014. Is there potential to adapt soybean (*Glycine max* Merr.) to future [CO_2]? An analysis of the yield response of 18 genotypes in free-air CO_2 enrichment. Plant Cell Environ. doi: 10.1111/pce.12443.

Blackman, B., 2013. Interacting duplications, fluctuating selection, and convergence: the complex dynamics of flowering time evolution during sunflower domestication. J. Exp. Bot. 64, 421–431.

Blackshaw, R., Johnson, E., Gan, Y., May, W., McAndrew, D., Barthet, V., McDonald, T., Wispinski, D., 2011. Alternative oilseed crops for biodiesel feedstock on the Canadian prairies. Can. J. Plant Sci 91, 889–896.

Bloem, E., Haneklaus, S., Schnug, E., 2014. Milestones in plant sulfur research on sulfur-induced-resistance (SIR) in Europe. Front. Plant Sci. 5, 779.

Bloomfield, J., Rose, T., King, G., 2014. Sustainable harvest: managing plasticity for resilient crops. Plant Biotechnol. J. 12, 517–533.

Brown, T., Cheng, R., Sirault, X., Rungrat, T., Murray, K., Trtilek, M., Furbank, R., Badger, M., Pogson, B., Borevitz, J., 2014. TraitCapture: genomic and environmental modelling of plant phenomic data. Curr. Opin. Plant Bio. 18, 73–79.

Bruce, T., 2014. Glucosinolates in oilseed rape: secondary metabolites that influence interactions with herbivores and their natural enemies. Ann. Appl. Biol. 164, 348–353.

Cartagena, J., Seki, M., Tanaka, M., Yamauchi, T., Sato, S., Hirakawa, H., Tsuge, T., 2014. Gene expression profiles in Jatropha under drought stress and during recovery. Plant Mol. Biol. Rep. doi: 10.1007/s11105-014-0815-0.

Chardon, F., Jasinski, S., Durandet, M., Lécureuil, A., Soulay, F., Bedu, M., Guerche, P., Masclaux-Daubresse, C., 2014. QTL meta-analysis in *Arabidopsis* reveals an interaction between leaf senescence and resource allocation to seeds. J. Exp. Bot. 65 (14), 3949–3962.

Cheng, J., Salentijn, E.M.J., Huang, B., Denneboom, C., Qi, W., Dechesne, A.C., Krens, F.A., Visser, R.G.F., van Loo, E.N., 2014. Detection of induced mutations in CaFAD2 genes by next-generation sequencing leading to the production of improved oil composition in *Crambe abyssinica*. PlantBiotechnol. J. 13 (4), 471–481.

Choudhury, S., Riesselman, A., Pandey, S., 2014. Constitutive or seed-specific overexpression of Arabidopsis G-protein c subunit 3 (AGG3) results in increased seed and oil production and improved stress tolerance in *Camelina sativa*. Plant Biotechnol. J. 12, 49–59.

Colombini, S., Broderick, G., Galasso, I., Martinelli, T., Rapetti, L., Russo, R., Reggiani, R., 2014. Evaluation of *Camelina sativa* (L.) Crantz meal as an alternative protein source in ruminant rations. J. Sci. Food Agric. 94, 736–743.

Cooper, M., Messina, C., Podlich, D., Radu Totir, L., Baumgarten, B., Hausmann, N., Wright, D., Graham, G., 2014. Predicting the future of plant breeding: complementing empirical evaluation with genetic prediction. Crop Pasture Sci. 65, 311–336.

Cowling, W., 2013. Sustainable plant breeding. Plant Breed. 132, 1–9.

Crane, T., Roncoli, C., Hoogenboom, G., 2011. Adaptation to climate change and climate variability: the importance of understanding agriculture as performance. NJAS- Wageningen J. Life Sci. 57, 179–185.

De Jong, T., Rong, J., 2013. Crop to wild gene flow: does more sophisticated research provide better risk assessment? Environmental Sci. Policy 27, 135–140.

De Storme, N., Geelen, D., 2014. The impact of environmental stress on male reproductive development in plants: biological processes and molecular mechanisms. Plant Cell Environ. 37, 1–18.

Dechaine, J., Brock, M., Weinig, C., 2014. QTL architecture of reproductive fitness characters in *Brassica rapa*. BMC Plant Biol. 14, 66. http://www.biomedcentral.com/1471–2229/14/66.

Dempewolf, H., Eastwood, R., Guarino, L., Khoury, C., Müller, J., Toll, J., 2014. Adapting agriculture to climate change: a global initiative to collect, conserve, and use crop wild relatives. Agroecol. Sust. Food Systems 38, 369–377.

Dias, 2014. Guiding strategies for breeding vegetable cultivars. Agric. Sci. 5, 9–32.

Diez, C., Trujillo, I., Martinez-Urdiroz, N., Barranco, D., Rallo, L., Marfil, P., Gaut, B., 2014. Olive domestication and diversification in the Mediterranean Basin. New Phytol. 206 (1), 436–447.

Elbeyrouthy, M., Abirizk, A., Wakim, H., 2014. Genomic approaches of crops genetic diversity. Adv. Crop Sci. Tech. 2, 1. http://dx.doi.org/10.4172/2329-8863.1000e110.

Eldakak, M., Milad, S., Nawar, A., Rohila, J., 2013. Proteomics: a biotechnology tool for crop improvement. Front. Plant Sci. 4doi: 10.3389/fpls.2013.00035, Article 35.

Eriksson, D., Stymne, S., Schjoerring, J., 2014. The slippery slope of cisgenesis. Nature Biotechnology 32, 727.

FAO, 2014. Oil crops, oils and meals market assessment – http://www.fao.org/ economic/est/est-commodities/oilcrops/en/ (accessed 12 10 14).

FAOSTAT, 2012. FAO Statistical Database. FAO, Rome. Available at: http://faostat.fao.org. (accessed 12 10 14).

Finnan, J., Styles, D., 2013. Hemp: a more sustainable annual energy crop for climate and energy policy. Energy Policy 58, 152–162.

Fu, Y.-B., 2011. Genetic evidence for early flax domestication with capsular dehiscence. Genet. Resour. Crop Evol. 58, 1119–1128.

Gan, Y., Liang, C., Huang, G., Malhi, S., Brandt, S., Katepa-Mupondwa, F., 2012. Carbon footprint of canola and mustard is a function of the rate of N fertilizer. Int. J. Life Cycle Assess. 17, 58–68.

Gao, G., Li, J., Li, H., Li, F., Xu, K., Yan, G., Chen, B., Qiao, J., Wu, X., 2014. Comparison of the heat stress induced variations in DNA methylation between heat-tolerant and heat-sensitive rapeseed seedlings. Breed. Sci. 64, 125–133.

García-López, J., Lorite, I., Garcia-Ruiz, R., Dominguez, J., 2013. Evaluation of three simulation approaches for assessing yield of rainfed sunflower in a Mediterranean environment for climate change impact modeling. Climate Change 124, 147–162.

Geleta, M., Ortiz, R., 2013. The importance of *Guizotia abyssinca* (niger) for sustainable food security in Ethiopia. Genet. Resour. Crop Evol. 60 (5), 1763–1770.

George, T., Hawes, C., Newton, A., McKenzie, B., Hallett, P., Valentine, T., 2014. Field phenotyping and long-term platforms to characterize how crop genotypes interact with soil processes and the environment. Agronomy 4, 242–278.

Gepts, P., 2002. A comparison between crop domestication, classical plant breeding, and genetic engineering. Crop Sci. 42, 1780–1790.

Ghamkar, K., Croser, J., Aryamanesh, N., Campbell, M., Francis, C., 2010. Camelina (*Camelina sativa* (L.) Crantz) as an alternative oilseed: molecular and ecogeographic analyses. Genome 53, 558–567.

Gilchrist, E., Sidebottom, C., Koh, C., MacInnes, T., Sharpe, A., et al.,2013. A mutant *Brassica napus* (Canola) population for the identification of new genetic diversity via TILLING and next generation sequencing. PLoS ONE 8 (12), e84303.

Gill, B., Raupp, W., Friebe, B., 2014. Dual threats of imperiled native agroecosystems and climate change to world food security: genomic perspectives. J. Crop Improvement 28, 88–98.

Gouache, D., Le Bris, X., Bogard, M., Deudon, O., Pagé, C., Gate, P., 2012. Evaluating agronomic adaptation options to increasing heat stress under climate change during wheat grain filling in France. Eur. J. Agron. 39, 62–70.

Greene, E., Codomo, C., Taylor, N., Henikoff, J., Till, B., Reynolds, S., Enns, L., Burtner, C., Johnson, J., Odden, A., Comai, L., Henikoff, S., 2003. Spectrum of chemically induced mutations from a large-scale reverse-genetic screen in Arabidopsis. Genetics 164, 731–740.

Gressel, J., 2014. Dealing with transgene flow of crop protection traits from crops to their relatives. Pest Manag. Sci. 71 (5), 658–667.

Haile, T., Holzapfel, C., Shirtliffe, S., 2014. Canola genotypes and harvest methods affect seedbank addition. Agron. J. 106, 236–242.

Haililu, A., Ogunwole, J., 2013. Genetic variability, genetic gain and relationships of yield and yield components in castor (*Ricinus communis* L.). RRBS 7, 181–186.

Hannoufa, A., Pillai, B., Chellamma, S., 2013. Genetic enhancement of *Brassica napus* seed quality. Transgenic Res. 23 (1), 39–52.

Harwood, J., Ramli, U., Tang, M., Quant, P., Weselake, R., Fawcett, T., Guschina, I., 2013. Regulation and enhancement of lipid accumulation in oil crops: the use of metabolic control analysis for informed genetic manipulation. Eur. J. Lipid Sci. Technol. 115, 1239–1246.

Hatzig, S., Zaharia, I., Abrams, S., Hohmann, M., Legoahec, L., Bouchereau, A., Nesi, N., Snowdon, R., 2014. Early osmotic adjustment responses in drought resistant and drought-sensitive oilseed rape. JIPB 56, 797–809.

Heslot, N., Akdemir, D., Sorrells, M., Jannink, J.-L., 2014. Integrating environmental covariates and crop modeling into the genomic selection framework to predict genotype by environment interactions. Theor. Appl. Genet. 127, 463–480.

Himelblau, E., Gilchrist, E., Buono, K., Bizzell, C., Mentzer, L., Vogelzang, R., Osborn, T., Amasino, R., Parkin, I., Haughn, G., 2009. Forward and reverse genetics of rapid-cycling *Brassica oleracea*. Theor. Appl. Genet. 118, 953–961.

Hou, H., Atlihan, N., Lu, Z.-X., 2014. New biotechnology enhances the application of cisgenesis in plant breeding. Frontiers Plant Sci. 5, 389.

Hyun, T., Kumar, D., Cho, Y.-Y., Hyun, H.-N., Kim, J.-S., 2013. Computational identification and phylogenetic analysis of the oil-body structural proteins, oleosin and caleosin, in castor bean and flax. Gene 515, 454–460.

Iha, O., Alves, F., Suzrez, P., Silva, C., Meneghetti, M., Meneghetti, S., 2014. Potential application of *Terminalia catappa* L. and *Carapa guianensis* oils for biofuel production: Physical-chemical properties of neat vegetable oils, their methyl-esters and bio-oils (hydrocarbons). Ind. Crops Prod. 52, 95–98.

IPCC, 2013. Climate change 2013: impact, adaptation, and vulnerability. Contribution of working group ii to the fifth assessment report of the intergovernmental panel on climate change. In: Stocker, T.F., Qin, D., Plattner, G.-K., Tignor, M., Allen, S.K., Boschung, J., Nauels, A., Xia, Y., Bex, V., Midgley, P.M. (Eds.), Contribution of working group I to the fifth assessment report of the intergovernmental panel on climate change. Cambridge University Press, Cambridge.

Jaggard, K., Qi, A., Ober, E., 2010. Possible changes to arable crop yields by 2050. Phil. Trans. R. Soc. B 365, 2835–2851.

Jain, R., 2014. Importance of flowering performance and pattern in Linum seed yield. DRJAFS 2, 82–97.

Jang, H.-Y., Rhee, J., Carlson, J., Ahn, S.-J., 2014. The Camelina aquaporin CsPIP21 is regulated by phosphorylation at Ser273, but not at Ser277, of the C-terminus and is involved in salt- and drought-stress responses. J. Plant Physiol. 171, 1401–1412.

Jaradat, A.A., 2010. Genetic resources of energy crops: biological systems to combat climate change. Austr. J. Crop Sci. 4, 309–323.

Jaradat, A.A., 2014. Can carbon in bioenergy crops mitigate global climate change? Tuteja, N., Gill, S.S. (Eds.), Climate Change and Plant Abiotic Stress Tolerance, 2, Wiley Blackwell Weinheim, Germany, pp. 345–420.

Jaradat, A.A., 2015. Beyond biodiversity: Ecosystem services of crop wild relatives. In: Redden, R., Yadav, S., Maxted, N., Dulloo, M., Guarino, L., Smith, P. (Eds.), Crop Wild Relatives and Climate Change. John Wiley & Sons, Inc, pp. 247–334.

Javed, N., Tahir, M., Geng, J., Li, G., McVetty, P., 2014. Identification of Brassica genotypes and molecular markers for increased seed oil content. Can. J. Plant Sci. 94, 1103–1108.

Johnson, T., Eswaran, N., Sujatha, M., 2011. Molecular approaches to improvement of *Jatropha curcas* Linn. as a sustainable energy crop. Plant Cell Rep. 30, 1573–1591.

Jung, H.-J., Dong, X., Park, J.-I., Thamilarasan, S.K., Lee, S.S., et al.,2014. Genome-wide transcriptome nalysis of two contrasting *Brassica rapa* doubled haploid lines under cold-stresses using Br135K oligomeric chip. PLoS ONE 9 (8), e106069.

Kang, F., Cournede, P-H., Lecoeur, J., Letort, V., 2014. SUNLAB: A functional–structural model for genotypic and phenotypic characterization of the sunflower crop. Ecological Modeling http://dx.doi.org/10.1016/j.ecolmodel.2014.02.006.

Kant, P., Wu, S., 2011. The extraordinary collapse of *Jatropha* as aglobal biofuel. Environ. Sci. Technol. 45, 7114–7115.

Kantar, M., Betts, K., Michno, J., Lubyb, J., Morrell, P., Hulke, B., Stupar, R., Wyse, D., 2014. Evaluating an interspecific *Helianthus annuus* × *Helianthus tuberosus* population for use in a perennial sunflower breeding program. Field Crops Res. 155, 254–264.

Karim, M., Siddika, A., Tonu, N., Hossain, D., Meah, M., Kawanabe, T., Fujimoto, R., Okazaki, K., 2014. Production of high yield short duration *Brassica napus* by interspecific hybridization between *B. oleracea* and *B. rapa*. Breed. Sci. 63, 495–502.

Kebede, B., Rahman, H., 2014. Quantitative trait loci (QTL) mapping of silique length and petal colour in *Brassica rapa*. Plant Breed. 133 (5), 609–614.

Keneni, G., Bekele, E., Imtiaz, M., Dagne, K., 2012. Genetic vulnerability of modern crop cultivars: causes, mechanism and remedies. Intl. J. Plant Res. 2, 69–89.

Kim, M., Yang, S., Mao, H-Z., Veena, S., Yin, J-L., Chua, N-H., 2014. Gene silencing of Sugar-dependent 1 (JcSDP1), encoding a patatin-domain triacylglycerol lipase, enhances seed oil accumulation in *Jatropha curcas*. Biotechnology for Biofuels 2014, 7:36http://www.biotechnologyforbiofuels.com/content/7/1/36.

Kovalchuk, I., Weselake, R., 2014. Trait introduction methods and innovation platforms in plant biotechnology. Biocatal. Agric. Biotechnol. 3, 1.

Kumar, A., Boualem, A., Bhattacharya, A., Parikh, S., Desai, N., Zambelli, A., Leon, A., Chatterjee, M., Bendahmane, A., 2013. SMART-Sunflower mutant population and reverse genetic tool for crop improvement. BMC Plant Biology 13, 38. http://www.biomedcentral.com/1471-2229/13/38.

Lai, K., Lorence, M., Edwards, D., 2012. Genomic databases for crop improvement. Agronomy 2, 62–73.

Laviola, B., Rocha, R., Kobayashi, A., Rosando, T., Bhering, L., 2010. Genetic improvement of Jatropha for biodiesel production. Ceiba 51, 1–10.

Lee, S., Kim, H., Kim, R., Suh, M., 2014. Overexpression of Arabidopsis *MYB96* confers drought resistance in *Camelina sativa* via cuticular wax accumulation. Plant Cell Rep. 33, 1535–1546.

Lei, X., Xiao, Y., Xia, W., Mason, A., Yang, Y., Ma, Z., Peng, M., 2014. RNA-seq analysis of oil palm under cold stress reveals a different C-repeat binding factor (CBF) mediated gene expression pattern in *Elaeis guineensis* compared to other species. PLoS ONE 9 (12), e114482.

Lenser, T., Theißen, G., Molecular mechanisms involved in convergent crop domestication. Trends Plant Sci. http://dx.doi.org/10.1016/j.tplants.2013.08.007.

Leonardos, E., Rauf, S., Weraduwage, S., Marilli, E.-F., Taylor, D., Micallef, B., Grodzinski, B., 2014. Photosynthetic capacity of the inflorescence is a major contributor to daily-C-gain and the responsiveness of growth to elevated CO_2 in *Arabidopsis thaliana* with repressed expression of mitochondrial- pyruvate-dehydrogenase-kinase. Environ. Exp. Bot. 107, 84–97.

Li, X., Mupondwa, E., 2014. Life cycle assessment of camelina oil derived biodiesel and jet fuel in the Canadian Prairies. Sci. Total Environ. 481, 17–26.

Li, X., van Loo, N., Gruber, J., Fan, J., Guan, R., Frentzen, M., Stymne, S., Zhu, L.-H., 2012. Development of ultra-high erucic acid oil in the industrial oil crop *Crambe abyssinica*. Plant Biotechnol. J. 10, 862–870.

Li, F., Chen, B., Xu, K., Wu, J., Song, W., Bancroft, I., Harper, A., Trick, M., Liu, S., Gao, U., Wang, N., Yan, G., Qiao, J., Li, J., Li, H., Xiao, X., Zhang, T., Wu, X., 2014a. Genome-wide association study dissects the genetic architecture of seed weight and seed quality in rapeseed (*Brassica napus* L.). DNA Research 21 (4), 355–367.

Li, J., Liu, B., Cheng, F., Wan, X., Aarts, M., Wu, J., 2014b. Expression profiling reveals functionally redundant multiple-copy genes related to zinc, iron and cadmium responses in *Brassica rapa*. New Phytol. 203, 182–194.

Li, N., Shi, J., Wang, X., Liu, G., Wang, H., 2014c. A combined linkage and regional association mapping validation and fine mapping of two major pleiotropic QTLs for seed weight and silique length in rapeseed (*Brassica napus* L.). BMC Plant Biology 2014, 14:114. http://www.biomedcentral.com/1471-2229/14/114.

Liu, Y., Wei, W., Ma, K., Li, J., Liang, Y., Darmency, H., 2013. Consequences of gene flow between oilseed rape (*Brassica napus* L.) and its relative. Plant Sci. 211, 42–51.

Liu, Y-F., Li, Q-T., Lu, X., Song, Q-X., Lam, S-M., Zhang, W-K., Ma, B., Lin, Q., Man, W-Q., Du, W-G., Shui, G-H., Chen, S-Y., Zhang, J-S., 2014. Soybean *GmMYB73* promotes lipid accumulation in transgenic plants. BMC Plant Biol. 14:73 http://www.biomedcentral.com/1471-2229/14/73.

Lusser, M., Parisi, C., Plan, D., Rodreguez-Cerezo, E., 2012. Deployment of new biotechnologies in plant breeding. Nat. Biotechnol. 30, 231–239.

Maghuly, F., Laimer, M., 2013. *Jatropha curcas*, a biofuel crop: functional genomics for understanding metabolic pathways and genetic improvement. Biotechnol. J. 8, 1172–1182.

Maheshwari, P., Kovalchuk, I., 2014. Genetic engineering of oilseed crops. Biocatal. Agric. Biotechnol. 3, 31–37.

Matthews, R., Rivington, M., Muhammed, S., Newton, A., Hallett, P., 2013. Adapting crops and cropping systems to future climates to ensure food security: the role of crop modelling. Global Food Security 2, 24–28.

Merilä, J., Hendry, A., 2014. Climate change, adaptation, and phenotypic plasticity; the problem and evidence. Evol. Appl. 7 (1), 1–14.

Meyer, R., Purugganan, M., 2013. Evolution of crop species: genetics of domestication and diversification. Nat. Rev. 14, 840 852.

Murphy, D., 2014. The future of oil palm as a major global crop: opportunities and challenges. J. Oil Palm Res. 26, 1–24.

Nayeri, F., Tarizade, K., 2014. Bioinformatic study of delta-12 fatty acid desaturase 2 (FAD2) gene in oilseed. Mol. Biol. Rep. 41, 5077-50-87.

Nelissen, H., Moloney, M., Inze, D., 2014. Translational research: from pot to plot. Plant Biotechnol. J. 12, 277–285.

Newton, A., Johnson, S., Gregory, P., 2011. Implications of climate change for diseases, crop yields and food security. Euphytica 197, 3–18.

Nicotra, A., Atkin, O., Bonser, S., Davidson, E., Finnegan, E., Mathesius, U., Poot, P., Purugganan, M., Richards, C., Valladares, F., van Kleunen, M., 2010. Plant phenotypic plasticity in a changing climate. Trends Plant Sci. 15, 684–692.

Nogueira, L., 2011. Does biodiesel make sense? Energy 36, 3659–3666.

Olesen, J., Trnka, M., Kersebaum, K., Skjelvåg, A., Seguin, B., Peltonen-Sainio, P., Rossi, F., Kozyra, J., Micale, F., 2011. Impacts and adaptation of European crop production systems to climate change. Europ. J. Agron. 34, 96–112.

Ortiz, R., Jarvis, A., Fox, P., Aggarwal, P., Campbell, B., 2014. Plant genetic engineering, climate change and food security. CCAFS Working Paper no. 72. CGIAR Research Program on Climate Change, Agriculture and Food Security (CCAFS). Copenhagen, Denmark. Available online at: www.ccafs.cgiar.org.

Pal, M., Chaturvedi, A., Pandey, S., Bahuguna, R., Khetarpal, S., Anand, A., 2014. Rising atmospheric CO_2 may affect oil quality and seed yield of sunflower (*Helianthus annua* L.). Acta Physiol. Plant 36, 2853–2861.

Palta, J., 2014. Merging physiological and genetic approaches to improve abiotic stress resistance. J. Crop Improvement 28, 260–304.

Pandey, M., Upadhyaya, H., Rathore, A., Vadez, V., Sheshshayee, M., Sriswathi, M., Govil, M., Kumar, A., Gowda, M., Sharma, S., Hamidou, F., Kumar, V., Khera, P., Bhat, R., Khan, A., Singh, S., Li, H., Monyo, E., Nadaf, H., Mukri, G., Jackson, S., Guo, B., Liang, X., Varshney, R., 2014. Genome-wide association studies for 50 agronomic traits in peanut using the 'Reference Set' comprising 300 genotypes from 48 countries of the semi-arid tropics of the World. PLoS ONE 9, e105228.

Parent, B., Tardieu, F., 2014. Can current crop models be used in the phenotyping era for predicting the genetic variability of yield of plants subjected to drought or high temperature? J. Exp. Bot. 65, 6179–6189.

Pearl, S., Burke, J., 2014. Genetic diversity in *Carthamus tinctorius* (Asteraceae; Safflower), an underutilized oilseed crop. American J. Bot 101 (10), 1640–1650.

Pecchia, P., Russo, R., Brambilla, I., Reggiani, R., Mapelli, S., 2014. Biochemical seed traits of *Camelina sativa* – an emerging oilseed crop for biofuel: environmental and genetic influences. J. Crop Improvement 28, 465–483.

Peleg, Z., Blumwald, E., 2011. Hormone balance and abiotic stress tolerance in crop plants. Curr. Opin. Plant Biol. 14, 290–295.

Peltonen-Sainio, P., Jauhiainen, L., Hyöveläb, M., Nissilä, E., 2011. Trade-off between oil and protein in rapeseed at high latitudes: means to consolidate protein crop status? Field Crops Res. 121, 248–255.

Podevin, N., Davies, H., Hartung, F., Nogué, F., Casacuberta, J., 2013. Site-directed nucleases: a paradigm shift in predictable, knowledge-based plant breeding. Trends Biotechnol. 31, 375–383.

Pugesgaard, S., Schelde, K., Larsen, S., Eriklaerke, P., Jorgensen, U., 2014. Comparing annual and perennial crops for bioenergy production – influence on nitrate leaching and energy balance. GCB Bioenergy. doi: 10.1111/gcbb.12215.

Rahman, H., 2013. Review: breeding spring canola (*Brassica napus* L.) by the use of exotic germplasm. Can. J. Plant Sci. 93, 363–373.

Rahman, H., Kebede, B., 2012. Improvement of spring canola *Brassica napus* by use of winter canola. J. Oilseed Brassica 3, 1–17.

Rahman, H., Harwood, J., Weselake, R., 2013. Increasing seed oil content in Brassica species through breeding and biotechnology. Lipid Technol. 25, 182–185.

Ram, B., Meena, H., Singh, V., Singh, B., Nanjundan, N., Kumar, A., Singh, S., Bhogal, N., Singh, D., 2014. High temperature stress tolerance in Indian mustard (*Brassica juncea*) germplasm as evaluated by membrane stability index and excised-leaf water loss techniques. J. Oilseed Brassica 5, 149–157.

Redden, R., 2013. New approaches for crop genetic adaptation to the abiotic stresses predicted with climate change. Agronomy 3, 419–432.

Rodríguez, V., Soengas, P., Cartea, E., Sotelo, T., Velasco, P., 2014. Suitability of a European nuclear collection of *Brassica oleracea* L. landraces to grow at high temperatures. J. Agron. Crop Sci. 14, 183–190.

Ruiz-López, N., Sayanova, O., Napier, J., Haslam, R., 2012. Metabolic engineering of the omega-3 long chain polyunsaturated fatty acid biosynthetic pathway into transgenic plants. J. Exp. Bot. 63, 2397–2410.

Runck, B., Kantar, M., Jordan, N., Anderson, J., Wyse, D., Eckberg, J., Barnes, R., Lehman, C., DeHaan, L., Stupar, R., Sheaffer, C., Porter, P., 2014. The reflective plant breeding paradigm: a robust system of germplasm development to support strategic diversification of agroecosystems. Crop Sci. 54, 1939–1948.

Schiessl, S., Samans, B., Huttel, B., Reinhard, R., Snowdon, R.m, 2014. Capturing sequence variation among flowering-time regulatory gene homologs in the allopolyploid crop species *Brassica napus*. Front. Plant Sci. 5, 404.

Schulman, A., 2012. Genomics tools for accelerating plant breeding. Maataloustieteen Päivät 2012. www.smts.fi.

Schulte, L., Ballard, T., Samarakoon, T., Yao, L., Vadlani, P., Staggenborg, S., Rezac, M., 2013. Increased growing temperature reduces content of polyunsaturated fatty acids in four oilseed crop. Ind. Crops Prod. 51, 212–219.

Sedbrook, J., Phippen, W., Marks, M., 2014. New approaches to facilitate rapid domestication of a wild plant to an oilseed crop: example pennycress (*Thlaspi arvense* L.). Plant Sci. 227, 122–132.

Séguin-Swartz, G., Beckie, H., Warwick, S., Roslinsky, V., Nettleton, J., Johnson, E., Falk, K., 2013. Pollen-mediated gene flow between glyphosate-resistant *Brassica napus* canola and *B. juncea* and *B. carinata* mustard crops under large-scale field conditions in Saskatchewan. Can. J. Plant Sci. 93, 1083–1087.

Sharma, S., Upadhyaya, H., Varshney, R., Gowda, C., 2013. Pre-breeding for diversification of primary gene pool and genetic enhancement of grain legumes. Front. Plant Sci. 4, 309.

Sheoran, P., Sheoran, O., Sardana, V., 2013. Modeling sunflower productivity and profitability in relation to adequate and limited sulphur availability under semiarid irrigated conditions. Int. J. Agron., vol. 2013, Article ID 738263, 4 pages. http://dx.doi.org/10.1155/2013/738263.

Shrestha, B., Desjardins, R., McConkey, B., Worth, D., Dyer, J., Cerkowniak, D., 2013. Change in carbon footprint of canola production in the Canadian Prairies from 1986 to 2006. Renew. Energ. 63, 634–641.

Singh, S., Dubey, S., Ali, M., Nigam, S., Srivastava, R., Saxena, S., Yadav, Y., Kumar, A., 2012. Development and promotion of an informal and formal seed system through farmer participatory seed production of pigeonpea (*Cajanus cajan* L.) in Uttar Pradesh, India. Agroecol. Sust. Food Syst. 37, 531–549.

Snowdon, R., Luy, F., 2012. Potential to improve oilseed rape and canola breeding in the genomics era. Plant Breed. 131, 351–360.

Sønderby, I., Geu-Flores, F., Halkier, B., 2010. Biosynthesis of glucosinolates – gene discovery and beyond. Trends Plant Sci. 15, 283–290.

Soto-Cerda, B., Duguid, S., Booker, H., Rowland, G., Diederichsen, A., Cloutier, S., 2014. Genomic regions underlying agronomic traits in linseed (*Linum usitatissimum* L.) as revealed by association mapping. JIPB 56, 75–87.

Sun, L., Hu, R., Shen, G., Zhang, H., 2013. Genetic engineering peanut for higher drought- and salt-tolerance. Food Nutr. Sci. 4, 1–7.

Sweetlove, L., Obata, T., Fernie, A., 2014. Systems analysis of metabolic phenotypes: what have we learnt? Trends Plant Sci. 19, 222–230.

Takashima, N., Rondanini, D., Puhl, L., Miralles, D., 2013. Environmental factors affecting yield variability in spring and winter rapeseed genotypes cultivated in the southeastern Argentine Pampas. Eur. J. Agron. 48, 88–100.

Tardieu, F., Tuberosa, R., 2010. Dissection and modelling of abiotic stress tolerance in plants. Curr. Opin. Plant Biol. 13, 206–212.

Tsuda, M., Ohsawa, R., Tabei, Y., 2014. Possibilities of direct introgression from *Brassica napus* to *B. juncea* and indirect introgression from *B. napus* to related Brassicaceae through *B. juncea*. Breed. Sci. 64, 74–82.

Urano, K., Kurihara, Y., Seki, M., Shinozaki, K., 2010. 'Omics' analyses of regulatory networks in plant abiotic stress responses. Curr. Opin. Plant Biol. 13, 132–138.

Uusitalo, V., Väisänen, S., Havukainen, J., Havukainen, M., Soukka, R., Luoranen, M., 2014. Carbon footprint of renewable diesel from palm oil, jatropha oil and rapeseed oil. Renew. Energ. 69, 103–113.

Vasilakoglou, I., Dhima, K., 2014. Potential of two cardoon varieties to produce biomass and oil under reduced irrigation and weed control inputs. Biomass Bioenerg. 63, 177–186.

Vindhiyavarman, P., Manivannan, N., Muraldharan, V., 2010. Farmers' participatory varietal selection in groundnut: a case study from tamilnadu. India. Electronic J. Plant Breed. 1, 878–881.

von Maltitz, G., Gasparatos, A., Fabricius, C., 2014. The Rise, fall and potential resilience benefits of Jatropha in Southern Africa. Sustainability 6, 3615–3643.

Wang, X., Liu, A., 2014. Expression of genes controlling unsaturated fatty acids biosynthesis and oil deposition in developing seeds of Sacha Inchi (*Plukenetia volubilis* L.). Lipids 49, 1019–1031.

Warwick, S., Martin, S., 2013. Gene flow from transgenic oilseed *Brassica juncea* (L.) Czern. into weedy *Sinapis arvensis* L. (wild mustard). Plant Breed. 132, 688–693.

Wells, R., Trick, M., Soumpourou, E., Clissold, L., Morgan, C., Werner, P., Gibbard, C., Clarke, M., Jennaway, R., Bancroft, I., 2014. The control of seed oil polyunsaturate content in the polyploid crop species *Brassica napus*. Mol. Breed. 33, 349–362.

Weselake, R., Taylor, D., Rahman, H., Shah, S., Laroche, A., McVetty, B., Harwood, J., 2009. Increasing the flow of carbon into seed oil. Biotechnol. Adv. 27, 866–878.

Wilkes, M., Takei, I., Caldwell, R., Trethowan, R., 2013. The effect of genotype and environment on biodiesel quality prepared from Indian mustard (*Brassica juncea*) grown in Australia. Ind. Crops Prod. 48, 124–132.

Wood, C., 2014. Leafy biofactories: producing industrial oils in non-seed biomass. EMBO Rep 15, 201–202.

Wu, X.-L., Liu, Z.-H., Hu, Z.-H., Huang, R.-Z., 2014. BnWRI1 coordinates fatty acid biosynthesis and photosynthesis pathways during oil accumulation in rapeseed. J Integr. Plant Biol. 56, 582–593.

Xing, N., Fan, C., Zhou, Y., 2014. Parental selection of hybrid breeding based on maternal and paternal inheritance of traits in rapeseed (*Brassica napus* L.). PLoS ONE 9 (7), e103165.

Yu, E., Fan, C., Yang, Q., Li, X., Wan, B., Dong, Y., Wang, X., Zhou, Y., 2014. Identification of heat responsive genes in *Brassica napus* siliques at the seed-filling stage through transcriptional profiling. PLoS ONE 9 (7), e101914.

Zhang, X., Lu, G., Long, W., Zou, X., Li, F., Nishio, T., 2014. Recent progress in drought and salt tolerance studies in Brassica crops. Breed. Sci. 64, 60–73.

Zhang, J.-L., Zhang, S.-B., Zhang, Y.-P., Kitajima, K., 2015. Effects of phylogeny and climate on seed oil fatty acid composition across 747 plant species in China. Ind. Crops Prod. 63, 1–8.

Zhou, Q.-H., Fu, D.-H., Mason, A., Zeng, Y.-J., Zhao, C.-X., Huang, Y.-J., 2014. *In silico* integration of quantitative trait loci for seed yield and yield-related traits in *Brassica napus*. Mol. Breed. 33, 881–894.

Zou, J., Raman, H., Guo, S., Hu, D., Wei, Z., Luo, Z., Long, Y., Shi, W., Fu, Z., Du, D., Meng, J., 2014. Constructing a dense genetic linkage map and mapping QTL for the traits of flower development in *Brassica carinata*. Theor. Appl. Genet. 127, 1593–1605.

CHAPTER

19

Possibilities of Sustainable Oil Processing

Bertrand Matthäus

Max Rubner-Institut, Federal Research Institute for Nutrition and Food, Department for Quality and Safety of Cereals, Working Group for Lipid Research, Detmold, Germany

INTRODUCTION

The production and consumption of vegetable oils nearly tripled in the period 1992–2014 with an increase from 59 Mt to 161 Mt. This increase in oil consumption was accompanied by a more or less linear increase in the world's population from about 5.5 billion to more than 7 billion (Fig. 19.1). This increasing demand is expected to continue in the coming decades. One reason for this growing demand for vegetable oils is the increasing incomes of the world's population, resulting in increasing demand for healthier vegetable oils of higher quality in processed foods. Moreover, the increasing use of oils in technical applications is another reason for the strong increase. In the last 20 years palm oil and soybean oil showed a remarkable increase from 12 Mt to 52 Mt and 17 Mt to 42 Mt, respectively. The third important oil crop is rapeseed with world production of 24 Mt in 2012.

The world's increasing population and the increasing demand of vegetable oils for food, animal nutrition, and a number of technical applications – such as fuel, lubricants, or raw material for the production of chemical products – means sustainable oil production is increasingly becoming a major concern. Today, it is clear that the survival of humans is only possible if the biological systems used for the production of food and technical raw materials remain diverse and productive. Moreover, protecting the environment and conservation of natural resources should have prominent relevance. Negative human impacts within the production chain need to be replaced with the aim "to meet the needs and aspirations of the present without compromising the ability to meet those of the future" (Anon, 1987). Since it is clear that land available for oilseed or fruit cultivation is limited, but the demand for oil is increasing, oil-processing technology has to be optimized with respect to oil yield and oil quality.

The term "sustainability" is derived from the Latin word *sustinere* which is equivalent to "maintain," "support," or "endure." Today, it describes the principle that consumption should be restricted to that which can be replenished or regenerated. The extensive production of

Breeding Oilseed Crops for Sustainable Production. http://dx.doi.org/10.1016/B978-0-12-801309-0.00019-7

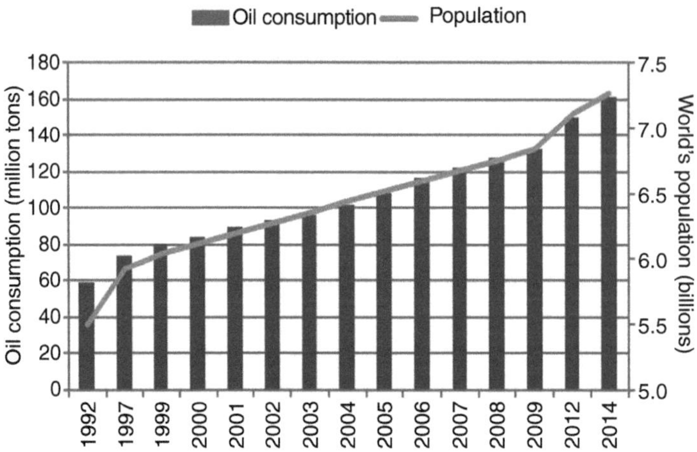

FIGURE 19.1 **Increase of vegetable oil consumption in comparison to world's population from 1992 to 2014.**

vegetable oils necessary to satisfy worldwide needs means that it is not only the cultivation of oil crops that has to be sustainable but so do the following processing steps and marketing of the oil.

Today, modern oil processing is a complex process encompassing a number of steps to increase oil yield, on the one hand, and to improve oil quality after extraction, on the other hand. There are a number of ways to make the whole process more sustainable regarding not only higher output but also environmental aspects. Ways that result in higher sustainability of oil processing include reduction of water and energy necessary for extraction and purification of the oil, reduction of fumes and waste, and improving oil yield or quality of the resulting oil.

However, sustainability in oil processing involves not only increasing efficiency regarding oil yield, but also extracting valuable byproducts from processing, and making them available for further use. Using only oil and extraction meal for further use without isolating valuable byproducts means wasting important resources. In the case of oilseed processing, the aim of sustainable production has to be to partition the whole crop into as many valuable products as possible. Separating oil and solid material is the main step, but – before or after oil extraction or during the refining process – different constituents can be isolated, too.

This concept of a biorefinery entails using various fractions of the oilseed crop to produce various products including fuel, power, heat, food, chemicals, and materials (Dunford, 2012). Examples are valuable tocopherols or phytosterols enriched in the deodistillate during deodorization which can be recovered and purified, solid residue which contains valuable compounds such as proteins or phenolic compounds, or glucosinolates usable as natural pesticides and defense against herbivores in plants.

Another approach to more sustainable oil production is the use of raw materials other than oilseed crops. Certain raw materials contain more oil and their cultivation is faster, easier, and needs fewer limited resources than oilseed crops. Microalgae are an example of such a new source for oil processing. The photosynthetic efficiency of microalgae is often higher than that of terrestrial plants (Pirt, 1986) and, as powerful biochemical factories in miniature format, microalgae sometimes contain more than 80% oil on a dry weight basis (Chisti, 2007). Taking into account the high net annual harvest of microalgae in subtropical areas of 40 tons ha^{-1} (Klass, 1998), clearly microalgae could be an interesting alternative for sustainable oil production to traditional oilseed crops.

OIL PROCESSING

The oldest and easiest way to separate oil from oil-bearing seeds is extraction by hot-water flotation. Ground seed material is placed in boiling water and simmered for several hours. After removal from the heat and cooling, the oil floats to the surface and is skimmed off. In general, the oil is then heated in a shallow pan to drive off the last traces of water (FAO, 1992). Although such an oil extraction process is environmentally friendly in that it does not use organic solvents and requires low-energy input, it is obvious that this method is not up to satisfying the world's huge demand for vegetable oils. The process depends on a heat source that is sustainable from an environmental point of view, oil yield is poor, and the efficiency of the process is very low.

An example of this ancient process is the traditional preparation of argan oil from fruits of the argan tree (*Argania spinosa*) endemic to southwestern Morocco. After removing argan nuts from their shells and gentle roasting of the white kernels, the roasted kernels are crushed using a millstone. Then, the brownish viscous liquid is mixed with water and the dough is hand-malaxed for several minutes, slowly getting solid and releasing an emulsion from which argan oil is finally decanted.

Traditional oil processing carried out in large facilities is divided into a number of steps: pretreatment of seeds or fruits to improve extractability and oil yield, extraction to separate oil from oil-bearing solid material, and finally purification to make the oil suitable for human consumption or other uses and to achieve adequate oxidative and thermal stability. This process is optimized for high oil yield during solvent extraction resulting in maximum amounts of residual oil in the extraction meal between 1% and 2%. The high efficiency of the extraction process makes it possible for oilseed processing to be economical. On the other hand, all over the world small oil mills extract oil from seeds or fruits by using a screw press with sedimentation, filtration, or centrifugation only used for purification purposes. In developed countries, this process is mainly used for the production of high-quality edible niche products such as virgin olive oil, rapeseed oil, or cold-pressed linseed oil with the aim to provide products that are less processed. Such products are expected to contain a high amount of valuable bioactive compounds. They are characterized by the typical taste and smell of the seed or fruit material from which the oil is derived. For example, rapeseed-like and sometimes pungent mustard-like in the case of rapeseed oil, or sunflowerseed-like or fruity in the case of sunflower oil. These virgin or cold-pressed oils belong to the high-price segment of edible oils. The processing itself is easy to handle and equipment-related requirements are comparatively low. However, for the production of high-quality virgin oils the choice of proper raw material is decisive because oil quality is directly affected. Small oil mills are unable to improve oil quality by further processing, while large facilities can compensate for inferiority of the raw material in the refining process. In rural areas of developing countries, this easy-to-handle technique of oil extraction is often the only possibility to produce oil from available raw materials.

Preparation of Seeds for Oil Extraction

Economical oil extraction requires the pretreatment of seeds or fruits before extraction to improve extractability of the oil and to lower the energy input of the process. Both result in more economical processing since oil yield increases and energy costs declines. Therefore, the main objective in preparing the seeds before oil extraction is to have feed material that is easily extractable for the extraction process. Any failure during seed preparation leads to deterioration of oil yield, and it is not possible to correct mistakes by further processing. Another issue that needs consideration is the demand for reduced hexane consumption during extraction and fewer losses during desolventization. For this to happen, seed material is needed that ensures fast and easy penetration of the material by solvents as well as easy, fast, and comprehensive removal of the solvent from the extracted material. To achieve economical and environmental oil processing, comprehensive preparation of the seed material comprises three steps: preconditioning, flaking, and cooking/tempering.

Preconditioning

Depending on climate conditions and raw material, oil processing starts with preconditioning the material at 30–40°C by direct contact of the material with hot air or indirect heating. This makes the seeds softer. Heating avoids shattering during flaking and improves screw-pressing capacity, cake formation, extractability, and hexane recovery from extracted oilseed flakes (Unger, 1990).

Size Reduction

For economical and comprehensive extraction, it is necessary to reduce the size of seeds to a specific thickness and length for better oil recovery. Additionally, the recovery of solvent after solvent extraction is made easier. Size reduction is achieved either by flaking or by extrusion depending on the material; sometimes pelletizing is used.

FLAKING

Flaking has two effects on the extraction of oil. On the one hand, it serves to improve the extractability of the oil by destroying the cellular structure and as many cell walls as possible and, on the other hand, lipid particles can more easily come into contact with each other resulting in larger oil droplets that are easier to extract. This process results in about 80% of membranes being destroyed, which allows unhindered oil drainage from the seed material. Moreover, the accessibility of oil droplets for the extraction solvent will be improved and the contact between solvent and seed material increases. From an economical point of view, flaking leads to the need for reduced thermal energy during cooking, less mechanical energy during screw pressing, and better oil extractability during solvent extraction.

The general design of a flaking mill entails two cast iron rollers with smooth surfaces turning in opposite directions (Fig. 19.2). In general, the size of the rollers is between 0.5 m and 0.8 m in diameter and 0.9–2.0 m long, depending on the throughput target. The capacity of such flaking mills can be between 300 t $day^{-1}$and 800 t day^{-1}. The thickness of the flakes is set by the gap between the rollers. During running, one roller revolves 2–5 rpm faster than the other, with the result that the roller surfaces wipe each other. Distribution of seeds along the whole length of the rollers is achieved by using vibrating feeders that result in even spreading.

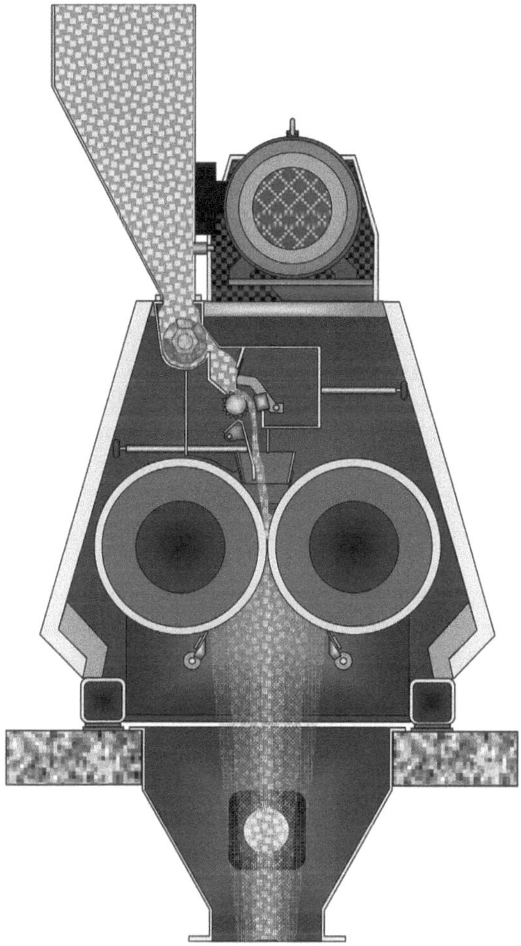

FIGURE 19.2 **Schematic drawing of a flaking mill.** *Source: Courtesy of Bühler AG, Uzwil, Switzerland.*

For big industrial facilities using screw pressing and solvent extraction, the aim of flaking is to achieve a particle size of 0.3–0.4 mm thickness. A higher level of fines should be avoided, because flakes thinner than 0.2 mm are very fragile, while flakes thicker than 0.4 mm result in lower oil yield (Carr, 1995). Fines can also get into the oil, resulting in problems during filtration of the oil. A lower particle size between 0.13 mm and 0.25 mm is sufficient for hydraulic pressing.

Important for the result of flaking and further processing is the temperature and moisture content of the oilseed during flaking. If the seeds are too cold or too dry many fines will be produced because the material is too brittle and the flaking process results more in shattering of the seeds rather than flaking. On the other hand, high moisture content in seeds can result in plastic deformation of the material without flaking. Hence, moisture content has to be adapted according to the seed material.

It is also important to take into consideration that, after flaking, oil processing has to be carried out without any delay. Since the protective seed coat has been destroyed and the cells are disrupted during flaking, the work of appropriate enzymes and attendant microorganisms can start immediately. The result is drastic impairment of flakes and subsequently of oil. Therefore, storage of the flakes before further processing is not advisable.

EXTRUSION

Extrusion is mainly used for pretreatment of oil-bearing materials with low oil content such as soybeans, cottonseeds, or rice brans. The principle behind extrusion is compression of the seed material in an extruder, which results in heat treatment of the particles at temperatures well above the boiling point of water. Directly when the particles leave the extruder, they expand due to the release of compressed water vapor leading to additional destruction of cell walls. Extruded seeds have some advantages over flaked material with respect to increased extractor capacity, better extraction percolation, improved drainage of solvent and reduced consumption of solvent, lower energy consumption, lower residual oil in the meal, and reduced operator attention (Chakrabarty, 2003).

Cooking/Tempering

Before the oil extraction process starts, cooking or tempering the flaked seed material further improves the extractability of oil which increases the efficiency of the process. After flaking, still intact cells are destroyed and oil is released. Additionally, the viscosity of oil droplets is reduced resulting in better drainage of the oil from the seed material and oil–protein emulsions, making separation of the oil more difficult. As a result of high-temperature proteins coagulating, this leads to better permeability of the seed material for floating oil and less affinity of the oil for the solid surface of seeds. A further effect of cooking/tempering is the confluence of smaller oil droplets to larger ones, which also accelerates the drainage of oil from the solid material during extraction. The success of the cooking process and the confluence of oil droplets can be monitored by the oily appearance of the flakes. In addition to better extractability of oil from the solid material during extraction, cooking or tempering enables sterilization of the seed material and destroys undesirable seed constituents, such as gossypol in cotton seed.

Optimal cooking temperatures and durations are in the range between 105°C and 130°C and 30–120 min, respectively, depending on initial moisture content, chemical composition, cooking method and equipment, as well as the extraction method used. During cooking/tempering, the moisture content of the seed material is reduced to a level optimal for subsequent screw pressing which needs moisture contents between 3% and 6%. If the moisture content is too high during screw pressing, the material is slippery resulting in bad buildup of pressure in the screw press. Moreover, during solvent extraction, high moisture content results in improper diffusion of the solvent into the oilseeds and creates nonpercolation problems. Both lead to bad results in oil yield during the extraction process, making the process less economical. For oilseeds with lower oil content, such as soybeans which can only be solvent extracted, cooking/tempering is not necessary. In that case pretreatment of the seeds before extraction is limited to the conditioning and flaking steps.

For oil quality, it is necessary for the seed to pass through the temperature optimum of the enzymes present as fast as possible to avoid degradation of seed ingredients. Examples are lipases which lead to the degradation of triacylglycerols and phospholipids resulting in higher amounts of free fatty acid or myrosinase in rapeseed breaking down glucosinolates to isothiocyanates or nitriles which remain in small amounts in the solid residue after the extraction process and have adverse effects on animals when the meal is used as feed.

Another important point is that temperature and duration of cooking/tempering should not be too high because overcooking results in dark color of the seed material and consequently dark color of the oil and press cake. Hence, the need for more sophisticated refining conditions for the oil.

In general, conditioning of flakes takes place using drum or stack-type cookers. The stack-type cooker is a vertical vessel consisting of several steam-heated floors. The flaked material is placed on the first floor on the top of the cooker and a stirrer mixes the material for the desired time on the floor to ensure that uniform uptake of temperature takes place by the flakes. The temperature run of the plate is adjusted so that the desired temperature is achieved as quickly as possible, to avoid enzymatic degradation reactions in the material. The time the flakes remain on each plate depends on the properties of the flakes before cooking, but in most cases a time between 20 and 40 min is used. After this time, the material drops to the next floor and so on. In this procedure the flakes run progressively through each plate of the cooker. As a result of temperature and time control as well as mixing of the material on each plate, uniform heating is possible. During the process all enzymes are deactivated. The use of hot steam or hot air makes it possible to optimize the final water content of the seeds for the pressing process.

Drum cookers consist of a number of individual horizontal tubes that can be heated. Each tube is equipped with a stirrer to enable equal motion of the flakes in the cooker. Screw conveyers fill the first tube with the material and, depending on the properties of the flakes, a certain number of tubes are used for cooking.

The temperature used for cooking should be higher than 80°C to deactivate enzymes, but at temperatures higher than 100°C there are changes in the aroma and taste of the resulting oil. Additionally, higher cooking temperatures lead to an increase of the amount of phosphorus and free fatty acids in the oil. For example, rise in temperature from 80°C to 100°C results in an increase of phosphorus from 125 mg kg^{-1} to more than 300 mg kg^{-1} and at the same time the amount of free fatty acids increases from 0.5% to about 0.6% (Unger, 1990). To ensure high oil quality at the end of the refining process as a result of using higher contents of phosphorus and free fatty acids, higher costs and higher losses of oil have to be considered. A further aspect of the use of high temperatures during cooking over a longer period of time is the damage to protein components, which can impair the protein quality of the residue.

High Temperature Short Time (HTST) Conditioning

An alternative method of conditioning oilseeds before pressing has been described by Eggers. This is high-temperature short-time (HTST) conditioning (Eggers, 2004). The basis of this method is short preheating of the raw material to temperatures up to 145°C with the aim of enriching valuable minor components in the oil such as phenolic compounds. The material is treated for less than 20 s with heated steam, which is the heat carrier at temperatures between 120°C and 145°C. This results in temperatures higher than 100°C in the seeds. Another effect is the direct deactivation of enzymes.

A number of valuable compounds are released from the cell structure by the hydrolysis effect of steam pretreatment which enrich the oil during pressing. Eggers (2004) found that the effect was stronger for heat-treated press cake compared with oil directly produced by the screw press. The antioxidant effect of the oil increased remarkably due to the higher content of phenolic compounds, while the content of tocopherols was unaffected by the treatment (Eggers, 2008).

Oil Extraction

The objectives of oil extraction are obtaining (i) high-quality oil with minimal undesirable components to ensure low costs of purification and (ii) high-quality cake or meal for animal feed or extraction of valuable proteins.

Today, oilseed and fruit are mainly extracted using a screw press or elution with a solvent; a combination of both processes is also widespread. In general, oil-rich seed materials are partially deoiled by a screw press to reduce the oil content, before using a solvent to wash out as much oil as economically reasonable. Seeds with lower oil content, such as soybeans, are extracted by solvent directly without further prepressing; as a result of the low oil content, solvent extraction is sufficient and more economical. Application of a screw press alone is used for the preparation of specialty oils such as virgin oils, oils produced in rural areas with no access to large oil-producing facilities, and palm oil. The investment cost for a screw press is much lower than for a solvent extraction plant. Moreover, the safety requirements for screw-pressing facilities are comparatively manageable.

Production facilities can vary in size from a few dozen kilograms per day up to several thousand tons. Depending on processing capacity the area covered by marketing ranges from a small region within a country to one or a few countries; a handful of companies sell oil worldwide. Very small companies supplying oil only to a limited regional market are often connected with a farm. They process their own raw material and produce cold-pressed oils which they sell in small markets either because this is the only way of processing oil in the region or because they can skim off the added value of a high-price specialty product. Depending on the quality of the raw material used and the extent of pressing, these cold-pressed oils have higher oil quality than the crude oil from solvent extraction, but the efficiency of the process is significantly poorer than that of conventional extraction.

Pressing

Pressing oil-bearing materials for the production of oil is a very ancient process (known for more than 2000 years). Only the tools used for pressing the material have changed over the years. Originally, simple wooden screw presses with small capacity and low efficiency were used (Fig. 19.3); today, metallic presses are optimized for high throughput and concomitant high efficiency. The first successful continuously working screw press (Expeller® press) was developed in 1900 by Valerius D. Anderson (Fig. 19.4).

Two approaches are used for the extraction of oil by a screw press. The first is to extract as much oil from the oilseed as possible under the effect of high pressure in a single pressing resulting in 5–7% residual oil in the press cake. Although costs per ton seed for full pressing are higher than for solvent extraction, this method has some advantages if the oilseed has high oil content and it is not problematic to leave a higher amount of oil in the press cake. Investment and maintenance costs are lower and oil quality is higher than those for solvent extraction (Niewiadomski, 1990). Thus, in areas with a poor infrastructure

FIGURE 19.3 **Wooden screw press.** *Source: Courtesy of the Ethnographic Museum, Spittal an der Drau, Austria.*

FIGURE 19.4 **Anderson No. 1 Oil Expeller®.** *Source: Courtesy of Anderson International Corp, Stow, OH, USA.*

FIGURE 19.5 **Screw press without barrel cage to show worm sections.**

the application of full pressing is not only an interesting option, it is advisable for a flexible production line of different products. It is much easier to switch between different raw materials in a short period of time since cleaning of the screw press is faster than cleaning combined extraction machinery in big facilities. In this way it is possible to produce small amounts of high-price products that meet the increasing demand of consumers for food that is less processed. More environmentally friendly production, minus the use of solvents, short routes of transport, and economical activities in regional material cycles make these products especially acceptable to consumers. Thus, these plants are able to supply a manageable and quantitatively limited market with high-quality pressed vegetable oils.

As a result of high pressure within the screw press (up to 1600 bar in large plants during full pressing) and frictional resistance of the seed material, temperatures up to 170°C can occur, which can adversely affect the quality of the oil and press cake.

The second approach involves partial deoiling of the oilseed or fruit under moderate pressure, which leaves about 15–20% oil in the press cake before the residual oil is comprehensively extracted by solvent. For seeds high in oil content, such as rapeseed or sunflower seed, combined extraction is the method of choice because direct solvent extraction would be both ineffective and uneconomical. To achieve the same high oil yield using direct solvent extraction of the oilseed, the amount of solvent and the extraction time needed would increase dramatically.

From an economical point of view, the combination of screw pressing and subsequent solvent extraction has the advantages of lower costs than using a screw press and higher oil yield than using solvent extraction.

The general construction is very similar for most screw presses on the market regardless of the manufacturer and envisaged capacity. The principle is similar to that of a meat mincer with a rotating screw shaft within a horizontal barrel (Fig. 19.5). The main difference is that the screw shaft transports the raw material through the screw flight in which volume decreases from the beginning to the end because the radius of the screw flight increases. The further the material passes through the screw flight the higher the pressure on the material, since the volume of the screw increases as the diameter of the screw flight decreases in the direction of seed flow. This results in less space for and increasing pressure on the material and final release of

FIGURE 19.6 **Barrel of a screw press surrounded by flat bars during pressing of rapeseed.**

oil from the seeds. It is possible to adapt the slope of the rotating screw to specific needs for oilseeds to improve the efficiency of the press.

The extraction process can achieve better results by changing the space between the bars that surround the barrel of the screw press (Fig. 19.6). These bars form a sort of cave around the screw allowing the oil to be expelled while retaining the solid material. Reducing the space between the bars leads to higher pressure within the barrel. At the end of the screw flight the seed material has to pass through a small press cake exit the size of which strongly influences pressure within the barrel. By adjusting the spaces between the bars and the size of the press cake exit, it is possible to optimize the pressing process regarding oil yield, energy consumption, and oil quality. Moreover, as a result of further development of screw presses by manufacturers, energy consumption has been significantly reduced in recent decades making the process more economical from the same throughput.

Today, the trend is increasingly toward very large facilities that are able to process 2000 tons day^{-1} or more, which makes processing more economical as a result of high throughput and little difference in investment compared with smaller facilities. Screw presses with capacities of 200–500 tons day^{-1} are typical of these large facilities.

Cold Pressing

For some years now the production of cold-pressed oils has been growing in popularity. The consumer connects with this type of oil healthier food because of reduced processing. According to the Codex Alimentarius Standard for Named and Vegetable Oils, cold-pressed oils have to be obtained only, without altering the oil, by mechanical procedures (e.g., expelling or pressing) without the application of heat. They can only be purified by washing with water, settling, filtering, and centrifuging (Anon, 2001). Hence, extensive extraction of the oilseed is not permissible because it results in frictional heat. The composition and properties of the oil are not allowed to be altered by the processing conditions.

In general, cold pressing is a very simple process similar to the abovementioned screw pressing of oil-bearing material in which the resulting oil undergoes nothing more than filtration or sedimentation. The only difference from screw pressing in large facilities is that the seeds are normally not pretreated before pressing. There are a few seeds such as linseeds in which the raw material is preheated at 40–60°C for better release of oil, but this should have no effect on the composition and properties of the oil.

Depending on the type of press and the settings, about 10–15% oil remains in the press cake, making the process not very efficient. On the other hand, cold pressing is environmentally friendly because energy is only needed for the screw press and to pump the oil through the chamber filter.

Since no further purification of the oil other than filtration or sedimentation is envisaged, very gentle processing is recommended. Extensive extraction results in the release of more free fatty acids, chlorophyll, and phosphorus into the oil. These components influence stability negatively. However, phosphorus is an important parameter for usability of the oil as fuel.

Among the most important features of cold-pressed oils are the typical taste and smell of oils according to the raw material used. Typical attributes are seed-like and nutty for virgin rapeseed oils; fruity, bitter, and pungent for virgin extra olive oil; sunflower seed–like, nutty, and sometimes fruity for virgin sunflower oil; or vinous and fruity, reminiscent of raisins for virgin grape seed oil. The use of improper raw material can result in attributes such as rancid, straw-like, wood-like, fried, burnt, fusty, musty, bitter, and astringent, which are typical off-flavors. These attributes result from the formation of volatile degradation products or the development of aroma-active compounds during faulty processing.

Solvent Extraction

For comprehensive deoiling of seed material, extraction by solvents is necessary. In this way it is possible to reduce the amount of oil in the residue to less than 0.5% depending on the number of extraction cycles, extraction time, and extent to which seed material is prepared before extraction. Complete extraction of oil from seed material is only limited by economical aspects. The more oil has to be extracted from the material the greater the expense.

Oilseed extraction by means of nonpolar solvents involves solid–liquid extraction, in which diffusion is the main driving and limiting factor for transfer of oil from the oil-bearing material to the solvent: (i) diffusion of solvent into the solid material, (ii) dissolution of oil droplets into the solvent, and (iii) diffusion of oil from solid particles to the liquid. For comprehensive and fast extraction of oilseed, it is important that there is a high gradient of oil concentration in the desired direction from the seed material to the oil–solvent mixture (miscella). Extraction capacity of the miscella is reached when the concentration of oil in free miscella is equal to the concentration of oil within the pores of the solid. To optimize this, it is necessary to consider the time necessary for oil within the particles to diffuse out into the solvent. Here each facility has to strike a balance between time necessary for extraction and oil yield necessary for economical work. The highest amount of oil is extracted within the first minutes of the process (75% in the first 25 min), whereas total oil yield depends on a number of further extraction steps and total extraction time. Extraction of the remaining 25% oil from the press cake takes disproportionately more time.

Although different solvents have been evaluated for the extraction of oil from oilseeds, such as ethane, propane, carbon dioxide, hydrogen sulfide, or trichloroethylene, the solvent of choice for solvent extraction is hexane. It is cheap, has good oil solubility even at lower temperatures, the boiling point of 70°C is not too high (which ensures easy and almost complete removal of solvent from the residue with low energy input and without impairment of the raw oil), it is inert to metal from the extraction apparatus and the extracted oil, is stable under process conditions, and is not mixable with water. Additionally, it shows high solubility of the oil at elevated temperatures and low solubility at ambient temperature making the recovery of oil and meal easier. Hence, the physical properties of hexane – such as specific heat, latent heat of vaporization, boiling point, oil solubility, viscosity, specific gravity, and polarity – favor this solvent for oil extraction. Solvents that are more polar have the disadvantage that they are partly absorbed by proteins, making it difficult to remove them after extraction.

Nevertheless, hexane has some disadvantages. Safety requirements for the use of this solvent are high due to its easy flammability and the risk of mixtures of hexane and air exploding. A number of requirements apply to the use of hexane, such as the use of explosion-proof electrical equipment, avoidance of hexane accumulation at deeper locations (since hexane is heavier than air), and avoidance of hexane coming into contact with hot components of the extraction apparatus.

For successful extraction, intensive washing of the oil-bearing press cake with solvent is necessary. The extraction time should be as brief as necessary and extraction results as good as possible. To achieve this, the solvent should be heated. This improves the solubility of oil in hexane, increases the diffusion of oil, and reduces the viscosity of the solvent, but at the same time results in a higher extraction rate of minor components such as phospholipids, chlorophyll, free fatty acids, color pigments, and degradation products of oxidative reactions the removal of which during refining is costly. Kock (1981) showed that an increase in solvent temperature from 40°C to 58°C results in increasing phospholipids in the oil from 0.2% to 0.8%. On the other hand, higher operating temperatures result in lower energy requirements for the extraction process as a result of faster extraction and better extraction results. Additionally, lower extraction temperatures result in higher content of air and oxygen within the extractor, which leads to corrosion of the extractor material and as a consequence repairs to the apparatus.

The investment for a solvent extraction plant is much higher than for a screw-pressing plant because the requirements concerning the construction and safety aspects are much more complex. A solvent extraction plant not only comprises extraction apparatus but also needs a facility for the removal of solvents from the meal including toasting, drying, and cooling as well as a plant for distillation of the solvent–oil mixture (miscella) and solvent recovery (Fig. 19.7).

In general, two types of solvent extractor operations are possible (i.e., batch vs. continuous). When solvent extraction of oilseeds began in the 1800s, batch-type extractors were very common, but today this method is only used for specialty and high-value oils or when oil is reextracted from bleaching earth. The construction of these batch extractors is very easy since they comprise little more than either a rotating drum or a cylindrical kettle in which crushed and pretreated oilseed is agitated together with the solvent by a low-speed agitator. After separating the solid material, the solvent is evaporated. This process is slow and time-consuming, and the extraction rate of oil from the oilseed is only moderate as a result of the oil extraction in one-step saturating the solvent very quickly .

For continuous solvent extraction from oil-bearing material, the solvent can come into contact with the material either by percolating through a bed of the material or by conveying material along with the solvent. Moreover, a combination of both systems is possible. The technique used by percolation-type extractors is comparable with oil extraction using the Twisselmann technique, in which the solvent drops from the cooler onto a bed consisting of

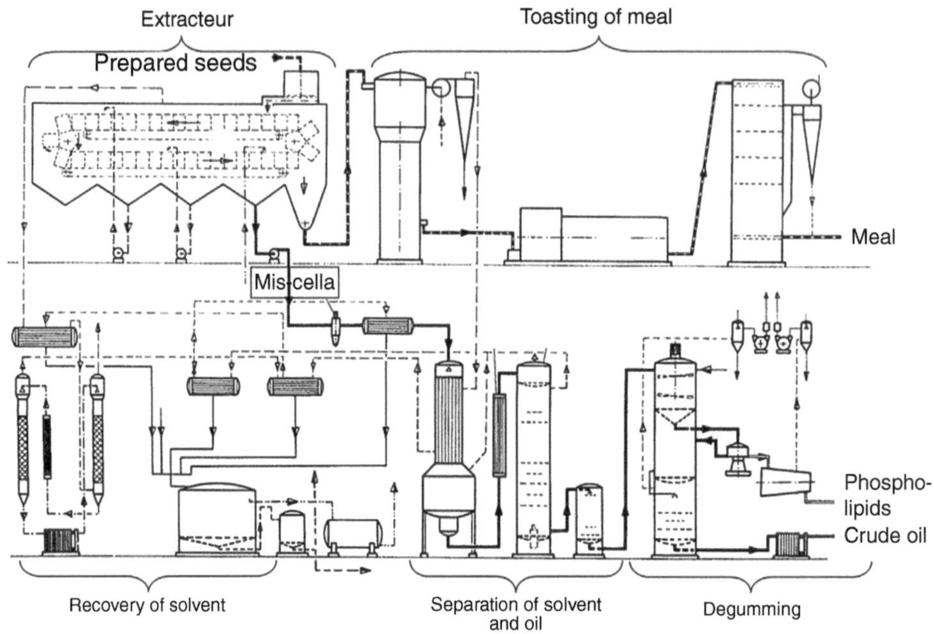

FIGURE 19.7 **Extraction plant (Fa. Lurgi).** *Source: Reprinted from Gupta (2007), p. 501 with permission from Elsevier.*

oil-bearing material, drains through the bed while the oil is dissolved in the solvent, and is then transported together with the solvent into a flask, before the solvent is evaporated again. The only difference is that the number of extraction cycles is determined by the number of baskets used by the percolation-type extractor and by the time taken to operate the Twisselmann apparatus. The principle behind the immersion-type extractor is similar to Soxhlet extraction, in which solid material is immersed in the solvent before the solvent–oil mixture is sucked into the flask.

For immersion-type extraction, the particle size of the solid material is decisive because there needs to be the greatest possible contact between the solvent and the material. Therefore, particle size should be as small as possible; nevertheless, particles that are too small result in difficulties when separating the miscella from the solid material. Hence, immersion-type extractors are mainly used in small and medium-sized plants while large facilities prefer percolation-type extractors that are more efficient regarding energy consumption and space requirements. It is necessary here for flakes to have an appropriate size to ensure a reasonable flow rate of the solvent through the bed of seed material, which enables comprehensive deoiling of the material. From this it can be seen that flaking and subsequent prepressing are important factors when considering the efforts and costs necessary for further solvent extraction. For an oil extraction process that is sustainable, all steps must be attuned and optimized for final solvent extraction. Small particles during percolation-type extraction worsen the surface/volume ratio resulting in higher residual oil content in the press cake (Unger, 1990) and high levels of solvent in the meal. Both lead to more extraction steps, higher solvent requirement, and more energy for solvent removal from the press cake.

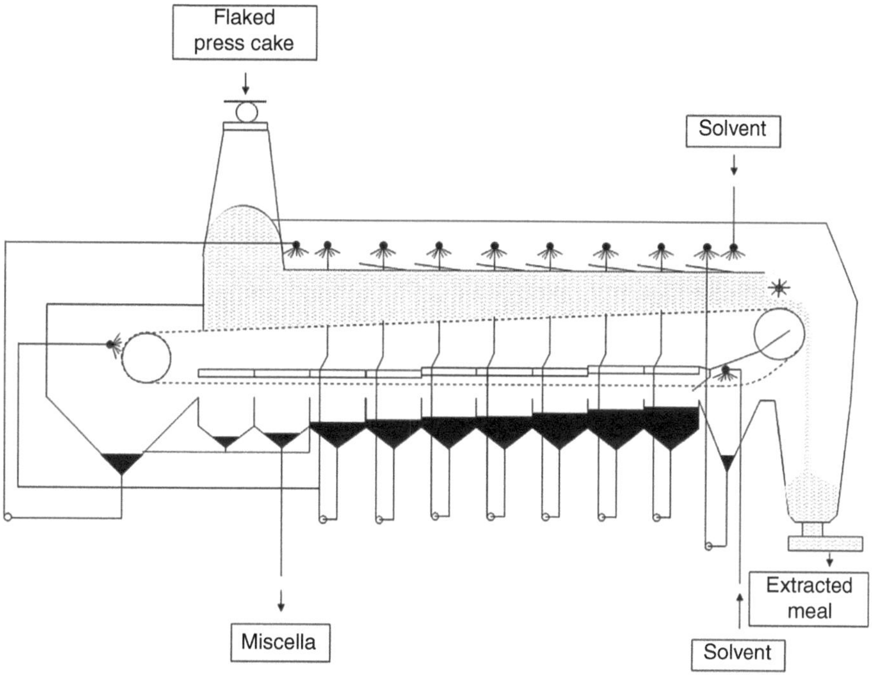

FIGURE 19.8 **Solvent extractor.** *Source: Reprinted from Gupta (2007), p. 500 with permission from Elsevier.*

For solvent extraction by percolation, a number of different extractors are on the market including rotary, shallow bed or deep bed, horizontal belt, and continuous loop. All of these extractors are countercurrent extractors (Fig. 19.8). The main difference between them is the way they transport solid material through the extractor. The design of different extractors is very similar and comprises a different number of extraction steps during which the oil-bearing material loses stepwise as much oil as possible. The number of extraction steps depends on the performance of the extractor and the arrangement of the baskets (most often 7–10 successive wash stages are necessary).

The basic element of most percolation extractors is the number of percolation units that contain oil-bearing material (flakes or press cake). Within the percolation unit, the material forms a fixed bed, the solvent is pumped or sprayed over the material and then drained through the material, dissolving some of the oil. The solvent–oil mixture is collected at the bottom and used for the next extraction step before reaching a final oil-in-solvent concentration of about 20–30%. At the beginning of the extraction process, the material percolates through a miscella that is nearly oil-saturated, while with each further extraction step the concentration of oil in the solvent decreases and in the final step before the meal leaves the extractor pure solvent is used.

In addition to size of the oil-bearing material, depth of the cake bed is crucial to outcomes of the extraction process. Depending on the type of extractor, depth of the cake can vary from 20–50 cm, and extractors today offer the possibility of modifying the height of the cake bed so that the different percolation parameters of different materials can be considered.

The first percolation extractors were vertical basket-type extractors (called "Bollman extractors") which had a number of baskets filled with flaked seed material by an automatic feed hopper at the top. The baskets of this extractor were directed by endless chains like gondolas in a big wheel in a sealed housing, and descending baskets were sprayed with solvent. At the end of the extraction cycle, deoiled material was discharged into a hopper. This type of extractor was problematical in that it was difficult to maintain.

Today the design of percolation extractors is horizontal. In general, three different types of extractors are in use: (i) the Crown or vertical loop extractor with a bed depth of 0.3–1 m, (ii) belt-type extractor with a bed depth of 0.9–1.8 m, and (iii) round extractors with a bed depth of 1.6–2.9 m. Which type of extractor to choose depends on different factors such as space available, target throughput, or company philosophy.

A very fast extraction technique that has only two or three extraction steps has been described by Beach (1983). He mentioned that extensive extraction with fewer steps is possible if the cake is saturated by solvent allowing it to soak over a sufficient time followed by flushing the enriched miscella from the surface of particles and replacing it with a leaner miscella.

Modified Oil Recovery Techniques

ENZYMATIC PRETREATMENT

Enzymatic pretreatment of raw materials to enhance oil recovery from seeds or fruits is increasingly becoming the focus of interest due to improving the oil recovery process regarding oil yield and sustainability. The aim of this technique is to modify the seed coat and cell walls for faster and easier release of oil under milder processing conditions (Fig. 19.9). In this way it is possible to save energy during the extraction process, to shorten the extraction time, to have fewer losses of hexane when using solvent extraction, and to improve the quality of the final product regarding such parameters as content of free fatty acids and pigments as well as oxidative stability. Enzymatic pretreatment also results in higher oil yield and protection of susceptible and valuable substances as a result of the mild conditions (H. Olsen, 1988). Thus, enzymatic pretreatment is considered to be an environmentally friendly technique that reduces energy consumption during oil processing. In addition to enhanced release of oil during extraction, hexane release from the meal during desolventizing is improved. In the case of rapeseed, less temperature is necessary to remove the solvent after extraction which results in reduced heat treatment of the meal, less strenuous treatment of the meal, and better availability of nutrients.

For the production of specialty oils, enzymatic pretreatment also allows more comprehensive extraction of oil without use of solvents, which makes the process more economical. One example is the application of enzymatic pretreatment on an industrial scale to process extra virgin olive oil to increase oil yield (Christensen, 1991). This method uses oil-containing material that has been prepared with an enzyme that degrades cell walls to liberate oil in the plant cell under very mild processing conditions. This facilitates separation of cell components – such as protein, oil, polysaccharides – by centrifuge processes. When no solvent is used for the recovery of oil, enzymatic pretreatment also enables simultaneous recovery of good-quality oil and protein from oilseeds (Rosenthal et al., 1996).

The enzymatic process is based on hydrolysis of structural polysaccharides forming cell walls of oilseeds or fruits and of proteins that form cell and lipid body membranes (Rosenthal

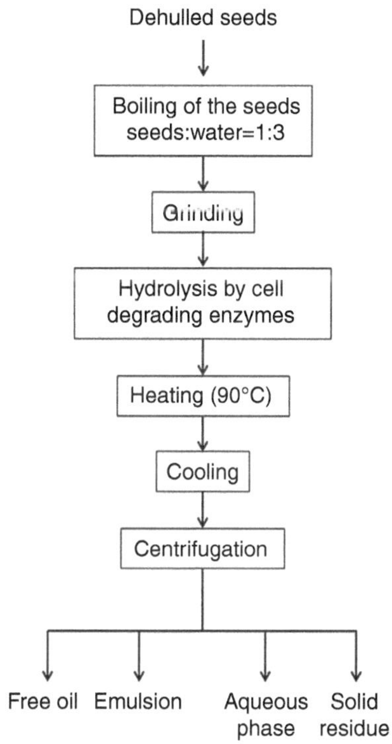

FIGURE 19.9 **Enzymatic pretreatment of oilseeds for oil and protein processing.**

et al., 1996). The main factors that affect the hydrolytic process are particle size, moisture, hydrolysis time, and enzyme/substrate mass ratio (Couri and Freitas, 2001).

For enzymatic pretreatment of seeds or fruits, a broad spectrum of enzymatic activities is necessary because the cell walls feature a complex structure (Christensen, 1991). Within the cell the oil droplets are surrounded by several barriers that make the use of different types of enzymes necessary for the release of oil. Dominguez et al. (1994) showed that cell wall polysaccharides of rapeseed consist of 39% pectins, 29% hemicelluloses, 22% celluloses, and arabinogalactans which amount to 20–28% of total seeds.

Major challenge of this process is knowing what type of enzyme is necessary for the oilseed in question and what are the optimal conditions necessary to obtaining optimal destruction of cell walls and optimal release of oil. Ricochon and Muniglia (2010) give a nice overview of different cell wall constituents and specific enzymes necessary to destroy the cell walls of oilseeds. Enzymes for the degradation of cell wall material mainly include different types of carbohydrolases – such as amylase, β-glucanase, pectinase, hemicellulase –mixtures of these enzymes are also in use. Sosulski et al. (1988) showed the effectiveness of these enzymes in the following order: mixed enzyme > β-glucanase > pectinase > hemicellulase > cellulase. As the number of effective enzymes in a mixture increases, the destruction of cell walls becomes quicker and more complete. On the other hand, it is not possible to define an optimal mixture of enzymes because the composition of cell walls is different for different oilseed species and varieties.

Depending on the optimal requirements of the enzymes used, temperature and pH value strongly influence the result of enzymatic pretreatment of seeds or fruits. The temperature and pH value of the system should be in such a range that allows optimal activity of enzymes to get the best possible result from enzymatic pretreatment in as short a time as possible with less degradation of valuable substances. Often temperature programs are in use, especially when different types of enzymes are needed to destroy cell walls, all of which have different temperature optimums. For rapeseed and soybeans, Fullbrook (1983) suggested 60 min at 50°C and 120 min at 63°C for comprehensive treatment of cell walls before inactivation of the enzymes at 80°C for 13 min. Moreover, a stepwise increase in temperature (40, 50, and 65°C) over a longer period of time (3 h) has been published (Lanzani et al., 1975).

Other important factors for enzymatic pretreatment are number of enzymes, enzyme/substrate ratio, and hydrolysis time. All these factors strongly depend on the type of oilseed, thus general recommendations are not possible. However, in general it can be said that the higher the number of enzymes and the longer the time of hydrolysis the better the destruction of cell walls, although both parameters reach an optimal value at which oil yield does not increase further (Zhang et al., 2007). On the other hand, it has be taken into consideration that different factors also have an effect on oil and protein quality and on the formation of emulsions obtained after treatment, which strongly affect separation of the different phases. Therefore, it is important to optimize enzymatic pretreatment regarding oil yield, on the one hand, and costs for separation of the oil and oil quality, on the other hand.

Enzymatic oil extraction involves taking into account the activity of endogenous enzymes which can result in degradation of seed ingredients and deterioration of oil quality. Therefore, it is necessary to inactivate such undesirable enzymes at the outset to optimize the conditions for desirable enzymes. Examples of such enzymes are myrosinases, present in rapeseed or mustard seeds and responsible for the release of isothiocyanates or nitriles from glucosinolates, or lipases which degrade triacylglycerols. Van Eylen et al. (2006) showed that, for thermal inactivation of myrosinase from mustard seeds in intact tissue, temperatures of 72.5°C for 30 min or 75°C for 10 min would be sufficient. For crushed seeds the time of treatment can be lower.

Different investigations have shown that the throughput of screw presses and oil yield increase as a result of enzymatic pretreatment before pressing. Sosulski and Sosulski (1993) showed that material throughput for enzyme-treated seeds increased by 30–50% over screw-pressed material without enzyme treatment, depending on the canola variety, while oil recovery increased from 72% for control samples to 90–93% for enzyme-treated samples. They also found that residual oil content decreases from 14% to 18% in control samples to 6.4–8.0% in press cakes from enzyme-treated seeds. The quality of oil from enzyme-treated seeds was worse than the quality of cold-pressed oil but much better than solvent-extracted oil regarding the content of free fatty acids (0.61 vs. 0.4%) and peroxide value (0.4 vs. 0.29 nM kg^{-1}). According to Bocevska et al. (1993) the quality of corn germ oil extracted by aqueous enzymatic pretreatment with 1.5% free fatty acids revealed a peroxide value of 0.0 mmol kg^{-1}, anisidine value 8.1, 1350 mg kg^{-1} tocopherols, and oxidative stability of 14.6 h at 100°C. This was better than for Expeller crude oil or degummed oil especially regarding total oxidation (TOTOX) value and total content of tocopherols. Sosulski and Sosulski (1988) showed that enzymatic pretreatment of rapeseed with cellulase, protease, β-glucanase, and two mixed activity enzymes improved quality regarding digestibility and available lysine content.

AQUEOUS ENZYMATIC OIL EXTRACTION

The idea behind extracting oilseeds by means of an aqueous extraction process first arose in the 1950s, with the aim of making oil extraction cheaper and less dangerous. Moreover, aqueous enzymatic oil extraction is a promising green alternative to hexane extraction that avoids environmental emission of hexane. In the United States, solvent emission in oilseeds has been identified as a significant source of air pollution justifying the search for alternative, more environmentally friendly techniques. Thus, the advantages of aqueous enzymatic oil extraction are less capital investment due to fewer safety requirements, environmentally friendly process conditions, and simultaneous extraction of oil and protein of higher quality. Despite its obvious advantages regarding environmental process conditions and gentle processing, this type of process is only economically applicable today for the production of specialty oils in which high quality gives an added value.

The object of the process is the destruction of cell walls by enzymatic pretreatment using water, followed by centrifugation of the slurry and separation into oil, emulsion, aqueous, and solid phases. The water phase dissolves degraded cellular material and facilitates oil entering the liquid phase. The process also allows separation of higher quality proteins in the aqueous phase or the solid phase depending on treatment conditions (Cater et al., 1974). In this way it is possible to release the oil from the cells and separate it from the solid material without using organic solvents. The different steps of aqueous enzymatic oil extraction are (i) grinding the seed or fruit to reduce the size of the material making it better accessible to enzymes, (ii) treatment with enzymes and extraction, (iii) separation of liquids and solids, (iv) separation of the oil-enriched phase, (v) deemulsification to obtain more oil, and (vi) drying to remove moisture from the oil.

A major disadvantage of aqueous enzymatic oil extraction is low oil yield as a result of formation of an emulsion. This creamy emulsion amounts to almost two thirds of the oil content extracted. For oil recovery from extruded soybean flakes, Lamsal and Johnson (2007) showed that the total amount of oil consisted of three fractions: free oil (16%), skim (13%), and cream (60%). One reason for formation of the strong emulsion is the presence of large protein molecules. Different investigations have shown that the use of proteases for the degradation of large protein molecules can be a promising way to reduce or avoid formation of emulsions (H. A. S. Olsen, 1981). Complete release of the oil from the emulsion is not possible, resulting in remarkable loss of oil which can amount to approximately 10–15% of the oil separated from the solid material. Thus, the real challenge to make aqueous enzymatic oil extraction more economical is deemulsification of the cream to obtain free oil that can easily be separated.

Similar to enzymatic pretreatment before solvent extraction, such factors as solid/liquid ratio, extraction time, temperature, pH, and agitation rate also strongly influence the result of aqueous enzymatic oil extraction (Campbell and Glatz, 2009; Nikiforidis and Kiosseoglou, 2009; Zhang et al., 2007).

Research has shown that oil composition is very different from that obtained by solvent extraction, but in general the quality is not worse. For oil extraction of three olive varieties, Ranalli et al. (1998) concluded that enzymatic treatment resulted in (i) higher contents of total phenols, o-diphenols, tyrosol and hydroxytyrosol-aglycones, tocopherols, trans-2-hexenal and other pleasant aromatic volatile compounds, and triterpene alcohols; (ii) higher values of oxidative stability and carotenoid color index; (iii) higher ratios of 1,2-diglycerols/1,3-diglycerols, campesterol/stigmasterol, trans-2-hexeal/hexanal, and trans-2-hexenal/total aroma; (iv) lower values of turbidity; (v) similar values of acidity, peroxide value, carbonyl value, UV (ultraviolet) spectrophotometric indices and alcoholic index; (vi) similar contents of saturated fatty acids, unsaturated fatty acids, triglycerols, diglycerols, sterols, and green lipochromes; and (vii) similar sensory score. No stigmastadienes or trans-isomeres were detected which would indicate the use of heat during processing.

EXTRACTION BY SUPERCRITICAL CARBON DIOXIDE

A very gentle technique for the extraction of oil-bearing seeds or fruits is the application of carbon dioxide under supercritical conditions. This technique was first investigated in the early 1980s as an alternative to conventional solvent extraction of oil-bearing oilseeds and fruits (Friedrich et al., 1982). Similar to conventional extraction by hexane, supercritical carbon dioxide washes out the oil from the solid material. Under supercritical conditions with temperature >31.1°C and pressure >73 atm the physical properties of carbon dioxide are neither those of a liquid nor those of a gas, with higher diffusivity than a liquid. This leads to faster penetration of the solid material by supercritical fluids than in a liquid resulting in higher mass transfer and faster extraction. The extraction properties of carbon dioxide are very similar to those of hexane.

The main advantages of this technique are that carbon dioxide is not toxic, is readily available, is easy to remove from the oil and the meal, and is nonflammable. Treatment of the material is very gentle since the extraction temperature can be very low. The gentle process conditions minimize deteriorations in the product during extraction and the amount of minor compounds coextracted with the oil from the seeds is reduced since carbon dioxide is highly selective for triacylglycerols.

Although the extraction method has some obvious advantages, the technique is not yet in widespread use and there is no commercial application for commodity oils. The reasons for this are the relative high costs of high-pressure equipment and technology necessary for the application of carbon dioxide and difficulties in using the method for continuous extraction. Thus, extraction of oil-bearing seeds and fruits is limited to specialty products produced in low amounts and sold at a high price, when application of this technique is justified. Examples are the extraction of blackcurrant seed oil, borage seed oil, or evening primrose seed oil for application in cosmetics, where even traces of solvents are undesired.

Recently, screw pressing combined with the supercritical carbon dioxide technique have shown promising results in laboratory-scale application (Voges et al., 2007) and in industrial -scale application (Müller and Eggers, 2014). During pressing, the gas is forced into contact with the oilseed to reduce the level of residual oil. The carbon dioxide technique, in particular, showed good results regarding oil yield. For rapeseed, a considerable increase in oil yield from 27% to 71% was found (Voges et al., 2007) both for laboratory-scale and industrial-scale application. Müller and Eggers (2014) found that residual oil levels were reduced to 7.7 wt% by gas assistance at an average CO_2 pressure of 12.5 MPa compared with 9.9 wt% achieved by conventional pressing with the same process. The authors indicated that oil quality was improved and the costs of investment would be recouped within 3 years.

The idea behind this process is to inject supercritical carbon dioxide into the modified cage of a screw press to lower the viscosity of the oil, which leads to easier drainage of the oil from the material. A consequence of this is the liquid–volume effect, which can be imagined as a form of displacement within the screw press when the oil and gas compete for the limited space as a result of massive use of carbon dioxide. Müller and Eggers (2014) described this effect as comparable to oil spilling out of a container.

The high-pressure liquid extraction (HIPLEX) system patented by Harburg-Freudenberger Maschinenbau GmbH is a commercially available system that uses the described technique (Homann et al., 2006). The system has been divided into a prepressing zone, a gas-tight extruder zone, and a final pressing zone.

REMOVAL OF THE SOLVENT

Removal of Solvent from the Oil

The concentration of oil in the miscella as a result of solvent extraction is between 20–30%. In addition to the oil, the miscella comprises hexane and fine particles of seed material which were formed mainly as result of screw pressing. However, transportation of the seed material in baskets during solvent extraction also results in fine particles. While direct solvent extraction without prepressing results in comparably higher numbers of particles, the amount is much lower if prepressing is used before solvent extraction.

For further oil processing, it is necessary to remove solvents and fine particles because they interfere with the refining process. For economical solvent removal, it is important to obtain as much oil from the miscella as possible, consume as little energy as possible, and simultaneously recover as much hexane as possible.

Prior to solvent distillation, particles have to be removed because they can influence heat transfer to the solvent resulting in higher energy demand during distillation. The recovery of solvent from the miscella is achieved by column distillation using an additional stripping column to remove residual solvent and water from the oil. In general, a two or three-stage process is carried out. In the first stage, hexane and steam vapour from the desolventizer–toaster, used to remove hexane from the meal, are applied as the heat source. During this stage the mixture reaches temperatures of about 50°C. In the second stage, the resulting oil-enriched liquid is treated in a steam-heated exchanger, where the concentration of hexane decreases to about 5% of the miscella mass at a temperature of about 80°C. The solvent is recovered by condensation and then used for solvent extraction again. In the third stage, hexane is reduced to less than 800 ppm by passing the oil through an oil stripper tower under a pressure of 13.3 kPa and temperature of about 100°C. High temperatures during solvent removal deeply affect the quality of the oil. Therefore, the miscella is passed through an oil stripper tower under low pressure with steam injected from the bottom to increase the effectiveness of solvent removal. The aim of oil stripping is to reduce the concentration of hexane to an amount necessary for a flash point higher than 250°C. From an economical and environmental point of view, the loss of hexane during solvent recovery should be as low as possible. For a well-operated extraction plant, it is possible to maintain hexane losses during the process lower than 3.0 L per metric ton (Unger, 1990).

Removal of Solvent from the Meal

The meal retains 25–35% of the solvent after extraction, which has to be recovered for economical, environmental, and safety reasons. In addition to regulations for the use of meal in animal feedstuff, limits for hexane are also regulated. Solvent is recovered from the meal in a meal desolventizer or toaster by heating. The main aim of the toasting process is to recover as much solvent from the meal as possible, which can be fed back into the extraction process. Many other effects can be observed during heating: (i) drying and crisping of the meal, (ii) degradation of antinutritive compounds, such as glucosinolates in rapeseed meal or protease inhibitors in soybean meal (trypsin and chymotrypsin inhibitors), and the inactivation of enzymes such as myrosinase (rapeseed) or urease (soybeans), and (iii) improving the digestibility of proteins.

During desolventizing, it is important to control temperature, moisture, and retention time to avoid degradation of valuable amino acids, which would lower the nutritional value of the meal. On the other hand, temperature and retention time need to be adequate to degrade unwanted substances such as glucosinolates or protease inhibitors.

The most widely used of today's desolventizer–toasters consist of an enclosed vertical vessel equipped with several steam-heated horizontal floors. After extraction, the solvent-containing meal is transported to the first floor by a screw conveyer. A rotating sweep arm ensures uniform distribution of the material on the floor, provides uniform heat transfer into the material by mixing, and transports the cake down to the next floor. The cake enters the apparatus at about 60°C and is heated during the process to 105°C. The efficiency of the process can be improved by direct steam injection. This ensures the recovery of hexane absorbed by proteins and trapped within the cell walls. Steam also acts as a carrier to transport the solvent through the bed of extracted cake.

At the end of the process, meal moisture is about 15–18%. For storage and delivery to feed manufacturers, it is necessary to dry the meal to a moisture content of 8–10%. Afterward, the meal is cooled down and milled.

Alternative Desolventizer

Heat treatment of the meal during desolventizing can lead to deterioration of the protein quality of the meal. In the production of protein for human consumption, the quality and technofunctional properties of protein are critical to further use. Meal is often used for animal nutrition with a relatively low added value. Increasing demand for plant protein for human nutrition has aroused great interest in the use of protein from residual meal of oil processing . Protein from soybeans is here the focus of interest; however, rapeseed comprises a very valuable protein that has an interesting amino acid composition.

Flash Desolventizer

A flash desolventizer can be used to desolventize soybean meal. It allows high heat transfer into the material *via* thin flakes using superheated hexane at about 85°C. The result is rapid solvent removal while subjecting solids to elevated temperature for only a few seconds, thus keeping damage to heat-sensitive proteins to a minimum. The final desolventized meal is light-colored, has a protein dispersibility index (PDI) ranging from 10 to 85, and contains a minimum of fines. Due to the absence of noncondensable gas in the system, solvent loss is negligible (Becker, 1983).

Very fast heat and mass transfer from superheated hexane to the meal and from the meal to superheated hexane is partly due to the very small size of the flaked meal, which results in an extensive surface. Vavlitis and Milligan (1993) described the active surface in a flash desolventizer as being about 6700 m^2 for flakes with a thickness of 0.23 mm, but the active surface is drastically reduced if the flakes are only a little thicker. Spherical particles led to a very low active surface of only 1500 m^2, resulting in poor removal of the solvent. Since hexane-wetted rapeseed meal flakes have a more spherical shape than that of flat blanks, the application of a flash desolventizer is not suitable for rapeseed meal.

Fluidized Bed Desolventizer

Another approach to reducing the temperature load of hexane-containing meal during desolventizing is use of a batch fluidized bed desolventizer system. This was developed in a joint project between PPM Pilot Pflanzenöltechnologie Magdeburg e.V. (Germany), Dr. Weigel Anlagenbau Magdeburg (Germany), and Otto-von-Guericke-University, Magdeburg (Germany) (Pudel, 2012). The principle behind fluidized bed desolventizers is that meal material is brought *via* a fluid (superheated hexane) into a state such that the solid/fluid mixture behaves as a fluid. Under these conditions, very high heat and mass transfer is ensured and, hence, low temperature load of the material is guaranteed.

The velocity of the fluid must be higher than minimum fluidized bed velocity; below that, a fixed bed occurs. At the upper end, velocity has to be lower than fluctuation velocity; above that, pneumatic transport begins. Minimum fluidized bed velocity and fluctuation velocity depend on particle size. Therefore, the operating range of a stable fluidized bed is defined by the minimum fluidized bed velocity of the largest particles and the fluctuation velocity of the smallest particles. Particles with a size less than 0.4 mm would begin to leave the apparatus if fluid velocity is just high enough that particles of about 5 mm can be fluidized.

Superheated hexane is fed up from the bottom and distributed by a perforated plate. It leaves the separation chamber on the top. The meal is fed in from the top and fluidized by the fluid. After treatment the distributor plate is turned and the desolventized meal can be removed from the equipment. After filtration, the fluid is partially condensed, while hexane and water are separated and fed back into the system.

REMOVAL OF SUSPENDED MATERIAL

After the extraction process, the crude oil contains many impurities in suspended form which adversely affect the quality of the oil. These impurities include water (which improves the conditions for enzymes, fungi, and bacteria) and seed material (which contains enzyme systems that not only degrade the oil, but are also settled by fungi and bacteria). Lengthy storage of oil with these impurities leads to an increase in free fatty acids (degradation products of triacylglycerols) and finally to the formation of bad-smelling compounds. For the crude oil from solvent extraction, the higher content of free fatty acids means losses of neutral oil during refining while bad-smelling compounds can be removed easily during deodorization. Oils from cold pressing have an even greater problem. Refining steps are not allowed and, once formed, compounds such as free fatty acids or bad-smelling compounds cannot be removed by sedimentation, filtration, or centrifugation. Therefore, it is strongly recommended to remove suspended impurities as fast after extraction as possible. In general, three different methods can be used to remove suspended impurities from the oil: sedimentation, filtration, and centrifugation.

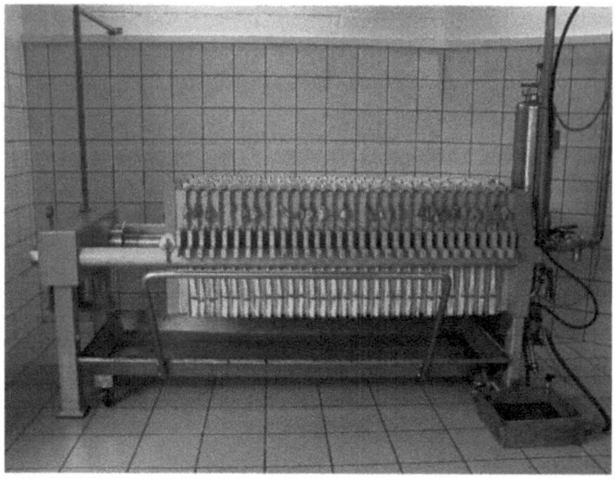

FIGURE 19.10 **Chamber filter press.**

The easiest way with the lowest need for sophisticated technical equipment is sedimentation. This uses the different specific densities of the liquid and solid phase and brings about slow settling of the suspended material as a result of gravity. The main drawback with the method is it is very time consuming; the longer the time the oil is in contact with water and seed material the greater the previously described degradation of the oil.

Today, filtration is the method of choice for small and medium-sized facilities, because it is fast, effective, and comparatively economical. Filtration involves mechanical separation of the solid and liquid phase, which allows removal of insoluble solids or suspended material from a liquid by passing it through a porous medium that is only permeable for the liquid phase. In small and medium-sized facilities, this is mostly achieved by forming a filter cake from the seed particles between porous filter material within a pressure gradient. The filter cake improves the results of filtration. Moreover, filter materials can be used to improve the effectiveness of the process. These materials – such as inert cellulose and cotton –improve the formation of a filter cake for the filtration process. Filtration is carried out by chamber filter presses (Fig. 19.10) or vertical pressure plate filters (Fig. 19.11) which enable much higher throughput than sedimentation.

The most sophisticated and effective way of removing suspended impurities is centrifugation. Centrifuges can work in continuous mode and remove water and gums in addition to solid particles from seed material.

REFINING PROCESS

The oil obtained from the extraction process generally comprises desirable triacylglycerols. However, depending on the process conditions, undesirable compounds can also be coextracted impairing the oil quality with respect to taste, smell, oxidative stability, and safety. These undesirabe compounds are metals, free fatty acids, oxidation products, contaminants, pigments, and hydrocarbons. On the other hand, the oil also contains minor compounds such

FIGURE 19.11 **Vertical pressure plate filter.**

as phytosterols, phenolic compounds, carotenoids, and tocopherols which are described to be beneficial nutritionally; the more extensive the extraction conditions the higher the number of minor compounds.

Solvent extraction is followed by a refining process consisting of a number of different steps. This enables comprehensive purification of the oil so that it is suitable for human consumption (Fig. 19.12). The result is a bland-tasting, light-colored product, without odor or flavor, which meets the requirements of consumers in being a universally applicable and safe oil. Thus, refining is a balancing act between removal of undesired compounds that impair oil quality and retaining minor compounds with known positive impacts on health.

Depending on the type of oil, refining is carried out in one of two ways: physical refining (with degumming, bleaching, and deodorization stages) or chemical refining (with an additional neutralization stage to remove most free fatty acids, phospholipids, and other impurities before deodorization) (Fig. 19.13). Neutralization is carried out at the deodorization stage in physical refining. Chemical refining has the advantage of lower temperature demand during deodorization but the disadvantage of more waste as a result of neutralization. In the physical refining process most free fatty acids are removed during deodorization by steam distillation. Physical refining is of interest for oils with high content of free fatty acids. With respect to the formation of unwanted contaminants such as *trans* fatty acids or

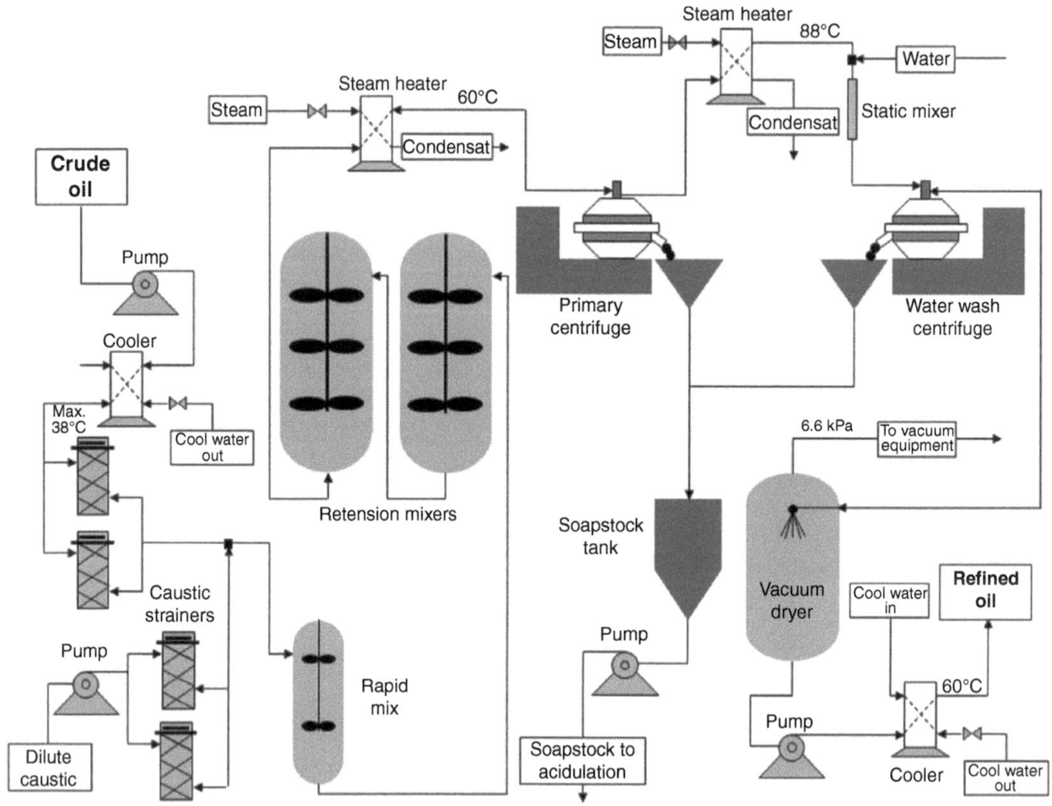

FIGURE 19.12 **Refining.** *Source: Reprinted from Gupta (2007), p.511 with permission from Elsevier.*

3-monochloropropane-1,2-diol (3-MCPD) and glycidyl fatty acid esters, chemical refining is recommended over physical refining.

The main advantages of physical refining are reduced demand for chemicals (such as phosphoric acid) or alkaline solutions and water, reduced waste production, and higher oil yield, all of which result in improved overall economical and environmental assessment of the process. On the other hand, the process is more sophisticated and is not applicable to all types and qualities of crude oil. Chemical refining is strongly recommended for cottonseed oil since gossypol, a phenolic aldehyde that acts as an inhibitor of several dehydrogenase enzymes, can only be removed by alkali treatment. From an economical point of view, physical refining makes sense for more expensive oils that have higher content of free fatty acids because of the lower loss of neutral oil, while the higher oil yield of physical refining is of less importance for cheaper oils.

Degumming

Degumming is the first stage of the refining process. It is often carried out in the oil mill before storing the crude oil for further purification. The aim is to reduce the content

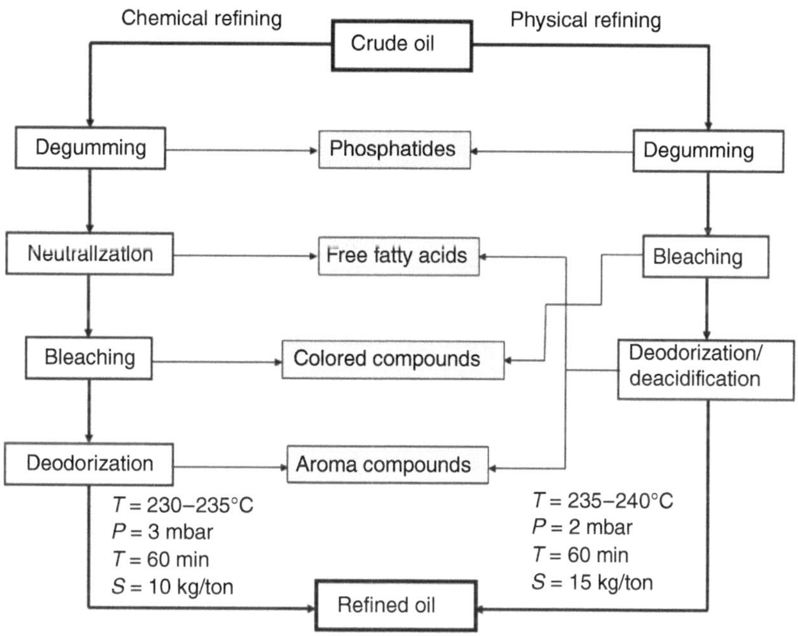

FIGURE 19.13 **Different steps of the refining process.** *Source: Reprinted from Gupta (2007), p.508 with permission from Elsevier.*

of phospholipids, which mainly consist of phosphatidylcholine (PC), phosphatidylethanolamine (PE), phosphatidylinositol (PI) and phosphatidic acid (PA). Such compounds belong to the phosphoglyerols which contain two ester-bound fatty acid molecules in position one and two of the glycerol molecule and one ester-bound phosphoric acid in position three. Different types of alcohol can be bound to the phosphoric acid (Fig. 19.14). Special properties of phospholipids arise from this structure which has a polar hydrophilic head and a nonpolar hydrophobic tail. Phospholipid content depends on oil-processing conditions and varies widely with type of oil. Soybean and rapeseed oil contain up to 3% phospholipids, while only 0.1% has been found in peanut oil.

Although dietary phospholipids are described as most likely having a positive effect on health (Wehrmüller, 2008), their presence will be detrimental to oil yield and quality. On the one hand, they act as strong emulsifiers during purification resulting in losses of neutral oil; on the other hand, they lead to deterioration during storage of oil (Young et al., 1994). It is suggested that amino groups from phospholipids can react during condensation with carbonyl compounds developed in the course of autoxidation of rapeseed oils. This results in colored compounds called "melanophosphatides" (Bratkowska, 1978). Moreover, phospholipids can cause difficulties during further processing (e.g., hydrogenation).

The simplest form of degumming involves the addition of water, which leads to enrichment of water-hydratable phospholipids in the water phase since they have greater affinity to the water than to the oil. Water degumming involves mixing the oil intensively with about 2% water at 80°C. This higher temperature lowers the viscosity of the oil and

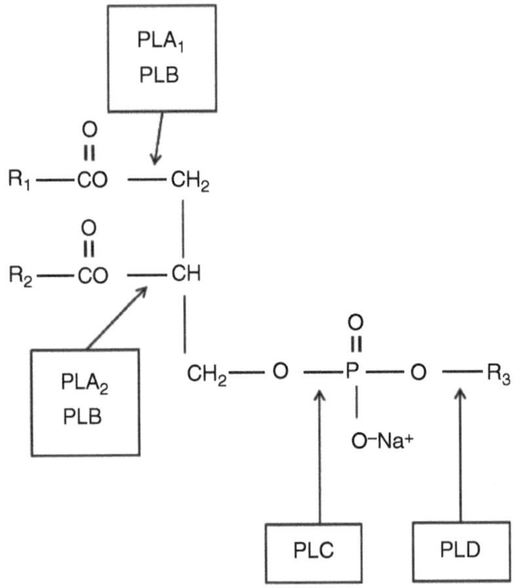

FIGURE 19.14 **Types of phospholipases PL (A_1, A_2, B, C, and D) and their effect on different bonds in the phospholipid molecule.**

accelerates the transfer of hydratable phospholipids into the water phase forming an emulsion called "gums" or "wet gums." Hydratable phospholipids become insoluble in the oil within 5–45 min and form a precipitate. Gentle agitation keeps the precipitate in suspension where it agglomerates to larger particles that are easier to separate. Finally, the suspension is centrifuged to separate as much oil from the hydrated phospholipids as possible, but 35–40% of the separated material is oil, making degumming a costly stage within the refining process (Fig. 19.15). After degumming, the phosphorus content is reduced to between 100 mg kg^{-1} and 200 mg kg^{-1} depending on the oilseed, the properties and quality of the seed, and oil-processing conditions. It is also possible to use hot steam instead of hot water, which improves the contact between oil and water resulting in better separation of phospholipids. Hydrated gums can easily be removed by centrifugation at a later stage. Depending on the type of oil, hydrated gums are used in human nutrition for the production of natural emulsifiers or stabilizers in various food applications as in the case of soybean oil, or phospholipids are readded to the meal to improve the value of feed or to reduce dustiness during the use of meal since gums are too dark to be used in food or pharmaceuticals as in the case of rapeseed oil.

Water degumming is not suitable for physical refining since the crude oils contain remarkable amounts of phospholipids that have no affinity to water. These so-called "nonhydratable phospholipids" (NHP) occur as salts of Fe, Ca, and Mg of phosphatidic acid or phosphatidyl ethanolamine and are not hydratable under water-degumming conditions due to their very low polarity making them highly oil soluble. The higher temperatures used during deodorization within physical refining lead to degradation of phospholipids resulting in darkening

FIGURE 19.15 **Centrifuge for the separation of oil/water mixtures.** *Source: Courtesy of MUEZHEST Process Technologies Pvt. Ltd., Mumbai, India. Reprinted from Gupta (2007), p. 509 with permission from Elsevier.*

of the oils. Therefore, the degumming process plays a critical role in the physical refining of edible oils.

Depending on the oil quality, the amount of NHPs can vary from only 10% for soybean oil from fresh good-quality beans to 50% which indicates bad oil quality (Rossel and Pritchard, 1991). Process conditions during oil processing can help to avoid the formation of NHPs by controlling the moisture content of flakes or inactivating the enzymes.

The removal of NHPs is more difficult since a more sophisticated treatment is necessary when mild organic acids are used. Examples are the application of 0.1–0.3% of 85% phosphoric acid or 0.1–1.0% of 30% citric acid solution at temperatures between 60°C and 70°C. Malic acid can also be used. By adding strong acids to the oil, the metal–phospholipid complexes which have a very strong binding at pH = 7 are dissociated in oil-insoluble metal, salts, and hydratable phospholipids, which can easily be removed by adding about 2% water. During the process, intensive mixing is necessary to ensure close contact between the oil and the acidic solution influencing the reaction time. Then, the mixture is further mixed for 30 min to 3 h to allow the hydratable phospholipids that have formed to agglomerate to larger groups which can be separated more easily with reduced loss of neutral oil (Fig. 19.16). While hydratable phospholipids are separated for further use,

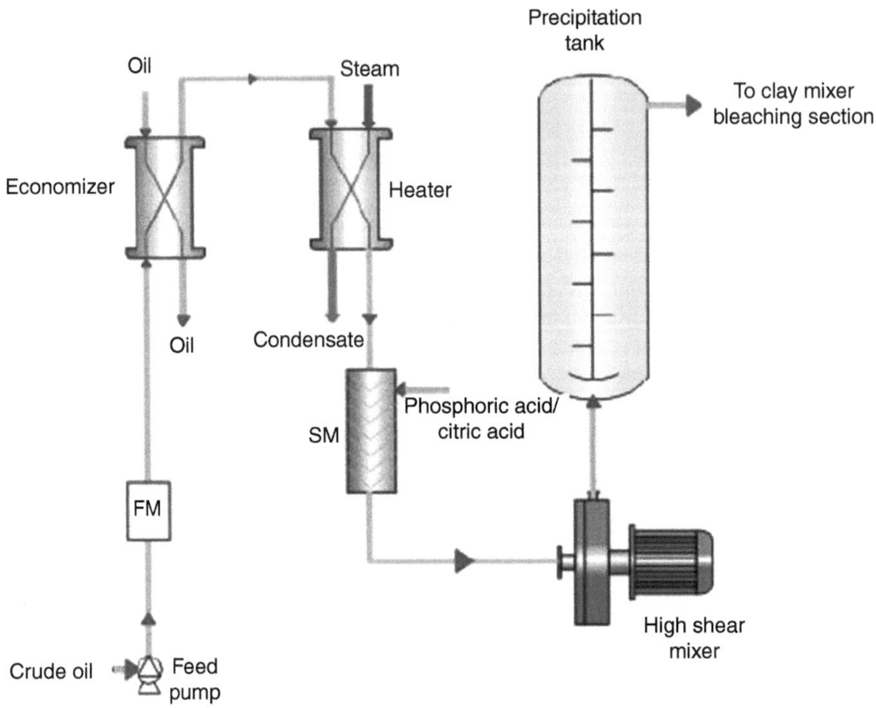

FIGURE 19.16 **Degumming.** *Source: Courtesy of MUEZ-HEST Process Technologies Pvt. Ltd., Mumbai, India. Reprinted from Gupta (2007), p. 500 with permission from Elsevier.*

NHPs are generally not recovered. The final oil should contain less than 25 mg kg^{-1} phospholipids calculated as phosphorus. Degumming is also a major source for the production of lecithin.

Degumming by EDTA

The so-called "soft degumming process" involves using chelating chemicals such as ethylenediaminetetraacetic acid (EDTA) in the presence of an emulsifier to effectively remove metal ions from the oil (Choukri et al., 2001). Depending on total content of phospholipids, the crude or water-degummed oil is treated with an aqueous solution of the chelating agents. Due to the high pK values that phosphatidic acid has with calcium and magnesium, these compounds form stable complexes that are not soluble in water. By the addition of EDTA, the metal ions are easily displaced from the phosphatidic acid complex to the EDTA complex whose pK value is much higher resulting in a more stable complex. For adequate contact between oil-soluble ions and water-soluble EDTA, the mixture has to be thoroughly mixed. During soft degumming, the NHPs are transferred into hydratable phospholipids as a result of loss of the metal and then the hydratable phospholipids can be enriched in the aqueous phase. Finally, the oil is recovered from the emulsion by decanting or centrifugation.

Dry Degumming

Oils with low phospholipid content (<20 mg kg^{-1} of phosphorus) such as palm oil, lauric oils (like coconut oil), palm kernel oil, and animal fats can be acid degummed using a small amount of concentrated acid such as phosphoric or citric acid followed by mixing with bleaching earth. For oils high in phospholipids this process would be uneconomical since a lot of bleaching earth would be needed.

Enzymatic Degumming

Conventional degumming is costly when it comes to equipment but worth consideration from the environmental point of view since much acidic waste is produced. Moreover, the level of phosphorus in physical refining needs to be reduced to 3 mg kg^{-1} before bleaching which involves a great deal of work.

An alternative technique is enzymatic degumming which was developed in the 1990s by Lurgi Öl Gas Chemie in Frankfurt (Germany); it is commonly known as the "EnzyMax® process" (Aalrust et al., 1992). By using such enzymes as phospholipases, the emulsifying nature of the phospholipids is changed, which results in reduced loss of neutral lipids during separation. The degumming conditions are much milder than for conventional refining and the process is more efficient.

Phospholipid degradation involves the use of different types of phospholipases which attack the phospholipid molecule at different positions (Abb. 5): phospholipase A_1 (PLA_1) is specific and cleaves the binding of the fatty acid at position 1 in the triacylglycerol molecule, while phospholipase A_2 (PLA_2) cleaves the fatty acid in position 2; phospholipase B (PLB) is more unspecific and separates both fatty acids; phospholipase C (PLC) results in the formation of a diacylglycerol by cleaving the phosphate ester from the glycerol side; and phospholipase D (PLD) leads to elimination of the alcohol from the phosphate ester.

The EnzyMax® process is based on the transfer of nonhydratable phospholipids into a hydratable form, with the advantages of reduced waste water, increased oil yield, and the reduced operational costs (Dahlke, 1998).

In general the process is divided into three steps: (i) adjusting the pH value of the oil using a buffer, (ii) enzyme reaction in a tank, and (iii) removal of gums and sludge from the oil. Although in small-scale processes several phospholipases from different sources have been used successfully, for industrial applications only three enzymes are in use, PLA_2 from porcine pancreas and PLA_1 from *Fusarium oxysporum*, and *Thermomyces lanuginosus/Fusarium oxysporum* (Clausen, 2001). When the pH was kept in an optimal range, less than 10 mg kg^{-1} of phosphorous in the oil was obtained. Gums and oil were easily separated after centrifugation and the oil loss was minimal under the process conditions. Enzymatic degumming is suitable for oils with relatively low levels of phospholipids while for phospholipid-rich oils a water degumming step should be inserted upstream.

The different steps of the process are (i) heating of the raw oil to 70–75°C, (ii) adding 0.5–0.6 kg citric acid per metric ton of raw oil, (iii) keeping the mixture in a residence container for 15–20 min after intensive mixing, (iv) adding caustic soda to buffer the water phase to achieve the optimal pH value for the enzyme, (v) adding demineralized water as well as 30–50 g enzyme per metric pound of raw oil, (vi) intensive mixing for 4–6 h, (vii) heating of the mixture to 80°C for a better separation, and (viii) centrifugation of the mixture and drying of the oil.

The enzymatic reaction takes place only at the interface between oil and enzyme-containing water. Therefore, a continuous mixing of the oil with the enzymatic solution is important for a successful and quick enzymatic degumming process. This mixing with a high-shear mixer results in an enlarged reaction surface.

One important aspect of enzymatic degumming is the remarkable reduction of waste and the avoidance of chemicals in comparison with conventional degumming. The waste water arising from conventional degumming is acidic and has to be neutralized using caustic chemicals. Additionally, a lot of water is necessary to separate the gums from the oil. For enzymatic degumming only 1–2% water is added which after degumming only contains lysophospholipids recyclable by mixing with oilseed meal or recoverable by evaporation (Dixit and Kanakraj, 2010).

With the discovery and development of commercial and cost-effective enzymes the process of enzymatic degumming becomes more interesting to the industry. The efficiency of the process has been further increased by the use of different combinations of enzymes.

Membrane Degumming

Another process for the removal of impurities from the oil is the use of a membrane technology that allows the selective removal of undesired compounds. The advantage of this technology is a higher efficiency, a mimimization of the processing steps, lower energy consumption, lower waste production, and no use of chemicals. The principle of membrane degumming is relatively simple since the oil is pumped through a membrane that allows the passage of molecules of a certain size, while larger compounds are retained. Although the size of triacylglycerols and phospholipids is very similar the separation is possible since phospholipids tend to form reverse micelles in nonpolar media which are large enough to be retained (Ribeiro et al., 2008).

Alkaline Neutralization

Alkaline neutralization by use of an alkaline solution in combination with heat treatment is used to neutralize the free fatty acids in vegetable oils by saponification. This is the key step of the chemical refining and this removal of free fatty acids from the crude oil is different from the physical refining. Additionally to free fatty acids other undesirable compounds are removed mainly by adsorption on the formed soaps such as phospholipids, metal ions, color pigments, or oxidation products. The process of neutralization also results in some losses of neutral oil that is saponified and finally also removed during the washing process.

The advantage of chemical refining in comparison to physical refining is that this process is less susceptible to specific process conditions. While for physical refining the amount of phospholipids should be very low, chemical refining tolerates remarkably higher contents since this process will remove most of the phospholipids. If no degumming is carried out the purification of the oil starts with the alkaline neutralization of the oil which removes impurities that would result in a darkening of the oil when it becomes heat treated further on in the process of refining.

The main factors affecting the result of the neutralization step and the quality of the oil are: (i) the type of alkali; (ii) the concentration of the solution; (iii) the temperature during treatment; and (iv) the excess of alkaline solution over the stoichiometrically required amount.

When alkaline neutralization was first adopted, batch or kettle refining was used and today this procedure is still popular for small and medium size plants. The main advantages are that processing is cheap, it is easy to handle, and also the associated investment costs are comparably low. The principle of this batch process is that oil and alkaline solution are mixed and then the oil and the resulting soapstock are allowed to settle before the two phases are separated. This process is slow, inefficient, affects oil quality, and results in remarkably high losses of neutral oil. Today, in large facilities, continuous neutralization is possible by the application of centrifuges that have increased the speed of the process and increased the throughput remarkably.

Alkaline neutralization starts with the addition of 0.05–0.1% concentrated phosphoric acid to the oil to precipitate the phospholipids, chlorophyll, and oxidant traces of metals. For a better contact between acid and oil intensive mixing is necessary, reducing the processing time. Further into the process this treatment results in reduced losses of neutral oil, reduced need for bleaching earth, and better flavor, and flavour stability, of oil after deodorization. In general phosphoric acid is used for this step since it is more effective than other organic acids (Ohlsen and Svensson, 1976).

Most commonly the neutralization of the free fatty acids is carried out by 2–3 N sodium hydroxide solution under continuous mixing to ensure close contact between oil and alkaline solution and heating. It is worth noting that other alkaline solutions can be used such as sodium carbonate or ammonium hydroxide. The resulting soapstock is removed by continuous centrifugation. As a result of alkaline neutralization the amount of free fatty acids is reduced to $< 0.05\%$, phosphorous < 5 mg kg^{-1}, chlorophyll < 30 mg kg^{-1}, and sulfur < 10 mg kg^{-1}.

The amount of alkaline solution necessary for a comprehensive neutralization of crude oil can be calculated from the amount of free fatty acids. The theoretical amount of alkaline solution necessary for the neutralization of free fatty acids, plus an excess, is necessary to also remove other impurities as well as to neutralize the phosphoric acid added at the beginning of the process to precipitate phospholipids, chlorophyll, and oxidant traces of metals. The preparation of the alkaline solution can be controlled by using a specific gravity hydrometer calibrated in degrees Baumé (Be), measuring the relative density of the solution. For most oils an alkaline solution with 16–24 Be is sufficient for successful neutralization.

The saponification of the free fatty acids takes place immediately when caustic acid and oil come into contact. But the process takes more time, necessary for the hydratization of phospholipids or the reaction with colored pigments.

The main drawback of chemical refining is the great loss of neutral oil making the process less economical. Additionally, during the process greater quantities of alkaline contaminated water arise which have a strong impact on the environment. This water also contains higher amounts of fatty matter from the oil and other organic and inorganic materials which are considered as harmful to the environment. Thus, before the water can be disposed to municipal sewers its pH value has to be neutralized to pH 6–9 and the main part of the oil phase must be separated. On the other side of the discussion, alkali refining allows the refining of almost all types, and qualities, of oil without any specific requirements to produce high-quality oil.

Physical Refining

Today, physical refining of vegetable oils is the favored method for oil purification since it has some environmental and economical benefits, namely, less waste water and discharge produced.

One disadvantage is that the requirements concerning the content of phosphorus are much stronger than for chemical refining. For physical refining the oil should have less than 15 mg kg^{-1}, preferably less than 10 mg kg^{-1}, and in some cases less than 5 mg kg^{-1} (Narayana et al., 2002). The efficiency of physical refining is directly correlated to a proper degumming and influences the quality of the resulting oils.

Bleaching

Bleaching removes colored compounds such as chlorophyll and carotinoids, soaps, oxidation products, phospholipids, and other minor compounds such as metals or contaminants such as polycyclic aromatic hydrocarbons or pesticides which bind to the surface of the bleaching earth. The removal of chlorophyll is important to improve the stability of the oil during storage because it acts as photosensitizer catalyzing the oxidation of oil in light that is a problem for bottled, edible oils. Chlorophyll also results in an undesired dark green colour to the oil. This is especially the case for rapeseed oil that often contains higher amounts of chlorophyll that have to be removed during bleaching. Carotinoids are removed from the oil during bleaching only to a lesser extent. More successful is the so-called heat bleaching during deodorization at higher temperatures when the compounds are destroyed. Bleaching is also the last chance to remove phospholipids from the oil, which affect the oil quality during deodorization. Soaps can form polymers in the deodorizer, which affect not only the processing but also the quality of the oil. During bleaching higher amounts of both phosphorous and soaps should be avoided since these compounds inactivate the active centers of the clays resulting in worse results for the bleaching process with respect to decolorization and the need for a greater amount of clay. As catalysts for the oxidation process, the removal of metal ions as early as possible within the bleaching process is recommended although metal ions are also removed during deodorization by chelation when citric acid has been added. The implementation of a bleaching procedure is particularly important for further physical refining or hydrogenation as low levels of impurities are a basic requirement for a successful processing.

The following parameters should be taken into consideration for a successful and economical bleaching process: in the case of alkali refining free fatty acids $< 0.05\%$, soaps < 50 mg kg^{-1}, phosphorous < 5 mg kg^{-1}, and water $< 0.1\%$ are recommended, while soaps and phosphorous should be below 50 mg kg^{-1}, and 20 mg kg^{-1}, respectively for oil provided for physical refining (Zschau, 1999).

The principle of the bleaching process is the adsorption of compounds at an adsorptive surface which means a mass transport from the oil to the bleaching earth. The mechanisms that take place at the surface are either physical adsorption as a result of van der Waals forces or chemisorptions, with a chemical binding between compound and adsorbents (Khoo et al., 1979). The aim of the bleaching process is to adsorb as much undesired compounds as possible from the oil at the surface of the bleaching earth. The adsorption capacity depends on the adsorption material, the adsorbed compound, but also on other parameters such

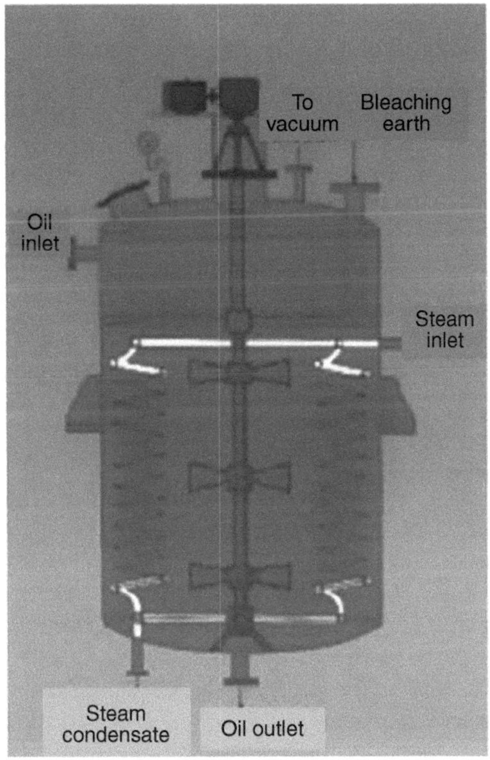

FIGURE 19.17 **Bleacher.** *Source: Courtesy of MUEZ-HEST Process Technologies Pvt. Ltd., Mumbai, India. Reprinted from Gupta (2007), p. 515 with permission from Elsevier.*

as temperature, viscosity, or pH value which enables a shifting of the balance significantly toward a better adsorption capacity.

Most commonly used in big facilities is continuous bleaching, which avoids unwanted oxidation and allows the processing of huge quantities of oil, while batch bleaching is more suitable for small and middle size plants and has high flexibility (Fig. 19.17). It is important to avoid oxidative processes occurring on the large surfaces of the adsorbent material. Therefore, it is recommended to carry out bleaching under vacuum conditions (26.7–40.0 hPa) with deaerated oil dried to result in moisture contents less than 0.1% in order to reduce the presence of oxidative compounds and oxidation reactions.

Bleaching starts by preheating the oil to 70–90°C. This decreases the viscosity of the oil and simplifies further handling. Then the preheated oil is pumped, together with the clay adsorbent material, into a clay slurry tank and after intensive mixing the slurry is transferred into the bleaching vessel. Oils with low amounts of phospholipids are degummed simultaneously with the bleaching process by adding the acid into the hot oil before adding the adsorbent. This combined degumming and bleaching saves the individual step of degumming which results in a more environmental friendly processing.

After adding the bleaching clay the mixture is heated to 95–105°C for 15–30 min under vacuum conditions. For bleaching of palm oil, temperatures of 100–130°C are in use to get the optimum bleaching result within 30 min. A higher temperature is not recommended due to its adverse effects on oil quality. On the other side of the argument, the temperature should be as high as necessary to ensure a low viscosity of the oil which improves the diffusion within the oil and the mass transfer of impurities to the adsorbent material. This mass transfer is also supported by continuously stirring the slurry which enables an intense contact between oil and clay. This helps to move the equilibrium between impurities adsorbed by the clay and impurities remaining in the oil in the direction of a removal of impurities from the oil. On the other hand a too rapid agitation of the slurry leads to poorer bleaching results due to incorporation of air into the slurry (Brimberg, 1982). The authors found the best results for bleaching of rapeseed and palm oil were with the slowest agitation rates – just enough to keep the clay dispersed.

Another possible way to improve the adsorbent capacity of the added material is the addition of clay in two steps. The weight of impurities adsorbed per unit weight of adsorbent decreases as lower impurity levels are achieved. Therefore, a lower adsorbent dosage results in a higher equilibrium adsorption capacity. On the other hand, to achieve lower impurity levels in the oil disproportionately higher amounts of adsorbent are required. Using this two-step approach it is possible to reduce the amount of impurities more effectively with the same amount of clay or it is possible to reduce the amount of clay while still achieving the same quality of oil.

The separation of clay and oil takes place by horizontal or vertical leaf filters. The mixture is pumped through the filter in a cycle in order to build up a filter cake. At the end of filtration the cake is dried by steam blowing or nitrogen at 2–3 bar and 140–150°C to lower the loss of neutral oil. Nevertheless 20–40% of the cake consists of oil. For batch units it is more economical to run two or more bleaching reactors in parallel with two independent filter systems in order to ensure a semicontinuous bleaching in one reactor while the other one is emptied.

In the process of bleaching different types of clays are used depending on the clay material, the activation process, or other specific properties. In most cases the adsorbent materials are acid-activated clays, which are obtained from inactive aluminium silicates (bentonite, montmorillonite, attapulgite), activated by mineral acids like sulphuric or hydrochloric acid and then washed and dried to a moisture content of 10–15%. The aim of the activation process is to enlarge the surface of the clay and to increase the number of active centers, which increases the mass transfer from the oil to the adsorbents improving the quality of the bleaching process. While acid-activated clays have surfaces between 150 $m^2 g^{-1}$ and 350 $m^2 g^{-1}$ the surfaces of natural clays have half the size or even less. Nevertheless in most cases the pore volume and the pore size of natural clays are sufficient for a good bleaching result (Zschau, 2001).

Another important factor is the reduction of the metal content of the clay, which also improves the adsorptive activity of the material. An important characteristic feature of the activated clay is the total acidity that seems to be directly correlated to the bleaching power of the clay. Therefore, clays with a high acidity are preferred while clays which have been washed completely free of residual acid show a reduced bleaching power. However, different investigations have shown that the acidity of bleaching earths is well correlated with the content of 3-MCPD esters in refined oils, which should be avoided for toxicological reasons (Ramli et al., 2011). These studies showed that a higher acidity of bleaching earth resulted in higher formation of 3-MCPD esters and they recommended the use of natural bleaching earths and acid-activated ones with more neutral pH value as options to reduce the formation of 3-MCPD esters.

Another important factor influencing the properties of the clay is the moisture content which normally is between 10% and 15%. Only when there is a certain amount of moisture in the structure of the clay is it able to absorb impurities from the oil. If the clay is too dry the bleaching power is reduced remarkably, due to a reduced surface area.

Modified Bleaching Techniques

The bleaching process creates a huge amount of waste in the form of oil containing clays that also contain impurities absorbed from the oil. Therefore, some efforts have been undertaken to improve the efficiency of such clays with the aim to reduce the total consumption and the resulting waste. Additionally, some approaches that change the conventional bleaching process have been described.

MODIFIED CAUSTIC REFINING™

Most interesting is Modified Caustic Refining™ which uses synthetic silicas from an aqueous solution of alkali-silicates, acidified with diluted sulfuric acid that have not the same adsorptive properties as conventional clays. This synthetic silica material contains a high amount of functional water (up to 65%) that serves as a polar medium to transfer polar impurities such as soaps or phospholipids from the oil through the pores onto the internal surface of the silica. Because of its composition silica has a high activity and selectivity for the adsorption of soaps and phospholipids, however, they are not suitable for the removal of chlorophyll. Thus, these materials are used to adsorb metals, soaps, and phospholipids from the oil, which allows the operation of caustic refining without water washing. This all results in a more cost-effective and environmentally friendly removal of these impurities before the bleaching process starts. Additionally, Modified Caustic Refining™ reduces oil losses during water washing and the need for clay during bleaching. High investment, operating costs, and maintenance costs for the water-washing centrifuges can, therefore, be avoided.

Another important point is that the final bleaching process can "concentrate" on the removal of pigments and trace metals, instead of being inactivated by soaps and phospholipids. This enables the development of bleaching earths with a higher affinity and specificity to these compounds resulting in a reduced need for clay during bleaching.

PACKED-BED BLEACHING

A more efficient bleaching technique is the counter-current adsorption of the impurities on a packed bed of clay. The principle of this approach is that higher impurity levels have a higher driving force for mass transfer onto partially spent adsorbents, while fresh and fully active adsorbents have a stronger affinity for the last trace amounts of impurities compared with partially spent adsorbents (Huddleston, 2014). During bleaching the front end of the packed bed removes most of the impurities and the amount is further reduced when passing the oil through the bed. With ongoing bleaching the saturated zone at the front end of the bed grows through the bed.

DOUBLE-PASS BLEACHING SYSTEM

Another approach has been described by the Crown Iron Work Company (Minneapolis, USA) using the filtration of silica pretreated oil through a bed of once-used bleaching earth as a first bleaching step (Anon, 2015). During silica treatment soaps and phospholipids are removed from the oil. Then the silica material is removed from the oil passing the mixture through a filter bed of once-used bleaching earth. A large part of the color is adsorbed in this step by the still-active bleaching earth. The actual bleaching is then carried out by fresh bleaching earth that removes residual pigments and other impurities from the oil. At the end, the bleaching earth is removed by filtration and can be used as a filter aid for the removal of the silica (Fig. 19.18). Using this technique the need for bleaching earth can be reduced remarkably in comparison to conventional bleaching as a result of the separate filtration of spent silica before adding bleaching earth and the prebleaching of oil with once-used bleaching earth.

SYSTEM-RELATED OPPORTUNITIES

For a sustainable and economical bleaching process the developments regarding centrifugation and system monitoring are very important too. These allow a more efficient and targeted removal of phospholipids and soaps from the oil as well as a more optimized processing with respect to the addition of bleaching earths. The better centrifugation reduces

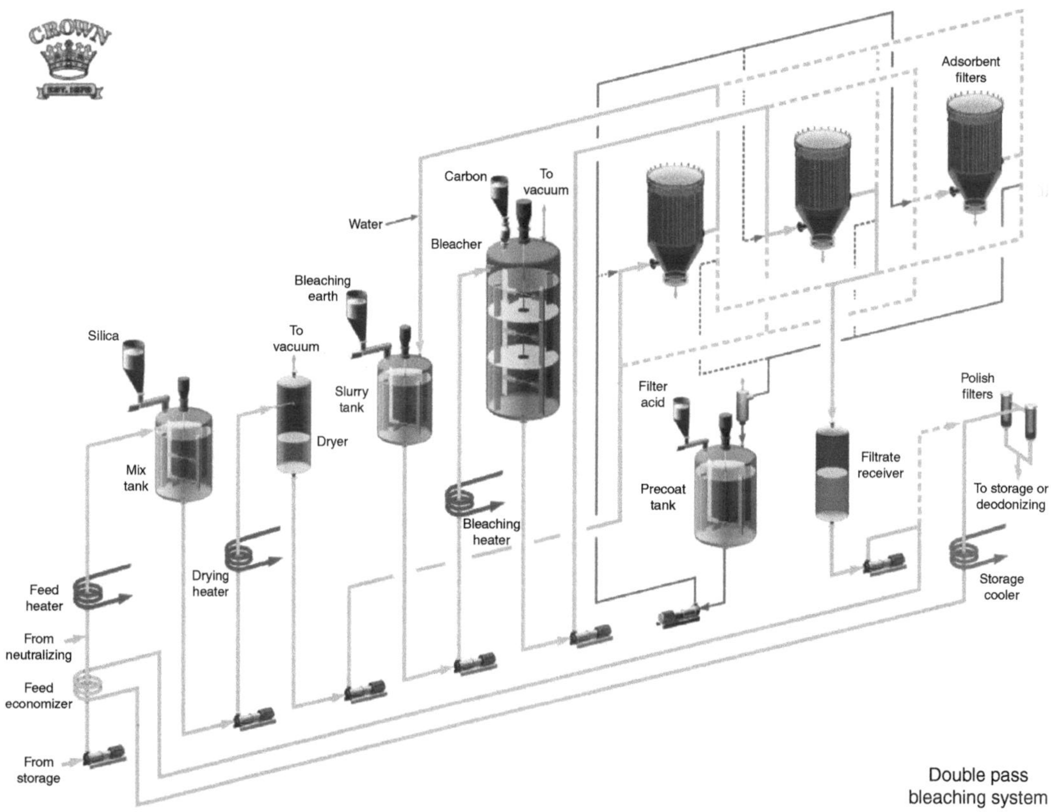

FIGURE 19.18 Flowchart of the Double pass bleaching system. *Source: Courtesy of Crown Iron Works, Minneapolis, USA.*

remarkably the losses of neutral oil during processing, making the whole process more economical and reducing waste. On the other hand, a more sophisticated monitoring system of the impurities allows for better control of the process and enables the dosage of the bleaching earth to be more precisely adjusted in order to meet the necessities of the oil.

Deodorization

In the final stage of the purification process the degummed and bleached oil is treated by hot steam at temperatures between 220°C and 270°C. The aim of this last step is to remove the aroma and other minor compounds that adversely affect the sensory quality and the shelf life of the oil, such as flavor components, oxidation products, or residual free fatty acids. The high temperature during deodorization also leads to a heat bleaching of carotenoids that are destroyed to colorless compounds. Additionally, contaminants are removed. The result is oil, neutral in taste and smell, light yellow in color, and low in free fatty acids with good stability.

The principle of deodorization is steam distillation at high temperature (220–250°C and 270°C for physical refining) and low pressure (0.3–0.6 kPa).

The relevant parameter of a given compound that describes volatility is vapor pressure. Compounds with high vapor pressure under given conditions will leave the oil more easily than compounds with a low vapor pressure. Vacuum conditions and temperature improve the volatility of compounds and they are the driving forces for the transfer of these compounds from the oil into the so-called "deodistillate." Since the unwanted substances have a remarkably higher vapor pressure than triacylglycerols it is possible to remove these compounds while triacylglycerols remain left behind. Unfortunately also some of the desirable compounds such as tocopherols or phytosterols have a relatively high vapor pressure that results in a distinctive reduction of these compounds during deodorization. What is decisive for a successful deodorization is there being sufficient contact between the oil and steam which ensures the most comprehensive removal of undesired compounds.

Parameters affecting the result of deodorization are temperature, time, pressure, and volume of stripping gas. Generally speaking the higher the temperature, the longer the time, the lower the pressure, and the higher the amount of stripping gas the more comprehensive is the removal of volatile compounds. However, other desirable compounds are also removed, *trans* fatty acids are formed, and the process becomes costly. Therefore, it is important to find the optimal conditions for these different parameters in order to get an optimal result with respect to oil quality and cost. What is decisive for the removal of volatile compounds is the amount of stripping steam which is indirectly proportional to the pressure and directly proportional to the temperature. It is important to note that the effect of temperature is not as great as the effect of pressure.

The removal of volatile aroma compounds originally from the oilseed can easily and rapidly be achieved by stripping. However, oil also contains nonvolatile flavor compounds whose removal requires some time. The presence of these compounds can adversely affect the storage stability of the oils when they decompose and form aroma relevant volatile substances. Additionally, they are able to accelerate oxidation during storage. During heat treatment in the process of deodorization these compounds are degraded and the formed volatile fragments can be removed from the oil. Therefore, it is important to give the nonvolatile compounds sufficient time for decomposition.

In big facilities deodorization is carried out in a continuous process that has some advantages with respect to efficiency and oil quality. In smaller sized plants batch or semicontinuous processes are in use too.

The deodorizer vessel in the batch process is a cylindrical tank. Some 10–20 tons of oil are pumped into the vessel and after deaeration it is heated up by heating coils. Deaeration is necessary to ensure a low concentration of oxygen in the vessel to reduce the risk of oxidation and polymerization during the treatment. Hot steam, necessary for the process, is supplied near the bottom of the vessel and a vacuum system provides the required low pressure. The process takes about 3–5 hr and 3–5% of the steam is used. During this time the temperature is between 180°C and 240°C, whereas the effect of deodorization starts at about 160°C.

Afterwards the vessel is cooled down to 60–80°C under vacuum conditions and the oil is sparged with nitrogen before storing in tanks or bottles.

For continuous working processes different types of deodorizers are available. One common used type consists of a vertical stainless steel vessel in which 4–7 trays are stacked (Fig. 19.19). At the beginning of the process the oil is pumped under vacuum conditions into the top tray of the deodorizer, which is heated. Here the oil is deaerated and heated to about 160°C for 15–20 min. After that the oil moves down stepwise from one tray to the next, counter-current, to a super-heated, stripping steam flow. During this process the oil is further heated to 220–270°C depending on the conditions and it is agitated through sparging steam at about 0.5% per hour. In the deodorizing section of the vessel about 1.5% of the sparging steam is used, before the oil reaches final cooling at about 50–60°C. The advantage of this process is that the oil bed is much smaller resulting in shorter processing time (50–80 min), reduced labor, and less heat treatment of the oil, resulting in a smaller

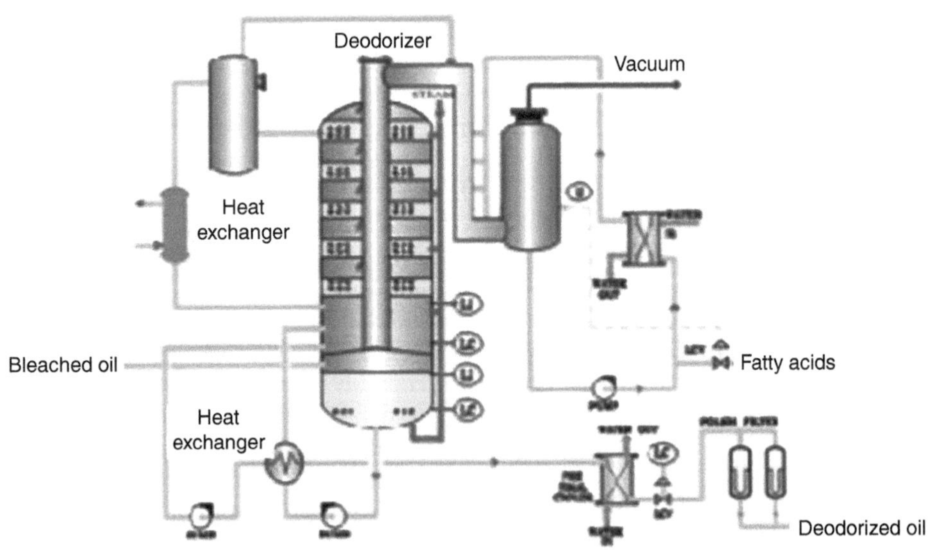

FIGURE 19.19 **Deodorization.** *Source: Courtesy of MUEZ-HEST Process Technologies Pvt. Ltd., Mumbai, India. Reprinted from Gupta (2007), p. 519 with permission from Elsevier.*

formation of degradation products. Finally, the oil is transferred to a storage tank by passing it through a polishing filter.

Another type of deodorization is based on a series of horizontal, single-shell, cylindrical vessels. In this apparatus hot steam is injected by several pipe distributors running the length of the vessels. The main advantage of this arrangement is that it enables a maximum surface for increased contact between the oil and hot steam. The cylindrical vessels are combined to form a plant, whereas each individual vessel is about 1 m in diameter and 4.5–12.5 m in length.

Within the refining process deodorization is one of the most sensitive steps. The high temperature required runs the risk of degrading oil quality. The input of oxygen via the steam could result in very rapid oxidation of the oil. Therefore, some authors suggest substituting steam with nitrogen as the stripping gas to avoid oxidation of valuable products and the formation of an emulsion with water (Decap et al., 2004).

Additionally, deodorization time and temperature is crucial for the formation of unwanted 3-MCPD and glycidyl esters since their formation is attributed to heat-induced reactions. While the formation of 3-MCPD esters starts at a temperature of 180°C in a very fast reaction, the formation of glycidyl esters is favored at temperatures higher than 240°C. The content of 3-MCPD esters is relatively independent from the temperature and time with values between 2 mg kg^{-1} and 4 mg kg^{-1}, the formation of glycidyl esters is favored and increases with increasing temperature and time up to 53 mg kg^{-1} under deodorization conditions with 290°C for 120 min (Pudel et al., 2011).

Removal of Valuable Compounds from the Deodistillate

During deodorization not only do the impurities leave the oil but also valuable compounds such as tocopherols, phenolic compounds, or phytosterols which are described to have some positive effects on health. During the refining process the amount of tocopherols is reduced by 10–60% and phytosterols aby 20–50%, depending on the type of oil and the refining conditions (Ferrari et al., 1996; Jung et al., 1989; Ramamurthi et al., 1998). A considerable amount of these compounds can be found in the deodistillate which also consists of steryl esters, sterol hydrocarbons, free fatty acids, mono-, di- and triacylglycerols, hydrocarbons, aldehydes, ketones, and different kinds of contaminants. The content of tocopherols is between 1 wt% and 20 wt% and phytosterols can be found in amounts between 5 wt% and 30 wt% (Maria and Miguel, 2010; Verleyen et al., 2001). Therefore, from an economical point of view and with respect to sustainable oil processing it makes sense to recover these valuable products from the deodistillate. However, using the deodistillate for the isolation of valuable compounds requires consideration that deodistillates from chemical refining comprise of 9.2–15% tocopherols and 9.0–17.6% phytosterols with low levels of free fatty acids (10–24.5%), while the amount of free fatty acids in physical refined oil is between 76.2% and 83.6% with only 1.4–4.3% tocopherols and 1.8–6.9% phytosterols (Verhé et al., 2006). In addition to the type of refining used, the nature of the oil, design of deodorizer, column, deodorization time, the operating parameters of the deodorizer, and the ratio of reduced pressure and temperature influences the composition of the deodistillate (De Greyt and Kellens, 1996; Kellens and De Greyt, 2000).

Different approaches for recovering the valuable compounds from the undesired substances have been developed depending on the compounds of interest. In addition to classical methods such as hydrolysis, distillation, and crystallization there are pretreatments such as crystallization, washing with different solvents, or a combination of esterificaton and transesterification. Torres et al. (2007) described an enzymatic two-step procedure which enzymatically modified the deodistillate to obtain a product mixture comprised mainly of sterol esters, tocopherols, and fatty acid ethyl esters. A semicontinuous column fractionation of canola oil (Guclu-Ustundag, and Temelli, 2007) and sunflower oil (Vázquez et al., 2006) deodistillate produced using supercritical carbon dioxide was carried out for the recovery of bioactive components such as sterols and tocopherols. In a patent Sun et al. (1998) defined a multistep process that comprised of esterification, water bathing, alcoholysis, cooling precipitation, and distillation for the efficient separation of tocopherols, sterols, and fatty acid methyl esters from the deodistillate. Carmona et al. (2010) achieved the enrichment of phytosterols from the deodistillate of sunflower oil refining it to 92% purity after esterification of its free fatty acids and transesterification of acylglycerols with methanol in the presence of an acid catalyst at 65°C for 4 h followed by cooling at −5°C for 24 h. Czuppon et al. (2008) achieved the removal of free fatty acids from the deodistillate by vacuum distillation or by continuation solvent saponification. The residue was reacted with an aromatic carboxylic acid anhydride at a temperature of 50–150°C under reduced pressure and afterwards tocopherols were removed from the mixture and crystalline free sterols were recovered from the distillation residue.

Harbaum-Piayda et al. (2010) described the finding of 2,6-dimethoxy-4-vinylphenol (canolol) and its dimers in the deodistillate of crude rapeseed oil and Aachary et al. (2014) found that the use of a deodistillate extract protected frying oil against deterioration with a considerable reduction in the hydroperoxides present after the first day of frying. The TOTOX and conjugated diene/triene values of oils fried with deodistillate extract were significantly lower than for butylated hydroxytoluene (BHT).

NEW CONCEPTS OF SEED PROCESSING

Production of Canolol from Fluidized Bed–treated Rapeseed Meal and Cake

Rapeseed contains comparably high amounts of phenolic compounds in free, esterified, or insoluble bound form, mainly as derivatives of sinapic acid. During oil extraction most of the phenolic compounds remain in the press cake or meal, but Koski et al. (2003) and Wakamatsu et al. (2005) showed that in crude rapeseed oil the decarboxylation product of sinapic acid, 2,6-dimethoxy-4-vinylphenol (4-vinylsyringol or canolol), can be found initiated by heat treatment. In contrast to the hardly oil-soluble sinapic acid canolol is soluble in oil. For canolol a high-antioxidant activity is described in the literature (Matthäus, 2012; Moltke Sørensen et al., 2013), making isolation of this compound promising for use in pharmaceuticals or food.

Pudel et al. (2014) developed a two-step method to isolate canolol-enriched extract from rapeseed meal or cake comprising heat treatment of the material at 165°C in a fluidized bed followed by extraction with supercritical carbon dioxide. The advantage of this approach is that the conventional oil mill process remains untouched for the isolation of the canolol-enriched extract. The fluidized bed treatment allows a very high and consistent heat and mass transfer to the meal or cake, which results in a temperature load for the material which is as low as possible.

The optimal temperature for such treatment was found to be 165°C, after which the fluidized bed treatment has to be interrupted immediately by cooling down the roasted material. A longer heating period reduced the canolol content of the extract. Fluidized bed treatment achieves about 500 mg canolol kg^{-1} rapeseed meal, while in cake more than 700 mg kg^{-1} was achieved. On the other hand, the use of meal for the fluidized bed treatment has the advantage of less oil being finally extracted by supercritical carbon dioxide, leading to higher canolol contents in the oily extract.

The canolol concentration in the canolol enriched extract was about 3%, while the extract additionally comprised of 68% triacylglycerols, 11% diacylglycerols, 10% free fatty acids, 2.5% phytosterols, and about 5% other components.

It was found that canolol-enriched extracts obtained from smaller heat-treated particles (<0.8 mm) showed a higher content of canolol due to an averaged shorter residence time within the fluidized bed reactor and the improved heat and mass transfer due to faster drying (Pudel et al., 2014). It was not possible to increase the canolol formation and yield of the extract by use of additional heat exchangers immersing in the fluidized bed or by applying super-heated steam as a fluidization medium, although these treatments improved the effectiveness of the process.

In a frying experiment over a period of 30 h, oil fortified with canolol enriched extract containing 200, 500, or 750 mg canolol per kilogram frying oil showed a two- to threefold better frying stability in comparison with oils fortified with commonly used antioxidants such as tert-butylhydroquinone (TBHQ) (200 mg kg^{-1}) or rosemary extract (40 or 200 mg kg^{-1}) (Matthäus et al., 2014).

Biorefinery Concept

The aim of the biorefinery concept is an environmental friendly processing of oilseeds with the comprehensive fractionation of the pretreated crop into oil, protein, and valuable bioactive compounds at the end of the process usable in the food, nonfood, and feed industry. Another characteristic feature of the biorefinery concept is economic activity in closed circles, for example, short routes of transport. A key step is the enzymatic treatment of the raw material, which allows a gentle extraction and fractionation of the different components without the use of organic solvents and without changing the functional properties of the resulting products.

The biorefinery process has been well investigated for rapeseed, which initially was developed by Novo Nordisk and the chemistry department of the Royal Veterinary and Agricultural University, Copenhagen, Denmark in the late 1980s. In the case of rapeseed the concept involves (i) the inactivation of enzymes such as lipoxygenase, myrosinase, and lipases which would adversely affect the final products, (ii) degradation of cell wall constituents by enzymes in milled seed material suspended in water, and (iii) separation of the different fractions, namely, oil, protein-rich meal, syrup, and hulls by centrifugation. The physicochemical properties of this extraction system and the good water solubility of many high-value compounds such as glucosinolates (usable as a natural pesticide) and some proteins, permit the simultaneous extraction of oil and these products from cruciferous oilseed meals (Ugolini et al., 2008).

After cleaning the seeds, and milling in a hammer mill, inactivation of enzymes takes place by heating at 85–90°C for 20 min with water (Jensen et al., 1990). After this heating process, cold water has to be added before treatment with cell wall–degrading enzymes, based on Aspergillus niger, performed for about 4 h at 50°C. Subsequently, the hulls were separated by decantation, oil, protein, and other valuable compounds were obtained by three washing and centrifugation steps, and finally the fractions with protein-rich meal and syrup were spray dried.

The oil yield from a biorefinery is about 35% seed dry matter, which is distinctively lower than for solvent extraction (C. Bagger et al., 2007). Due to the higher content of antioxidants the oil shows a better oxidative stability and during biorefinery processing less phospholipids go into the oil resulting in amounts of 0.03% compared with 1.8% for conventional processing (Anon, 1998). In addition, the content of other unwanted substances is very low, thus further oil purification by refining is not necessary (C. L. Bagger et al., 2003).

Another advantage of the biorefining concept is the mild treatment of the raw material which allows the isolation of valuable bioactive compounds from the meal. This may improve the quality of the meal with regard to its suitability as animal feed. An example is the isolation of pure glucosinolates which adversely affect the usability of meal and limit the use of rapeseed meal as feed. The isolated glucosinolates can be used as natural pesticides in crop management (Bellostas et al., 2007). In the same way pure myrosinases can be isolated directly from the crop (C. Bagger et al., 2007).

WASTE TREATMENT

During oil processing a huge amount of waste is formed in the different steps of refining. It starts with dust arising during cleaning of the seed material followed by the air necessary for drying seeds or desolventizing, to water used for neutralization or degumming, and also residual bleaching earths. The disposal and management of this huge amount of waste is a real challenge to sustainable oil processing today. The legal regulations for air and water pollution are becoming ever stricter and therefore the disposal of waste becomes more and more expensive. Thus, the aim of oil processing has to be to produce as low a level of waste as possible.

One reason why handling waste is very difficult is the number of places waste can arise, but also the influence different types of oil have on it, making it necessary to develop individual concepts of waste disposal for each oil-processing facility. Sources, among others, include heat exchangers, barometric condensors, gas scrubbers, cleaning of greasy floors and equipment, leakages, processed water from the refining plant, and soapstock splitting effluents (Chipasa, 2001). Another problem is that the composition of the waste varies widely depending on the processed oil or oilseed and the processing conditions. This means that air or water needs a comprehensive purification before it can be channeled back into the material cycle, because it can contain pesticides, phenols, metals from bleaching earths, oxidizable substances, and also fats and oils.

On average a refinery generates 700 m^3 of waste water daily, with 80–120 m^3 acid waste water and 470–600 m^2 technological waste water (Chipasa, 2001). Data for the wastewater pollutant concentrations are available only for the chemical oxygen demand (COD), which is in the range of 700–2000 mg L^{-1} (Rosenwinkel et al., 2004). Although different techniques for the removal of oil and other components from the waste water are known, there is no standard method for the treatment of water. In general the treatment comprises a set of methods such as floatation, gravitational methods, biological and chemical treatment, dissolved air floatation, the use of membranes, filtration, and neutralization which, taken step by step, free the water from unwanted and noxious substances.

The installation of fat separators is the minimum required pretreatment of waste water although the installation of a flotation system using additives is a more efficient way to eliminate emulsified, mainly lipophilic substances. Often an additional biological wastewater treatment system, with aerobic activated microorganisms, is used to reduce the biochemical oxygen demand (BOD) and COD. Niewiadomski and Szczepanska (1995) gave a nice overview of the different methods available for the protection of the natural environment by pretreatment and disposal of waste. They give significant attention to the different approaches to pretreatment of liquid waste.

FINAL CONCLUSIONS

With a further rise in the world's population and a subsequent increase in the demand for vegetable oils and fats for nutritional and technical purposes the challenges for a sustainable oil processing industry is increasing. Sustainable oil processing means that the available resources are used without consuming more than can be regenerated from them. To achieve this objective is difficult when an increase in growing areas is strongly limited since the further exploitation of land is not recommended. Thus, a sustainable increase in oil production is only possible by improving the agronomic performance of oilseeds and fruits, by discovery of new oil crops, or by improving the oil-processing technology with regard to higher oil yield during extraction, reduced losses of neutral oil during purification, and a more comprehensive use of the whole oilseed or fruit for the production of valuable bioactive compounds.

Another point which makes sustainable oil production difficult is that this industry has a high energy consumption and high generation of waste (dust, contaminated air, waste water, and contaminated residues such as bleaching earths). Stricter legal regulations have resulted in increased efforts by the oil-processing industry to reduce this waste by purification, reextraction, and recovery of air, water, and hexane. Several measures have been introduced into the process.

For a better, more comprehensive and environmental friendly extraction of oilseeds and fruits different techniques have been developed which are partially used on industrial scales. Unfortunately, most of them are too costly for large-scale applications at the moment but the industrial feasibility of these innovations has been shown to work for small-scale plants. Presently, there is interest in the byproducts of oil processing such as press cake as well as extraction meal increases for isolation of bioactive compounds which are usable in human nutrition, plant protection, or technical applications.

In summary it can be stated that the development of traditional oil processing toward a modern sustainable production is ongoing. Many different approaches are available, and already used in practice, but further efforts have to be made toward increasing the sustainability and environmental compatibility in connection with the economic usefulness of the process.

References

Aachary, A.A., Chen, Y., Eskin, N.M.A., Thiyam-Holländer, T., 2014. Effect of crude canolol and canola distillate extracts on the stability of refined canola oil during deep-fat frying. Eur. J Lipid. Sci. Technol. 116, 1467–1476.

Aalrust, E., Beyer, W., Ottofrickenstein, H., 1992. Enzymatic method for reducing the amount of phosphorous-containing components in vegetable and animal oils. Europe Patent 0,513,709.

Anon, 1987. Report of the World Commission on Environment and Development: Our Common Future. United Nations General Assembly. Transmitted to the General Assembly as an Annex to document A/42/427 – Development and International Co-operation: Environment. Retrieved on: 2009-02-15.

Anon., 1998. Aqueous enzymatic extraction of oil from rapeseeds. Manufacture of food products and beverages. Environmental management centre, international cleaner production information clearinghouse, Denmark, 1991–1994.

Anon., 2001. Codex standard for named vegetable oils; CODEX STAN 210 (amended 2003, 2005).

Anon, 2015. The "DoublePass" Bleaching System. http://www.crowniron.com/userimages/bledouble.pdf (03.08.2015).

Bagger, C., Bellostas, N., Jensen, S.K., Sørensen, H., Sørensen, J.C., Sørensen, S., 2007. Processing – bioprocessing of oilseed rape in bioenergy production and value added utilization of remaining seed components. Tingdon, F.U., Chunyun, G. (Eds.), Proceedings of the 12th International Rapeseed Congress, vol. 5, Science Press USA, Inc., Wuhan, China.

Bagger, C.L., Sorensen, H., Sorensen, J.C., Sorensen, S., 2003. Biorefining, the soft processing alternative. Proceedings of the 11th GCIRC International Rapeseed Congress, Copenhagen, Denmark, 650.

Beach, C.D.H., 1983. Rapeseed crushing and extraction. Academic Press, Toronto.

Becker, K.W., 1983. Current trends in meal desolventizing. J. Amer. Oil Chem. Soc. 60, 216–218.

Bellostas, N., Sorensen, J.C., Sorensen, H., 2007. Biofumigation: from the "classical" approach to the use of biorefined glucosinolates as natural plant protection agents. GCIRC Bulletin n°24. http://gcirc.org/fileadmin/documents/Bulletins/B24/B24_Biofumigation.pdf.

Bocevska, M., Karlovic, D., Turkulov, J., Pericin, D., 1993. Quality of corn germ oil obtained by aqueous enzymatic extraction. J. Am. Oil Chem. Soc. 70 (12), 1273–1277.

Bratkowska, I., 1978. Effect of autoxidation of rapeseed oil on development of melanophosphatides. Acta Alimentaria Polonica 4 (3), 255–262.

Brimberg, I.U., 1982. Kinetics of bleaching of vegetable oils. J. Am. Oil Chem. Soc. 59 (2), 74–78.

Campbell, K.A., Glatz, C.E., 2009. Mechanisms of aqueous extraction of soybean oil. J. Agric. Food Chem. 57, 10904–10912.

Carmona, M.A., Jiménez, C., Jiménez-Sanchidrián, C., Peña, F., Ruiz, J.R., 2010. Isolation of sterols from sunflower oil deodorizer distillate. J. Food Eng. 101, 210–213.

Carr, R.A., 1995. Processing the seed and oil. CAB International, Oxon, UK.

Cater, M.C., Rhee, C.K., Hagenmaier, D.R., Mattil, F.K., 1974. Aqueous extraction – an alternative oilseed milling process. *J. Am. Oil Chem. Soc.*, 51 (4), 137–141.

Chakrabarty, M.M., 2003. Extraction of Fats and Oils. Allied Publishers Pvt. Ltd., Mumbai, India.

Chipasa, K.B., 2001. Limits of physicochemical treatment of wastewater in the vegetable oil refining industry. Pol. J. Environ. Stud. 10, 141–147.

Chisti, Y., 2007. Biodiesel from Microalgae. Biotechnol. Adv. 25, 294–306.

Choukri, A., Kinany, M.A., Gibon, V., Tirtiaux, A., Jamil, S., 2001. Improved oil treatment conditions for soft degumming. J. Amer. Oil Chem. Soc. 78, 1157–1160.

Christensen, M.F., 1991. Extraction by aqueous enzymatic processes. Inform 2 (11), 984–987.

Clausen, K., 2001. Enzymatic oil-degumming by a novel microbial phospholipase. Eur. J. Lipid Sci. Technol. 103, 333–340.

Couri, S., Freitas, S.P., 2001. Aplicação de enzimas na extração aquosa de óleos vegetais. UNICAMP, Campinas.

Czuppon, T., Kemeny, Z., Kovari, E., Rescseg, K., 2008. Process for recovery of plant sterols from by-product of vegetable oil refining, US Patent US 7,368,583 B2.

Dahlke, K., 1998. An enzymatic process for the physical refining of seed oils. Chem. Eng. Technol. 21, 278–281.

Decap, P., Danthine, S.B., Vanbrabant, B., Greyt, W.D., Deroanne, C., 2004. Comparison of steam and nitrogen in the physical deacidification of soybean oil. J. Amer. Oil Chem. Soc. 81, 611–617.

De Greyt, W., Kellens, M., 1996. Refining Practice. Sheffield Academic Press, Sheffield, UK.

Dixit, S., Kanakraj, S., 2010. Enzymatic degumming of feedstock's (vegetable oil) for biodiesel – a review. J. Eng. Sci. Manag. Edu. 3, 57–59.

Dominguez, H., Nunez, J.M., Lema, J.M., 1994. Enzymatic pretreatment to enhance oil extraction from fruits and oilseeds: a review. Food Chem. 49, 271–286.

Dunford, N.T., 2012. Co-products from cereal and oilseed biorefinery systems. In: Bergeron, C., Carrier, D.J., Ramaswamy, S. (Eds.), Biorefinery Co-Products: Phytochemicals, Primary Metabolites and Value-Added Biomass Processing. John Wiley & Sons, Ltd., The Atrium, Southern Gate, Chichester, West Sussex, UK.

Eggers, R., 2004. Processing of oilseeds with increase of minor components. Refining: handling by-products and minor components. March 17/18, 2014, Leipzig, Germany.

Eggers, R., 2008. Innovative Verfahrensansätze in der Speiseölprozesstechnik. Chemie Ingenieur Technik 80, 1059–1068.

FAO, 1992. Principles of oil extraction. FAO Agribultural Services Bulletin No. 94.

Ferrari, R.A., Schulte, E., Esteves, W., Brühl, L., Mukherjee, K.D., 1996. Minor constituents of vegetable oils during industrial processing. J. Amer. Oil Chem. Soc. 73 (5), 587–592.

Friedrich, J.P., List, G.R., Heakin, A.J., 1982. Petroleum-free extraction of oil from soybeans with supercritical CO_2. J. Amer. Oil Chem. Soc. 59, 288–292.

Fullbrook, D.P., 1983. The use of enzymes in the processing of oilseeds. J. Am. Oil Chem. Soc. 60 (2), 476–478.

Guclu-Ustundag, O., Temelli, F., 2007. Column fractionation of canola oil deodorizer distillate using supercritical carbon dioxide. J. Amer. Oil. Chem. Soc. 84, 953–961.

Gupta, S.K., 2007. Rapeseed breeding, Oil Technology, vol. 45, Elsevier, p. 501.

Harbaum-Piayda, B., Oehlke, K., Sönnichsen, F.D., Zacchi, P., Eggers, R., Schwarz, K., 2010. New polyphenolic compounds in commercial deodistillate and rapeseed oil. Food Chem. 123, 607–615.

Homann, T., Schulz, J., Zmudzinski, R., 2006. Verfahren und Vorrichtung zum Pressen. Europe Patent EP1717014.

Huddleston, S., 2014. Adsorption solutions for triglyceride oil refining & pretreatment applications. IOMSA Summer Convention, June 21–24, 2014, Austin, Texas.

Jensen, S.K., Olsen, H.S., Sorensen, H., 1990. Aqueous enzymatic processing of rapeseed for production of high quality products. In: Shahidi, F. (Ed.), Canola and Rapeseed – Production, Chemistry, Nutrition and processing Technology. Van Nostrand Reinhold, New York, pp. 331–343.

Jung, M.Y., Yoon, S.H., Min, D.B., 1989. Effects of processing steps on the contents of minor compounds and oxidation of soybean oil. J. Amer. Oil Chem. Soc. 66, 118–120.

Kellens, M., De Greyt, W., 2000. Deodorization. AOCS Press, Champaign, IL, USA.

Khoo, L.E., Morsingh, F., Liew, K.Y., 1979. The adsorption of β-carotene by bleaching earths. J. Amer. Oil Chem. Soc. 56, 672–675.

Klass, D.L., 1998. Biomass for renewable energy, fuels, and chemicals. Academic Press, San Diego, CA.

Kock, M., 1981. Practical experience with a process for encyme deactivation of soybean flakes before extraction and its influence on the oil quality. American Soybean Association Congress, Antwerpen.

Koski, A., Pekkarinen, S., Hopia, A., Wähälä, K., Heinonen, M., 2003. Processing of rapeseed oil: effects on sinapinic acid derivative content and oxidative stability. Eur. Food Res. Technol. 217, 110–114.

Lamsal, B.P., Johnson, L.A., 2007. Separating oil from aqueous extraction fractions of soybean. J. Amer. Oil Chem. Soc. 84, 785–792.

Lanzani, A., Petrini, C.M., Cozzoli, O., Gallavresi, P., Carola, C., Jacini, G., 1975. On the use of enymes for vegetable oil extraction. A preliminary report. Rivista Italiana delle Sostanze Grasse 52 (7), 226–229.

Maria, A.P., Miguel, A.B., 2010. Production of natural antioxidants from deodorizer distillates: effect of catalytic hydrogenation. Bioresour. Technol. 101, 1369–1376.

Matthäus, B., 2012. Effect of Canolol on Oxidation of Edible Oils. CRC Press, Boca Raton, USA.

Matthäus, B., Pudel, F., Chen, Y., Achary, A., Thiyam-Holländer, T., 2014. Impact of canolol-enriched extract from heat-treated canola meal to enhance oil quality parameters in deep-frying: a comparison with rosemary extract and TBHQ-fortified oil systems. J. Amer. Oil Chem. Soc. 91 (12), 2065–2076.

Moltke Sørensen, A.D., Friel, J., Winkler-Moser, J.K., Jacobsen, C., Huidrom, D., Reddy, N., Thiyam-Holländer, U., 2013. Impact of endogenous canola phenolics on the oxidative stability of oil-in-water emulsions. Eur. J. Lipid Sci. Technol. 115, 501–512.

Müller, M., Eggers, R., 2014. Gas-assisted oilseed pressing on an industrial scale. J. Amer. Oil Chem. Soc. 91, 1633–1641.

Narayana, T., Kaimal, B., Vail, S.R., Surya, B.V., Rao, K., Chakrabarti, P.P., Vijayalakshmi, P., Kale, V., Narayana, K., Ran, P., Rajamm, I.O., Bhaskar, P.S., Rao, T.C., 2002. Origin of problems encountered in rice bran oil processing. Eur. J. Lipid Sci. Technol. 104, 203–211.

Niewiadomski, H., 1990. Rapeseed - Chemistry and Technology. Amsterdam: Elsevier.

Niewiadomski, H., Szczepanska, H., 1995. Protection of the Natural Environment. Polish Scientific Publishers PWN Ltd., Warsaw, Poland.

Nikiforidis, C.V., Kiosseoglou, V., 2009. Aqueous extraction of oil Bodies from maize germ (*Zea mays*) and characterization of the resulting natural oil-in-water emulsion. J. Agric. Food Chem. 57, 5591–5596.

Ohlsen, R., Svensson, C., 1976. Comparison of oxalic acid and phosphoric acid as degumming agents for vegetable oils. J. Amer. Oil Chem. Soc. 53, 8–11.

Olsen, H., 1988. Aqueous extraction of oil from seeds. Asian food conference, Bangkok.

Olsen, H.A.S., 1981. Method of producing soy protein hydrolysate from fat-containing soy material, and soy protein hydrolysate. US Patent 4,324,805.

Pirt, S.J., 1986. The thermodynamic efficiency (quantum demand) and dynamics of photosynthetic growth. New Phytol. 102, 3–37.

Pudel, F., 2012. Rapeseed Proteins – Recent Results on Extraction and Application. CRC Press, Boca Raton, USA.

Pudel, F., Benecke, P., Fehling, P., Freudenstein, A., Matthäus, B., Schwaf, A., 2011. On the necessity of edible oil refining and possible sources of 3-MCPD and glycidyl esters. Eur. J. Lipid Sci. Technol. 113, 368–373.

Pudel, F., Habicht, V., Piofczyk, T., Matthäus, B., Quirin, K.W., Cawelius, A., 2014. Fluidized bed treatment of rapeseed meal and cake as possibility for the production of canolol. OCL Oilseeds Fats Crops Lipids 21, Article number D103.

Ramamurthi, S., McCurdy, A.R., Tyler, R.T., 1998. Deodorizer Distillate: A Valuable Byproductvol. 1AOCS Press, Champaign, IL.

Ramli, M.R., Siew, W.L., Ibrahim, N.A., Hussein, R., Kuntom, A., Razak, R.A.A., Nesaretnam, K., 2011. Effects of degumming and bleaching on 3-MCPD esters formation during physical refining. J Am. Oil Chem. Soc. 88 (11), 1839–1844.

Ranalli, A., de Mattia, G., Ferrante, M.L., 1998. The characteristics of percolation olive oils produced with a new processing enzyme aid. Int. J. Food Sci. Technol. 32, 247–258.

Ribeiro, A.P.B., Bei, N., Goncalves, L.A.G., Petrus, J.C.C., Viotto, L.A., 2008. The optimisation of soybean oil degumming on a pilot plant scale using a ceramic membrane. J. Food Eng. 87, 514–521.

Ricochon, G., Muniglia, L., 2010. Influence of enzymes on the oil extraction processes in aqueous media. OCL 17 (356-359), .

Rosenthal, A., Pyle, L.D., Niranjan, K., 1996. Aqueous and enzymic processes for edible oil extraction. Enzyme Microb. Technol. 19 (6), 402–420.

Rosenwinkel, K.-H., Austermann-Haun, U., Meyer, H., 2004. Industrial Wastewater Sources and Treatment Strategies. Wiley-VCH Verlag GmbH & Co. KGaA, Weinheim, Germany.

Rossel, J.B., Pritchard, J.L.R., 1991. Analysis of Oilseeds, Fats and Fatty Foods. Elsevier Applied Science, London.

Sosulski, K., Sosulski, W.F., 1993. Enzyme-aided vs. two-stage processing of canola: technology, product quality and cost evaluation. J. Am. Oil Chem. Soc. 70 (9), 825–829.

Sosulski, K., Sosulski, W.F., Coxworth, E., 1988. Carbohydrase hydrolysis of canola to enhance oil extraction with hexane. J. Am. Oil Chem. Soc. 65 (3), 357–361.

Sun, D., Qu, D., Lui, F., Lu, Z., 1998. A new process to extract tocopherols and sterols from by-product of refined vegetable oil, CN Patent ZL 9310025.

Torres, C.F.G.T., Seňorans, F.J., Reglero, G., 2007. A two steps enzymatic procedure to obtain sterol esters, tocopherols and fatty acid ethyl esters from soybean oil deodorizer distillate. Process Biochem. 42, 1335–1341.

Ugolini, L., De Nicola, G., Palmieri, S., 2008. Use of reverse micelles for the simultaneous extraction of oil, proteins, and glucosinolates from cruciferous oilseeds. J. Agric. Food Chem. 56, 1595–1601.

Unger, E.H., 1990. Commercial Processing of Canola and Rapeseed: Crushing and Oil Extraction. van Nostrand Reinhold, New York.

Van Eylen, D., Indrawati, M., Hendrickx, M., Van Loey, A., 2006. Temperature and pressure stability of mustard seed (*Sinapis alba* L.) myrosinase. Food Chem. 97, 263–271.

Vavlitis, A., Milligan, E.D., 1993. Flash desolventizings. AOCS Press, Champaign, Illinois, USA.

Vázquez, L., Torres, C.F., Fornari, T., Grigelmo, N., Señoráns, F.J., Reglero, G., 2006. Supercritical fluid extraction of minor lipids from pretreated sunflower oil deodorizer distillates. Eur. J. Lipid Sci. Technol. 108, 659.

Verhé, R.V.T., Van Hoed, V., De Greyt, W., 2006. Influence of refining of vegetable oils on minor components. J. Oil Palm Res. 18, 168–179.

Verleyen, T., Verhe, R., Garcial, L., Huyghebaert, A., de Greyt, W., 2001. Gas chromatographic characterization of vegetable oil deodorization distillate. J. Chromatogr. A921, 277–285.

Voges, S., Eggers, R., Pietsch, A., 2007. Gas assisted oilseed pressing. Sep. Purif. Technol. 63, 1–14.

Wakamatsu, D., Morimura, S., Sawa, T., Kida, K., Nakai, C., Maeda, H., 2005. Isolation, identification, and structure of a potent alkyl-peroxyl radical scavenger in crude canola oil, canolol. Biosci. Biotechnol Biochem 69 (8), 1568–1574.

Wehrmüller, K., 2008. Impact of dietary phospholipids on human health – technical-scientific information. *ALP science*, vol. 524. Forschungsanstalt Agroscope Liebefeld-Posieux (ALP), Bern, pp. 1-15.

Young, F.V.K., Poot, C., Biernoth, E., Krog, N., Davidson, N.G.J., Gunstone, F.D., 1994. Processing of Fats and Oils. In: Gunstone, F.D., Harwood, J.L., P.F.B., (Eds.), The Lipid Handbook, second ed. Chapman and Hall, London, pp. 249–318.

Zhang, S.B., Zhang, W., Xu, S.Y., 2007. Optimization of the aqueous enzymatic extraction of rapeseed oil and protein hydrolysates. J. Amer. Oil Chem. Soc. 84, 97–105.

Zschau, W., 1999. Die Bleichung von Speisefetten und Ölen .lf.VII. Qualitätskontrolle der Bleichung. Fett/Lipid 101 (1), 117–119.

Zschau, W., 2001. Bleaching of edible fats and oils. Cooperative work of the German Society for Fat Science (DGF). Eur. J. Lipid Sci. Technol. 103, 505–551.

CHAPTER

20

Integrated Pest Management

Dharam P. Abrol, Uma Shankar

Faculty of Agriculture, Division of Entomology, Sher-e-Kashmir University of Agricultural Sciences & Technology Chatha, Jammu, Shalimar, J&K, India

INTRODUCTION

Oilseed crops have been the backbone of several agricultural economies from antiquity and play a prominent role in agricultural industries and trade throughout the world. A major amount of edible oil comes from these crop resources. The quantum increase in production through various methods of improvement and technological intervention has also offered a congenial and suitable environment for insect pests. Among several factors, insect pests are a significant constraint, limiting the production of oilseeds. Oil from the seeds of plants belonging to the genus *Brassica*, family Cruciferae, have been utilized by man for thousands of years (Prakash and Hinata, 1980), but it is only during the last 30 years that oilseed *Brassica* crops have become internationally important (Lamb, 1989). The two most important commercial oilseed crops are *Brassica campestris* var. *oleifera*, which has a worldwide distribution, and *B. napus* var. *oleifera*, which tends to be restricted to Europe and North Africa (Weiss, 1983). The other oilseed crops are groundnut, sunflower, soybean, safflower, and sesame yielding high-grade, edible oil. The nonedible oil-producing species include castor and linseed.

SCENARIO OF OILSEED CROPS THROUGHOUT THE WORLD

There are about 40 different oilseeds crops in the world whose oil can be consumed (Lennerts, 1983), but only a few are significant. Oilseed crops are of three distinct categories: the annual or biennial group, which includes soybeans, sunflower, groundnuts, and rapeseed; the perennial tree crops including coconuts and oil palms. Cotton and corn germ belong to the third group in which the embryo is a by-product. Other oilseeds are of minor importance in world trade but they do play a significant role in local markets or serve as raw materials for

Breeding Oilseed Crops for Sustainable Production. http://dx.doi.org/10.1016/B978-0-12-801309-0.00020-3

special products (Sharma et al., 2012). *Brassica* species are the second largest oilseed crop after soybean (*Glycine max* (L.) Merr.) in world oilseed production, surpassing peanut (*Arachis hypogaea* L.), sunflower (*Helianthus annuus* L.), and cottonseed (*Gossypium hirsutum* L.) over the last two decades (FAO, 2010; Raymer, 2002; Sharma et al., 2014).

Of the 37 species in the *Brassica* genus, the 4 most widely cultivated species for oilseed and vegetable production are *B. rapa* L., *B. juncea* (L.) Czernj and Cosson, *B. napus* L., and *B. carinata* A. Braun (Raymer, 2002; Rakow, 2004; Sovero, 1993). Rapeseed is the most favored vegetable oil. The leading producers are Canada, China, India, Germany, and Ukraine along with Poland, the United Kingdom, and Australia, which contribute about 77% of the total rapeseed mustard production of the world (FAO, 2010). Germany has the highest productivity of rapeseed (37.60 q ha^{-1}) followed by the United Kingdom (32.98 q ha^{-1}), and the Czech Republic (29.38 q ha^{-1}).

Oil palm is an alternative and potential long-term source of edible oil, which is expected to contribute significantly toward meeting the growing edible oil demand. It can yield, on average, 4 tons of vegetable oil per hectare as compared with less than half a ton from other annual oilseed crops. Palms are grown predominantly in the tropical areas of the world as perennial trees and are an important source of vegetable oil. Production of palm kernels was around 11.74 MMt; however, with respect to oil production, palm oil dominates the world vegetable oil economy with its production touching 42.41 MMt during 2008–2009. It contributes about 36% of the total world oil production. The major palm oil-producing countries are Indonesia, Malaysia, Thailand, Nigeria, and Colombia which together contributed about 93% of total world palm oil production in 2008–2009. Cotton is cultivated in 70 countries worldwide with a total coverage of about 34 Mha and production of around 41.26 MMt of cottonseed and 4.72 MMt of cotton seed oil in 2008–2009 (FAO, 2010). Over a quarter of the world cotton area is in India, followed by the United States (16%), China (14%), and Pakistan (8%). The remaining production comes from Turkey, Australia, Greece, Brazil, and Egypt. Cotton is a dual-purpose crop, producing both seed and fiber as valuable primary agriculture products. Groundnut (*Arachis hypogaea* L.) is a leading oilseed crop in India and an important oilseed crop found in tropical and subtropical regions of the world. The seeds are a rich source of edible oil (43–45%) and protein (25–28%). Groundnut (*A. hypogaea* L.) is the most important oilseed and a highly explored agricultural commodity. Groundnut is grown in about 84 counties over an area of 24.52 Mha with a production of 34.15 MMt and average productivity of 13.92 q ha^{-1}. Among the groundnut producing nations, India sows the greatest area (around 28% of total world's area) under this crop and is the second largest producer, next only to China. In India, groundnut occupies an area of about 6.85 Mha with a production of about 7.34 MMt. China, India, Nigeria, the United States, and Myanmar contributed about 85% of the total world's production of groundnut in 2008. China (31.02 q ha^{-1}), the United States (38.29 q ha^{-1}), Nicaragua (36.10 q ha^{-1}), Turkey (34.33 q ha^{-1}), Egypt (33.99 q ha^{-1}), and Syrian Arab Republic (31.87 q ha^{-1}) are countries having very high average yields and thus contribute significantly to the world's pool of groundnut crops.

Among the oilseed crops, sunflower (*H. annuus* L.) occupies fourth position in area and production in the world after soybean, rapeseed mustard, and groundnut. The world harvest of sunflower was 33.03 MMt from an area of 18.98 Mha during 2008. The Russian Federation, Ukraine, Argentina, India, and China are the top five countries in the world with respect to production (21.49 MMt) and area under crop (15.93 Mha). Europe alone accounts for about

50% of world's area and total production. Asia and Africa share about 21% and 6% of the total sunflower area in the world, respectively. The average yield of the crop is significantly high in countries like Crotia (31.03 q ha^{-1}), Austria (29.73 q ha^{-1}), and Switzerland (29.00 q ha^{-1}). The United States, Romania, Bulgaria, and Hungary are the other countries, which have high potential for this crop. Olive, *Olea europaea* L., is a subtropical, evergreen tree and one of the world's oldest cultivated crops originating in the Mediterranean region. The olive tree is the sixth most important oil crop in the world. World olive oil production is around 2.9 MMt, the Mediterranean countries being its major contributors (International Olive Council 2009, http://www.internationaloliveoil.org/downloads/production1_ang.PDF). Spain has the largest area (2.60 Mha) and highest production (6.22 MMt) followed in production by Italy, Greece, Turkey, Tunisia, and Greece. These five countries accounted for 80.84% of the total world's production (18.32 MMt) in 2008. Egypt has the highest productivity (63.60 q ha^{-1}) followed by Peru (55.46 q ha^{-1}) and Australia (54.40 q ha^{-1}). The United States, Chile, Mexico, Argentina, and Slovenia are some of the other countries having high-productivity levels. Sesame (*Sesamum indicum* L.) is known as the "queen of oilseeds" and one of oldest oilseed crops grown in India. India ranks first in area under cultivation representing 30% of the world's production. Castor (*Ricinus communis* L.) is one of the industrially important, nonedible oilseed crops of the world. India ranks first among the major castor producing countries (as well as Brazil and China) of the world, with 68% of the area for the castor crops being in India, which also contributes to 76% of world's production.

THE SCENARIO OF OILSEED CROPS IN INDIA

India is among the largest oil economies in the southeast Asian region and occupies a distinct position in terms of diversity in annual oilseed crops. The prevailing agroecological conditions have been favorable for growing several important annual oilseeds, including edible (namely, groundnut, rapeseed mustard, soybean, sunflower, safflower, and sesame) and nonedible oilseeds (namely, castor and linseed). In addition, a wide range of other minor oilseeds and oil-bearing tree species add to both the diversity and production of oilseed in the country. On the world map, India occupies a prominent position in acreage and production of oilseed crops. It constitutes the second largest agricultural commodity next to cereals in India, sharing 14% of the country's gross cropped area, accounting for nearly 5% of the gross national product, and 10% of the value of all agricultural products. Today, oilseed crops are cultivated on about 16.5 Mha, with a total production of 10 MMt, which constitutes approximately one tenth of the total cultivated area in India. The major oilseed crop–growing states are Madhya Pradesh (24.1%), Gujarat (11.6%), Andhra Pradesh (11.5%), Rajasthan (11.4%), Maharashtra (10.9%), Karnataka (9.5%), Uttar Pradesh (6%), and Tamil Nadu (4.5%). Approximately 15 million people are engaged in the production and processing of oilseeds (Hegde, 2000). In India oilseed crop importance is ranked thus: groundnut, rapeseed mustard, sesame, linseed, safflower, castor, sunflower, and niger. More than half of the world's dietary fats and oils come from oilseeds: coconuts, cotton, oil palm, olives, peanuts, rape, soybeans, and sunflower (Guidry, 1964).

India contributes a large share of the global castor production (76.9%) and also makes a substantial contribution to the production of sesame (31.2%) and groundnut (25.1%). The country is the largest producer of castor and sesame and the second largest producer of groundnut and rapeseed mustard (second only next to China).

Castor productivity in India is more than the world's average mainly due to high productivity in Gujarat, India, where castor is cultivated under irrigated conditions (it is otherwise grown in semiarid and arid regions). Castor is grown in almost all states of the country, Gujarat, Andhra Pradesh, Tamil Nadu, Rajasthan, and Karnataka account for about 51, 29, 6, 5, and 3% of the total area under the crop, respectively. Gujarat alone contributes 83% of the total country's production.

CONSTRAINTS IN OILSEED CROP PRODUCTION

Oilseed crops are affected by a large number of insect pests of which, about a dozen, are of serious concern. These include leaf miner, thrips, and tobacco caterpillar. Specifically, head borer in sunflower; aphids and *Helicoverpa armigera* in safflower; leaf webber and capsule borer in sesame; aphids, sawfly, cabbage butterfly, and painted bug in rapeseed mustard; semilooper and Bihar hairy caterpillar in castor; and bud fly and *H. armigera* in linseed. These crops are often attacked by polyphagous pests like red hairy caterpillar, Bihar hairy caterpillar, root grubs, termites, cutworms, and serpentine leafminer. Yield losses due to these pests can range from 50–90% across oilseed growing areas. Soybean crops are often severely attacked by about a dozen insect pests causing damage ranging from 20–100%. However, with implementation of an integrated pest management (IPM) schedule, 30–35% additional yield can be obtained (Sharma et al., 2014). In order to develop high-yielding genotypes having resistance to insect pests, detailed knowledge of the insects, the plant, and their interaction is important. In general, the available oilseed crop cultivars are susceptible to major insect pests, however, some have shown resistance to some insect pests. Although the use of insecticides is quite effective, their excessive and injudicious use has resulted in an imbalance in agroecosystems. Therefore, the answer lies in the development of resistant/tolerant varieties, exploitation of biological control agents, and use of biorational approaches to combat the major insect pests facing oilseed crops. One viable option is the adoption of IPM involving several approaches like seed treatment with *Trichoderma* sp., release of *Trichogramma,* use of pheromone traps, and use of measured applications of pesticides based on specific needs, etc. New technology has made significant impacts in the field of IPM.

Climate change, food insecurity, and energy demand are major concerns for modern agriculture and their impact is increasing rapidly. A rapidly growing human population has resulted in a demand for increased food production. As land resources become limited, efforts have been made to increase productivity by combatting losses inflicted by insects, weeds, and plant pathogens through pesticide intervention. At the same time, pesticides have proved to be dangerous due partly to their indiscriminate and excessive use, contaminating food and the environment, and resulting in pest resistance, pest resurgence, and pest outbreaks. Consequently, the need arose for ecofriendly, IPM strategies to produce food safe from the negative impacts of pesticide residues (Abrol and Shankar, 2014). The IPM principle does not preclude chemical pesticide use, but rather uses it as one of the tools in the management package, to be used prudently whilst being integrated with other tools. Thus, the concept of the IPM of insects contains three basic elements: (i) maintaining insect populations below levels that cause economic damage;

(ii) the use of multiple tactics to manage insect populations; and (iii) the conservation of environmental quality. Excessive and indiscriminate use of pesticides endangers the health of farm workers and consumers of agricultural products worldwide (Goodell, 1984). Many of the steps of IPM depend upon accurate and readily available information for proper decision-making. Approaches to be used within IPM strategies include: (i) the greater use of monitoring to determine pest abundance more accurately; (ii) the use of trap crops to attract and to concentrate pests thereby reducing pest density and damage to the main crop; and (iii) the promotion of conservation biological control with particular emphasis on minimum tillage and avoidance of insecticides during flowering (Cook and Denholm, 2008). The important insect pests and their management strategies, for different oilseed crops, are briefly discussed in the sections below.

IMPORTANT INSECT PESTS OF OILSEED CROPS

Insect Pests of *Brassica*

Many species of the insect pests that attack the oilseed *Brassica* crops also attack other cruciferous vegetables such as cabbage and radish. The common pests of crucifers are described by Bonnemaison (1965) prior to oilseed rape becoming an important crop in Europe and North America. Crucifers crops are susceptible to attack by insect pests in the early seedling stages, that is, from emergence until they have 4–5 leaves. The number of insect pests associated with rapeseed and mustard crops has increased from 24 (Rai, 1976) to about 40 (Bakhetia and Sekhon, 1989; Sachan and Gangwar, 1980) owing to changed cropping patterns and the introduction of new crops like *B. napus* and *B. carinata*. The mustard aphid, *Lipaphis erysimi* (Kaltenbach), is the key pest responsible for the bulk of yield losses. The most serious pests of the oilseed *Brassica* crops are members of the order coleoptera, particularly in Europe and North America. On the Indian subcontinent and in New Zealand, aphids cause severe damage (Bakhetia and Sekhon, 1989; Lammerink and Banfield, 1980; Singhvi et al., 1973), although in India insect pests from other orders are also important (Bakhetia and Labana, 1978). In Asia, insecticides are applied to control aphids and the diamondback moth, *Plutella xylostella*, in oilseed *Brassica* crops (Lee, 1968), but relatively little on the subject has been published. Similarly, little has been published on the pests of oilseed *Brassica* crops in South America, Australia, and Africa. It is important to correctly identify the pest causing damage so appropriate control measures may then be taken in good time.

The flea beetle, *Phyllotreta cruciferae*, is not usually a significant pest in Europe (Augustin et al., 1986), although it is the dominant flea beetle pest of oilseed *Brassica* crops in Canada (Burgess, 1977) and in India (Bakhetia and Labana, 1978). Root maggots, *Delia* spp., are significant pests of rape only in parts of Alberta, Canada (Griffiths, 1986) and in Finland (Tiittanen and Varis, 1960), although *Delia radicum* is a widespread pest of other crucifers throughout most of Europe and North America (Bonnemaison, 1965). The externalities triggered by the chemical control of insect pests have necessitated the search for alternative control methods. Among these, the role of resistant varieties comprises the foremost strategy because of its well-documented advantage. In India, in the recent past, extensive work has been done on the various aspects of host plant resistance (HPR) on the key pest (*Lipaphis erysimi*), however, work on the other major insect pests of rapeseed and mustard has been negligible.

***Mustard Aphid,* Lipaphis erysimi *Kalt., and* Brevicoryne brassicae**

This aphid is active throughout the year, however, its main activity period coincides with the growing period of cruciferous crops from September until March. During the lean season the aphid shifts to wild or cultivated off-season crucifers in damp places in orchards and kitchen gardens (Sidhu and Singh, 1964). Different off-season hosts have been recorded for this pest like Asvagandha (*Withania somnifera*) from May to July, cauliflower (*B. oleracea*) from August to March, cabbage (*Brassica oleracea* var. capitata), Indian mustard (*Brassica juncea*) and radish (*Raphanus sativus*) from March to April, (Chandra and Kushwaha, 1987). Mustard aphid (*Lipaphis erysimi* Kalt and *Brevicoryne brassicae* L.) is a key pest of rapeseed mustard (Figures 20.1–20.3), causing a 35–75% reduction in yield (Rohilla et al., 1987; Rana, 2005) and a 6% reduction in oil content (Singh et al., 1987). A minimum temperature in the range of 4–13.8°C, a mean maximum temperature ranging between 14.0°C and 30.0°C, and 48–80% relative humidity, favor quick aphid multiplication (Sekhon and Bakhetia, 1994; Bishnoi et al., 1992). Rainfall has been reported to have a profound effect on aphids with mild rainfall helping with aphid

FIGURE 20.1 **Mustard aphid, *Lipaphis erysimi*, on flowers.** *Courtesy: Dr Uma Shankar, SKUAST-Jammu.*

FIGURE 20.2 ***Lipaphis erysimi* infestation on mustard siliqua.** *Courtesy: Dr Uma Shankar, SKUAST-Jammu.*

FIGURE 20.3 ***Brevicorny brassicae* infestation on mustard pods/crucifers.** *Courtesy: Dr Uma Shankar, SKUAST-Jammu.*

multiplication when crops are flowering (Singh et al., 1990). Rainfall totally eradicates aphids at the pod stage of this crop (Mahal et al., 1988).

Flea Beetle, *Phyllotreta cruciferae Gorge*

This pest sometimes becomes very serious causing economic damage to a crop. It has been reported to cause serious damage to the rapeseed mustard crops in Uttar Pradesh and West Bengal while moderate incidence has been reported in all areas of India (Rai, 1976) and other parts of the world (Burgess, 1977; Tansey et al., 2009). Adult beetles are present in the fields all year round except from November–January, when they hibernate in the soil or plant debris. Furthermore, Makhmoor and Shinde (1986) had observed a positive correlation between flea beetle incidence and nitrogen application.

Painted Bug, *Bagrada hilaris, B. cruciferarum*

This pest causes damage to rapeseed and mustard crops at the seedling and maturity/harvesting stage of the crop (Fig. 20.4). Both adults and nymphs suck the sap from the leaves, young plants, or even from the seed. The seedling stage of the crop is the most sensitive and

FIGURE 20.4 **Painted bug, *Bagrada* on mustard.** *Courtesy: Dr Uma Shankar, SKUAST-Jammu.*

may even be killed if heavy incidence of this pest occurs at this time (Vora et al., 1985). On full-grown plants the nymphs and adults congregate on the pods and suck the sap.

Mustard Sawfly, *Athalia lungens Klug*

This pest preferably feeds on mustard plants if given the choice, otherwise it is considered a serious pest to all the cruciferous crops at seedling stages (Vora et al., 1985). The grub of the sawfly causes the most damage, feeding on the leaves (Figs 20.5 and 20.6), causing shot holes and sometimes riddling the whole leaf in severe situations. In the case of less damage, the plant recovers quickly and the shot holes are of no economic importance. In some months this pest aestivates as a grub/pupa and again becomes active when the temperature starts to decline. Mustard sawfly completes three overlapping generations from September–January.

Pea Leafminer, *Chromatomyia horticola*

This pest is a polyphagous pest and may cause damage of 15.2% to crop yield (Singh, 1991; Singh et al., 1984). It also affects plants of compositae and leguminoseae in addition to the family Cruciferae. The maggots of this pest mine the leaves with typical gallery formation. The larvae bore the mesophyll and spongy parenchymatous tissues of the leaves. They do not usually consume the entire leaf surface. The mines run both on the upper and lower leaf

FIGURE 20.5 **Sawfly larva feeding on mustard foliage.** *Courtesy: Dr Uma Shankar, SKUAST-Jammu.*

FIGURE 20.6 **Adult sawfly foraging on mustard flowers.** *Courtesy: Dr Uma Shankar, SKUAST-Jammu.*

surface. During summer and autumn months, it aestivates and a fly emerges in December with its peak activity period being February–April. It has 4–5 generations in a year. The population of this pest generally cannot sustain temperature conditions beyond 30°C.

Management Strategies

CULTURAL CONTROL

Fewer nitrogen applications normally restrict fecundity, faster development, and multiplication of aphids.

NATURAL ENEMIES

There are many natural enemies, which play role in the management of rapeseed and mustard insect pests at one or another stage of crop growth. Mustard aphid is one of the most problematic pests. It is parasitized by *Diaretiella rapae* up to 93% (Chandra and Kushwaha, 1987) (Fig. 20.7). Parasitization is quite high in January and February. Pea leafminer has been reported to be parasitized by *Chrymotomyia formosa, Diglyphus isaea, Opius exigua,* and Sohegion sp. by up to 86% (Kumar, 1985). Many coccinellids like *Coccinella septempunctata* (Figs 20.8 and 20.9), *Macaria sexmaculata, Brumoides suturalis, Hippodamia variegata, Scymnus nubilus,* and *Verania viacta* have been found to be active in aphid infested fields (Atwal et al., 1971; Kalra, 1988; Rai et al., 2010). *Ischiodon scutellaris* (Fig. 20.10) is the only syrphid feeding on mustard aphids (Cherian, 1934; Kumar et al., 2011) (Fig. 20.9). In Punjab, *Diaeretiella rapae* parasitized only 8.2% of the aphids in February, increasing to 100% in April (Atwal et al., 1969). A bacteria, *Serratia marcescens,* has been reported to cause disease to *Athalia proxima* larvae (Bogawat et al., 1966). On the other hand *Entomophthora* sp. and *Cephalosporium* sp. have been reported to cause disease in aphids, *Isipaphis erysimi* (Singh, 1982). The mustard aphid is known to be preyed upon or parasitized by a large number a predators and parasites like coccinellids, syrphids, *Chrysoperla carnea, Aphidius* sp., etc. There are only a few reports on the biology and efficacy of syrphids and coccinellids in the laboratory and the population dynamics of the aphid in the field. In India, the rate of parasitism of mustard aphid, *Lipaphis erysimi* (Kaltenbach) on rapeseed mustard crops varied from 20% (Thakur et al., 1989) to 51% (Dogra et al., 2003) in Himachal Pradesh to 60–97% in Maharashtra (Katole et al., 1998)

FIGURE 20.7 **Parasitoids probing for aphid parasitization.** *Courtesy: Dr Uma Shankar, SKUAST-Jammu.*

FIGURE 20.8 ***Coccinella septumpunctata*** **feeding on mustard aphids.** *Courtesy: Dr Uma Shankar, SKUAST-Jammu.*

FIGURE 20.9 ***Coccinella*** **grub feeding on mustard aphids.** *Courtesy: Dr Uma Shankar, SKUAST-Jammu.*

FIGURE 20.10 **Adult syrphid fly foraging on a mustard flower for nectar.** *Courtesy: Dr Uma Shankar, SKUAST-Jammu.*

MECHANICAL CONTROL

Mechanical removal of twigs and leaves harboring the hairy caterpillar pest is very effective and is advised on a community basis. In the case of the mustard aphids, heavily infested twigs removed without disturbing the whole plant can check their proliferation.

Insect Pests of Groundnut

Insect pests form a major constraint in the production of groundnut. Red hairy caterpillars: *Amsacta albistriga* Walker, *Ambulyx moorei*; leafminer: *Aproaerema modicella* Deventer; bihar hairy caterpillar: *Spilosoma obliqua* Walker; tobacco caterpillar: *Spodoptera litura* (Fabricius); aphids: *Aphis craccivora* Koch; and leaf hopper: *Empoasca kerri* Pruthi, are problematic insect pests of groundnut crops in India. In recent years the incidence of thrips on groundnut crops has increased and it is known to cause yield losses of 14–40% (Ghewande, 1987) and also is responsible for spreading a viral disease called bud necrosis in groundnut. The leafminer is considered the most important insect pest of groundnut in South India, particularly under rainfed situations (Ayyar, 1963; Nair, 1975; Reddy, 1988). The pest initially appears as a leaf miner causing short, blister-like mines. Older larvae fold the leaflets and feed within them. As a result, the leaflets turn brown, shrivel, and dry up. A severely infested crop gives a burnt up appearance and yield losses can reach up to 76% (Anonymous, 1986).

Management Strategies

CULTURAL CONTROL

Mixed sowing of cereals with groundnut is common practice and is known to reduce the incidence of insect pests like thrips, jassids, and leafminer. Mulches and summer ploughing can also protect crops in many different ways. Summer ploughing is known to reduce grub population significantly (>70%).

RESISTANT VARIETIES

It has been reported that the bunch varieties have a higher population of, and subsequently showed more injury owing to, leaf webber, aphid, and grey weevil, whereas the spreading and semispreading types harbored lower populations (Brar and Sandhu, 1976). This difference may be due to more foliage and more succulence of bunch types, which provide conditions favorable for better shelter and food. A high level of resistance to jassids has been noticed in EC 5240. Successful trasfer of jassid resistance from EC 5040 and EC 5043 into suitable lines, having good agronomic traits and high yields, has been achieved.

BIOLOGICAL CONTROL

More than 30 species of hymenopterous parasitoids attack the eggs, larvae, and pupae of *Aproaerema modicella* on groundnut and soyabean in India. Gadad and Hegde (2014) evaluated the efficacy of different biopesticides and biorationals against groundnut thrips and leaf miner at Dharwad, India. Results revealed that spinosad 45 EC at 0.20 ml L^{-1}, *Lecanicillium lecanii* at 6 g L^{-1}, and *L. lecanii* at 4 g L^{-1} was found to be effective in reducing the thrips population, whereas spinosad 45 EC at 0.20 ml L^{-1}, buprofezin 25 EC at 1.0 ml L^{-1}, *Beauveria bassiana* at 6 g L^{-1}, and *Metarrhizium anisopliae* at 6 g L^{-1} was found to be effective against leaf miner. Considering the cost of cultivation and the gross profit in different treatments, the benefit cost ratio (BCR) and net profit was calculated. Among the different treatments, spinosad and buprofezin recorded a higher BCR and net profit and was found to be superior to all other treatments in the trial. Among the biopesticides, *L. lecanii* at 6 g L^{-1} recorded a higher BCR and net profit compared with other biopesticide treatments.

Trichogramma evanescens minutum has often been recorded parasitizing the eggs of *Amsacta albistriga* and *Diacrisia obliqua* (Krishnamurti and Usman, 1954). *A. albistriga* eggs are frequently parasitised by *T. eyanescens minulum* with an effective recovery of 59.4% (Sundaramurthy et al., 1976; Gunathilagaraj and Babu, 1987). However, only 10.3% parasitization has been recorded in *Aproerema modicella* eggs in this crop (Muthiah and Kareem, 2000). Leaf miner, *Aproerema modicella,* have been parasitized by an eulophid, *Stenomesius japonicum,* and a chalcid, *Brachymeria wittei* for the first time by Muthiah (1991). *Goniozus indicus* and *B. wittei* have been found parasitizing 16.7% and 20% of the larvae of *A. modicella* in Tamil Nadu (Muthiah and Kareem, 2000). It was found that 26% of *Helicoverpa armigera* pupae were found parasitized by *Carcelia illota* and *Chelonus* sp. in Tamil Nadu (Muthiah and Kareem, 2000).

Coccinellids, *Menochiles sexmaculatus* and *Coccilnella septempunctata* have been found responsible for a 10–97% decrease in the population of *Aphis craccivors* in the groundnut grown in Gujrat (Patel et al., 1976). Patel and Vyas (1984), while studying the feeding potential of *M. sexmaculata* reported them to be consuming 474 aphids during their 6 days of larval development. *Ischiodon scutellaris, Xanthogramma scutellaris,* and *Chrysoperla carnea* were also recorded predating upon *A. craccivora* in Gujrat. Thrips on the other hand were being preyed upon by *C. septempunctata, Orius* sp. and *M. sexmaculata.* In another very important observation, Sahayaraj (1999) reported a 92.73% and 94.61% reduction in the population of *Helicoverpa armigera* and *Spodoptera litura,* respectively by *Rhinicoris marginatus.*

Bacillus thuringiensis have been found to be effectively managing the population of *Aproerema modicella* up to 36.4% (Brar and Sandhu, 1982; Chandla et al., 1988). *Holotrichia consanguinea,* a very important pest of groundnut in the subtropics, was effectively managed by *Bacillus popillae* var. holotrichiae at 5 billion spores per square meter (Vyas et al., 1991). A fungi, *Metarrhizium anisopliae,* was found infesting *Helicoverpa armigera* by Kalra and Kulshreshtha (1961). *A modicella* was found diseased by *Beauveria bassiana* by up to 30.80%. Many caterpillars infesting groundnut crops have been reported to be infected by nuclear polyhedrosis virus (NPV). SNPV (subgenus of NPV) was found to be causing 100% infection in *Spodoptera litura* larvae on castor followed by groundnut with 98.17% infection under laboratory conditions (Gururaj et al., 1999). A protozoon, *Nosema* sp. was parasitizing *S. litura* to the extent of 13.8% in Andhra Pradesh (Sridhar and Prasad, 1996).

BOTANICALS

Neem seed extract is found to be as effective against *Aproerema modicella* and *Aphis craccivora* as phosphamidon.

Insect Pests of Sesame

Sesame crops are attacked by several insect pests, which cause considerable damage (Satyagopal et al., 2014). The main insect pests include sesame shoot webber and pod borer, *Antigastra catalaunalis;* sesame gall fly, *Asphondylia sesame;* sphingid moth, *Acherontia* styx, and hairy caterpillar, *Spilosoma obliqua.*

FIGURE 20.11 **Sesame pod borer.** *Courtesy: Dr Uma Shankar, SKUAST-Jammu.*

Sesame Shoot Webber and Pod Borer, **Antigastis catalaunalis**

This pest is distributed throughout the sesame growing area of India. It causes terminal destortion of plants through rolling and webbing of a few top leaves. Early stage infestation (10–15 day old plants) results in plant death without producing any branches or shoots. Sometimes larvae enter the shoots and pods (Fig. 20.11). One larva can destroy 2–3 young plants and 4–6 flower buds. This leads to 10–70% plant infestation, 1–43.5% pod infestation, and 8.86–71.53% yield lose. This pest completes 14 generations in a year and is active from March–December.

Sesame Gallfly, **Asphondylia sesami**

The female lays its eggs in flower buds. The incubation period is 2–4 days. Its maggots feed on floral parts, especially the gynaeceum. The plant transforms into malformed gall-like buds, which fail to turn into seed capsules. The infested buds wither and fall into ground. The pest is active from the flower initiation stage of the plant. The intensity of the pest is 1–21%.

Sphingid Moth, **Acherontia styx**

This is massive caterpillar (7.5–10 cm) feeds voraciously on foliage. Robust moths are known as honey robbers as they steal honey from beehives. The pest is active from mid March to early December but maximum infestation is in September. Full-grown larvae are green with oblique yellow streaks along their sides. Larvae can be recognized by the presence of a curved horn on their dorsal sides. Three colors of larvae have been identified, green, yellow, and brown.

Hairy Caterpillars, **Diacrisia obliqua,** *and* **Amsacta moorei**

This pest is highly polyphagous and defoliates a number of kharif crops (Fig. 20.12). The larvae have long hairs all over their bodies. Young caterpillars are gregarious. The insects feed on all parts of the crop except the stem. There is only one generation of *A. moorei* each year and 3–8 generations of *D. obliqua* per year.

Pod-sucking Bug, **Nezara virudula, Aphanus sordidus,** *and Thrips*

These bugs suck the sap from pods of standing crops (Fig. 20.13) and even from the crops stored on thrashing floors, while thrip infestation results in malformation of the stamens,

FIGURE 20.12 **Bihar hairy caterpillar.** *Courtesy: Dr Uma Shankar, SKUAST-Jammu.*

FIGURE 20.13 **Green Stink bug nymphs infesting sesame pods.** *Courtesy: Dr Uma Shankar, SKUAST-Jammu.*

injury to the ovary, and development of a dark pigment on the fruit wall instead of the usual green coloring.

Management Strategies

CULTURAL CONTROL

Sowing the crop during June and July will avoid leaf webber damage. Delayed sowing resulted in a significantly higher level of damage to leaves, flowers, pods, and also poor yield (Patra, 2001). Intercropping with pigeon pea (Nath et al., 2003) and black gram, green gram, cluster bean, sorghum, and pearl millet (Ahirwar et al., 2009; Ahuja et al., 2009) was found to significantly reduce leaf webber damage.

MECHANICAL CONTROL

Collection and destruction of infested parts of plants further minimizes the damage caused by caterpillars. If possible manual collection and destruction of larvae should be undertaken in order to reduce the population build up.

RESISTANCE TO SESAME LEAF-ROLLER

It has been reported that the branching pattern and maturity of the plants are positively correlated with their resistance to leafroller. Profuse branching and early maturing varieties provide conditions, which are more conducive to pest development. The late maturing varieties NP 29, C 1036, and Chanda 3 have been reported to be less susceptible to the pest.

BIOLOGICAL CONTROL

Conservation of existing natural enemies (spiders, coccinellid beetles, predatory stink bugs, preying mantids, black ants) and parasitoids (Braconids and Ichneumonids) through economic threshold level (ETL) based (2 webbed leaves m^{-2} or 10% damage) applications of botanical insecticides and safer chemicals is encouraged. Augmented release of parasitoids, namely, *Trathala flavoorbitalis* (Behera, 2011) and *Apanteles* sp., and predators like *Chrysoperla carnea* also reduce help to reduce population build up. Sesame leaf webber and pod borer, *Antigastra catalaunalis* is parasitized by five different parasitoids. Cheema and Singh (1987) have recorded parasitization up to 20% in Punjab, India whereas, Kalra (1989) recorded parasitism up to 42%. In Gujrat, the *Diadegma* sp. has been reported as parasitizing leaf webber and pod borer up to 50% (Patel and Bhalani, 1990). In earlier studies, Khan and Hussain (1964) found the Lucern caterpillar, *Laphygina exigua* parasitized by the Mermethid nematode, *Mermis indica* up to 92.8% and Rao et al. (1980) recorded parasitization up to 93% of *Archerontia styx* by *T. australiacum.*

BIOPESTICIDES

It is suggested to spray neem oil at 1% or Neem Seed Kernel Extract at 5% at the early stage of infestation. Application of bioinoculants (*Azospirillum*) induces insect resistance among the treated plants and recorded a minimum leaf damage (Anandh et al., 2010) by increasing levels of phosphorus and potassium in the plants (Selvanarayanan, 2013). A chemical control spray of carbaryl at 0.1% or dusting carbary at 15% reduces damage. Two sprays of quinalphos at 0.05% 30 and 45 days after sowing will control the pest effectively.

Insect Pests of Sunflower

Besides, other factors, attack from insect pests is a major constraint in sunflower seed production, which causes considerable economic loss (Atwal and Dhaliwal, 2009). A large number of insect pests have been reported causing extensive damage to sunflower. In India, more than 50 insect species have been found feeding on sunflower, of which seedling pests such as cutworm (*Agrotis* spp.) and grasshopper (*Atractomorpha crenulata*), soil insects (termites and whitegrubs), sucking pests such as leaf hopper (*Amrasca biguttula biguttula*), whitefly (*Bemesia tabaci*), thrips (*Scirtothrips dorsalis*), defoliators like Bihar hairy caterpillar (*Spilosoma obliqua*), tobacco caterpillar (*Spodoptera litura*), green semilooper (*Plusia orichalcea*), cabbage semilooper (*Trichoplusia sppi*), and capitulum borer (*Helicoverpa armigera*) (Fig. 20.14) are of economic importance (Abrol et al., 2007; Basappa, 2004). In addition, non-insect pests such as rabbits, parakeets, doves, house sparrows, crows, rats, etc., have been reported to cause severe damage to sunflower crops (Goel and Kumar, 1990; Sandhu et al., 1973). A survey of insect pest attacks on sunflower was conducted in different locations of the Jammu region. For this purpose, 20 plants were selected randomly in the crop and the number of

FIGURE 20.14 ***Helicoverpa armigera* larva feeding on a sunflower head/capitulum.** *Courtesy: Dr Uma Shankar, SKUAST-Jammu.*

insect pests/plant of each species was recorded. The mean of these observations constituted the reading for each pest. The studies revealed the occurrence of several insect pests, which have been identified as *Altica cyanea* Weber (Coleoptera: Chrysomelidae); *Agonoscelis nubila* Fabricius (Heteroptera: Pentatomidae); *Eysarcoris guttger* Thunberg; *Eurydemaventralis* Westwood (Heteroptera: Pentatomidae); *Reptortus linearis* Fabricius (Heteroptera: Alydidae); *Monolepta signata* Olivier (Coleoptera: Chrysomelidae); and *Dolycoris indicus* Stal (Heteroptera: Pentatomidae). All these insect pests were found to cause considerable damage during different growth stages of the sunflower crop with up to 30% being seedling pests, 46% sucking pests, and 50% inflorescence pests. The number of insect pests attacking sunflower/plant (n = 20) on an average was 3.0 ± 1.2 (*A. cyanea*); 3.6 ± 0.8 (*A. nubila*); 2.6 ± 1.0 (*E. guttger*); 4.2 + 0.8 (*R. linearis*); 6.0 ± 1.2 (*M. signata*); 3.2 ± 1.0 (*Eysarcoris ventralis*); 1.8 ± 0.8 (*D. indicus*). Perusal of the literature reveals that *A. nubila, E. guttger; R. linearis,* and *M. signata* have not been recorded attacking sunflower in India and elsewhere. However, several other species of the genus *Agonoscelis, Eysarcoris, Riptortus*, and *Monolepta* have been recorded attacking several other crops in different parts of the world (Goel and Kumar, 1990). Evidently, their occurrence with sunflower constitutes new recorded evidence from India and elsewhere. Though *A. cyanea* has earlier been recorded from different parts of the world attacking several other crops (Goel and Kumar, 1990; Sandhu et al., 1973; Sethi et al., 1978) its occurrence with sunflower in the present study constitutes the first record of its association with the crop in India and elsewhere. In earlier studies, *D. indicus* has been recorded attacking sunflower in India (Atwal and Dhaliwal, 2004). However, its occurrence with sunflower in the present study constitutes the first record from Jammu and Kashmir.

Management Strategies

CULTURAL PRACTICES

Mixed cropping of sunflower with red gram reduced the population of *Helicoverpa armigera* on sunflower to some extent. In BSH-I or the modern variety of sunflower at 4:2 or 2:4 ratio, the number of borers were low compared with that in a solo crop of hybrid plants (Singh et al., 2003). For the control of cutworms, growing sunflower crops on ridges reduced the damage to seedlings to a very significant extent. In the case of sucking pests, increased spacing encouraged the jassid build up, while increased dosage of nitrogen fertilization favored heavy multiplication of jassids.

RESISTANCE SOURCES

More than 100 germplasm lines and several cultivars have been evaluated in the field against *Helicoverpa armigera.* The lines were categorized on the basis of egg or larval counts and damage to the capitulum.

BIOLOGICAL CONTROL

Various biological control agents attack pests. *Spilosoma obliqua* larvae have been reported to be parasitized by *Charops* sp. by up to 38% and up to 15.12% by *Apanteles obliquaei* and *A. ruides* (Shetgar et al., 1990). *Helicoverpa armigera,* a dreaded pest of sunflower is parasitized by two species of tachinid, namely, *Carcelia illata* and *Exorista xanthaspis* by up to 24.54% in Gujrat (Patel and Talati, 1987). Timely spray of Ha*NPV* at 450 LE ha^{-1} twice can protect sunflower crops from head borer. Two species of *B. thuringiensis* when evaluated against the fourth instar larvae of *Spilosoma obliqua* under laboratory conditions gave 100% mortality within 24 hrs. (Chahal et al., 2001). On the basis of an extensive study in the northern districts of Karnataka, various treatments with NPV, *Chysoperla carnea*, and endosulfan alone or in various combinations were found to hold a great promise for effective management of this pest on sunflower.

BOTANICALS

Many plant products like neem and sweet flag, etc. have been found effective against *Spodoptera litura* and *Helicoverpa armigera.*

Insect Pests of Safflower

This crop is mainly attacked by Safflower aphid, *Uroleucon carthemi* and pod borer, *Helicoverpa armigera.* There are a few natural enemies, which have been reported feeding on these insect pests.

Management Strategies

CULTURAL PRACTICES

Early sowing can be a suitable component in integrated aphid management since it escapes the peak incidence of pests. In Karnataka and Maharasthra, September sown crops experienced less damage.

RESISTANCE VARIETIES

Woody, spiny, and early maturing types were mostly tolerant while late, nonspiny types were more susceptible.

MECHANICAL

Collection and destruction of larvae of *Helicoverpa* as well as severely affected plants.

BIOLOGICAL CONTROL

Ladybird beetles (coccinellids), *Chrysoperla* sp., and syrphids are the main predators of the safflower aphid. Pawar et al. (1985) have recorded *Chrysoperla* sp. feeding on safflower aphids. Another parasitoid, *Endaphis aphidimyra*, has been reported parasitizing aphids by up to 32%. A cecidomyiid, *Pseudendaphis* sp., has been reported to cause 60% parasitization in Karnataka during the period February–May (Bhumannavar and Thontadarya, 1979a,b).

CHEMICAL CONTROL

Among several seed dressers evaluated, carbosulfan 25 DS at 20 g kg^{-1} of seed, was found to check the initial build up – control of the population later on was effectively managed by peripheral applications (1.8 m all around the field) with NSKE (5%) or dimethoate (0.05%). Several plant products, either aqueous petroleum ether or absolute alcohol extract, were found to be effective against aphids under laboratory conditions. Kernel extract from neem seed (5.7%) and custard apple seed were effective in managing aphids. Three years of experimentation with spraying NSKE (5%) 45 days after sowing followed by dimethoate (0.05%) spray at 15 days, in Karnataka, has confirmed its supremacy over 2 sprays of dimethoate (Singh et al., 2003).

Insect Pests of Olive

The main bottle-neck in olive fruit production is the crop being ravaged by several insects, diseases, nematodes, and weed pests throughout the world. Among the insect pests, olive fruit fly (*Bactrocera oleae*), olive psyllid (*Euphyllura olivine* Costa) (Fig. 20.15), olive moth (*Prays oleae*), olive black-tinged bug (Fig. 20.16), and black scale (*Saissetia oleae* Bern) have been reported to be of major concern (Daku et al., 2000) in addition to green stink bug (*Nezara virudula* L.) in India (Kaul et al., 2007a). While studying the pest complex in Himachal Pradesh, Thakur et al. (1997) observed 17 insects infesting cultivated olives, among the sap-suckers, olive psylla was found to be a regular pest. Farahbakhch and Moini (1975) presented

FIGURE 20.15 **Olive psylla infestation on olive crop.** *Courtesy: Dr Uma Shankar, SKUAST-Jammu.*

FIGURE 20.16 **Olive black-tinged bug.** *Courtesy: Dr Uma Shankar, SKUAST-Jammu.*

a list of all the known pests and fungus diseases reported from olive-growing countries in Iran, which include 95 species of insects and 6 species of mites.

Olive fruit fly is one of the key pests causing substantial damage in olive-growing regions all over the Mediterranean area. The major potential economic losses caused by this pest include preharvest fruit drop and reduction of olive fruit and oil quality due to increased acidity caused by fungi entering the fruit through the *Bactrocera olea* maggots exit holes. The olive psylla, also known as jumping plant lice, *Euphyllura olivine*, is considered as one of the most important flowering predators of olive trees causing 60–100% losses (Zouiten and Hadrami, 2001). This species is widely distributed in many parts of the world where olive crops are grown (Mathur, 1975). The first report of Olive psylla, *Euphyllura pakistanica*, from India was made by Thakur et al. (1989) on cultivated olive crops. In this connection, the incidence and management of olive psylla, *E. pakistanica*, was also discussed by Kaul et al. (2007b). Black scale damages olive trees in two ways: by sucking the sap and by secreting honeydew, which serves as a substrate for the development of sooty mould.

Management Strategies

Tilling the area under the trees and in the entire olive orchard after harvest helps to kill the pupae of the olive fruit fly. Pheromone traps can be used to monitor the fruit fly population. Furthermore, pruning crowded and bushy branches can help increase the mortality of black scale and olive psylla. The use of resistant varieties, and if necessary spraying with biopesticides during evening hours, will conserve the natural fauna available to predate olive pests. In addition to the abovementioned strategies, migrating and mobile pests can be suppressed by the collective action of various groups including the agrochemical industry, pesticide dealers, government institutions, NGOs, and groups of farmers (Knight and Norton, 1989).

Insect Pests of Castor

Castor is an industrially important oilseed crop, which is damaged by several insect pests out of which castor semilooper, *Achaea Janata;* tobacco caterpillar, *Spodoptera litura;* castor

FIGURE 20.17 **Bihar hairy caterpillar and castor spiny caterpillar on castor crops.** *Courtesy: Dr Uma Shankar, SKUAST-Jammu.*

shoot and capsule borer, *Conogathus punctiferalis;* leafhopper, *Empoasca flavescense;* whitefly, *Trialeurodes ricini,* and red hairy caterpillar, *Amsacta* spp. are of great economic importance. In recent years serpentine leaf miner, *Liriomyza trifolii* (Burgess) is also becoming a serious problem to castor crops. Additionally, *Spilosoma obliqua* (Walker), spiny caterpillar, *Ergollis merione* (Fig. 20.17), *Euproctis* spp., *Pericalia ricini* Fabricius, *Amsacta albistriga* Walker, *Amsacta moorei* Butl., castor slug caterpillar, *Parasa lepida,* castor gall fly, *Asphondylia ricini,* and red spider mite, *Tetranychus telarius* L., are also assumed regional importance and are sporadic pests. In recent years, castor inflorescence thrips (*Scirtothrips dorsalis* Hood.) also attained pest status by infesting crops at their flowering stage causing considerable loss to crops, especially in Gujarat. The polyphagous pest *Helicoverpa armigera* also caused considerable damage to castor crops by boring castor capsules in addition to simply feeding on foliage (Basappa, 2003).

Management Strategies

In the castor ecosystem, insect pests are attacked by a good number of natural enemies among them, egg parasitoid, *Trichogramma chilonis* Ishii, parasitized castor semilooper up to 100%, with another parasite, *Telenomus* spp. parasitizing this pest between 10.6–98.4% (Kulshreshta et al., 1967). Records of this reached 80% in endemic areas of Andhra Pradesh (Thobbi et al., 1976). A larval parasitoid, *Microplitis ophiusae,* has been reported parasitizing 70–80% of the population of *Achaea janata* in Coimbatore (Ayyar, 1927). An ectoparasite, *Eurytoma* sp., parasitized pupae of the castor gall fly, *A. licini* (Rao, 1952). Insect predators, insectivorous birds, and some microbial agents, such as *Bacillus thuringiensis,* cause mortality of up to 52.3% in castor semilooper (Kulshreshtha et al., 1965) and *Nosema rileyi* cause diseases in up to 100% of castor semilooper in Maharashtra (Phadke and Rao, 1978).

CULTURAL PRACTICES

Sowing cucumber crops in the borders, or interspersed within the main crop of castor, preferably earlier than the main crop itself, attracts migrating larvae, which can be destroyed mechanically. Similarly, wild castor and *Calotropis* twigs can be used to attract migrating larvae and the congregated larvae can be easily killed mechanically or otherwise by using insecticides.

RESISTANCE SOURCES

RC 1066, RC 488, and Aruna showed multiple resistances, thus, paving the way for evolving new varieties with higher levels of resistance in desirable agronomic backgrounds.

MECHANICAL CONTROL

Collection and destruction of eggs and gregarious larvae of the hairy caterpillar can be very effective.

BIOLOGICAL CONTROL

Among the 27 species parasitizing *Achaea janata*, *Trichogramma chilonis*, *T. achaeae*, and *Telenomus* spp., are egg parasitoids and are favored over others for inundative release because of their easy multiplication, low cost, and multiparasitic nature. It has been found that the releases of 50,000 parasitized eggs/ha/week during the egg-laying period could keep the pest population under check.

CHEMICAL CONTROL

Spraying of castor with pesticide should be done for controlling semilooper before adding 5 1arvae/plant for efficient plant protection.

Insect Pests of Linseed

Linseed crops are attacked by about a dozen insect pests among which linseed bud fly, *Dasyneura lini* (Fig. 20.18) and *Helicoverpa armigera* (Fig. 20.19) are key pests (Abrol and Shankar, 2011).

Linseed Bud Fly, Dasyneura lini

This insect appears as a serious pest of linseed in some parts of India including Andhra Pradesh, Madhya Pradesh, Bihar, Uttar Pradesh, Delhi, and Punjab. The adult is a small dipteran fly, which is a delicate, mosquito-like, orange-colored insect. The damage it causes

FIGURE 20.18 **Linseed pod damaged by linseed bud fly.** *Courtesy: Dr Uma Shankar, SKUAST-Jammu.*

FIGURE 20.19 ***Helicoverpa armigera* damaging linseed buds.** *Courtesy: Dr Uma Shankar, SKUAST-Jammu.*

comes from the maggot stage. They feed on the flower buds and prevent their proper opening affecting seed set. Due to their feeding on the crop, galls are produced but there is no pod formation. The incidence of this pest is up to 20%.

Management Strategies

Linseed bud fly is parasitized up to 50.7% in their larval stage during February. It is thought that 1 larva may parasitize 3–4 larvae in a bud (Ahmad and Mani, 1939). Coccinellids, *Coccinella septempunctata* and *Macaria sexmaculata*, are predators of midge and predation of up to 10–15% has been reported in Kanpur (Malik, 2000; Malik and Singh, 1997). Similarly, planting of resistant varieties is a promising option in order to mitigate linseed bud fly attacks. Other options include the installation of pheromone traps for early detection of pests and therefore mass destruction of adults of *Helicoverpa armigera* as well as the erection of bird perches encouraging black drongo (Fig. 20.20) to predate on *Helicoverpa* larvae. Biological control agents such as *Campoletes chloridae* are promising because they suppress *Helicoverpa* larval populations on linseed crops (Abrol and Shankar, 2011).

FIGURE 20.20 **Black drongo looking for larval predation in linseed.** *Courtesy: Dr Uma Shankar, SKUAST-Jammu.*

CONCLUSIONS AND FUTURE STRATEGIES

The multitude of insect pests occurring in different locations in different oilseed crops cause heavy losses in the production of oilseeds in an already oilseed-deficient world. The pests occurring at different stages of crop growth are being tackled using chemical control methods, which drastically effect the population of beneficial insects such as natural enemies, predators, parasitoids, and pollinators. It is therefore recommended that agriculture adopts the most widely acceptable tools of IPM in order to sustain the production of oilseed crops. This might be through regular monitoring of the pest situation along with weather conditions in a particular area, allowing timely protection from pests and applications of prophylactic measures, followed by the conservation and augmentation of beneficial species of parasites (Powell, 1986; Blande et al., 2008), predators, and microorganisms, by selectively using chemical pesticides and training farmers and farm workers via IPM demonstrations in their fields.

References

Abrol, D.P., Shankar, U., 2011. All India Co-ordinated Project on Linseed. Annual reports-2010–2011. SKUAST-Jammu.

Abrol, D.P., Shankar, U., 2014. Pesticides, food safety and integrated pest management. In: Pimental, D., Peshin, R. (Eds.), Integrated Pest Management – Pesticide Problems. Springer, New York, pp. 167–199.

Abrol, D.P., Kotwal, D.R., Hafeez, A., 2007. New insect pests attacking sunflower (*Helianthus Annuus* L.). J. Res. SKUAST-J 6 (1), 117–119.

Ahirwar, R.M., Banerjee, S., Gupta, M.P., 2009. Seasonal incidence of insect pests of sesame in relation to abiotic factors. Ann. Plant Prot. Sci. 17 (2), 351–356.

Ahmad, T., Mani, M.S., 1939. Two ne chalcidoid parasites of linseed midge, *Dasyneura lini*. Indian J. Agric. Sci. 9, 531–539.

Ahuja, D.B., Rajpurohit, T.S., Singh, M., Solanki, Z.S., Kalyan, R.K., Ahuja, U.R., Dhandapani, A., Sundria, M.M., 2009. Indian J. Agric. Sci. 79 (10), 808–812.

Anandh, G.V., Selvanarayanan, V., Tholkappian, P., 2010. Influence of arbuscular mycorrhizal fungi and bio-inoculants on host plant resistance *Antigastra catalaunalis* Duponchel in sesame *Sesamum indicum* Linn. J. Biopest. 3 (1), 152–154.

Anonymous., 1986. Proceedings of Seventh Annual Rabi-Summer Groundnut Workshop, Mahatma Phule Agricultural University Rahuri, Ahmednagar, pp. 4–5.

Atwal, A.S., Dhaliwal, G.S., 2004. Agricultural pests of South Asia and their management. Kalyani Publishers, New Delhi.

Atwal, A.S., Dhaliwal, G.S., 2009. Agricultural Pests of South Asia and Their Management. Kalyani Publishers, New Delhi.

Atwal, A.S., Chowdhary, J.P., Ramzan, M., 1969. Some preliminary studies in India on the bionomics and rate of the parasitism of *Diaeretiella rapae* (Braconidae: Hym.) a parasitoid of aphid. J. Res., PAU Ludhiana 6, 177–182.

Atwal, A.S., Chaudhary, J.P., Ramzan, M., 1971. Mortality factor on the natural population of cabbage aphid, *Lipaphis erysimi* (Kalt.) (Aphididae: Homoptera) in relation to parasites, predators and weather conditions. Ind. J. Agric. Sci. 41, 507–510.

Augustin, A., Tulisalo, U., Korpela, S., 1986. Flea beetles (Coleoptera, Chrysomelidae, Halticinae) on rapeseed and sugarbeet in Finland. J. Agric. Sci. Finl. 58, 69–82.

Ayyar, T.V.R., 1927. The parasitic hymenoptera of economic importance noted from South India. Bull. Entomol. Res. 18, 73.

Ayyar, T.V.R., 1963. Handbook of Economic Entomology for South India. Government Press, Madras, p. 516.

Bakhetia, D.R.C., Labana, K.S., 1978. Insect resistance in *Brassica* crops. Crop Improvement 5, 95–103.

Bakhetia, D.R.C., Sekhon, B.S., 1989. Insect-pests and their management in rapeseed mustard. J. Oilseeds Res. 6 (2), 269–299.

Basappa, H., 2003. Integrated Pest Management in Castor. Directorate of Oil seeds Research, Rajendranagar, Hyderabad-500 030, India. 8, p. 52.

Basappa, H., 2004. Integrated pest management in sunflower: an Indian scenario. In: Proceedings of Sixteenth International Sunflower Conference, Fargo, ND USA. pp. 853–859. August 29– September 2, 2004.

Behera, P.K., 2011. Trathala flavo-orbitalis – natural enemy of sesame shoot webber and capsule borer in Coastal Odisha. Insect Environ. 17 (3), 133–134.

Bhumannavar, B.S., Thontadarya, T.S., 1979a. Estimation of safflower crop losses due to the aphid *D. caarthami* Theobold (Hemiptera:Aphididae) under rainfed conditions. Pestology 3 (6), 18–19.

Bhumannavar, B.S., Thontadarya, T.S., 1979b. Effectiveness of different insecticides against safflower aphid, *D. compoitae* Theobald (Hemiptera Aphididae) on safflower in Karnataka. Pesticides 13 (9), 28–30.

Bishnoi, O.P., Singh, H., Singh, R., 1992. Incidence and multiplication of mustard aphid (Lipaphis erysimi) in relation to meteorological variables. Indian J. Agric. Sci. 62 (10), 710–712.

Blande, J.D., Pickett, J.A., Poppy, G.M., 2008. Host foraging for differentially adapted brassica-feeding aphids by the braconid parasitoid Diaeretiella rapae. Plant Signal Behav. 3 (8), 580–582.

Bogawat, J.K., Rangarajan, M., Chakravarthi, B.P., 1966. Bacterial mortality of the grubs of saw fly. Athalia proxima Klug (Hymenoptera:Tenthrenididae). Indian J. Entomol. 28, 417–421.

Bonnemaison, L., 1965. Insect pests of crucifers and their control. Ann. Rev. Entomol. 10, 233–256.

Brar, K.S., Sandhu, G.S., 1976. Major pests of oilseed crops in Northern India and their control. Oilseeds J. 6, 16.

Brar, K.S., Sandhu, G.S., 1982. Field biology of white grub, *Holotrichia consanguinea* Blanchard (Scarabaeidae: Coleoptera) in Punjab. J. Soil. Biol. Eco. 2, 32–35.

Burgess, L., 1977. Flea beetles (Coleoptera: Chrysomelidae) attacking rape crops in the Canadian prairie provinces. Can. Entomol. 109, 21–32.

Chahal, V.P.S., Singh, G., Chahal, P.P.K., 2001. In: Biopesticides : Emerging Trends BET 2001, Chandigarh, India, pp.136.

Chandla, V.K., Misra, S.S., Bhala, O.P., Thakar, J.R., 1988. White grub *Brahmina coriacea* (Hope) infesting potato in Shimla hills. Seeds and Forms 14, 12–13.

Chandra, S., Kushwaha, K.S., 1987. Impact of environmental resistance on aphid complex of cruciferous crops under the agroclimatic conditions of Udaipur II. Biotic components. Ind. J. Entomol. 49, 86–113.

Cheema, J.S., Singh, G., 1987. Biology of sesame leaf webber and capsule borer, *Antigastra catalaunalis*(Duponchel) (Pyralidae: Lepidoptera) in Punjab. J. Res. Punjab Agric. Univ. 24 (1), 65–74.

Cherian, M.C., 1934. Notes on some south Indian Syrphids. J. Bombay Nat. Hist. Soc. 37, 697–699.

Cook, S.M., Denholm, I., 2008. Ecological approaches to the control of pollen beetles in oilseed rape. IOBC-WPRS Bull. 38, 110–113.

Daku, L., Norton, G.W., Pfeiffer, D.G., Luther, G.C., Pitts, C.W., Taylor, D.B., Tedeschini, J., Uka, R., 2000. "Farmers' knowledge and attitudes towards pesticide use and olive pest management practices in Vlora, Albania: A Baseline Survey." IPM CRSP Working Paper 001-1, Virginia Polytechnic Institute and State University.

Dogra, I., Devi, N., Raj, D., 2003. Parasitization of mustard aphid, *Lipaphis erysimi* (Kalt.) by *Diaeretiella rapae* in mid hill zone of Himachal Pradesh (India). J. Entomol. Res. 27 (2), 145–149.

FAO., 2010. FAO agricultural outlook 2010–2019.

Farahbakhch, G., Moini, M., 1975. Olive pests in Iran-Plant Pests and Diseases Research Institute, National Agricultural Research Organisation, Ministry of Agriculture and Natural Resources, Tehran, Iran, pp. 73.

Gadad, H., Hegde, M., 2014. Evaluation of biopesticides and biorationals against thrips and leafminer in *rabi*/summer groundnut. Res. Rev. J. Agric. Allied Sci. 3 (3), 25–31.

Ghewande, M.P., 1987. Virus diseases of groundnut and their management. In: Vyas, S.C.C., Sgroff, V.N. (Eds.), Symposium on Diseases of Oilseed and Pulse Crops and Their Control. Jawaharlal Nehru Krishi Vishwa Vidyalaya, Indore, MP, pp. 62–67.

Goel, S.C., Kumar, A., 1990. Insect pests and predators associated to sunflower in winter of northern India. Indian J. Entomol. 52, 39–45.

Goodell, G., 1984. Challenges to international pest management research and extension in the third world: do we really want IPM to work? Bull. Entomol. Soc. Am. 30, 18–26.

Griffiths, G.C.D., 1986. Relative abundance of the root maggots Delia radicum (L.) and D. floralis (Fallén) (Diptera: Anthomyiidae) as pests of canola in Alberta. Quaestiones Entomologicae 22, 253–260.

Guidry, N.P., 1964. Graphic summary of world agriculture. US Dept Agric Misc Publ, 705:64.

Gunathilagaraj, K., Babu, P.C.S., 1987. FAO Plant Prot. Bull. 35 (2), 63–64.

Gururaj, G., Kulkarni, Hugar, P.S., 1999. Effect of host plant on POBs yield and mortality of *Spodoptera litura* larvae infected with NPV. Karnataka J. Agric. Sci. 12 (1–4), 60–63.

Hegde, D.M., 2000. Oilseeds Situation. Directorate of Oilseeds Research, ICAR, Hyderabad.

Kalra, V.K., 1988. Population dynamics of various predators associated with mustard aphid, L. erysimi Kalt. J. Bio. Cont. 2, 77–79.

Kalra, V.K., 1989. Natural parasitism of sesame leaf webber and pod borer, *Antigastra catalaunalis* Duponchel. Indian J. Plant Prot. 17 (1), 9–11.

Kalra, A.N., Kulshreshtha, J.P., 1961. Studies on the biology and control of *Lachnosterna consanguinea* Bl. a pest of sugarcane in Bihar. Bull. Ent. Res. 52, 1577–1587.

Katole, S.R., Nimbalkar, S.A., Yadgirwar, P.V., Kadu, N.R., Kolhe, A.V., 1998. Note on parasitization of *Diaeretiella rapea* M'Intosh on mustard aphids, *Lipaphis erysimi* Kalt. PKV Res. J. 22, 214–216.

Kaul, V., Tiku, A.K., Shankar, U., Monobrullah, Md., 2007a. Green stink bug (Hemiptera: Pentatomidae) recorded as a new pest of olive in India. J. Asia-Pacific Entomol. 10 (1), 81–83.

Kaul, V., Malik, G.H., Shankar, U., Monobrullah, Md., 2007b. Incidence and management of olive psylla, *Euphyllura pakistanica*. Indian J. Entomol. 69 (4), 331–340.

Khan, M.Q., Hussain, M., 1964. Natural control of Lapbygma exigua by a Mermis sp. (Nematoda). Indian J. Entomol. 26, 124–125.

Knight, A.L., Norton, G.W., 1989. Economics of agricultural pesticide resistance in arthropods. Ann. Rev. Entomol. 34, 293–313.

Krishnamurti, B., Usman, S., 1954. Some insect parasites of economic importance noted in Mysore State. Indian J. Entomol. 16, 327–344.

Kulshreshta, J.P., Sanghi, P.K., Thobbi, V.V., Kulkarni, L.G., 1967. Mass multiplication of Telenomus sp. (Hymenoptera: Scelionidae) an egg parasite imported in India for biological control of castor semilooper, *Achea janata* L. and some observations on its biology. Indian J. Entomol. 29, 290–294.

Kulshreshtha, J.P., Sanghi, P.K., Ravindranath, V., 1965. Microbial control of castor semilooper *Achea janata*. Indian J. Entomol. 27, 353–354.

Kumar, A., 1985. Estimation of damage caused to *Chromatomyia horticola* population on *Brassica campestris* by hymenopteran parasites. Ann. Entomol. 10, 49–53.

Kumar, G., Prasad, D.K., Pavan, S., 2011. Hover-flies (Diptera : Syrphidae) of the Coromandel Coast in Andhra Carnatic, Peninsular India. Bionotes 13 (2), 78–86.

Lamb, R.J., 1989. Entomology of oilseed *Brassica* crops. Ann. Rev. Entomol. 34, 211–229.

Lammerink, J., Banfield, R.A., 1980. Effect of aphid control by disulfoton on seed yield components and seed quality of oilseed rape. NZ J. Exp. Agric. 8, 45–48.

Lee, H.S., 1968. Evaluation of some granulated insecticides against the aphids and diamond-back moth on rapeseed by soil treatment. Plant Prot. Bull. 10, 69–70, (In Chinese).

Lennerts, L., 1983. Oelschrote, oelkuchen, pfl anzliche Oele und Fette, Herkunft, Gewinning, Verwendung, Bonn 1983. Alfred Strothe, Hannover.

Mahal, M.S., Singh, R., Singh, B., 1988. Dispersion of *Lipaphis erysimi* (Kalt) and development of a sequential sampling plan for its supervised control on raya, *Brassica juncea* Coss. J. Insect Sci. 1, 27–38.

Makhmoor, H.D., Shinde, C.D., 1986. Preliminary study on the effect of different doses of N. P. and K. and their combination in mustard in on the incidence of *Athalia proxima* Klug., *Crocidolomia binotalis* Zell. and *Phytomyza atriconis*. Res. Devel. Rep. 3, 66–71.

Malik, Y.P., 2000. Assessment of yield loss due to host pest interaction in linseed, *Linum usitatissimum*. Indian J. Agric. Sci. 70, 53–54.

Malik, Y.P., Singh, S.V., 1997. Evaluation of linseed germplasm against bud fly, *Dasyneura lini* Barnes and some important disease. J. Insect Sci. 10, 70–71.

Mathur, R.N., 1975. Psyllidae of the Indian Subcontinent. Indian Council of Agricultural Research, New Delhi, pp. 429.

Muthiah, C., 1991. Some new parasites of groundnut leaf miner (*Aproaerema modicella* Dev.). Indian J. Plant Prot. 19 (1), 105.

Muthiah, C., Kareem, A.A., 2000. Survey of groundnut leafminer and its natural enemies in Tamil Nadu, India. Int. Arachis Newsl. 20, 62–63.

Nair, M.R.G.K., 1975. Insect and Mites of Crops in India. ICAR, New Delhi, pp. 404.

Nath, P., Bhushan, S., Singh, A.K., 2003. Effect of intercropping on the population of insect pests of sesamum. J. Appl. Zool. Res. 14 (1), 76–79.

Patel, A.A., Bhalani, P.A., 1990. A new record of *Diadegma* sp. as a larval parasite of the sesamum leafroller in Gujarat. Indian J. Entomol. 51 (4), 474.

Patel, A.J., Talati, G.M., 1987. Biology of *H. armigera* (Hb.) as a pest of sunflower. Gujarat Agric. Univ. Res. J. 12 (2), 54.

Patel, A.G., Vyas, H.N., 1984. Studies on predatory capacity of ladybird beetle, *Menochilus sexmaculatus* (Fab.) against *Aphis craccivora* Koch under laboratory condition. Pesticides 18 (11), 8–9.

Patel, R.C., Yadav, D.N., Patel, J.R., 1976. Natural control of groundnut aphid, *Aphis craccivora* Koch in Central Gujrat. Curr. Sci. 45 (1), 34–35.

Patra, A.K., 2001. Response of sesame (Sesamum indicum) varieties to dates of sowing during rainy season. Indian J. Agric. Sci. 71 (8), 523–524.

Pawar, C.S., Bhatnagar, U.S., Jadhav, D.R., 1985. Some predatory insects and their parasites. Sci. Cult. 51 (1), 101–102.

Phadke, C.H., Rao, V.G., 1978. Studies on entomogenous fungus, *Nomuraea rilyei* (Farlow). Curr. Sci. 47, 511–512.

Powell, W., 1986. Enhancing parasitoid activity in crops. In: Insect, Parasitoids, Thirteenth Symposiumof the Royal Entomological Societyof London, September 18–19, 1985 (Waage, J., Greathead, D. (eds.), Academic Press, London, 19–340.

Prakash, S., Hinata, K., 1980. Taxonomy, cytogenetics and origin of crop brassicas. Rev. Opera Bot. 55, 1–57.

Rai, B.K., 1976. Pests of Oilseed Crops in India and Their Control. Indian Council of Agricultural Research, New Delhi, pp. 121.

Rai, S., Singh, N.N., Shankar, U., 2010. Functional response of grub and adult of *Coccinella septempunctata* (L.) and *Coccinella transversalis* (Fab.) on mustard aphid, *Lipaphis erysimi* (Kalt). Arch. Plant Prot. Phytopathol. 43 (18), 1829–1835.

Rakow, G., 2004. Species origin and economic importance of Brassica. Pua, E.C., Douglas, C.J. (Eds.), Biotechnology in Agriculture and Forestry, vol. 54, Springer, Berlin, pp. 3–11.

Rana, J.S., 2005. Performance of *Lipaphis erysimi* (Homoptera: Aphididae) on different *Brassica* species in a tropical environment. J. Pest Sci. 78, 155–160.

Rao, A.S., 1952. A gall midge pest of castor in India. Curr. Sci. 21, 347–348.

Rao, J.G., Kameswara Rao, P., Satyanarayan, M.M., 1980. Note on natural egg parasitization of Archerontia styx-Westw. at Hyderabad. Indian J. Entomol. 42 (2), 283.

Raymer, P.L., 2002. Canola: an emerging oilseed crop. In: Janick, J., Whipkey, A. (Eds.), Trends in New Crops and New Uses. ASHS, Alexandria, VA, pp. 122–126.

Reddy, L.J., 1988. The Groundnut. ICAR Publication, New Delhi, India, pp. 572.

Rohilla, H.R., Singh, H., Kalra, V.K., Kharub, S.S. 1987. Losses caused by mustard aphid, *lipaphis erysimi* (Kalt.) in different brassica genotypes. Proceedings of Seventh International Rapeseed Congress Poland. 5, pp. 1077–1083.

Sachan, J.N., Gangwar, S.K., 1980. Vertical distribution of important insect-pest of cole crops in Meghalaya as influenced by the environment factors. Indian J. Entomol. 42, 414–421.

Sahayaraj, K., 1999. Field evaluation of *Rhynocoris marginatus* (Fab.) against two groundnut defoliators. Int. Arachis Newsl. 19, 41–42.

Sandhu, G.S., Brar, K.S., Bakhetia, D.R.C., 1973. Soil systemic insecticides for the control of leafminer on mustard. Pesticides 11 (7), 19–21.

Satyagopal, K., Sushil, S.N., Jeyakumar, P., Shankar, G., Sharma, O.P., Boina, D., Sain, S.K., Chattopadhyay, D., Asre, R., Kapoor, K.S., Arya, S., Kumar, S., Patni, C.S., Chattopadhyay, C., Pandey, A., Pachori, R., Thakare, A.Y., Basavanagoud, K., Halepyati, A.S., Patil, M.B., Sreenivas, A.G., 2014. AESA based IPM package for Mustard/Rapeseed. 49.

Sekhon, B.S., Bakhetia, D.R.C., 1994. Relationship between population of mustard aphid, *Lipaphis erysimi* (Kalt.) and yield losses in *Brassica napus*. J. Insect Sci. 7, 67–69.

Selvanarayanan, V., 2013. Bio-inoculants to boost resistance in crop plants against insects: field success and struggles in India. IOBC-WPRS Bull. 89, 373–378.

Sethi, M.A., Singh, K.M., Prasad, H., 1978. Incidence of insect pests on different varieties of sunflower. Indian J. Ent. 40, 101–103.

Sharma, P., Saini, M.K., Deep, S., Kumar, V., 2012. Biological control of groundnut root rot in farmer's field. J. Agric. Sci. 4 (8), 48–59.

Sharma, A.N., Gupta, G.K., Verma, R.K., Sharma, O.P., Bhagat, S., Amaresan, N., Saini, M.R., Chattopadhyay, C., Sushil, S.N., Asre, R., Kapoor, K.S., Satyagopal, K., Jeyakumar, P., 2014. Integrated Pest Management for Soyabean. Directorate of Plant Protection, Quarantine & Storage, CGO Complex, NH IV, Faridabad – 121001, pp. 41.

Shetgar, S.S., Bilapate, G.G., Patel, V.V., Londhe, G.M., 1990. A note on the natural enemies of Bihar hairy caterpillar, *Spilosoma oblique* (Walker). Indian J. Entomol. 52 (1), 158.

Sidhu, H.S., Singh, S., 1964. Biology of the Mustard Aphid, *Lipaphis erysimi* in the Punjab. Indian Oilseeds J. 8 (4), 348–359.

Singh, H., 1982. Studies on insect-pest complex in *Brassica campestris* L. var. brown 'sarson'. Thesis, Doctor of Philosophy, Entomology, submitted to Haryana Agricultural University, Hisar, pp. 192.

Singh, H., 1991. A Success Story of Five-Fold Increase of Oilseeds Production in Haryana. Haryana Agricultural University, Hisar, India.

Singh, H., Rohilla, H.R., Kalra, V.K., Yadava, T.P., 1984. Response of Brassica cultivars sown on different dates to attack of mustard aphid, *Lipaphis erysimi* (Kalt.). J. Oilseeds Res. 1, 49–56.

Singh, H., Singh, Z., Yadav, J.P., 1987. Post harvest losses in rapeseed caused by aphid pest. Proceedings of Seventh International Congress Poland, 5, pp. 1138–1142.

Singh, H., Singh, Z., Yadava, T.P., 1990. Influence of abiotic factors on alate production in mustard aphid *Lipaphis erysimi* (Kaltenbach). J. Aphidol. 4, 71–74.

Singh, S., Singh, N., Bhat, N. S., Singh, S.B., 2003. Development and validation of IPM in oilseed based production systems. In: Proceedings and recommendations of the NATP interactive Workshop on Integrated Pest Management, February 26–28, 2003. Recent Advances in Integrated Pest Management. Singh, A., Trivedi, T.P, Sardana, H.R., Sharma,O.P., Sabir, N. (eds.), pp. 91–98.

Singhvi, S.M., Verma, N.D., Yadava, T.P., 1973. Estimation of losses in rapeseed (*Brassica campestris L. var. toria)* and mustard *(Brassica juncea* Coss.) due to mustard aphid (*Lipaphis erysimi* Kalt.). Haryana Agric. Univ. J. Res. 3, 5–7.

Sovero, M., 1993. Rapeseed, a new oilseed crop for the United States. In: Janick, J., Simon, J.E. (Eds.), New Crops. Wiley, New York, pp. 302–307.

Sridhar, V., Prasad, V.D., 1996. Life table studies on natural population of *Spodoptera litura* on groundnut. Ann. Pl. Protec. Sci. 4, 142–147.

Sundaramurthy, V.T., Jayraj, S., Swamiappan, M., 1976. *Telenomus malonus* Nixon, A potential egg parasite of groundnut red hairy caterpillar, *Amsacta albigastra* Walker. Curr. Sci. 45 (21), 775.

Tansey, J., Dosdall, L.M., Keddie, B.A., 2009. *Phyllotreta crucifera* and *Phyllotreta striolata* responses to insecticidal seed treatments with different models of action. J. Appl. Entomol. 133, 201–209.

Thakur, J.R., Sharma, P.C., Gupta, P.R., 1989. Occurrence of olive psylla, *Euphyllura pakistanica* on olive in India: a new record. Trop. Pest Manag. 35 (3), 331.

Thakur, J.R., Atwal, R.K., Gupta, P.R., 1997. Insect pests of olive in Himachal Pradesh. Pest Manag. Econ. Zool. 5 (1), 75–79.

Thobbi, V.V., Singh, B.U., Amin-ul-Hussain, Sarves, S.O.H., Rao, N.G.P., 1976. Biological control of castor semilooper. Indian Farm. 26 (6), 15–17.

Tiittanen, K., Varis, A.L., 1960. The treatment of seeds of swede, turnip, and turnip rape in the control of flea beetles (*Phyllotreta* spp.) and cabbage root flies (*Hylemyia brassicae* Bouch6 and *H. floralis* Fall.). Valt. Maatalouskoetoiminnan Julk. 181, 1–11.

Vora, V.J., Bharodia, R.K., Kapadia, M.N., 1985. Pests of oilseed crops and their control on rapeseed and mustard. Pesticides 19, 38–40.

Vyas, R.V., Yadav, D.N., Patel, R.J., 1991. Studies on mass production of *Bacillus popillae* var. *holotrichiae,* a promising pathogen of white grub in Gujarat. Gujarat Agric. Univ. Res. J. 17, 30–37.

Weiss, E.A., 1983. Oilseed processing and products, In: Oilseed crops, pp. 528–596, ISBN 0-632-05259-7, London: Longman.

Zouiten, N., Hadrami, I.E., 2001. Olive tree psyllid: state of knowledge and perspective of control. Agricultures 4 (10), 225–232.

Subject Index

C

D

Q

R

S

T

Kolibri 360 GmbH
Ferdinand-Jühlke-Straße 7
99095 Erfurt, Deutschland
produktsicherheit@kolibri360.de